# Biotechnology

## Progress and Applications

## The Editors

**Dr. Saif Hameed** is currently Assistant Professor and Research Coordinator at Amity Institute of Biotechnology, Amity University Haryana (AUH). He is also an active member of Directorate of Research and Publications (DRP) and University Science Instrumentation Center (USIC) committee in addition to being Member Secretary, Research Ethics Committee at AUH. Dr. Hameed did his Bachelor from University of Delhi and Master degree from Jamia Hamdard with distinction in 2003 and 2005 respectively. He did his doctoral studies in Life Sciences from Jawaharlal Nehru University in 2010 where he also received CSIR Research Associateship for his one year postdoctoral work. He also worked as Visiting Scholar in Institut für Mikrobiologie, Heinrich-Heine-Universität, Düsseldorf, Germany in 2008. Dr. Hameed has received Young Scientist award under Fast Track Scheme from Science and Engineering Research Board, Department of Science and Technology, New Delhi in the year 2012. He is member of International Society of Infectious Diseases (ISID) and member of editorial board and reviewer panel of E3 Journal of Biotechnology and Pharmaceutical Research (EJBPR) and International Journal of Scientific and Research Publication (IJSRP) respectively. Dr. Hameed is actively engaged in research in the field of infectious diseases related to multidrug resistance (MDR) in pathogenic fungi. He has around 30 peer-reviewed papers to his credit in both international and national journals of repute, organized various conferences. He has successfully convened four national conference as Organizing Secretary in 2014. He is currently supervising 5 Ph.D. students and guided 5 UG and 3 PG level students for their research projects. He has bagged regular project for funding from Board of Research in Nuclear Sciences (BRNS), Mumbai.

**Dr. Zeeshan Fatima** is currently working as an Assistant Professor at Amity Institute of Biotechnology (AIB), Amity University Haryana (AUH). She is Research Coordinator at AIB and also an active member of Directorate of Research and Publications (DRP) and University Science Instrumentation Center (USIC) committee and Research Ethics Committee at AUH. Dr. Fatima did her Bachelors and Masters from Banaras Hindu University in 2000 and 2002 respectively. She earned her doctoral degree in Biochemistry from Aligarh Muslim University in 2008 during this period she availed University Scholarship for Research. She has held research positions under nationally and internationally funded research projects which also include her Research Associateship at B.H.U. and J.N.U. and postdoctoral training from University of Cincinnati, Ohio, USA in 2010. Dr. Fatima has received two Young Scientist awards under Fast Track and Women Scientist Schemes respectively from Science and Engineering Research Board, Department of Science and Technology, New Delhi in the year 2012. She has also bagged regular project for funding from Board of Research in Nuclear Sciences (BRNS), BARC Mumbai. She is actively engaged in research in the field of infectious diseases and particularly on the aspect of Multidrug resistant in human pathogen *Mycobacterium tuberculosis* and *Candida albicans*. She has 2 books and around 30 peer-reviewed papers to her credit in both international and national journals of repute. She participated in several international and national conferences and received various accolade in the form of best paper award. She has successfully convened three national conference as Organizing Secretary in 2014 and organized various guest lectures. She is currently supervising 4 Ph.D. students and guided 10 UG and PG level students for their research projects.

# Biotechnology

## Progress and Applications

— *Editors* —

**Saif Hameed**

**Zeeshan Fatima**

*Amity Institute of Biotechnology,*
*Amity University Haryana,*
*Manesar, Gurgaon-122413, Haryana, India*

2016

**Daya Publishing House®**

*A Division of*

**Astral International Pvt. Ltd.**

**New Delhi – 110 002**

**Cataloging in Publication Data--DK**
Courtesy: D.K. Agencies (P) Ltd. <docinfo@dkagencies.com>

**Biotechnology** : progress and applications / editors, Saif Hameed, Zeeshan Fatima.
pages cm
Contributed articles.
Includes index.

ISBN 978-93-5130-950-5 (Internationa Edition)

1. Biotechnology. I. Hameed, Saif, editor. II. Fatima, Zeeshan, editor.

TP248.2.C66 2016 DDC 660.6 23

*Published by* : **Daya Publishing House®**
*A Division of*
**Astral International Pvt. Ltd.**
– ISO 9001:2008 Certified Company –
4760-61/23, Ansari Road, Darya Ganj
New Delhi-110 002
Ph. 011-43549197, 23278134
E-mail: info@astralint.com
Website: www.astralint.com

*Laser Typesetting* : **Classic Computer Services,** Delhi - 110 035

*Digitally Printed at* : **Replika Press Pvt. Ltd.**

*"Dedicated with Honor to the Living Legend"*

**Prof. S.M. Paul KHURANA**

*for his*

*Impeccable & Path Breaking Research Achievements,*
*Excellence in Higher Education Management*
*&*
*Democratization of Knowledge with Human Values*

# Foreword

It gives me immense pleasure to note that the book *"Biotechnology: Progress & Applications"* is being published to widely disseminate the current research topics and frontier areas in the era of biotechnology being pursued at global level. I am also happy to note that this book is carrying the latest research and developments with focus on the recent advancements and modern trends in the field of biotechnology.

In the previous centuries, technology was primarily driven by basic sciences, however, with the change in millennium, we have entered into a revolutionary transformation in biology drastically improving our quality of life and impacting our economy. This has made biotechnology indispensible for studies and research. I am indeed happy to note that in the book diverse R&D areas of vital interest to the industry and society have been touched upon. It is also a matter of great satisfaction that there is a fine blend of research and innovation culture in this book.

The book is primarily meant for the teachers and researchers in the field, nevertheless it will also prove useful to biotechnology students of UG and PG levels. I am hopeful that this book will be highly useful in creating an interdisciplinary mindset among students and young faculty for biotechnology research and innovations. It will also provide an opportunity to the students, faculty and research scholars to gain the knowledge and plan their prospective research. The innovative and creative minds of the faculty and inspired students may engage themselves in accelerating the pace of society development from relevant research and innovations discussed in the book. I wish the book a great success in the field.

**_Prof. P. B. Sharma_**

*Vice Chancellor,*
*Amity University Haryana, Gurgaon*

# Preface

*"Information wants to be free"*. Information is like fragrance that could not be kept concealed and reaches to all on its own. This book *"Biotechnology: Progress & Applications"* affirms to dwell upon the vast dimension of vibrant areas of biotechnology and to update the current research and developments. This archive is an effort to pair classic with contemporary treasure of natural sciences to foster the new horizon of frontier futuristic research. The chapters in the book are contributed by the authorities in their own right from reputed Universities of India *viz.* JNU, DU, AIIMS, JMI, BARC etc. providing holistic integration of current biotechnological advancements to solve complex problems encountered by real world.

We are entering a zone where focal research ideas often transcend the scope of mono-disciplinary study and pushing forward for accelerating interdisciplinary research. Current researches are wrestling hard against complex set of equations from a variety of perspectives. Thus the book attempts to connect scientific advancement, intellectual adventure and human awareness through threading by spreading the knowledge across varied disciplines within natural sciences intercalating biotechnology.

Collectively twenty two chapters have been divided into 9 thematic sections integrating vast areas of biotechnology. The chapters are presented in an impeccable manner to update and provide comprehensive knowledge to the scientists, teachers, UG/PG research and students scholars and also shall serve some interest for the public.

Genomics and Proteomics taking the lead through upholding direct impact on environment, society and economy followed by Virology, a subject of great concern and challenge for researches since last two decades. The next theme covers Plant Biotechnology, being the prerequisite in order to address the growth of socio-economic condition of our agriculture based economy. Building a bridge between academia and industry, the following theme covers Microbial Technology and Environmental Biotechnology as a unique blend of science and business associated with real world biotechnology implicating individual, community and industry. A vital theme touching human health includes Infectious Diseases and Pharmaceutical Biotechnology taking the centre stage to create synergy among researches, health care workers and industry. Subsequent topics on Cancer, Stem Cells & Nanotechnology are incorporated to register the speed of scientific breakthroughs of vital biotechnology research.

We feel privileged and proud to dedicate this piece of work to our true mentor Prof. S.M. Paul Khurana, a scholar of rare distinctions with unique amalgamation of scientific aptitude, administrative qualities and amiable disposition. We record our intense gratitude to him for making us realize the need of such a compilation and support throughout which made this venture possible. We are sincerely grateful to him for his magnanimity and guidance both professionally and even personally.

*Dr. Saif Hameed*

*Dr. Zeeshan Fatima*

# Contents

## Section VIII: Cancer



## Section IX: Stem Cells and Nanotechnology

# List of Contributors

## EDITORS

**Dr. Saif Hameed**

*Assistant Professor and Research Coordinator, Amity Institute of Biotechnology,*

*Member Secretary, Research Ethics Committee, Member, Directorate of Research and Publications, Amity University Haryana, Gurgaon (Manesar)- 122413*

*Email: saifhameed@yahoo.co.in, shameed@ggn.amity.edu*

**Dr. Zeeshan Fatima**

*Assistant Professor and Research Coordinator, Amity Institute of Biotechnology,*

*Member, Research Ethics Committee, Member, Directorate of Research and Publications, Member, Directorate of International Affairs, Amity University Haryana, Gurgaon (Manesar)- 122413*

*Email: drzeeshanfatima@gmail.com, zfatima@ggn.amity.edu*

## CONTRIBUTORS

**Anchal Singh;** Amity Institute of Biotechnology, Amity University Uttar Pradesh, Noida- 201313,India

**Anil Kumar;** Amity Institute of Biotechnology, Amity University Haryana, Gurgaon (Manesar)-122413, India.

**Anupam Dikshit;** Biological Product Lab, Department of Botany, University of Allahabad, Allahabad : 211002, India.

**Anurag Jyoti;** Amity Institute of Biotechnology, Amity University Madhya Pradesh, Gwalior -474005, India.

**Anurag Sharma;** Department of Chemistry, Amity School of Applied Sciences, Amity University Haryana, Gurgaon (Manesar)-122413, India.

**Ashish S Verma;** Pro-Vice Chancellor, Jadavpur University, Kolkata- 700032, India.

**Ashish Sharma;** Department of Botany, School of Biotechnology and Biosciences, Lovely Professional University, Punjab -144411, India.

**Ashutosh Pathak;** Biological Product Lab, Department of Botany, University of Allahabad, Allahabad : 211002, India.

**Chiranjib Banerjee;** Department of Bio-Engineering, Birla Institute of Technology, Mesra, Ranchi-835215, India.

**Digvijay Verma;** Department of Microbiology, University of Delhi South Campus, New Delhi-11002, India.

**Harsh Kumar Agrawal;** Department of Bio-Engineering, Birla Institute of Technology, Mesra, Ranchi-835215, India.

**Jawaid A. Khan;** Department of Biosciences, Jamia Millia Islamia, New Delhi- 110025, India.

**Leena Parihar;** Department of Botany, School of Biotechnology and Biosciences, Lovely Professional University, Punjab -144411, India.

**Madan Kumar;** School of Environmental Sciences, Jawaharlal Nehru University, New Delhi-10067, India.

**Malik Zainul Abdin;** Department of Biotechnology, Jamia Hamdard, New Delhi-110062, India.

**Manju Sharma;** Amity Institute of Biotechnology, Amity University Haryana, Gurgaon (Manesar)-122413, India.

**Maria Habib;** Department of Biosciences, Jamia Millia Islamia, New Delhi-110025, India.

**Moiz A. Ansari;** Amity Institute of Biotechnology, Amity University Haryana, Gurgaon (Manesar)-122413, India.

**Monika Vats;** Department of Chemistry, Amity School of Applied Sciences, Amity University Haryana, Gurgaon (Manesar)-122413, India.

**Munindra Ruwali;** Amity Institute of Biotechnology, Amity University Haryana, Gurgaon (Manesar)-122413, India.

**Narender Kumar;** Amity Institute of Biotechnology, Amity University Haryana, Gurgaon (Manesar)-122413, India.

**Nitai Debnath;** Amity Institute of Biotechnology, Amity University Haryana, Gurgaon (Manesar)-122413, India.

**Pathak A.N;** Amity Institute of Biotechnology, Amity University Rajasthan, Jaipur-302006, India.

**Pramod D. Sherkhane;** Nuclear Agriculture and Biotechnology Division, Bhabha Atomic Research Centre, Trombay, Mumbai- 400085, India.

**Prasun K. Mukherjee;** Nuclear Agriculture and Biotechnology Division, Bhabha Atomic Research Centre, Trombay, Mumbai- 400085, India.

**Pratyoosh Shukla;** Enzyme Technology and Protein Bioinformatics Laboratory, Department of Microbiology, Maharshi Dayanand University, Rohtak-124001, India.

**Pravej Alam;** Department of Biosciences, Jamia Millia Islamia, New Delhi- 110025, India.

**Puneet Kumar Singh;** Enzyme Technology and Protein Bioinformatics Laboratory, Department of Microbiology, Maharshi Dayanand University, Rohtak-124001, India.

**Rajesh Kumar;** Biological Product Lab, Department of Botany, University of Allahabad, Allahabad : 211002, India.

**Rajesh Singh Tomar;** Amity Institute of Biotechnology, Amity University Madhya Pradesh, Gwalior -474005, India.

**Rajib Bandopadhyay;** Department of Bio-Engineering, Birla Institute of Technology, Mesra, Ranchi-835215, India.

**Rakesh Kumar;** Department of Chemistry, Amity School of Applied Sciences, Amity University Haryana, Gurgaon (Manesar)-122413, India.

**Ravindra Bansal;** Nuclear Agriculture and Biotechnology Division, Bhabha Atomic Research Centre, Trombay, Mumbai- 400085, India.

**Rekha Kansal;** NRC on Plant Biotechnology, IARI, New Delhi-110012, India.

**Ruby Bansal;** Crosslay Wellness Program, Pushpanjali Crosslay Hospital, Ghaziabad-201012, India.

**Sangeeta Kumari;** Amity Institute of Biotechnology, Amity University Haryana, Gurgaon (Manesar)-122413, India.

**Seema R. Pathak;** Department of Chemistry, Amity School of Applied Sciences, Amity University Haryana, Gurgaon (Manesar)-122413, India.

**Shaili Srivastava;** Amity School of Earth and Environmental Sciences, Amity University Haryana, Gurgaon (Manesar)-122413, India.

**Shashi Kant Shukla;** Biological Product Lab, Department of Botany, University of Allahabad, Allahabad-211002, India.

**Shuchi Kaushik;** Amity Institute of Biotechnology, Amity University Madhya Pradesh, Gwalior -474005, India.

**Shweta;** Department of Biosciences, Jamia Millia Islamia, New Delhi- 110025, India.

**Simpi Mehta;** Department of Chemistry, Amity School of Applied Sciences, Amity University Haryana, Gurgaon (Manesar)-122413, India.

**Somesh Gupta;** Department of Dermatology, All India Institute of Medical Sciences, New Delhi-110029, India.

**Sujata Mohanty;** Centre of Excellence for Stem Cell Research, All India Institute of Medical Sciences, New Delhi-110029, India.

**Sumistha Das;** Amity Institute of Biotechnology, Amity University Haryana, Gurgaon (Manesar)-122413, India.

**Supriya Chakraborty;** Molecular Virology Laboratory, School of Life Sciences, Jawaharlal Nehru University, New Delhi 110067, India

**Syed Akhtar Husain;** Department of Biosciences, Jamia Millia Islamia, New Delhi-110025, India.

**T. Satyanarayana;** Department of Microbiology, University of Delhi South Campus,New Delhi-11002, India.

**Veerendra Kumar Sharma;** Molecular Virology Laboratory, School of Life Sciences, Jawaharlal Nehru University, New Delhi 110067, India

**Vinod Singh;** Department of Microbiology, Barkatullah University, Bhopal-462026, India.

**Zahid Ali;** Department of Biotechnology, Jamia Hamdard, New Delhi- 110062, India.

*Section I*

# Genomics and Proteomics

2016, **Biotechnology: Progress and Applications** *Pages* **3–17**
*Editors:* **Saif Hameed and Zeeshan Fatima**
*Published by:* **DAYA PUBLISHING HOUSE, NEW DELHI**

# *Chapter 1*

# Proteomic Approaches in Microalgal Research: Challenges and Opportunities

***Chiranjib Banerjee*[1], *Harsh Kumar Agrawal*[2], *Puneet Kumar Singh*[c], *Rajib Bandopadhyay*[1,2] *and Pratyoosh Shukla**[1,3]**

[1]*Department of Bio-Engineering, Birla Institute of Technology, Mesra, Ranchi – 835 215, Jharkhand, India*
[2]*Department of Botany, The University of Burdwan, Golapbag, Bardhaman – 713 109, W.B., India*
[3]*Enzyme Technology and Protein Bioinformatics Laboratory, Department of Microbiology, Maharshi Dayanand University, Rohtak – 124 001, Haryana, India*

## ABSTRACT

Microalgae have a capacity to entrap $CO_2$ which increases their biomass through photosynthesis and are important for the value added product including biofuel. It is evident that proteins expressed in such cells play decisive role at molecular level and their investigation through proteomics approach gives a clear indication and directives for interlinking of genotypic and phenotypic properties of microalgae. Despite the fact that proteomic approaches are currently applied to numerous fields of biotechnology especially in plants and this appraisal

* Corresponding Author: E-mail: pratyoosh.shukla@gmail.com

compiles the up to date advances on such techniques. Furthermore, it will also emphasizes on modern techniques including shotgun proteomics which helps in identifying proteins in complex mixtures using a combination of high performance liquid chromatography (HPLC) combined with mass spectrometry (MS) tools including details of software and databases.

## Introduction

Over last decades, aquatic ecosystem has shown a diverse growth with an emerging production and supply of various useful products. The aquatic organisms have been found as an important source to fulfill the requirement from the delivered products. The sustainable production through aquatic organisms has enhanced the importance of using proteomic approach for aquatic research and application. Proteomics, the study of protein structure and function, has emerged as a powerful tool to deliver workflow of production of these products from aquatic organisms among which microalgae has been found to be a vital source. Proteins are the effectors of almost all biological research. Hence, its analysis is important and complex, however, the proteome of several organisms are still unexplored. Proteomics contribute in providing the molecular, physiological and taxonomic data to explore the unraveling facts about protein. Different methods have been developed through proteomic study to identify, characterize and compare large protein sequences. Some of the important techniques used for microalgae research using proteomic approach are Shotgun Proteomic Analysis, Identification of Protein using Mass Spectrophotometer, electrospray analysis, Stiochiometric analysis of proteome etc. Microalgae having a huge diversity and wide range provide a useful source to extract many important products. From a long time proteomics had devoted in the findings of micro algal proteome profile. A list of microalgae in which proteomic research had been done recently is shown in Table 1.1. The basic steps involved in proteomics approach are depicted in Figure 1.1.

There are numbers of microalgae species which increases the complexity of proteome as more in number, more data to analyze, more complex. Modification in proteome also increases the complex nature of the protein. Shotgun proteomics involves techniques used for the identification of single protein to generate a global protein profile within the mixture of protein. Identifying protein is a very challenging task which can be resolved using shotgun identification. A basic process for the identification should involve three steps: identification of complement protein, detection of modifications and comparison among stored database (Valcu and Kempenaers 2014). Individual bands of protein generated from two-dimensional gel electrophoresis, can be analyzed using Mass Spectrophotometer and then comparison can be done using various bioinformatics software and protein databases. A list of some software and databases are shown in Tables 1.2 and 1.3. Microalgae have been found as one of the important renewable sources for the production of various useful products such as, biofuel, methyl ester, nutrients, astaxanthin, exopolysaccarides, vegetable oils, etc. (Staats *et al.*, 2000; Fuentes *et al.*, 2000; Milne *et al.*, 1990; Sawayama *et al.*, 1999; Lorenz and Cysewski 2000; Chand *et al.*, 2000; Nagle and Lemke 1990; de la Noue and de Pauw 1988; Tapie and Bernard 1988). As microalgae belong to a

**Table 1.1: List of Proteomic Approaches Adopted in Microalgae in Recent Years**

| Sl.No. | Algae Used | Field Studied | Sequencing Identification Method | Reference |
|---|---|---|---|---|
| 1. | *Chlamydomonas reinhardtii* | Expression and translation profile | LC-triple quadrupole-MS | Recuenco-Munoz ***et al.***, 2015 |
| 2. | *Synechocystis* sp. | Acid tolerance | iTRAQ-LC-MS/MS | Ren ***et al.***, 2014 |
| 3. | *Tisochrysis lutea* | Lipid accumulation | LC–MS/MS | Garnier ***et al.***, 2014 |
| 4. | *Ostreococcus tauri* | Protein response to daylight to night transition | Capillary-HPLC-MS/MS | Le Bihan ***et al.***, 2011 |
| 5. | *Chlorella vulgaris* | Examination of Triacylglycerol Biosynthetic Pathways | LC-MS/MS | Guarnieri ***et al.***, 2011 |
| 6. | *Haematococcus lacustris* | Astaxanthin accumulation | MALDI-TOF-MS | Tran ***et al.***, 2009 |
| 7. | *Haematococcus pluvialis* | Molecular defense mechanisms occur in response to oxidative stress | MALDI–TOF-MS | Wang ***et al.***, 2004 |
| 8. | *Stramenopiles* | Cell motility, signal transduction and other metabolic activities | LC-MS/MS | Fu ***et al.***, 2014 |
| 9. | *Chlamydomonas reinhardtii* | Identification of novel mitochondrial protein | Microsequencer | van Lis ***et al.***, 2003 |
| 10. | *Chlamydomonas reinhardtii* | Down regulation of heat shock proteins | LC-MS | Mahong ***et al.***, 2012 |
| 11. | *Aureoumbra lagunensis* | Identification of differentially expressed proteins under N and P deficiency. | QqTOF-MS/MS | Sun ***et al.***, 2012 |
| 12. | *Synechocystis* sp. *PCC 6803* | Express ability of different proteins in thermotolerance | 4-plex iTRAQ | Rowland ***et al.***, 2010 |
| 13. | *Bigelowiella natans* | characterize the diversity of nucleus-encoded proteins | HPLC-tandem MS | Hopkins ***et al.***, 2012 |
| 14. | *Gracilaria changii* | Identification of pigment proteins, metabolic enzymes, ion transporter etc. | 2-DE-MS | Wong ***et al.***, 2006 |
| 15. | *Prorocentrum micans* | Identification of proteins in different trophic modes | MALDI-TOF-MS | Shim ***et al.***, 2011 |
| 16. | *Dunaliella salina* | Plasma membrane protein analysis with respect to changes in salt concentration | 2-DE- LC-MS/MS | Katz ***et al.***, 2007 |
| 17. | *Phaeodactylum tricornutum* | Enzymes involve in carbon flux towards maximization of lipid accumulation | 2-DE-MS | Yang ***et al.***, 2014 |

***Note***: 2DE: Two-Dimensional gel electrophoresis; MS: Mass spectrometry; LC: Liquid chromatography; HPLC: High-performance liquid chromatography; MALDI-TOF: Matrix-assisted laser desorption/ionization-Time of Flight; iTRAQ: Isobaric Tag for Relative and Absolute Quantitation; Qq-TOF: Quadrupole/Time-of-Flight.

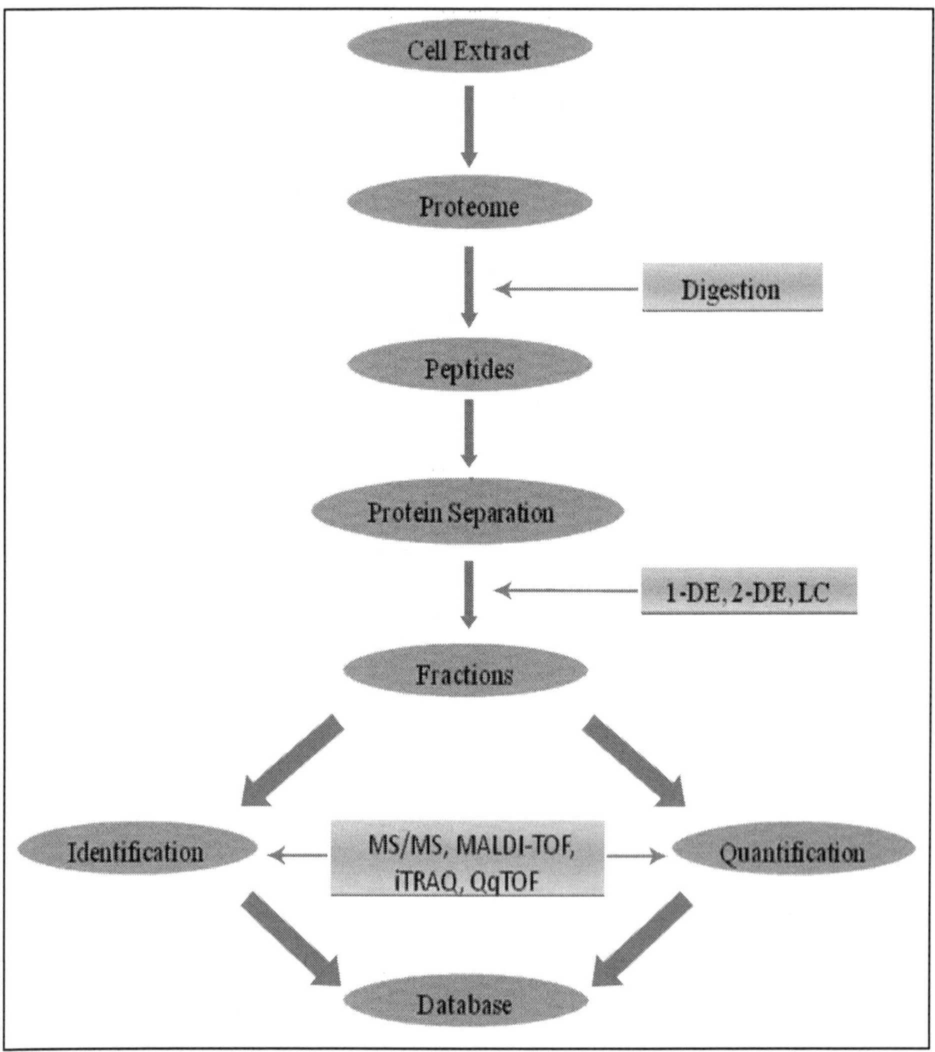


**Figure 1.1: Schematic Diagram Showing Steps Involved in Proteomics Approach.**

diverse range of eukaryotic organisms its proteome is much more complex and its analysis can be done easily by Shotgun Proteomic approach. Some of the advancements done in microalgae using shotgun proteomic are sequencing of the whole genome of *Chlorella protothecoides* sp. 0710 using shotgun approach to find out the oil accumulation mechanisms in *Chlorella protothecoides* sp. 0710 (Gao *et al.*, 2014), D/H fractionation during protein biosynthesis in *Leptospirillum* spp. was done by shotgun proteomics (Fischer *et al.*, 2013), identification of protein involved in the cell signaling and cell cycle regulation in *Chlorella vulgaris* (Guarnieri *et al.*, 2013), study of the metaproteome of the GF/F-preltered fraction of a microbial community from Shantou coast summer surface waters (Dong *et al.*, 2014), profiling of molecular process involved in the cell cycle regulation at proteomic level in *Prorocentrum donghaiense* (Wang *et al.*, 2013), a nano-scale shotgun proteomic approach to determine the changes in the composition of Proteome of *Emiliania huxleyi* (McKew *et al.*, 2013), a shotgun proteomic approach to monitor proteome in a heat stress dynamics in *Chlamydomonas reinhardtii* (Muhlhaus *et al.*, 2011), global proteomic study in *ostreococcus tauri* (Le Bihan *et al.*, 2011) etc. Hence, a shotgun proteomic approach provides a good platform to explore the vast use of microalgae.

**Table 1.2: List of Software Used for Proteomics Study**

| *Sl.No.* | *Softwares* | *Links* |
|---|---|---|
| 1. | Mascot | http://www.matrixscience.com/server.html |
| 2. | SEQUEST | http://fields.scripps.edu/sequest/ |
| 3. | X!Tandem | http://www.thegpm.org/tandem/ |
| 4. | SIMS | http://emililab.med.utoronto.ca/ |
| 5. | SIM Tandem | http://siret.ms.mff.cuni.cz/novak/simtandem/ |
| 6. | Maxquant | http://www.maxquant.org/ |
| 7. | MasschroQ | http://pappso.inra.fr/bioinfo/masschroq/ |
| 8. | PIQMIe | http://piqmie.semiqprot-emc.cloudlet.sara.nl/ |
| 9. | Proteowizard | http://proteowizard.sourceforge.net/ |
| 10. | PEAKS | http://www.bioinfor.com/ |
| 11. | Spectrolyzer | http://spectrolyzer.com/ |
| 12. | Proteomatic | http://www.proteomatic.org/ |
| 13. | ProteolQ | http://www.premierbiosoft.com/protein_quantification_software/index.html |

**Table 1.3: List of Protein Databases**

| *Sl.No.* | *Protein Databases* | *Links* |
|---|---|---|
| 1. | NCBI | http://www.ncbi.nlm.nih.gov/ |
| 2. | UniProt | http://www.uniprot.org/ |
| 3. | Protein Information Resource | http://pir.georgetown.edu/ |
| 4. | PDB | http://www.rcsb.org/pdb/home/home.do |
| 5. | PRIDE | http://www.ebi.ac.uk/pride/archive/ |
| 6. | MitoMiner | http://mitominer.mrc-mbu.cam.ac.uk/release-3.1/begin.do |
| 7. | Protein Data Bank in Europe | http://www.ebi.ac.uk/pdbe/ |
| 8. | GelMap | https://gelmap.de/ |

## Proteomics and Mass Spectrometry (MS)

Proteomics involves the complete identification of protein from the genome which is challenging because as changes occur at phenotypic level, there is respective change at genomic level. These phenotypic changes may occur due to evolution of species. Protein modifications involve pre and post translational modifications, due to which the metabolism changes causing a change in proteome. As, proteome comprises of a large number of genes its analysis is based on genomic level. There is a transition of techniques involved in proteomics from findings phase to securing phase. Finding phase involves techniques to explore the proteome while securing phase identifies the genes involved. At present, proteomic study with MS techniques has become the most powerful tool to analyze even the proteome of higher organisms (Kuster *et al.*, 2005). According to Zhang *et al.*, 2014 proteomic approach coupled with MS had provided information about human proteome. Numbers of MS based proteomic techniques had been developed so far and experiments are still going on (Aebersold and Mann 2003; Pereira *et al.*, 2014; Wang *et al.*, 2014; Zhang *et al.*, 2013). These techniques contribute to develop platform for the quantitative and qualitative analysis of proteome.

In MS based proteomic approach the protein is first digested with some enzymes as a result of which peptides are generated which are then separated by chromatographic techniques. The separated fractions thus obtained are subjected to MS for further analysis. However there are several limitations of these techniques as thousands of peptides are generated due to which fragmentation may not occur properly, it is difficult to separate very similar profile of peptides when there is a huge number of peptides and selection of peptide is very much important which is a difficult task. But, difficult doesn't mean impossible, despite of these limitations proteomic approach with MS analysis is a breakthrough in proteomic research.

An additional technique for the quantitative analysis of complex protein mixtures is the use of Isotope-Coded Affinity Tags (ICAT) with Tandem MS (Gygi *et al.*, 1999). Several microalgae research have been done in analyzing algal proteome using proteomic approach with MS. Global changes in protein expression of *Synechocystis* sp. PCC6803 was studied using MS based proteomic to find out the changes in protein expression under stress conditions (Hong *et al.*, 2014), a comparative proteomic study with MS was done for the find out the protein functioning of thylakoids in various microalgae (Tomizioli *et al.*, 2014), screening of different microalgae species (*Chlorella protothecoides, Haematococcus pluvialis* etc.) for biofuel and bioactive products was done by Liu *et al.*, 2014 using proteomic level two-dimensional gel electrophoresis and mass spectrometry, complete isotope labeling of blue green algae (*Spirulina platensis*) was done to utilize it as dietary nitrogen source for mice (Price and Ghaemmaghami 2014), the wild type strain of *Tisochrysis lutea* was studied for the lipid accumulation at the proteomic level followed by identification of proteins through MS (Garnier *et al.*, 2014), a proteomic data pipeline was adopted to combine multiple proteomic search engines results for information about biofuel production from microalgae *Chlorella vulgaris* along with identification through MS (Longworth J 2013) LC-MS/MS based proteomic analysis was done to study response of *Chlamydomonas reinhardtii* to ROS, light and oxygen (Barth *et al.*, 2014). In addition, these techniques are worldwide accepted for proteomic research in microalgae leading to explore the nature of microalgae species.

## Proteomics: A Way Towards Biofuel

Increase in the consumption of energy from fossil fuels has created a chaos among the scientists to save the sources left for energy production. Various petroleum, diesel, coal mines are being shut down due to the failure of providing sufficient production which is a very important issue for today and for the future. Looking to the solution to this problem comes, biofuel. Biofuels are the most prominent option to protect the left fossil fuels for our future for which alga culture is the leading field involved in the research of producing renewable bioenergy carriers and designing new technologies for better production. Several microalgae species are being monitored for producing biofuel. Algae can grow under diverse conditions easily while other autotrophs cannot. Algal biomass is produced by the mechanism of photosynthesis, due to which solar energy is converted into biomass. (Beer *et al.*, 2009). Proteomic approach can be done to understand the metabolic pathways involved in the synthesis. According to Briens *et al.*, 2008 algal biomass can provide 25 per cent of the total energy requirement. Lipid accumulation and the proteins expressed can be understood using different proteomic approaches. The rate of lipid synthesis drives the accumulation of oil in the starved cells, which generally varies as the climatic stress conditions change. On harvesting microalgae for the production of biofuel long chain fatty acids are generated such as FAMEs (Fatty Acid Methyl Esters) and FAEE (Fatty Acid Ethyl Esters) (Meng *et al.*, 2009). For better and economic production of biofuel from microalgae the mechanism involved in lipid accumulation and other important metabolisms are to be monitored for which Omics approach is one of the most effective methods.

Strategies had been developed in designing various methods and techniques using proteomic approach for the monitoring the pathways involved in biofuel production from microalgae. A proteomic approach was developed by Yu *et al.*, 2011 which identified a major lipid droplet protein that affects the size of lipid droplets from different microalgae species. In *Chlamydomonas reinhardtii*, a proteomics based methodology was adopted to perform profiling of oil bodies to identify the proteins involved in lipids metabolism when grown under nitrogen deprivation (Nguyen *et al.*, 2011).

## Tools for Proteomic Analysis

Biological systems involve various cellular processes which includes other small molecules which are regulated continuously in different metabolic pathways. The proteome is the core of the entire metabolism taking place which creates a link between the genome and the phenotype. Varying environmental and stress conditions causes a variation in the proteome, so it is very much important to do the genomic level analysis including the proteome study to learn about the in depth of these pathways. The previous conventional methods for proteomic approach were two-dimensional gel electrophoresis, by which the analysis of only the most abundant protein can be done (Kumar and Mann, 2009). For understanding the in depth knowledge of the proteome modifications, the gel electrophoresis method is not sufficient. To overcome these limitations, new experimental strategies were adopted which includes proteomic analysis using MS, different bioinformatics tools and other computational methods (Colinge *et al.*, 2007). Some of the important tools for proteomic analysis are entitled here:

- ☆ Proteomic approach through bioinformatics tools: As discussed earlier, for a proteomic analysis of the proteome, it is first digested with enzymes to result large number of peptides. These peptides can be bioinformatically analyzed to understand the physicochemical and other characteristics of the peptides. Some of the bioinformatics tools are Expasy (a bioinformatic resource portal, www.expasy.org), COBALT, CD-Search etc. Multiple reaction monitoring (MRM) of the peptides can be done to analyze the monitoring levels of the selected proteins (Lange *et al.*, 2008). A comparative proteogenomics discipline was adopted by Gupta *et al.*, 2008 to predict the genes involved in the expression of different proteins.
- ☆ Shotgun and Mass Spectrometry based proteomic tools: The Shotgun tools and MS based proteomic methods are discussed earlier. Detection and identification of a specific peptide in a mixture of hundreds and thousands of peptides is a tough task which can be made possible using MS instrumentations. SRM co-ordinates (Selected Reaction Monitoring) derived from SRM Atlas provides detection more easily. Further, the analysis can be subjected to nano scale triple quadruple MS techniques which provide a quantitative analysis of the selected proteins (Hood *et al.*, 2012).
- ☆ Use of *in vivo* molecular Imaging: To visualize the expressed protein in different tissues, protein-capture agents can be designed which provides the information about the complexes in the protein up to a single molecular level. This can be done using fluorescent recovery after photo bleaching (FRAP) (Bancaud *et al.*, 2010). The visualization of protein interactions and differentiation of expressed proteins can also be done through molecular imaging.
- ☆ De novo peptide sequencing: The results of MS based methods and the data generated are computational analysis and depends upon the availability of different databases in which the whole genome sequences of proteins must be reported. In case, there is no such database or the whole genome sequence of the specific protein is not available, chances of getting false match increases. This could be avoided by the use of De novo peptide sequencing, which provides the predication of the peptide sequence directly from the MS spectrum (Colinge *et al.*, 2007). Various sequence tags could be designed to get the sequence of the peptides which makes it easier for the identification of the specific protein.

## Opportunities

The emergence of new technologies in proteomics for the quantitative and qualitative analysis of the complete proteome has created opportunities for those who are seeking to gain in depth knowledge of the proteome. High throughput technologies in proteomics combined with other related techniques are extensively applied to identify smaller molecules responsible for the phenotype evolution. One of the most important issues now a day is human health, methods are being adopted to detect the insights of diseases and proteomics provides a way for discovery of biomarkers whose aim is to develop a simple differential test for clinical evaluation tools (Wasinger *et al.,* 2013). Over the last decades, these biomarkers have been significantly applied for developing tests to indicate cancer risk, detection of cancer, classification of tumors for the diagnosis of cancer (Mesri 2014). Numbers of results have been published that has explored the use of protein as cancer biomarkers (Hanash *et al.,* 2008). For the discovery of new biological markers, proteomic approaches are collaborated at nano level, which is including nanotechnology in proteomics, nanoproteomics. Nanoprotemics provides an analytical platform for discovery of new biomarkers which are highly sensitive, selective and provides the low cost identification of low abundance proteins along with low volume sample consumption (Agrawal *et al.,* 2013). Membrane proteins are involved in the transfer of fluids from cell to cell which are very much difficult to analyze as the results are confound. Proteomics enables the opportunities to develop extraction procedures of these membrane proteins so that the specific properties can be explored (Bertone and Snyder 2005).

New generation vaccines designing for cattle diseases are another opportunity in the field of proteomics (Marcelino *et al.,* 2012). Along with this, it has a wide application in drug discovery (Burbaum and Tobal, 2002) which includes the study of expression of proteins and changes in expression at different stages in diseases affected conditions so that an appropriate drug could be developed to diagnose various harmful diseases. The current toxicology studies lack measurement at molecular levels which is removed using proteomic based approaches (Titz *et al.,* 2014). The integral study of productive toxicology assessment could be done to understand the biology of toxicology in a simpler way. Apart from this, we all know that three fourth of earth's surface is surrounded by oceans; proteomics has also speeded legs in solving the problems in marine ecosystem. Dinoflagellates are the essential component of food chain in marine ecosystem and they also produce harmful algal blooms (Wang *et al.,* 2014). Proteomic study helps in understanding the physiological and metabolic characteristics of these dinofagellates so that the secrets behind their metabolism could be revealed.

## Challenges

Proteomics involves the study of the complete proteome, which includes expression of proteins, protein-protein interactions, pre and post translational modifications, molecular study of different complexes of protein, identification of genes etc. As a whole, the analysis is based on the genomic level and involves huge data interpretation of numbers of genes, peptides and proteins. The genomic level study of protein or proteomic techniques are not as easy as to eat served food on the table. It means there are lots of challenges to tackle when using proteomic approach. It is because proteomic study involves the throughout analysis of smaller molecules, genes. It was initially believed that proteomics is a global technique for the characterization of all the components in a proteome. But, the challenge arise from this instance only as genes expressed varies from cell to cell, organisms to organisms, it is difficult to design different protocols which involves functioning of different unknown genes. It is also seen that sometimes, the expressed gene is not related to the protein content (Rogers *et al.,* 2008). Apart from this, another challenge is that not all protein is expressed at similar levels of the proteorne. The most abundant protein will be large in number and they may be expressed at each and every level, which causes a problem in identification from MS if the protein of interest is less abundant may not be detected as they are less in number. Further, coming to the basic step of proteomics *i.e.* two-dimensional gel electrophoresis, it provides the separation of protein segments on the basis of isoelectric point and molecular point. The bands arising from electrophoresis could be a band of two or more protein segments whose molecular weight or isoelectric point is nearly equal. Further, proteomics has a wide application in designing biomarkers to develop clinical tools; the main challenge for designing a suitable biomarker is to identify the specific protein which is expressed during the time when diseases occurred, sometimes the modifications in expression due to foreign pathogen has also seen. Blue biotechnology comprises of aquatic environment is well known for the production of various useful products. MS based proteomic efforts in 16S sequencing to discover new organisms, in depth profiling of the complete proteome and metaproteome (Hartmann *et al.*, 2014). It also helps in comparing the wide variety of species present in the marine ecosystem.

## Acknowledgement

The Authors gratefully acknowledge the Senior Research Fellowship (SRF) to Mr. Chiranjib Banerjee from Council of Scientific and Industrial Research (CSIR), Government of India [Grant No. 09/554/(0036)/2013 EMR-I] and Junior Research Fellowship(JRF) to Mr. Puneet Kumar Singh, (Grant No. 42-457/2013(SR)), from University Grants Commission (UGC), New Delhi, India.

## References

1. Aebersold R and Mann M (2003). Mass spectrometry-based proteomics. *Nature* 422: 198-207.
2. Agrawal GK, Timperio AM, Zolla L *et al.* (2013). Biomarker discovery and applications for foods and beverages: Proteomics to nanoproteomics. *Journal of Proteomics* 93: 74-92.
3. Bancaud A, Huet S, Rabut G *et al.* (2010). Fluorescence perturbation techniques to study mobility and molecular dynamics of proteins in live cells: FRAP, photoactivation, photoconversion, and FLIP. Cold Spring Harbor Protocols DOI: 10.1101/pdb.top90.
4. Barth J, Bergner SV, Jaeger D *et al.* (2014). Reactive oxygen stress response of *Chlamydomonas Reinhardtii* dissected by quantitative mass spectrometry. *Molecular and Cellular Proteomics* 13: 969-989.
5. Beer LL, Boyd ES, Peters JW *et al.* (2009). Engineering algae for biohydrogen and biofuel production. *Current Opinion in Biotechnology* 20: 264-271.
6. Bertone P and Snyder M (2005). Prospects and Challenges in Proteomics. *Plant Physiology* 138: 560-562.
7. Briens C, Piskorz J, Berruti F (2008). Biomass valorization for fuel and chemicals production -a review. *International Journal of Chemical Reactor Engineering* 6: 1-49.

8. Burbaum J and Tobal GM (2002). Proteomics in drug discovery. *Current Opinion in Chemical Biology* 6: 427-433.

9. Chand RL, Kumar HD, Mohn FH *et al.* (2000). Services of algae to the environment. *Journal of Microbiology and Biotechnology* 10: 119-136.

10. Colinge J and Bennett KL (2007). Introduction to computational proteomics. *PLoS Computional Biology* 3(7): e114.

11. de la Noue J and de Pauw N. (1988). The potential of microalgal biotechnology: A review of production and uses of microalgae. *Biotechnology Advances* 6: 725-770.

12. Dong HP, Hong YG, Lu S *et al.* (2014). Metaproteomics reveals the major microbial players and their biogeochemical functions in a productive coastal system in the northern South China Sea. *Environmental Microbiology Reports* DOI: 10.1111/1758-2229.12188.

13. Fischer CR, Bowen BP, Pan C *et al.* (2013). Stable-isotope probing reveals that hydrogen isotope fractionation in proteins and lipids in a microbial community are different and species-specific. *American Chemical Society* 8: 1755-1763.

14. Fu G, Nagasatoa C, Okab S *et al.* (2014). Proteomics analysis of heterogeneous flagella in brown algae (*Stramenopiles*). *Protist* 165: 662-675.

15. Fuentes MMR, Fernandez GGA, Perez JAS *et al.* (2000). Biomass nutrient profiles of the microalga *Porphyridium cruentum*. *Food Chemistry* 70: 345-353.

16. Gao C, Wang Y, Shen Y *et al.* (2014). Oil accumulation mechanisms of the oleaginous microalga *Chlorella protothecoides* revealed through its genome, transcriptomes, and proteomes. *BMC Genomics* 15: 582.

17. Garnier M, Carrier G, Rogniaux H *et al.* (2014). Comparative proteomics reveals proteins impacted by nitrogen deprivation in wild-type and high lipid-accumulating mutant strains of *Tisochrysis lutea*. *Journal of Proteomics* 105: 107-120.

18. Guarnieri MT, Nag A, Smolinski SL *et al.* (2011). Examination of triacylglycerol biosynthetic pathways via de novo transcriptomic and proteomic analyses in an unsequenced microalga. *PLoS ONE* 6: e25851.

19. Guarnieri MT, Nag A, Yang S *et al.* (2013). Proteomic analysis of *Chlorella Vulgaris*: potential targets for enhanced lipid accumulation. *Journal of Proteomics* 93: 245-253.

20. Gupta N, Benhamida J, Bhargava V *et al.* (2008). Comparative proteogenomics: combining mass spectrometry and comparative genomics to analyze multiple genomes. *Genome Research* 18: 1133-1142.

21. Gygi SP, Rist B, Gerber SA *et al.* (1999). Quantitative analysis of complex protein mixtures using isotope-coded affinity tags. *Nature Biotechnology* 17: 994-999.

22. Hanash SM, Pitteri SJ, Faca VM (2008). Mining the plasma proteome for cancer biomarkers. *Nature* 452: 571-579.

23. Hartmann EM, Durighello E, Pible O *et al.* (2014). Proteomics meets blue biotechnology: A wealth of novelties and opportunities. *Marine Genomics* 17: 35-42.

24. Hong S, Kim HJ, Jang JH *et al.* (2014). Proteomic analysis of *Synechocystis* sp. $PCC68_{03}$ responses to low-temperature and high light conditions. *Biotechnology and Bioprocess Engineering* 19: 629-640.

25. Hood LE, Omenn GS, Moritz RL *et al.* (2012). New and improved proteomics technologies for understanding complex biological systems: Addressing a grand challenge in the life sciences. *Proteomics* 12: 2773-2783.

26. Hopkins JF, Spencer DF, Laboissiere S *et al.* (2012). Proteomics reveals plastid- and periplastid-targeted proteins in the chlorarachniophyte alga *Bigelowiella natans*. *Genome Biology and Evolution* 4: 1391-1406.

27. Katz A, Waridel P, Shevchenko A *et al.* (2007). Salt-induced changes in the plasma membrane proteome of the halotolerant alga *Dunaliella Salina* as revealed by blue native gel electrophoresis and nano-LC-MS/MS analysis. *Molecular and Cellular Proteomics* 6: 1459-1472.

28. Kumar C and Mann M (2009). Bioinformatics analysis of mass spectrometry-based proteomics data sets. *FEBS Letters* 583: 1703-1712.

29. Kuster B, Schirle M, Mallick P *et al.* (2005). Scoring proteomes with proteotypic peptide probes. *Nature Reviews Molecular Cell Biology* 6: 577-583.

30. Lange V, Picotti P, Domon B *et al.* (2008). Selected reaction monitoring for quantitative proteomics: a tutorial. *Molecular System Biology* 4: 222.

31. Le Bihan T, Martina SF, Chirnsidea ES *et al.* (2011). Shotgun proteomic analysis of the unicellular alga *Ostreococcus tauri*. *Journal of Proteomics* 74: 2060-2070.

32. Liu J, Li Q, Liu Q *et al.* (2014). Screening of unicellular microalgae for biofuels and bioactive products and development of a pilot platform. *Algological Studies* 145: 99-117.

33. Longworth J (2013). Proteomics in microalgae: a postgenomic approach for improved biofuel production. *Ph.D. Thesis*, University of Sheffield.

34. Lorenz RT and Cysewski GR (2000). Commercial potential for *Haematococcus* microalgae as a natural source of astaxanthin. *Trends in Biotechnology* 18: 160-167.

35. Mahong B, Roytrakul S, Phaonaklop N *et al.* (2012). Proteomic analysis of a model unicellular green alga, *Chlamydomonas reinhardtii*, during short-term exposure to irradiance stress reveals significant down regulation of several heat-shock proteins. *Planta* 235: 499-511.

36. Marcelino I, de Almeida AM, Ventosa M *et al.* (2012). Tick-borne diseases in cattle: Applications of proteomics to develop new generation vaccines. *Journal of Proteomics* 75: 4232-4250.

37. McKew BA, Lefebvre SC, Achterberg EP *et al.* (2013). Plasticity in the proteome of *Emiliania huxleyi* CCMP 1516 to extremes of light is highly targeted. *New Phytologist* 200: 61-73.

38. Meng X, Yang J, Xu X *et al.* (2009). Biodiesel production from oleaginous microorganisms. *Renewable Energy* 34: 1-5.

39. Mesri M (2014). Advances in proteomic technologies and its contribution to the field of cancer. *Advances in Medicine* DOI: 10.1155/2014/238045.

40. Milne TA, Evans RJ, Nagle N (1990). Catalytic conversion of microalgae and vegetable oils to premium gasoline, with shape-selective zeolites. *Biomass* 21: 219-232.

41. Muhlhaus T, Weiss J, Hemme D *et al.* (2011). Quantitative shotgun proteomics using a uniform $^{15}N$-labeled standard to monitor proteome dynamics in time course experiments reveals new insights into the heat stress response of *Chlamydomonas reinhardtii. Molecular and Cellular Proteomics* DOI: 10.1074/mcp.M110.004739.

42. Nagle N and Lemke P (1990). Production of methyl ester fuel from microalgae. *Applied Biochemistry and Biotechnology* 25: 355-361.

43. Nguyen HM, Baudet M, Cuine S *et al.* (2011). Proteomic profiling of oil bodies isolated from the unicellular green microalga *Chlamydomonas reinhardtii*: With focus on proteins involved in lipid metabolism. *Proteomics* 11: 4266-4273.

44. Pereira M, Bartolome CM, Sanchez-Fortun S (2014). Photosynthetic activity and protein overexpression found in Cr(III)-tolerant cells of the green algae *Dictyosphaerium chlorelloides. Chemosphere* 108: 274-280.

45. Price JC and Ghaemmaghami S (2014). Analysis of proteome dynamics in mice by isotopic labeling. *Methods in Molecular Biology* 1156: 111-131.

46. Recuenco-Munoz L, Offre P, Valledor L *et al.* (2015). Targeted quantitative analysis of a diurnal RuBisCO subunit expression and translation profile in *Chlamydomonas reinhardtii* introducing a novel Mass Western approach. *Journal of Proteomics* 113: 143-153.

47. Ren Q, Shi M, Chen L *et al.* (2014). Integrated proteomic and metabolomic characterization of a novel two-component response regulator Slr1909 involved in acid tolerance in *Synechocystis* sp. PCC 6803. *Journal of Proteomics* 109: 76-89.

48. Rogers S, Girolami M, Kolch W *et al.* (2008). Investigating the correspondence between transcriptomic and proteomic expression profiles using coupled cluster models. *Bioinformatics* 24: 2894-2900.

49. Rowland JG, Simon WJ, Nishiyama Y *et al.* (2010). Differential proteomic analysis using iTRAQ reveals changes in thylakoids associated with Photosystem II-acquired thermotolerance in *Synechocystis* sp. PCC 6803. *Proteomics* 10: 1917-1929.

50. Sawayama S, Minowa T, Yokoyama SY (1999). Possibility of renewable energy production and $CO_2$ mitigation by thermochemical liquefaction of microalgae. *Biomass and Bioenergy* 17: 33-39.

51. Shim J, Klochkova TA, Han JW *et al.* (2011). Comparative proteomics of the mixotrophic dinoflagellate *Prorocentrum micans* growing in different trophic modes. *Algae* 26: 87-96.

52. Staats N, Stal LJ, Mur LR (2000). Exopolysaccharide production by the epipelic diatom *Cylindrotheca closterium*: effects of nutrient conditions. Journal of *Experimental Marine Biology and Ecology* 249: 13-27.

53. Sun MM, Sun J, Qiu JW *et al.* (2012). Characterization of the proteomic profiles of the brown tide alga *Aureoumbra Lagunensis* under phosphate- and nitrogen-limiting conditions and of its phosphate limitation-specific protein with alkaline phosphatase activity. *Applied and Environmental Microbiology* 78: 2025-2033.

54. Tapie P and Bernard A (1988). Microalgae production: Technical and economic evaluations. *Biotechnology and Bioengineering* 32: 873-885.

55. Titz B, Elamin A, Martin F *et al.* (2014). Proteomics for systems toxicology. *Computational and Structural Biotechnology Journal* 11: 73-90.

56. Tomizioli M, Lazar C, Brugiere S *et al.* (2014). Deciphering thylakoid sub-compartments using a mass spectrometry-based approach. *Molecular and Cellular Proteomics* 13: 2147-2167.

57. Tran NP, Park JK, Hong SJ *et al.* (2009). Proteomics of proteins associated with astaxanthin accumulation in the green algae *Haematococcus lacustris* under the influence of sodium orthovanadate. *Biotechnology Letters* 31: 1917-1922.

58. Valcu CM and Kempenaers B (2014). Proteomics in behavioral ecology. *Behaviour Ecology* 00: 1-15.

59. van Lis R, Atteia A, Mendoza-Hernandez G, *et al.* (2003). Identification of novel mitochondrial protein components of *Chlamydomonas reinhardtii.* A proteomic approach. *Plant Physiology* 132: 318-330.

60. Wang DZ, Zhang H, Zhang Y *et al.* (2014). Marine dinoflagellate proteomics: Current status and future perspectives. *Journal of Proteomics* 105: 121-132.

61. Wang DZ, Zhang YJ, Zhang SF *et al.* (2013). Quantitative proteomic analysis of cell cycle of the dinoflagellate *Prorocentrum Donghaiense* (Dinophyceae). *PLoS ONE* 8: e63659.

62. Wang SB, Chen F, Sommerfeld M *et al.* (2004). Proteomic analysis of molecular response to oxidative stress by the green alga *Haematococcus pluvialis* (Chlorophyceae). *Planta* 220: 17-29.

63. Wasinger VC, Zeng M, Yau Y (2013). Current status and advances in quantitative proteomic mass spectrometry. *International Journal of Proteomics* DOI: 10.1155/2013/180605.

64. Wong PF, Tan LJ, Nawi H *et al.* (2006). Proteomics of The Red Alga, *Gracilaria Changii (Gracilariales, Rhodophyta)*. *Journal of Phycology* 42: 113-120.

65. Yang ZK, Ma YH, Zheng JW *et al.* (2014). Proteomics to reveal metabolic network shifts towards lipid accumulation following nitrogen deprivation in the diatom *Phaeodactylum tricornutum*. *Journal of Applied Phycology* 26: 73-82.

66. Yu WL, Ansari W, Schoepp NG *et al.* (2011). Modifications of the metabolic pathways of lipid and triacylglycerol production in microalgae. *Microbial Cell Factories* 10: 91.

67. Zhang W, Tan NG, Li SF (2013). NMR-based metabolomics and LCMS/MS quantification reveal metalspecific tolerance and redox homeostasis in *Chlorella vulgaris*. *Molecular Biosystems* 10: 149-160.

68. Zhang W, Wei Y, Ignatchenko V *et al.* (2014). Proteomic profiles of human lung adeno and squamous cell carcinoma using super-SILAC and label-free quantification approaches. *Proteomics* 14: 795-803.

**2016, Biotechnology: Progress and Applications** *Pages* **19–34**
*Editors:* **Saif Hameed and Zeeshan Fatima**
*Published by:* **DAYA PUBLISHING HOUSE, NEW DELHI**

## *Chapter 2*

# Retrieval of Xylanase Genes from Environmental Metagenomes by Metagenomic Approaches

***Digvijay Verma and T. Satyanarayana****

*Department of Microbiology, University of Delhi South Campus, Benito Juarez Road, New Delhi – 110 021, India*

## ABSTRACT

The thermo-alkali-stable xylanases are in high demand for pulp bleaching in paper industry and generating lower xylooligosaccharides by hydrolyzing xylan component of agro-residues. The community DNA from hot environmental samples could be a rich source of novel biocatalysts. The compost-soil samples are rich in microbes with high possibilities for retrieving thermostable enzyme encoding genes. While screening metagenomic library constructed from the compost-soil metagenome in the p18GFP vector, a clone (TSDV-MX1) was detected that exhibited clear zone of xylan hydrolysis on RBB xylan plate. The sequencing of 6.321 kb DNA insert and its BLAST analysis detected the presence of xylanase encoding gene that comprised 1077 bp. The deduced protein sequence (358 amino acids) displayed homology with glycosyl hydrolase (GH) family 11 xylanases. The gene was subcloned into pET28a vector and expressed in *E. coli* BL21(DE3). The recombinant xylanase (rMxyl) exhibited activity over a broad range of pH and temperature with optima at pH 9.0 and 80 °C. The recombinant xylanase is highly thermostable having $T_{1/2}$ of 2 h at 80 °C and 15 min at 90 °C. This is the first report on the retrieval of xylanase gene by metagenomic approach that encodes

* Corresponding Author: E-mail: tsnarayana@gmail.com

an enzyme with alkalistability and thermostability. The recombinant xylanase has potential applications in paper and pulp industry in pulp bleaching and generating lower xylooligosaccharides from the abundantly available agro-residues.

***Keywords:*** *Thermo-alkalistable xylanase, Metagenomics, Biobleaching, Saccharification.*

## Introduction

Hemicellulose represents the second most abundant renewable polymer of plant cell walls after cellulose, and xylan is the main constituent in the hemicellulose of component of agro-residues. Xylan is composed of 1,4 β-D xylopyranosyl residues (Figure 2.1). Various attached groups (acetyl, arabinosyl and glucuronosyl) of the backbone of xylan make it a heteropolysaccharide (Whistler and Richards, 1970). While homoxylans exclusively constitute xylosyl residues and chiefly present in esparto grass (Chanda *et al.*, 1950) and tobacco stalks (Eda *et al.*, 1976). These heterogeneous polysaccharides play a critical role in maintaining the cell wall integrity by making covalent and non-covalent bonds with cellulosic fibres and lignins (Coughlan and Hazlewood, 1993). The heterogeneity of xylans necessitates multiplicity in the xylanolytic enzyme system that comprises endo-xylanase (1,4-b-D-xylan xylanohydrolase; EC 3.2.1.8), β-xylosidase (1,4 β-D-xylan xylohydrolase; EC 3.2.1.37) α-glucuronidase, α-L-arabinofuranosidase and acetyl xylan esterase. The CAZY database (http://www.cazy.org/fam/acc_GH.html) categorized xylanases into six glycosyl hydrolase families GH5, GH8, GH10, GH11, GH30 and GH43 (Collins *et al.*, 2005; Paes *et al.*, 2012). Among these, family 10 and 11 xylanases are widely distributed where GH10 xylanases share versatile substrate specificity with higher molecular weight, while GH11 xylanases are more stringent in their substrate specificity with low molecular weight and are considered as true xylanases (Henrissat, 1991; Verma and Satyanarayana, 2012a). The hydrolysis of xylans by xylanases is of interest to many industries like ramie fibre degumming, food processing, textile, biofuel, feed, and paper and pulp industries (Khandeparkar and Bhosle, 2007; Menon *et al.*, 2010; Woldesenbet *et al.*, 2012). The xylanases that withstand extreme conditions prevailing in the industrial processes are in high demand in paper pulp processing and feed industries. Although several xylanases have been reported from a number of microorganisms, most of them do not have adequate thermostability and alkalistability for their application in paper and pulp industries.

Majority of xylanases have been obtained from the culturable 0.1 to 1 per cent of the total microbial diversity existing in natural environments. The culture-independent metagenomic approaches permit retrieval of genes encoding useful enzymes from environmental samples without involving laborious and elaborate methods for the cultivation of microbes. The immense demand for alkalistable and thermostable xylanases encouraged us to adapt this innovative strategy for retrieving genes that encode thermo-alkali-stable xylanases. A metagenomic library was constructed and screened for clones having xylanase activity. Xylanase encoding gene (*Mxyl*) was sub-cloned and expressed, and the recombinant xylanase was

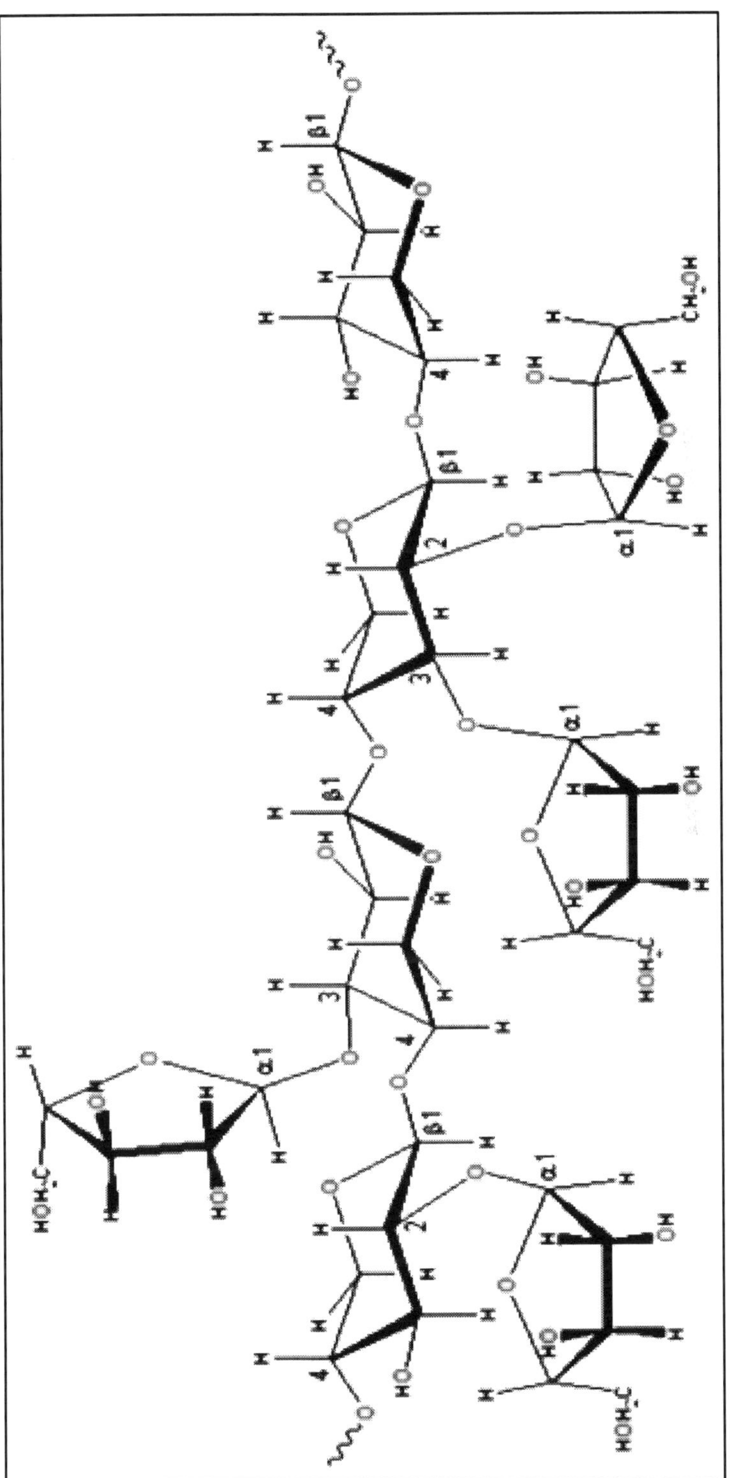

**Figure 2.1: Structure of Heteroploysaccharide Form of Xylan.**

purified and characterized. This is the first report on retrieving thermo-alkali-stable xylanase by a metagenomics approach (Verma *et al.*, 2013).

## Methods for Retrieving Xylanase Encoding Genes from Environmental Metagenomes

Several methods have been developed for isolating humic acid free metagenomic DNA from environmental samples (Rajendran and Gunasekran, 2007). The majority of the methods, however, need extraction and purification steps in order to obtain high quality DNA. Humic acid free DNA could be extracted from compost-soil samples in a single step by the protocol developed by Verma and Satyanarayana (2011) that could be processed for constructing the plasmid based metagenomic library. Five µg of DNA was partially digested with Sau3AI to achieve the fragments of 3–12 kb. Hundred nanogram (ng) of gel extracted purified insert DNA fragments were ligated with 300 ng of BamHI digested and dephosphorylated p18 GFP vector by using T4 DNA ligase at 16 °C overnight followed by transformation into competent *E. coli* DH10B cells using heat shock method. The metagenomic library was spread and screened for xylanase activity on 0.3 per cent (w/v) RBB-xylan (4-O-methyl-D-glucurono-D-xylan-remazol brilliant blue R, procured from Sigma, USA) - LB-ampicillin agar plates. The transformants were grown at 37 °C overnight and observed for the zone of xylan hydrolysis. Raplica plating method frequently used for identifying xylanase encoding genes was followed by Congo red staining. This method is labour intensive and time consuming. Direct incorporation of RBB-xylan into the medium enhances the possibilities of selecting xylanase producing clones. Dye based substrates, therefore, are highly useful for screening whole metagenomic libraries. Several xylanases have been obtained using RBB-xylan as the substrate.

## Xylanases Genes Recovered by Metagenomic Approaches

DNA cloning is a powerful tool that empowers the workers to achieve the production of xylanases in different hosts, and also facilitates their over expression by coupling their structural genes with their efficient promoters. Despite the tremendous developments, there is a scarcity of recombinant microbial strains that produce xylanase with the desired properties for application in extreme industries. Although the enzymes can be stabilized conventionally by various strategies like introduction of clusters of aromatic-aromatic interactions such as salt bridges, helix capping and entropic stabilization (Matthews *et al.*, 1987; Marshall *et al.*, 2002; Puchkaev *et al.*, 2003), a long way to go in generating the enzymes with desirable properties. Industrially relevant xylanases often require stability at alkaline pH and elevated temperatures especially in pulp and paper industry. Extremozymes can withstand the extremes and serve as pre-bleaching agents for reducing hazardous chlorine requirement for pulp bleaching. Several reports are available where xylanase genes have been cloned and expressed in a suitable host to overproduce xylanases of interest. Cloning demands the availability of full length gene of xylanase which can be pulled out from xylanase producing microorganisms and environmental metagenomes (Raha *et al.*, 2006; Hu *et al.*, 2008) using PCR based approaches (Zhang *et al.*, 2007; Guo *et al.*, 2009). Several xylanases have been reported till date using metagenomics approaches. Gut microbiota of insects like Lepidoptera and Isoptera have been used for recovering several xylanases. Among all, four xylanases were extensively characterized; one of them belongs to GH10 family while three shared homology with GH8 family. All the xylanases showed maximum specificity towards arabinoxylan having optimum activity in acidic conditions (Bertram *et al.*, 2008). Similarly full-length xylanase encoding gene was obtained from vermiform appendix of insects and horses using pre-amplified PCR (PAI-PCR) approach (Yamada *et al.*, 2008). A cold active xylanase (xyn8) encoding gene with high molecular weight (45.9 kDa) was isolated from an environmental metagenomic library. However, on sequencing, it neither shared homology with family 10 or family11 xylanases rather it shared identity with GH8 family xylanase (Lee *et al.*, 2006). Another GH10 xylanase (XynH) recovered from a soil based metagenomic library also showed activity at lower temperature under slightly alkaline conditions and exhibited activity in the presence of majority of the divalent metal ions (Hu *et al.*, 2010). On the contrary, xylanase obtained from a manure library showed complete inhibition of activity in the presence of 10 mM $Cu^{2+}$ (Li *et al.*, 2009). Several xylanases belong to GH10/GH11 family have been successfully amplified using degenerate primers of GH10/GH11 xylanases from soil metagenome of Alpine Tundra soil that exhibited less identity with known sequences available in the Genbank (Wang *et al.*, 2010). A xylanase obtained from compost soil library exhibited optimum pH in the acidic range (5.5-7.0) [Kwon *et al.*, 2010]. As the compost soil is acidic in nature, enzymes obtained from such habitats will reveal their habitant characteristics. This myth has, however, been broken, when an alkaline xylanase was reported by metagenomics approach. This xylanase (Mxyl) exhibits optimally activity at pH 9.0 and 80 °C with $T_{1/2}$ of 120 min under optimal conditions (Verma *et al.*, 2013). This is the only report of a metagenomic xylanase that exhibits stability at higher pH and temperature. The translated xylanase gene (*Mxyl*) comprises 358 amino acids, of which 21 are basic and 16 acidic. The predicted molecular weight, pI and instability index of recombinant xylanase are, 40 kDa, 8.8 and 33.44 respectively. The xylanase contained a long leader of 43 amino acids at the N terminal region followed by a catalytic domain ($44^{th}$–$212^{th}$) of GH11 family interrupted by a stretch of non-catalytic region ($WSVRQ_2R_2TG_2TIT_2$) rich in threonine and arginine (Figure 2.2). Furthermore, serine rich Q linker region ($S_2GS_2DITVG_2TS_2G_2TS_2G_2S_3G_2S_{10}G_4$) has also been detected from amino acid 213 to 248 just after catalytic domain. Linker regions generally discriminate catalytic domain from the carbohydrate binding domain (Gilkes *et al.*, 1991; Li *et al.*, 2009; Brennan *et al.*, 2004). The linkers have also been proposed to function in stabilizing the protein that connects surface layer homology domains, thermo-stabilizing domains, and dockerin domains which play a role in stabilizing the protein (Winterhalter and Liebl, 1995; Notenboom *et al.*, 2002; Li *et al.*, 2009; Shibuya *et al.*, 2009). Hydrophobic cluster analysis followed by amino acid homology grouped this high molecular weight xylanase into GH11 family, which includes low molecular weight xylanases. Xylanases of *Clostridium stercorium* also had been classified as a GH11 xylanase irrespective of its high molecular weight (56 kDa) [Sakka *et al.*, 1993]. Lower homology, higher molecular weight (30 kDa), metagenomic origin and distinct characteristics, make this a novel xylanase. Despite several attempts, metagenomics has yet to deliver an ideal xylanase for pulp bleaching.

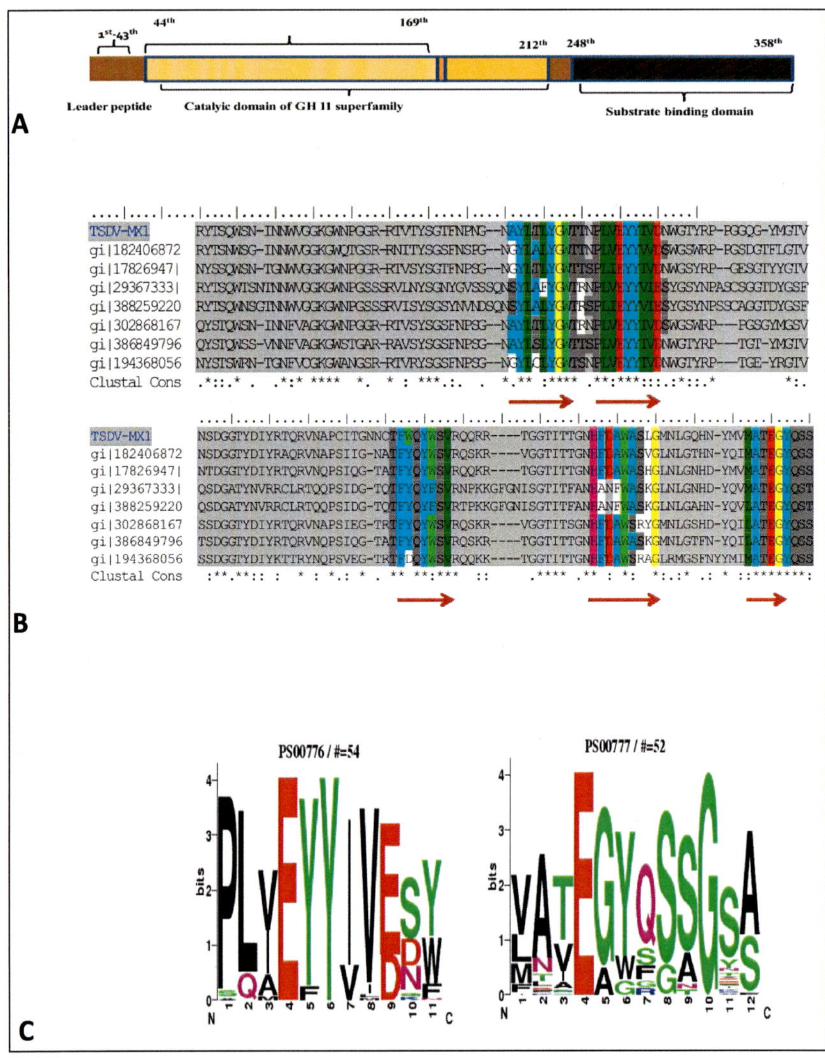


**Figure 2.2**

**A. Pictorial description of xylanase shows catalytically important region and substrate binding domain. B. Multiple sequence alignment of xylanase with other xylanases available in database. GenBank accession number and source of microorganisms were given as follows: 182406872 (glycosyl hydrolase family 11 precursor [uncultured bacterium]), 17826947 (*Pseudomonas* sp. ND137), 29367333 (uncultured *Cellvibrio* sp.), 388259220 (*Cellvibrio* sp. BR), 302868167 (*Micromonospora aurantiaca* ATCC 27029), 386849796 (*Actinoplanes* sp. SE50/110), 194368056 (*Streptomyces* sp. S27). Five signature sequences: I (AYLTLYGW) II (VEYYIVDN), III (FWQYWSV), IV (HFDAWASLG) and V (MATEGY) of GH11 family are coloured. The two catalytically important residues (Glu117 and Glu209) are marked with black circle. C. Logos retrieved from the prosite analysis. Signature sequences II (VEYYIVDN) and V (MATEGY) reveal that glutamate (E) is 100 per cent conserved residue at position 4 in both the respective logos.**

## Characteristics of Metagenomic Xylanases

The rMxyl showed optimum activity at elevated temperature at alkaline pH, which is similar to xylanases of *Geobacillus thermoleovorans, Dictyoglomus thermolacticum, Bacillus acidicaldarius, Thermotoga maritima* having optimal activity at or above 80 °C (Uchino and Fukuda, 1983; Mathrani and Ahring, 1992; Yoon *et al.*, 2004; Verma and Satyanarayana, 2012b). Xylanases of *Bacillus firmus, G. thermoleovorans* and *B. stearothermophilus* T-6 exhibited optimum activity at pH at 8.0 or above (Khasin *et al.*, 1993; Zhao *et al.*, 2011; Verma and Satyanarayana, 2012b). The shelf-life of rMxyl is more than 3 months at 4 °C, which retains greater than 90 per cent activity. The recombinant xylanase is optimally active at pH 9.0 and 80 °C that distinguishes it from the existing xylanases. The xylanase produced from *Thermotoga maritima* exhibits Topt of 90 °C, but it gets inactivated fast at pH 6.0 (Yoon *et al.*, 2004). The highly thermostable xylanase with optimum temperature in the range between 80 and 105 °C are available, but these xylanases show optimum activity either at acidic or neutral pH (Mathrani and Ahring, 1992; Yoon *et al.*, 2004). Similarly the stability of xylanases at high pH has been reported that are active at lower temperatures (Khasin *et al.*, 1993; Chang et. 2004). The recombinant GH10 family xylanase of *Bacillus halodurans* showed both the properties having optimum pH 9.0 and 75 °C, but it gets inactivated very fast at 80 °C and losses 50 per cent activity at 65 °C after 4 h (Mamo *et al.*, 2006). In contrast, Mxyl has good thermostability at 60, 70 and 80 °C with only 20–30 per cent loss after 3 h exposure. The most significant aspect of this investigation is obtaining a highly thermostable and alkalistable xylanase by metagenomic approach (Figure 2.3). The enzyme is stimulated by $Mg^{2+}$, $Sn^{2+}$ and $Fe^{2+}$, while drastic inhibition was recorded at even 1 mM of $Hg^{2+}$ and $Mn^{2+}$. Similar stimulation has been reported in xylanase of *B. subtilis* AMX-4 and an uncultured microbe (Yoon *et al.*, 2009; Zhao *et al.*, 2011). The inhibition of xylanase activity by $Hg^{2+}$ suggests the presence of tryptophan residue that is gets oxidized (Zhang *et al.*, 2007; Maalej *et al.*, 2009; Zhao *et al.*, 2011; Liu *et al.*, 2011). Xylanases from *Streptomyces* sp. S27, *Streptomyces olivaceoviridis* A1 and *Bacillus subtilis* strain R5 had also been reported to be stimulated by $Fe^{2+}$ and $Mg^{2+}$ (Wang *et al.*, 2007; Hu *et al.*, 2008; Li *et al.*, 2008; Jalal *et al.*, 2009; Li *et al.*, 2009), and complete loss of the enzyme in presence of $Hg^{2+}$ and $Mn^{2+}$ was reported in the xylanases from an uncultured microbe (Hu *et al.*, 2008; Li *et al.*, 2009), *Penicillium* sp. (Liu *et al.*, 2010) and *Plectosphaerella cucumerina* (Zhang *et al.*, 2007). The inhibition of xylanase activity by $Cu^{2+}$ is very similar to the majority of the xylanases (Li et. 2008; Son-Ng *et al.*, 2009; Zhao *et al.*, 2011; Matteotti *et al.*, 2012). A stimulatory effect EDTA was recorded on xylanase activity which suggests that Mxyl is not a metallozyme.

## Potential Applications of Xylanases

Saccharification of agroresidues using xylanase liberates sugars which are useful in producing products like lower xylooligosaccharides, ethanol, organic acids, SCP and hydrogen production (Barnard *et al.*, 2010; Lo, *et al.*, 2010; Li *et al.*, 2011; Shi *et al.*, 2013). The recombinant xylanase (rMxyl) hydrolyzed xylan of birchwood and beechwood efficiently due to their structural similarity (Guo *et al.*, 2009). The enzyme showed almost similar activities on xylan of arabinoxylan and oat spelt. Xylan from

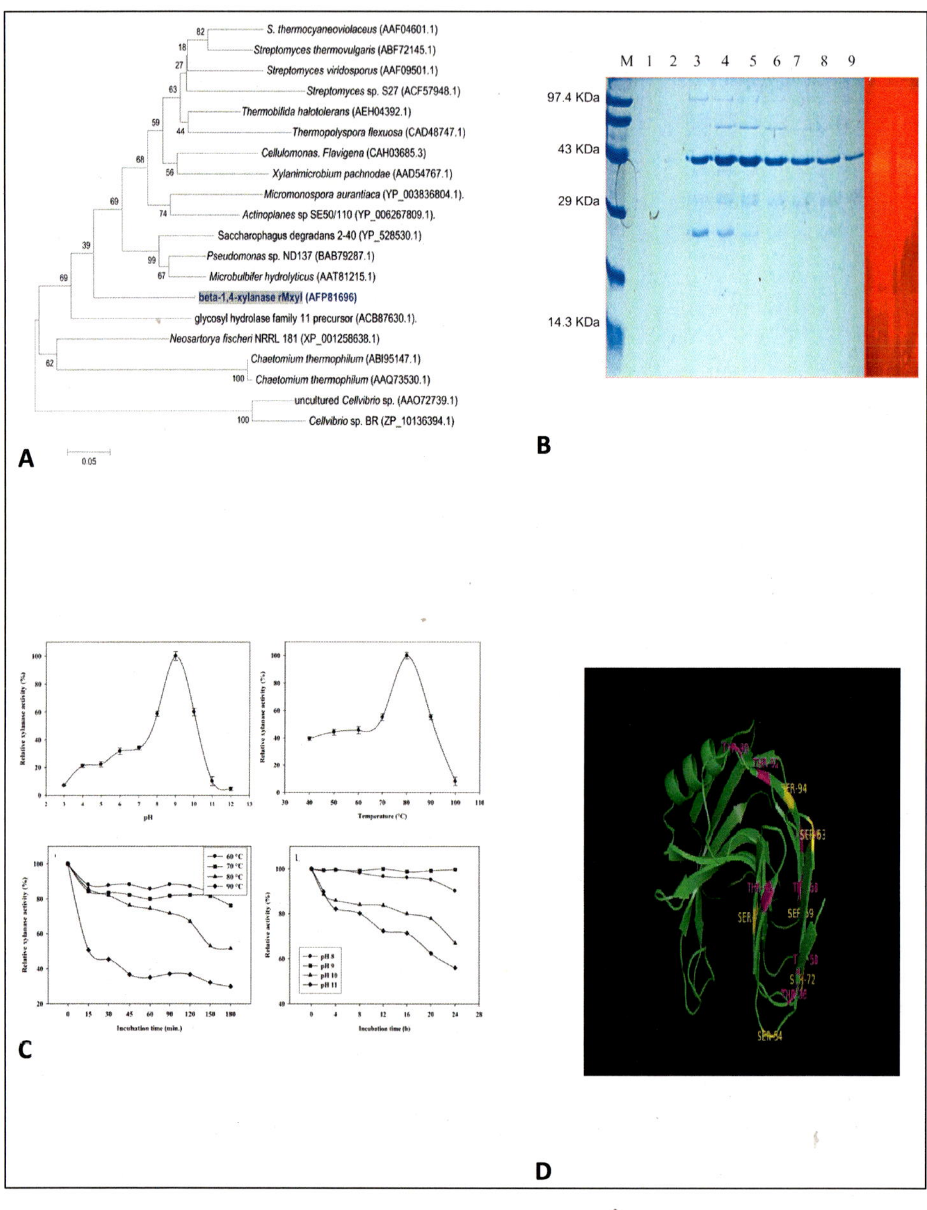
S. thermocyaneoviolaceus (AAF04601.1)
Streptomyces thermovulgaris (ABF72145.1)
Streptomyces viridosporus (AAF09501.1)
Streptomyces sp. S27 (ACF57948.1)
Thermobifida halotolerans (AEH04392.1)
Thermopolyspora flexuosa (CAD48747.1)
Cellulomonas. Flavigena (CAH03685.3)
Xylanimicrobium pachnodae (AAD54767.1)
Micromonospora aurantiaca (YP_003836804.1).
Actinoplanes sp SE50/110 (YP_006267809.1).
Saccharophagus degradans 2-40 (YP_528530.1)
Pseudomonas sp. ND137 (BAB79287.1)
Microbulbifer hydrolyticus (AAT81215.1)
beta-1,4-xylanase rMxyl (AFP81696)
glycosyl hydrolase family 11 precursor (ACB87630.1).
Neosartorya fischeri NRRL 181 (XP_001258638.1)
Chaetomium thermophilum (ABI95147.1)
Chaetomium thermophilum (AAQ73530.1)
uncultured Cellvibrio sp. (AAO72739.1)
Cellvibrio sp. BR (ZP_10136394.1)
0.05
A
M 1 2 3 4 5 6 7 8 9
97.4 KDa
43 KDa
29 KDa
14.3 KDa
B
C
D

**Figure 2.3**

**A. Phylogenetic tree of recombinant xylanase. Mxyl showed highest homology with xylanase of *Cellulomonas fimi* ATCC 484 followed by uncultured microbial GH11 xylanase. Neighbour Joining (NJ) Tree is constructed by using MEGA 4.0 software. Bootstrap values (n = 1000 replicates) are represented as percentage. The scale bar depicts the allowed changes per amino acid position. B. Analysis of rMxyl using SDS–PAGE (15 per cent polyacrylamide gel). Lane 1 protein marker, Lane 2 and 3 are washes with 20 and 30 mM imidazole. Recombinanat xylanase was eluted using different concentrations of imidazole (100, 200, 250, 300, 400, 450, 500 mM). Purified xylanase showed molecular mass of ~40 kDa on staining Coomassie Brilliant Blue R-250. Zymogram analysis of the purified xylanase using Congo red staining. C. Effect of pH and temperature on the activity and stability of rMxyl: The recombinant xylanase incubated in various buffers (pH 3.0–12.0) and temperatures (40–100 ϵC) and assayed for xylanase activity. Recombinant xylanase was incubated in glycine-NaOH buffer without substrate and kept at various temperatures. Aliquots were collected at various time interval and store at 0 ϵC for calculating residual activity. Similarly enzyme was incubated in various buffers (pH 8.0–11.0) and aliquots of different time intervals were used xylanase assays. D. Typical three dimensional structure of xylanase (rMxyl) showing ser/thr rich surface.**

oat spelt comprises short chain of arabinoxylan highly rich in arabinose (xylose/ arabinose = 66:34) (Gruppen *et al.*, 1992; Kormelink and Voragen, 1993). Interestingly, the rMxyl liberated lower xylooligosaccaharides within 5 min and this was sustained on prolonged incubation. Although several xylanases have been reported from various microorganisms for liberating xylooligosaccharides following xylan hydrolysis, this is the only report from metagenomic approach that shows applications in saccharification. Alkaline xylanases enhance the solubility of hemicellulosic materials and act efficiently on agro-residues by lowering the steric hindrance caused by cellulose (Gruppen *et al.*, 1992). Xylanases from *Bacillus halodurans* (Mamo *et al.*, 2007), *Thermomonospora fusca* (Sun *et al.*, 2007) and *G. thermoleovorans* (Verma and Satyanarayan, 2012b) have been reported to generate a similar profile of lower xylooligosaccharide. The recombinant rMxyl has a potential application in food industry for the production of xylooligosaccharides as prebiotics (Vazquez *et al.*, 2007). Another very significant application of xylanase is in paper and pulp industry. Vikarii *et al.* (1986) reported its utility as a bleaching agent. Xylanases hydrolyze β-1, 4 liked xylosyl residues and facilitates the release of xylan and lignin (polyphenolic compounds responsible for brown colour of paper) to make paper bright and white. In addition, xylanases enhance retention of water and pulp fibrillation (Viikari *et al.*, 1986). This way xylanases can trim down the chlorine consumption up to 25-30 per cent and make the process eco-friendly and cost-effective. Due to hot and alkaline nature of the pulp, industry demands thermo-alkalistable xylanases. There are several reports on the successful application of xylanases to reduce chlorine consumption (Ratto *et al.*, 1994; Azeri *et al.*, 2010). Low quality jute fibres, flex, hemp and ramie are rich in hemicellulosic content, and therefore, needs pre-treatment with xylanase to cope with the problems encountered during spinning in the mills. Thermostable xylanases remove hemicellulosic content like xylan without compromising the strength, and thus, soften them up to the required level (Patra *et al.*, 2010; Bhalla *et al.*, 2013).

Thermostable xylanases find applications in the food industry. Pre-treatment of cereals and millets using xylanases not only improve the quality of the processed food products but also reduce the processing time during fermentation. The treatment of wheat flour with thermostable xylanases changes the texture of loaf as well as significantly reduces the dough water absorption. Xylanases are also used in improving nutritional properties, clarifying fruit juices and wines, of silage and to achieve the desirable texture and loaf volume of bread in baking industry (Subramaniyam and Prema, 2002). The production of lower xylooligosaccharides (XOs) from agro-residues using xylanase has significantly enhanced the application of xylanases. There are several reports on liberation of XOs from agro-residues. Alkalistable and thermostable xylanase from *Bacillus halodurans*, *G. thermoleovorans*, and *Streptomyces thermocyaneoviolaceus* have successfully been employed in generating XOs from wheat bran as well as birchwood xylan (Shin *et al.*, 2009; Kumar and Satyanarayana, 2011; Verma and Satyanarayana, 2012a). Recently, an endo-xylanase (SCxyl) has been reported from sugarcane soil metagenome that finds application in the generation of XOs (Alvarez *et al.*, 2013).The lower XOs are useful as prebiotics which are fermented by *Bacterioides* and *Bifidobacterium* spp. to propionate and butyrate. Butyrate has been shown to have anti-tumour effect in colon cancer. The xylanase for feed industries should be thermostable, but alkalistability is not an issue.

## Conclusions

Undoubtedly metagenomics has emerged as a useful tool in recovering novel genes from those microorganisms that are not yet culturable. Xylanase, a promising industrial enzyme, has been recovered from various environmental metagenomes. The discovery of thermo-alkali-stable xylanase (Mxyl) is the first enzyme that has been reported to have thermostability as well as alkalistability. Most of the xylanases retrieved by culture-dependent and culture-independent approaches exhibit optimum activity in the pH and temperature range of 6.0–8.0 and 40–60 °C, respectively. A very low similarity in the amino acid sequence of Mxyl with other known xylanases makes it a novel xylanase. The possibility of obtaining thermo-alkali-stable xylanase encoding genes from composts may stimulate further search for similar enzymes in this niche. Metagenomics is being peeled but extensive efforts are needed to understand the relations of microorganisms with their habitats and their non-culturability on plates. Till date more than 150 xylanase genes have been retrieved using metagenomic approaches, where GH10 xylanases dominate indicating their abundance in the environment. GH11 xylanases recovered by metagenomic approaches are very few. It is worthwhile to retrieve xylanases from the extreme environmental metagenomes which withstand extreme conditions.

## References

1. Alvarez TM, Goldbeck R, Santos CRd, Paixao DAA, Goncalves TA, *et al.* (2013). Development and biotechnological application of a novel endo-xylanase family GH10 identified from sugarcane soil metagenome. *PLoS One* 8(7): e70014.
2. Azeri C, Tamer AU, Oskay M (2010). Thermoactive cellulase free xylanase production from alkaliphilic *Bacillus* strains using various agro–residues and their potential in biobleaching of kraft pulp. *Afr J Biotechnol* 9: 63–72.
3. Barnard D, Casanueva A, Tuffin M, Cowan D (2010). Extremophiles in biofuel synthesis. *Environ Technol* 31: 871–888.
4. Bertram M, Hildebrandt P, Weiner DP, Patel JS, Bartnek F, *et al.* (2008). Characterization of lipases and esterases from metagenomes for lipid modification. *J Am Oil Chem Soc* 85: 47–53.
5. Bhalla A, Bansal N, Kumar S, Bischoff KM, Sani RK (2013). Improved lignocellulose conversion to biofuels with thermophilic bacteria and thermostable enzymes. *Biores Technol* 128: 751–759.
6. Brennan Y, Callen WN, Christoffersen L, Dupree P, Goubet F, *et al.* (2004). Unusual microbial xylanases from insect guts. *Appl Environ Microbiol* 70: 3609–3617.
7. Chanda SK, Hirst EL, Jones, JKN, Percival EGV (1950). The constitution of xylan from esparto grass (*Stipa tenacissima* L.). *J Chem Soc* 1289–1297.

8. Chang WS, Tsai CL, Tseng MJ (2004). Cloning and characterization of two thermostable xylanases from an alkaliphilic *Bacillus firmus*. *Biochem Biophys Res Commun* 319: 1017–1025.
9. Collins T, Gerday C, Feller G (2005). Xylanases, xylanase families and extremophilic xylanases. *FEMS Microbiol Rev* 29: 3–23.
10. Coughlan MP, Hazlewood GP (1993). β–1,4 D xylan degrading enzyme systems: Biochemistry, molecular biology and applications. *Biotechnol Appl Biochem* 17: 259–289.
11. Eda S, Ohnishi A, Kato L (1976). Xylan isolated from the stalks of *Nicotiana tobaccum*. *Agri Biol Chem* 40: 359–36.
12. Fontes CM, Hazlewood GP, Morag E, Hall J, Hirst BH, *et al.* (1995). Evidence for a general role for non–catalytic thermostabilizing domains in xylanases from thermophilic bacteria. *Biochem J* 307: 151–158.
13. Gilkes NR, Henrissat B, Kilburn DG, Miller RCJr, Warren RA (1991). Domains in microbial β-1,4-glycanases: sequence conservation function and enzyme families. *Microbiol Rev* 55: 303–315.
14. Gruppen H, Hamer RJ, Voragen AGJ (1992). Water unextractable cell wall material from wheat flour. 2. Fractionation of alkali extracted polymers and comparison with water extractable arabinoxylans. *J Cereal Sci* 16: 53–67.
15. Guo B, Chen X, Sun C, Zhou B, Zhang Y (2009). Gene cloning expression and characterization of a new cold–active and salt tolerant endo-β-1,4 xylanase from marine *Glaciecola mesophila* KMM 241. *Appl Microbiol Biotechnol* 84: 1107–1115.
16. Henrissat B (1991). A classification of glycosyl hydrolases based on amino acid sequence similarities. *Biochem J* 280: 309–316.
17. Hu Y, Zhang G, Li A, Chen J, Ma L (2008). Cloning and enzymatic characterization of xylanase gene from a soil derived metagenomic library with an efficient approach. *Appl Microbiol Biotechnol* 80: 823–830.
18. Jalal A, Rashid N, Rasool N, Akhtar M (2009). Gene cloning and characterization of a xylanase from a newly isolated *Bacillus subtilis* strain R5. *J Biosci Bioeng* 107: 360–365.
19. Khandeparkar R, Bhosle NB (2007). Application of thermoalkalophilic xylanase from *Arthrobacter* sp. MTCC 5214 in biobleaching of kraft pulp. *Biores Technol* 98: 897–903.
20. Khasin A, Alchanati I, Shoham Y (1993). Purification and characterization of a thermostable xylanase from *Bacillus stearothermophilus* T–6. *Appl Environ Microbiol* 59: 1725–1730.
21. Kormelink FJM, Voragen AGJ (1993). Degradation of different (glucurono)arabino xylans by a combination of purified xylan-degrading enzymes. *Appl Microbiol Biotechnol* 38: 688–695.

22. Kumar V, Satyanarayana T (2011). Applicability of thermo-alkali-stable and cellulose-free xylanase from a novel thermo-halo-alkaliphilic *Bacillus halodurans* in producing xylooligosaccharides. *Biotechnol Lett* 33: 2279–2285.

23. Kwon EU, Jeong YS, Kim YH, Kim SK, Na HB, *et al.* (2010). Construction of a metagenomic library from compost and screening of cellulase and xylanase-positive clones. *J Korean Soc Appl Biol Chem* 53: 702–708.

24. Lee CC, Kibblewhite-Accinelli RE, Wagschal K, Robertson GH, Wong DWS (2006). Cloning and characterization of a cold–active xylanase enzyme from an environmental DNA library. *Extremophiles* 10: 295–300.

25. Li N, Meng K, Wang Y, Shi P, Luo H *et al.* (2008). Cloning expression and characterization of a new xylanase with broad temperature adaptability from *Streptomyces* sp. S9. *Appl Microbiol Biotechnol* 80: 231–240.

26. Li PP, Wang XJ, Yuan XF, Wang XF, Cao YZ, Cui ZJ (2011). Screening of a composite microbial system and its characteristics of wheat straw degradation. *Agric. Sci. China* 10: 1586–1594.

27. Li R, Kibblewhite R, Orts WJ, Lee CC (2009). Molecular cloning and characterization of multidomain xylanase from manure library. *World J Microbiol Biotechnol* 25: 2071–2078.

28. Lo YC, Lu WC, Chen CY, Chang JS (2010). Dark fermentative hydrogen production from enzymatic hydrolysate of xylan and pretreated rice straw by *Clostridium butyricum* CGS5. *Biores Technol* 101: 5885–5891.

29. Liu W, Shi P, Chen Q, Yang P, Wang G *et al.* (2010). Gene cloning over expression characterization of a xylanase from *Penicillium* sp. CGMCC 1699. *Appl Biochem Biotechnol* 162: 1–12.

30. Maalej I, Belhaj I, Masmoudi NF, Belghith H (2009). Highly Thermostable xylanase of the thermophilic fngus *Talaromyces thermophilus*: Purification and characterization. *Appl Biochem Biotechnol* 158: 200–212.

31. Mamo G, Delgado O, Martinez A, Mattiasson B, Kaul RJ (2006). Cloning, sequencing, analysis and expression of a gene encoding an endo-xylanase from *Bacillus halodurans* S7. *Mol Biotechnol* 33: 149–159.

32. Mamo G, Hatti–Kaul R, Mattiasson B (2007). Fusion of carbohydrate binding modules from Thermotoga neapolitana with a family 10 xylanase from *Bacillus halodurans* S7. *Extremophiles* 11: 169–177.

33. Marshall SA, Morgan CS, Mayo SL (2002). Electrostatics significantly affects the stability designed homeodomain variants. *J Mol Biol* 316: 189–199.

34. Mathrani IM, Ahring BK (1992). Thermophilic and alkaliphilic xylanase from several *Dictyoglomus* isolates. *Appl Microbiol Biotechnol* 38: 23–27.

35. Matteotti C, Bauwens J, Brasseur C, Tarayre C, Thonart P, *et al.* (2012). Identification and characterization of a new xylanase from Gram–positive bacteria isolated from termite gut (*Reticulitermes santonensis*). *Protein Expr Purif* 83: 117–127.

36. Matthews BW, Nicholson H, Becktel WJ (1987). Enhanced protein thermostability from site directed mutations that decrease the entropy of unfolding. *Proc Natl Acad Sci* 84: 6663–6667.

37. Menon V, Prakash G, Prabhune A, Rao M (2010). Biocatalytic approach for the utilization of hemicellulose for ethanol production from agricultural residue using thermostable xylanase and thermotolerant yeast. *Biores Technol* 101: 5366–5373.

38. Notenboom V, Boraston AB, Williams SJ, Kilburn DG, Rose DR (2002). High resolution crystal structures of the lectin–like xylan binding domain from *Streptomyces lividans* xylanase 10A with bound substrates reveal a novel mode of xylan binding. *Biochemistry* 41: 4246–4254.

39. Paes G, Berrin JG, Beaugrand J (2012). GH11 xylanases: Structure/function/ properties relationships and applications. *Biotechnol Adv* 30: 564–592.

40. Patra AK, Madhu A (2010). Studies on enzymatic pretreatment of linen. *Ind J Fibre Text* 35: 337–341.

41. Puchkaev AV, Koo LS, Ortiz de Montellano PR (2003). Aromatic stacking as determinant of the thermal stability of CYP119 from *Sulfolobus solfataricus*. Arch *Biochem Biophys* 409: 52–58.

42. Raha AR, Chang LY, Sipat A, Yusoff K Haryanti T (2006). Expression of a thermostable xylanase gene from *Bacillus coagulans* ST–6 in *Lactococcus lactis*. *Lett Appl Microbiol* 42: 210–214.

43. Rajendhran J, Gunasekaran P (2007). Strategies for accessing soil metagenome for desired applications. Biotechnol Adv 26: 576–590.

44. Ratto M, Poutanen K, Viikari L (1992). Production of xylanolytic enzymes by an alkalitolerant *Bacillus circulans* strain. *Appl Microbiol Biotechnol* 37: 470–473.

45. Sakka K, Kojima Y, Kondo T, Karita S, Ohmiya K, *et al.* (1993). Nucleotide sequence of *Clostridium stercorarium* xynA gene encoding xylanase A: Identification of catalytic and cellulose binding domains. *Biosci Biotechnol Biochem* 57: 273–277.

46. Shi H, Zhang Y, Li X, Huang Y, Wang L, Wang Y, Ding H, *et al.* (2013). A novel highly thermostable xylanase stimulated by $Ca^{2+}$ from *Thermotoga thermarum*: Cloning expression and characterization. *Biotechnol Biof* 6: 26.

47. Shibuya H, Kaneko S, Hayashi K (2000). Enhancement of the thermostability and hydrolytic activity of xylanase by random gene shuffling. *Biochem J* 349: 651–656.

48. Shin JH, Choi JH, Lee OS, Kim YM, Lee DS, Kwak YY *et al.* (2009). Thermostable xylanase from *Streptomyces thermocyaneoviolaceus* for optimal production of xylooligosaccharides. *Biotechnol Bioproc Engg* 14: 391–399.

49. Son-Ng I, Li CW, Yeh Y, Chen PT, Chir JL *et al.*(2009). A novel endoglucanase from the thermophilic bacterium *Geobacillus* sp. 70PC53 with high activity and stability over a broad range of temperatures. *Extremophiles* 13: 425–435.

50. Subramaniyan S, Prema P (2002). Biotechnology of microbial xylanases: Molecular biology and applications. *Crit Rev Biotechnol* 22: 33–64.

51. Sun JY, Lu MQ, Weng XY, Qian LC, Gu SH (2007). Expression of recombinant *Thermomonospora fusca* xylanase A in *Pichia pastoris* and xylooligosaccharides released from xylan by it. *Food Chem* 104: 1055–1064.

52. Uchino F, Fukuda O (1983). Taxonomic characteristics of an acidophilic strain of *Bacillus* producing thermophilic acidophilic amylase and thermostable xylanase. *Agric Biol Chem* 47: 965–967.

53. Vazquez MJ, Alonso JL, Dominguez H, Parajo JC (2000). Xylooligosaccharides: manufacture and applications. *Trends Food Sci Technol* 11: 387–393.

54. Verma D, Satyanarayana T (2011). An improved protocol for DNA extraction from alkaline soil and sediment samples for constructing metagenomic libraries. *Appl Biochem Biotechnol* 165: 454–464.

55. Verma D, Satyanarayana T (2012a). Molecular approaches for ameliorating microbial xylanases. *Biores Technol* 117: 360–367.

56. Verma D, Satyanarayana T (2012b). Cloning expression and applicability of thermo–alkalistable xylanase of *Geobacillus thermoleovorans* in generating xylooligosaccharides from agro-residues. *Biores Technol* 107: 333–338.

57. Verma D, Kawarabayasi Y, Miyazaki K, Satyanarayana T (2013). Cloning, expression and characteristics of a novel alkalistable and thermostable xylanase encoding gene (*Mxyl*). retrieved from compost-soil metagenome. *PLoS One*; 8: e52459.

58. Viikari L, Ranua M, Kantelinen A, Sundquist J, Linko M (1986). Bleaching with enzymes. *Proc. 3rd Int. Conf. Biotechnol. Pulp and Paper Industry* Stockholm pp. 67–69.

59. Wang G, Wang Y, Yang P, Luo H, Huang H, *et al.* (2010). Molecular detection and diversity of xylanase genes in Alpine tundra soil. *Appl Microbiol Biotechnol* 87: 1383–1393.

60. Wang Y, Zhang H, He Y, Luo H, Yao B (2007). Characterization, gene cloning and expression of a novel xylanase XYNB from *Streptomyces olivaceoviridis* A1. *Aquaculture* 267: 328–334.

61. Whistler RL, Richards EL (1970). Hemicelluloses. In: *The Carbohydrates*. eds. (Pigman W, Horton D), Academic Press, New York. pp. 447–469.

62. Winterhalter C, Liebl W (1995). Two extremely thermostable xylanases of the hyperthermophilic bacterium *Thermotoga maritima* MSB8. *Appl Environ Microbiol* 61: 1810–1815.

63. Woldesenbet F, Puneet A, Virk, Gupta N, Sharma P (2012). Effect of microwave irradiation on xylanase production from wheat bran and biobleaching of eucalyptus kraft pulp. *Appl Biochem Biotechnol* 167: 100–108.

64. Yamada K, Terahara T, Kurata S, Yokomaku T, Tsuneda S, Harayama S (2008). Retrieval of entire genes from environmental DNA by inverse PCR with pre-amplification of target genes using primers containing locked nucleic acids. *Environ Microbiol* 10: 978–987.

65. Yoon HS, Han NS, Kim CH (2004). Expression of *Thermotoga maritima* endo-β-1,4 xylanase gene in *E. coli* and characterization of the recombinant enzyme. *Agric Chem Biotechnol* 47: 157–160.

66. Yoon KH (2009). Cloning of the *Bacillus subtilis* AMX-4 xylanase gene and characterization of the gene product. *J Microbiol Biotechnol* 19: 1514–9.

67. Zhang GM, Huang J, Huang GR, Ma LX, Zhang XE (2007). Molecular cloning and heterologous expression of a new xylanase gene from *Plectosphaerella cucumerina*. *Appl Microbiol Biotechnol* 74: 339–346.

68. Zhao Y, Luo H, Meng K, Shi P, Wang G, *et al.* (2011). A xylanase gene directly cloned from the genomic DNA of alkaline wastewater sludge showing application potential in the paper industry. *Appl Biochem Biotechnol* 165: 35–46.

**2016, Biotechnology: Progress and Applications** *Pages* **35–65**
*Editors:* **Saif Hameed and Zeeshan Fatima**
*Published by:* **DAYA PUBLISHING HOUSE, NEW DELHI**

*Chapter 3*

# Detection of Bacterial Pathogens: Advances in the Post Genomic Era

***Rajesh Singh Tomar* and Anurag Jyoti***

*Amity Institute of Biotechnology, Amity University, Maharajpur, Gwalior – 474 005, Madhya Pradesh, India*

## ABSTRACT

The consumption of unsafe water and food in developing countries causes a number of disease outbreaks. Existing methods for detection of pathogens prevalent in water and food are highly diverse. Conventional methods for detection rely on the cultivation of specific bacteria in appropriate culture media and on their biochemical characterization. These methods are laborious, non-specific and time consuming. With the advancement in microbial genome sequencing, there are possibilities of mining gene sequence and the signature sequences of particular bacteria can be deciphered. The molecular approach, Polymerase Chain Reaction (PCR) is now used for testing of water and food borne pathogens. The sensitivity of the PCR is often not sufficient and post PCR processing is needed. Real Time PCR, more sensitive than conventional PCR, is being applied in the environmental risk assessment. Although the method is highly sensitive, expensive instrumentation is required. One promising and currently underexplored area is the field of nanotechnology in relevance to bio-diagnostics, where nanoparticle based assays has been developed for specific detection of bioanalytes of clinical interest. Gold nanoparticles have unique optical properties and high surface areas which make them ideal candidates for developing an ideal detection system. The gold nanoparticles can be modified by thiolated

* Corresponding Author: E-mail: rstomar33@hotmail.com

single stranded DNA probes. Depending on their size, shape and degree of aggregation the colloidal solution of gold nanoparticles can become red and blue/purple (aggregated form).

## Introduction

The presence of pathogens in water and food in rural and urban environment is among the most serious public health concerns due to the associated morbidity and mortality. In a developing country like India, water is used for many purposes including drinking, cooking, personal hygiene, food production and recreational activities. Microbial pathogens, causative agents for morbidity and mortality, are discharged into the city sewage that enters untreated in receiving surface waters (Tomar *et al.*, 2014).

Despite ongo-ing efforts, the 2002 Millennium Development Goal of halving the population without clean water or sanitation by 2015 is unlikely to be achieved. Therefore, infectious waterborne diseases acquired through these uses are of supreme importance. Recent reports indicate that waterborne diseases of faecal-oral origin (mostly acute gastroenteritis) are the most common based on the number of people affected (WHO, 2008; Stevens *et al.*, 2009). The outbreaks have an enormous impact on the burden of diseases and sustainable development in developing countries. In contrast, the populations of developed countries are generally protected by legislation and regulations regarding water quality usage (Agarwal *et al.*, 2014).

Globally, the source of almost the two third of the drinking water consumption is surface water, which may easily be contaminated by sewage discharges, animals defecation, faulty infected channels, municipal wastes and industrial effluents (Basualdo *et al.*, 2000; Ahmed *et al.*, 2005; Hamelin *et al.*, 2006; Hamner *et al.*, 2006; Orsi *et al.*, 2007). These surface waters harbor a variety of pathogens such as Shiga Toxin producing *Escherichia coli* (STEC), Enterotoxigenic *Escherichia coli* (ETEC) and *Salmonella* (Belanger *et al.*, 2002; Shelton *et al.*, 2006; Lothigious *et al.*, 2007; Sur *et al.*, 2007; Fong *et al.*, 2007, Jyoti *et al.*, 2010a). Presence of *Salmonella* in temperate and tropical waters has been well demonstrated (Alcaide *et al.*, 1984; Jiménez *et al.*, 1989; Moganedi *et al.*, 2007). Once introduced in tropical waters *S. typhimurium* and other serovars could represent a serious public health hazard due to its long term survival (Lehloesa *et al.*, 2000; Moganedi *et al.*, 2007).

Diseases caused by food-borne microbial hazards constitute a world-wide public health concern, despite the enormous volume of food produced and consumed safely throughout the world on a daily basis. During the past decades, the incidence of certain food-borne diseases has increased in many parts of the world despite the many improvements made to public health and food safety systems. Statistics indicate that even in industrialized countries one out of every three people has a food-borne microbial illness event every year (WHO, 2003). It is recognized that food safety is not an absolute but rather a continuum of more or less safety and also that food-borne microbial hazards are likely to remain or (re-)emerge over time for a number of reasons. These include microbial adaptation, changes in the food production systems, changes in human demographics and behaviour, international travel and trade.

Global data indicate that the epidemiology of food-borne diseases is changing and that an increased range of microorganisms and foods are the leading cause of food-borne illness. A better understanding of the distribution, epidemiology and threat posed by emerging and uncharacterized pathogens is needed due to rapid, poorly controlled global spread through the food chain and ultimately and communities. The detection, reporting and characterization of food- and water-borne illnesses play an important role in identifying the origins and incidence of disease when links can be made to the causative agents and the foods involved.

## Prevalent Water and Food Borne Pathogenic Bacteria

Water and food-borne bacterial agents have been the most well investigated and monitored causes of intestinal infectious disease. In developed countries, the risk assessment is being regulated, but still the disease burden is substantial. For the past two decades, *Salmonella* spp. and *E. coli* are among the three most persistant food borne pathogens. The serotypes of both these pathogens are responsible for majority of burdens of food and water borne illnesses (Newell *et al.*, 2010).

### *Salmonella*

*Salmonella* spp. is one of the most prevalent food and water borne pathogens, which causes morbidity and mortality in developing and developed world. The pathogen causes salmonellosis, leads to significant medical and economic burdens throughout the world (Voetsch *et al.*, 2004). The causative organism can pass through the food chain from primary production to ultimate consumers and causes gasteroenterititis.

*Salmonellae* are gram-negative, rod-shaped, non-spore forming, mainly motile enterobacteria. They are facultative anaerobes; belong to *Enterobacteriaceae* family having dimensions around 0.7 to 1.5 µm (diameters) and 2 to 5 µm (lengths) (Madigan *et al.*, 2009). *Salmonellae* can grow at optimum temperature between 35 to 37°C, but growth has also been observed at temperatures ranging from 5°C to 47 °C. The optimum pH for growth is near 7.0, but growth may occur between pH 4.0 and pH 9.0 (Lanata *et al.*, 2003). *Salmonellae* produce hydrogen sulfide, decarboxylate lysine and ornithine, and do not hydrolyze urea. They are oxidase negative and catalase positive and can grow on citrate as the sole carbon source.

The CDC distinguishes two *Salmonella* species (or subgenera): *S. enterica* and *S. bongori*. *S. enterica* is further divided into six subspecies, of which *S. enterica* subsp. *enterica* is the most clinically significant, causing 99 per cent of *Salmonella* infections (Brenner *et al.*, 2000). WHO report confirms there are 2,579 serovars of *Salmonella*, out of which approximately 59 per cent belong to *S. enterica* subsp. enterica (Table 3.1) (Grimont *et al.*, 2007).

The genome sequence of *Salmonella enterica* serovar Typhimurium has been envisaged. The genome size of S. Typhimurium has been determined to be 4.9 mega base pairs (McClelland *et al.*, 2001). *Salmonella* present in the environment is usually expressed as concentrations of genomic nucleic acids (genome equivalents) or the number of copies of a specific target DNA or RNA sequence. The total number of *Salmonella* genomic equivalent (GE) in purified DNA is calculated as follows: $m = n \times$

(1.013 x $10^{-21}$ g/bp), where *m* is the mass and *n* is the number of base pairs. Consequently, according to the aforesaid equation, one *Salmonella* genome is about 5.0 fg (Malorny *et al.*, 2003).

**Table 3.1: Salmonella Species and Serovars**

| | *Salmonella Species and Subspecies* | *Number of Serovars* |
|---|---|---|
| | *S. enterica* | 2,557 |
| I | *S. enterica* subsp. *enterica* | 1,531 |
| II | *S. enterica* subsp. *salamae* | 505 |
| III | *S. enterica* subsp. *arizonae* | 99 |
| IV | *S. enterica* subsp. *diarizonae* | 336 |
| V | *S. enterica* subsp. *houtenae* | 73 |
| VI | *S. enterica* subsp. *indica* | 13 |
| | *S. bongori* | 22 |
| | Total (genus *Salmonella*) | 2,579 |

The pathogenicity of *Salmonella* serotypes vary widely. Most *Salmonella* virulence factors are clustered on chromosome referred as *Salmonella* pathogenicity islands (Shea *et al.*, 1996). So far five *Salmonella* pathogenicity islands have been identified which encode different secretion systems required by pathogenesis (Schmidt *et al.*, 2004). Typhimurium, Enteritis, Newport, and Heidelberg were among the top four most commonly isolated serotypes from human infections. These are also the most frequently isolated serotypes from food samples (CDC, 2003). The pathogenicity of *Salmonella* requires the expression of virulent genes. Serotypes, not possessing the *invABC* genes are unable to invade mammalian cells (Galan *et al.*, 1992). Virulent chromosomal genes such as *invA* (Malorny *et al.*, 2003; Zahraei *et al.*, 2007) sequences unique to this genus and has been proved to be a suitable PCR target with potential diagnostic application (Jamshidi *et al.*, 2008). This gene is recognized as an international standard for detection of *Salmonella* genus (Malorny *et al.*, 2003).

Infection of *Salmonella* in immune-compromised persons causes Salmonellosis. The symptoms including diarrhea, fever and abdominal cramps develop within 3 days after infection, the illness may last 4 to 7 days (FDA/CFSAN, 2008). In severe cases, the *Salmonella* infection may become systemic and spread from intestines to other body sites through blood stream and can cause death. The type of *Salmonella* usually associated with infections in humans is non-typhoidal *Salmonella*. It is usually spread from sources such as: infected poultry, eggs, egg products, cattle, milk and contaminated fruits and vegetables. Another form of *Salmonella*, typhoidal *Salmonella* can lead to typhoid fever. It is carried only by humans and is usually spread through direct contact with the infected person and its fecal matter. The infectious dose of *Salmonella* varies and depends upon serovars, source of infection and host susceptibility. In general, the minimum infective dose of *Salmonella* is $10^6$ and *Salmonella typhimurium* 10-100 cells (Lanata *et al.*, 2003).

*Salmonella* is constantly released into the environment from infected humans, farm animals, pets, and wildlife (Baudart *et al.*, 2000). The organism is frequently isolated from surface and potable water (Chao *et al.*, 1987; Cherry *et al.*, 1972; Jyoti *et al.*, 2010a), which serve as bacterial reservoirs. *Salmonella* has the ability to grow and survive in harsh conditions too as it can survive for 10 to 15 days in a septic system (Parker *et al.*, 1982).

Sediments and soils are the potential habitats of *Salmonella* as the water currents and runoff carries contaminated materials (Monem *et al.*, 1990). *Salmonella* has been detected in soil samples collected from both agricultural and recreational areas (Thomason *et al.*, 1975a and 1977). *Salmonella* can survive and multiply for at least 1 year in this ecosystem (Davies *et al.*, 1996; Thomason *et al.*, 1977). The ability of *Salmonella* to grow and survive in diverse environments and ecosystem represent its cyclic lifestyle in host and nonhost environments.

## Diarrheagenic *Escherichia coli* (DEC)

*E. coli* is gram-negative, facultative anaerobic and non-sporulating bacterium. It is generally rod-shaped, and dimensions are about 2.0 µm (length) and 0.5 µm (diameter) (Kubitschek, 1990). *E. coli* generally grows at 37°C but some strains can grow upto 49°C (Fotadar *et al.*, 2005). Diarrheagenic *E. coli* causes the diarrhea in infants, children and immuno-compromised people. *E. coli* colonise the gastrointestinal tract of infants just after birth and is the predominant throughout lives. The bacterium is responsible for a variety of diseases like gastroenteritis, urinary tract infections (UTI), and meningitis. Infections due to pathogenic *E. coli* may be limited to the mucosal surfaces or can disseminate throughout the body. DEC can be categorized into six different groups (pathovars) of pathogenic *E. coli* that cause diarrheal disease (Kaper *et al.*, 1998). The evolution of these pathovars have made through the attainment of virulence factors and pathogenicity islands by horizontal gene transfer. Compared to non-pathogenic *E. coli* some of these pathovars differ (more than 30 per cent) in their genetic composition (Pallen *et al.*, 2007; Welch, *et al.*, 2002).

The diffusely adherent *Escherichia coli* (DAEC), causing watery diarrhea, uniformly covers the entire epithelium cell surface of the small intestine and stimulates elongation of the microvilli (Scaletsky *et al.*, 2002). Another pathovar Enteroaggregative *Escherichia coli* (EAEC) produces toxin, which causes mucosal damage resulting in watery mucoid diarrhea (Weintraub, 2007). The Enteroinvasive *Escherichia coli* (EIEC) pathovar produces an enterotoxin which invades and destroys the epithelium in the large intestine, resulting in watery diarrhea and in severe cases dysentery (Kaper *et al.*, 2004). The Enteropathogenic *Escherichia coli* (EPEC) infects by attaching and effacing lesion in epithelial cell membrane leading to fever, dehydration and weight loss. It causes infant diarrhea (mostly watery diarrhea) in developing countries (Rosas *et al.*, 2006). Enterotoxigenic *Escherichia coli* (ETEC) is the most common pathovar of *E. coli* in developing countries. ETEC causes watery diarrhea. The pathovar secretes enterotoxins: heat labile (LT) and/or heat stable (ST), causing diarrhea in children and traveler's diarrhea. Enterohemorrhagic *Escherichia coli* (EHEC) is one of the most lethal pathovars of *E. coli*. It produces cytotoxins, called as shiga toxin, hence also called as Shiga toxin producing *E. coli* (STEC). EHEC causes abdominal pain followed by diarrhea and vomiting, later on becomes bloody. The toxins produced can cause severe diseases like hemorrhagic colitis and hemolytic uremic syndrome (Sears *et al.*, 1996). EHEC releases its toxins in the cytoplasm of host cells which inhibits protein synthesis leading to cell death.

## Enterotoxigenic *E. coli* (ETEC)

ETEC is the most common diarrheal agent among the six recognized pathovars of DEC, particularly in the developing world. ETEC produces at least one member of two defined groups of enterotoxins: ST (heat-stable enterotoxin) and LT (heat-labile enterotoxin) and colonization factors (CFs) which allow the organisms to colonize the small intestine causes diarrhea in infants as well as adult human beings (Levine, 1987; Qadri *et al.*, 2005).

Genome sequencing of *E. coli* has revealed that chromosome of ETEC isolate (E24377A) is ~4.9 Mb, smaller than those of other pathogenic *E. coli* isolates examined to date. One genome of ETEC is about 5.3 fg (Rasko *et al.*, 2008).

The heat labile enterotoxin (LT) is oligomeric toxin which is closely related in structure and function to the cholera enterotoxin (CT) of *Vibrio cholerae* (Sixma *et al.*, 1993). The toxin exists in two forms, namely LT-I and LT-II (Nataro *et al.*, 1998). LT-I is pathogenic for both humans and animals. LT-II has only been found in ETEC strains infecting animals. LT-I consists of one A subunit and five identical B subunits (Nataro *et al.*, 1998; Nagy *et al.*, 2005). After the colonization of the small intestine by ETEC the LTB subunits bind to ganglioside GM1, and the A subunit activates for the enzymatic activity of the toxin by activating adenylate cyclase which results in increases in cyclic adenosine monophosphate (cAMP) concentrations. This inhibits neutral sodium chloride in the villus tips and stimulates chloride secretion in the crypt cells thereby resulting in osmotic diarrhea (Nataro *et al.*, 1998, Kaper *et al.*, 2004).

Heat stable enterotoxin (ST) is small, monomeric toxin in comparison to LT. ST can be categorised into two groups; STa (or STI) and STb (or STII). STa is produced by ETEC strains infecting both humans and animals whereas STb is primarily associated with disease in animals, and not oftenly human disease (Nataro *et al.*, 1998). The receptor for the ST toxins is guanylate cyclase in the small intestine. Binding to the receptor results in increased intracellular cyclic guanylate monophosphate (cGMP) levels which stimulate chloride secretion and inhibit NaCl absorption, resulting in subsequent net intestinal fluid secretion leading to diarrhea (Qadri *et al.*, 2005). ETEC has great importance for developing countries due to its association with surface and potable water borne diarrheal disease outbreaks.

## Enterohemorrhagic *E. coli* (EHEC)

Enterohemorrhagic *E. coli* (EHEC), Verotoxin producing *E. coli* (VTEC) or Shiga-toxin producing *E. coli* (STEC) are the synonyms for the same group of organism. EHEC infections in humans can result in bloody diarrhea and hemorrhagic colitis (Nataro *et al.*, 1998). In severe infections the hemorrhagic colitis progresses to hemolytic uremic syndrome (HUS), an important cause of acute renal failure in children and morbidity and mortality in adults (fatality rate in elders: 50 per cent).

*E. coli* O157:H7 (EHEC O157:H7) has been recognized as the most commonly occurring member of this group. EHEC O157:H7 infections in humans can be acquired by direct contact with animal carriers, their feces, ingestion of undercooked ground beef and contaminated water, vegetables and fruits. The low infectious dose (1-100 CFU) enhances the risk of occurrence of disease (Paton *et al.*, 1998; Sharma *et al.*, 2006).

The genome size of the EHEC strain (O157 Sakai) is 5,594,477 bp. One copy of the genome is about 5.7 fg (Hayashi *et al.*, 2001). The pathogenicity of EHEC is determined by pathogenic factors. The most important factors are the phage-encoded Shiga toxins (stx) elements on the EHEC chromosome which inhibit protein synthesis (Nataro *et al.*, 1998). Shiga toxin contains two major toxins, called Stx1 and Stx2, which share approximately 55 per cent amino acid homology, these are similar to the one produced by *Shigella dysenteriae* serotype 1, (Nakao *et al.*, 2000; Gourmelon *et al.*, 2006). The gene *stx1* is highly conserved, *stx2* has 11 distinct variants, with *stx2c, stx2d, stx2e,* and *stx2f* being the most frequently reported (Nakao *et al.*, 2000). Epidemiological studies have revealed that Stx2 is 1000 times more cytotoxic than Stx1 towards human renal endothelial cells. EHEC producing Stx2 are more associated with serious diseases than isolates producing Stx1 or Stx1 and Stx2 (Louise *et al.*, 1995; Boerlin *et al.*, 1999; Paton *et al.*, 1998). Shiga toxins are proteins formed by AB holotoxins comprising one catalytic subunit (A) of 32 kDa and five B subunits. The A subunit cleaves a single adenine residue from the 28S rRNA component of the eukaryotic ribosomal 60S subunit. The B subunits mediate binding to cells through interaction with globotriaosylceramide (Gb3) expressed on epithelial and endothelial cell membranes. A cluster of genes involved in the generation of attaching and effacing lesions are chromosomally carried on a "pathogenicity island" named the locus of enterocyte effacement (McDaniel *et al.*, 1995). This pathogenicity island also carries the *eaeA* gene, which secretes an outer membrane protein called intimin that is necessary for attachment to enterocytes. All strains of O157:H7 contain a highly conserved plasmid, designated pO157, which varies in size from 93.6 to 104 kb encoding virulence genes which secrete different virulent factors, such as the enterohemolysin E-Hly, the catalase-peroxidase KatP etc. (Schmidt *et al.*, 1999).

EHEC enters the human body through contaminated food or water. The ability of bacterium to survive at lower pH (up to 2-3) facilitates survival through the acidic condition of the stomach. In the colon, the virulence genes are turned on by environmental signals. Secretion of Shiga toxins leads to local damage blood vessels in the colon result in bloody diarrhea. Absorbance of sufficient Stx in circulation causes damaging and impaired function of vascular endothelial sites. The kidneys and central nervous system are frequently affected, and hemolytic uremic syndrome (HUS) may develop (Gyles, 2007).

Environmental reservoirs are the potential sources of pathogenic *E. coli*. These are ideal habitat for survival, growth, and proliferation of *E. coli* as they are rich in nutrients. A wide range of different animals harbor pathogenic *E. coli* (Chapman, 1994), which are consistently associated with human and animal diarrheal infection. Infected cattle typically excrete (>$10^3$ CFU/g of feces) pathogenic *E. coli* (Robinson *et al.*, 2004; Topping *et al.*, 2008). These discharges may be released directly into the aquatic environment or indirectly through storm water run off. The surface water carries a number of pathogenic bacteria when used for the drinking and bathing purposes. Apart from these, the water is treated in municipal treatment plants and supplied through pipelines for drinking purposes (Figure 3.1). The processed clean water sometimes also gets contaminated due to leakage of nearby sewer lines (Ram *et al.*, 2008a).

*E. coli* can also survive in estuarine sediment environments (Brettar *et al.*, 1992). Two highly prevalent diarrheagenic *E. coli* variants have been reported in surface and potable waters as well as in macrophytes in India. The multiple antibiotic resistant Enterotoxigenic *E. coli* (ETEC) and Shiga toxin producing *E. coli* (STEC) have been reported in surface waters of river Ganga and its tributaries Gomti and Saryu (Ram *et al.*, 2007, 2008a, b and c). Potentially pathogenic isolates of *Escherichia coli* serotype O157:H7 have been isolated from the Ganges river near Varanasi (Hamner *et al.*, 2007).

## Detection of Bacterial Pathogens

The detection of bacterial pathogens has been a major issue since decades. Initially the culture based colony count methods have been used (Figure 3.2). The culture plate method is the oldest bacterial detection technique and remains the standard detection methodology.

However, culture methods are excessively time-consuming and generally take 5-8 days to obtain result (Brooks *et al.*, 2004). Different selective media are used to detect particular bacterial species.

### Conventional Methods for Detection of *Salmonellae*

The detection of *Salmonellae* faces several challenges, such as the low concentration of target bacteria in foods, high amount of other bacteria and inhibitory components in samples. The traditional confirmation and identification of *Salmonella* spp. are usually based on the colony appearance on chromogenic and other selective agar media, followed by confirmation using classical biochemical and serological testing. The biochemical tests include fermentation of glucose, negative urease reaction, lysine decarboxylase activity and $H_2S$ production. In the European Union, the reference detection methods are published by the International Organization for Standardization (ISO). The procedure for the detection of food borne *Salmonella* includes non-selective enrichment in buffered peptone water (BPW) broth for 16–20 hours followed by selective enrichment cultivation in selective media and incubation on two different selective agar plates for isolation of colonies (Anonymous 2002). The colonies are then identified by means of biochemical tests. This procedure takes at least 72 hours to complete. In Bacteriological Analytical Manual (BAM), published from the United States, the procedure for the detection of *Salmonella* includes non-selective enrichment in nutrient broth (NB) for 16 hours and followed by another 16 hours selective enrichment cultivation in either Rappaport–Vassiliadis (RV) or tetrathionate brilliant green (TBG) broth (Andrews *et al.*, 2007). The colonies are isolated by using selective agar plates and identified biochemically. In general, the immunoassays use antisera for detection of flagellar (H) and somatic (O) antigens. Isolates with a typical biochemical profile, which agglutinate with both H and O

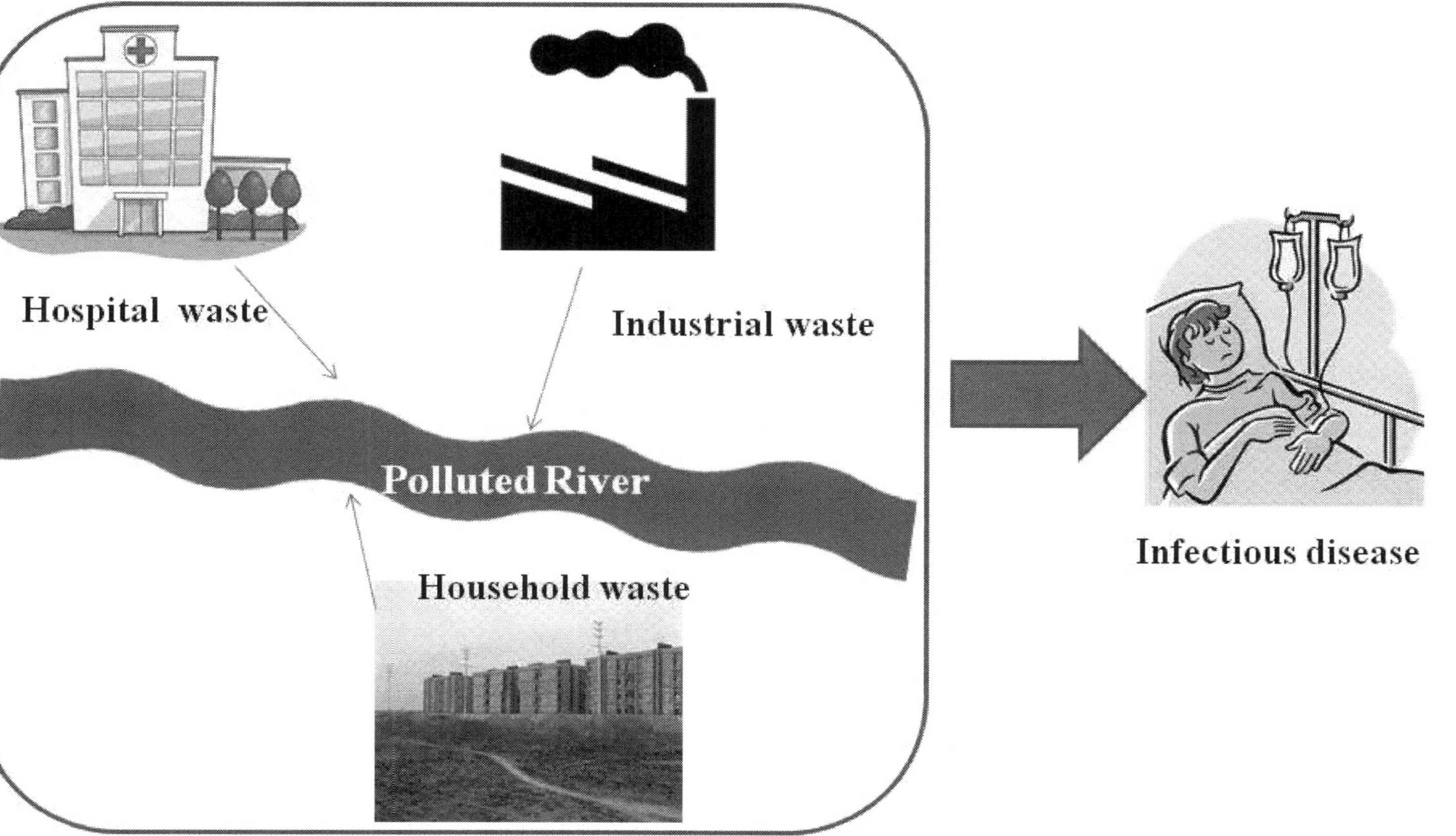


**Figure 3.1: Schematic Representation of the Contamination of Environmental Reservoirs by Pathogenic Microorganisms.**

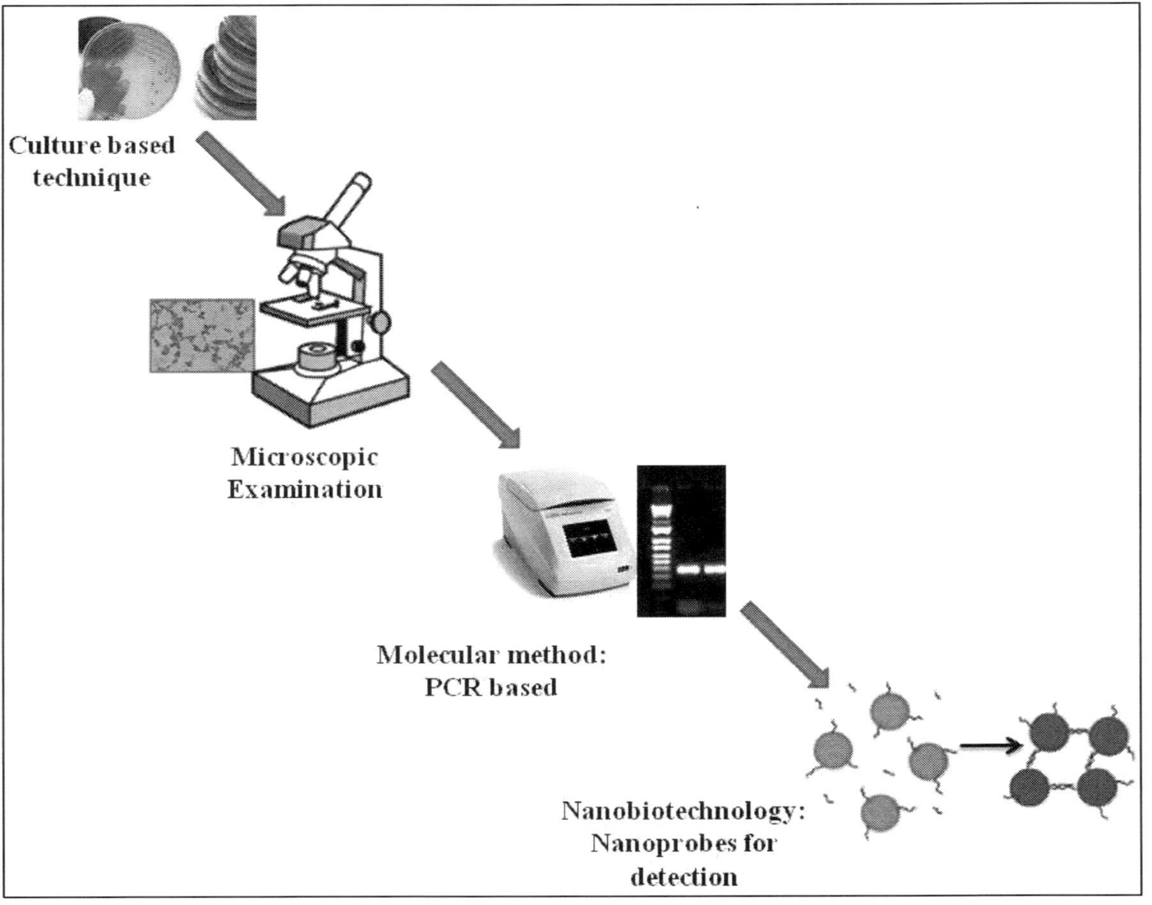


**Figure 3.2: Schematic Representation of the Development of Pathogen Detection Systems.**

antisera, are identified as *Salmonella* spp. (Cudjoe *et al.*, 1995; Valdivieso *et al.*, 2001; Magliulo *et al.*, 2007). Further, the positive isolates are often confirmed by serotyping and using techniques such a phage typing and pulsed-field gel electrophoresis (PFGE). The immunoassays are capable of detecting $10^4$ to$10^6$ cells of *Salmonella* in foods per assay. The antibodies may cross-react with antigens in closely related bacteria, while showing low reactivity with some *Salmonella* serotypes.

Polymerase Chain Reaction (PCR), a molecular technique has been used most frequently for the detection of *Salmonella* (Malorny *et al.*, 2007; Chen *et al.*, 2010; O'Regan *et al.*, 2008; McGuinness *et al.*, 2009). More than 30 *Salmonella* specific genes have been used for different PCR applications (Levin, 2009). These target genes include rRNA genes, virulent genes and antibiotic resistance genes. The virulence *invA* gene has been extensively used for the detection of *Salmonella* in water and food. Most of the PCR methods which are in routine use provide results in approximately 24 hours, and the detection limits are in the range of $10^3$–$10^4$ cells/ml.

## Conventional Methods for Detection of *E. coli* Pathotypes

The traditional method of detection of ETEC involves cultivation of bacteria in selective media. This includes the cultivation in MacConkey or Eosin Methylene Blue (EMB) agar followed by overnight incubation. The characteristic colonies are typical pink to red color for *E. coli*. Further confirmation is done by IMViC (Indole, Methyl red, Voges-Proskauer, and Citrate) test which examines the ability to produce indole and sufficient acid to change the color of a methyl red indicator (Todar, 2008). These assays are laborious and time consuming. ELISA has been widely used for detection of LT using microtiter GM1 ganglioside methods. It requires culturing of the bacteria before testing for presence of enterotoxins. This technique has also been further evolved into the inhibition-ELISA for detection of ST (Svennerholm *et al.*, 1986). Both ELISA and inhibition ELISA, based on monoclonal antibodies against LT and ST, are today used in many laboratories for detection and identification of ETEC. Although, the immunoassays, based upon the antigen-antibody interactions are established techniques, they often lack specificity. Serotyping has also been used to identify and characterize ETEC strains. The ETEC has more than 78 O groups and 34 H groups identified; therefore the determination of O serogroups associated with the lipopolysaccharides in the cell wall and H serogroups of the flagella is difficult (Qadri, *et al.*, 2005). A huge number of combinations of O and H groups make serotyping less suitable for identification of ETEC.

The identification of EHEC in accurate manner is still a major issue. In general, most of the agars for EHEC detect *E. coli* O157. The inability of most *E. coli* O157 to ferment sorbitol is exploited in sorbitol McConkey agar (SMAC). The ISO standard protocol (ISO 16654) recommends addition of Cexifime and potassium tellurite to SMAC (CT-SMAC) may increase the selectivity in samples. *E. coli* O157 generally produces colourless colonies when cultured on this media, thus distinguishing it from other EHEC serogroups. EHEC O157:H7 colonies are confirmed with biochemical tests and immunoassays having the O157 somatic antigen and H7 flagellar antigen. A variety of tests including enzyme-linked immunosorbent assays (ELISAs), agglutination, PCR, immunoblotting or Vero cell assay can be used to detect the verocytotoxin or its genes.

PCR, a sensitive and specific method can be used for differentiating diarrheagenic strains from nonpathogenic strains in both clinical and environmental samples (Sjöling *et al.*, 2005). This approach to ETEC detection relies on the presence of the genes encoding LT and/or ST enterotoxins. The *E. coli* O157:H7 specific virulent genes such as *stx1* and *stx2* are used for the reliable and accurate detection (Ram *et al.*, 2011).

## Challenges in Pathogen Detection

Detection of microbial pathogens relies upon predominant conventional microbiology monitoring approaches. These methods suffer from a number of considerable drawbacks. Well established standard culture and susceptibility tests allow pathogen detection but are laborious, time-consuming and expensive. These tests are unable to differentiate between pathogenic and non-pathogenic strains.

Bacterial pathogens, after prolonged exposure to water, might enter a viable but non-culturable stage, in which they cannot be detected by culture, although they retain their infective potential (Cenciarini *et al.*, 2009). Apart from these, the recovery of microbes and concentration re-concentration of samples are the major roadblocks in pathogen detection. Although, PCR is a standard and most frequent method for detection, the post PCR analysis, unable to quantify viable and dead pathogens are the major drawbacks (Figure 3.3).

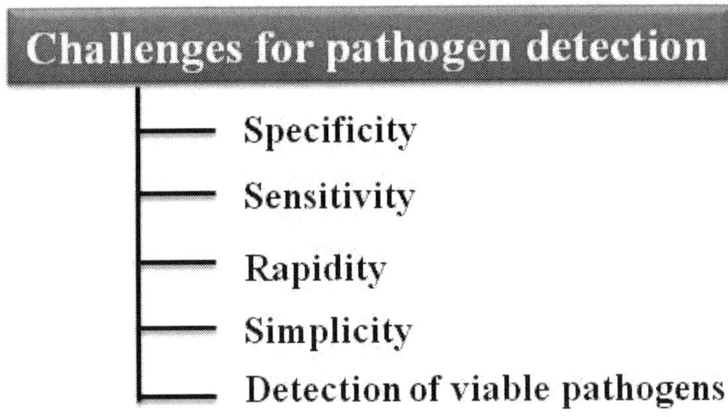


**Figure 3.3: Certain Hurdles in Detection and Monitoring of Water-borne Pathogens.**

## PCR in Real Time: Real Time PCR

Real Time PCR, a recent advancement in PCR technology uses fluorescent detection strategies for the quantitation of nucleic acid in a reaction without post-PCR processing. The addition of an optical module to conventional PCR machines, real time PCR allows the capture of fluorescent signals from amplified PCR products. This method of detecting the amplicon is advantageous over the conventional PCR technology, as the instrument detects the intensity of the fluorescent signal during each replication cycle of the PCR (Mackay 2004). Computer software records and displays the amount of fluorescence in relative fluorescence units (RFU). The amplification cycle at which the fluorescence exceeds the cutoff level is known as threshold cycle ($C_T$) (Corless *et al.*, 2000). This threshold is calculated as a function of the amount of background fluorescence and is plotted at a point in which the signal generated from a sample is significantly greater than background fluorescence (Kubista *et al.*, 2006; Espy *et al.*, 2006). Since the amplification is monitored during the reaction, real time PCR can be used for simultaneous detection and quantification of the target. With increasing copy number of target DNA, the threshold of the sample fluorescence will be achieved earlier. The amount of initial copy number of the target in unknown sample can be calculated by comparing the $C_T$ obtained of that sample with a standard curve, generated from serial dilutions of a known amount of the target DNA (Ibekwe *et al.*, 2002).

The possibility of cross-contamination between analyses is reduced due to closed tube format, and provides rapid, continuous data collection and reduction in time required to obtain results. These salient features have made real time PCR an appealing alternative to conventional culture-based or immunoassay-based testing methods used for diagnosing infectious diseases. The advantage and limitations of real time PCR are given (Table 3.2).

**Table 3.2: Advantages and Limitations of Real Time PCR**

| *Advantages* | *Limitations* |
|---|---|
| ☆ Wide dynamic range of quantification (7-8 logarithmic decades) | ☆ PCR product increases exponentially cannot monitor amplicon size. |
| ☆ High analytical sensitivity (<5 copies; 1fg to 10 pg bacterial DNA per PCR reaction) | ☆ Variation increase with cycle number |
| ☆ Better precision (<2 per cent standard deviation) | ☆ Maximum of four simultaneous reactions |
| ☆ Closed system to reduce risk of contamination | ☆ Overlap of emission spectra |
| ☆ No post PCR processing | ☆ Increased risk of false negative in pathogen detection (particularly for new emerging or highly variable pathogens) |
| ☆ Lower turnaround time | ☆ Non-specific binding (SYBR green analysis) |
| ☆ Increased throughput | |
| ☆ Multiplexing capabilities | |

## Detection Chemistries

The fluorescent reporters can be classified on the basis of nucleic acid dyes and molecular probes. Further, probes can also divided into hydrolysis and hybridization probe. The most relevant hydrolysis probes are TaqMan, MGB TaqMan and hybridization probes are FRET, Molecular beacon and Scorpion probe etc.

### Nucleic Acid Dye: SYBR Green I

SYBR Green I is an asymmetrical cyanine dye used to detect the accumulation of any double-stranded DNA (dsDNA) product. The resulting DNA-dye-complex absorbs blue light ($\lambda_{max}$ = 498 nm) and emits green light ($\lambda_{max}$ = 522 nm) (Zipper *et al.*, 2004; Bengtsson *et al.*, 2003). After binding, the dye emits a strong fluorescent signal (depending on amplified product), which is more than 1000 times greater than that emitted by unbound dye. Advantages of SYBR Green include the relative ease with which it can be applied to existing PCR assays, relatively lower cost, and ease of assay design. Since the dye binds to dsDNA irrespective of sequence it may also bind to nonspecific dsDNA sequences (primer dimmers, mispriming) which generate false positive signals (Missel *et al.*, 2001; Ramos *et al.*, 2003). After the amplification, the melting curve analysis is run to verify the fragment obtained is the correct sequence. The melting temperature of dsDNA increases with longer length and higher GC content. By analyzing the melting temperatures a distinction can be made between the desired amplicons and any nonspecific products, or between different amplicons in a multiplexed reaction (Wang *et al.*, 2004). SYBR Green I has been applied successfully in real time PCR (Jothikumar *et al.*, 2002; Knutsson *et al.*, 2002, McCrea *et al.*, 2007). *Escherichia coli* O157:H7, *Listeria monocytogenes*, and *Salmonella* strains in fresh produce have been detected simultaneously using SYBR Green I and melting curve analysis (Bhagwat, 2003).

**Real Time Fluorescent Probes: Sequence-Specific**

**Hydrolysis Probe: TaqMan Probe**

The hydrolysis probes are cleaved during the elongation phase of primers by the proof reading activity of DNA polymerase (Holland *et al.*, 1991). This results in a real time measurable fluorescence emission. The best-known and most frequently used hydrolysis probe is TaqMan. This is the first real time fluorescent probe developed. A TaqMan probe is short oligonucleotide (20-30 bp) which contains reporter fluorophore at the 5'- end and a quencher fluorophore at the 3'- end. Initially, the probe binds to a complementary strand of DNA at 60°C. This is followed by the cleavage of 5′- end of the TaqMan probe by the 5′ nuclease activity of Taq polymerase at same temperature during the extension phase. This result in separating the fluorescent dye from the quenching dye and signal is generated. The TaqMan probe is cleaved only if DNA amplification occurs, separating the reporter and quencher molecules. PCR products are monitored during increasing fluorescent signal exponentially with successive PCR amplification cycles (Kubista *et al.*, 2006). TaqMan probe is highly specific towards the target. A number of studies have been reported which detect pathogens using TaqMan probe (Malorny *et al.*, 2007; Hein *et al.*, 2006).

**Hybridization Probes**

**Molecular Beacon Probe (MBP)**

Molecular beacon is a short single-stranded oligonucleotide molecule (between 25 and 35 bases in length) that forms a hairpin with a loop and stem. The loop, also termed as probe, is designed to hybridize to a 15-20 nucleotide fragments of the target DNA sequence while the stems are composed of 5-7 self complementary nucleotides. A fluorescent reporter molecule is attached at the 5′- end of the MB and a dark quencher at 3′- end. In the native state, the MB is closed and both the reporter and quencher are in close proximity of each other. The signal released by reporter is masked by the quencher. During annealing step of PCR, the binding of MB to its target leads to increased fluorescence, due to separation of reporter and quencher, which directly

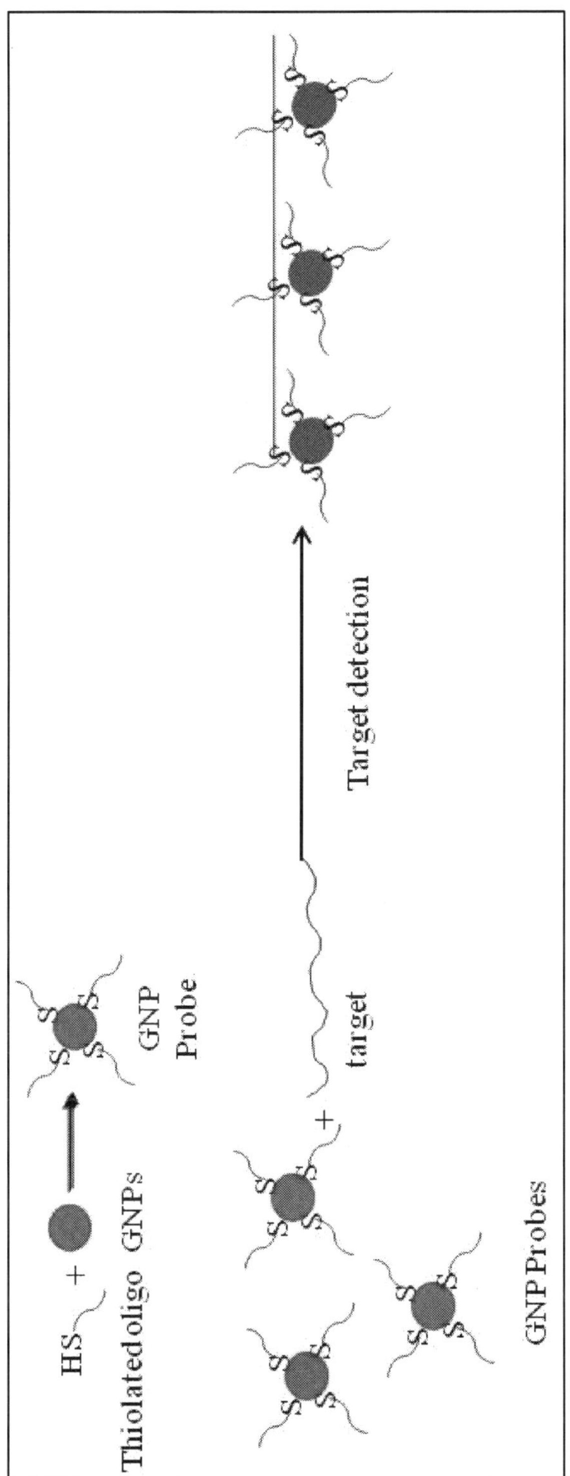


**Figure 3.4: Mechanism of Target Detection using Gold Nanoparticle Probes.**

corresponds to accumulation of product (Figure 3.4) (Tyagi *et al.*, 1996; Vet *et al.*, 1999; Belanger *et al.*, 2002). Molecular Beacon has an advantage over conventional nucleic acid probes because they have a higher degree of specificity with better signal-to-noise ratios (Park *et al.*, 2000). The probe has been used for detection of a number of pathogens (Ram *et al.*, 2008b, 2011; Lata *et al.*, 2009; Hadjinicolaou *et al.*, 2009; Jyoti *et al.*, 2010a).

**Fluorescence Resonance Energy Transfer Probes (FRET Probes)**

FRET hybridization probes, labeled with different fluorescent dyes are two DNA probes designed to anneal in a head-to-tail orientation on the PCR product. Each probe has a single label either a donor fluorophore at the 3′- end or an acceptor fluorophore at 5′- end. Fluorophores, having the emission spectrum one overlapping with the excitation spectrum of the other are taken into account. During the annealing of probes to the target PCR product, fluorescence from the 3′dye is absorbed by the adjacent acceptor dye on the 5′end of the second probe. The second dye is excited and emits light at a different wavelength which is detected. The FRET probes must be blocked at their 3′- end using phosphate moiety to prevent the extension during the annealing step (Bernard *et al.*, 2002).

Molecular beacons and FRET hybridization probes, unlike TaqMan probes, are both recycled in each round of PCR temperature cycle.

**Scorpion Probe**

Scorpion probe is bi-functional molecule, having a primer covalently linked to the probe. It consists of a reporter dye attached to the 5′ end, followed by a complementary stem-and-loop structure (the loop contains specific probe sequence), a quencher dye, a DNA polymerase blocker, and a PCR primer on the 3′ end. The quencher attached to the 3′- end of the loop quenches the fluorophore attached to the 5′ - end. The 3′- end of loop is linked to the 5′- end of primer via non-amplifiable stopper moiety such as hexa ethylene glycol (HEG). In addition, the probe sequence is complementary to an internal region of the sequence extended by the adjacent primer (Whitcombe *et al.*, 1998, Whitcombe *et al.*, 1999). During the annealing, the primers anneal to the specific target region. After extension of the primer, the specific probe sequence is able to bind to target sequences. This hybridization event opens the hairpin loop so that fluorescence is no longer quenched and an increase signal is observed (similar to Molecular Beacons). In the absence of a target, the quencher absorbs the fluorescence emitted by the reporter dye via contact quenching. As the probe is physically attached with the primer, the reaction generates signal via uni-molecular kinetics. The blocker is present on the Scorpion probe to limit non-specific amplification.

Scorpion probe is used in a fluorescence based method for real time PCR analysis, which is ideal for diagnostic applications such as pathogen detection (Marco *et al.*, 2007; Stroup *et al.*, 2006; Stricker *et al.*, 2008).

## Computation of Primers and Probes for Real Time PCR Based Pathogen Detection

In the recent years, with the advancements of genome sequencing technologies and due to the relatively smaller size, about 1660 number of microbial genomes have been sequenced (Up to June 2011). The majority of sequenced genomes are freely available in public databases such as the National Center for Biotechnology Information Genome Web site (http://www.ncbi.nlm.nih.gov/sites/entrez). This includes the sequencing of numerous water and food borne pathogens and their strains which include *Escherichia coli* (Hayashi *et al.*, 2001, Perna *et al.*, 2001), *L. monocytogenes* (Glaser *et al.*, 2001) and *Salmonella enterica* serovar Typhimurium LT2 (McClelland *et al.*, 2001) etc.

The genomics-driven studies have direct applications in the development of molecular tools for rapid detection of pathogens. The traditional methods are based on biochemical and antigenic properties of bacteria which give variable and unreliable results and have several lacunae (Liu, 2006). In recent years, with the advancement of molecular biology in combination with genomics studies, real time PCR a molecular-based diagnostic method has evolved. Genomic sequence analysis enables the identification of unique (based on existing sequence data) and conserved (across different strains of pathogen) regions to individual pathogens and permits the computation of primers and probes with increased specificity. The oligonucleotides used to detect target genes in the real time PCR require on bench validation, which is time consuming and expensive (Patel *et al.*, 2011). Computational exercises of the oligonucleotides prior to real time PCR are useful in minimizing the time and effort taken in optimization and validation.

### Steps Involved in Designing and Computation of Primers and Probes

1. Retrieving of sequences from database. Nucleotide sequences (genomic islands, conserved sequences and virulence determinant genes) of target genes of selected pathogens are retrieved from the NCBI GenBank (*www.ncbi.nlm.nih.gov/genbank/*), National Centre for Biotechnology Information, Bethesda, MD, U.S.A.
2. Multiple sequence alignment. Clusters having two or more sequences are used to construct conserved regions. The multiple sequence alignments are created using ClustalW programme (www.ebi.ac.uk/clustalW) to determine conserved sequences.
3. Primer design. The conserved regions, unique to a target pathogen, obtained from ClustalW are used to design primers and probes using either of Lasergene (DNASTAR, Madison, WI), Primer3, or Beacon Designer (Premier Biosoft. International, Palo Alto, California, USA). Primer and probe designing parameters such as melting temperature, GC content, amplicon length, etc are taken into consideration. The parameters can be set depending upon the type of probe designing.
4. Specificity of designed primers and probes. The specificity of computed primers and probes is determined against the known microbial genomes

and sequences by BLAST (Basic Local Alignment Search Tool) programme (http://www.ncbi.nlm.nih.gov/BLAST) of NCBI to ensure no homology is found in other genera or species.

5. *In-Silico* analysis of primers. The primer-pairs will then be used for *in-silico* analysis using, *in-silico* PCR (http://insilico.ehu.es/PCR/). The designed primers are subjected to amplification to ensure that each primer pair amplifies in the conserved region of the selected target gene.
6. Analysis of folding and secondary structure formation. The computed probes are analysed for correct stem loop configuration and no secondary structure formation in mFold server (http://mfold.rna.albany.edu/?q=mfold/DNA-Folding-Form). The reaction conditions used for calculation of secondary structure are: 0.25 mM nucleic acid, 50 mM monovalent ion, 3 mM free $Mg^{++}$ ion and 269 mM total $Na^{+}$ equivalent for MB. The free energy calculation temperature for MB was 55 and 25°C for all primers and other probes. Free energy values range obtained with Zuker program should usually range between - 3 to + 0.5 kcal/mol.

## Criteria for Designing Primers and Probes

The parameters for designing of primers and probes are as follows: length of primer 18-20 bp for Molecular beacon/FRET and 22-24 bp for TaqMan, primer annealing temperature in range between 50-60 °C. The other primer search parameters are as follows: no ambiguous base in amplicon, maximum primer $T_M$ mismatch was 3 °C for MB and 2 °C for TaqMan and FRET. Cross dimer maximum ΔG (internal/3′end) is -6/-5 kcal/mol. The hairpin maximum ΔG (internal/3′end) = -3/-3 kcal/mol for TaqMan and Molecular beacon, while -2/-3 kcal/mol for FRET, self-dimer ΔG (internal/3′end) and 3′end maximum ΔG = -6/-5 kcal/mol and -10 kcal/mol for all primers, respectively.

The lengths of probes are 20-24, 18-30 and 25-35 bases for TaqMan/Molecular beacon and FRET, respectively. The $T_m$ of TaqMan and FRET probes were 8-10 °C and that of Molecular beacon 4-14 °C higher than $T_m$ of primers respectively. The other search parameters are: hairpin maximum ΔG= -3 kcal/mol, self dimer ΔG= -6 kcal/mol, run repeat not more than 4 bp, GC clamp consecutive C/G at 3′end is 2 and multiplex ΔG is -6 kcal/mol for all robes. The melting temperature of Molecular beacon stem can be calculated using Zuker hairpin folding formula; it is used to estimate the free energy of the formation of the stem hybrid from which its melting temperature can be predicted.

This article is not included in your organization's subscription. However, you may be able to access this article under your organization's agreement with Elsevier.

## Detection of Viable Pathogens

The presence of viable pathogens in water and food is well established. However, well established detection methods are still unable to detect accurate viable pathogens in samples. Molecular methods are adapted to assess viable cells, and not dead cells or DNA, in the samples. A possible solution is to target mRNA, as it tends to degrade relatively rapidly after cell death. However, in many cases, expression and persistence of mRNA depends on the genes targeted and the conditions in which the cells were inactivated. Apart from this, some RNA molecules can also persist in cells after loss of viability. Moreover, some mRNA molecules are not transcribed in the viable but non-culturable (VBNC) state. All these may lead to false positive or negative results in the real time PCR assay. Therefore, on the basis of reverse transcription real time PCR the detection and quantification of viable pathogens is inaccurate (Cenciarini *et al.*, 2009).

The use of DNA intercalating dyes like ethidium monoazide (EMA) and propidium monoazide (PMA) opens new avenues in pathogen detection. These dyes enter the membrane compromised dead cells and intercalate the DNA, which on further is unable to amplify in PCR and give signal in real time PCR. The EMA has certain drawbacks, as it has been found that EMA treatment of live *E. coli* 0157:H7 cells leads to a loss of more than 60 per cent of the genomic DNA during the DNA extraction procedure (Nocker *et al.*, 2006). PMA has the advantage over EMA in terms of increased selectivity. The use of real time PCR with PMA treatment may also enable quantization of viable pathogenic Cryptosporidium oocysts in environmental samples (Brescia *et al.*, 2009).

## Nanoparticle Based Probes for Detection of DNA

Nanobiotechnology is evolving with the simultaneous developments in nanoscience, biology and medicine. The potential applications of nanotechnology are becoming increasingly important for the development of ultrasensitive DNA detection systems. The molecular recognition is fundamental for the development of diagnostic tools. Various organic molecules, possessing unique properties, have been used to achieve the recognition of different targets. Gold nanoparticles (GNPs) have been explored for bio-diagnostics due to unique optical properties (*i.e.* surface plasma resonance absorption and resonance light scattering), a variety of surface coatings and great biocompatibility (Mirkin *et al.*, 1996; Storhoff *et al.*, 2004; Sato *et al.*, 2007). GNP-based DNA detection has high sensitivity as compared to conventional fluorescence based assays due to the extremely high molar absorptivity of GNPs (1000 times higher than that of organic dyes) (Zhao *et al.*, 2008a).

The optical properties of gold nanoparticles are governed by collective oscillation of electrons at surfaces (known as surface plasmon resonance, (SPR). The resonance frequency of this oscillation lies in the visible region of the electromagnetic spectrum (Sayed, 2001). The GNPs have a high surface to volume ratio, therefore the surface electrons are sensitive to change in the dielectric (refractive index) of the medium. Any changes to the environment of these particles (surface modification, aggregation, medium refractive index, etc.) lead to colorimetric changes of the dispersions (Daniel *et al.*, 2004; Rosi *et al.*, 2005; Murphy, 2002; Kelly *et al.*, 2003; Burda *et al.*, 2005; Yguerabide *et al.*, 1998a,b; Niemeyer *et al.*, 2005). The aggregation behaviour of GNPs has been widely explored for different applications (Katz *et al.*, 2004; Stewart *et al.*, 2008). This has further facilitated the application in biodetection via numerous detection methods (Boyer *et al.*, 2002; Niemeyer 2001). The versatile surface chemistry of GNPs can be achieved by attaching various biofunctional groups, such as nucleic acids, sugars, proteins via the strong affinity of gold surface with thiol ligands (Schofield *et al.*, 2007).

The colloidal solution of GNPs is monodisperse red and exhibit a narrow surface plasmon absorption band centered on 520 nm (depends upon particle size) in the UV-Visible spectrum. In contrast, when the particles aggregate, the solution appears purple/blue, corresponding to a characteristic red shift in the surface plasmon resonance (SPR) towards higher wavelength. A number of factors such as size and shape of the nanoparticle, refractive index of the surrounding media and inter-particle distance are taken into account for use in colorimetric detection of DNA (Baptista *et al.*, 2008). The GNPs obtained by citrate-reduction present in solution are charged particles and they are sensitive to changes in solution dielectrics. Hence, with the addition of NaCl the surface charge is shielded leading to a decrease in inter-particle distance and particle aggregation (Niemeyer *et al.*, 2005).

## GNP Probes for DNA Detection

The use of GNP-based colorimetric methods for detection of DNA has been widely reported and reviewed (Daniel *et al.*, 2004; Rosi *et al.*, 2005). Bio-functionalization of GNPs with the thiol modified single stranded DNA (ssDNA) also called GNP probe, has been synthesized for the DNA detection (Mirkin *et al.*, 1996; Lu *et al.*, 2006). The ssDNA probe strand is designed to be complementary to a target of interest and is attached to the GNPs through chemisorption of the thiol group onto the surface of the GNPs. After hybridization with the target, the GNP probes come in the close proximity, which leads to change in color.

Mirkin and co-workers reported the colorimetric detection of DNA targets based on the cross-linking mechanism use of GNP probes. The two different batches of probes are designed to target the DNA. Thus, upon the addition of target DNA, a polymeric network of GNP probes is generated due to aggregation, turning the solution from red to blue (Mirkin *et al.*, 1996). This aggregation mechanism is mainly applied to detect small sized targets. The GNP aggregation induced by interparticle crosslinking is relatively a slower process (Elghanian *et al.*, 1997). The relatively slow aggregation is due to the nature of the interparticle crosslinking aggregation mechanism. In general, the aggregation is driven by random collisions between nanoparticles with relatively slow Brownian motion (Sato *et al.*, 2003).

Similarly, the DNA detection based on a non-cross-linking mechanism has been well documented. Sato and co-workers described non-cross linking mechanism of GNP aggregation for DNA detection. The single stranded DNA can be immobilized on GNPs above the physiological temperature. The GNP probes aggregate together at considerably high salt concentration when the target DNA is perfectly complementary to the probe (Sato *et al.*, 2007). Other detection techniques have also been applied in the GNPs-based DNA assay for improving detection sensitivity and selectivity (Dai *et al.*, 2008; Zhang *et al.*, 2007; Chiu *et al.*, 2008). In noncrosslinking aggregation systems, the van der Waals force of attraction dominates. Many parameters, including surface charge properties (*e.g.*, charge density, the amount of associated counter ions) and entropy factors are involved in aggregation.

Compared to interparticle crosslinking aggregation systems, the noncrosslinking aggregation mechanism has some attractive features. Aggregation induced by the noncrosslinking process is very rapid and assays can often be completed in a few minutes (Sato *et al.*, 2003). This is presumably due to the mechanism of non-crosslinking aggregation. The interparticle attractive forces (van der Waals) dominate over the interparticle repulsive forces, which results in rapid aggregation (Zhao *et al.*, 2008b; Sato *et al.*, 2003).

The use of GNP probes for the colorimetric detection of DNA targets represents an inexpensive and simple workable alternative to fluorescence or radioactivity based assays (Storhoff *et al.*, 2004).

In view of these, there is an urgent need to develop an ideal detection system which should be based on simple, rapid, highly sensitive and specific and overcome drawbacks in existing methods for detection of water and food-borne pathogens and risk assessment. The use of molecular methods can improve the sensitivity and specificity to the detection system. Optical properties of gold nanoparticles can lead to development of a portable and robust hand-held device for on-site diagnostics of pathogens.

## References

1. Ahmed W, Neller R, Katouli M. Host species-specific metabolic fingerprint database for *Enterococci* and *Escherichia coli* and its application to identify sources of fecal contamination in surface waters. *Appl. Environ. Microbiol.* 2005, *71*, 4461-4468.
2. Alcaide E, Garay E. R-Plasmid transfer in *Salmonella* spp. isolated from wastewater and sewage-contaminated surface waters. *Appl. Environ. Microbiol.* 1984, *48*, 435-438.
3. Anonymous. ISO 6579: 2002 Microbiology of food and animal feeding stuffs-Horizontal method for the detection of *Salmonella* spp. International Organization for Standardization. 2002.
4. Antony B, Dias M, Shetty AK, Rekha B. Food poisoning due to *Salmonella enterica* serotype Weltevreden in Mangalore. *Ind. J. Med. Microbiol.* 2009, *27*, 257-258.
5. Banatvala N, Griffin PM, Greene KD, Barrett TJ, Bibb WF, Green JH, Wells JG, Hemolytic Uremic Syndrome Study Collaborators. The United States national prospective hemolytic uremic syndrome study: microbiologic, serologic, clinical, and epidemiologic findings. *J. Infect. Dis.* 2001, *183*, 1063-1070.
6. Basualdo J, Pezzani B, De Luca M, Córdoba A, Apezteguía M. Screening of the municipal water system of La Plata, Argentina, for human intestinal parasites, *Int. J. Hyg. Environ. Health.* 2000, *203*, 177-182.
7. Baudart, J, Lemarchand K, Brisabois A, Lebaron P. Diversity of *Salmonella* strains isolated from the aquatic environment as determined by serotyping and amplification of the ribosomal DNA spacer regions. *Appl. Environ. Microbiol.* 2000, *66*, 1544–1552.

8. Belanger SD, Boissinot M, Menard C, Picard FJ, Bergeron MG. Rapid detection of shiga toxin producing bacteria in feces by multiplex PCR with molecular beacons on the smart cycler. *J. Clin. Microbiol.* 2002, *40,* 1436 -1440.

9. Bengtsson M, Karlsson JH, Westman G, Kubista M. A new minor groove binding asymmetric cyanine reporter dye for real-time PCR. *Nucleic Acids Res.* 2003, *31,* e45.

10. Bernard PS, Wittwer CT. Real-time PCR technology for cancer diagnostics. *Clin. Chem.* 2002, *48,* 1178-1185.

11. Bhagwat AA. Simultaneous detection of *Escherichia coli* O157: H7, *Listeria monocytogenes* and *Salmonella* strains by real-time PCR. *Int. J. Food Microbiol.* 2003, *84,* 217– 224.

12. Bhunia R, Hutin Y, Ramakrishnan R, Pal N, Sen T, Murhekar M. A typhoid fever outbreak in a slum of South Dumdum municipality, West Bengal, India, 2007: Evidence for food borne and waterborne transmission. *BMC Public Health* 2009, *9,* 115.

13. Boyer D, Tamarat P, Maali A, Lounis B, Orrit M. Photothermal imaging of nanometer-sized metal particles among scatterers. *Science.* 2002, *297,* 1160-1163.

14. Brenner FW, Villar RG, Angulo FJ, Tauxe R, Swaminathan B. *Salmonella* nomenclature. *J. Clin. Microbiol.* 2000, *38,* 2465-2467.

15. Brescia CC, Griffin SM, Ware MW, Varughese EA, Egorov AI, Villegas EN. Cryptosporidium propidium monoazide-PCR, a molecular biology-based technique for genotyping ofviable Cryptosporidium oocysts. *Appl. Environ. Microbiol.* 2009, 75, 6856-6863.

16. Brooks WA, Hossain A, Goswami D, Nahar K, Alam K, Ahmed N. Bacteremic typhoid fever in children in an urban slum, Bangladesh. *Emerg. Infect. Dis.* 2005, *11,* 326-329.

17. Burda C, Chen X, Narayanan R, Sayed MAE. Chemistry and properties of nanocrystals of different shapes. *Chem. Rev.* 2005, *105,* 1025-1102.

18. CDC. Centers for Disease Control and Prevention (CDC) *Morbid. Mortal. Weekly Rep.* 2010, 59.

19. Cenciarini-Borde C, Courtois S, La Scola B. Nucleic acids as viability markers for bacteria detection using molecular tools. *Fut. Microbiol.* 2009, *4,* 45-64.

20. Chao W, Ding R, Chen R. Survival of pathogenic bacteria in environmental microcosms. *Chin. J. Microbial Immunol.* (Taipei). 1987, *20,* 339–348.

21. Chapman PA, Wright DJ, Siddons CA. A comparison of immuno-magnetic separation and direct culture for the isolation of verocytotoxin-producing *Escherichia coli* O157 from bovine feces. *J. Med. Microbiol.* 1994, *40,* 424-427.

22. Chen J, Zhang LD, Paoli GC, Shi CL, Tu SI, Shi XM. A real-time PCR method for the detection of *Salmonella enterica* from food using a target sequence identified by comparative genomic analysis. *Int. J. Food Microbiol.* 2010, *137,* 168–174.

23. Cherry WB, Hanks JB, Thomason BM, Murlin AM, Biddle JW, Croom JM. *Salmonellae* as an index of pollution of surface waters. *Appl. Microbiol.* 1972, *24*, 334–340.

24. Corless CE, Guiver M, Borrow R, Edwards JV, Kaczmarski EB, Fox AJ. Contamination and sensitivity issues with a real-time universal 16S rRNA PCR. *J. Clin. Microbiol.* 2000, *38*, 1747–1752.

25. Cudjoe KS, Hagtvedt T, Dainty R. Immunomagnetic separation of *Salmonella* from foods and their detection using immuno magnetic particle (IMP)-ELISA. *Int. J. Food Microbiol.* 1995, *27*, 11-25.

26. Daniel MC, Astruc D. Gold nanoparticles: assembly, supramolecular chemistry, quantum-size-related properties, and applications toward biology, catalysis, and nanotechnology. Chem. Rev. 2004, 104, 293-346.

27. Das S, Saha R, Singhal S. Enteric pathogens in North Indian patients with diarrhea. *Ind. J. Comm. Med.* 2007, *32*, 1–3.

28. Espy MJ, Uhl JR, Sloan LM, Buckwalter SP, Jones MF, Vetter EA, Yao JDC, Wengenack NL, Rosenblatt JE, Cockerill III FR, Smith TF. Real-Time PCR in clinical microbiology: applications for routine laboratory testing. *Clin. Microbiol. Rev.* 2006, *19*, 165–256.

29. FDA/CFSAN - Food Safety A to Z Reference Guide-Salmonella. *FDA - Center for Food Safety and Applied Nutrition.* 2008-07-03.

30. Fong T, Mansfield LS, Wilson DL, Schwab DJ, Molloy SL, Rose JB. Massive microbiological groundwater contamination associated with a waterborne outbreak in Lake Erie, South Bass Island, Ohio. *Environ. Health Perspect.* 2007, *115*, 856-864.

31. Fotadar U, Zaveloff P, Terracio L. Growth of *Escherichia coli* at elevated temperatures. *J. Bas. Microbiol.* 2005, *45*, 403-404.

32. Grimont PAD, Weill FX. Antigenic formulae of the *Salmonella* serovars. WHO Collaborating Centre for Reference and Research on *Salmonella* 2007, 9th edition.

33. Gould LH, Demma L, Jones TF, Hurd S, Vugia DJ, Smith K, Shiferaw B, Segler S, Palmer A, Zansky S, Griffin PM. Hemolytic uremic syndrome and death in persons with *Escherichia coli* O157: H7 infection: Foodborne diseases active surveillance network sites, 2000-2006. *Clin. Infect. Dis.* 2009, *49*, 1480-1485.

34. Gourmelon M, Montet MP, Lozach S, Le Mennec C, Pommepuy M, Beutin L, Vernozy-Rozand C. First isolation of shiga toxin 1d producing *Escherichia coli* variant strains in shellfish from coastal areas in France. *J. Appl. Microbiol.* 2006, *100*, 85-97.

35. Gupta SN, Gupta N. Outbreak of Gastroenteritis in Tibetan transit school, Dharamshala, Himachal Pradesh, India, 2006. *Ind. J. Comm. Med.* 2009, *34*, 97-101.

36. Hadjinicolaou AV, Demetriou VL, Emmanuel MA, Kakoyiannis CK, Kostrikis LG. Molecular beacon-based real-time PCR detection of primary isolates of *Salmonella typhimurium* and *Salmonella enteritidis* in environmental and clinical samples. *BMC Microbiol.* 2009, *9*, 97.

37. Hamelin K, Guillaum B, El-Shaarawi A, Hill S, Edge T.A, Bekal S, Fairbrother JM, Harel J, Maynard C, Masson L, Brousseau RA. Virulence and antimicrobial resistance DNA microarray detects a high frequency of virulence genes in *Escherichia coli* isolates from great lakes recreational waters. *Appl. Environ. Microbiol.* 2006, *72*, 4200-4206.

38. Hamner S, Broadaway SC, Mishra VB, Tripathi A, Mishra RK, Pulcini E, Pyle BH, Ford TE. Isolation of Potentially Pathogenic *Escherichia coli* O157: H7 from the Ganges River *Appl. Environ. Microbiol.* 2007, *73*, 2369–2372.

39. Hayashi T, Makino K, Ohnishi M, Kurokawa K, Ishii K, Yokoyama K, Han CG, Ohtsubo E, Nakayama K, Murata T, Tanaka M, Tobe T, Iida T, Takami H, Honda T, Sasakawa C, Ogasawara N, Yasunaga T, Kuhara S, Shiba T, Hattori M, Shinagawa H. Complete genome sequence of enterohemorragic *Escherichia coli* O157: H7 and genomic comparison with a laboratory strain K-12. *DNA Res.* 2001, *8*, 11–22.

40. Hein I, Flekna G, Krassnig M, Wagner, M. Real-time PCR for the detection of Salmonella spp. in food: An alternative approach to a conventional PCR system suggested by the FOOD-PCR project. *J. Microbiol. Method.* 2006, *66*, 538–547.

41. Holland PM, Abramson RD, Watson R, Gelfand DH. Detection of specific polymerase chain reaction product by utilizing the 5′-3′ exonuclease activity of *Thermus aquaticus* DNA polymerase. *Proc. Natl. Acad. Sci.* 1991, *88*, 7276-7280.

42. Ibekwe AM, Watt PM, Grieve CM, Sharma VK, Lyons SR. Multiplex fluorogenic real-time PCR for detection and quantification of *Escherichia coli* O157: H7 in dairy wastewater wetlands. *Appl. Environ. Microbiol.* 2002, *68*, 4853–4862.

43. Jamshidi A, Bassami MR, Afshari NS. Identification of *Salmonella* spp. and *Salmonella typhimurium* by multiplex PCR-based assay from poultry caracasses in Mashhad, Iran. *Int. J. Vet. Res.* 2008, *3*: 43-48.

44. Jiménez L, Muñiz I, Toranzos GA, Hazen TC. Survival and activity of *Salmonella typhimurium* and *Escherichia coli* in tropical freshwater. *J. Appl. Bacteriol.* 1989, *67*, 61-69

45. Joshi R, Janagama H, Dwivedi HP, Kumar TMAS, Jaykus, LA, Schefers J, Sreevatsan S. Selection, characterization, and application of DNA aptamers for the capture and detection of *Salmonella enterica* serovars. *Mol. Cell. Prob.* 2009, *23*, 20–28.

46. Jothikumar N, Griffiths MW. Rapid detection of *Escherichia coli* O157: H7 with multiplex real-time PCR assays. *Appl. Environ. Microbiol.* 2002. *68*, 3169-3171.

47. Jyoti A, Vajpayee P, Singh G, Patel CB, Gupta KC, Shanker R. Identification of environmental reservoirs of non-typhoidal salmonellosis: aptamer assisted bio-concentration and subsequent detection of *Salmonella typhimurium* by quantitative

Polymerase Chain Reaction. *Environmental Science and Technology* 2011, *45*, 8996-9002.

48. Jyoti A, Ram S, Vajpayee P, Singh G, Dwivedi PD, Jain SK, Shanker R. Contamination of surface and potable water in South Asia by *Salmonellae:* Culture-Independent quantification with Molecular Beacon real-time PCR. *Sci. Tot. Environ.* 2010a, *408*, 1256–1263.
49. Katz E, Willner I. Integrated nanoparticle–biomolecule hybrid systems: synthesis, properties, and applications. *Angew. Chem. Int. Ed.* 2004, *43*, 6042-6108.
50. Kelly KL, Coronado E, Zhao LL, Schatz GC. The optical properties of metal nanoparticles: the influence of size, shape, and dielectric environment. *J. Phys. Chem.* B. 2003, 107, 668-677.
51. Knutsson R, Lofstrom C, Grage H, Hoofar J, Radstrom P. Modeling of 5′ nuclease real-time responses for optimization of a high throughput enrichment PCR procedure for *Salmonella enterica. J. Clin. Microbiol.* 2002, *40*, 52-60.
52. Kubista M, Andrade JM, Bengtsson M, Forootan A, Jonak J, Lind K, Sindelka R, Sjoback R, Sjogreen B, Strombom L, Stahlberg A, Zoric N. The real-time polymerase chain reaction. *Mol. Asp. Med.* 2006, *27*, 95-125.
53. Lanata CF. Studies of food hygiene and diarrhoeal disease. *Int. J. Environ. Heal. Res.* 2003, *13*, 175-183.
54. Lehloesa LJ, Muyima NYO. Evaluation of the impact of household treatment procedures on the quality of groundwater supplies in the rural community of the Victoria district, Eastern Cape. *Water SA.* 2000, *26*, 285-290.
55. Lata P, Ram S, Agrawal M, Shanker R. Real Time PCR for the rapid detection of *vanA* gene in surface waters and aquatic macrophyte by molecular beacon probe. *Environ. Sci. Technol.* 2009, *43*, 3343-3348.
56. Levin RE. The Use of Molecular Methods for Detecting and Discriminating *Salmonella* Associated with Foods - A Review. *Food Biotechnol.* 2009, *23*, 313–367.
57. Levine MM. *Escherichia coli* that cause diarrhea: enterotoxigenic, enteropathogenic, enteroinvasive, enterohemorrhagic, and enteroadherent. *J. Infect. Dis.* 1987, *155*, 377-389.
58. Lothigius Å, Sjöling Å, Svennerholm AM, Bölin I. Survival and gene expression of enterotoxigenic *Escherichia coli* during long-term incubation in sea water and freshwater. *J. Appl. Microbiol.* 2010, *108*, 1441-1449.
59. Mackay I. Real-time PCR in the microbiology laboratory. *Clin Microbiol. Infect.* 2004, *10*, 190–212.
60. Magliulo M, Simoni P, Guardigli M, Michelini E, Luciani M, Lelli R, Roda A. A rapid multiplexed chemiluminescent immunoassay for the detection of *Escherichia coli* O157: H7, *Yersinia enterocolitica, Salmonella typhimurium,* and *Listeria monocytogenes* pathogen bacteria. *J. Agr. Food Chem.* 2007, *55*, 4933–4939.
61. Madigan M *et al.*, Brock Biology of Microorganisms. 12 ed. San Francisco, Pearson Education. 2009.

62. Mairal T, Özalp VC, Sánchez PL, Mir M, Katakis I, O'Sullivan CK. Aptamers: molecular tools for analytical applications. *Anal. Bioanal. Chem.* 2008, *390*, 989–1007.

63. Malorny B, Bunge C, Helmuth R. A real-time PCR for the detection of *Salmonella* Enteritidis in poultry meat and consumption eggs. *J. Microbiol. Meth.* 2007 *70*, 245–251.

64. Malorny B, Hoorfar J, Bunge C, Helmuth R. Multicenter Validation of the Analytical Accuracy of *Salmonella* PCR: towards an International Standard. *Appl. Environ. Microbiol.* 2003, *69*, 290–296.

65. Manna SK, Brahmane MP, Manna C, Batabyal K, Das R. Occurrence, virulence characteristics and antimicrobial resistance of *Escherichia Coli* O157 in slaughtered cattle and diarrhoeic calves in West Bengal, India. *Lett. Appl. Microbiol.* 2006, *43*, 405-409.

66. Marco ED, Cangemi G, Filippetti M, Melioli G, Biassoni R. Development and clinical validation of a real-time PCR using a uni-molecular Scorpion-based probe for the detection of *Mycoplasma pneumoniae* in clinical isolates *New Microbiologica,* 2007, *30*, 415-421.

67. McClelland M, Sanderson KE, Spieth J, Clifton SW, Latreille P, Courtney L, Porwollik S, Ali J, Dante M, Du F, Hou S, Layman D, Leonard S, Nguyen C, Scott K, Holmes A, Grewal N, Mulvaney E, Ryan E, Sun H, Florea L, Miller W, Stoneking T, Nhan M, Waterston R, Wilson RK. Complete genome sequence of *Salmonella enterica* serovar Typhumurium LT2. *Nature* 2001, *413*, 852–56.

68. McCrea JK, Liu C, Ng LK, Wang G. Detection of the *Escherichia coli* pathogenic gene *eae* with three real-time polymerase chain reaction methods. *Can. J. Microbiol.* 2007, *53*, 398-403.

69. McGuinness S, McCabe E, O'Regan E, Dolan A, Duffy G, Burgess C, Fanning S, Barry T, O'Grady J. Development and validation of a rapid real-time PCR based method for the specific detection of *Salmonella* on fresh meat. *Meat Sci.* 2009, *83*, 555–562.

70. Mead PS, Shuster L, Dietz V, McCaig LF, Bresee JS, Shapiro C, Griffin PM, Tauxe RV. Food-related illness and death in the United States. *Emerg. Infect. Dis.* 1999, *5*, 607-625.

71. Mirkin CA, Letsinger RL, Mucic RC, Storhoff JJ. A DNA-based method for rationally assembling nanoparticles into macroscopic materials. *Nature.* 1996, *382*, 607-609.

72. Missel A, Decker K, Loeffert D, Mueller S, Dwyer K, Kang J. Improvements in quantitative PCR and RT-PCR using SYBR Green. Hilden, Germany: QIAGEN. Available from: http: //www1.qiagen.com/literature/posters/PDF/ PCR_RTPCR_RNAamp/1016700_SPOS_RTPCR_0201.pdf. 2001.

73. Moganedi KLM, Goyvaerts EMA, Venter SN, Sibara MM. Optimisation of the PCR-invA primers for the detection of *Salmonella* in drinking and surface waters following a pre-cultivation step. *Water.SA.* 2007, *33*, 195–202.

74. Mohit Agrawal, Rajesh Singh Tomar and Anurag Jyoti. Detection of Water-borne Pathogenic Bacteria: Where Molecular Methods Rule. *Int. Jour. Multidisc. Curr. Res.* 2014, *2*, 351-358.

75. Monem AMHAA, Dowidar A. Recoveries of *Salmonella* from soil in eastern region of Saudi Arabia Kingdom. *J. Egypt. Public Health Assoc.* 1990, *65*, 61–75.

76. Munoz FX, Lazcka O, Campo FJD. Pathogen detection: A perspective of traditional methods and biosensors. *Biosens. Bioelectron.* 2007, *22*, 1205–1217.

77. Murphy CJ. Nanocubes and nanoboxes. *Science.* 2002, *298*, 2139-2141.

78. Nakao H, Takeda T. *Escherichia coli* shiga toxin. *J. Nat. Toxins.* 2000, *9*, 299-313.

79. Nataro JP, Kaper JB. Diarrheagenic *Escherichia coli. Clin. Microbiol. Rev.*, 1998 *1*, 142-201.

80. Newell DG, Koopmans M, Verhoef L, Duizer E, Kane AA, Sprong H, Opsteegh M, Langelaar M, Threfall J, Scheutz F, van der Giessen JV, Kruse H. Food-borne diseases–The challenges of 20 years ago still persist while new ones continue to emerge. *Int. J. Food Microbiol.* 2010, *139*, 3-15.

81. Niemeyer CM. Nanoparticles, proteins, and nucleic acids: biotechnology meets materials science. *Angew. Chem. Int. Ed.* 2001, *40*, 4128-4158.

82. Niemeyer CM, Simon U. DNA-based assembly of metal nanoparticles. *Eur. J. Inorg.* 2005, *18*, 3641-3655.

83. Nocker A, Cheung CY, Camper AK. Comparison of propidium monoazide with ethidium monoazide for differentiation of live vs. dead bacteria by selective removal of DNA from dead cells. *J. Microbiological Method.* 2006, *67*, 310–320.

84. O'Regan E, McCabe E, Burgess C, McGuinness S, Barry T, Duffy G, Whyte P, Fanning S. Development of a real-time multiplex PCR assay for the detection of multiple *Salmonella* serotypes in chicken samples. *BMC Microbiol.* 2008, 8.

85. Orsi RH, Stoppe NC, Sato MIZ, Gomes TAT, Prado PI, Manfio GP, Ottoboni LMM. Genetic variability and pathogenicity potential of *Escherichia coli* isolated from recreational water reservoirs, *Res. Microbiol.* 2007,*158*, 420-427.

86. Park S, Wong M, Marras SAE, Cross EW, Kiehn TE, Chaturvedi V, Tyagi S, Perlin DS. Rapid identification of *Candida dubliniensis* using a species-specific Molecular beacon. *J. Clin. Microbiol.* 2000, *38*, 2829-2836.

87. Patel CB, Vajpayee P, Singh G, Jyoti A, Upadhyay RS, Shanker R. Computation and *in-silico* validation of a Real-Time PCR array for quantitative detection of fecal coliforms in surface and potable water. *Int. J. Biosci. Biotechnol.* 2011, *1*, 184-195.

88. Paton JC, Paton AW. Pathogenesis and diagnosis of shiga toxin-producing *Escherichia coli* infections. *Clin. Microbiol. Rev.* 1998, *11*, 450-479.

89. Perna NT, Plunkett G 3rd, Burland V, Mau B, Glasner JD, Rose DJ, Mayhew GF, Evans PS, Gregor J, Kirkpatrick HA, Pósfai G, Hackett J, Klink S, Boutin A, Shao Y, Miller L, Grotbeck EJ, Davis NW, Lim A, Dimalanta ET, Potamousis KD,

Apodaca J, Anantharaman TS, Lin J, Yen G, Schwartz DC, Welch RA, Blattner FR. Genome sequence of enterohaemorrhagic *Escherichia coli* O157: H7. *Nature* 2001, *409*, 529-533.

90. Qadri F, Svennerholm AM, Faruque AS, Sack RB. Enterotoxigenic *Escherichia coli* in developing countries: epidemiology, microbiology, clinical features, treatment, and prevention. *Clin Microbiol. Rev.* 2005, *18*, 465-483.

91. Ram S, Vajpayee P, Shanker R. Contamination of potable water distribution systems by multiantimicrobial-resistant enterohemorrhagic *Escherichia coli. Environ. Heal. Perspect.* 2008a, *116*, 448-452.

92. Ram S, Vajpayee P, Shanker R. Rapid culture-independent quantitative detection of enterotoxigenic *Escherichia coli* in surface waters by Real-Time PCR with molecular beacon. *Environ. Sci. Technol.* 2008b, *42*, 4577–4582.

93. Ram S, Vajpayee P, Tripathi U, Singh RL, Seth PK, Shanker R. Determination of antimicrobial resistance and virulence gene signatures in surface water isolates of *Escherichia coli. J. Appl. Microbiol.* 2008c, *105*, 1899–1908.

94. Ram S, Vajpayee P, Dwivedi PD, Shanker R. Culture-free detection and enumeration of STEC in water. *Ecotoxicol. Environ. Saf.* 2011, *74*, 551-557.

95. Ramos-Payan R, Aguilar-Medina M, Estrada-Parra S, Gonzalez-Y-Merchand JA, Favila-Castillo L, Monroy-Ostria A, Estrada-Garcia IC. Quantification of cytokine gene expression using an economical real-time polymerase chain reaction method based on SYBR Green I. Scand. *J. Immunol.* 2003, *57*, 439–45.

96. Ramteke PW, Tewari S. Serogroups of *Escherichia coli* from drinking water, *Environ. Monit. Assess.* 2007, *130*, 215-220.

97. Rasko DA, Rosovitz MJ, Myers GSA., Mongodin EF, Fricke WF, Gajer P, Crabtree J, Sperandio V, Ravel J. The pan-genome structure of *Escherichia coli:* comparative genomic analysis of *E. coli* commensal and pathogenic isolates. *J. Bacteriol.* 2008, *190*, 6881-6893.

98. Robinson SE, Wright EJ, Hart CA, Bennett M, French NP. Intermittent and persistent shedding of *Escherchia coli* O157 in cohorts of naturally infected calves. *J. Appl. Microbiol.* 2004, *94*, 1045-1053.

99. Rosi NL, Mirkin CA. Nanostructures in biodiagnostics. Chem. Rev. 2005, 105, 1547-1562.

100. Sato K, Hosokawa K, Maeda M. Rapid aggregation of gold nanoparticles induced by non-cross-linking DNA hybridization. J. Am. Chem. Soc. 2003, 125, 8102–8103.

101. Sato K, Hosokawa K, Maeda M. Colorimetric biosensors based on DNA-nanoparticle conjugates. *Anal. Sci.* 2007, 23, 17-20.

102. Sayed MAE. Some interesting properties of metals confined in time and nanometer space of different shapes. *Acc. Chem. Res.* 2001, 34, 257-264.

103. Stewart ME, Anderton CR, Thompson LB, Maria J, Gray SK, Rogers JA, Nuzzo RG. Nanostructured plasmonic sensors. *Chem. Rev.* 2008, 108, 494-521.

104. Schmidt H, Hensel M. Pathogenicity islands in bacterial pathogenesis. *Clin. Microbiol. Rev.* 2004, *17*, 14-56.

105. Shea JE, Hensel M, Gleeson C, Holden DW. Identification of a virulence locus encoding a second type III secretion system in *Salmonella typhimurium*. *Proc. Natl. Acad. Sci. USA*. 1996, *93*, 2593-2597.

106. Sixma TK, Kalk KH, van Zanten BAM, Duater Z, Kingma J, Witholt B, Hol WGJ. Refined structure of *Escherichia coli* heat-labile enterotoxin, a close relative of cholera toxin. *J. Mol. Biol.* 1993, *230*, 890-918.

107. Sharma V. Real-time reverse transcription-multiplex PCR for simultaneous: Detection of *rfbE* and *eae* genes for *Escherichia coli* O157: H7. *Mol. Cell. Probes* 2006, *20*, 298-306.

108. Shelton DR, Karns JS, Higgins JA, VanKesel JA, Perdue ML, Belt KT, Anelli JR, C. Debroy. Impact of microbial diversity on rapid detection of enterohemorraghic *Escherichia coli* in surface waters. *FEMS Microbiol. Lett.* 2006, *261*, 95-1011.

109. Sjöling Å, Wiklund G, Savarino SJ, Cohen DI, Svennerholm, AM. Comparative analyses of phenotypic and genotypic methods for detection of enterotoxigenic *Escherichia coli* (ETEC) toxins and colonisation factors. *J. Clin. Microbiol.* 2007, *45*, 3295-3301.

110. Stevens G, Mascarenhas M, Mathers C. Global health risks: progress and challenges. *Bull. World Heal. Organ.* 2009, *87*, 646-646.

111. Stroup S E, Roy S, Mchele J, Maro V, Ntabaguzi S, Siddique A, Kang G, Guerrant R L, Kirkpatrick BD, Fayer R, Herbein J, Ward H, Haque R, Houpt E R. Real-time PCR detection and speciation of Cryptosporidium infection using Scorpion probes *J.Med. Microbiol.* 2006, *55*, 1217–1222.

112. Stricker AR, Wilhartitz I, Farnleitner AH, Mach RL. Development of a Scorpion probe-based real-time PCR for the sensitive quantification of *Bacteroides* sp. ribosomal DNA from human and cattle origin and evaluation in spring water matrices. *Microbiol. Res.* 2008, *163*, 140-147.

113. Storhoff JJ, Lucas AD, Garimella V, Bao YP, Muller UR. Homogeneous detection of unamplified genomic DNA sequences based on colorimetric scatter of gold nanoparticle probes. *Nat. Biotechnol.* 2004, *22*, 883–887.

114. Sur D, Dutta S, Sarkar BL, Manna B, Bhattacharya MK, Datta KK, Saha A, Dutta B, Pazhani GP, Choudhuri AR, Bhattacharya SK. Occurrence, significance and molecular epidemiology of cholera outbreaks in West Bengal. *Indian J. Med. Res.* 2007, 125, 772-776.

115. Todar K. Pathogenic *Escherichia coli*. In: Todar's Online Textbook on Bacteriology. University of Wisconsin, Madison, Dept. of Bacteriology, 2008.

116. Tomar R S, Jyoti A, Mishra R K, Shrivastava V, Kaushik S. *In-silico* designing of SYBR Green based Real-Time PCR array for the quantification of Salmonellae and Enterotoxigenic *Escherichia coli* in water. *Europ. Acad. Res.* 2014, *1*, 5945-5958.

117. Topping MC, Gally D, Low C, Matthews L, Woolhouse M. Super-shedding and the link between human infection and livestock carriage of *Escherichia coli* O157. *Nat. Rev. Microbiol.* 2008, *6*, 904-912.

118. Turner M, German *E. coli* outbreak leads to drug trial. *Nat. News.* 2011, doi: 10.1038/news.2011.332

119. Tyagi S, Kramer FR. Molecular beacons: Probes that fluoresce upon hybridization. *Nat. Biotech.* 1996, *14*, 303-308.

120. Valdivieso-Garcia A, Riche E, Abubakar O, Waddell TE, Brooks BW. A double antibody sandwich enzyme-linked immunosorbent assay for the detection of *Salmonella* using biotinylated monoclonal antibodies. *J. Food Protect.* 2001, *64*, 1166–1171.

121. Vikesland PJ, Wigginton KR. Nanomaterial enabled biosensors for pathogen monitoring - a review. *Environ Sci Technol.* 2010, *44*, 3656-69.

122. UNDP-World Bank. Water for India's poor - Who pays the price for broken promises? UNDP-World Bank Water and Sanitation Program Report: South Asia Region Report, 2002.

123. Vet JAM, Majithia AR, Marras SAE, Tyagi S, Dube S, Poiesz BJ, Kramer FR. Multiplex detection of four pathogenic retroviruses using molecular beacons on smart cycler. *Proc. Nat. Acad. Sci.* 1999, *96*, 6394-6399.

124. Wang X, Jothikumar N, Griffiths M. Enrichment and DNA extraction protocols for the simultaneous detection of *Salmonella* and *Listeria monocytogenes* in raw sausage meat with multiplex real-time PCR. *J. Food Prot.* 2004, *67*,189–92.

125. Whitcombe D, Brownie J, Gillard HL, McKechnie D, Theaker J, Newton CR, Little S. A homogeneous fluorescence assay for PCR amplicons: Its application to real-time, single-tube genotyping. *Clin. Chem.* 1998, *44*, 918-923.

126. Whitcombe D, Theaker J, Guy SP, Brown T, Little S. Detection of PCR products using self-probing amplicons and fluorescence. *Nat. Biotechnol.* 1999, *17*, 804 - 807.

127. Wang S, Singh AK., Senapati D, Neely A, Yu H, Ray PC. Rapid colorimetric identification and targeted photothermal lysis of *Salmonella* bacteria by using bioconjugated oval-shaped Gold nanoparticles. *Chem. Eur. J.* 2010, *16*, 5600–5606.

128. WHO: Water Quality Interventions to Prevent Diarrhoea: Cost and Cost-Effectiveness. 2008.

129. WHO: Quantifying Selected Major Risks to Health: The World Health Report, 2002 (Chapter 4). World Health Organization, Geneva, 2003.

130. Yguerabide J, Yguerabide EE. Light-scattering submicroscopic particles as highly fluorescent analogs and their use as tracer labels in clinical and biological applications: i. theory. *Anal. Biochem.* 1998a, *262*, 137-156.

131. Yguerabide J, Yguerabide EE. Light-scattering submicroscopic particles as highly fluorescent analogs and their use as tracer labels in clinical and biological

applications: ii. experimental characterization. *Anal. Biochem.* 1998b, *262*, 157-176.

132. Zahraei ST, Tadjbakhsh H, Atashparvar N, Nadalian MG, Mahzounieh MR. Detection and identification of *Salmonella typhimurium* in bovine diarrhoeic fecal samples by immunomagnetic separation and multiplex PCR assay. *Zoon. Pub. Heal.* 2007, *54*, 231-236.

133. Zhao W, Chiuman W, Lam J, McManus SA, Chen W, Cui Y, Pelton R, Brook MA, Li Y. DNA aptamer folding on gold nanoparticles: from colloid chemistry to biosensors. *J. Am. Chem. Soc.* 2008a, *130*, 3610–3618.

134. Zhao W, Brook MA, Li YF. Design of Gold nanoparticle-based colorimetric biosensing assays. *Chem. Bio. Chem.* 2008b, *9*, 2363–2371.

135. Zipper H, Brunner H, Bernhagen J, Vitzthum F. Investigations on DNA intercalation and surface binding by SYBR Green I, its structure determination and methodological implications. *Nucleic Acids Res.* 2004, *32*, e103.

*Section II*

# Virology

**2016, Biotechnology: Progress and Applications** *Pages* **69–106**
*Editors:* **Saif Hameed and Zeeshan Fatima**
*Published by:* **DAYA PUBLISHING HOUSE, NEW DELHI**

# *Chapter 4*

# *In silico* RNA Interference: A Powerful Weapon against Virus Defense in Plants

***Shweta and Jawaid A. Khan****

*Plant Virus Laboratory, Department of Biosciences,*
*Jamia Millia Islamia, New Delhi – 110 025, India*

## ABSTRACT

Virus diseases are major threat for agricultural production. They cause severe damages to crops leading to huge economic losses. Upon virus infection, plant defence system triggers RNA interference (RNAi) pathway which is a homology-dependent gene silencing mechanism implicating degradation of complementary RNA. Employing small RNA molecules such as miRNA and siRNA, RNAi has been widely manipulated to engineer resistance in plants against viruses. Small RNA species have been demonstrated to play primary roles in several regulatory pathways inside the host. To counteract host-mediated RNA silencing, viruses have evolved several silencing suppressor genes. A large number of experimental evidences validate that inhibition of cytoplasmic RNA silencing occurs via virus derived suppressor genes. In this review, several conventional and non-conventional control measures to combat virus diseases have been described. Progression of *In silico* biology, involving miRNA and siRNA, equip the plant with one more arm in repertoire of tools against viruses. It holds great promise in delivering immunity against virus infection. Future investigations will provide more information about small RNA(s) expression profiles in response to virus

* Corresponding Author: E-mail: jawaidkhan14@yahoo.co.in

infection and other biotic and abiotic stresses. It will lead to better understanding of pathways, intricate virus-host interactions and developing efficient combinatorial approach to unravel the complex signalling network of plants and viruses.

***Keywords:*** *RNA interference, miRNA, siRNA, Gene silencing, Virus resistance.*

## Introduction

RNA interference (RNAi) is a homology-dependent flexible mechanism that implicates degradation of cellular RNA by a complex of enzymes in biological systems. The phenomenon was first called as "post-transcriptional gene silencing" (PTGS; Hamilton and Baulcombe, 1999). During RNAi mechanism, primary role is played by small RNAs [known as short interfering RNA (siRNA) and microRNA (miRNA)], that in turn bring about effects of silencing (Brodersen and Voinnet, 2006). In plants, RNAi is reported to be involved in regulation of developmental processes as well as in defense against aberrant foreign nucleic acids, viruses and transposons. In viruses, specificity of RNAi is determined by the sequence complementarity between the viral genome itself, and the small RNA either siRNA/miRNA, that acts as a guide and results down-regulation of cognate gene expression as well as viral DNA accumulation (Vanitharani *et al.*, 2003).

During infection cycle in plant hosts, DNA and RNA viruses produce double stranded RNA (dsRNA) structures. In RNA viruses, dsRNA intermediates are formed during replication while in DNA viruses overlapping transcripts from opposite strands are generated. Plants recognize dsRNA as foreign/aberrant molecules which are cleaved into (21-26 nt) siRNAs and (21-22 nt) miRNAs by a ribonuclease III-like enzyme called "Dicers" (Bernstein *et al.*, 2001) (Figure 4.1).

Viruses are inducers as well as targets of RNAi machinery. One strand of the siRNA/miRNA is incorporated into a ribonuclease complex known as the RNA-induced silencing complex (RISC) and serves as the guide for sequence-specific degradation of homologous mRNAs (Hammond *et al.*, 2000). In addition to cleavage of target mRNA, plant miRNAs also inhibit the translation of target genes if there is any mismatch around the central complementary region (Brodersen *et al.*, 2008) (Figure 4.2).

siRNAs complementary to promoter regions of target genes induce transcriptional gene silencing (TGS) leading to methylation of promoter and consequently inhibition of transcription. siRNAs/miRNAs complementary to mRNAs induce PTGS, which results in sequence-specific RNA degradation. Both PTGS and TGS are correlated, but TGS functions in nucleus while PTGS occurs in cytoplasm (Vaucheret *et al.*, 2001).

The significance of RNA silencing in host defence against viruses is apparent from the finding that viruses, as a counter defence, have developed silencing suppressor genes which operate to defeat host RNA silencing by interfering with several regulatory components of silencing machinery and weakens the host immunity. Till now numerous viruses have been demonstrated to encode one or more

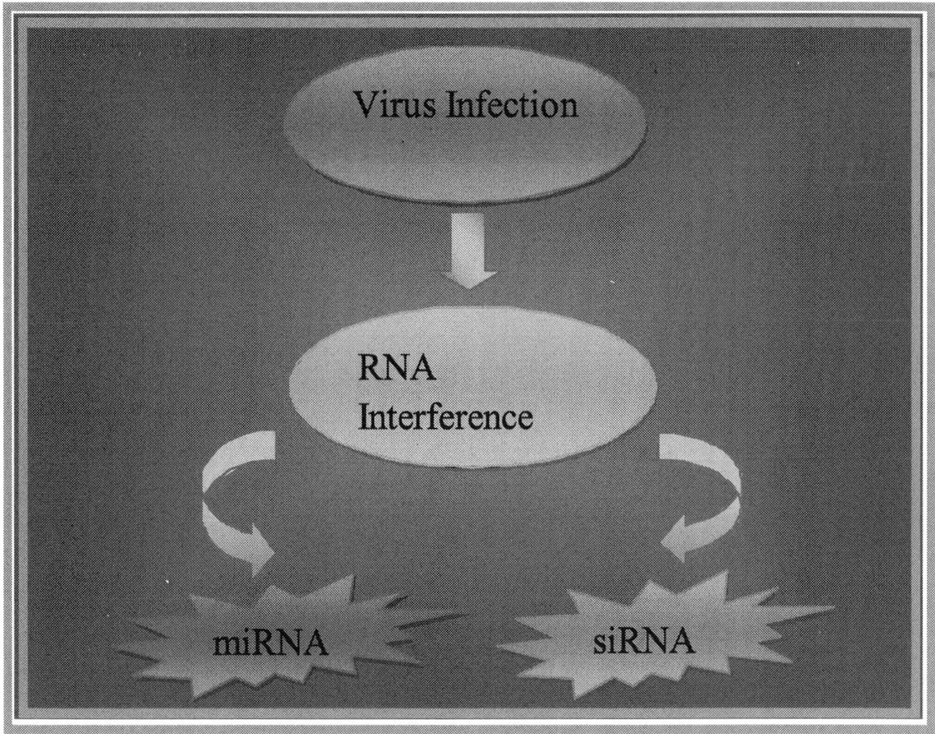


**Figure 4.1: Small RNA in Host Mediated RNAi Response. Upon virus infection, host mediated RNAi responses are induced. Virus infection triggers both short interfering RNA (siRNA) and microRNA pathway.**

silencing suppressor genes (Bivalkar- Mehla *et al.*, 2011; Voinnet *et al.*, 1999; Sharma *et al.*, 2012; Singh *et al.*, 2012).

P1/HC-Pro of the Potyvirus targets RISC assembly (Brigneti *et al.*, 1998; Anandalakshmi *et al.*, 1998; Kasschau *et al.*, 2003), 2b of *Cucumber mosaic virus* (CMV) binds to dsRNA and interferes with spread of silencing signal (Brigneti *et al.*, 1998; Qi *et al.*, 2004), p19 of tombusviruses binds to siRNAs (Voinnet *et al.*, 1999; Silhavy *et al.*, 2002; Qi *et al.*, 2004), p38 encoded by the *Turnip crinkle virus* binds AGO1 and inhibits the activity of DCL-4 (Thomas *et al.*, 2003; Azevedo *et al.*, 2010), P25 of *Potato Virus* X interferes with the spread of silencing signal (Voinnet *et al.*, 2000), P23, S, γb, P15, P0, P1, P30, P69, NS and coat protein of closterovirus (Reed *et al.*, 2003; Lu *et al.*, 2004), comovirus (Liu *et al.*, 2004), hordeivirus (Yelina *et al.*, 2002), pecluvirus (Dunoyer *et al.*, 2002), polerovirus (Pfeffer *et al.*, 2002), sobemovirus (Voinnet *et al.*, 1999), tobamovirus (Kubota *et al.*, 2003), tymovirus (Chen *et al.*, 2004), tospovirus (Bucher *et al.*, 2003) and carmovirus (Qi *et al.*, 2004), respectively, inhibit processing of dsRNA and serve as pathogenicity determinant. Similiarly *V2*, *C2*, *C4* (encoded by DNA-A) and the β*C1* genes of begomoviruses have been shown to have suppressor activity (Voinnet *et al.*, 1999; Van Wezel *et al.*, 2002; Amin *et al.*, 2011a) (Table 4.1).

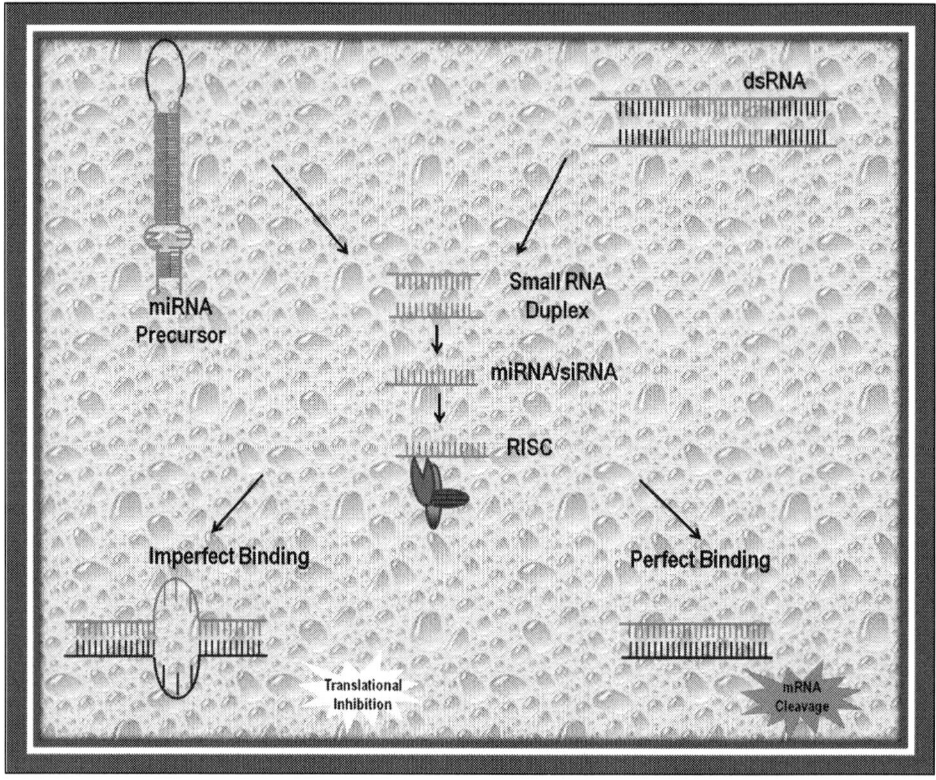


**Figure 4.2: RNA Interference Pathway and its Components.**

**Small RNA duplex gets converted into single-stranded miRNA/siRNA. miRNA/siRNA conjoin with RNA induced silencing complex (RISC) and bind to target transcripts. Depending upon the degree of complementarity between target and miRNA/siRNA complex; target is either cleaved or undergoes translation inhibition.**

DNA viruses like geminiviruses, replicate in the nucleus and within their replication cycle have no dsRNA phase, are known to trigger PTGS in plants with the production of virus-specific siRNA (Lucioli *et al.*, 2003; Chellappan *et al.*, 2004). Contrary to RNA viruses that can be controlled only by PTGS, geminiviruses may be manipulated by both PTGS and TGS. In a transient assay, TGS was reported to be effective against the begomovirus *Mungbean yellow mosaic virus* (Pooggin *et al.*, 2003). TGS is introduced when siRNAs corresponding to the promoter regions are produced that direct methylation of the promoter followed by inhibition of transcription (Mette *et al.*, 2000).

Viral invasiveness is reported to be promoted by the ability of *C1* to suppress PTGS (Cui *et al.*, 2005). It has been shown that *C4* gene of *Cotton leaf curl Multan virus* and β*C1*gene of Cotton leaf curl Multan betasatellite bind short RNAs, with a preference for the double stranded and single stranded forms, respectively, implying that *C4* and β*C1*genes sequester siRNAs and prevent their incorporation into the RISC involved in sequence specific mRNA degradation (Hammond *et al.*, 2000).

**Table 4.1: Virus Encoded Silencing Suppressor Proteins in Plant**

| *Virus Genus* | *Virus Species* | *Viral Suppressor* | *Function* | *References* |
|---|---|---|---|---|
| | | **Positive-strand RNA viruses** | | |
| *Carmovirus* | *Turnip Crinkle virus* | P38 | Coat protein, Binds Ago | Azevedo ***et al.*** (2010), Thomas ***et al.*** (2003) |
| *Cucumovirus* | *Cucumber mosaic virus, Tomato aspermy virus* | 2b | Host-specific movement, Binds dsRNA | Brigneti ***et al.*** (1998), Qi ***et al.*** (2004) |
| *Closterovirus* | *Beet yellows virus, Citrus tristeza virus* | P21, P20, P23, CP | Replication enhancer Nucleic-acid binding, Coat protein | Reed ***et al.*** (2003), Lu ***et al.*** (2004) |
| *Comovirus* | *Cowpea mosaic virus* | S protein | Small coat protein | Liu ***et al.*** (2004) |
| *Hordeivirus* | *Barley yellow mosaic virus* | γb | Replication enhancer, movement, seed transmission, pathogenicity determinant | Yelina ***et al.*** (2002) |
| *Pecluvirus* | *Peanut clump virus* | P15 | Movement | Dunoyer ***et al.*** (2002) |
| *Polerovirus* | *Beet western yellows virus, Cucurbit aphid-borne yellows virus* | P0 | Pathogenicity determinant | Pfeffer ***et al.*** (2002) |
| *Potexvirus* | Potato virus X | P25 | Interfere with spread of silencing signal, Movement | Voinnet ***et al.*** (2000) |
| *Potyvirus* | *Potato virus* Y, *Tobacco etch virus, Turnip yellow virus* | P1/HcPro | Movement, polyprotein processing, aphid transmission, pathogenicity | Brigneti ***et al.*** (1998) Anandalakshmi ***et al.*** (1998) |
| | | | determinant, target RISC | Kasschau ***et al.*** (1998) |
| *Sobemovirus* | *Rice yellow mottle virus* | P1 | Movement, pathogenicity determinant | Voinnet ***et al.*** (1999) |
| *Tombusvirus* | *Tomato bushy stunt virus, Cymbidium ringspot virus, Carnation Italian ringspot virus* | P19 | Movement, pathogenicity determinant, Binds siRNA | Silhavy ***et al.*** (2002), Qi ***et al.*** (2004), Voinnet ***et al.*** (1999) |
| *Tobamovirus* | *Tobacco mosaic virus, Tomato mosaic virus* | P30 | Replication | Kubota ***et al.*** (2003) |
| *Tymovirus* | *Turnip yellow mosaic virus* | P69 | Movement, pathogenicity determinant | Chen ***et al.*** (2004) |

*Contd...*

**Table 4.1**–*Contd...*

| *Virus Genus* | *Virus Species* | *Viral Suppressor* | *Function* | *References* |
|---|---|---|---|---|
| | | **Negative-strand RNA viruses** | | |
| *Tospovirus* | *Tomato spotted wilt virus* | NSs | Pathogenicity determinant | Bucher ***et al.*** (2003) |
| | | **Single-stranded DNA viruses** | | |
| *Begomovirus* | *African cassava mosaic virus* | AC2/C2 | Transcriptional activator protein (TrAP) | Voinnet ***et al.*** (1999), Van Wezel ***et al.*** (2002) |
| | *Tomato yellow leaf curl virus, Mungbean yellow mosaic virus* | | | |
| | *Cotton leaf curl Multan virus* | V2, C2, C4 βC1 | Pre-coat protein, TrAP | Amin ***et al.*** (2011a) |

V2, a unique protein of monopartite begomoviruses has been shown to suppress PTGS in transient assays (Zrachya *et al.*, 2007). It was demonstrated that ability of V2 to interact with SISG3, the tomato homologue of *Arabidopsis* SGS3 leads to suppression of PTGS (Glick *et al.*, 2008). Furthermore, it was shown that the Rep proteins encoded by two alphasatellites, Gossypium darwinii symptomless alphasatellite and Gossypium mustelinium symptomless alphasatellite possess suppressor activity (Nawaz-ul-Rehman *et al.*, 2010).

AC2 protein mediates inhibition of PTGS as evidenced by transcriptional profiling followed by transient expression of AC2 protein in *Arabidopsis* protoplasts. During this experiment several genes were found to be over expressed and one of them was a homologue of Werner syndrome-like exonuclease (WEX), Werner exonuclease-like 1 (WEL1) (Trinks *et al.*, 2005). WEX is required to direct PTGS (but not TGS) against transgenes (Glazov *et al.*, 2003), and it was speculated that WEL1 over-expression might compete for factors required for WEX function (Trinks *et al.*, 2005).

The *AC4* gene resides completely within the *Rep* coding region, but within different reading frame. Although *Rep* is most conserved, *AC4* is the one of the least conserved of all geminivirus genes. Silencing-active AC4 proteins interfere with RISC loading by acting downstream of small RNA biogenesis and duplex unwinding, thus facilitating the degradation of single-stranded miRNAs and siRNAs (Bisaro *et al.*, 2006).

Expression of *AC4* gene of *African cassava mosaic virus* (ACMV) causes developmental defects. These defects occurred due to reduced accumulation of specific miRNAs and over-accumulation of their mRNA targets. *In vitro* studies showed that single-stranded miRNA and siRNA binds to *AC4* gene of ACMV. Complementary miRNA oligonucleotide (miR159*) worked as bait to pull down *AC4* gene from the protoplast extracts of ACMV (Chellappan *et al.*, 2005).

It has been reported that mixed infection of cassava by ACMV (recovery-type, with relatively strong AC4 suppressor) and *East African cassava mosaic virus* (EACMV; non-recovery-type, with a relatively strong AC2 suppressor) caused very unusual severe disease symptoms in the field of cassava characterised as synergistic disease resulting from mixed infection. On further investigation, AC4 proteins of ACMV showed suppressor activity with increased accumulation of GFP mRNA and inhibition of GFP-specific siRNAs. While AC4 protein from EACMV showed little or no activity in this assay. Contrary to this, the AC2 protein of the non-recovery viruses (EACMV) turned up as effective silencing suppressors, while that from recovery-type viruses (ACMV) was found to be less effective (Vanitharani *et al.*, 2005).

All these findings validates that silencing suppressor genes obstructs cytoplasmic RNA silencing, and involuntarily the miRNA pathway, via binding to single-strand of siRNA and miRNA. One question is very prevalent in the context of these reports, that why the appearance of virus disease complex requires so many suppressors. There is a need of devising resistance considering all the suppressor genes of all the available isolates of all the species of specific viruses because DNA viruses like geminiviruses (unlike RNA viruses) employ multi protein strategy concept where more than one gene has silencing suppressor activity.

## Small RNA Molecules and their Role in RNA Interference

### miRNA

Till now several studies illustrating the role of miRNAs in mediating RNAi have been reported. A study showing downregulation in expression pattern of conserved miRNA families (miR319 and miR172) on microarray analysis of tomato plants agro-infected with *Tomato leaf curl New Delhi virus* (TLCNDV) was demonstrated (Naqvi *et al.*, 2010). In another experiment on begomovirus infection, miRNAs expressed in plant developmental processes were found to be up-regulated followed by suppression of corresponding endogeneous targets (squamosa promoter binding protein, cup-shaped cotyledon, phabulosa and phavoluta) leading to developmental defects. These findings suggest that miRNAs expression and the virus infection are manipulated in such a way that several viral proteins exclusively stimulate a common set of miRNAs which direct to the symptom development (Amin *et al.*, 2011).

Upon virus infection, RNA silencing is induced in plants (Burgya'n and Havelda 2011; Sharma *et al.*, 2012; Muthamilarasan and Prasad 2013). RNA silencing is regulated following the generation of host-derived miRNAs and virus-induced siRNAs. Several studies have demonstrated the role of artificial miRNA in post-transcriptional control against viral infection (Table 4.2).

**Table 4.2: Plant miRNA Precursors Manipulated for amiRNA Synthesis against Virus Species**

| *miRNA Precursor* | *Virus Species* | *Transformed Plant Species* | *Reference* |
|---|---|---|---|
| miR159 | *Turnip yellow mosaic virus, Turnip mosaic virus* | *Arabidopsis thaliana* | Niu *et al.* (2006) |
| miR171 | *Cucumber mosaic virus* | *Nicotiana tabacum* | Qu *et al.* (2007) |
| miR159a, miR167b and miR171a | *Potato virus X* | *N. tabacum* | Ai *et al.* (2011) |
| miR159a | *Tobacco mosaic virus, Tomato yellow leaf curl virus* or *Cucumber mosaic virus* | *Solanum lycopersicon* | Zhang *et al.* (2011) |
| miR319a | *Tomato leaf curl New Delhi virus* | *S. lycopersicon* | Yadava and Mukherjee, 2010 |
| miR169a | *Cotton leaf curl Burewala virus* and *Cotton leaf curl Kokhran virus* | *N. benthamiana* | Vu *et al.*, 2012 |

Niu *et al.* (2006) were among the first to successfully apply miRNA-based defense strategy against viruses. They showed that small miRNAs can be engineered to protect plants against the infection by tymoviruses and potyviruses. They engineered miRNA precursor of abundant natural plant miR159 a 273-nt backbone with complementary sequences of viral suppressors of RNA silencing (VSR), *viz.*, P69 of *Turnip yellow mosaic virus* and HC-Pro of *Turnip mosaic virus*. *Arabidopsis thaliana* plants transformed with these recombinant artificial miRNA precursors acquired immunity against infection of these viruses.

Transgenic *Nicotiana tabacum* expressing an amiRNA derived from miR171 precursor of *A. thaliana* targeting another VSR *i.e.* 2b gene of *Cucumber mosaic virus* (CMV), showed resistance against CMV (Qu *et al.*, 2007). This study also demonstrated that expression of miRNA was more effective than that of short hairpin sRNA.

In similar experiments, protection was achieved against *Potato virus X* (PVX) in *N. tabacum* expressing amiRNAs (developed from *A. thaliana* miR159a, miR167b and miR171a precursors) targeting TGBp1 D p25 domains of PVX (Ai *et al.*, 2011). Later, tomato plants expressing amiRNAs against highly conserved 3′ untranslated region and the overlapped coding sequence of RNA-dependent RNA polymerase, 2a protein and the suppressor 2b protein genes of CMV demonstrated resistance even following infection with *Tobacco mosaic virus*, *Tomato yellow leaf curl virus* or CMV (Zhang *et al.*, 2011).

Silencing suppressor genes, such as *AC2* or *AC4* impart pathogenic attributes to begomoviruses, appear to be the important targets for designing antiviral strategy. Transgenic tomato plants expressing amiRNA targeting *AC2* or *AC4* genes of TLCNDV led to the successful production of resistant transgenic tomato lines offering a great degree of resistance towards the challenge of *Tomato leaf curl virus* (Yadava and Mukherjee 2010).

Similarly, *N. benthamiana* transformed with cotton miRNA169a, sharing homology with the *AV2* gene of *Cotton leaf curl Burewala virus* showed good resistance against homologous as well as *Cotton leaf curl Kokhran virus* (Vu *et al.*, 2012).

An increasing number of experimental evidences draw attention towards competitive struggle for survival that takes place between hosts and viruses *in vivo* around the RNA silencing system. It seems that miRNA-mediated plant defence responses have emerged as relevant components of the innate immune system and some definite miRNA families that are evolutionary conserved and abundantly expressed in plants, appear to provide immunity to plants against viral infections (Ai *et al.*, 2010).

## siRNA

RNA-based antiviral immunity is controlled by the dsRNA–siRNA pathway. Production of virus derived small RNA is dependent on the Dicer protein that produces siRNAs from a viral dsRNA precursor. The first transgenic geminivirus resistant plant in the field was produced by manipulating dsRNA constructs designed to silence the *AC1* gene of *Bean Golden Mosaic Virus* in common bean (Bonfim *et al.*, 2007). Similarly, Vanderschuren, (2009) engineered resistant transgenic cassava against ACMV following expression of homologous hairpin dsRNAs of *AC1* gene of ACMV (Vanderschuren *et al.*, 2009).

In plants during RNA based antiviral immunity viral siRNAs are produced and these siRNAs have overlapping sequences (Wu *et al.*, 2010; Donaire *et al.*, 2009; Wang *et al.*, 2010). So, on cloning viral siRNAs quickly assembled back into large contigs and cover the complete genomic length of infecting viruses (Wu *et al.*, 2010; Kreuze *et al.*, 2009). In this way, virus-derived siRNAs are different from miRNAs, which are excised from single stem of a stem–loop precursor (Meyers *et al.*, 2008). However, siRNA based technique is oppressed with high risk factor of possible recombination between the transgene and other non targeted viruses.

Subsequent studies have demonstrated the role of virus-derived small RNAs and RNA-based antiviral immunity in plants against positive-strand RNA, negative-strand RNA, single-stranded DNA and double-stranded DNA viruses (Table 4.3).

**Table 4.3: Virus-derived siRNAs Targeting Plant Viruses**

| *Viral Genome* | *Virus Name* | *Viral Suppressor of RNA Silencing* | *Reference* |
|---|---|---|---|
| Positive-strand RNA | *Cucumber mosaic virus* | 2b | Wang XB *et al.* (2010), Diaz-Pendon *et al.* (2007) |
| | *Cymbidium ringspot virus* | P19 | Szittya *et al.* (2010), Szittya *et al.* (2002) |
| | *Potato virus X* | P25 | Hamilton and Baculombe (1999) |
| | *Tobacco rattle virus* | 16K | Deleris *et al.* (2006), Donaire *et al.* (2008,2009) |
| | *Turnip crinkle virus* | P38 | Deleris *et al.* (2006) |
| | *Turnip mosaic virus* | HC-Pro | Garcia-Ruiz *et al.* (2010) |
| Negative-strand RNA | *Tomato yellow ring virus* | NS(s) | Hassani-Mehraban *et al.* (2009) |
| Single-stranded DNA | *African cassava mosaic virus* | AC2 | Chellappan *et al.* (2004) |
| | *Cabbage leaf curl virus* | AL2 | Blevins *et al.* (2006) |
| Double-stranded DNA | *Cauliflower mosaic virus* | P6 | Blevins *et al.* (2006), Moissiard *et al.* (2006) |

Deep sequencing has validated that virus-derived siRNAs, characteristically cover every genomic position of many positive (+) strand RNA viruses and dsRNA viruses in plant, accompanied with those viruses that are bias for (+) strand to induce virus-derived small RNAs (Donaire *et al.*, 2009; Qi *et al.*, 2009; Wang *et al.*, 2010; Szittya *et al.*, 2010). *Tobacco rattle virus, Turnip mosaic virus* and *Cymbidium ringspot virus,* like (+) strand RNA viruses, produce one or more 3t co-terminal subgenomic RNAs and abundance of these RNAs is more than genomic RNA. But the density of virus-derived small RNAs targeting inside and outside of the subgenomic RNA-coding region of the viral RNA genome is independent to the ratio of subgenomic RNA to genomic RNA (Donaire *et al.*, 2009; Qi *et al.*, 2009; Wu *et al.*, 2010). *A. thaliana* plants infected with CMV mutant, deficient in viral suppressor of RNA silencing, showed plenty of 5t-terminal viral siRNAs. On the contrary to this, no 5t-terminal bias of viral siRNAs production was observed following infection of *N. benthamiana* with similar mutant of TuMV, CymRSV and various other (+) strand RNA viruses of plant (Donaire *et al.*, 2009; Wang *et al.*, 2010; Szittya *et al.*, 2010; Garcia-Ruiz *et al.*, 2010). All these findings established the role of siRNA in defense against different viruses in various plants.

Various reports validated that, among others, intergenic region that includes replication initiation site and bidirectional promoter of geminivirus genomes churn out siRNAs in response to resistance (Akbergenov *et al.*, 2006; Sahu *et al.*, 2010; Yadav and Chattopadhyay 2011; Sahu *et al.*, 2012b). Thus, genome-wide identification of regions that could generate significant amount of potential siRNAs is one of the potential areas for further progress in the gene silencing studies.

## Strategies to Combat Virus Infection

Plant viruses are responsible for huge economic losses. They are responsible for reduction of yield as well as quality of huge variety of crops. To protect crops from viral diseases, several control measures are employed. These control measures can be classified into two major classes: Conventional and Non-conventional.

### Conventional Control

Conventional management strategies to control viral diseases in crops include intercropping, planting of resistant cultivars, control of insect vectors, eradication of weeds and alternate hosts that serve as reservoirs of virus primary inoculum during off season (Khan and Ahmad, 2005).

Eradication of alternate host or weed is carried out to minimize the disease occurrence. In Sudan planting of okra is prohibited for at least 2 months before or 1 month after the sowing of cotton as okra is a favourite host to *Cotton leaf curl virus* (Sastry and Zitter, 2014). Occurring of tungro disease in rice plant caused by *Rice Tungro Virus* (RTV) in Philippines decreased significantly when planting of rice was done after a gap of 2 month (Chancellor *et al.*, 2006). Alternative host removal seems to lead decreased disease pressure. But the success of this conventional step is determined by the host range and mobility of viruses.

In several crops, intercropping has reduced the occurrence of virus disease contrary to cultivation of single or mono crop. Intercropping of cassava (*Manihot esculenta*) crop with maize, cowpea and peanut decreases both the attack of the whitefly (*Bemisia tabaci*), insect vector as well as incidence of cassava mosaic disease (Page *et al.*, 1999; Fondong *et al.*, 2002).The incidence of *Beet mild yellowing virus* (BMYV) in sugar beet (*Beta vulgaris*) is reduced when intercropped with mustard or barley (Sastry and Zitter, 2014). Page *et al.* (1999) reported that intercropping of beans and millets with maize crop reduced the prevalence of leafhoppers transmitted *Maize streak virus* in maize fields. Infection of lettuce with *Tomato spotted wilt virus* was reduced on planting of cabbage crop at 15 m-wide distance from lettuce (Coutts *et al.*, 2004). So, intercropping appear to be an effective measure against virus infection and dissemination.

It is a common practice to select the date for planting when the occurrences of viruliferous insect vectors are least and young plants get exposure to minimum infection. Disease incidence is reported less in many crops when plants are planted early as they become older and resistant by the time transmission vectors attacked the crop field. Younger infected plants suffers more than late flowering or fruiting stage plants so planting dates should be decided such that maximum vegetative growth phase never coincides with the maximum vector population period. There are several crops in which positive results were seen on employing proper planting dates such as for groundnut rosette, *Peanut mottle virus* in peanut, *barley yellow dwarf virus* in wheat (Sastry and Zitter, 2014). In Indian subcontinent, cotton plants showed decreased incidence of cotton leaf curl disease (CLCuD) when sowing was done between mid april to mid may as compared to delayed sowing from mid May to June (Farooq *et al.*, 2011). The severity of CLCuD reached to its peak 105 days followed early sowing. In case of late sown crop, *i.e.* (15 June or later) infestation became severe just within 45 days of sowing (Farooq *et al.*, 2011). When crops are sown near the old infected field during the season of peak vector population, viral diseases spread to the new crop very easily. And spread decreases with the increase of distance from the site of infection. The estimation of minimum isolation distance to minimize the virus spread is very difficult because it depends on various parameters such as, the type of virus, transmitting vector, vector density, direction of wind and several others. It is reported that extent of mosaic disease caused by aphid borne *Beet mosaic virus* and *Beet yellows virus* in sugar beet decreased parabolically with growing distance of sugar beet planting in fields of California (Sastry and Zitter, 2014). Increased plant spacing on early sowing and decreased plant spacing under late sown conditions is promising factor in management of CLCuD (Farooq *et al.*, 2011). 15 cm plant spacing during planting has also been recommended for late sowing in order to facilitate CLCuD management in the case of planting later than 15th of June. All these strategies involving management of insect vectors, appears to be an effective strategy against viruses. But control of vectors is not always practicable or successful to control virus transmission. Vectors like aphids and plant hoppers can travel distance of continents and oceans. Isolation strategy works in case of stylet-borne viruses because there is chance that they will lose virulence in long flights and suck the sap of non-susceptible host and transfer the virus to it. Isolation did not work in case of persistent viruses because once the viruliferous vectors acquire the viruses, it keeps on transmitting it through the entire life.

Effect of soil health in terms of nutrition found to play role in viral disease occurrence. In resistant cotton cultivars nitrogen concentration has hardly any affect, but in susceptible varieties nitrogen concentration plays a crucial role to combat disease severity. Similiarly, use of potassium in field upto 250kg/ha leads to the reduction of disease from 12 to 38 per cent (Farooq *et al.*, 2011). Therefore, strategies based on understanding of the physiological requirement of nutrition, could be designed to deal with viral diseases.

The use of molecular markers associated with virus disease resistance in plants has the potential to improve the efficiency of selection in breeding programs. Rice cultivar IR-64 containing *bph1*a dominant gene for resistance against planthopper, vector of *rice grassy stunt virus* showed better resistance against disease (Sastry and Zitter, 2014). Mapping of BPH-resistance gene-DNA markers is done and the markers are utilized in breeding programs (Sastry and Zitter, 2014). Similarly DNA marker-based assay utilizing DNA markers linked with CLCuV resistance gene is used for identification and selection of resistant cultivars (Farooq *et al.*, 2011). So, virus resistant cultivars could be selected using DNA markers. It minimizes possibility of virus spread into a new environment unlike other methods where infection with pathogen is required to identify the resistance.

Numerous resistant lines were developed utilizing wild resistance source through breeding programme such as P7 (*Abelmoschus esculentus* x *A. manihot* ssp *manihot*) and Parbhani Kranti (*A. esculentus* x *A. manihot*) like okra cultivars against *yellow vein mosaic virus* and many more (Sastry and Zitter, 2014). Resistance genes against *Tomato yellow leaf curl virus* (TYLCV), one of the major constraints for tomato production worldwide, were isolated and used in breeding programs from several wild species such as *Solanum pimpinellifolium*, *S. peruvianum*, *S. habrochaites*, and *S. chilense* (Sastry and Zitter, 2014). Barley germplasm (PI467884) was identified showing resistance against seed transmission of three isolates of *barley stripe mosaic virus* and 83 wild *Arachis* species were screened as resistance source against *Peanut bud necrosis virus* (Sastry and Zitter, 2014). CoPusa-3, N-2-1 and V-16 genotypes of *Vigna unguiculata* showed inherited seed transmission resistance against *Cucumber mosaic virus*, *Blackeye cowpea mosaic virus*, *Cow parsnip mosaic virus* and *Tobacco mosaic virus* (Sastry and Zitter, 2014). Wild *Gossypium* species (*G. thurberii*, *G. anomalum*, *G. raimondii*, *G. armourianum*, and *G. tomentosum*) have resistance genes to provide protection against insect pests such as boll worms, jassids, whitefly and mites and pathogen borne diseases (bacterial blight, and Verticillium wilt) (Farooq *et al.*, 2011). *G. arboreum*, resistant against CLCuD (Briddon and Markham, 2000) fungal and bacterial diseases (Farooq *et al.*, 2011), has been exploited for isolation of resistance genes and their incorporation into susceptible varieties via genetic transformation, such as two cotton lines, NIBGE-2 and NIBGE-115 were generated (Farooq *et al.*, 2011). During 1990s conventional breeding approaches involving indigeneous resistant varieties LRA5166 and CP-15/2 set back CLCuD to gloom. These varieties involving three genes, showed excellent resistance against 'Multan' strain of CLCuD. Out of these, two were responsible for resistance and one gene for resistance against suppressor gene (Rahman *et al.*, 2005). So for efficient management of viral disease, resistance inducing genes predominantly from the wild genotypes need to be incorporated into the commercial varieties.

But unfortunately, several instances of resistance failure has been recorded for many viruses like *cassava mosaic virus* in cassava, *tomato spotted wilt virus* in tomato, geminiviruses infecting legumes, cotton and tomato. For instance, second CLCuD epidemic came in the year 2001 after introduction of S12 cotton varieties of America in Pakistan due to emergence of *Cotton leaf curl Burewala virus* (CLCuBV) (Amrao *et al.*, 2010). Subsequently, CLCuD spread towards east out of Pakistan into the cotton-growing states of north-western India and several virus species like CLCuMV, CLCuKV and CLCuBV identified in Pakistan were identified from India along with *cotton leaf curl Rajasthan virus* (CLCuRV) (Rajagopalan *et al.*, 2012). Since then CLCuD is a limiting factor for cotton cultivation in Pakistan and north-western India. North-western region includes about 10 million hectares spread over the states of Punjab,

Haryana and Rajasthan, comprises 16 per cent of the total cultivated land where cotton is grown, and contributes 20 per cent to the total cotton production in India. More than a decade has been passed since the emergence of the very devastating "Burewala" strain of CLCuV, but still plant breeders are struggling to develop a resistant line of cotton against CLCuD infection (Farooq *et al.*, 2011). Some promising cultivars of cotton showing tolerance against CLCuV are available. But they still undergo virus infection, demonstrate mild symptoms and lead to systemic spread of virus. These tolerant cultivars may serve as pool of onward virus transmission and source of new destructive recombinant strain of virus (Rahman and Zafar, 2007). Thus, plant breeding appears to be failed in introducing disease resistance among crop cultivars that could provide a sustainable resistance for a longer period.

### Non-conventional Control

Conventional techniques have one major drawback; it varies from climate to climate and on available resistance sources. Introgression of resistance traits during breeding is also an issue and it takes very long time. With the advent of genetic engineering, modern molecular and transformation techniques, it has become easy to understand the functioning of viral genes. Presently, several crop plants having internal defense are developed using genetic engineering that are resistant against certain pathogens (Farooq *et al.*, 2011).

The genetic engineering strategies have resulted into minimal genetic diversity against pathogen resistance. It involves either pathogen-derived approach (sequence-derived from the pathogen is used to tailor the resistance) or non-pathogen-derived approach, where sequence derived from other heterologous sources is manipulated. Both of these strategies have been employed to develop transgenic resistance against viruses with a different level of achievements.

Pathogen derived resistance, based on RNA-mediated (sense and anti-sense RNA mediated) or protein-mediated resistance, is applied successfully for those plants which are susceptible against respective viruses (Vanderschuren *et al.*, 2007; Shepherd *et al.*, 2009; Raja *et al.*, 2010). For broad base resistance against viruses in plants, pyramiding of virus encoded genes is necessary rather than single one. Wide spectrums of genes are integrated in the genomes of plants to develop resistance against pathogens, which usually lack naturally (Farooq *et al.*, 2011).

## Approaches to Induce RNA Interference

Untill now, a wide range of antiviral resistance strategies have been examined like expression of viral mutated proteins (Lucioli *et al.*, 2003), antisense RNAi constructs of viral sequences (Shepherd 2009; Vanderschuren *et al.*, 2009), 21–25 nt-long siRNAs (Ramesh *et al.*, 2007), artificial micro-RNAs (amiRNAs) (Frizzi *et al.*, 2010) as well as transgenic constructs of chimeric origin (Lin *et al.*, 2011). Generally, these RNAi strategies have been utilized by employing entire, partial or mutated sequences of Rep or CP genes and the non-coding intergenic region (IR) of viruses to achieve resistance against concerned viral infection (Duan *et al.*, 2012).

## Transgene-mediated Resistance using Sense or Antisense Viral Sequences

In sense or antisense transgene-induced PTGS pathway, plant hosts recognize single- stranded RNA sequence and plant-encoded RNA-dependent RNA polymerase amplifies this exogenous aberrant transgenic sequence into dsRNA. DsRNA serves as a substrate to initiate RNA silencing following production of siRNA that binds to complementary pathogen gene and results in its cleavage (Dalmay *et al.*, 2000). The first report of homologous sequence-dependent RNA silencing mechanism came in year 1998. Well before this breakthrough, virologists had discovered that transgenic tobacco plants expressing viral coat protein (CP) of *Tobacco mosaic virus* (TMV) were resistant against infection by the homologous virus (Abel *et al.*, 1986). This type of protein-mediated resistance has been reported in diverse viruses such as tobamovirus, potexvirus, cucumovirus, tobravirus, carlavirus, potyvirus, alfalfa mosaic virus groups and luteovirus (Abel *et al.*, 1986; Beachy *et al.*, 1990; Kawchuk *et al.*, 1990; Lindbo *et al.*, 1992; Jan *et al.*, 1999).

### Coat Protein

One of the most favourite transgenes to produce virus resistant plants is coat protein (CP). Appreciable resistance level against virus infection has been reported in several transformed plants. A highly unpredictable resistance mechanism was reported on employment of CP gene-mediated defence against tospoviruses. When CP was used against similar isolates of tospoviruses it was RNA driven while in case of distant one, it was protein derived resistance (Pang *et al.*, 1994). In transgenic potato plants CP gene of *Potato mosaic virus* (PMV) strain N605 reported to provide resistance against both the N605 strain as well as 0803 strain (Malnoe *et al.*, 1994). Likewise, transgenic tobacco plants expressing a CP gene of TMV found to be resistant to TMV and other closely related species of TMV (Beachy, 1990). While contrary to this, transgenic papaya expressing CP gene of *Papaya ringspot virus* (PRSV) strain HA found to be resistant only for HA strain of PRSV (Tennant *et al.*, 1994). Even in geminiviruses, *AV1* gene was the first gene that is used to confer resistance against homologous closely related viral species infections and still remains one of the most widely used gene to confer resistance against geminiviruses. In some cases *AV1* gene had offered broad protection and found to be effective against several strains of the viruses (Prins *et al.*, 2008). So, as per these findings it could be concluded that viral coat protein mediates either broad or narrow resistance. Afterwards, several viral proteins have been used to engineer resistance against viruses, like movement protein (Sijen *et al.*, 1996), replication-associated protein (Canto *et al.*, 1998, Chellappan *et al.*, 2004), the potyvirus nuclear inclusion proteins (NIa and NIb) (Guo *et al.*, 1998), viral suppressor proteins of RNA silencing, and some other viral proteins (Fagoaga *et al.*, 2006).

### Replicase Protein

It has been proposed that replicase-protein mediated resistance repress virus replication by interfering with the functions of virus replicases like replication regulation and other viral gene expression (Beachy, 1997). Contrary to wild replicase, mutant replicase of *Potato virus Y* (Audy *et al.*, 1994), *Alfalfa mosaic virus* (AMV)

(Brederode *et al.*, 1995) and *Tomato yellow leaf curl virus* (Yang *et al.*, 2004) conferred better resistance against viral infection. Transformation of tobacco plants with replicase gene of CMV and TMV conferred resistance by both transgene mRNA mediated RNAi, and by its expressed protein mediated resistance (Goregaoker *et al.*, 2000). Transgenic tobacco plants expressing mutant replicases of CMV subgroup I conferred high levels of resistance against all subgroup I CMV strains but not against subgroup II strains of CMV or other viruses (Morroni *et al.*, 2008). Similarly, a mutant replicase conferred resistance to infecting *Potato mosaic virus* (Audy *et al.*, 1994) and AMV while wild one was failed in conferring resistance (Brederode *et al.*, 1995). Truncated replicase gene (*AC1*) of *Cotton leaf curl Kokhran virus* expressing its N-terminal (669nt) and C-terminal (783nt) were used in antisense orientation in *N. Tabacum* to engineer transcriptional control against CLCuV (Asad *et al.*, 2003). Similiar gene constructs were used to transform *G. hirsutum* which had shown delayed symptoms on exposure with viruliferous whitefly (Hashmi *et al.*, 2011). In geminiviruses, properties of *AC1* gene such as conservation, involvement in the modulation of geminiviral gene expression and ability to influence viral replication by decreasing viral DNA accumulation (Chellappan *et al.*, 2004a; Abhary *et al.*, 2006) has made *AC1* gene an outstanding target to engineer broad resistance against geminiviruses.

**Movement Protein**

Viral movement proteins (MPs) play role in spread of infection both cell to cell as well as systemic. Transgenic plants expressing movement protein MP of TMV demonstrate resistance against not only against several tobamoviruses but also towards AMV, *Cauliflower mosaic virus* and many other distantly related viruses (Cooper *et al.*, 1995). Mutated MP of *White clover mottle virus* comprised disordered nucleotide binding site, displayed resistance against various potexviruses (Cooper *et al.*, 1995). Transgenic *N. occidentalis* plants expressing a movement protein (P50) with partially functional deletion mutants (DeltaA and DeltaC) of the *Apple chlorotic leaf spot virus* showed resistance against *Grapevine berry inner necrosis virus* (Yoshikawa *et al.*, 2006). So, transgenic plants expressing mutant MPs efficiently inhibit movement of viruses and success of these experiments, encourage utilisation of MPs to generate antiviral resistance.

**Intergenic Region Sequences**

Intergenic region (IR), that carries promoters and viral origin of replication, reported to mediate antiviral defence. Transient expression of intron spliced hairpin IR sequences of the bipartite begomovirus *Mungbean yellow mosaic virus* (MYMV) provided complete recovery from diseased state in blackgram plants infected with MYMV via siRNA based RNAi (Pooggin *et al.*, 2003).

**Protease Protein**

Protease protein was also used to acquire resistance against virus. In one of the studies, ability of the gene to mediate DNA-directed RNAi against RNA virus, *Potato virus Y* in tobacco was investigated, by evaluation of the constructs expressing *protease* (Pro) gene transcripts of sense, antisense and both polarities (Waterhouse *et al.*, 1998). It was observed in both cases, that duplex RNA expressing both the polarities of protease (Pro) gene simultaneously was more effective than single expression of either sense or antisense RNA.

Earlier, it was supposed that resistance could be conferred only by the expression of transgene encoded viral proteins. However, later on several reports came where resistance against virus disease was acquired in plants even on expression of sense or antisense truncated viral protein sequence or the sense non-coding viral RNA sequences (Fagoaga *et al.*, 2006). And all these studies demonstrate that during viral infection resistance could be generated via RNAi pathway, by expressing viral gene in sense and/or antisense orientation.

## Transgene-Mediated Resistance using Sense or Antisense Non-pathogen Sequences

A wide variety of molecules derived either from the host plant or other non-pathogenic source have been utilized to generate resistance against virus infection.

Following expression of antisense RNA of uroporphyrinogen decarboxylase or coporphyrinogen oxidase in *N. tabacum*, a lowered accumulation of viral RNA was observed (Mock *et al.*, 1999). Transgenic tobacco plants expressing *catalase 1* (*Cat1*) or *catalase 2* (*Cat2*) genes in an antisense orientation reported to develop increased resistance against TMV (Takahashi *et al.*, 1997). Similarly, tobacco lines expressing the rice cysteine protease inhibitor gene, showed resistance against TEV and PVY infection (Gutierrez-Campos *et al.*, 1999). Further, ribosome inactivating protein, dianthin, isolated from *Dianthus caryophyllus*, has been utilized to engineer transgenic resistance against bipartite begomovirus ACMV in *N. benthamiana* (Hong *et al.*, 1997). Likewise, the RNAse barstar (Zhang *et al.*, 2003), an insect symbiont derived virus binding protein (GroEL) (Edelbaum *et al.*, 2009) and peptide aptamers (short peptides that interfere with enzyme activity) (Lopez-Ochoa *et al.*, 2006) have also been reported to generate resistance against geminiviruses. So, sequences of non pathogenic origin holds great possibility to generate antiviral resistance.

## Transgene-Mediated Resistance using hpRNA Derived from Viruses

Sense/antisense transgene approach has a critical drawback, wherein inefficient, variable and slow resistance is developed due to inappropriate amount of transgene-derived siRNA to provoke sense/antisense-PTGS. For successful RNA silencing higher accumulation of dsRNA is desired. But, sometimes during sense/antisense transgene approach the production of siRNA is not appropriate to generate resistance.

Afterwards, use of transgenes expressing self-complementary "hairpin" (hp) RNA was proposed for efficient induction of RNAi. The hpRNA transgene consists of inverted repeat sequence of the target gene (sense and anti-sense) interrupted with a spacer region or intron flanked by repeats sandwiched between plant promoter and terminator. After transcription of the very transgene, RNA forms that fold back to itself and forms a hairpin structure including a single-stranded loop region and a stem region with base pairing. Loop and stem regions are transcribed from the spacer region/intron and inverted repeats respectively. Base paired stem region serve as dsRNA structure and evokes RNAi (Smith *et al.*, 2000).

siRNAs are generated from stem and spacer region/intron provides stability to the construct and probably improves the efficiency of silencing on implementation against viruses (Zrachya *et al.*, 2007).

Transient expression of an intron spliced hairpin construct containing sequences of the IR conserved between the monopartite begomoviruses, *Tomato yellow leaf curl virus, Tomato yellow leaf curl Sardinia virus* and *Tomato yellow leaf curl Malaga virus* in tomato and *N. benthamiana* plants resulted a broad spectrum resistance against these viruses and the accumulation of TYLCV-specific siRNAs (the effector of the RNAi response) was observed in silenced plants (Abhary *et al.*, 2006).

On the basis of several experiments, it is recommended that transformation of plants should be done with multiple hpRNA constructs of different viral origin or with single hpRNA constructs containing several viral gene sequences for broad range and efficient resistance against viral infections (Bucher *et al.*, 2006).

Other parameters that ensure the success includes 100 to 800 nt long transgene sequence, more than one transgenic copies and hpRNA developed from multiple loci (Dalakouras *et al.*, 2011). Viruses having mutation frequency 10 per cent -20 per cent, smoothly overpowers the effect of transgene mediated RNAi resistance. So, amid of several factors that determine the success of transgene hpRNA construct driven RNAi, the most important is the sequence similarity among the transgene sequence and the infecting virus species sequence (de Haan *et al.*, 1992). Greater the homology between the hpRNA sequences and virus genes, more stable would be the resistance.

## RNAi Resistance Mediated by Artificial microRNA

Presently, RNA silencing in plants is mediated by artificial miRNA (amiRNA). It utilizes endogenous pri-miRNA backbone, original mature miRNA/miRNA* sequence of backbone is replaced accordingly with complementary target gene of virus. Mature miRNA comes into the cytoplasm utilizing natural miRNA biogenesis pathway. For smooth processing of miRNA biogenesis pathway, structural features of miRNA backbone like bulges, loop and mismatches is maintained. The silencing signal generated after processing is very specific and effective against infecting virus species. It has been turned up as a proven technology to carry out competent and precise RNA silencing of viruses, endogeneous genes, non-coding RNA and reporter genes both locally and system level (Eamens *et al.*, 2014). At the beginning it is used for functional annotation of host encoded genes (Schwab *et al.*, 2006) but presently it is among the most used technique that provide innate immunity against viruses both in plants as well as animals (Niu *et al.*, 2006; Qu *et al.*, 2007; Ai *et al.*, 2011).

In comparison to conventional RNAi strategies amiRNAs have many benefits: (1) It utilizes short nucleotide sequences of amiRNAs instead of long viral cDNA fragment; Contrary to siRNAs derived from long hairpin RNA, high sequence specificityof amiRNAs could be useful to target conserved domains and genes, reduces off target effects and satisfies biosafety concerns associated with transgenic crops; (2) Offers both systemic as well as tissue- or cell-specific silencing of genes of interest.

## Resistance Employing Transient RNA Silencing

The rising biosafety concerns have led the advent of transient RNA silencing. In this technique silencing molecules of RNAs are directly transferred into plant tissues. This strategy is implemented in plants through two modes: (a) mechanical inoculation, in vitro synthesized dsRNA molecules are injected to plant and (b) Agrobacterium-mediated, where transient expression of dsRNA is brought about. DsRNA molecules are synthesized specifically according to the target so efficient resistance was reported against homologous virus species (Tenllado *et al.*, 2001). However, the immunity attained against the viral infection was non heritable and did not confer long-term protection (Tenllado *et al.*, 2003). This methodology is very cost and labour intensive so it could not be used at field level. To overcome this issue, dsRNA were synthesized in bacteria and crude extract of bacteria containing dsRNA was sprayed over the plant (Gan *et al.*, 2010; Yin *et al.*, 2010).

## Need of RNA Interference to Engineer Resistance against Viruses?

Continuous evolution process of viruses needs exact detection of molecular source that confers capability to defeat host resistance in plants. Geminiviruses and associated betasatellites are continuously evolving to produce new virulent strains through point mutations and recombinations (Nawaz-ul-Rehman and Fauquet 2009). In the past, numerous plant genetic modification strategies comprising expression of coat proteins, replicase proteins, or nucleoproteins of virus origin have been utilized to generate resistance against these phytopathogenic viruses (Abel *et al.*, 1986; Jones *et al.*, 1998).

Efforts of conventional breeding and transgene-based methodologies for conveying tolerance to viruses did not turn up completely successful. It becomes urgent to explore the molecular interface between the host and virus elements to provide stable resistance against viruses in crop plants. It provokes the necessity of alternate strategies to combat economic losses due to viral diseases. Among the possible strategies, RNAi based techniques hold great promise (Mansoor *et al.*, 2006). RNA silencing can be triggered by inducing transgenes, RNA viruses or DNA sequences that are homologous to expressed genes. This phenomenon has already been verified as a device to obtain resistance against viruses in plants and animals.

As a counter defense, viruses have evolved "suppressor" proteins, capable to prevent or counteract RNAi (Levy *et al.*, 2008). Several studies exploring the utility of RNAi for obtaining resistance against viruses have been reported, with various levels of success (Ramesh *et al.*, 2014; Sahu *et al.*, 2013). RNA silencing process involving antisense suppression and co-suppression were recognized as efficient techniques to achieve gene silencing in plants (Tenllado *et al.*, 1999). siRNA performs gene silencing in a sequence-dependent manner (Vanitharani *et al.*, 2003; Ding *et al.*, 2007). Recently, one more powerful weapon is added in repertoire of PTGS mechanisms against viruses, termed as artificial miRNAs (amiRNAs). amiRNAs are substantiating as very effective tools because they make use of cellular miRNA pathway to develop antiviral resistance.

## *In silico* Analysis for Designing Efficient RNA Interference

A prime challenge for successful execution of RNA interference is the competency to find miRNA targets with high specificity. Even though several biological principles of miRNA-target pair have been revealed experimentally as well as computationally still, detection of true functional miRNA targets remains a challenging task. Even now, we are struggling to completely understand the rules defining interactions between plant–virus miRNAome and their respective targets. To decipher this uncertainity miRNA-target prediction algorithms come up with implementation of innovative determining factors and keep on progressing (Zuker, 2003; Rehmsmeier *et al.*, 2004; SaeTrom *et al.*, 2005; Sung-Kyu *et al.*, 2005; Wang *et al.*, 2005; Kruger *et al.*, 2006; Miranda *et al.*, 2006; Xiao *et al.*, 2009; Dai and Zhao, 2011).

So far, it has been observed that the degree of complementarity between a miRNA/siRNA and its target determines the stability of miRNA/siRNA: target duplex and fate of the miRNA. Several findings confirm that excessive complementarity favours mRNA destruction (Ameres *et al.*, 2010; Baig and Khan, 2013; Shweta and Khan, 2014). Due to Off-target effect loss of specificity of small RNA occur followed by irrelevant silencing of other genes. This undesirable off-targeting of other mRNA results non availability of these small RNAs to their original targets and loss of potency (Vert *et al.*, 2006).

For a miRNA, the complementarity between itself and its target site determines the stability of miRNA:target duplex. miRNA families illustrate same seed site, and are usually well-conserved among related species (Ding *et al.*, 2011).Therefore it has been utilized as a key feature for target gene analysis. The first 2–8 nucleotides originating from the 5' end to the 3' end of a miRNA is termed as seed sequence (Lewis *et al.*, 2003). Generally seed match is a Watson-Crick (WC) match between a miRNA and its target in the seed sequence. And WC match between a miRNA and mRNA nucleotide happens during pairing of adenosine (A) with uracil (U) and guanine (G) with cytosine (C). A perfect seed match between the miRNA and the mRNA target has no gaps in alignment within the WC matching (Lewis *et al.*, 2003 and 2005; Brennecke *et al.*, 2005; Krek *et al.*, 2005).

Similarly, on the basis of several experimental observations various position dependant preferences were identified for designing of siRNA that are as follows: (i) on antisense strand A/U at the 5′ end;(ii) on sense strand G/C at the 5′end; (iii) On one-third of the antisense strand not less than five A/U residues in the 5′ terminal; and (iv) not more than 9 nt long GC stretch (Ui-Tei *et al.*, 2004). (v) On sense strand, not less than 3 'A/U' bases at positions 15-19; (vi) no internal repeats; (vii) on sense strand, 'A' base at positions 3 and 19; 'U' at 10 position; 'G/C' at position 19; at position 13 all bases except 'G' (Reynolds *et al.*, 2004).

The advantages and disadvantages of using different sets of complementarities are that consideration of only stringent-pair type's increases specificity but it is quite possible that they could miss many potential targets, on the other hand consideration of both stringent and moderate-stringent-pair type's increases sensitivity but might also increase the number of false positives.

Additionally, 3′UTR and 5′UTR excluding SNPs, low GC content, common region of several exons transcribed from single mRNA and non palindromic regions are the most favoured regions for designing of the efficient siRNA (Birmingham *et al.*, 2007).

Perfect complementarity between miRNA and mRNA is not enough for target prediction. Accessibility in context of secondary structure is very important for any mRNA to qualify as target site and underestimation of this parameter could lead to false positive predictions (Dai *et al.*, 2011). Secondary structure is very important for prediction of small RNAs and their targets (Ding *et al.*, 2010). Site accessibility, parameter is calculated to estimate the ease with which a miRNA can locate and hybridize with an mRNA target. mRNA presumed to attain secondary structure after transcription (Mahen *et al.*, 2010) that interferes with a small RNA′s ability to bind to a target site. Small RNA:mRNA hybridization is a two-step process in which a miRNA binds first to a short accessible region of the mRNA (Mückstein *et al.*, 2006). Afterwards, mRNA secondary structure unfolds as the siRNA/miRNA completes binding to a target (Long *et al.*, 2007). Kiryu *et al.*, had carried out detailed investigation of accessibility of target sites (Kiryu *et al.*, 2011). They found that the efficacy of small RNAs depends strongly on the accessibility of both the very 5′ and 3′ end of their binding sites. Therefore, to evaluate the probability that candidate mRNA is the target of a small RNA, the predicted amount of energy is required to evaluate a site accessiblity of a small RNA:target pair. So, competent target site on mRNA should have minimal mRNA secondary structure, as it obstructs the miRNA/siRNA access to the target site.

Another parameter that is very important is multiplicity, since different miRNAs can cooperatively regulate individual targets, but miRNA expression patterns in vivo differs at several developmental stages, biotic and abiotic conditions (Voinnet, 2009). More number of putative miRNA sites per mRNA can significantly enhance target prediction (Ding *et al.*, 2012). The existence of multiple potential target sites on mRNA is sometimes viewed as a necessity for efficient translational inhibition. For example, dual target sites were reported on *AtTAS3*, a target gene of miR390. *AtTAS3* is the phasing precursor of ta-siRNA TAS3 and the biogenesis of the ta-siRNA have similar target recognition system like miRNA (Axtell *et al.*, 2006). In this case, two target sites (two hits) represents two cleavage events, which represents a stronger signal for degradation of target mRNA.

Apart from that, miRNA families have targets that are conserved among related species. This feature has been implemented in the target prediction (Ha *et al.*, 2008). But the major dispute of such a rule is that the expression pattern of mRNAs may be time- and space specific, thus it is quite possible that target sequences of miRNA may not be detected due to their low expression level in specific organisms (German *et al.*, 2008), leading to false negative predictions from conservative analysis. Apart from that, there are also species-specific miRNAs and targets. So, inclusion of conservation filters while identification of miRNA targets can decrease the false positive rate, but the side effect is also obvious, since it is only effective for conserved miRNAs. It is important to identify targets both with and without conservation when our interest is identification of species-specific miRNAs.

Today we have many small RNA-target prediction algorithms established on sequence complementarity (Wang *et al.*, 2005), thermodynamic properties (favouring energetically stable binding sites) (Zuker *et al.*, 2003; Rehmsmeier *et al.*, 2004; Kruger *et al.*, 2006), machine learning supervised techniques (utilizing patterns of experimentally verified miRNA–target pairs) (SaeTrom *et al.*, 2005; Sung-Kyu *et al.*, 2005; Miranda *et al.*, 2006; Xiao *et al.*, 2009) and plant miRNA target recognition rules (Dai and Zhao, 2011). Although there are so many questions that are need to be answered, recent progress in small RNA biology transends very clear verdict that combination of multiple strategies is required to obtain a comprehensive high-confidence description of miRNA targeting networks.

So, many conserved RNAs and unsorted target sequences are available that are yet to be confirmed and it is assumed that they can serve the purpose of devising immunity against viruses. In this context computational investigation would serve as powerful tool to eliminate junk data and to obtain the valuable lead.

## Significance of *In silico* Analysis

It has been documented that target gene selection and sequence identity were determining factor of RNAi success to achieve resistance against cotton leaf curl disease (Mubin *et al.*, 2011). This finding could be harnessed to mediate resistance against geminiviruses employing appropriate *in silico* studies.

Untill now numerous small ncRNAs have been reported that are involved in plant gene regulatory mechanisms. However, enough cheaper high throughput biological techniques are not available to evaluate their expression pattern; this has led dependency on *in silico* methods to find the way out. A computational approach for prediction of miRNA targets facilitates the process of narrowing down potential target sites for experimental validation.

Virus evolution is strongly contributed by recombination. Numerous examples are available on recombination directed generation of new DNA and RNA viruses from several groups such as bromoviruses (*Cowpea chlorotic mottle virus*) (Allison *et al.*, 1989), caulimoviruses (*Cauliflower mosaic virus*) (Chenault and Melcher 1994), luteoviruses (cucurbit aphid borne yellows luteovirus and pea enation mosaic RNA1) (Gibbs and Cooper 1995), nepoviruses (*Tomato black ring virus*) (Le Gall *et al.*, 1995), cucumoviruses (CMV and *tomato aspermy virus*) (Masuta *et al.*, 1998), geminiviruses (ACMV and EACMV) (Pita *et al.*, 2001). Among these, family *Luteoviridae* has undergone both inter and intrafamilial recombination (Moonan *et al.*, 2000). In the late 1990s, the widespread use of resistant varieties, ruled out the presence of CLCuD caused by geminiviruses from the fields of cotton in Pakistan and north- western India. However, during the cropping season in 2001, symptoms of CLCuD started appearing on all previously resistant cultivars at Burewala, district Vehari in Pakistan. By the year 2002, the disease attained the form of epidemic. This marked the emergence of a new resistance breaking strain of the virus (Mansoor *et al.*, 2003). Later on this begomovirus complex associated with resistance breakdown was characterized as *Cotton leaf curl Burewala virus*, a novel recombinant of *Cotton leaf curl Kokhran virus* and *Cotton leaf curl Multan virus* that lacks C2 gene, encoded by complementary strand of begomoviruses (Amrao *et al.*, 2010). Regrettably there are not lots of success stories of engineering resistance against begomoviruses in cotton. Presently there is no reliable source of resistance has been identified against the resistance breaking strain of the CLCuD. A family recombination analysis for a wide variety of geminivirus DNA-A components suggests that the common region (CR) is a hotspot of recombination irrespective of their geographical origin. On the other hand coding regions of geminiviruses are less susceptible to recombination (Padidam *et al.*, 1999). Furthermore, the CR of DNA-A component is not only exchanged between members of different geminivirus species, it is also exchanged with heterologous molecules such as betasatellites or DNA-B component (Tao *et al.*, 2008). So, like any other viruses, geminiviruses undergo very rapid mutation and recombination and it is part of their evolution. Gene silencing based technology is sequence specific and any small change in the targeted virus sequence can overcome the resistance. This emerged as major constraint in providing broad range resistance against viruses.

Thus, it is very important to identify those targets that are residing in conserved regions of genes among various viruses. And this is not an easy task, especially in case of viruses, where abundant discrete viruses can cause the disease. The only way out of this issue is to devise resistance either for single or cross species by considering all the available accessions of virus species and it could be possible with the aid of *in silico* study only.

A study conducted to identify antiviral characteristics of miRNAs obtained from six plant species (*Arabidopsis thaliana, Glycine max, Oryza sativa, Sorghum bicolor, Vitis vinifera*, and *Zea mays*) revealed that plant miRNAs have innate property to target genomes of respective invading viruses preferentially much better than random miRNAs or viral genomes of animal origin (Pérez-Quintero *et al.*, 2010) Similiarly, passenger miRNA strands (miRNA*) of *Solanum. Lycopersicon* showed tendency to bind with the genomes of ToLCNDV (Naqvi *et al.*, 2010).

These reports confirm that computational approaches are quite substantial for explaining the details of miRNA-mediated gene regulation and computationally parsed findings cannot be overlooked merely on the belief that they arose out just by chance pairings.

Artificially designed plant miRNAs (artificial miRNAs) produced following manipulations in backbone of corresponding endogeneous precursor miRNA complementary with viral genes have demonstrated successful defence against corresponding viral disease and validate the capability of predicted miRNA target sites to build up amiRNA centred antiviral defence in plants (Niu *et al.*, 2006; Qu *et al.*, 2007; Jelly *et al.*, 2012; Fahim *et al.*, 2013).

In a recent study it has been reported that amiRNA approach accompanied with computational analysis could bring about efficient resistance against *Cotton leaf curl Burewala virus* and degree of resistance increases with higher complementarity between the amiRNA and the target sequence (Ali *et al.*, 2013).Using accessible data sources computational prediction is also possible to investigate and discover the siRNAs from viral genomes.

## Conclusion

Our understanding on how plants adapt against virus attack has been greatly increased by deeper understanding of small RNA-mediated pathways. Both miRNA and siRNA, could be utilized concurrently as per the need of cell to mediate desired changes in the expression of transcriptome. To establish the virus infection, they have evolved viral suppressor proteins. Although lots of studies have been conducted to uncover this aspect of complex arms race between plant and viruses, still we are very far from the complete understanding of these adaptive responses in both plants and viruses. In this direction, a potential aspect is functional annotation of miRNA and siRNAs, so that existing information gaps of RNA interference pathway could be filled. Exploring plant-encoded miRNAs and siRNAs as well as virus-encoded miRNAs remain the potential areas to engineer defence against viral diseases.

A combinatorial approach, employing more than one computational program for predicting viral targets of plant host miRNA and siRNA and virus encoded miRNA set, need to be applied. Bioinformatics approaches should be explored to check the possibility whether miRNAs and siRNAs have any potential to target viral genomes. This is likely to provide an effective way for exploring complex host miRNA-virus target relationship and could serve as the basis to engineer resistance against viral infections in plants. It will further increase our understanding of the complex virus-host interactions. Unlike animals, plants cannot move from their respective positions. At a time, plants have to experience several form of stresses and understanding of endogeneous small RNA species reaction towards multiple stresses is very necessary to solve this riddle. In near future, we would like to have complete information of small RNA(s) in response to single or multiple stimuli. It will lead to better understanding of pathways and efficient strategies to unravel the complex signalling network of plants and viruses.

## References

1. Abel PP, Nelson RS, De B, Hoffmann N, Rogers SG, Fraley RT, Beachy RN (1986). Delay of disease development in transgenic plants that express the Tobacco mosaic virus coat protein gene. *Science* 232, 738–743.

2. Abhary MK, Anfoka GH, Nakhla MK, Maxwell DP (2006). Post-transcriptional gene silencing in controlling viruses of the tomato yellow leaf curl virus complex. *Arch Virology* 151, 2349–2363.

3. Ai T, Zhang L, Gao Z, Zhu CX, Guo X (2011). Highly efficient virus resistance mediated by artificial microRNAs that target the suppressor of PVX and PVY in plants. *Plant Biology* 13,304–316.

4. Akbergenov R, Si-Ammour A, Blevins T, Amin I, Kutter C, Vanderschuren H, Zhang P, Gruissem W, Meins F Jr, Hohn T, Pooggin MM (2006). Molecular characterization of geminivirus-derived small RNAs in different plant species. *Nucleic Acids Res*, 34,462–471.

5. Ali I, Amin I, Briddon RW, Mansoor S (2013). Artificial microRNA-mediated resistance against the monopartite begomovirus Cotton leaf curl Burewala virus. *Virol J*, 10, 231.

6. Allison R F, Janda M, Ahlquist P (1989). Sequence of cowpea chlorotic mottle virus RNAs 2 and 3 and evidence of a recombinant event during bromovirus evolution. *Virology* 172, 321-330.

7. Ameres SL, Horwich MD, Hung JH, Xu J, Ghildiyal M, Weng Z, Ameres SL, Horwich MD, Hung JH, Xu J, Ghildiyal M, Weng Z, Zamore PD (2010).Target RNA-directed trimming and tailing of small silencing RNAs. *Science* 328, 1534–1539.

8. Amin I, Basavaprabhu LP, Briddon RW, Mansoo S, Fauquet CM (2011). Common set of developmental miRNAs are upregulated in *Nicotiana benthamiana* by diverse begomoviruses. *Virol Journal* 8,143.

9. Amin I, Hussain K, Akbergenov R, Yadav JS, Qazi J, Mansoor S, Hohn T, Fauquet CM, Briddon RW (2011a). Suppressors of RNA silencing encoded by the components of the cotton leaf curl begomovirus-betasatellite complex. *Mol. Plant-Microbe Interaction* 24, 973–983.

10. Amrao L, Amin I, Shahid MS, Briddon RW, Mansoor S (2010). Cotton leaf curl disease in resistant cotton is associated with a single begomovirus that lacks an intact transcriptional activator protein. *Virus Research* 152, 153–163.

11. Anandalakshmi R, Pruss GJ, Ge X, Marathe R, Mallory AC, Smith TH, Vance VB (1998). A viral suppressor of gene silencing in plants. *Proc. Natl Acad. Sci. USA* 95, 13079–13084.

12. Asad S, Haris W A, Bashir A, Zafar Y, Malik K A, Malik N N, Lichtenstein C P (2003). Transgenic tobacco expressing geminiviral RNAs are resistant to the serious viral pathogen causing cotton leaf curl disease. *Arch. Virology* 148, 2341–2352.

13. Audy P, Palukaitis P, Slack S A, Zaitlin M (1994).Replicase mediated resistance to potato virus Y in transgenic tobacco plants. *Mol. Plant-Microbe Interaction* 7, 15–22.

14. Axtell MJ, Jan C, Rajagopalan R, Bartel DP (2006). A two-hit trigger for siRNA biogenesis in plants. *Cell*, 127,565–77.

15. Azevedo J, Garcia D, Pontier D, Ohnesorge S, Yu A, Garcia S, Braun L, Bergdoll M, Mohamed A H, Lagrange T, Voinnet O (2010). Argonaute quenching and global changes in Dicer homeostasis caused by a pathogen-encoded GW repeat protein. *Genes Dev* 24, 904–915.

16. Hamilton A J, Baulcombe D C (1999). A novel species of small antisense RNA in posttranscriptional gene silencing. *Science* 286, 950–952.

17. Beachy R N (1997). Mechanisms and applications of pathogen-derived resistance in transgenic plants. *Curr. Opin. Biotech* 8, 215-220.

18. Beachy RN, Loesch-Fries S, Tumer NE (1990). Coat Protein-Mediated Resistance Against Virus Infection. *Annu Rev Phytopathology* 28, 451–472.

19. Bernstein E, Caudy AA, Hammond SM, Hannon GJ (2001). Role for a bidentate ribonuclease in the initiation step of RNA interference. *Nature* 409, 363-366.

20. Birmingham A, Anderson E, Sullivan K, Reynolds A, Boese Q, Leake D, Karpilow J, Khvorova A (2007). A protocol for designing siRNAs with high functionality and specificity. *Nature Protocols*, 2(9), 2068-2078

21. Bisaro DM (2006). Silencing suppression by geminivirus proteins. *Virology* 344,158–68.

22. Bivalkar-Mehla S, Vakharia J, Mehla R, Abreha M, Kanwar J R, Tikoo A, Chauhan A (2011). Viral RNA silencing suppressors (RSS): novel strategy of viruses to ablate the host RNA interference (RNAi) defense system. *Virus Research* 155, 1–9.

23. Blevins T, Rajeswaran R, Shivaprasad PV, Beknazariants D, Si-Ammour A, Park HS, Vazquez F, Robertson D, Meins F Jr, Hohn T, Pooggin MM(2006). Four plant Dicers mediate viral smallRNA biogenesis and DNA virus induced silencing. *Nucleic Acids Research* 34, 6233–6246.

24. Bonfim K, Faria JC, Nogueira EOPL, Mendes ÉA, Aragão FJL (2007). RNAi-Mediated Resistance to Bean golden mosaic virus in Genetically Engineered Common Bean (Phaseolus vulgaris). *Mol. Plant Microbe Interaction* 20, 717–726.

25. Brederode FT, Taschner P E M, Posthumus E, Bol J F (1995). Replicase-mediated resistance to alfalfa mosaic virus. *Virology* 207, 467-474.

26. Brennecke J, Stark A, Russell RB, Cohen SM (2005). Principles of microRNA-target recognition. *PLoS Biol*, 3,e85.

27. Briddon RW, Markham PG (2000). Cotton leaf curl virus disease. *Virus Res* 71: 151-159.

28. Brigneti G, Voinnet O, Li WX, Ji LH, Ding SW, Baulcombe DC (1998). Viral pathogenicity determinants are suppressors of transgene silencing in *Nicotiana benthamiana. EMBO Journal* 17, 6739–6746.

29. Brodersen P, Sakvarelidze-Achard L, Bruun-Rasmussen M, Dunoyer P, Yamamoto YY Sieburth L, Voinnet O (2008). Widespread translational inhibition by plant miRNAs and siRNAs. *Science* 320,1185–90.

30. Brodersen P, Voinnet O (2006). The diversity of RNA silencing pathways in plants. *Trends Genet*, 22: 268-280.

31. Bucher E, Lohuis D, van Poppel PM, Geerts-Dimitriadou C, Goldbach R, Prins M (2006). Multiple virus resistance at a high frequency using a single transgene construct. *J Gen Virology* 87, 3697–3701.

32. Bucher E, Sijen T, De Haan P, Goldbach R, Prins M (2003).Negative-strand tospoviruses and tenuiviruses carry a gene for a suppressor of gene silencing at analogous genomic positions. *J. Virology* 77, 1329–1336.

33. Burgyán J, Havelda Z (2011). Viral suppressors of RNA silencing. *Trends Plant Science* 16, 265–272.

34. Canto T, Palukaitis P (1998). Transgenically expressed cucumber mosaic virus RNA 1 simultaneously complements replication of cucumber mosaic virus RNAs 2 and 3 and confers resistance to systemic infection. *Virology* 250, 325-336.

35. Chancellor TCB, Holt J, Villarreal S, Tiongco ER, Venn J (2006). Spread of plant virus diseases to new plantings: a case study of rice tungro disease. *Adv Virus Res* 66, 1–30.

36. Chellappan P, Masona MV, Vanitharani R, Taylor NJ, Fauquet CM (2004). Broad spectrum resistance to ssDNA viruses associated with transgene-induced gene silencing in cassava. *Plant Mol Biology* 56, 601–611.

37. Chellappan P, Vanitharani R, Fauquet CM (2005). MicroRNA binding viral protein interferes with Arabidopsis development. *Proc Natl Acad Sci USA* 102, 10381–10386.

38. Chellappan P, Vanitharani R, Pita J, Fauquet CM (2004a). Short interfering RNA accumulation correlates with host recovery in DNA virus-infected hosts, and gene silencing targets specific viral sequences. *J. Virology* 78, 7465–7477.

39. Chen J, Li WX, Xie D, Peng JR, Ding SW (2004). Viral virulence protein suppresses RNA silencing-mediated defense but upregulates the role of microRNA in host gene expression. *Plant Cell* 16, 1302–1313.

40. Chenault K D, Melcher U (1994).Patterns of nucleotide sequence variation among cauliflower mosaic virus isolates. *Biochimie* 76(1), 3-8.

41. Cooper B, Lapidot M, Heick, J A, Dodds, J A, Beachy, R N (1995). A defective movement protein of TMV in transgenic plants confers resistance to multiple viruses whereas the functional analog increases susceptibility. *Virology* 206, 307–313.

42. Coutts BA, Thomas-Carroll ML, Jones RAC (2004). Patterns of spread of Tomato spotted wilt virus in field crops of lettuce and pepper: spatial dynamics and validation of control measures. *Ann Appl Biology* 145,231–245.

43. Cui X, Li G, Wang D, Hu D, Zhou X (2005). A begomovirus DNAb-encoded protein binds DNA, functions as a suppressor of RNA silencing, and targets the cell nucleus. *J Virology* 79, 10764–10775.

44. Dai X, Zhao PX (2011). psRNATarget: a plant small RNA target analysis server. *Nucleic Acids Research* 39,W155–W159.

45. Dalakouras A, Tzanopoulou M, Tsagris M, Wassenegger M, Kalantidis K (2011). Hairpin transcription does not necessarily lead to efficient triggering of the RNAi pathway. *Transgenic Research* 20, 293–304.

46. Dalmay T, Hamilton A, Rudd S, Angell S, Baulcombe DC (2000). An RNA dependent RNA polymerase gene in Arabidopsis is required for posttranscriptional gene silencing mediated by a transgene but not by a virus. *Cell* 101, 543–553.

47. de Haan P, Gielen JJ, Prins M, Wijkamp IG, van Schepen A, Peters D, van Grinsven MQ, Goldbach R (1992). Characterization of RNA-mediated resistance to tomato spotted wilt virus in transgenic tobacco plants. *Biotechnology* (N Y) 10, 1133–1137.

48. Deleris A, Gallego-Bartolome J, Bao J, Kasschau KD, Carrington JC, Voinnet O (2006). Hierarchical action and inhibition of plant dicer-like proteins in antiviral defense. *Science* 313, 68–71.

49. Diaz-Pendon JA, Li F, Li WX, Ding SW (2007). Suppression of antiviral silencing by cucumber mosaic virus 2b protein in Arabidopsis is associated with drastically reduced accumulation of three classes of viral small interfering RNAs. *Plant Cell* 19, 2053–2063.

50. Ding J, Li D, Ohler U, Guan J, Zhou S (2012). Genome-wide search for miRNA-target interactions in *Arabidopsis thaliana* with an integrated approach. *BMC Genomics* 13,S3.

51. Ding J, Zhou S, Guan J (2010). MiRenSVM: towards better prediction of microRNA precursors using an ensemble SVM classifier with multiloop features. *BMC Bioinformatics* 11,S11.

52. Ding J, Zhou S, Guan J (2011). MiRFam: an effective automatic miRNA classification method based on n-grams and a multiclass SVM. *BMC Bioinformatics* 12, 216.

53. Ding SW, Voinnet O (2007). Antiviral immunity directed by small RNAs. *Cell* 130, 413–426.

54. Donaire L, Barajas D, Martínez-García B, Martínez-Priego L, Pagán I, Llave C (2008). Structural and genetic requirements for the biogenesis of tobacco rattle virus-derived small interfering RNAs. *J. Virology* 82, 5167–5177.

55. Donaire L, Wang Y, Gonzalez-Ibeas D, Mayer KF, Aranda MA, Llave C (2009). Deep-sequencing of plant viral small RNAs reveals effective and widespread targeting of viral genomes. *Virology* 392, 203–214.

56. Duan C, Wang C, Guo H (2012). Application of RNA silencing to plant disease resistance. *Silence* 3, 5.

57. Dunoyer P, Pfeffer S, Fritsch C, Hemmer O, Voinnet O, Richards KE (2002). Identification, subcellular localization and some properties of a cysteine-rich suppressor of gene silencing encoded by peanut clump virus. *Plant J.* 29, 555–567.

58. Eamens AL, McHale M, Waterhouse PM (2014). The use of artificial microRNA technology to control gene expression in Arabidopsis thaliana. *Methods Mol. Biology* 1062, 211–224.

59. Edelbaum D, Gorovits R, Sasaki S, Ikegami M, Czosnek H (2009). Expressing a whitefly GroEL protein in Nicotiana benthamiana plants confers tolerance to tomato yellow leaf curl virus and cucumber mosaic virus, but not to grapevine virus A or tobacco mosaic virus. *Arch Virology* 154(3), 399-407.

60. Fagoaga C, López C, Hermoso de Mendoza A H, Moreno P, Navarro L, Flores R, Peña L (2006). Post-transcriptional gene silencing of the p23 silencing suppressor of Citrus tristeza virus confers resistance to the virus in transgenic Mexican lime. *Plant Mol. Biol.* 66, 153–165.

61. Fahim M, Larkin PJ (2013). Designing effective amiRNA and multimeric amiRNA against plant viruses. *Methods Mol. Biol.* 942, 357–377.

62. Farooq A, Farooq J, Mahmood A, Shakeel A, Rehman A, Batool A, Riaz M, Shahid M T H, Mehboob S (2011). An overview of cotton leaf curl virus disease (CLCuD) a serious threat to cotton productivity. *Aust J Crop Sci* 5, 1823–1831.

63. Fondong VN, Thresh JM, Zok S (2002). Spatial and temporal spread of Cassava mosaic virus disease in cassava grown alone and when intercropped with maize and/or cowpea. *J Phytopathol* 150, 1–10.

64. Frizzi A, Huang S (2010). Tapping RNA silencing pathways for plant biotechnology. *Plant Biotechnol. Journal* 8, 655–677.

65. Gan D, Zhang J, Jiang H, Jiang T, Zhu S, Cheng B (2010). Bacterially expressed dsRNA protects maize against SCMV infection. *Plant Cell Rep* 29, 1261–1268.

66. Garcia-Ruiz H, Takeda A, Chapman EJ, Sullivan CM, Fahlgren N, Brempelis KJ, Carrington JC(2010). Arabidopsis RNA-dependent RNA polymerases and dicer-like proteins in antiviral defense and small interfering RNA biogenesis during turnip mosaic virus infection. *Plant Cell* 22, 481–496.

67. German MA, Pillay M, Jeong DH, Hetawal A, Luo S, Janardhanan P, Kannan V, Rymarquis LA, Nobuta K, German R, De Paoli E, Lu C, Schroth G, Meyers BC, Green PJ (2008). Global identification of microRNA-target RNA pairs by parallel analysis of RNA ends. *Nat Biotechnology* 26, 941–946.

68. Gibbs M J, Cooper J I (1995). A recombinational event in the history of luteoviruses probably induced by base-pairing between the genomes of two distinct viruses. *Virology* 206, 1129-1132.

69. Glazov E, Phillips K, Budziszewski GJ, Meins F, Levin J (2003). A gene encoding an RNase D exonuclease-like protein is required for post-transcriptional silencing in Arabidopsis. *Plant Journal* 35, 342–349.

70. Glick E, Zrachya A, Levy Y, Mett A, Gidoni D, Belausov E, Citovsky V, Gafni Y (2008). Interaction with host SGS3 is required for suppression of RNA silencing by tomato yellow leaf curl virus V2 protein, *Proc. Natl. Acad. Sci. U.S.A.* 105, 157–161.

71. Goregaoker SP, Eckhardt LG, Culver JN (2000). Tobacco mosaic virus replicase mediated cross-protection: contributions of RNA and protein-derived mechanisms. *Virology* 273, 267-275.

72. Guo HS, López-Moya JJ, García JA (1998). Susceptibility to recombination rearrangement of a chimeric Plum pox potyvirus genome after insertion of a foreign gene. *Virus Research* 57,183-195.

73. Gutierrez-Campos R, Torres-Acosta JA, Saucedo-Arias LJ, Gomez-Lim MA (1999). The use of cysteine proteinase inhibitors to engineer resistance against potyviruses in transgenic tobacco plants. *Nature Biotechnology* 17, 1223–1226.

74. Ha M, Pang M, Agarwal V, Chen ZJ (2008). Interspecies regulation of microRNAs and their targets. *Biochim Biophys Acta* 1779, 735–42.

75. Hamilton AJ, Baulcombe DC (1999). A species of small antisense RNA in posttranscriptional gene silencing in plants. *Science* 286, 950–952.

76. Hammond SM, Bernstein E, Beach D, Hannon GJ (2000). An RNA-directed nuclease mediates post-transcriptional gene silencing in Drosophila cells. *Nature* 404, 293–296.

77. Hashmi J A, Zafar Y, Arshad M, Mansoor S, Asad S (2011). Engineering cotton (Gossypium hirsutum L.) for resistance to cotton leaf curl disease using viral truncated ACI DNA sequences. *Virus Genes* 42, 286-296.

78. Hassani-Mehraban A, Brenkman AB, van den Broek NJ, Goldbach R, Kormelink R (2009). RNAi-mediated transgenic Tospovirus resistance broken by intraspecies silencing suppressor protein complementation. *Mol. Plant Microbe Interaction* 22, 1250–1257.

79. Hong Y, K Saunders, J Stanley (1997). Transactivation of dianthin transgene expression by African cassava mosaic virus AC2. *Virology* 228, 383-37.

80. Jan FJ, Pang SZ, Fagoaga C, Gonsalves D (1999). Turnip mosaic potyvirus resistance in Nicotiana benthamiana derived by post-transcriptional gene silencing. *Transgenic Research* 8, 203–213.

81. Jelly NS, Schellenbaum P, Walter B, Maillot P (2012). Transient expression of artificial microRNAs targeting Grapevine fanleaf virus and evidence for RNA silencing in grapevine somatic embryos. *Transgenic Research* 6, 1319–1327.

82. Jones AL, Johansen IE, Bean SJ, Bach I, Maule AJ (1998). Specificity of resistance to pea seed-borne mosaic potyvirus in transgenic peas expressing the viral replicase (NIb) gene. *J. Gen. Virology* 79, 3129–3137.

83. Kasschau KD, Carrington JC (1998). A counter defensive strategy of plant viruses: suppression of post-transcriptional gene silencing. *Cell* 95, 461–470.

84. Kasschau KD, Xie Z, Allen E, Llave C, Chapman EJ, Krizan KA, Carrington JC (2003). P1/HC-Pro, a viral suppressor of RNA silencing, interferes with Arabidopsis development and miRNA function. Dev. *Cell* 4, 205–217.

85. Kawchuk LM, Martin RR, McPherson J (1990). Resistance in transgenic potato expressing the potato leafroll virus coat protein gene. *Mol Plant Microbe Interaction* 3, 301–307.

86. Khan JA, Ahmad J (2005). Diagnosis, monitoring and transmission characters of Cotton leaf curl virus. *Current Science* 88, 1803-1809.

87. Kiryu H, Terai G, Imamura O, Yoneyama H, Suzuki K, Asai K (2011). A detailed investigation of accessibilities around target sites of siRNAs and miRNAs. *Bioinformatics* 27, 1788-1797.

88. Krek A, Grün D, Poy MN, Wolf R, Rosenberg L, Epstein EJ, MacMenamin P, da Piedade I, Gunsalus KC, Stoffel M, Rajewsky N (2005). Combinatorial microRNA target predictions. *Nat. Genetics* 37, 495–500.

89. Kreuze JF, Perez A, Untiveros M, Quispe D, Fuentes S, Barker I, Simon R (2009). Complete viral genome sequence and discovery of novel viruses by deep sequencing of small RNAs: a generic method for diagnosis, discovery and sequencing of viruses. *Virology* 388, 1–7.

90. Kruger J, Rehmsmeier M (2006). RNAhybrid: microRNA target prediction easy, fast and flexible. *Nucleic Acids Research* 34,W451–W454.

91. Kubota K, Tsuda S, Tamai A, Meshi T (2003). Tomato mosaic virus replication protein suppresses virus-targeted posttranscriptional gene silencing. *J. Virology* 77, 11016–11026.

92. Le Gall O, Lanneau M, Candresse T, Dunez J (1995). The nucleotide sequence of the RNA-2 of an isolate of the English serotype of tomato black ring virus: RNA recombination in the history of nepoviruses. *Journal of General Virology* 76, 1279-1283.

93. Levy A, Dafny-Yelin M, Tzfira T (2008). Attacking the defenders: plant viruses fight back. *Trends Microbiol* 16, 194–197.

94. Lewis BP, Burge CB, Bartel DP (2005). Conserved seed pairing, often flanked by adenosines, indicates that thousands of human genes are microRNA targets. *Cell* 120(1),15–20.

95. Lewis BP, Shih IH, Jones-Rhoades MW, Bartel DP, Burge CB (2003). Prediction of mammalian microRNA targets. *Cell* 115, 787–798.

96. Lin C, Ku H, Tsai W, Green S, Jan F (2011). Resistance to a DNA and a RNA virus in transgenic plants by using a single chimeric transgene construct. *Transgenic Research* 20, 261–270.

97. Lindbo JA, Dougherty WG (1992). Pathogen-derived resistance to a potyvirus immune and resistant phenotypes in transgenic tobacco expressing altered forms of a potyvirus coat protein nucleotide sequence. *Mol Plant Microbe Interaction* 5, 144–153.

98. Liu L, Grainger J, Canizares MC, Angell SM, Lomonossoff GP (2004). Cowpea mosaic virus RNA-1 acts as an amplicon whose effects can be counteracted by a RNA-2-encoded suppressor of silencing. *Virology* 323: 37–48.

99. Long D, Lee R, Williams P, Chan CY, Ambros V, Ding Y (2007). Potent effect of target structure on microRNA function. *Nat. Struct. Mol. Biology* 14, 287–294.

100. Lopez-Ochoa L, Ramirez-Prado J, Hanley-Bowdoin L (2006).Peptide aptamers that bind to a geminivirus replication protein interfere with viral replication in plant cells. *J Virology* 80(12), 5841-53.

101. Lu R, Folimonov A, Shintaku M, Li WX, Falk BW, Dawson WO, Ding SW (2004). Three distinct suppressors of RNA silencing encoded by a 20-kb viral genome. *Proc. Natl Acad. Sci. USA* 101, 15742–15747.

102. Lucioli A, Noris E, Brunetti A, Tavazza R, Ruzza V, Castillo A, Tavazza M (2003).Tomato yellow leaf curl Sardinia virus rep-derived resistance to homologous and heterologous geminiviruses occurs by different mechanisms

and is overcome if virus-mediated transgene silencing is activated. *J. Virology* 77, 6785–6798.

103. Mahen EM, Watson PY, Cottrell JW, Fedor MJ (2010). mRNA secondary structures fold sequentially but exchange rapidly in vivo. *PLoS Biology* 8,e1000307.

104. Malnoe P, Farinelli L, Collet G F, Reust W (1994). Small-scale field test with transgenic potato, cv. Bintje, to test resistance to primary and second infections with Potato virus Y. *Plant Mol. Biology* 25, 963-975.

105. Mansoor S, Amin I, Hussain M, Zafar Y, Briddon RW (2006). Engineering novel traits in plants through RNA interference. *Trends Plant Science* 11, 559–565.

106. Mansoor S, Amin I, Iram S, Hussain M, Zafar Y, Malik KA, Briddon RW (2003). Breakdown of resistance in cotton to cotton leaf curl disease in Pakistan. *Plant Pathology* 52, 784.

107. Masuta C, Ueda S, Suzuki M, Uyeda I (1998).Evolution of a quadripartite hybrid virus by interspecific exchange and recombination between replicase components of two related tripartite RNA viruses. *Proc Natl Acad Sci* 95(18), 10487–10492.

108. Mette MF, Aufsatz W, van der Winden J, Matzke MA, Matzke AJM (2000). Transcriptional silencing and promoter methylation triggered by double-stranded RNA. *EMBO Journal* 19, 5194-5201.

109. Meyers BC, Axtell MJ, Bartel B, Bartel DP, Baulcombe D, Bowman JL, Cao X, Carrington JC, Chen X, Green PJ, *et al.* (2008). Criteria for annotation of plant MicroRNAs. *Plant Cell* 20,3186-3190.

110. Miranda KC, Huynh T, Tay Y, Ang YS, Tam WL, Thomson AM, Lim B, Rigoutsos I (2006). A pattern-based method for the identification of microRNA binding sites and their corresponding heteroduplexes. *Cell* 126(6), 1203–1217.

111. Mock HP, Heller W, Molina A, Neubohn B, Sandermann H Jr, Grimm B (1999). Expression of uroporphyrinogen decarboxylase or coproporphyrinogen oxidase antisense RNA in tobacco induces pathogen defense responses conferring increased resistance to tobacco mosaic virus. *Journal of Biological Chemistry*, 274, 4231–4238.

112. Moissiard G, Voinnet O (2006).RNA silencing of host transcripts by cauliflower mosaic virus requires coordinated action of the four Arabidopsis Dicer-like proteins. *Proc. Natl Acad. Sci. USA* 103, 19593–19598.

113. Moonan F, Molina J, Mirkov TE (2000).Sugarcane yellow leaf virus: an emerging virus that has evolved by recombination between luteoviral and poleroviral ancestors. *Virology* 30; 269(1),156-71.

114. Morroni M, Thompson J R, Tepfer M (2008). Twenty years of transgenic plants resistant to Cucumber mosaic virus. *Mol Plant Microbe Interaction* 21, 675-684.

115. Mubin M, Hussain M, Briddon RW, Mansoor S (2011). Selection of target sequences as well as sequence identity determine the outcome of RNAi approach for resistance against cotton leaf curl geminivirus complex. *Virol Journal* 8,122.

116. Mückstein U, Tafer H, Hackermüller J, Bernhart SH, Stadler PF, Hofacker IL (2006). Thermodynamics of RNA-RNA binding. *Bioinformatics* 22,1177-1182.

117. Muthamilarasan M, Prasad M (2013). Plant innate immunity: an updated insight into defense mechanism. *J Bioscience* 38,433–449.

118. Naqvi AR, Haq QM, Mukherjee SK (2010). MicroRNA profiling of tomato leaf curl New Delhi virus (ToLCNDV) infected tomato leaves indicates that deregulation of mir159/319 and mir172 might be linked with leaf curl disease. *Virol Journal* 7, 281.

119. Nawaz-ul-Rehman MS, Fauquet CM (2009). Evolution of geminiviruses and their satellites. *FEBS Lett*, 583,1825–1832.

120. Nawaz-ul-Rehman MS, Nahid N, Mansoor S, Briddon RW, Fauquet CM (2010). Post-transcriptional gene silencing suppressor activity of two non-pathogenic alphasatellites associated with a begomovirus. *Virology* 405, 300–308.

121. Niu QW, Lin SS, Reyes JL, Chen KC, Wu HW, Yeh SD, Chua NH (2006) Expression of artificial microRNAs in transgenic Arabidopsis thaliana confers virus resistance. *Nature Biotechnology* 24,1420–1428.

122. Padidam M, Sawyer S, Fauquet CM (1999). Possible emergence of new geminiviruses by frequent recombination. *Virology* 265, 218–225.

123. Page WW, Smith MC, Holt J, Kyetere D (1999). Intercrops, Cicadulina spp., and maize streak virus disease. *Ann Appl Biol* 135,385–393.

124. Pang S Z, Bock, J H, Gonsalves C, Slightom J L, Gonsalves D (1994). Resistance of transgenic Nicotiana benthamiana plants to tomato spotted wilt and impatiens necrotic spot tospoviruses: evidence of involvement of the N protein and N gene RNA in resistance. *Phytopathology* 84, 243-249.

125. Perez-Quintero AL, Neme R, Zapata A, Lopez C (2010). Plant microRNAs and their role in defense against viruses: a bioinformatics approach. *BMC Plant Biology* 10,138.

126. Pfeffer S, Dunoyer P, Heim F, Richards KE, Jonard G, Ziegler-Graff V (2002). P0 of beet Western yellows virus is a suppressor of posttranscriptional gene silencing. *J. Virology* 13, 6815–6824.

127. Pita JS, Fondong VN, Sangaré A, Otim-Nape GW, Ogwal S, Fauquet CM (2001). Recombination, pseudorecombination and synergism of geminiviruses are determinant keys to the epidemic of severe cassava mosaic disease in Uganda. *J Gen Virol* 82, 655–665.

128. Pooggin M, Shivaprasad PV, Veluthambi K, Hohn T (2003). RNAi targeting of DNA virus in plants. *Nat. Biotechnol*, 21,131–132.

129. Prins M, Laimer M, Noris E, Schubert J, Wassenegger M, Tepfer M (2008). Strategies for antiviral resistance in transgenic plants. *Mol. Plant Pathology* 9, 73–83.

130. Qi X, Bao FS, Xie Z (2009). Small RNA deep sequencing reveals role for Arabidopsis thaliana RNA-dependent RNA polymerases in viral siRNA biogenesis. *PLoS One* 4, e4971.

131. Qi Y, Zhong X, Itaya A, Ding B (2004). Dissecting RNA silencing in protoplasts uncovers novel effects of viral suppressors on the silencing pathway at the cellular level. *Nucleic Acids Research* 32, e179.

132. Qu J, Ye J, Fang R (2007). Artificial microRNA-mediated virus resistance in plants. *J. Virology* 81, 6690–6699.

133. Rahman M, Hussain D, Malik T A, Zafar Y (2005). Genetics of resistance to cotton leaf curl disease in Gossypium hirsutum. *Plant Pathology* 54, 764–772.

134. Rahman M, Zafar Y (2007). Registration of NIBGE-115 cotton. *J Plant Reg* 1, 51–52.

135. Raja P, Wolf J, Bisaro D (2010). RNA silencing directed against geminiviruses: post-transcriptional and epigenetic components. BBA Gene Regul. *Mechanism* 1799, 337–351.

136. Rajagopalan P A, Naik A, Katturi P, Kurulekar M, Kankanallu RS, Anandalakshmi R (2012). Dominance of resistance-breaking cotton leaf curl Burewala virus (CLCuBuV) in northwestern India. *Arch Virol* 157, 855–868.

137. Ramesh S, Mishra A, Praveen S (2007). Hairpin RNA-mediated strategies for silencing of tomato leaf curl virus AC1 and AC4 genes for effective resistance in plants. *Oligonucleotides* 17, 251–257.

138. Ramesh SV, Ratnaparkhe MB, Kumawat G, Gupta GK, Husain SM (2014). Plant miRNAome and antiviral resistance: a retrospective view and prospective challenges. *Virus Genes* 48, 1–14 10.1007/s11262-014-1038-z.

139. Reed J C, Kasschaub KD, Prokhnevskya AI, Gopinatha K, Poguec GP, Carrington JC, Doljaa VV (2003). Suppressor of RNA silencing encoded by Beet yellows virus. *Virology* 306, 203–209.

140. Rehmsmeier M, Steffen P, Hochsmann M, Giegerich R (2004). Fast and effective prediction of microRNA/target duplexes. *RNA* 10(10), 1507–1517.

141. Reynolds A, Leake D, Boese Q, Scaringe S, Marshall WS, Khvorova A (2004). Rational siRNA design for RNA interference. *Nature Biotechnology*, 22(3),326-30.

142. SaeTrom OLA, Snove OJ, SaeTrom PAL (2005). Weighted sequence motifs as an improved seeding step in microRNA target prediction algorithms. *RNA* 11(7), 995–1003.

143. Sahu PP, Pandey G, Sharma N, Puranik S, Muthamilarasan M, Prasad M (2013). Epigenetic mechanisms of plant stress responses and adaptation. *Plant Cell Rep* 32,1151–1159.

144. Sahu PP, Puranik S, Khan M, Prasad M (2012b). Recent advances in tomato functional genomics: utilization of VIGS. *Protoplasma* 249, 1017–1027.

145. Sahu PP, Rai NK, Chakraborty S, SinghM, Ramesh B, Chattopadhyay D, Prasad M (2010). Tomato cultivar tolerant to Tomato leaf curl New Delhi virus infection induces virus-specific short interfering RNA accumulation and defence-associated host gene expression. *Mol Plant Pathology* 11, 531–544.

146. Sastry KS, Zitter TA (2014). Management of Virus and Viroid Diseases of Crops in the Tropics. In: *Plant Virus and Viroid Diseases in the Tropics, Volume 2: Epidemiology and Management;* Sastry KS, Zitter TA (eds); Springer Netherlands, pp. 149-480.

147. Schwab R, Ossowski S, Riester M, Weigel D (2006). Highly specific gene silencing by artificial microRNAs in Arabidopsis. *Plant Cell* 18, 1121-1133.

148. Sharma N, Sahu PP, Puranik S, Prasad M (2012). Recent advances in plant-virus interaction with emphasis on small interfering RNAs (siRNAs). *Mol Biotechnol,* doi: 10.1007/s12033-012-9615-9617.

149. Shepherd D, Martin D, Thomson J (2009). Transgenic strategies for developing crops resistant to geminiviruses. *Plant Science176,* 1–11.

150. Shweta, Khan JA (2014). *In silico* prediction of cotton (*Gossypium hirsutum*) encoded microRNAs targets in the genome of Cotton leaf curl Allahabad virus. *Bioinformation* (10(5): 251-255.

151. Sijen T, Wellink J, Hiriart J B, Kammen AV (1996). RNA-mediated virus resistance: role of repeated transgenes and delineation of targeted regions. *Plant Cell* 8, 2277-2294.

152. Silhavy D, Molnar A, Lucioli A, Szittya G, Hornyik C, Tavazza M, Burgyan J (2002). A viral protein suppresses RNA silencing and binds silencing-generated, 21- to 25-nucleotide double-stranded RNAs. *EMBO Journal* 21, 3070–3080.

153. Singh K, Talla A, Qiu W (2012). Small RNA profiling of virus-infected grapevines: evidences for virus infection-associated and variety specific miRNAs. *Funct Integr Genomics* 12,659–669.

154. Smith NA, SP S, Wang M-B, Stoutjesdijk PA, Green AG, Waterhouse PM (2000). Total silencing by intron-spliced hairpin RNAs. *Nature* 407, 319-320.

155. Sung-Kyu K, Jin-Wu N, Wha-Jin L, Byoung-Tak Z (2005). A Kernel method for microRNA target prediction using sensible data and position-based features in *Proceedings of the 2005 IEEE Symposium on Computational Intelligence in Bioinformatics and Computational Biology* (CIBCB, La Jolla, 2005) 46–52.

156. Szittya G, Molnar A, Silhavy D, Hornyik C, Burgyan J (2002). Short defective interfering RNAs of tombusviruses are not targeted but trigger posttranscriptional gene silencing against their helper virus. *Plant Cell* 14, 359–372.

157. Szittya G, Simon M, Vitantonio P, Gabor T, Rachel LRP, Vincent M, Jozsef B, Tamas D (2010). Structural and functional analysis of viral siRNAs. *PLoS Pathogy* 6, e1000838.

158. Takahashi H, Chen Z, Du H, Liu Y, Klessig DF (1997). Development of necrosis and activation of disease resistance in transgenic tobacco plants with severely reduced catalase levels. *Plant Journal* 11, 993–1005.

159. Tao X, Zhou X (2008). Pathogenicity of a naturally occurring recombinant DNA satellite associated with tomato yellow leaf curl China virus. *J. Gen. Virology* 89, 306–311.

160. Tenllado F, Dýaz-Ruýz JR (1999). Complete resistance to pepper mild mottle tobamovirus mediated by viral replicase sequences partially depends on transgene homozygosity and is based on a gene silencing mechanism. *Transgenic Res* 8, 83–93.

161. Tenllado F, Diaz-Ruiz JR (2001). Double-stranded RNA-mediated interference with plant virus infection. *J Virology* 75, 12288–12297.

162. Tenllado F, Martinez-Garcia B, Vargas M, Diaz-Ruiz JR (2003). Crude extracts of bacterially expressed dsRNA can be used to protect plants against virus infections. *BMC Biotechnology* 3, 3.

163. Tennant PF, Gonsaives C, Ling K S, Fitch M, Manshardt R, Slightom JL, Gonsalves D (1994). Differential protection against papaya ringspot virus isolates in coat protein gene transgenic papaya and classically cross-protected papaya. *Phytopathology* 84, 1359-1366.

164. Thomas CL, Leh V, Lederer C, Maule AJ (2003). Turnip crinkle virus coat protein mediates suppression of RNA silencing in *Nicotiana benthamiana*. *Virology* 306, 33–41.

165. Trinks D, Rajeswaran R, Shivaprasad PV, Akbergenov R, Oakeley EJ, Veluthambi K, Hohn T, Pooggin M (2005). Supression of RNA silencing by a geminivirus nuclear protein, AC2, correlates with transactivation of host genes. *J. Virology* 79, 2517–2527.

166. Ui-Tei K, Naito Y, Takahashi F, Haraguchi T, Ohki-Hamazaki H, Juni A, Ueda R, Saigo K (2004). Guidelines for the selection of highly effective siRNA sequences for mammalian and chick RNA interference. *Nucleic Acids Research* 32(3), 936-48.

167. Van Wezel R, Dong X, Liu H, Tien P, Stanley J, Hong Y (2002). Mutation of three cysteine residues in Tomato yellow leaf curl virus-China C2 protein causes dysfunction in pathogenesis and posttranscriptional genesilencing suppression. *Mol. Plant Microbe Interact.* 15, 203–208.

168. Vanderschuren H, Akbergenov R, Pooggin M, Hohn T, Gruissem W, Zhang P (2007). Transgenic cassava resistance to African cassava mosaic virus is enhanced by viral DNA-A bidirectional promoter-derived siRNAs. *Plant Molecular Biology* 64, 549-557.

169. Vanderschuren H, Alder A, Gruissem W, Zhang P (2009). Dose-dependent RNAi-mediated geminivirus resistance in the tropical root crop cassava. *Plant Molecular Biology* 70 (3), 265-272.

170. Vanitharani R, Chellappan P, Fauquet CM (2003). Short interfering RNA-mediated interference of gene expression and viral DNA accumulation in cultured plant cells. *Proc. Natl. Acad. Sci. U.S.A.*, 100: 9632–9636.

171. Vanitharani R, Chellappan P, Fauquet CM (2005). Geminiviruses and RNA silencing. *Trends Plant Science* 10, 144– 151.

172. Vaucheret H, Fagard M (2001). Transcriptional gene silencing in plants: targets, inducers and regulators. *Trends Genet* 17, 29-35.

173. Vert JP, Foveau N, Lajaunie C, Vandenbrouck Y (2006). An accurate and interpretable model for siRNA efficacy prediction. *BMC bioinformatics* 7(1), 520.

174. Voinnet O (2009). Origin, biogenesis, and activity of plant microRNAs. *Cell* 136, 669–687.

175. Voinnet O, Lederer C, Baulcombe DC (2000). A viral movement protein prevents systemic spread of the gene silencing signal in Nicotiana benthamiana. *Cell* 103, 157–167.

176. Voinnet O, Pinto YM, Baulcombe DC (1999).Suppression of gene silencing: a general strategy used by diverse DNA and RNA viruses. *Proc. Natl Acad. Sci. USA* 96, 14147–14152.

177. Vu TV, Roy Choudhury N, Mukherjee SK (2012). Transgenic tomato plants expressing artificial microRNAs for silencing the pre-coat and coat proteins of a begomovirus, Tomato leaf curl New Delhi virus, show tolerance to virus infection. *Virus Research* 172, 35–45.

178. Wang X, Zhang J, Li F, Jin G, Tao H, Zhang X, Yanda L (2005). MicroRNA identification based on sequence and structure alignment. *Bioinformatics* 21(18), 3610–3614.

179. Wang XB, Wu Q, Ito T, Cillo F, Li WX, Chen X, Yu JL, Ding SW(2010).RNAi-mediated viral immunity requires amplification of virus-derived siRNAs in Arabidopsis thaliana. *Proc. Natl Acad. Sci. USA* 107, 484–489.

180. Waterhouse P M, Graham M W, Wang M B (1998). Virus resistance and gene silencing in plants can be induced by simultaneous expression of sense and antisense RNA. *Proc. Natl Acad Sci USA* 95, 13959–13964.

181. Wu Q, Luo Y, Lu R, Lau N, Lai EC, Li WX, Ding SW (2010). Virus discovery by deep sequencing and assembly of virus-derived small silencing RNAs. *Proc. Natl. Acad. Sci. USA* 107, 1606–1611.

182. Xiao F, Zuo Z, Cai G, Kang S, Gao X, Li T (2009). miRecords: an integrated resource for microRNA–target interactions. *Nucleic Acids Research* 37(Suppl 1), 105–110.

183. Yadav RK, Chattopadhyay D (2011). Enhanced viral intergenic region specific siRNA accumulation and DNA methylation correlates with resistance against a geminivirus. *Mol Plant Microbe Interaction* 24, 1189–1197.

184. Yadava P, Mukherjee SK (2010). Artificial microRNA and its applications. In: BibekanandM, Zhumur G (eds) *Regulatory RNAs: basics, methods and applications.* Springer: Berlin, Heidelberg, Germany. pp. 505–521.

185. Yang Y, Sherwood T A, Patte C P, Hiebert E, Polston J E (2004). Use of Tomato yellow leaf curl virus (TYLCV) Rep gene sequences to engineer TYLCV resistance in tomato. *Phytopathology* 94, 490–496.

186. Yelina NE, Savenkov EI, Solovyev AG, Morozov SY, Valkonen JP (2002). Long-distance movement, virulence, and RNA silencing suppression controlled by a single protein in hordei- and potyviruses: complementary functions between virus families. *J. Virology* 76, 12981–12991.

187. Yin GH, Sun ZN, Song YZ, An HL, Zhu CX, Wen FJ (2010). Bacterially expressed double-stranded RNAs against hot-spot sequences of tobacco mosaic virus or potato virus Y genome have different ability to protect tobacco from viral infection. *Appl Biochem Biotechnology* 162, 1901–1914.

188. Yoshikawa N, Saitou Y, Kitajima A, Chida T, Sasaki N, Isogai M (2006). Interference of long-distance movement of Grapevine berry inner necrosis virus in transgenic plants expressing a defective movement protein of Apple chlorotic leaf spot virus. *Phytopathology* 96, 378–385.

189. Zhang P, Fütterer J, Frey P, Potrykus I, Puonti-Kaerlas J, *et al.* (2003) Engineering virus-induced African cassava mosaic virus resistance by mimicking a hypersensitive reaction in transgenic cassava. In: Vasil I, editor. *Plant Biotechnology 2002 and Beyond,* Proceedings of the 10th IAPTC and B Congress June 23–28, 2002 Orlando, Florida, U.S.A. Springer Netherlands. 143–145.

190. Zhang X, Li H, Zhang J, Zhang C, Gong P, Ziaf K, Xiao F, Ye Z (2011). Expression of artificial microRNAs in tomato confers efficient and stable virus resistance in a cell-autonomous manner. *Transgenic Research* 3, 1–13.

191. Zrachya A, Glick E, Levy Y, Arazi T, Citovsky V, Gafni Y (2007). Suppressor of RNA silencing encoded by Tomato yellow leaf curl virus-Israel, *Virology* 358, 159–165.

192. Zuker M (2003). Mfold web server for nucleic acid folding and hybridization prediction. *Nucleic Acids Research* 31(13), 3406–3415.

**2016, Biotechnology: Progress and Applications** *Pages* **107–129**
*Editors:* **Saif Hameed and Zeeshan Fatima**
*Published by:* **DAYA PUBLISHING HOUSE, NEW DELHI**

*Chapter 5*

# Biotechnology Approaches to HIV and AIDS

***Ashish S Verma[1]*, Vinod Singh[2], Ruby Bansal[3] and Anchal Singh[4]***

[1] *Jadavpur University, 188, Raja S.C. Mallik Road, Jadavpur, Kolkata, W.B., India*
[2]*Department of Microbiology, Barkatullah University, Bhopal, M.P., India*
[3]*Crosslay Wellness Program, Pushpanjali Crosslay Hospital, Ghaziabad, U.P., India*
[4]*Amity Institute of Biotechnology, Amity University, Sector-125, Noida, UP, India*

## SUMMARY

Human Immunodeficiency Virus (HIV) and Acquired Immunodeficiency Syndrome (AIDS) have been reported for the first time In USA. Since then it has achieved the status of an epidemic. This infection was reported among the people following a unique life-style, therefore various names were given to this disease. Finally these infections have been given official name as HIV and AIDS. Numerous medications are available to treat HIV infections, even though, HIV infections can be controlled only. There is no cure available for HIV and AIDS. Unfortunately, the worst affected area with HIV and AIDS is African continent. Almost 2/3rd of the HIV infected people live in this continent. Sincere efforts have been tried to control the spread of this disease. UNAIDS has played a key role for the monitoring of HIV and AIDS cases at global level. The good news is that there is a decline in the new infections of HIV, but a new disease has emerged among long-term HIV seropositives, which is known as NeuroAIDS. Although bone marrow transplantation with mutated CCR-5 has shown a possibility to cure HIV, but still this treatment have many limitations. There is no doubt more researches are needed to find a cure for HIV and AIDS.

There is no doubt that nowadays, anyone or everyone who is following any one type of media like print media or electronic media, must be aware of the words like HIV and AIDS. There is no doubt that increase in awareness about these two words is attributable to the sincere efforts of government as well as various non-government organizations (NGO), who are working in this area. HIV is abbreviated as Human Immunodeficiency Virus, while AIDS is abbreviated as Acquired Immune Deficiency Syndrome. Both of these words are associated with the same disease. There is no doubt that AIDS and HIV infections are closely associated with each other. AIDS is the final outcome of HIV infections. AIDS is the terminal stage of HIV infections. So, anyone who is infected with HIV will end up having AIDS, but all HIV seropositives cannot be considered as AIDS patients. So, it can be easily concluded that AIDS is the final chapter of HIV seropositives, because AIDS is the terminal stage of HIV infected patient. Certainly, confirmation of symptoms of AIDS is the start of the final chapter in the life of HIV patients and it is just a matter of time, when a HIV seropositive will get full blown AIDS? In the recent past, it has been observed that a large percentage of HIV seropositives show an emerging trend for new type of complications, which is related with neurological and neuropsychiatric health of HIV infected individuals. This new and emerging health problem is known as neuroAIDS (Verma *et al.*, 2014).

## Discovery of HIV

It is surprising to know that initially HIV was discovered as a mysterious disease with no known causative organism as well as no known medication for its treatment. At that time only thing that was known that it is a dangerous disease with fatal consequences for sure. At that time people were very skeptical about this new disease and every one was scared if the patients of this mysterious new disease would be around them.

For the 1$^{st}$ time, 5 cases of this mysterious disease were reported in 1981 by Dr. Gottlieb. These patients were reported from 3 different hospitals of Los Angeles, California, USA. At that time these patients were diagnosed with a very rare disease known as Pneumocystis Carinii Pneumonia (PCP). PCP is a very rare disease which affects only on those patients whose immune system is severely compromised. Usually, PCP is not reported in normal population. Later these 5 patients were confirmed to be infected with HIV and were suffering from AIDS. In 1981, neither the word HIV nor AIDS existed in medical dictionary. The only thing was noticed that this disease occurs among those people who follow a unique life-styles, *i.e.* Homosexuals. This kind of observation was taken as an alarming bell for the individuals who were following this unique life-style. A fatal disease was simply related with this unique life-style, which was one of the major reasons to create huge social pressure on the government. This pressure was the main reason that government had to take strong measures to find out that, What is happening with reference to this mysterious disease? What is the causative organism for this mysterious disease? How to treat this disease? Who is majorly getting infected with this disease? How this disease is getting transmitted? What is the final outcome of disease? Sincere efforts and approaches were put into place to find the answers for these haunting questions. At this moment it is very important to mention the name of Ms. Sandra Ford, a pharmacy technician working at Center for Disease Control and Prevention (CDC), Atlanta, USA. Ms. Ford was responsible for the dispension of a controlled drug known as Pentamidine. Pentamidine can only be dispensed by CDC. Pentamidine is a drug which is used in very rare cases. She was surprised to notice that a doctor asked for the refill of this drugs numerous times. This was a very rare and unique situation, so she brought this information to the attention of her superiors at CDC. Ms. Sandra was the first person to raise an alarm about the seriousness of the situation.

This concern prompted the formation of Task Force on Kaposi's Sarcoma and Opportunistic Infections (KSOI). KSOI started monitoring the cases of HIV infections in USA. In the mean-time different names were given to this mysterious disease. These names were somewhat indicative of a particular life style of most of the person reported to be affected with this disease. Some of the names for this new disease are Gay Compromise Syndrome (GCS), Gay Related Immune-deficiency Disease (GRID), *etc.* if you notice that either the gay word is associated with these names or the word immune-deficiency is associated with the names evolved for this disease. Soon it was realized that association of this fatal disease with life-style is not true, because HIV infections were reported even among the new born. Therefore, it was perceived that these names for this disease have no relevance, so CDC decided to rename this disease. In 1982, CDC has adopted a new name for this disease and now it is known as Acquired Immune-Deficiency Syndrome (AIDS). This name is certainly indicative of deficiency of immune-system, which is not natural but this immune-deficiency is acquired because of infection. This new term befitted with the conditions of the patients afflicted with this disorder and hence AIDS became an official name for this disease.

An ever increasing number of AIDS patients have created an environment for active researches for the search of its causative organism as well as a possible treatment for the disease. This was the reason to initiate intensive researches to find cures as well as treatments for AIDS. As a result of rigorous research programs the causative organism for AIDS was identified for the 1$^{st}$ time in 1983 by Dr. Luc Montagnier and his team from Pasteur Institute Paris, France. Their findings were reported in May 20$^{th}$, 1983 issue of Science magazine (Barre-Sinousi *et al.*, 1983). Initially Montagnier's team gave the name of this virus as Lymphadenopathy Associated Virus (LAV). Within a gap of few months Dr. Robert Gallo of National Cancer Institute (NCI), Bethesda, USA also reported the causative organism for the disease and but they gave the name of this new virus as Human T-Lymphotrophic Virus IIIB ($HTLV_{IIIB}$). The work of Dr. Gallo received unexpected attention from the scientific community. The virus identified by both the groups had shown >90 per cent similarity in their genetic sequences. Later, Dr. Gallo's group filed a patent for the diagnostic test of the disease at US Patient and Trade Mark Office, Alexandria, USA. This application lead to the major controversy regarding the credit for the discovery of this virus. Finally, it was decided that both of them have equal contribution towards the discovery of the causative organism for AIDS. Different names were used in literature for the same organism, which has created confusion. To avoid this confusion International Committee on Taxonomy of Viruses coined a new term for this virus as Human Immunodeficiency Virus (HIV). Later on it was confirmed that initial patients of PCP were also infected with HIV and they were suffering from AIDS at the time of initial observation (Gottlieb *et al.*, 1981).

## Disease Burden of HIV Infections

HIV infections are one of the most feared infections during initial days of its discovery. With time researches have provided a better understanding about HIV infections and its consequences. Still the fact is that HIV infections can only be controlled and can not be cured. Knowing this fact that HIV is an infectious disease with not many possibilities to cure, it became necessary to understand the disease burden of HIV infections. The knowledge of disease burden will help us to develop better policy to contain the HIV infections as well as to develop medications which can address the issues of HIV infections and it consequences among under-privileged people.

Although the initial cases of HIV infections were reported from USA, so it was initially believed that these infections are confined only in that nation. This fact turned out to be false, as the data regarding HIV infections started pouring from all over the world. Undoubtedly, our understanding about the HIV infections at global level has improved significantly. In view of these facts, a new organization was established to handle this new evolving and emerging epidemic. This organization was named as UNAIDS (United Nation AIDS) which works under the umbrella of United Nation, New York, USA. The headquarter of UNAIDS is located at Geneva, Switzerland. Since the establishment of UNAIDS, the quality of data collection about HIV infections and AIDS have improved significantly and it is considered as the most reliable and authentic source of data about HIV infection and AIDS. Soon it was realized the African Continents are the worst affected area for HIV infections and AIDS, while the oceanic region is least affected with HIV infections, therefore the prevalence of AIDS is also very low in this region. On the one hand UNAIDS has a mandate to monitor the status of HIV infections as well as AIDS at global level, while on the other hand UNAIDS provides support to all nations dealing with HIV infections and AIDS. UNAIDS started publishing annual report about the status of HIV and AIDS at global level.

As per UNAIDS (2012) report ~34 million people are living with HIV infections, which ranges from ~31.4-35.9 million. So far, it has been shown that HIV infections do not show any bioasness towards sex. That means males and females are equally vulnerable to HIV infections. One of the most important concerns about HIV is that children are also badly affected with HIV infections. As per UNAIDS reports ~2.3 million children are infected with HIV at global level who are below the age of 15 years. At global levels, African continent is the worst affected nation with HIV infections as well as incidences of AIDS. In fact ~2/3$^{rd}$ of total HIV infections are reported only from African continents. Out of ~33 millions of HIV infected people

~23.5 million HIV infected people live in African continent. The range of HIV infections in African continents is ~22.1-24.8 millions. There is no doubt that highest number of people with AIDS also die in this continent. Numerous initiatives have been taken by local government as well as various international organizations to bring a better control over HIV and AIDS. It seems like the world community has to relook their efforts regarding HIV infections and AIDS in African continents (Table 5.1).

**Table 5.1: HIV and AIDS: Infection and Death**

| | *New Infection (in millions)* | *Total Infection (in millions)* | *Death (in millions)* |
|---|---|---|---|
| A) Global | | | |
| Adults | **2.3**<br>(2.0-2.5) | **31.3**<br>(29.2-33.7) | **1.7**<br>(1.4-2.1) |
| Children | **0.43**<br>(0.24-0.61) | **2.1**<br>(1.2-2.9) | **0.28**<br>(0.15-0.41) |
| Total | **2.7**<br>(2.4-3.0) | **33.4**<br>(31.1-35.8) | **2.0**<br>(1.7-2.4) |

As per one of the calculations, ~7400 people get newly infected with HIV every day, while ~5,500 people die with HIV infections and AIDS. It means that ~1900 new HIV seropositive are added in existing pool of HIV seropositives. If we calculate these numbers on annual basis, it means we are adding ~0.7 million people every year in the existing pool of HIV seropositives. Although, there is a better control on HIV transmission among individuals now, even though, it is a real concern for an incurable infection like HIV. The only way we can emphasize the significance of these numbers is that ~0.7 million is almost equal to the total population of capital cities of European continents barring few large cities (UNAIDS, 2012).

There is no doubt that African continent is the worst affected geographical area in the world. Apart from African continents, other developing nations are also badly affected with HIV infections. There are multiple factors which are playing their role for such a high numbers of HIV infections in underdeveloped nations as well as developing nations. Some of those factors are 1) poor implementation of surveillance, 2) poor conditions of health care, 3) poor living conditions, 4) lack of resources, 5) social fabric of the society, 6) social sigma, *etc.*

In the present time, the situation of HIV infections and AIDS patients is not as grim as it was earlier. With time there is no doubt that world community did a commendable job to control the spread of HIV infections by increased awareness and development of new and effective drugs. Some of the salient feature of present report on HIV suggests that:

1. There is not a very significant increase in new HIV infections compared to its status in 1996,
2. The is no doubts that the long-term HIV seropositives as well as new HIV infections are the main reasons for the accumulation of HIV seropositives,

3. An encouraging fact is that some nations have shown a decline in new HIV infections,
4. There is certainly a decrease in AIDS related deaths due to significant improvements in anti-retroviral drugs,
5. In general it is a great news that new HIV infections are declining at global levels, although there is a variation in nation's status of infection.

No doubt the efforts that have been tried at global levels have given their fruit by bringing a better control over the new HIV infections. Now the global efforts are towards "0" new infections in future.

## HIV and its Structure

HIV taxonomically belongs to Lentivirus genus and Retroviridae family. Since HIV belongs to Retroviridae family, so it is also called as retrovirus. The main characteristic of HIV is that its genetic material is RNA instead of DNA. Being a retrovirus, HIV also has a unique enzyme known as reverse transcriptase (RT). The main function of the RT enzyme is to convert RNA into DNA *via* the process known as reverse transcription. Structurally, HIV is a spherical virus with diameter of ~120nm (Figure 5.1). The RNA which is present as the genomic nucleic acid is single stranded and a positive-sense RNA. The genome size of HIV is ~9.8 kb. The genome of HIV consists of 9 genes in total namely *tat, rev, pol, vif, vpr, env, nef, vpu* and *LTR*. LTR flanks both sides of HIV genome. These 9 genes of HIV produces 15 different proteins during the HIV replication (Abbas *et al.*, 2008, Verma *et al.*, 2014).

## Molecular Biology of HIV

HIV genome consists of 9 genes which produces only 15 proteins during the viral replication. Some of these proteins are produced as precursor proteins (Frankel and Young, 1998). On the basis of their functions these proteins are divided into 3 main categories, 1) Structural proteins are those proteins which are important for the structure of HIV particle and these are Tat and Rev, 2) Regulatory Proteins, are those proteins mainly responsible for various regulatory functions and these proteins are Gag, Env, and Pol, and 3) Accessory Proteins are those proteins which play role for accessory functions and these are Nef, Vpr, Vif and Vpu (Figure 5.2).

### Envelope (Env)

The name Env is derived from the word envelope. The genomic location of *env* is from nucleotide 5771 to 8341. The size of *env* gene of 2,570 nucleotides. The size of Env precursor is 160 kDa and this protein comprises of 856 amino acids (AA). Envelope is a glycoprotein in nature, which is synthesized as a poly-protein precursor. Env precursor is known as gp160. gp160 is processed by cellular proteases rather than viral proteases during the trafficking of Env to cell surface. Finally gp160 cleaves into two proteins namely gp120 and gp41. Both of these glycoprotein interact with T-cell receptors.

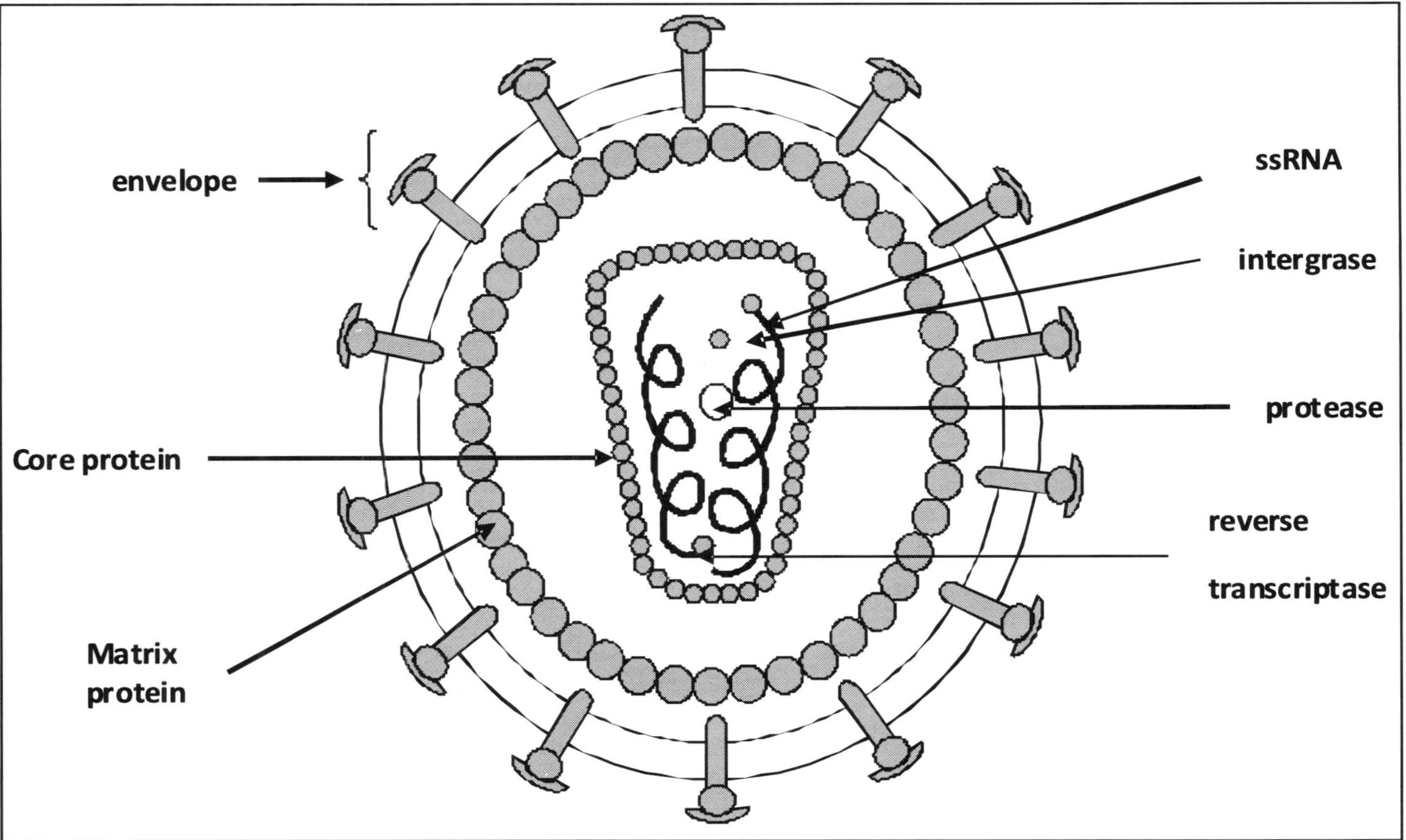


**Figure 5.1: Structure of HIV. This figure shows various component of HIV. The most unique feature of HIV structure is that it has single stranded RNA as a genetic material for HIV.**

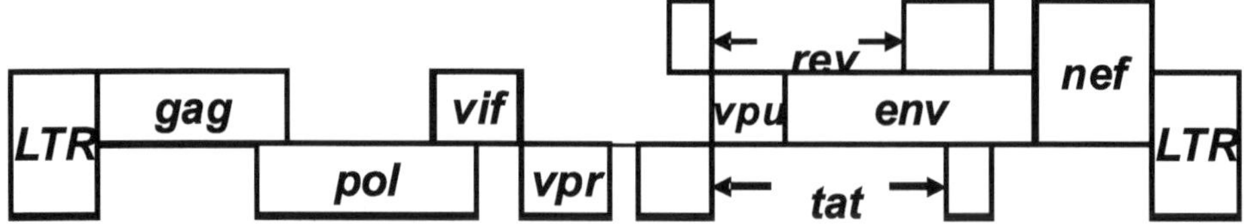


**Figure 5.2: HIV Genome.**

**The graphical representation of HIV genome. HIV genome is 9.8 kB in size and it consists of 9 genes. These 9 genes are *env*; envelope, *gag;* group specific antigen, *LTR;* Long Terminal Repeat, *nef*; negative factor, *pol;* polymerase, *rev*; regulator of expression of viral proteins, *tat*; transactivator of transcription, *vif;* viral infectivity factor, *vpr;* viral protein R, *vpu*; viral protein U.**

## Group Specific Antigen (Gag)

The gene location of *gag* is from nucleotide 336 to 1838. Gag gene comprises of total 1,502 nuceotides. *gag* gene produces a protein with its molecular weight of 55 kDa. Gag protein comprises of 550AA. Gag gene encodes for Pr55gag, which works as a polypeptide precursor. The polyprotein plays an important role for the completion of virion assembly at the end of replication. Viral proteases cleave this precursor peptide into 6 different peptides. Out of these 6 peptides two are known as spacer peptides *i.e.,* p1 and p2 while other 4 peptides are p6, p7, p17 and p24. Peptides p17 is known as matrix protein (MA), p24 is known as capsid protein (CA), p7 is known as nucleocapsid (NA), while p6 is considered as Vpr-binding protein. Matrix protein (MA) is mainly responsible for binding to the plasma membrane, which is the important step of viral assembly. Viral assembly takes place at the plasma membrane of HIV infected cells.

## Long-Term Repeats (LTRs)

Long Term Repeats (LTRs) are present on the either ends of HIV genome. LTR serves as the initiation site for the transcription of viral genome and it contains *cis*-acting elements. The *cis*-acting elements are required for RNA synthesis. LTR consists of 3 regions. 1) U3 which has unique sequence as 3′ end, 2) R is repeated sequences, and 3) U5, which has unique sequences from 5′end. The transcription starts at U3/R junction. Initiation of transcription at LTR leads to the formation of more than 30 copies of viral RNA. During the viral replication cycle newly synthesized RNA can be divided into 3 main categories which are 1) unspliced RNA, it acts as precursor for gag and gag-pol polyprotien, 2) partially spliced mRNA, it encodes for Env, Vif, Vpu and Vpr, while 3) multiple spliced RNA, it encodes for Rev, Tat and Nef. In general the basal transcription activity of LTR is low, which is enhanced significantly in the presence of Tat.

## Negative Factor (Nef)

The origin of word Nef has evolved from the word Negative Factor. The genomic location of *nef* is from nucleotide 8343 to 8963, so the total size of *nef* is 620 nucleotides. The molecular weight of nef protein is 27kDa and nef consists of 206 AA. Viral proteases cleave Nef protein into 20kDa by removing a 7kDa polypeptide chain. The main function of Nef is to downregulate the expression of major histocompatibility complex Class I and Class II and other cell receptors like CD4, CD8 and chemokine receptors. Earlier, Nef was considered as negative factor for viral replication, but with better understanding it was realized that Nef act as a positive factor for *in vivo* pathogenesis. Nef performs numerous biological functions to support viral replication; some of them are perturbation of endocytosis, modulation of signal transduction pathway in virally infected cells. In reality nef helps towards the enhancement of viral infectivity. Nef gets incorporated in HIV during the formation of a new viral particle.

## Polymerase (Pol)

The name Pol is derived from Polymerase. The genomic location of *pol* is from nucleotide 1839 to 4642. The total size of this gene is 2,803 nucleotides. Pol is a 112 kDa protein which comprised of 935AA. Pol coded enzymes are initially synthesized as a polypeptide precursor. The *pol* encoded enzymes are viral protease (PR or p10), reverse trancriptase (RT or p64) and integrase (IN or p32). Out of these one of the most important enzymes is RT. RT is responsible for reverse transcription of RNA into DNA. Integrase is a single polypeptide with molecular mass of 32kDa. Integrase is helpful for the integration of proviral DNA into host genome.

## Regulator of Expression of Viral Proteins (Rev)

The name for this gene is derived from the Regulator of Expression of Viral Proteins. Rev is encoded by two exons and both the exons are essential for the production of functional proteins. The genomic location of *rev* in genome is from nucleotide 5516 to 5591 and 7925 to 8199. Therefore *rev* comprises of two sizes *i.e.,* 75 nucleotides and 274 nucleotides. The collective length of *rev* gene is 349 nucleotides. The rev protein is 19kDa in size and it comprises of 116AA in total. The main function of rev protein is to regulate viral replication. Rev downregulates post-transcriptional splicing of viral mRNA. Rev contains two functional domains.

## Transactivator of Transcription (Tat)

The name Tat has been derived from the word Transactivator of Transcription. Tat is a regulatory protein and is essential for the regulation of viral replication. The genomic location of *tat* gene is from nucleotide 5377 to 5591 and from nucleotide 7925 to 7970. Tat is a 86-110AA long protein with a molecular mass of 16kDa. Tat is a non-structural protein which is the final product of two exons. Tat protein contains several functional domains. Out of these domains one is known as activation domain which comprised of 48AA at N-terminal, one is basic RNA-binding domain and an overlapping domain which is known as nuclear localization signal. Tat in secreted form is taken up by neighbouring cells. Tat also affects the HIV infected cells. This is the reason that tat affects both the infected and non-infected cells. Tat has been reported in peripheral circulation. Tat has also been reported from the brain of HIV infected individual with Central Nervous System (CNS) pathology. Tat protein has been reported to induce apoptosis of neurons both *in vivo* and *in vitro* conditions.

### Viral Infectivity Factor (Vif)

The Vif has been derived from the Viral Infectivity Factor. Vif is considered as one of the accessory proteins. Vif contains conserved sequences among lentiviruses. Genomic location of *vif* is from nucleotide 4587 to 5166, so *vif* comprised of 578 nucleotides in total. The Vif protein has molecular weight of 23kDa. Vif protein is comprised of 192AA. Vif is expressed during late stages of HIV replication and Vif localized in the cytoplasm of HIV infected cells. *vif* gene is responsible for the viral infectivity, but the exact mechanism of action of Vif is not known.

### Viral Protein Regulatory (Vpr)

The Vpr name has been derived from the Viral Protein R. The genomic location of *vpr* is from nucleotide 5105 to 5396. The total size of *vpr* gene is 292 nucleotides. The molecular weight is 12.7kDa. Vpr protein is comprised of 96AA. Vpr is an accessory protein of HIV. Vpr protein contains two parts one is trypsin resistant and other one is trypsin sensitive. Vpr is detected during early stages of HIV infection and Vpr is also released into the cells during infection. A higher quantity of Vpr can be detected during late stages of HIV replication. Vpr is known to affect various biological functions like transcription of new viral particle, apoptosis induction, disruption of cell cycle control, induction of defects in mitosis, *etc.*

### Viral Protein U (Vpu)

The name Vpu is derived from Viral Protein U. The genomic location of *vpu* is from nucleotide 5608 to 5856. *vpu* gene consists of 248 nucleotides in total and produces a Vpu protein whose molecular weight is 16kDa and contains 82AA. Vpu is not an essential protein of HIV, but it is considered to play an important role in enhancing the vial infectivity. Vpu protein cannot be detected in virions itself, because it is expressed during HIV replication in host cells. Vpu does not get packaged into virions. Vpu is known to play two important roles; on the one hand it enhances the release of viral particles while on the other hand it degrades CD4.

## HIV Replication

HIV is a retrovirus. HIV can be considered as a bag of nucleoprotein complex. It attains the ability to replicate only after infecting the host cells. HIV can replicate and produce new virion with the help and utilization of host cell's synthetic machinery (Freed EO, 2001; Verma *et al.*, 2014). After infection HIV leads into 2 different conditions for HIV infections 1) Productive infection, and 2) Restrictive infections. Therefore, infection of a cell by HIV may be:

### Permissive Infection

If cells are permissive for HIV replication in that case HIV replicate profoundly and release new progeny of HIV.

### Restrictive Infection

This is known as restrictive infection, this occurs to those cells which are permissive for HIV infection but restrictive for active HIV replication. In these cells

HIV replication is limited, so production of new progeny ceased, but HIV genome remains incorporated in the genome of persistently infected cells.

## HIV Replication: Different Steps

Various stages HIV replication starts after infection of HIV to the permissive cells. From infection of HIV to permissive cells to the production of new HIV progeny can be divided into 3 main steps. These steps are 1) Initiation, 2) Replication, and 3) Release.

### Initiation

This is the first phase in viral replication which can be subdivided into two steps which are 1) Attachment, and 2) Penetration. During initiation phase HIV attaches to the cell surface *via* specific receptors and co-receptors mainly on T-cells. After attachment fusion of HIV takes place with cells. After fusion HIV releases viral RNA and other essential proteins to the cells.

### Replication

This is the second stage of viral replication. Is the stage HIV RNA is reverse transcribed to synthesize pro-viral DNA. Conversion of viral RNA into DNA occurs due to the activity of Reverse Transcriptase enzyme. The replication of HIV can be divided into 3 main sub-stages. These stages are 1) Genome synthesis, 2) RNA production, and 3) Protein synthesis.

### Release

This is the last stage of HIV replication. After the complete synthesis of new HIV RNA and other protein in infected cells, virions get assembled and released as new virion particles from the cell. This step can be sub-divided further into 2 stages 1) Assembly, and 2) Release.

## HIV Infection and its Clinical Stages

HIV infection has been a major health concern. Initially, no one knew that what are the symptoms of infection that could be used for the disease diagnosis. The symptoms of the HIV infections vary at various stages, so it becomes quite difficult for clinicians to decide what is the stage of infection? and How to treat them? Clinical symptoms also become a major concern for those cases which has to be referred from one clinic to another or referred from one clinician to another. Therefore, it was realized that there was a dire need to evolve some kind of classification of different symptoms among HIV seropositives, which can be used by clinicians to treat the patients and to request for only mandatory or crucial diagnostic tests. If, someone can evolve a system to categorize or classify various stages of HIV infection that will be great service to the HIV patients as well as various communities engaged in various issues related with HIV infections like government, health policy makers, pharmaceutical companies, service providers, *etc.* Off course, the classification about different stages of HIV infection can be best determined with the availability of results from various diagnostic tests CDC, 1992).

The unfortunate part of HIV infections are that HIV infections exist among both the resourceful communities as well as resource-crunched communities. It is a reality that most of the diagnostic tests for HIV infections and HIV related diseases are quite expensive, therefore, only few can bear the financial cost for the diagnostic tests. The unfortunate fact about HIV epidemic is that the poor and developing nations are worst affected with this epidemic. African continents are worst affected with HIV infections. In this part of the world majority of people cannot afford the cost of diagnostic tests, therefore, a need for the development of classification for the stage HIV infections was realized. It will be better if classification could be based on clinical symptoms. This initiative was taken by World Health Organization (WHO), Geneva, Switzerland to develop guidelines for the classification of clinical stages for HIV infections. Still, WHO experts revise these guidelines time-to-time, so that this classification can serve its purpose effectively. In 2005, WHO has revised this classification, which is based on clinical symptoms. As per this classification, infection of HIV has been classified into 4 categories/or stages and those are 1) Stage I, 2) Stage II, 3) Stage III and 4) Stage IV (Figure 5.3). Main signs and symptoms of different stage of HIV infection are mentioned in (Table 5.2). A detailed description of each stage is discussed below:

**Table 5.2: Common Symptoms of Different Clinical Stages**

| *Stage* | *Name* | *Sign and Symptoms* |
|---|---|---|
| **Stage I** | Primary HIV Infection | Fever, Headache, Soar throat, *etc.* |
| **Stage II** | Asymptomatic Phase | Recurrent respiratory tract infections<br>Herpes zoster<br>Recurrent oral ulcerations<br>Fungal nail infections of fingers, *etc.* |
| **Stage III** | Persistent Generalized Lymphadenopathy | Severe weight loss<br>Unexplained chronic diarrhoea<br>Unexplained persistent fever<br>Pulmonary tuberculosis (TB), *etc.* |
| **Stage IV** | Symptomatic Phase | HIV wasting syndrome<br>*Pneumocystis* Pneumonia<br>Chronic herpes simplex infection<br>Extrapulmonary TB<br>Kaposi's sarcoma<br>Toxoplasmosis<br>HIV encephalopathy, *etc.* |

## Stage I: Primary HIV Infections/Seroconversion Stage

This is the 1$^{st}$ stage of HIV infection. As a matter of fact in a known case HIV infection, this stage is already established. But in unknown case of HIV infections this stage is determined on the basis of clinical symptoms presented by the patient. Stage I also called as Primary HIV Infection Stage or Seroconversion Stage. Stage I is also known as Acute HIV Infection Stage. This clinical stage can be observed with 2-4 weeks after the initial exposure to HIV. During this stage a large numbers of HIV viral particles are produced in the body due to the HIV replication. These HIV particles

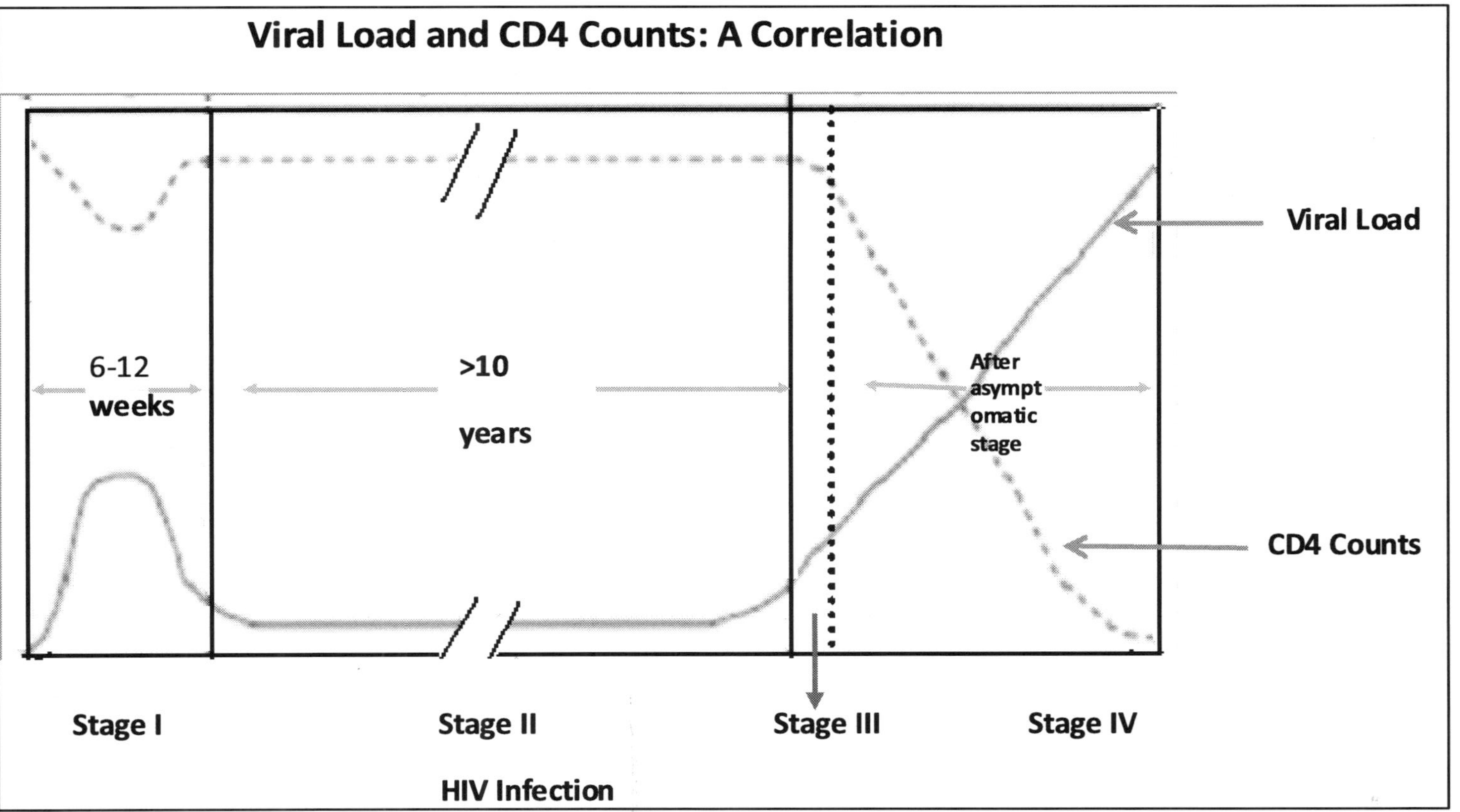


**Figure 5.3: Clinical Stages of HIV Infections.**
**This picture is depicting the different clinical stage commonly observed in a HIV patient.**

are present in peripheral blood or circulation of infected individuals. The most unique feature of Stage I of HIV infections is that at this stage immune-system of infected individual gets activated and as a result of activation of immune system, protective antibodies against HIV can be easily detected during this clinical stage. A cytotoxic T-cell response is also observed during Stage I of HIV infection. Some of the common symptoms at this stage are fever, diarrhoea, sore throat, headaches, *etc*. It has been observed that more than 65 per cent of HIV infected individuals show these symptoms. At this stage diagnosis of HIV is based on the detection of either HIV antigens present in circulation or presence of anti-HIV antibodies in peripheral circulation of the patient. At this stage, if these preliminary tests turn out to be negative in a patient with a background of high risk behavior, in that case other diagnostic tests for HIV are recommended to ascertain the infection of HIV.

## Stage II: Asymptomatic Stage

This is the 2$^{nd}$ clinical stage in case of HIV infected individuals and this stage is followed after seroconversion stage. This stage is commonly called as Asymptomatic Stage. At this clinical stage patients do not show any significant signs and symptoms of HIV infection that is the reason this stage is known as Asymptomatic Stage. At this stage viral replication takes place and patient's immune-system maintains a balance, however with time HIV replication increases with rapid rate leading to decline of immune-system, which leads to the next stage of clinical conditions. During this stage antibodies against HIV can be detected consistently. Usually CD4 counts are above 500 cells/mm$^3$. In some cases, the CD4 counts can be even lesser than 500 cells/mm$^3$. This stage could be of short span, if HIV seropositive is not on any anti-retroviral treatment. If someone is following the anti-retroviral drug regimen, in that case the asymptomatic stage lasts about ~10 years or more. Nowadays, with the improvement in anti-retroviral drugs this can be expanded upto 20 years or more post-HIV infection. Usually the main strategy for a clinician at this stage is to keep the viral replication minimal. A decrease in viral load prevents further deterioration of immune-system.

## Stage III: Persistent Generalized Lymphadenopathy

This is the 3$^{rd}$ Clinical Stage of HIV infection. This stage comes due to further deterioration of health during Stage II and this stage is called as Persistent Generalized Lymphadenopathy. This stage last for a very short span of time. It may last ~3 months. This clinical stage is an indication for further and rapid decline in health of HIV seropositives. This clinical stage is an indication for the worst. At this stage one of the most common symptoms is swelling of lymph-nodes. Most of the time the size of swollen lymph-nodes is >1cm. Except the swelling of lymph-nodes patient looks healthy even at this stage. In case of swelling of lymph-nodes, biopsy is not commonly recommended for diagnosis in cases of HIV seropositives.

## Stage IV: Symptomatic Stage

This is the 4$^{th}$ and certainly the last clinical stage in cases of HIV seropositives. Majority symptoms are present during this clinical stage. Since symptoms of HIV infections are clearly visible at this stage that is the reason this stage is known as

Symptomatic Stage. As a matter of fact the health of patient declines rapidly and choices of medications get limited due to serious side effects of therapeutic interventions. This stage finally leads to Acquired Immunodeficiency Syndrome (AIDS). At this stage higher incidences of opportunistic infections as well as AIDS associated cancers have been routinely observed. This stage of HIV infected individual is often characterized with high viral loads and low CD4 counts. Although some of the constitutional symptoms like fever, malaise, *etc.* are common in this clinical stage. Fortunately these symptoms can be treated easily, but the most unfortunate part of this stage is that it is almost impossible to stop further deterioration of health status of HIV patients' at this stage. Certainly it should be considered as the last stage of HIV patient, therefore counselling of HIV patient regarding the final outcome of the disease should be discussed. It is of no use to keep patient in dark about the consequences of such a significant decline in health.

With time more and more new symptoms have been reported among HIV seropositives, this is one of the reasons that WHO revise classification as and when it deemed necessary. In the recent past new symptoms have been reported among long-term HIV seropositives. These patients are presented with decline in neurological and neuropschyiatic health of HIV seropositives. These symptoms are grouped as NeuroAIDS, which is considered as a new emerging complication in long term HIV seropositives.

## Anti-Retroviral Treatment (ART)

The drugs that work against HIV are known as Anti-Retroviral Drugs and treatment against HIV is commonly known as Antiretroviral Treatment. Literally these drugs are good enough to treat any retroviral infections, but mainly they are considered as anti-HIV drugs. Initially, there was no drug available to treat HIV infections. The 1$^{st}$ drug that has been tried to treat HIV infections was Azidothymidine (AZT). This drug was approved by Food and Drug Administration (FDA), Silver Spring, USA in 1985 to treat HIV patients. After the initial success of AZT, the concerted efforts of scientists, various new drugs have been developed to treat HIV infections. The development of new drugs against HIV infections was possible due to the better understanding about HIV replication. A better understanding of HIV replication has provided clues for new drug targets which are used to develop new types of anti-retroviral drugs (Olender *et al.*, 2012; Verma *et al.*, 2012, 2014). At present we have numerous choices of anti-retroviral drugs, which are classified into the following categories: 1) NRTI, 2) NtRTI, 3) NNRTI, 4) PI and 5) InI (Table 5.3).

## Nucleoside Reverse Transcriptase Inhibitors (NRTI)

This is the 1$^{st}$ ever group of drugs developed against HIV infections. As a matter of fact AZT belongs to this group of anti-retroviral drugs. These drugs target the activity of reverse transcriptase enzyme, which is crucial for viral replication. These drugs act as "False Building Block". These drugs serve as an alternative substrate for RT enzyme. During the synthesis of pro-viral DNA, these drugs compete with physiological nucleosides and due to the competition further synthesis of pro-viral DNA is blocked. Therefore, NRTI acts as anti-retroviral drugs by preventing further replication of HIV in infected cells. Later other nucleosides were targeted for the development of new drugs which belong to this group. Some of the most important examples of NRTI are AZT, ddI, d4T, 3TC and FTC.

**Table 5.3: Anti-retroviral Drugs***

| *Class* | *Chemical Name* | *Generic Name* | *Trade Name* |
|---|---|---|---|
| **NRTI** | ZDV | Zidovudine | Retrovir |
| AZT | Azidothymidine | Azitidin | |
| ddI | Didanosine | Videx | |
| ddC | Zalcitabine | Hivid | |
| d4T | Stavudine | Zerit | |
| 3TC | Lamivudine | Epivir | |
| ABC | Abacavir | Ziagen | |
| **NtRTI** | TDF/PMPA | Tenfovir | Viread |
| **NNRTI** | NVR | Nevirapine | Viramune |
| DLV | Delavirdine | Rescriptor | |
| EFV | Efavirenz | Stocrin/Sustiva | |
| **PI** | SQV-hgc | Saquniavir | Invirase |
| RTV | Ritonavir | Norvir | |
| IDV | Indinavir | Crixivan | |
| NFV | Nelfinavir | Viracept | |
| APV | Amphrenavir | Agenerase | |
| LPV | Lponavir | Kaletra | |
| **Int. I** | RGV | Raltegravir | Isentress |

* It is not an exhaustive list of anti-HIV drugs.

NRTI: Nucleoside Reverse Transcriptase Inhibitors; NtRTI: Nucleotide Reverse Transcriptase Inhibitors; NNRTI: Non-nucleoside Reverse Transcriptase Inhibitors; PI: Protease Inhibitors; Int. I: Integrase Inhibitors.

## Nucleotide Reverse Transcriptase Inhibitors (NtRTI)

This is also a group of medication which affects Reverse Transcriptase enzyme. NtRTI also works on the principles of nucleoside reverse transcriptase inhibitors. NtRTI act as nucleotide competitors, therefore the medicines of this group are classified as Nucleotide Reverse Transcriptase Inhibitors (NtRTI). This drug has advantage over other NRTI because NtRTI drugs do not require activation of drug as in case of NRTI. This is the reason that toxicity of drug from this group is lesser compared to NRTI. But NtRTI is also known to have some serious side effects. TDF is one of the examples for NtRTI.

## Non-Nucleoside Reverse Transcriptase Inhibitors (NNRTI)

After the success of NRTI and NtRTI, a new target was tried in reverse transcriptase enzyme. So, NRTI, NtRTI and NNRTI group of drugs target reverse transcriptase enzyme, but the mode of action of NNRTI is different than NRTI and NtRTI. This is the reason this group of drugs is known as Non-Nucleoside Reverse Transcriptase Inhibitors (NNRTI). The drugs of NNRTI group bind with Reverse Transcriptase enzyme and makes RT enzyme incapable for the synthesis of proviral DNA. NNRTI binds with RT enzyme at a site closer to the binding site for nucleoside, therefore NNRTI blocks the binding of nucleosides with RT. This is mechanism NNRTI blocks the synthesis of proviral DNA (Boehringer Ingelheim, 2005). Some of the most common drugs of this group are Nevirapine, Delaviridine, Efavirenz, *etc.*

### Protease Inhibitors (PI)

This is another group of anti-retroviral drugs which is useful to treat HIV infections. The target of these drugs is entirely different than NRTI, NtRTI as well as NNRTI. As a matter of fact the drugs of PI group acts at the last step of viral replication. The drugs of PI group inhibit the activity of protease enzyme, therefore these drugs are known as Protease Inhibitors (PIs). Since these drugs block the activity of protease, therefore, these drugs prevent the formation of new viral particles. Initially these drugs have to face the criticism for its frequency to be taken by the patients. At present this criticism has died down due to significant improvements in these drugs. Some of the drugs of this group are Saquinavir, Ritonavir, Indinavir, Nelfinavor, *etc.*

### Integrase Inhibitors (InI)

This is another group of drugs to treat HIV infections. The target for this group of drugs is integrase enzyme of HIV. Drugs belonging to this group inhibit the activity of integrase enzyme which prevents integration of pro-viral DNA into the infected host or cells. The 1st drug to treat HIV infections from this group was approved by United States Federal Drug Administration (USFDA) in 2007, some of the drugs of this group are Raltegravir, Isentress, *etc.*

## New Target and New Drugs

There is no doubt that NRTI, NtRTI, NNRTI, PI and InI have lead to significant improvements in the management of HIV infections. It is one of the reasons that nowadays a HIV seropositive can lead a healthy life-style and average life-span has significantly improved among HIV seropositives post-HIV exposure. Continuous and serious researches on HIV for discovering newer targets to treat HIV infections are seriously warranted. Some of those targets that are used for drug development are at the level of HIV binding to the cells. We hope that some of these drugs would be in market in future and they may be of great help to change the scenario of HIV infections and HIV infected patients.

### Bone Marrow Transplantation: A New Hope for Treatment

In the recent past in Germany, an HIV seropositive has been shown to be free from HIV. This person is famously known as "Berlin Man", because this new treatment was tried in a hospital at Berlin, Germany. This HIV seropositive was treated by a team of doctors headed by Dr. Gero Hutter. As a matter of fact this HIV seropositive was tried with bone-marrow transplantation. Following the bone marrow transplantation, this patient did not show any signs and symptoms for the presence of HIV in his body even after 4 years post bone marrow transplantation. That is an amazing result in case of HIV infections. No doubt this kind of clinical approach has offered a new hope for the treatment of HIV patients.

It has been observed that some Caucasian people were found to be HIV negative, even though they are classified as High Risk individuals with reference to HIV infections. The most surprising fact about these cases was that they were never on any anti-HIV medications, even though they were following a High Risk Life-Style with reference to HIV infections. This observation lead to a curiosity among clinicians and scientists that what are the underlying reasons for these persons not getting HIV infection (Shearer and Clerici 1996). These patients were followed for their medical record which was studied thoroughly to find out the causes behind this medical mystery. Intensive researches in this direction revealed some of the surprising facts that these individuals have a mutation in their chemokine receptor. For HIV infections, two types of chemokine receptors *i.e.,* CXCR-4 and CCR-5 play important role. A deletion mutation in gene coding for CCR-5 receptor was identified in people who are resistant to HIV infections.

A deletion of 32bp in CCR-5 gene was found and hence it was called as deletion mutation for CCR-5. This mutation could be either homozygous or heterozygous in nature. Both of these mutations have been shown to reduce the possibility of HIV infections significantly. It has been shown that persons with homozygous mutation to be fully resistant against HIV infections. However, in case of individuals with heterozygous mutation progression of HIV infection is minimal or very slow. These mutations are written as Δ32/Δ32 CCR-5. This information proved useful to address the issues related with the treatment of HIV infections. But this information had to wait for the right opportunity to be tried for the treatment of HIV seropositive. For this treatment, an HIV seropositives of Caucasian origin was needed who may also be in need of bone marrow transplantation. Finally a patient turned up and clinicians believed that this patient could be tried with this new mode of treatment. This treatment truly turned out to be a success story and credit for the development of this new treatment for HIV seropositives goes to Dr. Gero Hutter and his team in Germany.

A 40 year male patient who was HIV positive for more than 10 years developed Acute Myelogenous Leukaemia (AML). This patient was under Highly Active Anti-Retroviral Treatment (HAART) and did not show any signs and symptoms either for HIV or for AIDS. Initially this patient was given chemotherapy for AML. During the course of AML treatment the patient continued with HAART regimen too but because of the severe toxicity of drugs clinicians had to stop HAART medication. Unfortunately, discontinuation of HAART treatment was the reason for the rebound of HIV replication. This was the reason HAART treatment was restarted which brought the HIV replication under control. This patient failed to show a significant improvement in AML. Therefore, the obvious choice for this patient was to go for bone marrow transplantation as a treatment strategy for AML.

Under these circumstances, it was decided to perform bone marrow transplantation with a bone marrow showing Δ32/Δ32 CCR-5. This strategy had two pronged benefits; on the one hand bone marrow transplantation will help to cure

AML in this patient, if the patient is transplanted with Δ32/Δ32 CCR-5 bone marrow cells in that case there is always a possibility that this patient may also develop resistance against HIV infection. With this strategy, a search for suitable donor was started among various bone marrow donor registries. Finally a matching donor was found and luckily, this bone marrow donor agreed to donate the bone marrow for this patient. Bone marrow transplantation was performed successfully, and the patient was cured from AML. All anti-HIV medications were stopped following transplantation. So far, routine tests for HIV turnout to be negative. Till date, this patient did not shown any signs for HIV replication. In other words we can say this patient has been cured from HIV infection, and this is the 1st case to cure HIV by bone marrow transplantation (Hutter *et al.*, 2009).

On the one hand Berlin Man has raised the hope for a possible cure for HIV infections, but this method has its own limitations like 1) so far, this mutation has been reported only in Caucasian populations, 2) the prevalence rate for this mutation is very very low *i.e.*, ~1-3 per cent in Caucasian populations, 3) till date, this mutation has not been reported in any other race, 4) infectivity of X-4 strain of HIV cannot be denied, 5) cost of bone-marrow transplantation, 5) benefits of bone marrow transplantation compared to anti-retroviral treatments, 6) maintenance of immunosuppression after transplantation could be an important reason for concern, *etc.* Apart from all these limitations a possibility for the improvement of this mode of treatment to be used more effectively and more commonly cannot be denied in near future.

Further development in stem cell research could open new avenues for the treatment of HIV infections.

## AIDS and Opportunistic Infections

There is always confusion about HIV and AIDS. The fact is that one can be infected with HIV but may not have AIDS. Acquired Immunodeficiency Syndrome (AIDS) is the terminal stage of HIV infected person. When HIV seropositives clinically reach to Stage IV *i.e.*, Symptomatic Phase of HIV infections, then they show signs and symptoms of AIDS. This stage of HIV infections is known as AIDS. AIDS defining diseases can be of two types which are grouped as 1) AIDS related cancers and 2) AIDS related infections or opportunistic infections.

As we know, if a HIV seropositive is on correct anti-HIV medication, in that case they can live for a long period as asymptomatic patients.

It means these patients do not show any clinical signs and symptoms of HIV infection, because anti-retroviral drugs keeps HIV replication to minimal or barely detectable levels and these HIV seropositive are able to maintain good counts for CD4 cell. Ultimately a situation arises, when these anti-HIV drugs are not as effective to stop HIV replication. As a result, these patients see a rapid increase in viral load as well as a significant drop in their CD4 counts. These clinical conditions of HIV seropositive suggest a decline in health of patient who is heading towards final outcome due to HIV infections, which is death. At this stage, there are limited choices for clinical interventions, but most of them can only provide some relief to the patients from the miseries, and may delay the final outcome a bit. So it is clear that all HIV seropositives cannot be called as AIDS patients, but it is for sure that all AIDS patients have HIV infections(Walker, 2006). Some of the common symptoms in AIDS patients are mentioned in (Table 5.4).

**Table 5.4: Clinical Symptoms of AIDS**

| |
|---|
| Thrush |
| Recurring vaginal yeast infections |
| Pelvic inflammatory disease (Chronic) |
| Severe and frequent infections |
| Unexplained tiredness |
| Quick loss of body weight |
| Frequent diarrhea |
| Frequent fevers |
| Swelling or hardening of glands located in the throat, armpit, or groin |
| Shortness of breath |
| Unexplained bleeding from mouth, nose, anus, or vagina |
| Frequent skin rashes |

A low counts of CD4 simply suggest that immune-system is badly compromised which could lead to the AIDS and AIDS related conditions. Most of the times the onset of AIDS is characterized with either AIDS related cancers or Opportunistic Infections.

## Opportunistic Infections

Stage IV of HIV seropositive is a state of immunosuppression due to very low CD4 counts. This kind of immune-status provides excellent opportunities for opportunistic infections to affect the host/patients. The opportunistic infections organisms are bacteria, virus, fungi, *etc.* In normal host these microorganisms do not survive for long periods because of intact immune-system. But in case of immune-deficient or immune-suppressed host these infections can grow easily, because the immune-system is not able to fight against the infection.

It is important to remind that the first cases of HIV infections were never reported to be positive for HIV or as AIDS patients, because at that point of time neither HIV was known nor AIDS. Even the term HIV and AIDS were not coined by that time. But Dr. Gottleib has noticed severe immune-compromised patients who were suffering from Pneumocystis carnii Pneumonia (PCP). Later on these initial patients were confirmed with HIV infection and AIDS patient. As time passed more information was gathered about HIV infections it was noticed that there is wide spectrum of opportunistic infections among AIDS patients (Table 5.5).

**Table 5.5: Opportunistic Infections**

| |
|---|
| Candidiasis |
| Coceidiodomyeosis |
| Cryptococeosis |
| Cryptosporidiosis |
| Cytomegalovirus disease |
| Histoplasmosis |
| Mycobacterium avium complex |
| Pneumocystis carinii pneumonia |
| Toxoplasmosis |

## AIDS Related Cancers

Another most common symptom in AIDS patients is cancers. These cancers have been commonly reported in AIDS patients. This is the reason these cancers are known as AIDS Related Cancers, if a HIV seropositive shows signs and symptoms of these cancers then it is an indication that this patient is suffering from AIDS. Some of the most commonly reported cancers among AIDS patients are Kaposi's sarcoma, Non-Hodgkin's Lymphoma and Cervical cancer. In literature, these cancers are also known as "AIDS Defining Cancer", because it has been observed that once a patient shows any of these cancers, they could be considered as AIDS patients. These cancers are commonly found in AIDS patients, apart from these cancers some less common cancer like cancer of mouth, throat, lung, colon, anus, rectum, *etc* have also been reported among AIDS patients (Table 5.6).

**Table 5.6: AIDS Related Cancers**

A) AIDS Related Cancers
  - i) Most Prevalent Cancers
    - Kaposi's Sarcoma
    - Non-Hodgkin's Lymphoma
  - ii) Less Prevalent Cancers
    - Hodgkin's Lymphoma
    - Angiosarcoma
    - Liver Cancer
    - Oral Cancer
    - Colorectal Cancer
    - Lung Cancer
    - Testicular Cancer

### Kaposi's Sarcoma

Kaposi's sarcoma is a type of skin cancer, which traditionally occurred in older men of Jewish or Mediterranean descent, as well as people who have received organ transplantation. But Kaposi's sarcoma is most commonly reported among homosexual men with HIV/AIDS. HIV/AIDS-related Kaposi's sarcoma causes lesions at multiple sites in the body, *e.g.*, skin, lymph nodes, and other organs.

**Non-Hodgkin Lymphoma**

After Kaposi's sarcoma, Non-Hodgkin Lymphoma is the second most common cancer associated with HIV infections. Non-Hodgkins Lymphomas are of different types. In case of HIV and AIDS patients, primary central nervous system lymphoma is very common. ~20 per cent of HIV and AIDS patients show primary effusion lymphoma which causes the accumulation of fluid around lungs or in abdomen. But more than 80 per cent of HIV and AIDS patients show high grade B-cell lymphoma. Surprisingly, this kind of lymphoma is reported ~10 per cent -15 per cent among cancer patients who are non-HIV and non-AIDS patients.

## Other Less Common Cancers in AIDS Patients

Some other cancers have also been reported in HIV/AIDS patients. Some of these cancers are Hodgkin's lymphoma, angiosarcoma, anal cancer, liver cancer, mouth cancer, colorectal cancer, lung cancer, testicular cancer and skin cancer to name a few.

## NeuroAIDS: A New and Emerging Health Concern

There is no doubt that after the discovery of HIV infections, there is significant improvements in the clinical conditions of HIV seropositives. Now a HIV seropositive lives a healthy life, if s/he is on anti-retroviral medications. This is one of the reasons that a HIV seropositive lives a lot longer after the initial HIV infections. It seems like that improvement in anti-retroviral medication worked as a double edged sword for HIV seropositives. One the one hand they live longer but they started showing signs and symptoms of neurological complications which is a matter of real concern. These neurological complications among HIV seropositives are grouped as NeuroAIDS. The most surprising fact about neuroAIDS is that HIV is not known to infect the brain cells, so how come patients are showing these symptoms? The present state of knowledge indicates that more aggressive researches are needed to address this new and other emerging health concerns among HIV infected individuals (Macombe *et al.*, 2009; Verma *et al.*, 2010, 2012, 2014).

There is a significant improvement in the management of HIV infected patients as well as AIDS patients. With time a better understanding has evolved about HIV and its replication. This is one of the most important factors for the development of better anti-retroviral drugs. In the recent past emergence of NeuroAIDS have been observed among long-term HIV-seropositives. Even though there are improvements towards the control of HIV and AIDS epidemic. Still more researches are warranted to find a cure for HIV.

## Acknowledgments

Authors are thankful to Prof. Chanderdeep Tandon, Director, AIB, AUUP, NOIDA for providing necessary resources to complete this book chapter. Mr. Dinesh Kumar's efforts for word processing and graphical work are duly appreciated by the authors.

## References

1. Abbas KA, Lichtman AH and Pillai S (2008). *Cellular and Molecular Immunology.* 6th edition, India: Elsevier Press.
2. Barre-Sinousi F, Chermann JC, Rey F, Nugeyre MT, Chamaret S, Gruest J, Dauget C, Axler-Blin C, Vezinet-Brun F, Rouzioux C, Rozenbaum W and Montagnier L (1983). Isolation of a T-Lymphotrophic retrovirus from a patient at risk for Acquired Immune Deficiency Symdrome. *Science* 220: 868-871.
3. Boehringer Ingelheim Pharmaceuticals Incorporated, (2005). *NNRTI Mode of Action (animation).* www.boehringer-ingelheim.com.
4. Centers for Disease Control and Prevention (1992). 1993 Revised classification system for HIV infection and expanded surveillance definition for AIDS among adolescents and adults. *Morbidity and Mortality Weekly Report* 41: 1-19.
5. Frankel AD and Young JA (1998). HIV-1: fifteen protein and an RNA. *Annual Review of Biochemistry* 67: 1-25.
6. Freed EO (2001). HIV-1 Replication. *Somatic Cell and Molecular Genetics* 26: 13-36.
7. Gottlieb MS, Schroff R, Schanker HM, Weisman J D, Fan P T, Wolf R A and Saxon A. (1981). *Pneumocystis carinii* Pneumonia and mucosal Candidiasis in previously healthy Homosexual men: Evidence of a New Acquired Cellular Immunodeficiency. *New England Journal of Medicine* 305: 1425-1431.
8. Hütter G, Nowak D, Mossner M, Ganepota S, Mubig A, Allers K, Schneider T, Hofmann J, Kücherer C, Blau O, Blau IW, Hofmann WK and Thiel E (2009). Long-term control of HIV by CCR5 delta32/delta32 Stem-cell transplantation. *New England Journal of Medicine* 360: 692-698.
9. McCombe JA, Noorbakhsh F, Buchholz C, Trew M and Power C (2009). NeuroAIDS: A watershed for mental health and nervous system disorders. *Journal of Psychiatry and Neurosciences* 34: 83-85.
10. Olender S, Wilkin TJ, Taylor BS and Hammer SM (2012). Advances in anti-retroviral therapy. *Topics in Antiviral Medicine* 20: 61-68.
11. Shearer GM, Clerici M (1996). Protective Immunity against HIV infections: has nature done experiment for us? *Immunology Today* 17: 21-24.
12. UNAIDS Report (2012). Report on the Global AIDS Epidemic.
13. Verma AS, Singh UP, Dwivedi PD and Singh A (2010). Contribution of CNS Cells in NeuroAIDS. *Journal of Pharmacy and Bioallied Sciences* 2: 300-306.
14. Verma AS, Singh UP, Mallick P, Dwivedi PD and Singh A (2012). NeuroAIDS and omics of HIV *vpr*. In: Barh D, Blum K and Madigan MA (eds.). *Omics Biomedical perspectives and applications.* pp. 477-511. USA: CRC Press.
15. Verma AS, Singh IG, Bansal R, and Singh A (2014). HIV and antiretroviral drugs. In: Verma AS and Singh A (eds.). Animal Biotechnology: Models in Discovery and Translations pp. 155-176, USA: Academic Press.
16. Walker BD (2006). AIDS and Secondary Immunodeficiency, In Male D, Brostoff J, WHO Guidelines, 2005, www.who.int/hiv/pub

**2016, Biotechnology: Progress and Applications** *Pages* **131–159**
*Editors:* **Saif Hameed and Zeeshan Fatima**
*Published by:* **DAYA PUBLISHING HOUSE, NEW DELHI**

*Chapter 6*

# Engineering Pathogen Derived Resistance against Plant Viruses: Current Scenario and Future Prospect

***Veerendra Kumar Sharma and Supriya Chakraborty****

*Molecular Virology Laboratory, School of Life Sciences, Jawaharlal Nehru University, New Delhi – 110 067, India*

## ABSTRACT

Around the globe, with increasing demand of food, viral diseases of crop plants are becoming major threat to agriculture. Classical methods of controlling viruses through chemicals, pesticides, cultivation of resistant varieties generated through conventional plant breeding have been of very limited success and attached with imminent concerns like adverse health and environmental consequences, a complex situation compounded with the emergence of resistant vector strains. In this context, development of durable and broad spectrum resistance strategies is the utmost requirement to meet the future food demands.

The advancement of new technologies to introduce foreign genes has opened up doors to apply novel pathogen derived resistance (PDR) strategies in developing resistance in plants. The current chapter focuses on the major developments made in the direction of generating antiviral resistance in plants based on PDR over the years.

* Corresponding Author: E-mail: supriyachakrasls@yahoo.com

## Introduction

Plant pathogens, continuously posing stress on plants for their survival, manifest their presence through abnormal morphology, physiology and productivity of the infected plants. The diverse nature and vast number of pathogens are serious threat to worldwide production and security of food as the population and demands are ever increasing.

As per the report of the International Committee on Taxonomy of Viruses (2012), 2619 virus species are known to exist. This number has increased upto 2828 virus species by 2013 and is continuously growing as new virus species are emerging due to several factors such as changes in environmental factors, high rate of recombination, pseudo - recombination, counter pressure of host defence machinery etc. Viral diseases of plants result in massive loss of productivity of economically important crops (Table 6.1). Agricultural practices in developing countries like India, where major concern is to maximize the production rather than to control pest and pathogens, further aggravate the complexity of the situation.

The classical methods used to combat virus diseases have not been proven sufficient and no strategy is known to counter viruses using chemical compound. The concept of pathogen derived resistance (PDR) (Sanford and Johnston 1985) refers to the expression of pathogen derived sequences in dysfunctional form or at a wrong stage, which would disrupt the normal life cycle of pathogen inside the host. Since its inception, PDR approach has been used to develop virus resistance in transgenic plants through the expression viral sequences (Baulcombe 1994b; Lomonossoff 1995; Lucioli *et al.*, 2008, Patil *et al.*, 2011; Wilson, 1993).

Here, in this book chapter, we focus on various PDR based strategies, their progress over the years, and possible mechanism of conferring resistance. The final section highlights the use of bioinformatics to design broad spectrum strategies against multiple viruses, the problems associated with the applicability of this technology to economically important crops and the future prospects.

## Methods of Generating PDR

### Protein Mediated Protection aginst Viruses

#### Coat Protein Mediated Resistance (CPMR)

The transgenic expression of *Tobacco mosaic virus* (TMV) coat protein gene in tobacco plant describes the first published report of expression of viral derived sequences in transgenic plants. However, the resistance in transgenic plants against TMV could not be detected, and the R1 progeny of transgenic plants, when inoculated with TMV, showed symptoms similar to non transgenic controls (Bevan *et al.*, 1985). The first demonstration of viral sequences mediated resistance in transgenic plants was shown by Beachy and colleagues (Powell Abel *et al.*, 1986). Transgenic tobacco plants expressing high levels of coat protein were found to be more resistant to TMV infection. Resistance, in this case, was manifested through delayed symptoms appearance as well as by reduced virus titer as compared to control plants. Since the first report of developing CPMR against TMV, there have been an increasing number

**Table 6.1: Important Virus Group Causing Crop Diseases in India**

| *Virus Group* | *Virus* | *Nature of Genetic Material* | *Disease* | *Yield Loss (per cent)* | *Crop* |
|---|---|---|---|---|---|
| Begomovirus | *Indian cassava mosaic virus* | ss DNA | mosaic | 20 - 25 | Cassava |
| | *Cotton leaf curl virus* | ss DNA | leaf curl | 65 - 70 | Cotton |
| | *Bhendi yellow vein mosaic* | ss DNA | mosaic | 80 - 96 | Bhendi |
| | *Tomato leaf curl* | ss DNA | leaf curl | 40 - 100 | Tomato |
| | *Mungbean yellow mosaic virus* | ss DNA | mosaic | 20 - 70 | Mungbean |
| | *Mesta yellow vein mosaic virus* | ss DNA | mosaic | 60 - 80 | Metsa |
| | *Chilli leaf curl virus* | ssDNA | leaf curl | 60 - 100 | Chilli |
| Potyvirus | *Rice necrosis mosaic virus* | +ve sensess RNA | necrosis | 12 - 100 | Rice |
| | *Potato virus Y* | +ve sensess RNA | mosaic | 85 | Potato |
| Badnavirus | *Rice tungro badna virus* | ds DNA | Rice tungro | 10 | Rice |
| | *Sugarcane bacilliform virus* | ds DNA | Leaf fleck | 5 - 100 | Sugarcane |
| | *Citrus yellow mosaic virus* | ds DNA | mosaic | 77 | Citrus |
| | *Piper yellowmottle virus* | ds DNA | mottle | - | Piper |
| Nanovirus | *Banana bunchytop virus* | ss DNA | bunchy | 70 - 90 | Banana |
| Tospovirus | *Groundnut bud necrosis virus* | –ve sensess RNA | necrosis | 50 | Groundnut |
| | *Watermelon bud necrosis virus* | –ve sensess RNA | necrosis | 50 - 100 | Watermelon |
| | *Capsicum chlorosis virus* | –ve sensess RNA | chlorosis | 50 - 80 | ***Capsicum*** |
| Cucumovirus | *Cucumber mosaic virus* | +ve senseRNA | Shoestring, Mosaic, Chlorosis | 70 - 80 | Cucurbits, Tomato, Chilli |

of examples of developing resistance through the expression of CP gene of different viruses in a number of transgenic dicots and monocots. Table 6.2 shows the examples where coat protein gene of viruses from different taxonomic groups, have been transgenically expressed in different crops to generate effective resistance.

Several lines of evidence support the fact that endogenous CP interfered with an early event in virus infection. Earlier studies conducted with TMV, *Alfalfa mosaic virus* (AlMV), and *Tobacco streak virus* (TSV) had shown that the resistance conferred by endogenous CP could be overcome by inoculation with viral RNA (Loesch Fries *et al.*, 1987; Nelson *et al.*, 1987). Based on these observations, it was suggested that transgenically expressed CP inhibited the disassembly of virions in the initially infected cells, while naked RNA does not require this step and can establish infection. To rule out the role of interaction between endogenous CP mRNA and virion RNA in resistance, Powell generated transgenic plants expressing untranslatable mRNA lacking initiation codon and plants expressing mRNA that would produce CP but lack the rep binding site. Upon infection with TMV, plants accumulating the CP rather than CP mRNA were found to be resistant to TMV infection (Powell *et al.*, 1990). The results obtained by Nejidat also supported the role of protein in resistance rather than RNA. They examined the level of TMV CP in transgenic tobacco and tomato plants under different temperature conditions and analyzed the effects of temperature on resistance. At high temperature the level of protection mediated by CP declined with levels of CP while the level of CP mRNA was constant however in transgenic tomato plants the level of protection was same at different temperatures indicating the role of host in conferring resistance (Nejidat and Beachy, 1990).

Several studies have shown the correlation between the level of protection and the potential of transgenically expressed wild type or mutant CP in forming aggregates or virus like particles (VLPs) in plants (Asurmendi *et al.*, 2007; Bendahmane *et al.*, 1997; Clark *et al.*, 1995). Based on the structure of TMV CP, mutants with enhanced subunit - subunit interaction were generated. Transgenic plants expressing these mutants showed higher level of protection while the transgenic plants expressing mutants that were unable to form aggregates were as susceptible as wild type plants (Bendahmane *et al.*, 1997, 2007; Register *et al.*, 1988). The expression of virus CP genes has provided one of the strongest examples of cross protection yet developed such as the coat protein of TMV provided effective levels of resistance against closely related strains of TMV and tobamoviruses but not against distantly related tobamoviruses (Nejidat and Beachy 1990). TMV CP expressing transgenic plants when challenged against TMV strains U1 and a severe strain PV230, developed reduced number of chlorotic local lesions correlating with reduced virus titer (Nelson *et al.*, 1987). Similar observations were also reported in transgenic plants expressing Potato virus X (PVX) CP and AlMV CP following inoculation with PVX and AlMV respectively (Hemenway *et al.*, 1988; Loesch *et al.*, 1987). Replication of PVX was reduced in transgenic plants and transgenic protoplasts expressing wild type or mutant ($CP^{T28W}$) TMV CP. Similarly transgenic protoplasts expressing PVX CP conferred transient protection against infection with TMV and TMV RNA. PVX CP conferred heterologous interference to PVX and TMV infection (Bazzini *et al.*, 2006). Several elegant experiments were performed that supported disassembly hypothesis. In one of the experiments TMV like pseudovirions containing β - glucuronidase (GUS) mRNA were produced *in vitro*. These pseudovirions when introduced in protoplast, the expression level of GUS mRNA was 100 fold lower in protoplasts expressing CP than in control protoplasts, while expression of unencapsidated GUS mRNA was lowered by 2.8 folds only. When TMV RNA was encapsidated with heterologous virions, the hybrid virions were found to be infectious against both CP expressing and control protoplasts. It was suggested that although endogenous CP inhibited the disassembly, it might interfere with a later step involving viral RNA (Osbourn *et al.*, 1989). The experiments conducted with AlMV also supported the disassembly hypothesis; however the protection conferred by endogenous CP was effective against both virions and viral RNA inocula. It was suggested that CPMP against AlMV acts on another step in addition to nucleocapsid disassembly (Taschner *et al.*, 1994). Recent work of Vasudevan proposed that the resistance developed in transgenic *Nicotiana benthamiana* plants expressing *Turnip crinkle mosaic virus* CP was the outcome of a combination of both protein mediated and RNA mediated resistance. Transgenic plants resistant to TCV infection accumulated CP 5 times higher than that in the susceptible transgenic line. The resistance was specific to TCV infection only and the progenies of resistant line showing milder symptoms were recovered in the later stages suggesting the role of RNA mediated resistance (Vasudevan *et al.*, 2008). Most of the studies conducted with TMV, AlMV, and TSV have suggested that endogenous CP is required for the resistance. However, studies conducted with cucumber mosaic virus (CMV), *Potato leaf roll virus* (PLRV), PVX, and *Potato virus Y* (PVY) showed a different mechanism. Plants transformed with truncated, antisense or nonexpressing CP genes provided measurable protection or even complete immunity against the corresponding viruses (Lindbo *et al.*, 1992a and 1992b). For instance, although the PVY coat protein was undetectable (Keniewski *et al.*, 1990) in the plant, transgenic plants expressing PVY CP was found to be resistant to PVY. CPMR has also been shown to greatly reduce the titer of PLRV in transgenic plants. In this case, the CP transcript was accumulated but the protein was not detected (Kawchuk *et al.*, 1991). Dougherty *et al.* (1994) observed that transgenic plants expressing full length *Tobacco etch virus* (TEV) CP sense RNA with frame shift mutations making it untranslatable, were highly resistant to Potyvirus infection.

Further, different stages of viral infection cycle at which transgenically expressed coat protein acts were shown to vary in different studies involving different virus groups. The actual mechanism of CPMR is not clear and several mechanisms may be involved in the specific resistance elicited via the expression of a particular CP in plants. These mechanisms could operate at the initial interaction between the transgenic CP and the virus, or if replication is initiated, resistance mechanism could involve the control of replication or a later stage of viral movement.

**Movement Protein Mediated Resistance (MPMR)**

The efficiency of cell to cell virus movement is important in determining pathogenicity, virulence and in some cases, the host range of a plant virus. Cell to cell and long - distance movement of viruses in plants are not passive events, and most of the, if not all, plant viruses have a gene or set of genes that mediate movement. Movement proteins (MP) encoded by the plant viruses are responsible for their local (cell to cell) and systemic (long distance) movement in the hosts enabling infection to reach distant sites. Movement proteins facilitate cell to cell transport of the viral genome by modifying plasmodesmata, the complex transwall channels that connect adjacent plant cells and provide symplastic movement between cells and tissues. The mechanism of transport of virus particles or virus nucleic acids through interaction of MPs with plasmodesmata is not clear. It is hypothesized that viral MPs increase the size exclusion limit of plasmodesmata by modifying their gating function to allow virus particles to move to adjacent cells. RNA viruses replicate in the cytoplasm and are known to encode one movement protein, while the DNA viruses like geminiviruses, replicate in the nucleus and require two types of MPs for their transport. One protein mediates viral transport from nucleus to cytoplasm, while the other transports viral nucleic acids to adjacent cells. Unlike most examples of CPMR, where transgenic expression of wild type CP confers resistance, resistance can also be engineered in plants through transgenic expression of dominant negative mutant forms of MPs. It is suggested that transgenically expressed dominant negative mutant forms or dysfunctional MP (dMP) competes with wild - type (WT) for plasmodesmata binding sites (Guy Kotlizky *et al.,* 2001; Lapidot *et al.,* 1993). Transgenic tobacco plants expressing defective TMV movement protein lacking three N - terminal amino acids (Cooper *et al.,* 1995) or a temperature sensitive mutant (Malyshenko *et al.,* 1993) of it showed delayed and attenuated symptoms. Chimeric MP constructs made with N - terminal sequences of *Tomato golden mosaic virus* (TGMV) BL1 and carboxy - terminal sequences of *African cassava mosaic virus* (ACMV) homologue BC1, inhibit the replication and symptom development by ACMV (Von Arnim and Stanley 1992). Resistance conferred by dMP was also effective against viruses of different groups and there were significant delays in appearance of disease symptoms and / or there was reduced systemic accumulation of virus in upper leaves of plants inoculated with *Tobacco rattle tobravirus, Tobacco ringspot nepovirus, Alfalfa mosaic alfamovirus, Peanut chlorotic streak caulimovirus,* and *Cucumber mosaic cucumovirus,* whereas the functional WT protein accelerated the symptoms appearance and infection (Cooper *et al.,* 1995). Fluorescent microscopic studies with GFP tagged MPs of different viruses and yeast two hybrid assay revealed the interaction of viral MPs with different cellular components like microtubules (Ashby *et al.,* 2006; Heinlein *et al.,* 1995), plasmodesmata (Atkins *et al.,* 1991) cell wall associated protein pectin methylesterases (Chen *et al.,* 2000), *Arabidopsis* Acetyltransferase AtNSI (Carvalho and Lazarowitz 2004). These interactions enable the viruses to exploit these host factors for their movement. Recent reports suggest the role of MPs as a suppressor of RNA silencing (Powers *et al.,* 2008) and systemic silencing through preventing the spread of silencing signal. The 25 kDa MP (p25) of PVX and 50 kDa MP (P50) of *Apple chlorotic leaf spot virus* have been shown to act as a suppressor of systemic silencing in *Nicotiana benthamiana* (Voinnet *et al.,* 2000, Yaegashi *et al.,* 2008). In contrast to the viruses encoding one or two MPs, Potexvirus, Carlavirus, Furovirus, and Hordeivirus contain a set of three partially overlapping genes known as the triple gene block (TGB) which encodes for nonvirion proteins of 26 kDa, 13 kDa and 7 kDa. Transgenic plants expressing the 13 kDa gene bearing mutation in a region conserved in all viruses were shown to be resistant to systemic infection by *White clover mosaic virus* strain (WClMV - O). Transgenic plant lines were found to be resistant to systemic infection with two other members of the Potexvirus group, PVX and *Narcissus mosaic virus*, and the Carlavirus but were susceptible to TMV (Beck *et al.*, 1994). Potato plants expressing mutant forms of triple gene block (TGB) protein were resistant to infection by PVX, *Potato aucuba mosaic potexvirus* and the carlaviruses potato virus M and S over a wide range of inoculum concentrations. However, as these plants were susceptible to PVY which lacks a TGB protein, it became apparent that resistance occurred due to interaction of modified and wild type MPs (Seppanen *et al.*, 1997).

Above studies suggest that expression of defective viral MPs from different virus groups offers a powerful strategy to generate broad spectrum transgenic resistance against multiple viruses. The mode of action of defective movement protein seems to compete with the wild type movement protein for host factors involved in virus trafficking and thus inhibiting virus movement in plant restricting the infection to a local site. The mechanism of MPMR appears to be protein mediated rather than RNA mediated.

**Replicase Protein Mediated Resistance (RPMR)**

RPMR is considered to be another effective strategy for developing virus resistance in transgenic plants. RPMR involves the expression of virus genes encoding full length, truncated or mutant form of virus replicase protein. RPMR in plants was first shown in case of transgenic tobacco plants transformed with TMV 54 - kDa gene, encoding a putative component of the replicase complex. The plants were resistant to infection when challenged with virions or viral RNA and the resistance was effective against closely related strains but not against distantly related viruses (Golemboski *et al.*, 1990). Expression of either the N - terminal portions or full length rep gene of PVX provided resistance to PVX infection in transgenic tobacco plants. The lines, when challenge inoculated with either viral RNA or virus inoculum, exhibited lower local lesion numbers and a reduction in virus levels in both the inoculated and upper leaves. On the other hand, transgenic tobacco lines expressing other portion of the ORF 1, which contained either NTP binding or GDD motif, were susceptible to PVX infection (Braun and Hemenway 1992). Expression of 54kDa full length rep gene of *Pea early browning virus* (PEBV) conferred high degree of resistance to *Broad bean yellow band virus* and an uncharacterized isolate of PEBV but not to other tobraviruses (MacFarlane and Davies 1992). The mechanism of resistance provided by the expression of Rep gene seems to be specific for different virus groups. Some reports of RPMR describe that the expression of full length, unmodified rep protein is required for virus resistance in transgenic plants, and the introduction of a mutation causing the premature termination of protein abolished resistance or the level of resistance was low (Braun and Hemenway 1992; Carr *et al.*, 1992; MacFarlane and Davies 1992; Thomas *et al.*, 2000; Shivaprasad *et al.*, 2006). Wintermantel and Zaitlin (2000) showed that transgene translatability is required for effective resistance against CMV. These reports support the hypothesis of protein mediated resistance of RPMR. While in few other cases, expression of truncated rep protein or N - terminal half of the rep protein provided high degree of resistance in transgenic plants and lines expressing full length protein and C - terminal half of the protein were susceptible (Anderson *et al.*,

1992; Braun and Hemenway 1992; Brederode *et al.*, 1995; Brunetti *et al.*, 1997; Carr *et al.*, 1994; Donson *et al.*, 1993; Lucioli *et al.*, 2003; Nguyen *et al.*, 1996).

RPMR has been shown to be effective against DNA viruses also. Expression of full length, mutated or truncated Rep gene derived from ssDNA viruses (geminiviruses) and dsDNA virus, *Rice tungro bacilliform virus* (RTBV) had been shown to provide resistance in transgenic plants. *N. benthamiana* transgenic plants expressing truncated or full length rep (AC1) gene from *Tomato yellow leaf curl virus* (TYLCV) and *African cassava mosaic virus* (ACMV) remained symptom free till 4 weeks post inoculation and accumulated 70 per cent less viral DNA in inoculated leaf discs and protoplasts as compared to infected control plants (Hong and Stanley 1996; Noris *et al.*, 1996,). Plants expressing ACMV rep gene (AC1) bearing mutation in putative NTP - binding domain were tolerant to ACMV and developed very mild symptoms and accumulated less viral DNA compared to non transgenic inoculated plants. Northern blot analysis of transgenic plants showed that a high level expression of mutated AC1 was essential for resistance (Sangare *et al.*, 1999). Interestingly, expression of truncated Rep (T - Rep; N - terminal 210 amino acids with the deletion of C - terminal NTP - binding site) of TYLCV was highly effective in tomato (Brunetti *et al.*, 1997). Expression of T - Rep inhibited viral - sense strand synthesis and repressed its own transcription (Brunetti *et al.*, 2001). It was suggested that in the case of homologous viruses, T - Rep - mediated resistance is through repression of Rep gene transcription while in the case of heterologous viruses the interference was due to Rep–Rep oligomerization (Brunetti *et al.*, 2001). A truncated Rep of TYLCV - Israel (mild) expressing the N - terminal 129 amino acids resulted in resistance against TYLCV - Is [mild] (Antignus *et al.*, 2004). The expression of the oligomerization domain of *Tomato leaf curl New Delhi virus* Rep interfered with DNA accumulation of not only the homologous virus, but also of ACMV, *Pepper huasteco virus* and *Potato yellow mosaic virus* (Chatterji *et al.*, 2001). Recently, to avoid virus induced gene silencing (VIGS) of transgene, Lucioli *et al.* (2008) generated synthetic rep130 transgene (Rep130syn) by introducing silent point mutation in such a way that the continuous homology between Rep130syn and Rep130wt was less than or equal to 5 nts. Transgenic plants accumulating high levels of Rep130syn showed ~80 per cent less viral DNA accumulation as compared to Rep130wt transgenic plants and the long lasting resistance in Rep130syn plants was linked to stable accumulation of Rep130syn. Furthermore, transgenic plants expressing Rep130wt showed late susceptibility to TYLCSV which was coupled to the reduction in Rep130wt.

The actual mechanism of Rep mediated resistance, whether it is protein or RNA mediated, is still not clear and based on the available literature the mechanism of resistance seems to be different in different virus groups, although the results of Lucioli *et al.* (2008) strongly suggest that higher amount of accumulation of Rep130syn was responsible for long lasting resistance.

## Nucleic Acid Mediated Protection against Viruses

In the previous section, we discussed the examples of protein mediated resistance where transgenically expressed protein was responsible for the resistance. But in several studies of protein mediated resistance, the protein encoded by the transgene was either failed to be detected or there was no correlation between the amount of accumulated protein and the level of resistance. These observations led researchers to suspect the possible role of transgene transcript in resistance. Further investigations into the role of RNA revealed that transgene transcript degraded the homologous RNA in a sequence specific manner. These findings played a major role in the discovery of post transcriptional gene silencing (PTGS), also called as RNA interference (RNAi) or RNA silencing (Voinnet 2005). Great amount of research has discovered the importance of PTGS as a natural defense mechanism in plants against viruses, transposons and foreign genetic material which involves sequence specific mRNA degradation that is triggered by double stranded RNA (dsRNA) (Bernstein *et al.*, 2001).

The role of transgene transcript in resistance was investigated when it was observed that transgenic plants showing high degree of resistance to corresponding virus did not accumulate any protein. Subsequently, resistance generation through the expression of untranslatable transgenes further supported the role of transgene transcript in resistance. The use of antisense RNAs to inhibit the expression of specific gene is known since early nineties where several groups reported specific inhibition of expression of endogenous gene by the expression of antisense RNA (Coleman *et al.*, 1984; Izant and Weintraub 1984; Mizuno *et al.*, 1984; Pestka *et al.*, 1984). In 1990, two different groups were attempting to make transgenic petunias that were expected to be purpler as a result of introduction of a transgene designed to overexpress the chalcone synthase enzyme. Surprisingly, over - expression of chalcone synthase transgenes did not increase colouration. Instead, some flowers were completely colorless (white), and others showed patterns such as loss of pigmentation along the veins. This phenomenon was then termed as cosuppression (Napoli *et al.*, 1990; Vander Krol *et al.*, 1990). Dougherty *et al.*, was the first to demonstrate the link between PTGS and cosuppression. They generated transgenic plants expressing *Tobacco etch virus* coat protein with control transgenics expressing non translatable sequence of coat protein. To their surprise, 30 per cent transgenic lines expressing non translatable sequences showed high degree of resistance to immunity against TEV infection (Lindbo *et al.*, 1993). In addition to immune lines some transgenic plants showed recovery to an early TEV infection and the newly emerged systemic leaves were symptom free. When inoculated with PVY, CMV, AlMV along with TEV, recovery phenotype was observed only in case of TEV. They suggested that the TEV CP RNA sequence, not the CP itself, was responsible for the virus resistance and the resistance was specific to TEV alone. TEV resistance and low transgene RNA levels in recovered plants were both due to a sequence - specic RNA - degrading activity in the cytoplasm. They referred this phenomenon as RNA mediated virus resistance (RMVR). They proposed that an RNA - directed sequence - specic RNA - degradation mechanism could be part of the mechanism of cross - protection between viruses or viroids as well as in the case of cosuppression that were post transcriptional in nature. Further, this proposal was confirmed and elaborated by using the viral vectors to silence endogenous plant genes or transgenes (Angell and Baulcombe 1997; Baulcombe *et al.*, 1995; Kumagai *et al.*, 1995; Ratcliff *et al.*, 2001). Later it was found that transgene derived small RNAs (21 - 24nts) are mediators of this sequence specific recognition and degradation of homologous RNA (Hamilton and Baulcombe 1999).

**RNA Silencing/RNA Interference/Post Transcriptional Gene Silencing**

RNA interference (RNAi) is an evolutionarily conserved gene regulatory phenomenon induced by double - stranded RNA (dsRNA) in which gene expression of target mRNA is inhibited through specific degradation of mRNA (Post transcriptional gene silencing), inhibition of transcription of transposons and repetitive DNA elements (Transcriptional gene silencing) and suppression of gene expression at translational level. The mechanism involves conversion of dsRNA into short RNAs that may direct ribonucleases to homologous mRNA targets and resulting in subsequent degradation, or binding to and methylation of the promoter of homologous gene causing post-transcriptional or translational gene silencing (Agrawal *et al.*, 2003; Bass 2000; Bosher and Labouesse 2000; Fire *et al.*, 1998; Fjose *et al.*, 2001). The basic mechanism and major players of RNAi machinery are common among various organisms (Cogoni and Machino, 1999a and b; Dalmay *et al.*, 2000; Hammond *et al.*, 2000; 2001; Kooter *et al.*, 1999; Tabara *et al.*, 1999). In plants, RNA silencing machinery is composed of at least three complexes and partially overlapping pathways: 1. Cytoplasmic RNA silencing mediated by siRNAs (PTGS), 2. Silencing mediated by miRNAs and 3. DNA methylation dependent silencing at transcriptional level (Baulcombe 2004). The major components of RNA silencing in plants are argonaute proteins (Ago), dicer like enzymes (DCLs), RNA dependent RNA polymerase (RDRs), HEN1 and HYL1. *Arabidopsis* genome encodes for ten Ago proteins, four DCLs, four DRBs and six RDRs, specifically involved in different silencing pathways with partially redundant functions (Baulcombe 2004). The Argonaute (Ago) proteins in plants, animals and fungi have been implicated in all three pathways of RNA silencing. AGO proteins are one of the key components of RISC and contain four functionally distinct domains to interact extensively with small RNA molecules while AGO1 is considered as the principal slicer (Bohmert *et al.*, 1998; Fagard *et al.*, 2000; Hunter *et al.*, 2003; Zilberman *et al.*, 2003).

Several attempts have been made to generate virus resistant transgenic plants exploiting RNAi mechanism against both RNA and DNA viruses. Based on the siRNAs - mediated RNA silencing (RNAi) mechanism, constructs were designed to target viral genes that play important roles in their infection cycle. Diverse approaches have been developed based on different features of precursor RNA for siRNA production, including sense/antisense RNA, small/long hairpin RNA and artificial miRNA precursors (Martinez de Alba *et al.*, 2002; Prins *et al.*, 2008; Simon - Mateo and Garcia 2011). The success of generating virus resistant transgenic plants depends upon following three criteria (1) approach used to induce RNAi; (2) selection of RNAi targets; and (3) pathogens targeted by RNAi.

***Sense or Antisense Viral Sequences in Transgene-Mediated Resistance***

Earlier, it was believed that transgenically expressed viral proteins are responsible for the resistant phenotype through their interference with target virus infection cycle (Abel *et al.*, 1986; Fagoaga *et al.*, 2006). However, in several cases of protein mediated resistance, the mechanism of resistance could not be explained as transgene encoded protein could not be detected and there was no correlation between the amount of protein and the level of resistance. Therefore, it was postulated that transgene transcript might be responsible for resistance. Further investigations into the role of RNA revealed that transgene transcript degraded the homologous RNA in a sequence specific manner. These findings played a major role in the discovery of post transcriptional gene silencing (PTGS) (Voinnet 2005). Furthermore, plants expressing antisense viral sequences also conferred high resistance. These results imply that the RNA sequence play a role in resistance, suggesting the involvement of other mechanisms now known as sense transgene - induced PTGS (S - PTGS) (Beclin *et al.*, 2002) S - PTGS has been well - documented in viral sequence - mediated resistance. Kawchuk (1991) developed resistance in potato against *Potato leaf roll virus* (PLRV) using CP derived sense and antisense constructs. Van Der Vlugt *et al.* (1992) generated transgenic tobacco plants resistant to *potato virus Y* (PVY$^{N)}$ infection using CP cistron. Several reports are available where virus resistance was developed using viral sequences in sense and antisense orientation (Lindbo and Dougherty 1992; Lindbo *et al.*, 1993; Goodwin *et al.*, 1996; Smith *et al.*, 1994; Waterhouse *et al.*, 1998).

***Virus-Derived Inverted Repeats (IRs)/HairpinRNAi Transgene-Mediated Resistance***

One of the major drawbacks of sense/antisense transgene constructs mediated resistance is instability and delayed resistance or resistance with low efficacy. This may be attributed to low level siRNA generation by S - PTGS construct. Moreover, several viruses, including potyviruses, cucumoviruses, and tobamoviruses, have developed suppressor to inhibit PTGS (Brigneti *et al.*, 1998; Voinnet *et al.*, 1999). With exploration of mechanism of RNA silencing it is now possible to design constructs with high efficacy to target viral pathogens (Wesley *et al.*, 2001). It has been reported by several studies that virus derived hpRNA constructs were 60 to 70 per cent more efficient in imparting resistance against viruses in transgenic plants as compared to sense/antisense constructs (Smith *et al.*, 2000; Waterhouse 1998; Waterhouse and Helliwell 2003). Using IRs, it was possible to generate effective RNA - mediated resistance against a wide range of RNA viruses, for which sense - RNA - mediated silencing was not effective (Chen *et al.*, 2004; Kalantidis *et al.*, 2002; Nomura *et al.*, 2004). Transgenic plants expressing intron hairpin RNAi (ihpRNAi) construct derived siRNAs contain precharged RISC complexes with sequence - specic antiviral recognition. Using RNAi mechanism, it is possible to design strategy to target multiple viruses with single construct containing sequences derived from multiple viruses. Based on this, Bucher *et al.* (2006) generated broad spectrum resistance, by pyramiding 150 - nt fragments of four tospoviruses in a single small chimeric IR construct. This strategy resulted in a high frequency of multiple virus resistant plants. It is proposed that by pyramiding the transgene construct with additional viral sequences, the spectrum of resistance could be broadened. Plants infected with RNA viruses produce virus - specic siRNAs, which are thought to be the breakdown product of dsRNA replicative forms or generated from secondary structures of the viral RNA (Hamilton and Baulcombe 1999; Molnar *et al.*, 2005). Interestingly, plant infecting DNA viruses like caulimoviruses and geminiviruses are also known to be targets of host endogenous RNA silencing pathway (Al - Khaff *et al.*, 1998; Chellappan *et al.*, 2004a and b; Lucioli *et al.*, 2003). During DNA virus infection, this response can lead to recovery from virus symptoms (Al - Khaff *et al.*, 1998; Chellappan *et al.*, 2005; Covey *et al.*, 1997), suggesting that RNA silencing is a remarkably robust natural defense mechanism that protects plants from viral infection (Covey *et al.*, 1997; Ratcliff 1997). Based on the above information, hpRNAi mediated resistance mechanism has been successfully extended against DNA viruses as well. Approaches based upon expressing sense and antisense RNA in transgenic plants have been employed successfully against *Tomato golden mosaic virus* (TGMV) (Day *et al.*, 1991), TYLCSV (Bendahmane and Gronenborn 1997) and TYLCV (Yang *et al.*, 2004), confirming the suggestion that RNA silencing can be exploited for defense against DNA viruses (Lapidot and Friedman 2002). Pooggin *et al.* (2003) generated IR construct targeting the common region of the begomovirus *Vigna mungo yellow mosaic virus* (VMYMV) and observed recovery from virus infection. Gafni and colleagues developed resistance in transgenic tomato plants resistant to TYLCV by targeting the CP gene with an IR construct (Zrachya *et al.*, 2007). Similarly, Noris *et al.* (2004) and Ribeiro *et al.* (2007) produced transgenic plants expressing siRNAs against TYLCSV and *Tomato chlorotic mottle virus* (ToCMoV), respectively. Fuentes *et al.* (2006) reported that the TYLCV resistant transgenic tomato line had a higher load of hairpin - derived small RNAs when compared to the other transgenic lines. During the last five years, several studies were carried out using different genes as hpRNAi constructs to generate transgenic resistance against geminiviruses (Bonfim *et al.*, 2007; Nahid *et al.*, 2011; Ramesh *et al.*, 2007; Vanderschuren *et al.*, 2009; Zrachya *et al.*, 2007).

***Artificial microRNA Mediated Resistance***

Plant micro RNA (miRNA) are encoded by the endogenous non protein coding genes and transcribed by RNA pol II to generate 21 nucleotide long small RNA that regulate the expression of endogenous target genes through cleavage at complementary region. Plant miRNA are important regulators of many developmental processes, metabolism and stress (Carrington and Ambros, 2003). These miRNAs are transcribed as long pre - miRNA (Bartel 2004) and processed by DCL - 1 (Xie *et al.*, 2004) to form mature ds miRNA. One of the strands called guide strand is loaded into RISC complex to perform the silencing of target mRNA by cleavage (mostly in plants) or by translational inhibition (mostly in animals). The finding that altering multiple nucleotides in 21 nt sequence of mature miRNA does not affect its maturation and processing, opened up the possibility of engineering artificial miRNA into plants for targeted silencing of gene or gene families (Vaucheret *et al.*, 2005). This technology was used to modify endogenous miRNA precursors to generate amiRNAs for specific gene silencing in different plants species (Khraiwesh *et al.*, 2008; Molnar *et al.*, 2009; Niu *et al.*, 2006; Ossowski *et al.*, 2008; Qu *et al.*, 2007; Schwab *et al.*, 2006; Warthmann *et al.*, 2008; Zhao *et al.*, 2009). Tang *et al.* (2010) developed miR - VIGS technology for virus based miRNA expression for gene functional analysis in plants. Using cabbage leaf - curl geminivirus vector system they silenced endogenous genes such as PDS, Su, CLA1, and SGT1. Niu *et al.* (2006) first exploited this strategy to generate resistance against viruses. They modified *Arabidopsis thaliana* miR159 precursor to express artificial miRNAs (amiRNAs) to target P69 of *Turnip yellow mosaic virus* (TYMV) and HC - Pro of *Turnip mosaic virus* (TuMV). Qu *et al.* (2006) compared siRNA and miRNA mediated approach against CMV. They designed ihpRNAi construct and miRNA construct targeting the same conserved site on 2b protein. They observed that miRNA construct was superior to ihpRNAi construct in anti - CMV mechanism in both transient and transgenic systems. The resistance provided by amiRNA construct was correlated with miRNA expression. Duan *et al.* (2008) studied the effect of local mRNA secondary structure on the efficiency of amiRNA mediated gene silencing. They observed that amiRNAs that were highly accessible to their targets sites conferred high resistance to the CMV challenge. Their study emphasized that while designing amiRNA against a viral sequence, local secondary structure of that target sequence must be studied to confer highly efficient gene silencing system. Another group developed high degree of resistance against PVX and PVY using amiRNA against the suppressors of PVX and PVY (Ai *et al.*, 2010). Expression of single and multiple amiRNA constructs targeting the conserved motif of replicase gene of *Watermelon silver mottle virus* (WSMoV) developed high degree of virus resistance (Kung *et al.*, 2012). In the first report of using amiRNA technique against geminivirus, transgenic *Nicotiana benthamiana* plants expressing two amiRNA constructs against V2 protein of *Cotton leaf curl Burewala virus* (CLCuBuV) based on the sequence of cotton miRNA169a were shown to be resistant to CLCuBuV. Resistance to heterologous viruses was lower and correlated with the numbers of sequence mismatches between the amiRNA and the V2 gene sequence (Ali *et al.*, 2014).

Table 6.2 lists some examples of PDR using either protein mediated or nucleic acid mediated approaches to generate transgenic resistance in plants against different viral pathogens.

***Comparison of siRNA and amiRNA Technology***

Relative efficiency of siRNA or amiRNA based approaches is a very pertinent aspect of designing antiviral strategy based on RNAi. Although both the approaches have their own pros and cons, amiRNA technology has some advantages over siRNA mediated one. In amiRNA technology probability of off target silencing of unintended region, an incident quite frequent in siRNA mediated silencing, is highly minimized due to the specificity of miRNA for their targets. Another advantage in amiRNA technology is that miRNAs are not direct targets of viral suppressors and thus may offer a durable resistance as compared to siRNA technology. On the other hand, siRNA technology is considered to be of broader impact as the generated secondary siRNAs may target additional sequences of pathogens. So far, siRNA technology has been the preferred choice for generating antiviral resistance because of the easier applicability of it compared to amiRNA technology and a number of reports are available for siRNA approach showing development of effective resistance against viruses.

## Role of *In silico* Analysis in PDR

Bioinformatics is one of the fastest emerging branches of biological science research. In the paradigm of generating antiviral resistance *in plants in silico* analysis adds a further layer of efficacy and may provide crucial breakthroughs in building strategies. Studies conducted over the years on different siRNA databases have helped to identify several characteristic features for functional siRNAs (Amarzguioui *et al.*, 2004; Chalk *et al.*, 2004; Elbashir *et al.*, 2001; Peek *et al.*, 2007; Reynolds *et al.*, 2004; Ui - Tei *et al.*, 2004;Vert *et al.*, 2006). Based on these patterns several online siRNA prediction programs were created time to time which are available freely and can predict all the possible siRNAs for a given target sequence and rank them based on

**Table 6.2: Examples of PDR Mediated Transgenic Resistance in different Plant Species**

| *Strategy of PDR* | *Plant Species* | *Viral Sequence* | *Target Virus* | *Reference* |
|---|---|---|---|---|
| CPMR | Tobacco | Coat protein | TMV | Powell Abel *et al.*, 1986 |
| | Tomato | Coat protein | TMV | Nelson *et al.*, 1987 |
| | *N. benthamiana* | Coat protein | TCV | Vasudevan *et al.*, 2008 |
| | *N. benthamiana* | Coat protein | TMV | Bendahmane *et al.*, 1997 |
| | Tobacco | Coat protein | TMV | Clark *et al.*, 1995 |
| | Tobacco | Coat protein | AlMV | Loesch Fries *et al.*, 1987 |
| | *Brassica napus* | Coat protein | TuMV | Lehmann *et al.*, 2003 |
| | Tobacco | Coat protein | TEV | Lindbo 1992b |
| | *N. benthamiana* | Coat protein | PVA | Savenkov and Valkonen 2001 |
| | *N. benthamiana* | Coat protein | PMTV | Barker *et al.*, 1993 |
| MPMR | *N. benthamiana* | BL1 and BR1 gene | ACMV | Von Arnim and Stanley 1992 |
| | Tobacco | Movement protein | BMV, TMV | Malyshenko *et al.*, 1993 |
| | Tobacco | Movement protein | TMV | Lapidot *et al.*, 1993 |
| | Tobacco | Movement protein | TMV | Cooper *et al.*, 1995 |
| | *N. benthamiana* | Triple gene block | WClMV | Beck *et al.*, 1994 |
| | Tobacco | BC1 gene | CabLCV | Duan *et al.*, 1997 |
| | Potato | Movement protein | PVX | Seppanen *et al.*, 1997 |
| | Tomato | Coat protein | | Kunik *et al.*, 1994 |
| RPMR | Tobacco | Replicase protein | TMV | Golemboski *et al.*, 1990 |
| | Tobacco | N - terminal and full Replicase gene | PVX | Braun and Hemenway 1992 |
| | *N. benthamiana* | Replicase gene | PEBV | MacFarlane and Davies 1992 |
| | Potato | Replicase gene | PLRV | Thomas *et al.*, 2000 |
| | Tobacco | Replicase gene | CMV | Wintermantel and Zaitlin 2000 |
| | Tobacco | Replicase gene | TMV | Carr *et al.*, 1992 |
| | *N. benthamiana* | Replicase gene | TYLCSV | Brunetti *et al.*, 2001 |
| | Tobacco | P1 and P2 genes | AlMV | Brederode *et al.*, 1995 |
| | *N. benthamiana* | Replicase gene | TYLCV | Noris *et al.*, 1996 |
| | Tomato | Replicase gene | TYLCV | Antignus *et al.*, 2004 |
| | Tobacco | Replicase gene | MYMV - Vigna | Shivaprasad *et al.*, 2006 |
| | *N. benthamiana* | Replicase gene | ToLCNDV | Chatterji *et al.*, 2001 |
| | Tomato | Replicase gene | TYLCSV | Lucioli *et al.*, 2008 |

*Contd...*

**Table 6.2**–*Contd...*

| *Strategy of PDR* | *Plant Species* | *Viral Sequence* | *Target Virus* | *Reference* |
|---|---|---|---|---|
| RNA interference | Tomato | CP region | TYLCV | Zrachya *et al.* (2003) |
| | Common bean | AC1 gene | BGMV | Bonfim *et al.*, 2007 |
| | Common bean | AC1, AC2, AC3 and BC1 ORFs of Bean Golden Mosaic virus (BGMV) | BGMV | Aragao and Faria 2009 |
| | *N. benthamiana* | N gene sequence (150 bp) derived from four Tospo viruses | Broad spectrum resistance to multiple Tospoviruses | Bucher *et al.*, 2006 |
| | Cassava | *African cassava mosaic virus* (ACMV) Rep gene sequence from nt 1690 to 1844 | ACMV | Vanderschuren 2009 |
| | *N. benthamiana* | Full length, N - terminal and C - terminal Coat protein gene | Cassava brown streak virus | Patil *et al.*, 2011 |
| | *N. benthamiana* | Intergenic region of *Ageratum yellow vein virus* (AYVV), Nucleocapsid protein of *Melon yellow spot virus* (MYSV) | AYVV, MYSV | Yang *et al.*, 2014 |
| Artificial miRNA | *Arabidopsis thaliana* | P69 and Hc - Pro gene | *Turnip yellow mosaic virus* (TYMV) | Niu *et al.*, 2006 |
| | *Arabidopsis thaliana* | CMV 2b gene | CMV | Qu *et al.*, 2006 |
| | *Arabidopsis thaliana*, Tobacco | CMV 3' UTR | CMV | Duan *et al.*, 2008 |
| | Tobacco | HC - Pro of *Potato virus Y* (PVY), TGBp1Dp25 (p25) of *Potato virus X* (PVX) | PVX and PVY | Ai *et al.*, 2011 |
| | *N. benthamiana* | Conserved motifs of replicase gene | *Watermelon silver mottle virus* (WSMoV) | Kung *et al.*, 2012 |
| | *Arabidopsis thaliana* | Hc - Pro gene of TuMV | TuMV | Martinez *et al.*, 2013 |
| | Tomato | 2a and 2b genes of CMV | CMV | Zhang *et al.*, 2011 |

their features (Table 6.3). Using these programs it is possible to identify the most potent siRNA generating sequence of viruses and these sequences can be used to generate resistance against single or multiple viruses. Effective siRNA generating sequences from viruses of different groups can be cloned as a single hairpin RNAi constructs and expressed in susceptible crops to generate broad spectrum resistance against multiple viruses. Further, by the use of *in silico* modeling of important viral proteins novel dominant negative mutant forms of these proteins can be predicted which may hinder the normal functioning of wild type viral protein and thus inhibiting viral life cycle. These studies are still in infancy and need more vigorous efforts to generate both broad spectrum and specific resistance in plants against viruses.

**Table 6.3: Online siRNA Designing Programs**

| *Program* | *Website* | *Reference* |
|---|---|---|
| siRNA at Whitehead | http://jura.wi.mit.edu/bioc/siRNAext/ | Yuan *et al.*, 2004 |
| siDirect | http://sidirect2.rnai.jp/ | Naito *et al.*, 2009 |
| BIOPREDsi | http://www.biopredsi.org/ | Huesken et. 2005 |
| DSIR | http://biodev.extra.cea.fr/DSIR/DSIR.html | Vert *et al.*, 2006 |
| AsiDesigner | http://sysbio.kribb.re.kr:8080/AsiDesigner/menu Designer.jsf | Park *et al.*, 2008 |
| siVirus | http://sivirus.rnai.jp/ | Naito *et al.*, 2006 |
| pssRNAit | http://plantgrn.noble.org/pssRNAit/ | - |

## Obstacles in Generating Virus Resistant Crops

A successful regeneration and transformation system is prerequisite for generating large number of transgenic plants using genetic engineering. So far, the concept of pathogen derived resistance is mostly limited to the model plants such as tobacco and *Arabidopsis* for which regeneration and transformation protocol is well standardized and a large number of transformants can be achieved. But in economically important crops this technology lags behind due to the non - availability of a well developed regeneration and transformation system. Food and spice crops are either recalcitrant in terms of regeneration and transformation or produce very few transformations. Serious efforts are required to transfer the knowledge of PDR and success of regeneration and transformation from model systems to economically important crops.

Simultaneous infection of a plant with more than one virus cause synergistic interaction between them, resulting in more severe symptoms and higher level accumulation of one or both the viruses as compared to either virus alone. Due to synergistic interaction, the host defense machinery is challenged by the pathogenicity factors of both the viruses. To overcome this problem, strategies should be designed to raise resistance simultaneously against multiple viruses making the crop better equipped to combat the synergistic viral threat.

## Future Prospects

The concept of PDR was originated more than two and half decades ago and till then there have been numerous developments in this field. The dilemma of using a particular PDR approach is still not over (protein or RNA mediated) and needs further efforts to prove which strategy is more effective. With the ever expanding nature of knowledge database in the concerned field and simultaneous advancement of genetic engineering to modify plant genome, the scope to generate transgenic plants in recalcitrant plant species has increased by many folds. Genetic engineering of plant species to develop crops with superior agronomic traits and resistance to different pathogens is considered as the most promising tool in the current global context. In the continuous battle between plants and viruses for survival and propagation, because of their small genome size and few numbers of proteins encoded by them, viruses always have an advantage. In this process of survival, viruses evolve and adapt very fast and develop factors in their genome to overcome the pressure exerted by the host plants. Thus in addition to the evolution of various defence mechanisms in plants, there is always a need for continuous efforts to upgrade existing strategies and developing new ones to combat pathogen attack.

With the advancement of genomics and increasing knowledge of plant pathogen interaction at molecular level, it is now possible to sequence the whole plant genomes within a very short span of time and the mechanism of resistance against a particular pathogen may be manipulated in a susceptible plant. The major point of concern with transgenic resistance is the presence of selectable marker genes, (antibiotic resistant or herbicide resistant) which may be detrimental to environment and human health. Moreover, these genes may be transferred to other crops by cross breeding. With the advancement of genetic engineering to develop marker free transgenic plants, it is now possible to address the issues raised by public concerns over the safety of biotechnology crops.

## References

1. Abel PP, Nelson RS, De B, Hoffmann N, Rogers SG, *et al.* (1986). Delay of disease development in transgenic plants that express the *Tobacco mosaic virus* coat protein gene. *Science* 232: 738-743.
2. Agrawal N, Dasaradhi PV, Mohmmed A, Malhotra P, Bhatnagar RK, *et al.* (2003). RNA interference: biology, mechanism, and applications. *Microbiol Mol Biol Rev* 67: 657-685.
3. Ai T, Zhang L, Gao Z, Zhu CX, Guo X (2011). Highly efficient virus resistance mediated by artificial microRNAs that target the suppressor of PVX and PVY in plants. *Plant Biology* 13: 304-316.
4. Ali I, Amin I, Briddon RW, Mansoor S (2013). Artificial microRNA-mediated resistance against the monopartite begomovirus *Cotton leaf curl Burewala virus*. *Virology J* 10: 231.
5. Al-Kaff NS, Covey SN, Kreike MM, Page AM, Pinder R, *et al.* (1998). Transcriptional and post transcriptional plant gene silencing in response to a pathogen. *Science* 279: 2113-2115.

6. Amarzguioui M, Prydz H (2004). An algorithm for selection of functional siRNA sequences. *Biochem Biophys Res Commun* 316: 1050–8.

7. Anderson JM, Palukaitis P, Zaitlin M (1992). A defective replicase gene induces resistance to *cucumber mosaic virus* in transgenic tobacco plants. *Proc Natl Acad Sci USA* 89: 8759-8763.

8. Angell SM, Baulcombe DC (1997). Consistent gene silencing in transgenic plants expressing a replicating potato virus X RNA. *EMBO J* 16: 3675-3684.

9. Antignus Y, Vunsh R, Lachman O, Pearlsman M, Maslenin L, *et al.* (2004). Truncated Rep gene originated from *Tomato yellow leaf curl virus*-Israel [Mild] confers strain-specific resistance in transgenic tomato. *Ann App Biol* 144: 39-44.

10. Aragao FJ, Faria JC (2009). First transgenic geminivirus-resistant plant in the field. *Nat Biotechnol* 27: 1086-1088.

11. Ashby J, Boutant E, Seemanpillai M, Groner A, Sambade A, *et al.* (2006). *Tobacco mosaic virus* movement protein functions as a structural microtubule-associated protein. *J Virol* 80: 8329-8344.

12. Asurmendi S, Berg RH, Smith TJ, Bendahmane M, Beachy RN (2007). Aggregation of TMV CP plays a role in CP functions and in coat-protein-mediated resistance. *Virology* 366: 98-106.

13. Atkins D, Hull R, Wells B, Roberts K, Moore P, *et al.* (1991). The *Tobacco mosaic virus* 30K movement protein in transgenic tobacco plants is localized to plasmodesmata. *J Gen Virol* 72 (Pt 1): 209-211.

14. Barker H, Reavy B, Webster KD, Jolly CA, Kumar A, *et al.* (1993). Relationship between transcript production and virus resistance in transgenic tobacco expressing the *Potato leaf roll virus* coat protein gene. *Plant Cell Rep* 13.

15. Bartel DP (2004). MicroRNAs: genomics, biogenesis, mechanism, and function. *Cell* 116: 281-297.

16. Bass BL (2000). Double-stranded RNA as a template for gene silencing. *Cell* 101: 235-238.

17. Baulcombe D (1994). Novel strategies for engineering virus resistance in plants. *Curr Opin Biotech* 5: 117-124.

18. Baulcombe D (2004). RNA silencing in plants. *Nature* 431: 356-363.

19. Baulcombe DC, Chapman S, Santa Cruz S (1995). Jellyfish green fluorescent protein as a reporter for virus infections. *Plant J* 7: 1045-1053.

20. Bazzini AA, Asurmendi S, Hopp HE, Beachy RN (2006). *Tobacco mosaic virus* (TMV). and potato virus X (PVX). coat proteins confer heterologous interference to PVX and TMV infection, respectively. *J Gen Virol* 87: 1005-1012.

21. Beck DL, Van Dolleweerd CJ, Lough TJ, Balmori E, Voot DM, *et al.* (1994). Disruption of virus movement confers broad-spectrum resistance against systemic infection by plant viruses with a triple gene block. *Proc Natl Acad Sci USA* 91: 10310-10314.

22. Beclin C, Boutet S, Waterhouse P, Vaucheret H (2002). A branched pathway for transgene-induced RNA silencing in plants. *Curr Biol* 12: 684-688.

23. Bendahmane M, Chen I, Asurmendi S, Bazzini AA, Szecsi J, *et al.* (2007). Coat protein-mediated resistance to TMV infection of *Nicotiana tabacum* involves multiple modes of interference by coat protein. *Virology* 366: 107-116.

24. Bendahmane M, Fitchen JH, Zhang G, Beachy RN (1997). Studies of coat protein-mediated resistance to tobacco mosaic Tobamovirus: correlation between assembly of mutant coat proteins and resistance. *J Virol* 71 (10): 7942–7950.

25. Bendahmane M, Gronenborn B (1997). Engineering resistance against *Tomato yellow leaf curl virus* (TYLCV). using antisense RNA. *Plant Mol Biol* 33: 351-357.

26. Bernstein E, Caudy AA, Hammond SM, Hannon GJ (2001). Role for a bidentate ribonuclease in the initiation step of RNA interference. *Nature* 409: 363-366.

27. Bevan MW, Mason SE, Goelet P (1985). Expression of *Tobacco mosaic virus* coat protein by a cauliflower mosaic virus promoter in plants transformed by *Agrobacterium. EMBO J* 4: 1921-1926.

28. Bohmert K, Camus I, Bellini C, Bouchez D, Caboche M, *et al.* (1998). AGO1 defines a novel locus of Arabidopsis controlling leaf development. *EMBO J* 17: 170-180.

29. Bonfim K, Faria JC, Nogueira EOPL, Mendes ÉA, Aragão FJL (2007). RNAi-mediated resistance to *Bean golden mosaic virus* in genetically engineered common bean (*Phaseolus vulgaris*). *Mol Plant Microbe Interact* 20: 717-726.

30. Bosher JM, Labouesse M (2000). RNA interference: genetic wand and genetic watchdog. *Nat Cell Biol* 2: E31-36.

31. Braun CJ, Hemenway CL (1992). Expression of amino-terminal portions or full-length viral replicase genes in transgenic plants confers resistance to *Potato Virus X* infection. *The Plant Cell* 4: 735-744.

32. Brederode FT, Taschner PE, Posthumus E, Bol JF (1995). Replicase-mediated resistance to *Alfalfa mosaic virus*. *Virology* 207: 467-474.

33. Brigneti G, Voinnet O, Li WX, Ji LH, Ding SW, *et al.* (1998). Viral pathogenicity determinants are suppressors of transgene silencing in *Nicotiana benthamiana*. *EMBO J* 17: 6739-6746.

34. Brunetti A, Tavazza M, Noris E, Tavazza R, Caciagli P, *et al.* (1997). High expression of truncated viral rep protein confers resistance to *Tomato yellow leaf curl virus* in transgenic tomato plants. *Mol Plant Microbe Interact* 10: 571-579.

35. Brunetti A, Tavazza R, Noris E, Lucioli A, Accotto GP, *et al.* (2001). Transgenically expressed T-Rep of *Tomato yellow leaf curl sardinia virus* acts as a trans-dominant-negative mutant, inhibiting viral transcription and replication. *J Virol* 75: 10573-10581.

36. Bucher E, Lohuis D, van Poppel PM, Geerts-Dimitriadou C, Goldbach R, *et al.* (2006). Multiple virus resistance at a high frequency using a single transgene construct. *J Gen Virol* 87: 3697-3701.

37. Carr JP, Gal-On A, Palukaitis P, Zaitlin M (1994). Replicase-mediated resistance to *Cucumber mosaic virus* in transgenic plants involves suppression of both virus replication in the inoculated leaves and long-distance movement. *Virology* 199: 439-447.

38. Carr JP, Marsh LE, Lomonossoff GP, Sekiya ME, Zaitlin M (1992). Resistance to *tTobacco mosaic virus* induced by the 54-kDa gene sequence requires expression of the 54-kDa protein. *Mol Plant Microbe Interact* 5: 397-404.

39. Carrington JC, Ambros V (2003). Role of microRNAs in plant and animal development. *Science* 301: 336-338.

40. Carvalho MF, Lazarowitz SG (2004). Interaction of the movement protein NSP and the *Arabidopsis* acetyltransferase AtNSI is necessary for Cabbage leaf curl geminivirus infection and pathogenicity. *J Virol* 78: 11161-11171.

41. Chalk AM, Wahlestedt C, Sonnhammer EL (2004). Improved and automated prediction of effective siRNA. *Biochem Biophys Res Commun* 319: 264–74.

42. Chatterji A, Beachy RN, Fauquet CM (2001). Expression of the oligomerization domain of the replication-associated protein (Rep). of *Tomato leaf curl New Delhi virus* interferes with DNA accumulation of heterologous geminiviruses. *J Biol Chem* 276: 25631-25638.

43. Chellappan P, Masona MV, Vanitharani R, Taylor NJ, Fauquet CM (2004). Broad spectrum resistance to ssDNA viruses associated with transgene-induced gene silencing in cassava. *Plant Mol Biol* 56: 601-611.

44. Chellappan P, Vanitharani R, Fauquet CM (2004). Short interfering RNA accumulation correlates with host recovery in DNA virus-infected hosts, and gene silencing targets specific viral sequences. *J Virol* 78: 7465-7477.

45. Chellappan P, Vanitharani R, Ogbe F, Fauquet CM (2005). Effect of temperature on geminivirus-induced RNA silencing in plants. *Plant Physiol* 138: 1828-1841.

46. Chen MH, Sheng J, Hind G, Handa AK, Citovsky V (2000). Interaction between the *Tobacco mosaic virus* movement protein and host cell pectin methylesterases is required for viral cell-to-cell movement. *EMBO J* 19: 913-920.

47. Chen Y-K, Lohuis D, Goldbach R, Prins M (2004). High frequency induction of RNA-mediated resistance against *Cucumber mosaic virus* using inverted repeat constructs. *Mole Breeding* 14: 215-226.

48. Clark WG, Fitchen JH, Beachy RN (1995). Studies of coat protein-mediated resistance to TMV. I. The PM2 assembly defective mutant confers resistance to TMV. *Virology* 208: 485-491.

49. Cogoni C, Macino G (1999). Gene silencing in *Neurospora crassa* requires a protein homologous to RNA-dependent RNA polymerase. *Nature* 399: 166-169.

50. Cogoni C, Macino G (1999). Posttranscriptional gene silencing in Neurospora by a RecQ DNA helicase. *Science* 286: 2342-2344.

51. Coleman J, Green PJ, Inouye M (1984). The use of RNAs complementary to specific mRNAs to regulate the expression of individual bacterial genes. *Cell* 37: 429-436.

52. Cooper B, Lapidot M, Heick JA, Dodds JA, Beachy RN (1995). A defective movement protein of TMV in transgenic plants confers resistance to multiple viruses whereas the functional analog increases susceptibility. *Virology* 206: 307-313.

53. Covey SN, Al-Kaff NS, Lángara A, Turner DS (1997). Plants combat infection by gene silencing. *Nature* 385: 781-782.

54. Dalmay T, Szittya G, Burgyan J (1995). Generation of defective interfering RNA dimers of cymbidium ringspot tombusvirus. *Virology* 207: 510-517.

55. Day AG, Bejarano ER, Buck KW, Burrell M, Lichtenstein CP (1991). Expression of an antisense viral gene in transgenic tobacco confers resistance to the DNA virus *Tomato golden mosaic virus*. *Proc Natl Acad Sci USA* 88: 6721-6725.

56. de Haan P, Gielen JJ, Prins M, Wijkamp IG, van Schepen A, *et al.* (1992). Characterization of RNA-mediated resistance to *Tomato spotted wilt virus* in transgenic tobacco plants. *Biotechnology (N Y)*. 10: 1133-1137.

57. Donson J, Kearney CM, Turpen TH, Khan IA, Kurath G, *et al.* (1993). Broad resistance to tobamoviruses is mediated by a modified *Tobacco mosaic virus* replicase transgene. *Mol Plant Microbe Interact* 6: 635-642.

58. Dougherty WG, Lindbo JA, Smith HA, Parks TD, Swaney S, *et al.* (1994). RNA-mediated virus resistance in transgenic plants: exploitation of a cellular pathway possibly involved in RNA degradation. *Mol Plant Microbe Interact* 7: 544-552.

59. Duan CG, Wang CH, Fang RX, Guo HS (2008). Artificial MicroRNAs highly accessible to targets confer efficient virus resistance in plants. *J Virol* 82: 11084-11095.

60. Duan YP, Powell CA, Webb SE, Purcifull DE, Hiebert E (1997). Geminivirus resistance in transgenic tobacco expressing mutated BC1 protein. *Mol Plant Microbe Interact* 10: 617-623.

61. Elbashir SM, Lendeckel W, Tuschl T (2001). RNA interference is mediated by 21- and 22-nucleotide RNAs. *Genes Dev* 15: 188–200.

62. Fagard M, Boutet S, Morel JB, Bellini C, Vaucheret H (2000). AGO1, QDE-2, and RDE-1 are related proteins required for post-transcriptional gene silencing in plants, quelling in fungi, and RNA interference in animals. *Proc Natl Acad Sci USA* 97: 11650-11654.

63. Fagoaga C, Lopez C, de Mendoza AH, Moreno P, Navarro L, *et al.* (2006). Post-transcriptional gene silencing of the p23 silencing suppressor of *Citrus tristeza virus* confers resistance to the virus in transgenic Mexican lime. *Plant Mol Biol* 60: 153-165.

64. Fire A, Xu S, Montgomery MK, Kostas SA, Driver SE, *et al.* (1998). Potent and specific genetic interference by double-stranded RNA in *Caenorhabditis elegans*. *Nature* 391: 806-811.

65. Fjose A, Ellingsen S, Wargelius A, Seo HC (2001). RNA interference: mechanisms and applications. *Biotechnol Annu Rev* 7: 31-57.

66. Fuentes A, Ramos PL, Fiallo E, Callard D, Sanchez Y, *et al.* (2006). Intron-hairpin RNA derived from replication associated protein C1 gene confers immunity to *Tomato yellow leaf curl virus* infection in transgenic tomato plants. *Transgenic Res* 15: 291-304.

67. Golemboski DB, Lomonossoff GP, Zaitlin M (1990). Plants transformed with a tobacco mosaic virus nonstructural gene sequence are resistant to the virus. *Proc Natl Acad Sci USA* 87: 6311-6315.

68. Goodwin J, Chapman K, Swaney S, Parks TD, Wernsman EA, *et al.* (1996). Genetic and biochemical dissection of transgenic RNA-mediated virus resistance. *Plant Cell* 8: 95-105.

69. Hamilton AJ, Baulcombe DC (1999). A species of small antisense RNA in posttranscriptional gene silencing in plants. *Science* 286: 950-952.

70. Hammond SM, Bernstein E, Beach D, Hannon GJ (2000). An RNA-directed nuclease mediates post-transcriptional gene silencing in *Drosophila* cells. *Nature* 404: 293-296.

71. Hammond SM, Caudy AA, Hannon GJ (2001). Post-transcriptional gene silencing by double-stranded RNA. *Nat Rev Genet* 2: 110-119.

72. Hannon GJ (2002). RNA interference. *Nature* 418: 244-251.

73. Heinlein M, Epel BL, Padgett HS, Beachy RN (1995). Interaction of tobamovirus movement proteins with the plant cytoskeleton. *Science* 270: 1983-1985.

74. Helliwell C, Waterhouse P (2003). Constructs and methods for high-throughput gene silencing in plants. *Methods* 30: 289-295.

75. Hemenway C, Fang RX, Kaniewski WK, Chua NH, Tumer NE (1988). Analysis of the mechanism of protection in transgenic plants expressing the *Potato virus* X coat protein or its antisense RNA. *EMBO J* 7: 1273-1280.

76. Hong Y, Stanley J (1995). Regulation of *African cassava mosaic virus* complementary-sense gene expression by N-terminal sequences of the replication-associated protein AC1. *J Gen Virol* 76 (Pt 10): 2415-2422.

77. Huesken D, Lange J, Mickanin C, Weiler J, Asselbergs F, *et al.* (2005). Design of a genome-wide siRNA library using an artificial neural network. *Nat Biotechnol* 23: 995-1001.

78. Hunter C, Sun H, Poethig RS (2003). The *Arabidopsis* heterochronic gene ZIPPY is an ARGONAUTE family member. *Curr Biol* 13: 1734-1739.

79. Izant JG, Weintraub H (1984). Inhibition of thymidine kinase gene expression by anti-sense RNA: a molecular approach to genetic analysis. *Cell* 36: 1007-1015.

80. Kalantidis K, Psaradakis S, Tabler M, Tsagris M (2002). The occurrence of CMV-specific short RNAs in transgenic tobacco expressing virus-derived double-stranded RNA is indicative of resistance to the virus. *Mol Plant Microbe Interact* 15: 826-833.

81. Kaniewski W, Lawson C, Sammons B, Haley L, Hart J, *et al.* (1990). Field Resistance of Transgenic Russeet Burbank Potato to Effects of Infection by Potato Virus X and Potato Virus Y. *Bio/Technology* 8: 750-754.

82. Kawchuk LM, Martin RR, McPherson J (1991). Sense and antisense RNA-mediated resistance to *Potato leafroll virus* in russet burbank potato plants. *Mol Plant Microbe Interact* 4: 247.

83. Khraiwesh B, Ossowski S, Weigel D, Reski R, Frank W (2008). Specific gene silencing by artificial MicroRNAs in *Physcomitrella patens*: an alternative to targeted gene knockouts. *Plant Physiol* 148: 684-693.

84. Kooter JM, Matzke MA, Meyer P (1999). Listening to the silent genes: transgene silencing, gene regulation and pathogen control. *Trends Plant Sci* 4: 340-347.

85. Kotlizky G, Katz A, van der Laak J, Boyko V, Lapidot M, *et al.* (2001). A dysfunctional movement protein of *Tobacco mosaic virus* interferes with targeting of wild-type movement protein to microtubules. *Mol Plant Microbe Interact* 14: 895-904.

86. Kumagai MH, Donson J, della-Cioppa G, Harvey D, Hanley K, *et al.* (1995). Cytoplasmic inhibition of carotenoid biosynthesis with virus-derived RNA. *Proc Natl Acad Sci USA* 92: 1679-1683.

87. Kung YJ, Lin SS, Huang YL, Chen TC, Harish SS, *et al.* (2012). Multiple artificial microRNAs targeting conserved motifs of the replicase gene confer robust transgenic resistance to negative-sense single-stranded RNA plant virus. *Mol Plant Pathol* 13: 303-317.

88. Lapidot M, Friedmann M (2002). Breeding for resistance to whitefly-transmitted geminiviruses. *Ann App Biol* 140: 109-127.

89. Lapidot M, Gafny R, Ding B, Wolf S, Lucas WJ, *et al.* (1993). A dysfunctional movement protein of *Tobacco mosaic virus* that partially modifies the plasmodesmata and limits virus spread in transgenic plants. *Plant J* 4: 959-970.

90. Lehmann P, Jenner CE, Kozubek E, Greenland AJ, Walsh JA (2003). Coat protein-mediated resistance to *Turnip mosaic virus* in oilseed rape (*Brassica napus*). *Mole Breeding* 11: 83-94.

91. Lindbo JA, Dougherty WG (1992). Pathogen-derived resistance to a potyvirus: immune and resistant phenotypes in transgenic tobacco expressing altered forms of a potyvirus coat protein nucleotide sequence. *Mol Plant Microbe Interact* 5: 144-153.

92. Lindbo JA, Dougherty WG (1992). Untranslatable transcripts of the tobacco etch virus coat protein gene sequence can interfere with *Tobacco etch virus* replication in transgenic plants and protoplasts. *Virology* 189: 725-733.

93. Lindbo JA, Silva-Rosales L, Proebsting WM, Dougherty WG (1993). Induction of a highly specific antiviral state in transgenic plants: implications for regulation of gene expression and virus resistance. *The Plant Cell Online* 5: 1749-1759.

94. Loesch-Fries LS, Merlo D, Zinnen T, Burhop L, Hill K, *et al.* (1987). Expression of *Alfalfa mosaic virus* RNA 4 in transgenic plants confers virus resistance. *EMBO J* 6: 1845-1851.

95. Lomonossoff GP (1995). Pathogen-Derived Resistance to Plant Viruses. Ann *Rev Phytopathol* 33: 323-343.

96. Lucioli A, Noris E, Brunetti A, Tavazza R, Ruzza V, *et al.* (2003). *Tomato yellow leaf curl Sardinia virus* rep-derived resistance to homologous and heterologous geminiviruses occurs by different mechanisms and is overcome if virus-mediated transgene silencing is activated. *J Virol* 77: 6785-6798.

97. Lucioli A, Sallustio DE, Barboni D, Berardi A, Papacchioli V, *et al.* (2008). A cautionary note on pathogen-derived sequences. *Nat Biotechnol* 26: 617-619.

98. MacFarlane SA, Davies JW (1992). Plants transformed with a region of the 201-kilodalton replicase gene from *Pea early browning virus* RNA1 are resistant to virus infection. *Proc Natl Acad Sci USA* 89: 5829-5833.

99. Malyshenko SI, Kondakova OA, Nazarova Ju V, Kaplan IB, Taliansky ME, *et al.* (1993). Reduction of *Tobacco mosaic virus* accumulation in transgenic plants producing non-functional viral transport proteins. *J Gen Virol* 74 (Pt 6): 1149-1156.

100. Martinez de Alba AE, Flores R, Hernandez C (2002). Two chloroplastic viroids induce the accumulation of small RNAs associated with posttranscriptional gene silencing. *J Virol* 76: 13094-13096.

101. Martínez F, Elena SF, Daròs J-A (2013). Fate of artificial microRNA-mediated resistance to plant viruses in mixed infections. *Phytopathology* 103: 870-876.

102. Mizuno T, Chou MY, Inouye M (1984). A unique mechanism regulating gene expression: translational inhibition by a complementary RNA transcript (micRNA). *Proc Natl Acad Sci USA* 81: 1966-1970.

103. Molnar A, Bassett A, Thuenemann E, Schwach F, Karkare S, *et al.* (2009). Highly specific gene silencing by artificial microRNAs in the unicellular alga *Chlamydomonas reinhardtii. Plant J* 58: 165-174.

104. Molnar A, Csorba T, Lakatos L, Varallyay E, Lacomme C, *et al.* (2005). Plant virus-derived small interfering RNAs originate predominantly from highly structured single-stranded viral RNAs. *J Virol* 79: 7812-7818.

105. Nahid N, Amin I, Briddon RW, Mansoor S (2011). RNA interference-based resistance against a legume mastrevirus. *Virol J* 8: 499.

106. Naito Y, Ui-Tei K, Nishikawa T, Takebe Y, Saigo K (2006). siVirus: web-based antiviral siRNA design software for highly divergent viral sequences. *Nucleic Acids Res* 34: W448-W450.

107. Naito Y, Yoshimura J, Morishita S, Ui-Tei K (2009). siDirect 2.0: updated software for designing functional siRNA with reduced seed-dependent off-target effect. *BMC Bioinformatics* 10: 392.

108. Napoli C, Lemieux C, Jorgensen R (1990). Introduction of a chimeric chalcone synthase gene into *petunia* results in reversible co-suppression of homologous genes in trans. *Plant Cell* 2: 279-289.

109. Nejidat A, Beachy RN (1990). Transgenic tobacco plants expressing a coat protein gene of *Tobacco mosaic virus* are resistant to some other tobamoviruses. *Mol Plant Microbe Interact* 3: 247-251.

110. Nelson A, Roth DA, Johnson JD (1993). *Tobacco mosaic virus* infection of transgenic *Nicotiana tabacum* plants is inhibited by antisense constructs directed at the 5' region of viral RNA. *Gene* 127: 227-232.

111. Nelson RS, Abel PP, Beachy RN (1987). Lesions and virus accumulation in inoculated transgenic tobacco plants expressing the coat protein gene of *Tobacco mosaic virus*. *Virology* 158: 126-132.

112. Nguyen L, Lucas WJ, Ding B, Zaitlin M (1996). Viral RNA trafficking is inhibited in replicase-mediated resistant transgenic tobacco plants. *Proc Natl Acad Sci USA* 93: 12643-12647.

113. Niu QW, Lin SS, Reyes JL, Chen KC, Wu HW, *et al.* (2006). Expression of artificial microRNAs in transgenic *Arabidopsis thaliana* confers virus resistance. *Nat Biotechnol* 24: 1420-1428.

114. Nomura K, Ohshima K, Anai T, Uekusa H, Kita N (2004). RNA silencing of the introduced coat protein gene of *Turnip mosaic virus* confers broad-spectrum resistance in transgenic *Arabidopsis*. *Phytopathology* 94: 730-736.

115. Noris E (2004). Tomato yellow leaf curl Sardinia virus can overcome transgene-mediated RNA silencing of two essential viral genes. *J Gen Virol* 85: 1745-1749.

116. Noris E, Accotto GP, Tavazza R, Brunetti A, Crespi S, *et al.* (1996). Resistance to tomato yellow leaf curl geminivirus in *Nicotiana benthamiana* plants transformed with a truncated viral C1 gene. *Virology* 224: 130-138.

117. Osbourn JK, Watts JW, Beachy RN, Wilson TM (1989). Evidence that nucleocapsid disassembly and a later step in virus replication are inhibited in transgenic tobacco protoplasts expressing TMV coat protein. *Virology* 172: 370-373.

118. Ossowski S, Schwab R, Weigel D (2008). Gene silencing in plants using artificial microRNAs and other small RNAs. *Plant J* 53: 674-690.

119. Park YK, Park SM, Choi YC, Lee D, Won M, *et al.* (2008). AsiDesigner: exon-based siRNA design server considering alternative splicing. *Nucleic Acids Res* 36: W97-W103.

120. Patil BL, Ogwok E, Wagaba H, Mohammed IU, Yadav JS, *et al.* (2011). RNAi-mediated resistance to diverse isolates belonging to two virus species involved in Cassava brown streak disease. *Mole Plant Pathol* 12: 31-41.

121. Peek AS (2007). Improving model predictions for RNA interference activities that use support vector machine regression by combining and filtering features. *BMC Bioinformatics* 8: 182.

122. Pestka S, Daugherty BL, Jung V, Hotta K, Pestka RK (1984). Anti-mRNA: specific inhibition of translation of single mRNA molecules. *Proc Natl Acad Sci USA* 81: 7525-7528.

123. Pooggin M, Shivaprasad PV, Veluthambi K, Hohn T (2003). RNAi targeting of DNA virus in plants. *Nat Biotechnol* 21: 131-132.

124. Powell PA, Sanders PR, Tumer N, Fraley RT, Beachy RN (1990). Protection against *Tobacco mosaic virus* infection in transgenic plants requires accumulation of coat protein rather than coat protein RNA sequences. *Virology* 175: 124-130.

125. Powers JG, Sit TL, Heinsohn C, George CG, Kim KH, *et al.* (2008). The *Red clover necrotic mosaic virus* RNA-2 encoded movement protein is a second suppressor of RNA silencing. *Virology* 381: 277-286.

126. Prins M, Laimer M, Noris E, Schubert J, Wassenegger M, *et al.* (2008). Strategies for antiviral resistance in transgenic plants. *Mole Plant Pathol* 0: 071127144754001.

127. Qu J, Ye J, Fang R (2007). Artificial MicroRNA-mediated virus resistance in plants. *J Virol* 81: 6690-6699.

128. Ramesh SV, Mishra AK, Praveen S (2007). Hairpin RNA-mediated strategies for silencing of *Tomato leaf curl virus* AC1 and AC4 genes for effective resistance in plants. *Oligonucleotides* 17: 251-257.

129. Ratcliff F (1997). A Similarity Between Viral Defense and Gene Silencing in Plants. *Science* 276: 1558-1560.

130. Ratcliff F, Martin-Hernandez AM, Baulcombe DC (2001). Technical Advance. *Tobacco rattle virus* as a vector for analysis of gene function by silencing. *Plant J* 25: 237-245.

131. Ratcliff FG, MacFarlane SA, Baulcombe DC (1999). Gene silencing without DNA. RNA-mediated cross-protection between viruses. *Plant Cell* 11: 1207-1216.

132. Register JC, 3rd, Beachy RN (1988). Resistance to TMV in transgenic plants results from interference with an early event in infection. *Virology* 166: 524-532.

133. Reynolds A, Leake D, Boese Q, Scaringe S, Marshall WS, Khvorova A (2004). Rational siRNA design for RNA interference. *Nat Biotechnol* 22: 326–30.

134. Ribeiro SG, Lohuis H, Goldbach R, Prins M (2006). *Tomato chlorotic mottle virus* is a target of RNA silencing but the presence of specific short interfering RNAs does not guarantee resistance in transgenic plants. *J Virol* 81: 1563-1573.

135. Sanford JC, Johnston SA (1985). The concept of parasite-derived resistance—Deriving resistance genes from the parasite's own genome. *J Theor Bio* 113: 395-405.

136. Sangaré A, Deng D, Fauquet CM, Beachy RN (1999). *Mole Breeding* 5: 95-102.

137. Savenkov EI, Valkonen JP (2001). Coat protein gene-mediated resistance to *Potato virus A* in transgenic plants is suppressed following infection with another potyvirus. *J Gen Virol* 82: 2275-2278.

138. Schwab R, Ossowski S, Riester M, Warthmann N, Weigel D (2006). Highly specific gene silencing by artificial microRNAs in *Arabidopsis*. *Plant Cell* 18: 1121-1133.

139. Seppanen P, Puska R, Honkanen J, Tyulkina LG, Fedorkin O, *et al.* (1997). Movement protein-derived resistance to triple gene block-containing plant viruses. *J Gen Virol* 78 (Pt 6): 1241-1246.

140. Shivaprasad PV, Thillaichidambaram P, Balaji V, Veluthambi K (2006). Expression of full-length and truncated Rep genes from *Mungbean yellow mosaic virus*-Vigna inhibits viral replication in transgenic tobacco. *Virus Genes* 33: 365-374.

141. Simon-Mateo C, Garcia JA (2011). Antiviral strategies in plants based on RNA silencing. *Biochim Biophys Acta* 1809: 722-731.

142. Smith HA, Swaney SL, Parks TD, Wernsman EA, Dougherty WG (1994). Transgenic plant virus resistance mediated by untranslatable sense RNAs: expression, regulation, and fate of nonessential RNAs. *Plant Cell* 6: 1441-1453.

143. Smith NA, Singh SP, Wang MB, Stoutjesdijk PA, Green AG, *et al.* (2000). Total silencing by intron-spliced hairpin RNAs. *Nature* 407: 319-320.

144. Tabara H, Yigit E, Siomi H, Mello CC (2002). The dsRNA binding protein RDE-4 interacts with RDE-1, DCR-1, and a DExH-box helicase to direct RNAi in C. elegans. *Cell* 109: 861-871.

145. Tang Y, Wang F, Zhao J, Xie K, Hong Y, *et al.* (2010). Virus-based microRNA expression for gene functional analysis in plants. *Plant Physiol* 153: 632-641.

146. Taschner PE, van Marle G, Brederode FT, Tumer NE, Bol JF (1994). Plants transformed with a mutant *Alfalfa mosaic virus* coat protein gene are resistant to the mutant but not to wild-type virus. *Virology* 203: 269-276.

147. Thomas PE, Lawson EC, Zalewski JC, Reed GL, Kaniewski WK (2000). Extreme resistance to *Potato leaf roll virus* in potato cv. Russet Burbank mediated by the viral replicase gene. *Virus Res* 71: 49-62.

148. Ui-Tei K, Naito Y, Takahashi F,et al. (2004). Guidelines for the selection of highly effective siRNA sequences for mammalian and chick RNA interference. *Nucleic Acids Res* 32: 936–48.

149. van der Krol AR, Mur LA, Beld M, Mol JN, Stuitje AR (1990). Flavonoid genes in *Petunia*: addition of a limited number of gene copies may lead to a suppression of gene expression. *Plant Cell* 2: 291-299.

150. van der Vlugt RA, Ruiter RK, Goldbach R (1992). Evidence for sense RNA-mediated protection to PVYN in tobacco plants transformed with the viral coat protein cistron. *Plant Mol Biol* 20: 631-639.

151. Vance V, Vaucheret H (2001). RNA silencing in plants—defense and counterdefense. *Science* 292: 2277-2280.

152. Vanderschuren H, Alder A, Zhang P, Gruissem W (2009). Dose-dependent RNAi-mediated geminivirus resistance in the tropical root crop cassava. *Plant Mol Biol* 70: 265-272.

153. Vasudevan A, Oh T-K, Park JS, Lakshmi SV, Choi BK, *et al.* (2008). Characterization of resistance mechanism in transgenic *Nicotiana benthamiana* containing *Turnip crinkle virus* coat protein. *Plant Cell Rep* 27: 1731-1740.

154. Vaucheret H, Vazquez F, Crete P, Bartel DP (2004). The action of ARGONAUTE1 in the miRNA pathway and its regulation by the miRNA pathway are crucial for plant development. *Genes Dev* 18: 1187-1197.

155. Vert JP, Foveau N, Lajaunie C, Vandenbrouck Y (2006). An accurate and interpretable model for siRNA efficacy prediction. *BMC Bioinformatics* 7: 520.

156. Voinnet O (2005). Induction and suppression of RNA silencing: insights from viral infections. *Nat Rev Genet* 6: 206-220.

157. Voinnet O, Lederer C, Baulcombe DC (2000). A viral movement protein prevents spread of the gene silencing signal in *Nicotiana benthamiana*. *Cell* 103: 157-167.

158. Voinnet O, Pinto YM, Baulcombe DC (1999). Suppression of gene silencing: a general strategy used by diverse DNA and RNA viruses of plants. *Proc Natl Acad Sci USA* 96: 14147-14152.

159. Von Arnim A, Stanley J (1992). Inhibition of *African cassava mosaic virus* systemic infection by a movement protein from the related geminivirus *Tomato golden mosaic virus*. *Virology* 187: 555-564.

160. Warthmann N, Chen H, Ossowski S, Weigel D, Herve P (2008). Highly specific gene silencing by artificial miRNAs in rice. *PLoS ONE* 3: e1829.

161. Waterhouse PM, Graham MW, Wang MB (1998). Virus resistance and gene silencing in plants can be induced by simultaneous expression of sense and antisense RNA. *Proc Natl Acad Sci USA* 95: 13959-13964.

162. Waterhouse PM, Helliwell CA (2003). Exploring plant genomes by RNA-induced gene silencing. *Nat Rev Genet* 4: 29-38.

163. Wesley SV, Helliwell CA, Smith NA, Wang MB, Rouse DT, *et al.* (2001). Construct design for efficient, effective and high-throughput gene silencing in plants. *Plant J* 27: 581-590.

164. Wilson TM (1993). Strategies to protect crop plants against viruses: pathogen-derived resistance blossoms. *Proc Natl Acad Sci USA* 90: 3134-3141.

165. Wintermantel WM, Zaitlin M (2000). Transgene translatability increases effectiveness of replicase-mediated resistance to *Cucumber mosaic virus*. *J Gen Virol* 81: 587-595.

166. Xie Z, Johansen LK, Gustafson AM, Kasschau KD, Lellis AD, *et al.* (2004). Genetic and functional diversification of small RNA pathways in plants. *PLoS Biol* 2: e104.

167. Yaegashi H, Tamura A, Isogai M, Yoshikawa N (2008). Inhibition of long-distance movement of RNA silencing signals in *Nicotiana benthamiana* by *Apple chlorotic leaf spot virus* 50 kDa movement protein. *Virology* 382: 199-206.

168. Yang C-F, Chen K-C, Cheng Y-H, Raja JAJ, Huang Y-L, *et al.* (2014). Generation of marker-free transgenic plants concurrently resistant to a DNA geminivirus and a RNA Tospovirus. *Sci Rep* 4.

169. Yang Y, Sherwood TA, Patte CP, Hiebert E, Polston JE (2004). Use of *Tomato yellow leaf curl virus* (TYLCV). rep gene sequences to engineer TYLCV resistance in tomato. *Phytopathology* 94: 490-496.

170. Yuan B, Latek R, Hossbach M, Tuschl T, Lewitter F (2004). siRNA Selection Server: an automated siRNA oligonucleotide prediction server. *Nucleic Acids Res* 32: W130-W134.

171. Zhang X, Li H, Zhang J, Zhang C, Gong P, *et al.* (2011). Expression of artificial microRNAs in tomato confers efficient and stable virus resistance in a cell-autonomous manner. *Transgenic Res* 20: 569-581.

172. Zhao T, Wang W, Bai X, Qi Y (2009). Gene silencing by artificial microRNAs in *Chlamydomonas*. *Plant J* 58: 157-164.

173. Zilberman D, Cao X, Jacobsen SE (2003). ARGONAUTE4 control of locus-specific siRNA accumulation and DNA and histone methylation. *Science* 299: 716-719.

174. Zrachya A, Kumar PP, Ramakrishnan U, Levy Y, Loyter A, *et al.* (2007). Production of siRNA targeted against TYLCV coat protein transcripts leads to silencing of its expression and resistance to the virus. *Transgenic Res* 16: 385-398.

*Section III*

# Plant Biotechnology

**2016, Biotechnology: Progress and Applications** *Pages* **163–180**
*Editors:* **Saif Hameed and Zeeshan Fatima**
*Published by:* **DAYA PUBLISHING HOUSE, NEW DELHI**

## *Chapter 7*

# Biological Control of Critical Fungal Phytopathogens of Cereal Crops

***Ashish Sharma and Leena Parihar****

*Department of Botany, School of Biotechnology and Biosciences, Lovely Professional University, Punjab – 144 411, India*

## ABSTRACT

A large quantity of food crops are lost every due to fungal infestation which can be controlled by application of chemical fungicides, a process which is costly as well as leads to the buildup of chemical residues in the environment which can be poisonous to the animals and can lead to development of resistance in the fungal pathogens. In this respect an environment friendly and more efficient process is the use of natural enemies of the fungal pathogens to control their growth and in turn disease severity. This process called as biological control or in short biocontrol uses this principle to control some critical fungal diseases of some important crop plants. One of the main benefits of this process is the reduced environmental stress as a result of reduction in the use of harmful chemicals, along with growth promotion and yield enhancement of the crop plants. Also the availability of more than one biocontrol agents for any pathogen reduces the chances of development of resistance. Therefore, in this manuscript we explore the phenomenon and mechanism of biocontrol which will prove beneficial for understanding the basic principles of the same.

* Corresponding Author: E-mail: leena.parihar@gmail.com

## Introduction

Plant diseases are major cause of grain yield loss, quality of seeds and other plant parts. The problem is managed up to some extent by using integrated disease management module, in which combination of physical, chemical and biological methods were applied in the field. Using so many approaches is also not so good for the plant health. Out of these methods biological mode of disease control is still under the process of development.

The term biological control was first coined by Harry Smith of the University of California in relation to the biological control of insects. As per Smith the biological control is the suppression of insect populations by the actions of their native or introduced enemies. The mode for Biocontrol of plant pathogens can be variable for different pathogens such as predation, parasitism, antagonism, pathogenicity, competition etc. and in these methods the population of one species will be suppressed or controlled by another species.

## Types of Biological Control

Biological control is of three types:

1. Classical biological control
2. Augmentation biological control
3. Conservation biological control

### Classical Biological Control

This generally refers to importation, introduction and establishment of a non-native natural enemy population for suppression of non-native or native organisms. Classical biological control is generally accepted to be the use of natural enemies against a host which is exotic in an area and has established without its full guild of natural enemies (Shah and Pell, 2003).

### Augmentation Biological Control

This approach refers to natural enemy populations that are increased through mass culture for periodic release (either inoculative or inundative) and colonization for suppression of native or exotic pests. Inoculative releases are intended to colonize natural enemies early in a crop season so that they and their offspring will provide pest suppression for an extended period (Wraight *et al.*, 2001).

### Conservation Biological Control

Conservation biological control refers to the modification of human influences to allow natural enemies to realize their potential to suppress pests. Conservation deals with resident enemy populations. It involves identification and remediation of negative influences that suppress natural enemies, and enhancement of systems as habitats for natural enemies. Classical biological control is long lasting and inexpensive. Other than the initial costs of collection, importation, and rearing, little expense is incurred. When a natural enemy is successfully established it rarely requires additional input and it continues to kill the pest with no direct help from humans and at no cost. This method does not always work. It is usually most effective against exotic pests and less so against native insect pests (Hoffmann and Frodsham, 2003).

**Table 7.1: Some Important Diseases of Major Cereals Caused by Fungi**

| *Cereal* | *Disease* | *Causal Fungi* |
|---|---|---|
| Rice | Black kernel | *Curvularia lunata* |
| | Leaf blast | *Pyricularia oryzae* |
| | Brown spot | *Cochliobolus miyabeanus* |
| | False smut | *Ustilaginoidea virens* |
| | Root rot | *Fusarium* spp.; *Pythium* spp. |
| | Sheath blight | *Rhizoctonia solani* |
| Wheat | Leaf blight | *Alternaria triticina* |
| | Leaf spot | *Ascochyta tritici* |
| | Ergot | *Claviceps purpurea* |
| | Karnal bunt | *Tilletia indica* |
| | Brown rust | *Puccinia recondite* |
| | Loose smut | *Ustilago tritici* |
| | Powdery mildew | *Eresyphae graminis* |
| | Black rust | *Puccinia graminis-tritici* |
| | Wheat blast | *Magnaporthe grisea* |
| Maize | Ear and kernel rot | *Aspergillus flavus* |
| | Black kernel rot | *Lasiodiplodia theobromae = Botryodiplodia theobromae* |
| | Brown spot | *Physoderma maydis* |
| | Leaf spot | *Curvularia clavata* |
| | Graminicola downy mildew | *Sclerospora graminicola* |
| | Fusarium ear and stalk rot | *Fusarium subglutinans = Fusarium moniliforme* |
| | Late wilt | *Cephalosporium maydis* |
| | Northern corn leaf blight | *Setosphaeria turcica* |
| Sorghum | Charcoal rot | *Macrophomina phaseolina* |
| | Damping-off and seed rot | *Aspergillus* spp. |
| | Gray leaf spot | *Cercospora sorghi* |
| | Leaf blight | *Setosphaeria turcica* |
| | Rust | *Puccinia purpurea* |
| | Smut, covered kernel | *Sporisorium sorghi* |
| | Smut, loose kernel | *Sporisorium cruentum* |
| Barley | Barley stripe | *Pyrenophora graminea* |
| | Crown rust | *Puccinia coronata var. hordei* |
| | Leaf rust | *Puccinia hordei* |
| | Stem rust | *Puccinia graminis f.sp. secalisPuccinia graminis f.*sp. *tritici* |
| | Stripe rust = yellow rust | *Puccinia striiformis f.* sp. *hordei* |
| | Covered smut | *Ustilago hordei* |

*Contd...*

**Table 7.1**–*Contd...*

| *Cereal* | *Disease* | *Causal Fungi* |
|---|---|---|
| | False loose smut | *Ustilago nigra*= Ustilago avenae |
| | Loose smut | *Ustilago nuda= Ustilago tritici* |
| Millets | Ergot | *Claviceps fusiformis* |
| | Exserohilum leaf blight | *Exserohilum rostratum* |
| | False mildew | *Beniowskia sphaeroidea* |
| | Rust | *Puccinia substriata var. indica* |
| | Smut | *Moesziomyces penicillariae* |
| | Southern blight | *Sclerotium rolfsii* |
| | Top rot | *Fusarium moniliforme* |
| | Zonate leaf spot | *Gleocercospora sorghi* |

**Table 7.2: Major Plant Diseases and their Biocontrol Agents**

| *Disease Name* | *Pathogen* | *Biocontrol Agents* | |
|---|---|---|---|
| Rice blast | *Pyricularia grisea* | *Pseudomonas fluorescens* | Vasudevan *et al.* (2002) |
| | *Pyricularia oryzae* | *Burkholderia Cepacia* | Homma *et al.* (1989) |
| Brown spot of rice | *Bipolaris oryzae* | *Pseudomonas* sp.<br>*P. aeruginosa*<br>*Bacillus* sp.<br>*B. subtilis* | Vasudevan *et al.* (2002) |
| Stem rot of rice | *Sclerotium oryzae* | *P. fluorescens*<br>*P. aeruginosa*<br>*B. subtilis*<br>*B. pumilus* | Vasudevan *et al.* (2002) |
| Sheath blight of rice | *Rhizoctonia solani* | *P. fluorescens*<br>*P. putida*<br>*Bacillus* sp.<br>*B. subtilis*<br>*B. laterosporus*<br>*B. pumilus*<br>*Serratia marcescens*<br>*Pseudomonas* sp.<br>*P. aeruginosa* | Vasudevan *et al.* (2002) |
| Sheath rot of rice | *Sarocladium oryzae* | *P. fluorescens*<br>*B. subtilis*<br>*P. aeruginosa*<br>*Pseudomonas* sp. | Vasudevan *et al.* (2002) |
| Bacterial blight of rice | *Xanthomonas oryzae pv. oryzae* | *Bacillus* sp. | Vasudevan *et al.* (2002) |
| Rice viral disease tungro | *Rice tungro virus* | *P. fluorescens* | Vasudevan *et al.* (2002) |

*Contd...*

**Table 7.2**–*Contd...*

| *Disease Name* | *Pathogen* | *Biocontrol Agents* | |
|---|---|---|---|
| Charcoal rot of sorghum | *Macrophomina phaseolina* | *Pseudomonas plecoglossicida, Brevibacterium antiquum, Bacillus altitudinis, Enterobacter ludwigii, E. ludwigii, Acinetobacter tandoii* and *P. monteilii* | Gopalakrishnan *et al.*, 2010 |
| Take all disease of wheat | *Gaeumannomyces graminisvar. tritici* | *Burkholderia glathei, B cepacia, Pseudomonas huttiensis, P aureofaciens, Bacillus pumilus, B. subtilis Xanthomonas sp.* | Kremer, RJ *et al.*, 2007 |
| | | *P. fluorescens 2-79* and *30-84* | Thomashow *et al.*, 1990 |
| Maize downy mildew | *Peronosclerospora sorghi* | *Trichoderma viride, Trichoderma harzianum, Gliocladium virens* and *Bacillus subtilis.* | Sadoma *et al.*, 2011 |
| Root Rot of Barley | *Cochliobolus sativus* | Vesicular arbuscular mycorrhiza | Arabi *et al.*, 2013 |
| Foot rot of fingermillet | *Sclerotium rolfsii* | *Trichoderma harzianum* | Manu *et al.*, 2012 |

## Chemical Control

Fungicides are biocidal chemical compounds or biological organisms used to kill or inhibit fungi or fungal spores. Fungi can cause serious damage in agriculture, resulting in critical losses of yield, quality, and profit. Fungicides are used both in agriculture and to fight fungal infections in animals. Chemicals used to control oomycetes, which are not fungi, are also referred to as fungicides as oomycetes use the same mechanisms as fungi to infect plants.

Classification of fungicides: Fungicides can be classified a number of different ways, including (1) mobility in the plant, (2) role in protection of plants, (3) breadth of activity, (4) mode of action, and (5) chemical group.

## Mobility in the Plant

### Contact Fungicide

A fungicide that remains on the surface where it is applied but does not go deeper; these fungicides have no after-infection activity. Repeated applications are needed to protect new growth of the plant and to replace material that has been washed off by rain or irrigation, or degraded by environmental factors such as sunlight.

### Systemic Fungicide

A fungicide that is absorbed into plant tissue and may offer some after-infection activity. Very few fungicides are truly systemic (*i.e.*, move freely throughout the plant);

however, some are upwardly systemic (*i.e.*, move only upward in the plant through xylem tissue), and some are locally systemic (*i.e.*, move into treated leaves and redistribute to some degree within the treated portion of the plant.

## Role in Protection

Some fungicides are showing the preventative activity for the host plant. It occurs when a fungicide is present on the plant as a protective barrier before the pathogen arrives or begins to develop so that it prevents infection from occurring. Early-infection activity occurs when the active ingredient of a fungicide can penetrate the plant and stop the pathogen in the plant tissues, usually most effective 24 to 72 hours after infection occurs, depending on the fungicide. Most fungicides that have early-infection activity also have preventative activity and are most effective when applied before infection occurs. The ability to stop disease development after symptoms has developed due to fungicide.

The chemical compound is able to show anti-sporulant activity in which it prevents spores from being produced. In this case, disease continues to develop (lesions continue to expand), but spores are not produced or released, so the amount of inoculum available to infect surrounding plants is reduced.

## Breadth of Activity

Single-site fungicide: fungicide active against only one point in one metabolic pathway in a fungus or against a single critical enzyme or protein needed by the fungus. These fungicides are less toxic to plants and tend to have systemic properties. Multi-site fungicide: fungicide that affects a number of different metabolic sites within the fungus. These fungicides are more toxic to plants and tend to have broad spectrum properties.

## Mode of Action

Mode of action: how a fungicide acts on a target fungus, which is the specific process in the metabolism of the target fungus that is affected by a fungicide. Examples are damaging cell membranes, inactivating critical enzymes or proteins, or interfering with key processes such as energy production or respiration.

## Chemical Group

The name given to a group of chemicals that share a common biochemical mode of action and may or may not have similar chemical structure. On the basis of chemical group the types are as follows: aliphatic nitrogen fungicides, amide fungicides, antibiotic fungicides, aromatic fungicides, arsenical fungicides, aryl phenyl ketone fungicides, benzimidazole fungicides, benzimidazole precursor fungicides, benzothiazole fungicides, botanical fungicides, bridged diphenyl fungicides, carbamate fungicides, conazole fungicides, copper fungicides, cyanoacrylate fungicides, dicarboximide fungicides, dinitrophenol fungicides, dithiocarbamate fungicides, dithiolane fungicides, fumigant fungicides, hydrazide fungicides, imidazole fungicides, inorganic fungicides, mercury fungicides, morpholine fungicides, organophosphorus fungicides, organotin fungicides, oxathiin fungicides, oxazole fungicides, polysulfide fungicides, pyrazole fungicides, pyridine fungicides, pyrimidine fungicides, pyrrole fungicides, quaternary ammonium fungicides, quinoline fungicides, quinone fungicides, quinoxaline fungicides, thiadiazole fungicides, thiazole fungicides, thiazolidine fungicides, thiocarbamate fungicides, thiophene fungicides, triazine fungicides, triazole fungicides, triazolopyrimidine fungicides, urea fungicides and zinc fungicides.

## Biocontrol

'Suppressive soils' soils that inhibit the growth of bacteria and fungi having phytotoxic effects on beneficial plant populations, contain rhizobacteria that are able to control plant diseases that are caused by these fungi or bacteria. Many different types of mechanisms have been identified for biocontrol activity which includes competition for nutrients, niche exclusion, induced systemic resistance (ISR), and the production of anti-fungal metabolites (AFMs). The biocontrol agents that are best-characterized at the molecular level belong to the genus *Pseudomonas.*

Many agricultural practices are also important for providing control of some plant diseases. In natural systems plant pathogens co-exist with host plants and other microorganisms, therefore, biological control means any reduction in the incidence and severity of the pathogen achieved through any biological mechanism. Disease control that is obtained through the action of naturally occurring or introduced antagonists, that occurs by manipulating the microenvironment to favor the activity of antagonists or that is achieved through the induction of plant resistance mechanisms by non-pathogenic or incompatible micro-organisms, are all examples of biological control.

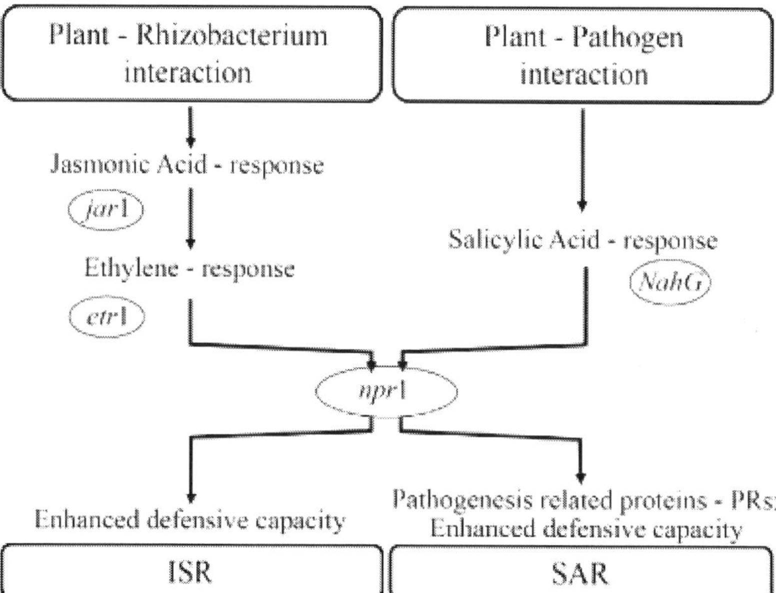


**Figure 7.1: Signal Transduction Pathway Leading to Pathogen-Induced Systemic Acquired Resistance (SAR) and Rhizobacteria Mediated Induced Systemic Resistance (ISR) in *Arabidopsis thaliana* (adopted from Beneduzi *et al.*, 2012).**

Microbial antagonism of plant pathogens occurs in several ways, the most common mechanisms being parasitism and predation, competition for nutrients or space, production of antimicrobial substances and induced resistance. As our knowledge of these mechanisms has increased, it has become apparent that antagonism often involves the synergistic action of several mechanisms.

## Mechanism of Biological Control

### Competition for Nutrients and Space

All organisms require nutrients and space for their comfortable growth, but many different types of microorganisms can occupy same space and will require same type of nutrients for their growth leading to competition. The organism which is able to utilize lesser space well, or is able to grow comfortably with less nutrients usually overgrow other microorganisms thus obliterating them from the space. A soil-borne pathogen which infects only certain parts of the root may therefore be limited by competition for suitable colonisable sites or space on the root surface. The rhizosphere is a region with high microbial activity, due to the production of many different types of metabolites as root exudates. This competition for nutrients also results in competition for oxygen. However, on the leaf surface, where nutrients are in short supply, competition for nutrients is thought to play a significant role in disease suppression. Competition for the same site is sometimes called 'site exclusion' and frequently takes place amongst organisms which are closely related taxonomically (*e.g.* the fireblight bacterium *Ertutnia amglouora* and the saprophytic species *E. herbicola*). In the case of competition for nutrients the situation is similar. Antagonists with the same nutrient requirements as the pathogen are its most effective competitors.

One of the best documented examples of nutrient competition in biological control involves competition for iron between fluorescent pseudomonads and soilborne fungal pathogens such as (*Fusarium oxysporum*). Bacteria like *Pseudomonas fluorescens* and *P. putida* produce siderophores, which bind iron and facilitate its transport from the environment into the microbial cell. The siderophores pyoverdine and pseudobactin have a high affinity for the soluble ferric iron ($Fe^{3+}$) and inhibit the growth of pathogens by making iron unavailable to them.

The applicability of nutrient competition for biological control depends on the type of pathogen that is targeted. It may not be useful in suppressing biotrophs (obligate parasites) such as rusts and powdery mildews because they do not require exogenous nutrients to infect the host. On the other hand, a nectrotrophic pathogen such as *Botrytis cinerea* is directly affected. Such pathogens require some exogenous nutrients during a definite saprophytic phase prior to infecting the host and are therefore vulnerable to nutrient competition.

### Parasitism and Predation

Parasitism is a type of relationship in which one organism depends on other organism for nutrients and growth and in turn harms the other organism. The parasitism of one fungus by another (hyperparasitism or mycoparasitism) is well documented and different steps involved are morphological disturbance, direct penetration of hyphae and hyphal lysis. Environmental factors like temperature, pH, light, nutrient status, C:N ratio have diverse effects on hyperparasitism. The genus *Trichoderrna* contains some of the most studied mycoparasites. Formulations of some species are available commercially and are used to control fungal pathogens in the soil and on aerial plant surfaces. This mycoparasite penetrates resting structures such as sclerotia or may parasitise growing hyphae by coiling round them. *Trtchoderma harzianum* degrades fungal cell walls by the lytic action of glucanases and chitinases, while other species also produce cellulase. *Trichoderma* species are amenable to genetic manipulation and their usefulness in integrated disease management programs has been enhanced by the production of pesticide-tolerant strains with enhanced hyperparasitic and lytic capabilities. Other widely studied mycoparasites include *Coniothgrium minitans* and *Sporidesmium sclerotiorum*, which are antagonists of sclerotial fungi, and *Gliocladium* spp., which parasitise a range of soil-borne pathogens. Mycoparasitic Pythium spp. appear to have some affinity for plant-parasitic members of the same genus. On the phylloplane, several fungi, including *Verticittium lecanii*, *Sphaerellopsis filum* and *Cladosporium* sp., are known to attack rust fungi.

## Cross Protection and Induced Resistance

It is difficult to distinguish clearly between cross protection and induced resistance. In cross protection, an organism which first arrives at an infection site acts directly or indirectly against a pathogen that arrives later. For example, symptomless strains of tobacco mosaic virus are used to protect tomatoes from virulent strains of the same virus. One of the best known examples of cross protection involves inoculation of young citrus trees in the nursery with a mild strain of tristeza virus (mild strain protection). Inoculated trees are protected from later infection by the more virulent strain and this protection will last for many years. Induced resistance which is a form of cross protection, is achieved by restricted inoculation of plants with pathogens, attenuated pathogens (*e.g.* pathogens inactivated by heat treatment) or selected non-pathogens or by treatment with chemical substances which stimulate an immune response. Host defence mechanisms recognise and respond to this perceived threat so that they are already operating when the pathogen subsequently attacks. The nature of the immune response is poorly understood, but the concept is not new and it provides protection against a broad spectrum of pathogens across many crop species including cereals, cucurbits, legumes, solanaceous plants and fruit trees. For instance, seedlings can be protected from wilt-inducing strains of *Fusarium oxysporum* by prior inoculation with non-pathogenic strains. However, the protection is not complete and impractically high quantities of the protectant are required for success. Inoculation of cucurbits with *Colletotrichum lagenarium* or tobacco necrosis virus systemically protects plants against disease caused by numerous pathogens. Infiltration of leaves with heat-killed cells of the bacterium *Ralstonia solanacearum* can give protection against the virulent pathogen. In this case some type of inhibitor is thought to be induced in response to the recognition of dead cells.

## Use of PGPRs

On the basis of effect of bacteria on the plants, plant-associated bacteria can be classified as beneficial, deleterious and neutral (Dobbelaere *et al.*, 2003). Beneficial free-living soil bacteria are usually referred to as plant growth-promoting rhizobacteria (PGPR, Kloepper *et al.*, 1989). PGPRs colonize the rhizosphere, the rhizoplane (root surface), or the root itself (ground and cortex tissues) (Gray and Smith, 2005). It is well established that only 1 to 2 per cent of bacteria promote plant growth in the rhizosphere (Antoun and Kloepper, 2001). Bacteria of diverse genera have been identified as PGPR, of which *Bacillus* and *Pseudomonas* spp. are predominant (Podile and Kishore, 2006). PGPR affect plant growth in two different ways, indirectly or directly. The direct promotion of plant growth by PGPR entails either providing the plant with a compound that is synthesized by the bacterium, for example phytohormones, or facilitating the uptake of certain nutrients from the environment (Glick, 1995). The indirect promotion of plant growth occurs when PGPR lessen or prevent the deleterious effects of one or more phytopathogenic organisms. This can happen by producing antagonistic substances or by inducing resistance to pathogens (Glick, 1995). A particular PGPR may affect plant growth and development by using any one, or more, of these mechanisms.

According to Beattie (2006), bacteria that reduce the incidence or severity of plant diseases are often referred to as biocontrol agents whereas those that exhibit antagonistic activity toward a pathogen are defined as antagonists. Some characteristics that provide antagonistic properties to the bacteria are as follows:

1. Synthesis of hydrolytic enzymes, such as chitinases, glucanases, proteases, and lipases, that can lyse pathogenic fungal cells (Neeraja *et al.*, 2010; Maksimov *et al.*, 2011). These enzymes are produced as extracellular secretary proteins, which normally remain inactive in their ambient environment. But when come in contact with their respective specific substrates specifically present in the fungal cell wall like chitin, etc. they degrade and destroy the fungal cell wall and in turn the fungal cells.
2. Competition for nutrients and suitable colonization of niches at the root surface (Stephens *et al.*, 1993; Kamilova *et al.*, 2005). Also called in short as competition for nutrition and niches, defined as bacteria and many other microorganisms compete within the rhizosphere for nutrients present in the root exudates and in turn for the space of growth. Therefore, the organisms which are able to grow well under stressful conditions of low available nutrition usually win this race and are able to colonize the rhizosphere of host plants and in turn obliterate the harmful pathogen population.
3. Regulation of plant ethylene levels through the ACC-deaminase enzyme, which can act to modulate the level of ethylene in a plant in response to stress imposed by the infection (Glick and Bashan, 1997; Van Loon, 2007), and
4. Production of siderophores and antibiotics.

## Production of Siderophores, Bacteriocins and Antibiotics

*Siderophores*: many microorganisms produce siderophores, which are low molecular weight iron chelators for satisfying the requirements for Fe nutritional

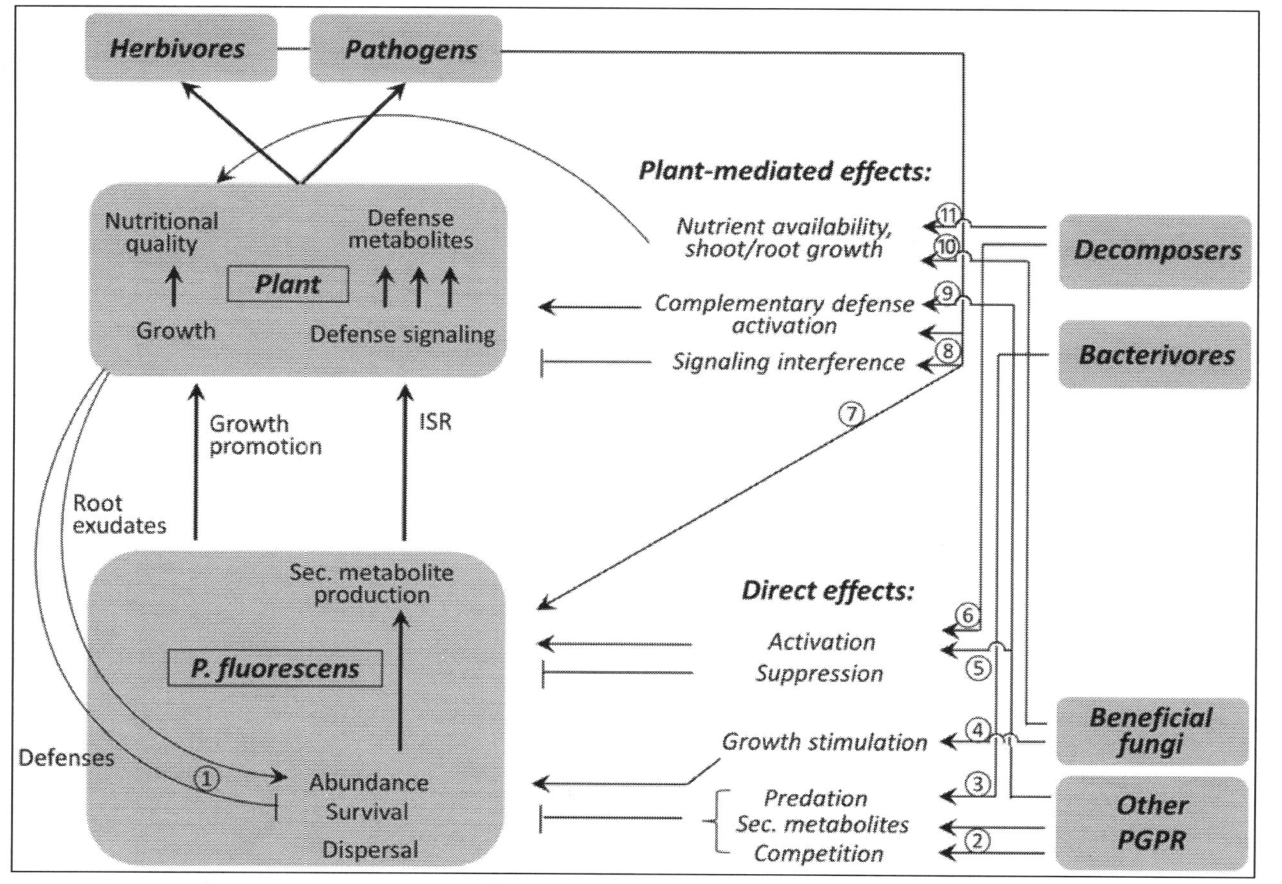


**Figure 7.2: Simplified Scheme of Direct and Indirect (Plant-mediated) Effects of Rhizosphere- and Plant-Associated Organisms on Interactions between *Pseudomonas fluorescens* and Host Plant Defenses (adopted from Hol *et al.*, 2013).**

deficiencies. These siderophores are secreted by the bacterial cells out into the soil which then solubilize iron from their surrounding environments, forming a complex ferric- siderophore that can move by diffusion and returns back to the cell surface where it is absorbed (Andrews *et al.*, 2003). Absorption of siderophore complexes require an active transport system that begins with the recognition of the ferric-siderophore by specific membrane receptors, which are different in Gram-negative and Gram-positive bacteria (Boukhalfa and Crumbliss, 2002). Siderophores can chelate ferric ion with high affinity, allowing its solubilization and extraction from most mineral or organic complexes (Wandersman and Delepelaire, 2004). In aerobic conditions at physiological pH, the reduced ferrous ($Fe^{2+}$) form is unstable and is readily oxidized to the oxidized ferric ($Fe^{3+}$) form, which normally occurs as a poorly soluble iron hydroxide basically unavailable to biological systems (Krewulak and Vogel, 2008; Osorio *et al.*, 2008). Hence, binding of these $Fe^{3+}$ with siderophores considerably improve their solubility and makes them available for absorption by plants.

Siderophores are defined as small peptidic molecules containing side chains and functional groups that have a high-affinity to form coordinate complexes with ferric ions (Crosa and Walsh, 2002). Based on their iron-coordinating functional groups, structural features and types of ligands present, bacterial siderophores have been classified into four main classes *i.e.* carboxylate, hydroxamates, phenol catecholates and pyoverdines (Crowley, 2006). Hundreds of siderophores have been identified from cultivable microorganisms, some of which are widely recognized and used by different microorganisms, while others are species- specific (Crowley, 2006; Sandy and Butler, 2009).

Siderophore production by bacteria is one of the major ability that determines whether a given microorganism will benefit the plant or not. Microbial siderophores enhance iron uptake by plants that are able to recognize the bacterial ferric-siderophore complex (Masalha *et al.*, 2000; Katiyar and Goel, 2004; Dimkpa *et al.*, 2009) and are also important in the iron uptake by plants in the presence of other metals such as nickel and cadmium which generally act as Fe antagonist and inhibits its uptake (Burd *et al.*, 1998; Dimkpa *et al.*, 2008). However, it is still unclear if bacterial siderophore complexes can significantly contribute to the iron requirements of the plant or not.

However, siderophore production also provides a competitive advantage to PGPR that can colonize roots, because it allows them to outgrow the other microorganisms growing in the rhizospheric zone (Haas and Defago, 2005). Under highly competitive conditions, the ability to acquire iron via siderophores may determine the outcome of competition, because many different types of C-sources are available in rhizosphere, due to root exudation. Siderophore producing bacteria generally use these diverse C-sources and grow easily while, other non-producers fail to colonize and grow (Crowley, 2006). The siderophores produced by *Pseudomonas* are most studied and have high affinity to the ferric ion. Pyoverdin, by far the most popular and potent siderophore, can inhibit the growth of bacteria and fungi that produce less potent siderophores in iron-depleted media *in vitro* (Kloepper *et al.*, 1980a). A pseudobactin siderophore produced by *P. putida* B10 strain was also able to suppress *Fusarium oxysporum* in soil deficient in iron; this suppression was lost when the soil was replenished with iron, a condition that represses the production of iron chelators by microorganisms (Kloepper *et al.*, 1980b). Recent studies have also demonstrated that iron chelation by siderophores produced by fluorescent pseudomonads makes it unavailable to other fungal pathogens inhabiting soil and hence the growth of these organisms (Loper, 1988; Paulitz and Loper, 1991; Dwivedi and Johri, 2003).

*Antibiotics*: Besides siderophore production, the biocontrol abilities of pseudomonad strains is also regulated by aggressive root colonization, induction of systemic resistance in the plant, and production of antifungal antibiotics (Haas and Keel, 2003). Many bacteria produce a plethora of antibiotics which is one of the major mechanism commonly associated with the ability of plant growth-promoting bacteria to act as antagonistic agents against phytopathogens (Glick *et al.*, 2007). The regulation of other microorganisms by antibiotics which forms the basis of biocontrol has been more clearly defined over the past decade and is now clear that it includes a secretion of molecules that interact with other microbes, inhibit their growth or kill them (Dowling and O'Gara, 1994; Whipps, 2001; Lugtenberg and Kamilova, 2009). Antibiotics encompass a heterogeneous group of organic, low-molecular-weight compounds that are deleterious to the growth or metabolic activities of other microorganisms (Duffy, 2003).

According to Haas and Defago (2005), six classes of antibiotic compounds related to the biocontrol of root diseases are: phenazines, phloroglucinols, pyoluteorin, pyrrolnitrin, cyclic lipopeptides (all of which are diffusible) and hydrogen cyanide (HCN; which is volatile). More recently, lipopeptide biosurfactants produced by *Pseudomonas* and *Bacillus* species have been implied in biocontrol due to their potential positive effect on competitive interactions with organisms including bacteria, fungi, oomycetes, protozoa, nematodes and plants (de Bruijn *et al.*, 2007; Raaijmakers *et al.*, 2010).

Many different bacterial and fungal strains produce a wide variety of antibiotics with a diverse mechanism of action including inhibition of pathogen cell walls synthesis, influence interaction and inhibition of membrane structures of cells and inhibit the formation of initiation complexes on the small subunit of the ribosome (Maksimov *et al.*, 2011). Pyrrolnitrin, the antibiotic produced by the *P. fluorescens* BL915 strain, is able to prevent the damage of *Rhizoctonia solani* during damping-off of cotton plants (Hill *et al.*, 1994). The 2,4-diacetylphloroglucinol (DAPG) produced by pseudomonads, an effective and extensively studied antibiotic, causes membrane damage to *Pythium* spp. and is particularly inhibitory to zoospores of this oomycete (de Souza *et al.*, 2003). Phenazine, also produced by pseudomonads, possesses redox activity and can suppress pathogens of plants such as *F. oxysporum* and *Gaeumannomyces graminis* (Chin-AWoeng *et al.*, 2003). The *P. chlororaphis* PCL1391 strain, isolated from roots of tomato plants, synthesizes phenazine-1-carboxamide, which is able to release soluble iron from insoluble ferric oxides at neutral pH, raising the possibility that phenazines might contribute to iron mobilization in soils (Hernandez *et al.*, 2004; Haas and Défago, 2005).

Antibiotics, such as polymyxin, circulin and colistin, produced by the majority of *Bacillus* ssp. Are active against Gram-positive and Gram-negative bacteria, as well as many pathogenic fungi (Maksimov *et al.*, 2011). The *B. cereus* UW85 strain, which suppresses oomycete pathogens and produces the antibiotics zwittermicin A (aminopolyol) and kanosamine (aminoglycoside), contributes to the biocontrol of alfalfa damping off (Silo-Suh *et al.*, 1994; He *et al.*, 1994). Regarding bacteria as biocontrol agents to act as a biological solution, some researchers have highlighted the use of sporulating Gram-positive species such as *Bacillus* and *Paenibacillus* spp., which can confer higher population stability during formulation and storage of inoculants products (Emmert and Handelsman, 1999; Kokalis-Burelle *et al.*, 2005).

Many antibiotic substances are only detectable when the producer organism is cultured in a nutrient rich medium. They are difficult to find in the nutrient poor environment that often exists on plant surfaces. The detection and extraction of antibiotics from the rhizosphere is hindered by adsorption to clay colloids and humus and also because only small amounts are produced. Antibiotics are also easily lost to the atmosphere, particularly from aerial plant surfaces. In addition, en4rrnatic breakdown of antibiotics can occur in all situations due to the activity of micro-organisms that are insensitive to these compounds.

*Bacteriocins*: After antibiotics, the most potent compounds used for microbial defense systems are the bacteriocins. According to Riley and Wertz (2002), bacteriocins are different from traditional antibiotics because they commonly have a relatively narrow killing spectrum and are only toxic to bacteria closely related to the producing strain. Almost all bacteria may make at least one bacteriocin, and many bacteriocins isolated from Gram-negative bacteria appear to have been created by recombination between existing bacteriocins (Riley, 1993). The colicins, proteins produced by some strains of *Escherichia coli* that are lethal for related strains, are the most representative bacteriocins produced by Gram-negative bacteria. Like colicin, a name derived from *E. coli*, other bacteriocins have been thus defined and named, such as pyocins from *P. pyogenes* strains, cloacins from *Enterobacter cloacae*, marcescins from *Serratia marcescens* and megacins from *B. megaterium* (Cascales *et al.*, 2007). Interestingly, bacteriocins from *Bacillus* spp. are increasingly becoming more important due to their sometimes broader spectra of inhibition (as compared with most lactic bacterial bacteriocins), which may include Gram-negative bacteria, yeasts or fungi, in addition to Gram-positive species, some of which are known to be pathogenic to humans and/or animals (Abriouel *et al.*, 2011).

## Conclusion

Biocontrol is the mechanism of control of harmful fungal pathogens by use of their natural enemies. It results in the regulation of pathogen population by many different mechanisms that include production of siderophores, antibiotics and bacteriocins. Besides these, biocontrol agents can also control the pathogen population by competing with them for the availability of nutrients and growth space, by parasitizing on them or by cross protecting the host species by inducing the resistance in them. Along with these mechanisms another very important mechanism of protection is the growth regulation and modification of the host plant by their growth promoting characteristics. As the discussion suggest, this approach is 100 per cent risk free to the host plants and environment friendly as it does not produce any harmful chemicals. Thus we can conclude that use of biocontrol can be the next revolutionary change in the world agriculture today.

## Future Prospects

The process of biocontrol holds immense potential to be developed in to a new field. Many numbers of new microbial species are yet to be found, which with the advancement of technologies and availability of better isolation and purification protocols will be found and may prove to be even more potential for the control of important plant pathogens. Moreover, we can engineer the microorganisms for improved biocontrol by over expressing the proteins underlying the biocontrol mechanism which will help us to enhance the efficiency of the biocontrol process, and also provide an improved plant growth and higher productivity.

## References

1. Abriouel H, Franz CM, Ben Omar N, Gálvez A (2011). Diversity and applications of *Bacillus* bacteriocins. *FEMS Microbiol Rev* 35: 201-232.
2. Andrews SC, Robinson AK, Rodríguez-Quiñones F (2003). Bacterial iron homeostasis. *FEMS Microbiol Rev* 27: 215- 237.
3. Antoun H, Kloepper JW (2001). Plant growth promoting rhizobacteria. In: Brenner S and Miller JH (eds). *Encyclopedia of Genetics*. Academic, New York, pp. 1477-1480.
4. Beattie GA (2006). Plant-associated bacteria: Survey, molecular phylogeny, genomics and recent advances. In: Gnanamanickam SS (ed). *Plant-Associated Bacteria*. Springer, Dordrecht, pp. 1-56.
5. Beneduzi A, Ambrosini A, Passaglia LMP (2012). Plant growth promoting rhizobacteria (PGPR): their potential as antagonists and biocontrol agents. *Genet Mol Biol.* 35: 1044-1051.
6. Boukhalfa H, Crumbliss AL (2002). Chemical aspects of siderophore mediated iron transport. *BioMetals* 15: 325-339.
7. Burd GI, Dixon DG, Glick BR (1998). A plant growthpromoting bacterium that decreases nickel toxicity in seedlings. *Appl Environ Microbiol* 64: 3663-3668.
8. Cascales E, Buchanan SK, Duché D, Kleanthous C, Lloubès R, Postle K, Riley M, Slatin S, Cavard D (2007). Colicin Biology. *Microbiol Mol Biol Rev* 71: 158-229.
9. Chin-A-Woeng TF, Bloemberg GV, Lugtenberg BJ (2003). Phenazines and their role in biocontrol by *Pseudomonas* bacteria. *New Phytol* 157: 503-523.
10. Crosa JH, Walsh CT (2002). Genetics and assembly line enzymology of siderophore biosynthesis in bacteria. *Microbiol Mol Biol Rev* 66: 223-249.
11. Crowley DE (2006). Microbial siderophores in the plant rhizospheric. In: Barton LL and Abadía J (eds). *Iron Nutrition in Plants and Rhizospheric Microorganisms*. Springer, Dordrecht, pp. 169-198.

12. de Bruijn I, de Kock MJD, Yang M, de Waard P, van Beek TA, Raaijmakers JM (2007). Genome-based discovery, structure prediction and functional analysis of cyclic lipopeptide antibiotics in *Pseudomonas* species. *Mol Microbiol* 63: 417-428.

13. de Souza JT, Arnould C, Deulvot C, Lemanceau P, Gianinazzi-Pearson V, Raaijmakers JM (2003). Effect of 2,4-diacetylphloroglucinol on *Pythium*: Cellular responses and variation in sensitivity among propagules and species. *Phytopathology* 93: 966-975.

14. Dimkpa C, Svatos A, Merten D, Büchel G, Kothe E (2008). Hydroxamate siderophores produced by *Streptomyces acidiscabies* E13 bind nickel and promote growth in cowpea (*Vigna unguiculata* L.). under nickel stress. *Can J Microbiol* 54: 163-172.

15. Dimkpa CO, Merten D, Svatos A, Büchel G, Kothe E (2009). Siderophores mediate reduced and increased uptake of cadmium by *Streptomyces tendae* F4 and sunflower (*Helianthus annuus*), respectively. *J Appl Microbiol* 107: 1687-1696.

16. Dobbelaere S, Vanderleyden J, Okon Y (2003). Plant growthpromoting effects of diazotrophs in the rhizosphere. *CRC Crit Rev Plant Sci* 22: 107-149.

17. Dowling DN, O'Gara F (1994). Metabolites of *Pseudomonas* involved in the biocontrol of plant disease. *Trends Biotechnol* 12: 133-141.

18. Dwivedi D, Johri BN (2003). Antifungals from fluorescent pseudomonads: Biosynthesis and regulation. *Curr Sci* 12: 1693-1703.

19. Emmert EAB, Handelsman J (1999). Biocontrol of plant disease: A (Gram-). positive perspective. *FEMS Microbiol Lett* 171: 1-9.

20. Glick BR (1995). The enhancement of plant growth by free-living bacteria. *Can J Microbiol* 41: 109-117.

21. Glick BR, Bashan Y (1997). Genetic manipulation of plant growth-promoting bacteria to enhance biocontrol of phytopathogens. *Biotechnol Adv* 15: 353-378.

22. Glick BR, Cheng Z, Czarny J, Duan J (2007). Promotion of plant growth by ACC deaminase-producingsoil bacteria. *Eur J Plant Pathol* 119: 329-39.

23. Gray EJ, Smith DL (2005). Intracellular and extracellular PGPR: Commonalities and distinctions in the plant-bacterium signaling processes. *Soil Biol Biochem* 37: 395-412.

24. Haas D, Défago G (2005). Biological control of soil-borne pathogens by fluorescent pseudomonads. *Nat Rev Microbiol* 3: 307-319.

25. Haas D, Keel C (2003). Regulation of antibiotic production in root-colonizing *Pseudomonas* spp. and relevance for biological control of plant disease. *Annu Rev Phytopathol* 41: 117-153.

26. He H, Silo-Suh LA, Handelsman J, Clardy J (1994). Zwittermicin A, an antifungal and plant protection agent from *Bacillus cereus*. *Tetrahedron Lett* 35: 2499-2502.

27. Hernandez ME, Kappler A, Newman DK (2004). Phenazines and other redox-active antibiotics promote microbial mineral reduction. *Appl Environ Microbiol* 70: 921-928.

28. Hill DS, Stein JI, Torkewitz NR, Morse AM, Howell CR, Pachlatko JP, Becker JO, Ligon JM (1994). Cloning of genes involved in the synthesis of pyrrolnitrin from Pseudomnas fluorescens and role of pyrrolnitrin synthesis in biological control of plant disease. *Appl Environ Microbiol* 60: 78-85.

29. Hoffmann MP, Frodsham AC (1993). *Natural Enemies of Vegetable Insect Pests.* Cooperative Extension, Cornell University, Ithaca, NY. 63 pp.

30. Homma Y, Kato Z, Hirayama F, Konno K, Shirahama H, Suzui T (1989). Production of antibiotics by Pseudomonas cepacia as an agent for biological control of soilborne plant pathogens. *Soil Biol Biochem.* 21: 723-728.

31. http: //www.ipm.iastate.edu/ipm/icm//ipm/icm/2006/5-15/ fungicides.html

32. Kamilova F, Validov S, Azarova T, Mulders I, Lugtenberg B (2005). Enrichment for enhanced competitive plant root tip colonizers selects for a new class of biocontrol bacteria. *Environ Microbiol* 7: 1809-1817.

33. Katiyar V, Goel R (2004). Siderophore-mediated plant growth promotion at low temperature by mutant of fluorescent pseudomonad. *Plant Growth Regul* 42: 239-244.

34. Kloepper JW, Leong J, Teintze M, Schroth MN (1980a). Enhancing plant growth by siderophores produced by plant growth-promoting rhizobacteria. *Nature* 286: 885-886.

35. Kloepper JW, Leong J, Teintze M, Schroth MN (1980b). *Pseudomonas* siderophores: A mechanism explaining disease suppressive soils. *Curr Microbiol* 4: 317-320.

36. Kloepper JW, Lifshitz R, Zablotowicz RM (1989). Free-living bacterial inocula for enhancing crop productity. *Trends Biotechnol* 7: 39-43.

37. Kokalis-Burelle N, Kloepper JW, Reddy MS (2005). Plant growth-promoting rhizobacteria as transplant amendments and their effects on indigenous rhizosphere microorganisms. *Appl Soil Ecol* 31: 91-100.

38. Krewulak KD, Vogel HJ (2008). Structural biology of bacterial iron uptake. *Biochim Biophys Acta* 1778: 1781-1804.

39. Loper JE (1988). Role of fluorescent siderophore production in biological control of *Pythium ultimum* by a *Pseudomonas fluorescens* strain. *Phytopathology* 78: 166-172.

40. Lugtenberg B, Kamilova F (2009). Plant-growth-promoting rhizobacteria. *Annu Rev Microbiol* 63: 541-55.

41. Maksimov IV, Abizgil'dina RR, Pusenkova LI (2011). Plant growth promoting rhizobacteria as alternative to chemical crop protectors from pathogens (Review). *Appl Biochem Microbiol* 47: 333-345.

42. Masalha J, Kosegarten H, Elmaci Ö, Mengel K (2000). The central role of microbial activity for iron acquisition in maize and sunflower. *Biol Fert Soils* 30: 433-439.

43. Neeraja C, Anil K, Purushotham P, Suma K, Sarma P, Moerschbacher BM, Podile AR (2010). Biotechnological approaches to develop bacterial chitinases as a bioshield against fungal diseases of plants. *Crit Rev Biotechnol* 30: 231-241.

44. Osório H, Martinez V, Nieto PA, Holmes DS, Quatrini R (2008). Microbial iron management mechanisms in extremely acidic environments: Comparative genomics evidence for diversity and versatility. *BMC Microbiology* 8: 203.

45. Paulitz TC, Loper JE (1991). Lack of a role for fluorescent siderophore production in the biological control of *Pythium* damping-off of cucumber by a strain of *Pseudomonas putida*. *Phytopathology* 81: 930-935.

46. Podile AR, Kishore GK (2006). Plant growth-promoting rhizobacteria. In: Gnanamanickam SS (ed). *Plant-Associated Bacteria*. Springer, Netherlands, pp. 195-230.

47. Raaijmakers JM, de Bruijn I, Nybroe O, Ongena M (2010). Natural functions of lipopeptides from *Bacillus* and *Pseudomonas*: More than surfactants and antibiotics. *FEMS Microbiol Rev* 34: 1037-1062.

48. Riley M (1993). Molecular mechanisms of colicin evolution. *Mol Biol Evol* 10: 1380-95.

49. Riley MA, Wertz JE (2002). Bacteriocins: Evolution, ecology, and application. *Annu Rev Microbiol* 56: 117-137.

50. Sandy M, Butler A (2009). Microbial iron acquisition: Marine and terrestrial siderophores. *Chem Rev* 109: 4580-4595.

51. Shah PA, Pell, JK (2003). Entomopathogenic fungi as biological control agents. *Appl Microbiol Biotechnol*. 61: 413–423.

52. Silo-Suh LA, Lethbridge BJ, Raffel SJ, He H, Clardy J, Handelsman J (1994). Biological activities of two fungistatic antibiotics produced by *Bacillus cereus* UW85. *Appl Environ Microbiol* 60: 2023-2030.

53. Stephens PM, Crowley JJ, O'Connell C (1993). Selection of pseudomonad strains inhibiting *Pythium ultimum* on sugarbeet seeds in soil. *Soil Biol Biochem* 25: 1283-1288.

54. Thomashow LS, Weller DM, Bonsall RF, Pierson LS (1990). Production of the antibiotic phenazine-1-carboxylic acid by fluorescent pseudomonas in the rhizosphere of wheat. *Appl Environ Microbiol*. 56: 908-912.

55. Van Loon LC (2007). Plant responses to plant growth-promoting rhizobacteria. *Eur J Plant Pathol* 119: 243-254.

56. Wandersman C, Delepelaire P (2004). Bacterial iron sources: From siderophores to hemophores. *Annu Rev Microbiol* 58: 611-647.

57. Whipps JM (2001). Microbial interactions and biocontrol in the rhizosphere. *J Exp Bot* 52: 487-511.

58. Wraight SP, Jackson MA, Kock SL de (2001). Production, stabilization and formulation of fungal biocontrol agents. In: Butt TM, Jackson C, Magan N (eds). *Fungi as biocontrol agents: progress, problems and potential*. CAB International, Wallingford, pp. 253–287.

**2016, Biotechnology: Progress and Applications** *Pages* **181–205**
*Editors:* **Saif Hameed and Zeeshan Fatima**
*Published by:* **DAYA PUBLISHING HOUSE, NEW DELHI**

## *Chapter 8*

# Crop Improvement Methods: A Walk Through

***Manju Sharma[1] and Rekha Kansal[2]***

[1]*Amity Institute of Biotechnology, Amity University Haryana, Manesar, Gurgaon, Haryana*
[2]*NRC on Plant Biotechnology, IARI, New Delhi*

## ABSTRACT

Developing improved varieties is a significant event for progressive agriculture which started way back in 1492 with the introduction of corn. Since then, crops were spread all over the world and flourished under different conditions. Before, Gregor Mendel's discoveries the entire agriculture was relying on selective breeding for crops but later with thorough understanding of the basis of genetics the scenario has started changing rapidly. Agriculturist started practicing induced mutation using mutagens *i.e.* chemicals and radiations as a part of classical breeding. Many varieties had been developed based on this technique having new desirable characteristics through modification of their genetic material. The most significant scientific work of Watson and Crick had made wonderful contribution in the field of biology with the understanding of gene. Introduction of transposons, tissue culture, embryo rescue, protoplast fusion, genetic engineering and marker assisted breeding widened the road to move forward in agricultural product development. Direct changes in the genetic makeup of the plant with predictable results in less than 10 years' time have popularized genetic engineering overnight. Transgenic breeding and molecular marker assisted

* Corresponding Author: E-mail: sharma.manju131@gmail.com

breeding has now taken up another plunge in this direction. But, judicious discrimination in use of these different tools needs to be assessed for associated risk with the technology. With all oppressive forces the extensive risk assessment and safety testing of crops developed through the use of genetic engineering has shown that there are no varieties in use that pose risks to consumers. Therefore, it is recommended to apply this technology with precautionary principle for sustainable development.

***Keywords:*** *Transposons, Tissue culture, Embryo rescue, Protoplast fusion, Genetic engineering, Marker assisted breeding, Transgenic breeding.*

## Introduction

Initially, as per the availability of coherent knowledge, crop improvement was based on conventional plant breeding methods which shifted to modern molecular approach with the development of knowledge and at present further resting on transgenic and molecular marker assisted breeding.

Conventional breeding methods can generally relying on selection, hybridization and increased number of chromosome sets (polyploidy) to manipulate plant chromosome combination. Green revolution has seen judicious use of these conventional methods to develop modern corn hybrids and wheat. In the genesis of these techniques spontaneous or artificially induced mutations were introduced as radiation breeding, to enhance the ability to develop new traits in crops (Poehlman 1979).

A decade between 1980's to 1990's has witnessed active role of genetic engineering techniques in crop improvement and made it most adopted technology in different parts of the world (USDA/APHIS 2002, James 2007). Also raised hope for developing countries (Herrera-Estrella 2014). Although, being more precise way of improving crops the involvement of inter and intra specific crops gene transfer techniques has raised questions about their safety to consumers and the environment (Harlander, 2002; Manshardt, 2004). Time span of 20 to 25 years to introduce new varieties was cut to half using the tools of genetic engineering. GM crop farming is expanding rapidly around the world but crops developed via biotechnology should undergo extensive safety testing. There are no human viral pathogens harboured by plants: thus, especially for pharmaceuticals, plants represent the safer production system (Lanfranco 2003). However, it is necessary for genetically engineered crops to undergo extensive food and environmental safety assessment as they may have unexpected secondary allergenic, toxic or anti nutritional properties.

Further with rapid development of molecular marker applications with conventional plant breeding practices is now on focus for molecular plant breeding. This technique is continue to evolve and a topic of intense interest among plant breeders and crop scientists (Cooper *et al.*, 2004; Nelson *et al.*, 2004; Lörz and Wenzel, 2005; Varshney *et al.*, 2006; Eathington *et al.*, 2007; Mumm, 2007). Molecular markers with high-throughput genome sequencing efforts, have dramatically increased knowledge of and ability to characterize genetic diversity in the germplasm pool for essentially any crop species.

Limitation of biotechnology to add only specific and limited traits encourage plant breeding to couple with transgenic improvement to address dual approach of both under the head of transgenic breeding to develop new viable produce (Visarada *et al.*, 2009). This could be done by backcross breeding with non-elite germplasm (Iftekharuddaula *et al.*, 2011; Zhong, 2001), pedigree method, and single seed descent can be used for forward breedingto develop inbred transgenic lines (Mumm, 2007), however, forward breeding is only favorable if the transgene donor also contains favorable allele(s) other than the transgenes per se. Sometimes, pyramiding transgenes at separate loci would favor backcrossing, although transgenes can also be pyramided as a single DNA cassette or multiple cassettes, which allows for easy breeding schemes, less crosses and potentially less cost in the regulation aspect of generating a transgenic cultivar, which is currently estimated at ≤100 million US dollars (Monsanto, 2005; Mumm, 2007; Naqvi *et al.*, 2010).

In recent years, biotechnology has developed rapidly as a practical means for accelerating success in plant breeding and improving economically important crops. Despite all gaps, biotechnology is predicated as an important tool which is being perfected day by day to produce an enormous food supply. But, the most important issues of safety concerns must be addressed on priority as safety assessment process required for these crops is contrasted with the lack of a formal process required for traditionally bred crops. It will be sustainable in the long run and meet the needs of increasing population worldwide (Serageldin, 1999; Borlaug 2000). These new techniques are supplementing and extending traditional breeding methods to enhance the production of food, fiber, and other agricultural products. Farmers in the United States, Canada, Argentina, and other countries have rapidly adopted transgenic crop varieties. Between 1996 and 2001, the global production of transgenic soybean, cotton, corn, and canola varieties increased from essentially none to over 125 million acres (50 million hectares). But there are many issues need critical analysis.

## Conventional Methods for Crop Improvement

It is well known that all traits of a plant are controlled by genes located on chromosomes; conventional plant breeding can be considered as the manipulation of the combination of chromosomes using various procedures and that can be done more preciously using modern techniques (Figure 8.1).

### Selection

This is the most ancient and basic procedure has 3 distinct steps;

1. Selection is done with purpose to identify a large number of genetically variants from original crop population on the basis of phenotype and performance.
2. Grow progeny from the individual plant selections for observational purposes. The superior performing material during selections is grown over several years after making elimination. The selections are now allowed to grow for stable performance under different environmental conditions.

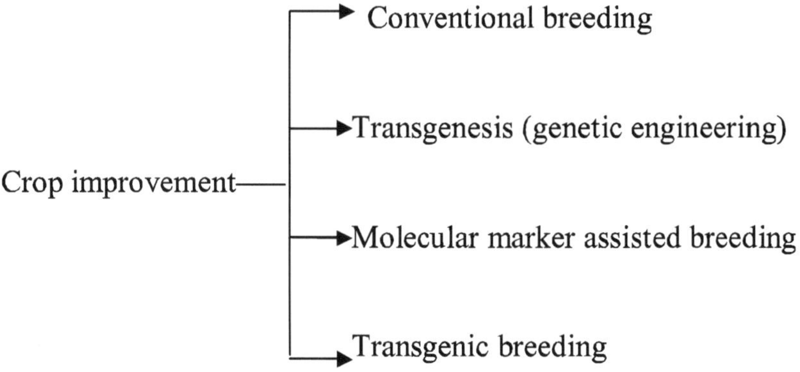


**Figure 8.1: Methods in Plant Breeding.**

The varieties don't perform up to the mark eliminated further.

3. The selected and inbred lines are subjected to compare with existing commercial varieties in yield performance and other aspects of agronomic importance.

**Hybridization**

This the most widely used procedure with aim to bring together desired traits found in different plant lines into one plant line via cross- pollination. It starts with generation of pure lines using self-pollination. Once after obtaining pure line it is combined with another inbred line although during this process the large quantity of undesired traits (low yield, low nutritional value, and unwanted taste) is also moved to the crop as well. These unfavorable traits must be removed by repeated back crossing with the crop parent. The conventional hybridization technique has limitation to not go beyond interspecific and intergeneric hybrids due to sexual incompatibility. Further limits the use of technique to transfer desirable traits from completely unrelated species.

> Do you know?
>
> The most popular and successful hybrid plant is - Hybrid Maize.
>
> Year of production- 1919 it was the first commercial hybrid
>
> First made available in - United States.
>
> Today, the maize varieties are in use all hybrid.

**Polyploidy**

Plants having more than two sets of the chromosome are referred as polyploidy. Generally, the main effect of polyploidy is enhancement in size and genetic variability. The induction of polyploidy is a frequent technique to overcome the sterility of a hybrid species during plant breeding. Polyploidy in crop plants is most commonly induced by treating seeds with the chemical colchicines, which leads to a doubling of the chromosome number. In fact, recent findings in genome research indicate that many species that are currently diploid, including humans, were derived from polyploid ancestors (Van de Peer and Meyer, 2005).

## Mutation Breeding

During 1970s to 1980s, breeders have started using mutations breeding for genetic improvement of plants. Mutations can be introduced using chemicals or radiations, to generate genetic variations from which desired mutants may be selected. Transposable elements can also cause gene disruptions, protein product alterations, or large-scale genome rearrangements. These will be discussed later elsewhere in the chapter. Using a cobalt 60 gamma source, seeds are irradiated to create mutations in the DNA. It offers the possibility of inducing desired attributes that either cannot be found in nature or have been lost during evolution (Navok and Bruner 1992). It is possible to adjust the dosage of mutagen treatment to the desired mutational response. High doses results in prominent mutation response with high lethality of plants. High survival rate can be achieved exposing plant to low doses but frequency of mutation will also be less.

Approximately more than 2540 mutagenic plant varietals have been released between 1930–2007 (Schouten and Jacobsen 2007) of these 70 per cent either as direct mutants or 30 per cent from their progenies (Maluszynsk 2000). The combination of *in vitro* culture and mutation techniques provides a simple, fast and highly efficient method to improve horticultural crops. For example, micro-cultured plants of chrysanthemum, potato, and streptocarpus were irradiated with 20 Gy gamma rays, and subcultured four times from single nodal cuttings (Ahloowalia 1996). Crop plants account for 75 per cent of released mutagenic species with the remaining 25 per cent ornamentals or decorative plants (Ahloowalia 2004).

In 1960s conventional plant breeding principles and agricultural techniques had been utilized to acclaim "Green Revolution" in Asia with the grand success of high-yielding wheat and rice lines. Many fold increase in productivity of these important crops triumphed agricultural revolution in the developing countries to describe the social, economic, and nutritional impact of the new high-yielding wheat and rice strains. However, the performance of high yielding varieties was dependent on fertilizers, irrigation and agrochemicals and required energy-intensive investments. Therefore, the "Green Revolution" has earned praise and criticism both. Hence, new bio-techniques, in combination to conventional plant breeding, are very much required to boost yields of the crops to feed the world.

## Biotechnology as Tool for Crop Improvement

Now, the techniques of biotechnology have become more widely adopted by plant breeders that make use of living organisms or their products for benefit of human beings. The careful choice biotechniques as well as going with conventional breeding hand in hand is required because (1) the elite cultivars will be the parents of the next generation of improved genotypes (2) field testing across locations or cropping systems and over the years will be needed to determine the best selections due to the genotype-by-environment interaction (Kang and Gauch 1996; Simmonds 1997). The different methods of biotechnology have great potential for plant improvement (Figure 8.2), provided plants have potential to regenerated readily in large numbers (Jain *et al.*, 1998) and also accommodate foreign DNA. Biotechnology has emerged as an indispensable new tool in aid of plant breeding to reach traditional goals and also in the production of industrial products in plants (Trevor *et al.*, 2002). The major areas worked on for genetic modification are insect-pest resistant plants, herbicide resistant, stress resistant, virus resistant, improved nutritional quality, increased disease resistance and male sterility.

## Plant Tissue Culture

Under carefully controlled laboratory conditions, tissues, organ or individual cells can be taken from a plant, to develop back into whole plants. Presence of the total potential is required in plant cells contribute to regenerate desired plants. The journey of plant tissue culture was initiated in 1902 which further developed different areas of this technique till 1950s and became popular in the 1960s. Today, *micro propagation, somaclonal variations, haploid production, embryo rescue and in vitro conservation* are standard techniques in most important crops.

1. Micropropagation is the production of multiple copies of a single plant using tissue culture techniques. Shoot meristem with two new leaves is excised and to grow into complete new plant. In many cases, the viruses that infect plant are not present in the growing tip, they can be eliminated from plants produced by micro propagation. This method has played an important role in the production of healthy, disease-free plants and in the rapid multiplication of scions and rootstocks with desirable trait fruits (Dobránszki and da Silva 2010), ornamentals (Podwyszyñska and Sochacki 2010), and vegetables (*Miassar et al.*, 2011) and other crops (Rout *et al.*, 2001, Sanger *et al.*, 2011). Micro-propagation has been the starting point for each new cycle of strawberry plants grown in California for over 20 years both in commercial fields and home gardens. Citrus, pineapples, avocados, mangoes, bananas, coffee and papaya have been multiplied by this technique.

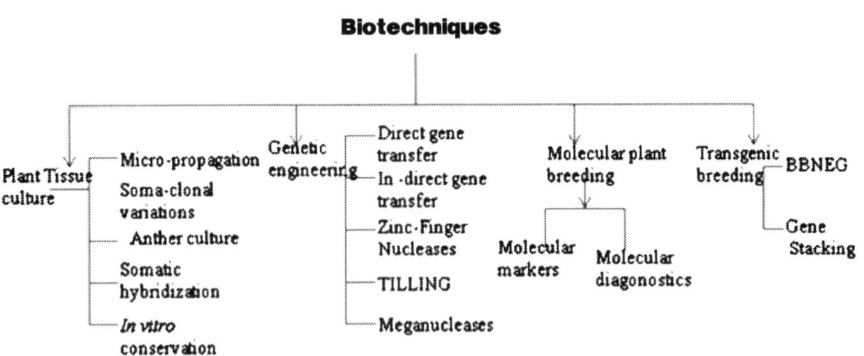


**Figure 8.2: Different Methods in Biotechnology for Crop Improvement.**

2. Embryo rescue technique has been used to grow hybrid embryos from wide crosses, which often fail to produce plant. To rescue them, the immature embryo tissue can be removed from the developing seeds and cultured in the laboratory to get hybrid plants to produce mature viable seeds. Embryo culture enables the breeder to successfully make wide crosses with a greater number of related species of wild plants and have access to a much wider range of genes that can be used for genetic improvement of crop plants. This has been valuable tool, especially for disease resistance genes transfer from wild relatives into crop plants. The development of a new rice plant type for West Africa (NERICA – New Rice for Africa) was a result of wide crosses between the Asian *Oryza sativa* and the African rice *Oryza glaberrima* (Aliou *et al.*, 2010). It employs embryo rescue in the initial breeding and in the successive back crossing work followed by anther culture to stabilize the breeding lines. The new plants had combined yield traits of the sativa parent with local adaptation traits from Glaberrima. Wild rice is a rich source of traits for resistance to pests and abiotic stresses. At the International Rice Research Institute, embryo rescue is utilized and facilitated the transfer of bacterial blight resistance genes from wild rice *Oryza longistaminata* to variety IR24 resulting to a bacterial blight resistant line (IRBB21).

3 Anther culture/microspore culture is a significant application of tissue culture being used to produce *haploid* plants, with single copy of each chromosome. The chromosome number of these haploid plants can be doubled by a chemical treatment, which quickly results in homozygous plants having two identical sets of chromosomes, or are completely inbred (*homozygous*). Pollen grain culture can dramatically reduce the time required to develop inbred parents for breeding of F1 hybrid varieties and facilitates the selection of recessive traits. Anther culture of F1 plants which are progenies in a specific breeding objective would allow many more different types of regenerates. The DH populations, which are similar in genetics to Recombinant Inbred Lines (RILs) generated by single seed descent approach, have been applied for mapping Quantitative Trait Loci (QTLs) for several desirable characters (Muñoz-Amatriaín *et al.*, 2008).

   Wheat cultivars developed from DH technology have been released for cultivation and have now turned out as dominant cultivars in several countries across the globe (Baenziger and DePauw, 2009; Touraev *et al.*, 2009). As the genetic constitution of the pollen will be more varied than those from the inbreds, thus breeders will have a wider range of traits to choose from. This technology has seen success in development of doubled haploid lines of rice, wheat, oats, sorghum, barley (Kiviharju *et al.*, 2005), and other field crops (Immonen and Anttila, 2000).

4. Somaclonal variations somewhat, limits the application of cell/tissue culture for plant propagation due to genetic variation in small fraction of the callus. Most plants regenerated from callus are genetically identical to the original plant from which the callus was derived, but a small fraction of

the resulting plants may exhibit genetic changes. While this is a problem for the *clonal propagation* of identical plants, it has been exploited as a new method to increase genetic variation. These changes are always not bad but many times can result in a range of genetically stable variation, useful in crop improvement (Skirvin *et al.*, 1993; Jain *et al.*, 1998). In long term cell and tissue cultures new genetic changes in the culture conditions become fixed, this supports the selection of quantitative and qualitative traits *in vitro*. Therefore *in vitro* cultures may be considered as mutagenic treatment and its user can be compared with mutation breeding (Buiatti 1993). Classical mutation breeding suffers from two serious drawbacks. Firstly, most mutations are recessive; therefore, it is difficult to improve character. Secondly, no easy method of the mutagenized materials is available.

On the other hand, most of the somaclonal variations are dominant but sometimes unnecessary changes are inevitable and finally selected variations are epigenetic (Roest and Gilisen,1989). Somaclonal variations can also be characterized by a variety of molecular methods (Henry, 1998) such as Isozymes (Martelli *et al.*, 1993), methylation sensitive amplified polymorphism (MSAP) (Xiong *et al.*, 1999) RFLP, RAPD (Piccioni *et al.*, 1997; Henry, 1998,) and SSR analysis. RAPD, SSR, and RFLP markers used to evaluate somaclonal variation in vegetative propagated chrysanthemum cultivars (Veilleux *et al.*, 1995; Wolff *et al.*, 1995).

All this notwithstanding, many encouraging results have been obtained (Buiatti 1993) with selection of genotype resistant to stress (Joshi and Rao, 2009). Pusa Jai Kisan, with bolder seeds and higher yield developed at Indian Agricultural Research Institute is another successful application of somaclonal variation of *Brassica juncea* variety 'Varuna'. Somaclonal variants in Lathyrus with low toxin level have also been developed at Indian Agricultural Research Institute (Rai, 2007). Recently, Khatab and Antar (2011) employed UV-C radiation to induce somaclonal variation in potato callus cultivar 'Cosima' and its detection using RAPD-PCR.

5. Protoplast fusion this technique works beyond the boundaries of texa/ species/genera. Protoplasts derived from genetically different plants allow DNA and other cellular contents to be combined. The fused cells can then be induced to regenerate into whole plants having traits from both parents. Some novel Chinese cabbage varieties have been developed through protoplast fusion for enhanced resistance to Blackleg (Hu *et al.*, 2002a, 2000b), introduction of the C3–C4 intermediate trait (Ishikawa *et al.*, 2003; Bang *et al.*, 2003). Citrus, members of solanaceae, different cereals and ornamental plants have been improved for various traits (Davey *et al.*, 2005). Protoplast fusion has also been used to transfer *male sterility* (lack of pollen viability) from radish into other *Brassica* species (cabbage, cauliflower, broccoli, canola), where it is used in the production of F1 hybrid seeds. The technique could not popularized more due to rapid regeneration of cell wall around isolated protoplast before fusion and extensive labour involved.

**Interesting Facts**

First plants regenerated from protoplasts (Takebe *et al.,* 1971)

Popular intergeneric somatic hybrid of tomato and potato was first produced by Melchers *et al.*1978 and named as Pomato.

## Genetic Engineering

Transferring genes from one organism to another is called genetic modification (GM) or genetic engineering (GE) or genetic improvement (GI). New DNA may be inserted in the host genome by first isolating and copying the genetic material of interest using molecular cloning methods to generate a DNA sequence, or by synthesizing the DNA, and then inserting this construct into the host organism. Different methods employed for transferring genes in plants are as follows:

### *Agrobacterium* Mediated Gene Transfer

*Agrobacterium tumefaciens* is the most commonly used transformation method employs the bacterium to transfer the desired DNA into the plant. *Agrobacterium* as natural genetic engineer is well known for its ability to transfer DNA between itself and plants, and for this reason it has become an important tool for genetic engineering. In early 1980s, scientists discovered use of these bacteria without affecting its ability to insert gene in plant genome. Desired genes can be spliced into the bacterial DNA, and *Agrobacterium,* transfers them into the plant's chromosomes. Soybean, cotton, sugar beet, corn, alfalfa, brassica, wheat, rice and finger millet (Chandler and Dunwell 2008; Sharma *et al.,* 2011) have been transformed using *Agrobacteria* for improving the crop quality. Initially it was limited to dicots only but use of acetosyringone during co-cultivation has made it applicable to monocots successfully.

### Biolistics or Gene Gun

A second extensively used approach is to transfer desired gene without using any vector applicable to all plant species. The desired DNA is coated onto microscopic metal beads and then shot them at high speed into pieces of plant tissue. The beads carry the DNA into the plant cells, some of which will incorporate the DNA into their chromosomes. The transgenic cells resulting from these processes can be selected and regenerated into whole plants by tissue culture. These plants will then express the characteristics encoded by the cloned genes. Biolistic mediated transformation method has been developed with various crops like wheat (Spark and Jones 2009) maize, pearl millets (O' Kennedy 2011), and finger millet (Jagga-Chugh *et al.,* 2012). Other methods of gene transfer like microinjection, electroporation and several others are not as popular as agrobacterium and gene gun.

Many agronomically beneficial traits are caused by reduction or elimination of gene expression (Gepts, 2002). Recent research in gene silencing has resulted in the production of transgene cassettes capable of silencing specific genes in a dominant fashion. This process is termed RNA interference, or RNAi, and results from the production of double-stranded RNA (dsRNA) in the plant cell. The dsRNA produced specifically prevents the production of protein from messenger RNA of the target gene (Bucher and Prins, 2006; Eamens *et al.*, 2010). These technologies circumvent the problems of generating knockout mutants described above, but are still functionally identical to them.

Although the mechanism was unknown at the time, the very first transgenic plants released for crop production were the product of gene silencing events. The FlavrSavr tomato was the result of silencing a gene responsible for fruit ripening, thereby extending the shelf-life of the transgenic tomato (Krieger *et al.*, 2008). Research is currently in progress to produce crops that are resistant to other diseases as well. Transgenic maize expressing a silencing vector was shown to be resistant to the coleopteran western corn rootworm (Baum *et al.*, 2007). Scientist still on the move for fungal (Nowara *et al.*, 2010), and nematode (Sindhu *et al.*, 2009) resistance.

**Zinc-Finger Nucleases**

Generally plant transformation technologies means random insertion of transgenes into plant genomes, sometimes can disrupt important genes or pathways. This disruption can lead to deleterious traits and transgene may link with undesirable alleles making so as to pose difficulty on achieving future breeding goals difficult. Monsanto's Roundup Ready® soybean lines are one of the best documented examples in transgenes' history which contain on average, a 5-10 per cent yield loss compared to isogenic lines that do not contain the insertion (Elmore *et al.*, 2001). Given these constraints, a large number of independent events must be produced in order to find one with high expression and good field performance.

Methods such as TILLING (Targeting Induced Local Lesions in Genomes), Zinc finger nuclease mediated mutagenesis, and the use of meganucleases, has allowed to produce targeted mutations in crop plants to delineate gene function as well as improve cultivars. These new and more specific methods are very promising. Zinc-finger nuclease technology allows for constructs to be designed to target transgene insertion into a specific locus within the plant genome, avoiding many of the problems addressed above. To employ zinc-finger nuclease technology researchers first engineer plants with zinc-finger nuclease constructs. The transformed plants produce zinc-finger nucleases that target loci specified in the construction of the zinc finger. The zinc fingers then cut both strands of DNA at the locus and then through recombination, inserts the transgene (Urnov *et al.*, 2010)). Zinc fingers can also be used to produce knockout mutations by disrupting the sequence of the target gene; *i.e.* causing a frame-shift mutation, elimination of the start codon or important motif, or a complete removal of the target gene (Zhang *et al.*, 2010). Also, researchers can target a known gene with zinc fingers and replace the sequence with a selectable maker (Shukla *et al.*, 2009). The transgene insertion destroys the function of the endogenous gene and the selectable maker allows one to select only transformed cells (Osakabe *et al.*, 2010). The zinc finger transgene, which is most likely inserted at a separate locus, can then be segregated out from the mutation it caused; thus resulting in a modified plant, without any transgenic sequences.

The benefit of this approach is the lack of recombinant protein although site directed gene insertion is quite profound, but the efficiency of the system is quite low. Osakabe *et al.* (2010) have studied that out of the transgenic arabidopsis lines recovered, only 0-3 per cent of the progeny had germinal mutations in the target gene. Furthermore the design of the zinc-finger nucleases is a limiting step. Townsend *et al.* (2009) designed for thirty-two zinc fingers, of these only three were able to bind DNA. The upfront screening assays required make zinc-finger nuclease technology unappealing for many researchers (Shukla *et al.*, 2009).

**Key Points**

In 1983, the first transgenic plant, a tobacco plant resistant to anti-biotics was created.

ATryn is "antithrombin " first transgenically produced protein approved for human treatment anywhere in the world. GTC Biotherapeutics has developed a way to produce a human anti-clotting protein, using genetically modified goats.

## Molecular Plant Breeding

### Markers

Molecular markers and other genomic applications have been highly successful in characterizing existing genetic variation within species; plant biotechnology generates new genetic diversity that often extends beyond species boundaries (Gepts, 2002; Johnson and McCuddin, 2008). The conventional method takes almost 20 years to develop pure homozygous plant while agricultural biotechnology has shortened time to develop 7-10 years. A *marker* is a "genetic tag" that identifies a particular location within a plant's DNA. *Marker-assisted* (or *molecular-assisted*) *breeding* promises a dramatic improvement where breeders can select plants with desirable combinations of genes. The need for genetic diversity must be balanced by elite performance, because choosing the best parents is key to maximizing the probability for successful improvement. Both DNA- and protein-based markers have been widely used in plant breeding, but DNA-based markers are expected to predominate in the future. Thus, molecular breeding is more precise and efficient than traditional breeding involves selection of individual plants based on visible or measurable traits. Another use of molecular markers is to identify undesirable genes that can be eliminated in future generations. Large number of DNA-based markers can be identified to cover entire regions of an organism's DNA, and they are not based on the developmental stage of the plant, as many protein based markers are. To visualize DNA-based markers rapid screening tests can be conducted on seeds or seedlings stage. Genomic information enables the discovery of beneficial alleles via QTL mapping and cloning, followed by the use of information learned from the molecular characterization of QTLs to design optimal transgenic strategies for crop improvement. Use of molecular markers increased breeding efficiency approximately 2-fold relative to phenotypic selection alone, with similar gains also observed in soybean (*Glycine max*) and sunflower (*Helianthus annuus*) populations (Eathington *et al.*, 2007).

The information generated using molecular markers' enriches investigations of plant evolution and comparative genomics, understanding of population structure, provides empirical measures of genetic responses to selection, increases efficiency of selection, action of favorable gene and also serves to identify and maintain reservoirs of genetic variability for future mining of beneficial alleles (McCouch, 2004; Slade *et al.*, 2005). In addition, knowledge of genetic relationships among germplasm sources may guide choice of parents for production of hybrids or improved populations (Collard and Mackill, 2008; Moose and Mumm2008).

## Molecular Diagnostics

Molecular diagnostics detect genes or gene products that are very precise and specific. Recent advances in Molecular biology and biotechnology are being focused on developing immunological and nucleic acid based hybridization methods for pathogen detection for crop systems (Weller *et al.*, 2006; Olmos *et al.*, 2007). Molecular methods have also been applied to the study of variation in plant pathogen populations, for example detection of different mating types or virulence types (McCartney et.al 2003). New variants of PCR, such as simple or multiplex nested PCR in a single closed tube, co-operative-PCR and real-time monitoring of amplicons or quantitative PCR, allow high sensitivity and precision in the detection of one or several pathogens in a single assay (Lopez *et al.*, 2003). PCR-based methods can provide new tools to monitor the exposure of a crop to pathogen inoculums that are more reliable and faster than conventional methods. This information can be used to improve disease control decision making.

## Transgenic Breeding

Developing transgenic line is no more dream for scientist though it has been difficult and expensive even after so much progressive growth of this technique. Yield which is main consideration for farmers so they would demand same performance on a genetically engineered variety to make decisions over non-genetically engineered varieties. Transgenes may or may not follow Mendelian segregation. Three important selection criteria have been identified for successful development of transgenic trait or product.

The first most important one is to confirm the presence of desired transgenic traits of interest in transformants.

The second important selection criterion is whether or not a transgenic trait has a predicted inheritance and expression pattern over generations, genetic background and environmental conditions.

Finally, it is significant to ensure that transgenes do not disrupt or have significantly negative impact on expression of endogenous genes that are critical to plant development. With more new transgenic traits generated, integration of multiple transgenic traits into base germplasm will be a substantial task in transgenic breeding. There are number of methods which could be employed to achieve goal in transgenic breeding.

## Backcross Breeding with Non-Elite Germplasm (BBNEG)

Yield drag and yield lag is major problem in crop performance. It is necessary to incessantly improve methods to transform and regenerate elite lines for each crop and expedite the backcrossing method.

Once a transgene has been backcrossed into an elite inbred background, it is no longer necessary for it to undergo backcrossing again. It can be used in the same manner as non-transgenic lines in breeding programs being mated to other lines and undergoing selection for improved qualities including yield.

The steps of backcross breeding are:

Mate a transgenic inbred with elite inbred. (F1 generation) All progeny will have 1 copy of the transgene and 50 per cent elite genes.

Harvest the progeny and mate them back to the elite inbred (BC1 generation). Half of the progeny will have 1 copy of the transgene. All will have 75 per cent elite genes.

Select the 50 per cent progeny with the transgene and mate back to the elite inbred. Again, half of the progeny will have 1 copy of the transgene. All progeny will have 87.25 per cent elite genes.

Continue selecting out progeny with the transgene and mating back to the elite inbred until the desired percentage of elite genes is reached (usually 98+ per cent) usually it takes about 5-7 generations (Patty and Straessue 2000). For each cross, the transgenic lines are typically the pollen donors. In order to reduce linkage drag from the donor parent, recurrent parent markers can be utilized for background selection and recombination markers can be used to verify the disruption of the donor parents' chromatin around the transgene due to crossover event(s) in meiosis (Young and Tanksley, 1989; Iftekharuddaula *et al.*, 2011).

An excellent use of MABC was demonstrated by integrating a gene from weedy rice (Oryza sativa L.) relative into elite varieties rapidly and efficiently using a non-transgenic approach (Iftekharuddaula *et al.*, 2011). About 116 simple sequence repeat (SSR) markers used for background selection of their elite parent, and six donor markers to select $BC_2F_2$ progeny (three crosses total) with their SUB1 allele for flood tolerance from the donor parent and 99.8 per cent of the recurrent parent's genome, where the average recovery of the recurrent genome is 93.75 per cent (Fehr, 1987). This same strategy can be used in transgenic breeding to rapidly and effectively recover a commercial line for deregulation.

As per Fehr (1987) the general guidelines for successful backcross breeding require at least more than five backcrosses. When applied to selecting transgenic lines the number of backcrosses is (1) based on the unwanted level of chromatin linked to the transgene, (2) the importance of the recurrent parent phenotype, (3) the amount of selection imposed during recurrent backcrosses, (4) and the level of eliteness of the donor parent.

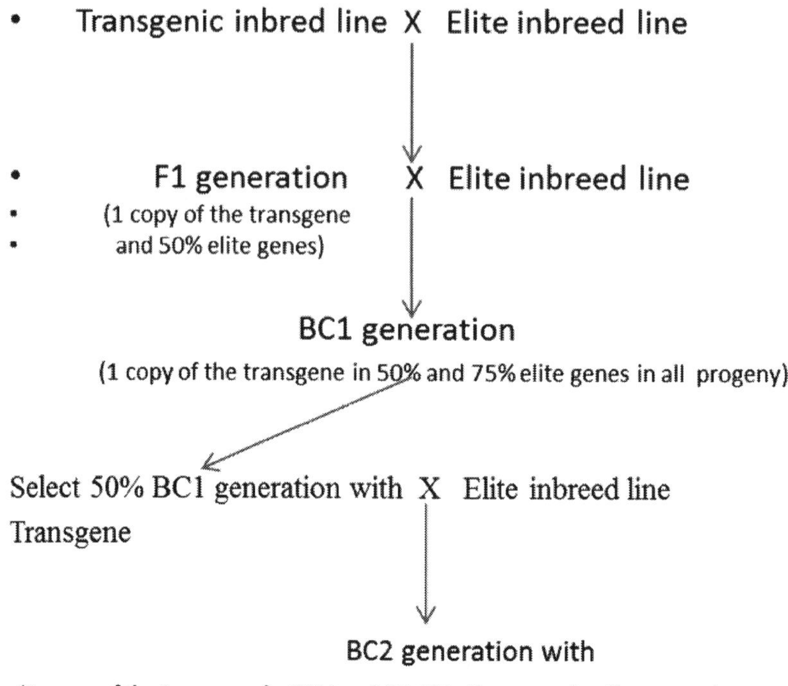


Continue selecting out progeny with the transgene and mating back to the elite inbred until the desired percentage of elite genes is reached (usually 98+%).

## Gene Stacking

Allowing crosses between parents each having a desired trait and then identifying offspring carrying both of these desired traits is one of the safest, quickest and easiest way to stack genes. Now a days, it is also possible transgenesis where more than one gene can be transferred in one or subsequent events. Here, the size of genes inserted into a chromosome location may limit the feasibility of stacking genes. The major effect of the transgenes on plant metabolism does also compelled to think that how many genes and what combinations of genes one can stack into a crop plant? In light of available facts we expect a few copies of one or two additional proteins from inserted genes which always have not been so accurate. As of now, the most suitably genes stacking can be performed via selected crosses between parents that possess the desired individual genes and identify progeny with the desired combination of traits.

Recently,Yield Guard VT Triple transgenic maize hybrids has been released, main traits combined in hybrid stacks are herbicide tolerance and multiple insect resistance integrated as one genomic locus or the combination of transgenes that simultaneously increase synthesis and decrease catabolism of Lys in maize seeds (Frizzi *et al.*, 2008). Stacked herbicide tolerance event 8264.44.06.1, includes a novel expression cassettes and transgenic inserts comprising multiple traits conferring resistance to glyphosate, aryloxyalkanoate, and glufosinate herbicides (Yunxing *et a.,l* 2013). In Brassica male sterility/fertility and multiherbicide tolerance genes, in Cotton insect and herbicide were stacked successfully (Taverniers *et al.*, 2008). But sometimes detection and quantification of stacked genes in crop is really challenging in comparison to non-transgenic. Still, it is needed to harmonize event specific methods to detect multiple genes in a single go.

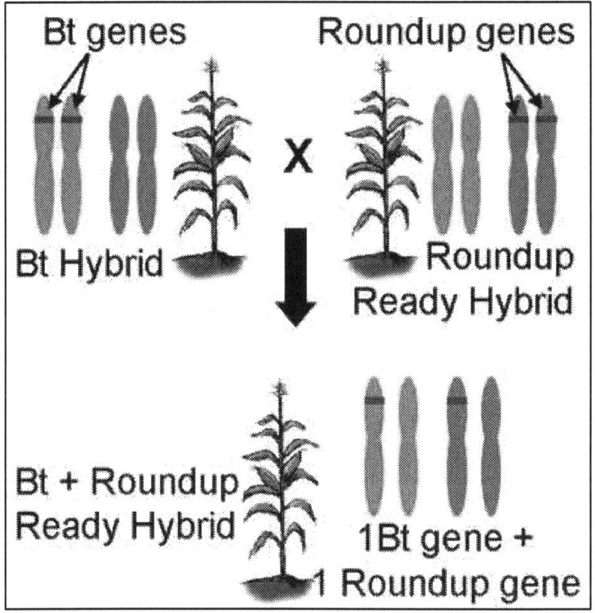


Mating a homozygous Roundup resistant plant with a homozygous Bt plant will result in progeny stacked for both traits. The progeny will have one copy of the Roundup resistance gene and the Bt gene (Source Monsanto, 2005).

With improved gene targeting technology (Ow, 2007) and the construction of meiotically transmissible plant minichromosomes (Carlson *et al.*, 2007; Yu *et al.*, 2007) it seems to introduce more traits with increasing complexity without much difficulty. With such advances, biotechnology is now poised to assemble useful genetic diversity from essentially any source into constructs that concentrate favorable gene action and maximize heritability for a greatly expanded set of traits.

**Classical Vs Transgenic**

| *Classical Breeding* | *Transgenic Breeding* |
|---|---|
| ☆ Desired gene is incorporated along with many other genes, which then have to be bred out | ☆ Desired gene only is transferred from an organism to another |
| ☆ Crosses limited by species barriers | ☆ Species barrier may be overcome |

## Biotechnology Vs safety

As far the different biotechnological techniques are concerned being perfected with time. Previously, it was completely relying on microbial gene but later the focus was shifted to plant genes confer disease resistant to insect pest (Ryan 1990). The presence of marker genes in commercialized transgenic crops has caused considerable public concern about the medical implications of GM food consumption and GM crop cultivation. A lot of research effort has been directed towards the development of marker-free transformation methods and selectable marker elimination strategies. These selectable markers only suppress the growth of non-transformed cells, in contrast to antibiotic and herbicide resistance markers. Likewise, use of phosphomannose isomerase gene (*pmi*) has been reported as a selectable marker for transformation of many plant species, including rice, wheat, millet, tobacco, sugarcane, apple and onion. The molecular cut and paste site specific recombination technique can eliminate the unwanted marker from GM plants, the most widely used of which is the Cre/*loxP* system from the bacteriophage P1.

The GM food items available in the market have undergone full review by regulatory agencies regarding safety and content relative to unmodified form. People's worry about ingestion of foreign DNA or modified protein products was addressed in July 2007 by European food safety authorities that it rapidly degrade on ingestion into small DNA or peptide fragment,so even no trace of them were detected in tissues, fluids or edible products of farm animals (Lemaux 2009). Currently, studies on impact assessment of transgenes moving into wild relatives and the potential to change ecosystem dynamics are requested in environmental impact statements before any GE plant is released. As far the moving of genes from GEP to soil bacteria is concerned could happen, but at very low frequencies. The economic benefits, time savings, and ease of agricultural practices using GM crops have already been realized worldwide by the farmers.

In 2009, countries that grew 95 per cent of the global transgenic crops were the United States (46 per cent), Brazil (16 per cent), Argentina (15 per cent), India (6 per cent), Canada (6 per cent), China (3 per cent), Paraguay (2 per cent) and South Africa (2 per cent). In the US, by 2009/2010, 93 per cent of the planted area of soybeans, 93 per cent of cotton, 86 per cent of corn and 95 per cent of the sugar beet were genetically modified varieties.

## Conclusion

Through the journey of agricultural development, the plant breeders and plant biologists has come up with "elite crop varieties" that promise to stand firm in wider range of climatic variability to tackle the food insecurities of future generations along with maintaining/exceeding quality parameters. Green revolution which was the major breakthrough in the 20th century was having the base in conventional plant breeding approaches. Now, in the 21st century modern plant breeders require precise gene modifications with a gene tracking system for the modified trait. Identifying the gene combinations that lead to significant crop improvement remains the challenges in plant biology. For accelerating efforts, it is better to integrate the different research disciplines and activities that form core components of molecular plant breeding as this integration requires knowledge of the genomic organization, genes function and understanding in statistical approaches to estimate genetic effects (Gepts and Hancock, 2006; Bliss, 2007). Progress made with the help of various approaches along with the creation of a wider gene pool by utilizing modern biotechnological tools is the best approach to improve crop productivity for meeting crop improvement goals.

## Glossary

**Embryo culture** —removal of a developing embryo from a seed for *in vitro* propagation

**Genetic engineering**— The manipulation of an organism's genetic endowment by introducing or eliminating specific genes through modern molecular biology techniques. A broad definition of genetic engineering also includes selective breeding and other means of artificial selection.

**Genetic marker**— A gene or group of genes used to "mark" or track the action of microbes.

**Hybridization**— The hydrogen bonding of complementary DNA and/or RNA sequences to form a duplex molecule. (See Northern hybridization, Southern hybridization.)

***In vitro***—To be grown in glass (Latin) Propagation of plants in a controlled, artificial environment using plastic or glass culture vessels, aseptic techniques, and a defined growing medium.

**Micropropagation**—*In vitro* Clonal propagation of plants from shoot tips or nodal explants, usually with an accelerated proliferation of shoots during subcultures.

**Molecular breeding**— It is the application of molecular biology tools, often in plant breeding.

**Plant Tissue Culture**—The growth or maintenance of plant cells, tissues, organs or whole plants *in vitro.*

**Somaclonal Variation**—Phenotypic variation, either genetic or epigenetic in origin, displayed among somaclones.

**Transgenic breeding**— An organism whose genome has been altered by the transfer of a gene and use this altered genome to breed with another organism.

**Yield drag**— represents a negative effect on grain yield associated with crop plants that have a specific gene or a specific trait.

**Yield lag**— is a relative reduction in yield observed in some hybrids or varieties compared with the yield observed in the most recently produced hybrids or varieties.

**Zinc fingers**— Configuration of a DNA-binding protein that resembles a finger with a base, usually cysteines and histidines, binding a zinc ion.

## References

1. *Agricultural Biotechnology* (A Lot More than Just GM Crops). 2010 ISAAA, URL: http: / /www.isaaa.org/
2. Ahloowalia, BS (2004). Global impact of mutation-derived varieties. *Euphytica* 135: 187–204.
3. Ahloowalia, BS(1996). improvement of horticultural plants through *in vitro* culture and induced mutations ishs Acta Horticulturae 447: *III International Symposium on In Vitro Culture and Horticultural Breeding* .
4. Aliou Diagne, Soul-Kifouly Gnonna Midingoyi, Marco Wopereis and Inoussa Akintayo(2010). The NERICA Success Story: Development, Achievements and Lessons Learned. (Ist draft The Africa Rice Center (Africa Rice).
5. Baenziger PS, DePauw RM (2009). Wheat breeding: Procedures and strategies. In: B.F. Carver (ed.). *Wheat: Science and Trade*. Wiley-Blackwell Publishing, Ames, IA.
6. Bang SW, Mizuno Y, Kaneko Y, Matsuzawa Y, Bang KS (2003). Production of intergeneric hybrids between the C-3–C- 4 intermediate species *Diplotaxis tenuifolia* (L). DC and *Raphanus sativus* L. *Breed Sci.*, 53: 231– 6.
7. Baum JA, Bogaert T, Clinton W, Heck GR, Feldmann P, Ilagan O, Johnson S, Plaetinck G, Munyikwa T, Pleau M, Vaughn T, Roberts J. (2007). Control of coleopteran insect pests through RNA interference. *Nature Biotechnology* 25: 1322-1326. DOI: 10.1038/nbt1359.
8. Bertrand C, Y Collard and David J Mackill (2008). Marker-assisted selection: an approach for precision plant breeding in the twenty-first century. *Phil. Trans. R. Soc.* B 363, 557–572.
9. Bliss FA (2007). Education and preparation of plant breeders for careers in global crop improvement. *Crop Sci* 47: S250–S261.
10. Borlaug NE (2000). Ending world hunger. The promise of biotechnology and the threat of antiscience zealotry. *Plant Physiol* 124: 487–490.
11. Boston, MA (2012). Biotechnology crop development in africa: Support biotechnology development in Africa-private sector urged (Press release).
12. Bucher E and Prins M (2006). *RNA silencing: A natural resistance mechanism in plants*. Springer.
13. Buiatti M and F Gimelli (1993). Somaclonal variation in ornamentals. In *Creating Genetic Variations on Ornamantals*, pp. 5-24 (eds T. Schiva and A. Merruri). Sanremo: Institute spermentale per la Forlicotura.

14. Carlson SR, Rudgers GW, Zieler H, Mach JM, Luo S, Grunden E, Krol C, Copenhaver GP, Preuss D (2007). Meiotic transmission of an *in vitro*-assembled autonomous maize minichromosome. *PLoS Genet* 3: 1965–1974.

15. Chandler, S, and JM Dunwell (2008). Gene flow, risk assessment and the environmental release of transgenic plants. *Critical Reviews in Plant Sciences* 27: 25-49.

16. Collard BCY, Mackill DJ (2008). Marker assisted selection: an approach for precision plant breeding in the twenty-first century. *Philos. Trans. R. Soc. Lond. Ser. B. Biol. Sci.* 363: 557–572.

17. Cooper M, Smith OS, Graham G, Arthur L, Feng L, Podlich DW (2004). Genomics, genetics, and plant breeding: a private sector perspective. *Crop Sci* 44 1907–1913.

18. Daveya MR, Paul Anthonya, J. Brian Powera, Kenneth C. Loweb (2005). Plant protoplasts: status and biotechnological perspectives. *Biotechnology Advances* 23L 131–171.

19. Dobránszki J, da Silva JA (2010). Micropropagation of apple: A review. *Biotechnol. Adv.* 28(4): 462-88.

20. Eamens A., Curtin S.J., Waterhouse P.M. (2010). *RNA Silencing in Plants.* Springer.

21. Eathington SR, Crosbie TM, Edwards MD, Reiter RD, Bull JK (2007). Molecular markers in a commercial breeding program. *Crop Sci* 47 S154–S163.

22. Eicher CK, Karim Maredia, Idah Sithole-Niang(2006). Crop biotechnology and the African farmer. *Food Policy* 31: 504–527.

23. Elmore R.W., Roeth F.W., Nelson L.A., Shapiro C.A., Klein R.N., Knezevic S.Z., Martin A. (2001a). Glyphosate-resistant soybean cultivar yields compared with sister lines. *Agronomy Journal* 93: 408-412.

24. Fehr WR (1987). *Principles of Cultivar Development: Theory and Technique.* Macmillan Publishing Company, Macmillan, New York.

25. Frizzi A, Huang S, Gilbertson LA, Armstrong TA, Luethy MH, Malvar TM (2008). Modifying lysine biosynthesis and catabolism in corn with a single bifunctional expression/silencing transgene cassette. *Plant Biotechnol J* 6: 13–21.

26. Gepts P (2002). A comparison between crop domestication, classical plant breeding, and genetic engineering. *Crop Sci* 42 1780–1790.

27. Gepts P, Hancock J (2006). The future of plant breeding. *Crop Sci* 46: 1630–1634.

28. Harlander Susan K (2002). The Evolution of Modern Agriculture and Its Future with Biotechnology. *Journal of the American College of Nutrition,* Vol. 21, No. 3, 161S–165S.

29. Henry RJ (1998). 'Molecular and biochemical characterization of somaclonal variation', in SM Jain, DS Brar and BS Ahloowalia (eds), *Somaclonal Variation and Induced Mutations in Crop Improvement,* Kluwer Academic Publishers, Dordrecht, Germany pp. 485-499.

30. Henry RJ (1998). 'Molecular and biochemical characterization of somaclonal variation', in SM Jain, DS Brar and BS Ahloowalia (eds), *Somaclonal Variation and Induced Mutations in Crop Improvement*, Kluwer Academic Publishers, Dordrecht, Germany pp. 485-499.

31. Herrera-Estrella L(1999). Transgenic plants for tropical regions: some considerations about their development and their transfer to the small farmer. *Proc Natl Acad Sci USA* 96: 5978–5981.

32. Hu Q, Andersen SB, Dixelius C, Hansen LN (2002a). Production of fertile intergeneric somatic hybrids between *Brassica napus* and *Sinapsis arvensis* for the enrichment of the rapeseed gene pool. *Plant Cell Rep.*, 21: 147–52.

33. Hu Q, Hansen LN, Laursen J, Dixelius C, Andersen SB(2002b). Intergeneric hybrids between *Brassica napus* and *Orychophragmus violaceus* containing traits of agronomic importance for oilseed rape breeding. *Theor Appl Genet.*, 105: 834–40.

34. Iftekharuddaula KM, Newaz MA, Salam MA, Ahmed HU, Mahbub MAA, Septiningsih EM, Collard BCY, Sanchez DL, Pamplona AM, Mackill DJ (2011). Rapid and high-precision marker assisted backcrossing to introgress the *SUB1* QTL into BR11, the rainfed lowland rice mega variety of Bangladesh. *Euphytica* 178: 83–97.

35. Immonen S, Anntila H (2000). Media composition and anther plating for production of androgenic green plants from cultivated rye (*Secale cereale* L.). *J. Plant Physiol.* 156: 204-210.

36. Ishikawa S, Bang SW, Kaneko Y, Matsuzawa Y (2003). Production and characterization of intergeneric somatic hybrids between *Moricandia arvensis* and *Brassica oleracea*. *Plant Breed.*, 122: 233–8.

37. Jagga-Chugh S, Kachhwaha S, Sharma M, Kothari-Chajer A, Kothari SL (2012). Optimization of factors influencing microprojectile bombardment-mediated genetic transformation of seed-derived callus and regeneration of transgenic plants in *Eleusine coracana* (L.). Gaertn. *Plant Cell Tiss. Org. Cult.*, 109: 401–410.

38. Jain PK, Lalwani AK, Li XC, Singleton TL, Smith TN, Chen A, Deshmukh D, Verma I, Smith RJ, Wilcox ER (1998). A gene for recessive nonsyndromic sensorineural deafness (DFNB18). maps to the chromosomal region 11p14-p15.1 containing the Usher syndrome type 1C gene. *Genomics* 1;50 (2): 290-2.

39. James C (2007). Global status of commercialized biotech/GM crops: In ISAAA Briefs No. 37. *International Service for the Acquisition of Agri-Biotech Applications*, Ithaca, NY.

40. Johnson GR, McCuddin ZP (2008). Maize and the biotech industry. In: JL Bennetzen, SC Hake, eds, *Handbook of Maize: Its Biology*. Springer, Berlin.

41. Joshi RK and Rao GJN (2009). Somaclonal variation in submergence tolerant rice cultivars and induced diversity evaluation by PCR markers. *International Journal of Genetics and Molecular Biology* Vol. 1 (5). 80-88.

42. Kang, M.S. and Gauch Jr., H.G. (eds.)(1996). *Genotype by Environment Interaction*. CRC Press, Inc., Boca Raton. 416 pp.

43. Khatab IA and El-Banna Antar N (2011). Detection of somaclonal variations in potato using RAPD markers. *Egypt. J. Genet. Cytol*, 40: 227-238.

44. Kiviharju E, Moisander S, Laurila J (2005). Improved green plant regeneration rates from oat anther culture and the agronomic performance of some DH lines. *Plant Cell Tissue Organ Cult*. 81: 1-9.

45. Krieger EK, Allen E, Gilbertson LA, Roberts JK, Hiatt W, Sanders RA (2008). The Flavr Savr tomato, an early example of RNAi technology. *Hortscience* 43: 962-964.

46. Lanfranco L (2003). Engineering crops, a deserving venture. *Riv Biol*. 96(1): 31-54.

47. Lemaux, Peggy G (2009). Genetically Engineered Plants and Foods: A Scientist's Analysis of the Issues (Part 11). *Annual Review of Plant Biology*. Vol. 60: 511-559.

48. Lifen Gao, Yinghao Cao, Zhihui Xia, Guanghuai Jiang, Guozhen Liu, Weixiong Zhang,corresponding and Wenxue Zhai (2013). Do transgenesis and marker-assisted backcross breeding produce substantially equivalent plants? - A comparative study of transgenic and backcross rice carrying bacterial blight resistant gene Xa21. *BMC Genomics*, 14: 738.

49. López MM, Bertolini, E, Olmos A, Caruso P, Gorris MT, Llop P, Penyalver R, and Cambra M (2003). Innovative tools for detection of plant pathogenic viruses and bacteria. *Int. Microbiol. 6*, 233-243.

50. Lörz H and Wenzel G (2005). *Molecular Marker Systems in Plant Breeding and Crop Improvement*. Springer, New York.

51. Maluszynsk MK, K Nichterlein, L. van Zanten and BS Ahloowalia (2000). "Officially released mutant varieties – the FAO/IAEA Database". *Mutation Breeding Review* (12): 1–84.

52. Manshardt Richard (2004). Crop Improvement by Conventional Breeding or Genetic Engineering: How Different Are They? *Biotechnology BIO*-5; 1-3.

53. Martelli GI, Greco B, Mezzetti P, Rosatti (1993). Isozymic analysis of somaclonal variation among regenerants from apple root stock leaf tissue. *Acta Hort*. 336: 381–387.

54. McCartney HA, Foster SJ, Fraaije BA(2003). Molecular diagnostics for fungal plant pathogens. *Pest Manag Sci.*, 59(2). : 129-42.

55. McCouch S (2004). Diversifying selection in plant breeding. *PloS. Biol*. 2: e347.

56. Melchers G, Sacristan MD and Holder AA (1978). *Carlberg Res Commen.*, 43 203-208.

57. Miassar M. Al-Taleb, Dhia S. Hassawi and Saeid M. Abu-Romman (2011). Production of Virus Free Potato Plants Using Meristem Culture from Cultivars Grown under Jordanian Environment. *American-Eurasian J. Agric. and Environ. Sci.*, 11 (4): 467-472.

58. Michael R. Daveya, Paul Anthonya, J. Brian Powera, Kenneth C. Loweb (2005). Plant protoplasts: status and biotechnological perspectives. *Biotechnology Advances* 23 (2005). 131–171.

59. Monsanto. (2005). Annual Report for fiscal year ended Aug. 31, Form 10-K, in: U. S. S. a. E. Commission (Ed.), Washington, D.C.

60. Moose SP, Mumm RH (2008). Molecular plant breeding as the foundation for 21st century crop improvement. *Plant Physiol.* 147: 969–977.

61. Mumm RH (2007). Backcross versus forward breeding in the development of transgenic maize hybrids: theory and practice. *Crop Sci* (Suppl 3). 47 S164–S171.

62. Muñoz-Amatriaín M, Castillo AM, Chen XW, Cistué L, Vallés MP (2008). Identifi cation and validation of QTLs for green plant percentage in barley (*Hordeum vulgare* L.). anther culture. *Mol. Breed.* 22: 119-129.

63. Naqvi S., Farre G., Sanahuja G., Capell T., Zhu C.F., Christou P. (2010). When more is better: multigene engineering in plants. *Trends in Plant Science* 15: 48-56.

64. Navok FJ and Brunner H(1992). Plant breeding: induced mutation technology for crop improvement. *IAEA Bull* 4; 25-33.

65. Nelson RJ, Naylor RL, Jahn MM (2004). The role of genomics research in improvement of "orphan" crops. *Crop Sci* 44 1901–1904.

66. Nowara D., Gay A., Lacomme C., Shaw J., Ridout C., Douchkov D., Hensel G., Kumlehn J., Schweizer P. (2010). HIGS: Host-Induced Gene Silencing in the Obligate Biotrophic Fungal Pathogen Blumeria graminis. *Plant Cell* 22: 3130-3141.

67. O'kennedy MM, Crampton BG, Ozias-akins P (2009). Pearl millet. In *Com-pendium of Transgenic Crop Plants, Volume 1: Transgenic Cereals and Forage Grasses* (Eds C Kole, TC Hall). John Wiley and Sons, Chichester, pp. 177-190.

68. O'Kennedy MM, Stark HC, Dube N.(2011). Biolistic-mediated transformation protocols for maize and pearl millet using pre-cultured immature zygotic embryos and embryogenic tissue. *Methods Mol Biol.* 710: 343-54.

69. Olmos, A. Capote, N Bertolini, E. Cambra, M (2007). *Molecular Diagnostic Methods for Plant Viruses.* Chapter: 10 (PageNo: 227). CABI (H ISBN 9781845932886).

70. Osakabe K., Osakabe Y., Toki S. (2010). Site-directed mutagenesis in *Arabidopsis* using custom-designed zinc finger nucleases. Proceedings of the National Academy of Sciences of the United States of America 107: 12034-12039. DOI: 10.1073/pnas.1000234107.

71. Ow DW (2007). GM maize from site-specific recombination technology, what next? *Curr Opin Biotechnol* 18: 115–120.

72. Patty Hain and Carda Strauss(2000). *Backcross Breeding*. University of Nebraska (Plant and soil Sciences, elibraray).

73. Piccioni E, Barcaccia G, Falcinelli M, Standardi A (1997). Estimating alfalfa somaclonal variation in axillary branching propagation and indirect somatic embryogenesis by RAPD fingerprinting. *Int. J. Plant Sci.* 158: 556–562.

74. Podwyszyñska M, Sochacki D (2010). Micropropagation of tulip: production of virus-free stock plants. *Methods Mol Biol.* 589: 243-56.

75. Poehlman JM (1979). Genetics and plant breeding: mutation. In "*Breeding Field Crops*," 2nd ed. Westport CT: AVI Publishing Company, Inc, pp. 73–87.

76. Rai Rhitu (2007). Introduction to Plant Biotechnology. In: *Plant Breeding and Genetic*. pp1-42.

77. Roest, S and Gilissen, LJW (1989). Plant regeneration from protoplasts: a literature review. *Acta Bot. Neerl.* 38: 1–23.

78. Rout GR, Das G, Samantaray S, Das P (2001). *In vitro* micropropagation of Lawsonia inermis (Lythraceae). *Rev Biol Trop.* 49(3-4): 957-63.

79. Ryan CA (1990). Protease inhibitors in plants: genes for improving defenses against insects and pathogens. *Annual Review of Phytopathology*, Vol. 28; 425-449.

80. Schouten, HJ, Jacobsen E (2007). "Are Mutations in Genetically Modified Plants Dangerous?". *Journal of Biomedicine and Biotechnology* 82612.

81. Sengar RS, Kalpana Sengar and Sanjay Kumar Garg (2011). Biotechnological approaches for high sugarcane yield. *Plant Sciences Feed* - 1 (7): 101-111.

82. Serageldin I (1999). Biotechnology and food security in the 21st century. *Science* 285(5426): 387–389.

83. Sharma M, Kothari-Chajer A, Jagga-Chugh S, Kothari SL (2011). Factors influencing *Agrobacterium tumefasciens*-mediated genetic transformation of *Eleucine coracana* (L.). Gaertn. *Plant Cell. Tiss. Organ. Cult*, 105: 93-104.

84. Shukla VK, Doyon Y, Miller JC, DeKelver RC, Moehle EA, Worden SE, Mitchell JC, Arnold NL, Gopalan S, Meng XD, Choi VM, Rock JM, Wu YY, Katibah GE, Zhifang G, McCaskill D., Simpson MA, Blakeslee B, Greenwalt SA, Butler HJ, Hinkley SJ, Zhang L, Rebar EJ, Gregory PD, Urnov FD (2009). Precise genome modification in the crop species Zea mays using zinc-finger nucleases. *Nature* 459: 437-U156. DOI: 10.1038/nature07992.

85. Simmonds NW (1997). *Pie in the sky*. Tropical Agriculture Association June 1997: 1-5.

86. Skirvin, RM, Norton, M, Mcpheeters, KD (1993). Somaclonal variation: has it proved useful for plant improvement? *Acta Horticulturae*, v.336, p.333-340.

87. Slade AJ, Fuerstenberg SI, Loeffler D, Steine MN, Facciotti D. A reverse genetic, nontransgenic approach to wheat crop improvement by TILLING. *Nat Biotechnol.* 2005 Jan;23(1): 75-81. Epub 2004 Dec 5. PubMed PMID: 15580263.

88. Sparks CA and Jones HD (2009). Biolistics transformation of wheat. *Methods Mol Biol.* 478: 71-92.

89. Takebe I, Labib G, Melchers G (1971). Regeneration of whole plants from isolated mesophyll protoplasts of tobacco. *Naturwissenschaften* 58: 318–320.

90. Taverniers I, Papazova N, Bertheau Y, De Loose M, Holst-Jensen A. (2008). Gene stacking in transgenic plants: towards complience between definitions, terminology and detection within the EU regulatory framework.

91. Touraev A, Forster BP, Jain SM (eds). (2009). *Advances in haploid production in higher plants*. Springer Science Business Media B.V., The Netherlands. P. 1-208.

92. Townsend J.A., Wright D.A., Winfrey R.J., Fu F.L., Maeder M.L., Joung J.K., Voytas D.F. (2009). High-frequency modification of plant genes using engineered zinc-finger nucleases. *Nature* 459: 442-U161. DOI: 10.1038/nature07845.

93. Trevor V. Suslow, UC Davis, Bruce R. Thomas Kent J. Bradford (2002). Biotechnology Provides New Tools for Plant Breeding. Agricultural Biotechnology in California series Publication 8043 by the Regents of the University of California, Division of Agriculture and Natural Sciences.

94. Tyson P (2001). *Should we Grow GM crops?* Retrieved from http: //www.pbs.org/wgbh/harvest/exist/.

95. Urnov F.D., Rebar E.J., Holmes M.C., Zhang H.S., Gregory P.D. (2010). Genome editing with engineered zinc finger nucleases. *Nature Reviews Genetics* 11: 636-646. DOI: 10.1038/nrg2842.

96. USDA/APHIS (2002). Lists for Field Test Releases in the U.S., http: //www.isb.vt.edu/cfdocs/ISBlists1.cfm.

97. Van de Peer, Y.; Meyer, A. (2005). "Large-scale gene and ancient genome duplications". In Gregory, T.R. *The Evolution of the Genome*. San Diego: Elsevier. pp. 329–68.

98. Varshney RV, Hoisington DA, Tyagi AK (2006). Advances in cereal genomics and applications in crop breeding. *Trends Biotechnol* 24 490–499.

99. Veilleux RE, Shen LY, Paz MM (1995). Analysis of the genetic composition of anther-derived potato by randomly amplified polymorphic DNA and simple sequence repeats. *Genome* 38: 1153–1162.

100. Visarada KBRS, Kanti Meena, C Aruna, S Srujana, N Saikishore and N Seetharama (2009). *Transgenic Breeding: Perspectives and Prospects*. National Research Centre for Sorghum (NRCS), Rajendranagar, Hyderabad 500 030, A.P., India.

101. Weller, SA, JG Elphinstone, N Parkinson and R Thwaites (2006). Molecular Diagnosis of Plant Pathogenic Bacteria. *Arab J. Pl. Prot*. 24: 143-146.

102. Wolff K, Zietkiewicz E, Hofstra H (1995). Identification of Chrysanthemum cultvars and stability of DNA fingerprint patterns. *Theor. Appl. Genet*. 91: 439–447.

103. Xiong LZ, Xu CG, Maroff MAS, Zhang QF (1999). Patterns of cytosine methylation in an elite rice hybrid and its parental lines, detected by amethylation-sensitive amplification polymorphism technique. *Mol.Gen. Genet*. 261: 439–446.

104. Young ND (1999). A cautiously optimistic vision for marker-assisted breeding. *Mol Breed* 5: 505–510.

105. Yu W, Han F, Gao Z, Vega JM, Birchler JA (2007). Construction and behavior of engineered minichromosomes in maize. *Proc Natl Acad Sci USA* 104: 8924–8929.

106. Yunxing Cui, Thom Hoffman, Ning Zhou, Stephen N. Novak, Julissa Colon, Dawn Parkhurst, Sandra Toledo, Terry Wright, Sean Russell, Bruce Held, Vaithilingam SekarStacked herbicide tolerance event 8264.44.06.1 related transgenic soybean lines, and detection.(patent).

107. Zhong GY (2001). Genetic issues and pitfalls in transgenic plant breeding. *Euphytica* 118: 137-144.

**2016, Biotechnology: Progress and Applications** *Pages* **207–221**
*Editors:* **Saif Hameed and Zeeshan Fatima**
*Published by:* **DAYA PUBLISHING HOUSE, NEW DELHI**

# *Chapter 9*

# *Trichoderma*: Biotechnological Applications

***Ravindra Bansal, Pramod D. Sherkhane and Prasun K. Mukherjee****

*Nuclear Agriculture and Biotechnology Division, Bhabha Atomic Research Centre, Trombay, Mumbai – 400 085*

## ABSTRACT

*Trichoderma* is among the few fungal genera that has affected humanity in a big way. These filamentous, basically saprophytic fungi are widely used in industry and agriculture, and have immense potential in other areas like bioremediation and health science. *Trichoderma reesei* is the major industrial source of cellulases and hemicellulases that are increasingly becoming significant with the advent of biofuel research and development. Several species of *Trichoderma* are used in agriculture as suppressor of plant diseases and plant growth enhancers. Several reports on enhancement of plant immunity due to *Trichoderma* applications have ushered in new areas of research and applications. Recent research reports also suggest that *Trichoderma* spp. can ameliorate abiotic stresses like drought and salinity, enhancing the scope of applications in crop production. Several genes from *Trichoderma* spp. have been transferred to plants for enhancing biotic and abiotic stress tolerance. These fungi are prolific producers of secondary metabolites and hence are sources of novel bioactive compounds. The genomes of seven species have been sequenced and these advancements will aid greatly in further exploiting these fungi in human welfare.

* Corresponding Author: E-mail: prasunmukherjee1@gmail.com

## Introduction

*Trichoderma* species are a group of fungi which have had astounding impact on human welfare, be it in agriculture, industry, environment, or in the field of health sciences (Mukherjee *et al.*, 2013a). These are the most popular and perhaps the most effective biopesticides used in the management of a range of plant diseases, both soil-borne and foliar (Verma *et al.*, 2007). These fungi also promote plant growth, boost plant immunity and make them somewhat tolerant to biotic and abiotic stresses (Mukherjee *et al.*, 2013b). *Trichoderma* spp. also produce a range of secondary metabolites, some having beneficial effects on plants, and some are of pharmaceutical significance (Mukherjee *et al.*, 2012a). Several genes from *Trichoderma* have been transferred to plants that have shown improved tolerance to biotic and abiotic stresses (Nicolas *et al.*, 2014). The impact of these fungi in industry is tremendous- *Trichoderma reesei* being the major source of cellulases and hemicellulases. On the other side, some species are opportunistic human pathogens while a few are enemies of the cultivated mushrooms. This chapter summarizes these multiple facets of *Trichoderma* applications and its existing and potential biotechnological applications.

## *Trichoderma* in Agriculture

The first report of biological control of plant pathogens was in early 1930s, when Weindling (1932) recorded reduction of *Rhizoctonia solani* infection in citrus seedlings through the application of *Trichoderma virens* (then identified as *T. lignorum*). However, even after a series of papers published by them on the mechanisms of biocontrol, including mycoparasitism and antibiotics production, the interest in biocontrol as such declined due to the development of synthetic crop protection chemicals. Later, after demonstration of biocontrol in the field by Wells *et al.* (1972), who achieved control of *Sclerotium rolfsii* through the application of *Trichoderma harzianum*, interest in *Trichoderma* was revived and still continues to dominate research in the area of plant disease biocontrol.

### Biopesticides

*Trichoderma* spp. constitute more than 60 per cent of the total biofungicides market the world over, showing the dominance of these fungi as crop protection biologicals (Verma *et al.*, 2007). In India alone, more than 300 formulation products are registered and sold for use in agriculture (Singh *et al.*, 2012). Several species of *Trichoderma* are used as biocontrol agents, *e.g.*, *Trichoderma harzianum, T. viride, T. koningii, T. virens, T. asperellum* and *T. asperelloides*. There is no generalized rule as to what species being the most effective, but this is determined by factors like relative abundance in soil, interest of research groups and also the availability of toxicological data that are mandatory for registration. Similarly, there is no delineation on what pathogens can be controlled by a particular species/strain. In general, majority of the data are available against soil-borne pathogens like *Sclerotium rolfsii, Rhizoctonia solani, Pythium* spp. and against the foliar pathogen *Botrytris cinerea*. This does not exclude many other pathogens that are controlled by *Trichoderma* spp. The direct antagonistic action of *Trichoderma* spp. on fungal and oomycete pathogens include mycoparasitism and/or antibiosis. Mycoparasitism is mediated by a series of hydrolytic enzymes like chitinases, glucanases, cellulases and proteases (Viterbo and Horwitz, 2010). The antibiosis, where *Trichoderma* inhibits/kills pathogens is mediated by the production of secondary metabolites that are lethal/inhibitory to the cells often facilitates competitive exclusion of the pathogens from the shared ecological niche, *e.g.*, the rhizosphere. Secondary metabolism will be discussed in details in section 4. *Trichoderma* species are primarily saprophytes, and hence can utilize a large number of natural substrates. Basically it can grow and sporulate on any organic substrate (as well as inorganic synthetic media) including grains, crop residues, composts and even fresh cow dung (Zaidi *et al.*, 2013). This attribute has been a major driving force behind commercial success of these fungi. Commercial scale multiplication is done both in solid state fermentation (like on sorghum/millet grains) or liquid fermentation (on semi-synthetic medium like potato dextrose medium, PDA). Many commercial companies also grow *Trichoderma* on standing broth medium, without shaking. At the end of the cultivation process, the biomass (comprising of mycelia and conidia, or mycelia and/or chlamydospores in a fermenter) is mixed with a neutral carrier like talcum powder. A sticker like carboxy -methyl cellulose and some detergent is used as a spreader/dispersing agent. The formulation thus developed is recommended as seed dressing, drenching, foliar spray, dip or application along with compost. For best results, a combination of seed dressing and soil application with pre-colonized compost is recommended (Zaidi *et al.*, 2013).

## Biofertilizers

Many *Trichoderma* strains are reported to have plant-growth promoting activities (Contreas-Cornejo *et al.*, 2013). Many different mechanisms, based on incidental observations were proposed for such effects on plants. Suppression of minor pathogens, increased solubilization of micronutrients, enhanced nutrient use efficiency, especially of nitrogen, enhanced photosynthesis, and release of plant hormones are some examples (Harman 2006, Harman *et al.*, 2004a). Many *Trichoderma* strains are known to promote root growth (Kleifeld and Chet 1992; Windham *et al.*, 1986; Harman 2000; Harman *et al.*, 2004; Contreas-Cornejo *et al.*, 2014; Martinez-Medina *et al.*, 2014; Reglinski *et al.*, 2012; Masunaka *et al.*, 2011). However, first conclusive evidence on the production of auxins came from the studies of Contreas-Cornejo *et al.* (2009). They demonstrated that *T. atroviride* and *T. virens* produce auxins and promote root growth in Arabidopsis. Subsequently, many *Trichoderma* spp. have been demonstrated to produce auxins that are responsible for plant growth promotion (*e.g.*, Kotasthane *et al.*, 2014). Some *Trichoderma* secondary metabolites like harzianic acid also promote plant growth (Vinale *et al.*, 2013). It is interesting that among the 200 *Trichoderma* spp. classified so far, the members belonging to "harzianum" group are known to promote plant growth.

## Bioherbicides

Though not popular and perhaps not economically feasible yet, there have been attempts to use some *Trichoderma* strains as bioherbicides. Notable among them is *Trichoderma virens* that produces a secondary metabolite viridiol (reduced form of antifungal compound viridin). *T. virens* grown on chicken manure produced plenty of viridiol and suppressed weeds (Hutchinson 1999; Heraux *et al.*, 2005a,b). Several secondary metabolites produced by *Trichoderma* spp. (like 6-pentyl pyrone) has plant growth regulator-like activities (Vinale *et al.*, 2012) and hence has a potential to be deployed as biological suppressor of weeds.

## Induced Resistance

After the landmark discovery that *Trichoderma* spp. internally colonize roots and induce immunity in the plants, this area has received considerable attention culminating in an in-depth understanding of how these fungi interact with plants (Yedidia *et al.*, 1999; Viterbo *et al.*, 2005). There are several reports on the penetration of plant roots by *Trichoderma* spp. and the transfer of sucrose from plants to *Trichoderma* has been reported (Vargas *et al.*, 2009, 2011). It was proposed that in this type of mutualistic interaction, *Trichoderma* obtains sucrose from the roots while, in turn, promotes photosynthesis in plants and also boosts immunity against invading root- and foliar- pathogens. An abundantly secreted elicitor protein, named Sm1 was purified and demonstrated to induce defense responses in plants (Djonovic *et al.*, 2006). In a subsequent study, deletion of the corresponding gene resulted in reduced ability of *T. virens* to induce resistance in maize against the foliar pathogen *Colletotrichum* sp. (Djonovic *et al.*, 2007; Vargas *et al.*, 2008). This protein was also expressed in yeast and the recombinant protein was active as an elicitor (Buensentai *et al.*, 2010). Several Sm1-like small secreted cysteine-rich proteins were subsequently identified in the genome of *Trichoderma* and an investigation into their biological activities is awaited (Horwitz *et al.*, 2013; Pazzagli *et al.*, 2014). Apart from proteins, some secondary metabolites are also reported to be involved in induced resistance. The trichovorin-type 18-residue peptaibiotics of *Trichoderma virens* are involved in induced resistance in cucumber against a bacterial pathogen (Viterbo *et al.*, 2007). Mukherjee *et al.* (2012b) by gene deletion, identified a gene cluster involved in biosynthesis of an unknown PKS/NRPS, to be involved in induced resistance in maize. The beauty of induced resistance is that this works at a location away from application site and is effective against bacterial, fungal and even viral pathogens (Shoresh *et al.*, 2010). There are reports on the reduction of insect pest infestation upon seed inoculation with endophytic *Trichoderma* (Muvea *et al.*, 2014). On the other hand, induced resistance alone may not be very effective, and there might be a need to combine with other measures of crop protection.

## Abiotic Stress Management

*Trichoderma* spp., having positive effects on growth and development of plants also helps plants in combating abiotic stresses (Harman 2000, 2006; Harman *et al.*, 2004). The most studied of these is drought or moisture stress tolerance. Many strains of *Trichoderma* promote root growth. This in turn helps in better water harvesting and hence delays the onset of drought stress syndromes. *Trichoderma* spp. have also been reported to help plants overcome physiological stresses, including breaking of seed dormancy and improved germination of ageing seeds (Matsouri *et al.*, 2010, 2012). Inoculation with *Trichoderma* also alleviates salinity stress in plants (Rawat *et al.*, 2011, 2012, 2013; Contreas-Cornejo *et al.*, 2014; Qi and Zhao 2013). Apart from the soil- and root-residing *Trichoderma*, many strains and new species have been reported to be typical endophytes (Bailey and Melnick 2013). Inoculation with these endophytic

*Trichoderma* helps in amelioration of biotic as well as abiotic stresses (Bailey *et al.*, 2008; Bae *et al.*, 2009; Hanada *et al.*, 2008; Bae *et al.*, 2011; Druzhinina *et al.*, 2011).

## Sources of Transgenes

After the pioneering work of Lorito *et al.* (1998), who expressed a *Trichoderma* endochitinase gene in plants to combat infection caused by fungi, there have been several successful attempts to introduce *Trichoderma* genes in plants for enhancing tolerance to biotic and abiotic stresses (reviewed by Nicolas *et al.*, 2014). Most popular candidates for improving biotic stresses have been the fungal cell-wall degrading enzymes chitinases and glucanases alone or in various combinations (Nicolas *et al.*, 2014). Several *Trichoderma* genes have also been transferred to transgenic plants for imparting tolerance to abiotic stresses. For example, a heat shock protein from *T. harzianum* imparted tolerance to high temperature, salinity, oxidative and osmotic stresses in Arabidopsis (Montero-Barrientos *et al.*, 2010). A *T. harzianum* kelch repeat protein also imparted tolerance to salinity and osmotic stresses in Arabidopsis (Hermosa *et al.*, 2011). Dixit *et al.* (2011a,b) transferred a GST gene from *T. virens* in tobacco. The plants showed enhanced tolerance to heavy metal cadmium and degraded the polyaromatic hydrocarbon anthracene to non-toxic naphthalenes.

## Industrial Applications

*Trichoderma reesei* is one species that is leading the industrial production of cellulases and hemicellulases. The wild type was isolated originally from the Solomon Islands during World War II, where it destroyed army canvas and other cellulosic materials. The original strain QM6a underwent several round of genetic manipulations and a mutant strain RutC 30 is currently the major source of industrial cellulases (Le Crom *et al.*, 2009). The interest in *Trichoderma* as sources of cellulases got a major boost with the advent of the biofuels industry (Martinez *et al.*, 2008). Apart from cellulases and hemicellulases, *Trichoderma* spp. are also used as a cell factory for production of value added compounds and novel pharmaceuticals (Steiger *et al.*, 2011, Gorshe *et al.*, 2013; Jovanovic *et al.*, 2013). Some strains of *Trichoderma* are also used in food and feed industry, *e.g.*, a food flavoring agent 6-pentyl pyrone with coconut aroma as a food additive (Fadel *et al.*, 2014). Some *Trichoderma* strains are also being used for synthesis of metal nanoparticles with medical applications (Vahabi *et al.*, 2011; Mishra *et al.*, 2014; Ahluwalia *et al.*, 2014; Sundaravadivelan *et al.*, 2014; Tripathi *et al.*, 2013; Kethirabalan and Annadurai 2013; Mukherjee *et al.*, 2008; Vivek *et al.*, 2014). The advantage of using *Trichoderma* for the production of enzymes or other value-added products is that the biomass, which is a byproduct, can be used as biofungicides. This of course requires strains that can have dual applications in industry as well as biofungicides.

## Sources of Bioactive Compounds/Secondary Metabolites

*Trichoderma* spp. are prolific producers of secondary metabolites, with more than a hundred already reported and chemically characterized (Reino *et al.*, 2008; Sivasithamparam and Ghisalberti 1998). *Trichoderma* secondary metabolites can be grouped into non-ribosomal peptides (NRPs), polyketides, terpenes, sterols and hybrid compounds (Ruiz *et al.*, 2013; Zeilinger and Schuhmacher 2013). The NRPs can further be sub-grouped into high molecular weight peptaibols and low molecular weight compounds. The peptaibols are linear peptides of approximately 6-20 amino acids (both proteinogenic and non-proteinogenic), and act on the membrane by forming voltage-gated pores, thus have immense potential as new class of antibiotics *i.e.*, peptaibiotics (Strieker *et al.*, 2010). These peptaibiotics have immense therapeutic potential both in humans and plants, being antifungal, antibacterial and even anti-viral (Mukherjee *et al.*, 2012b). Some peptaibols like alamethicin and trichovorins are also inducers of systemic resistance in plants (Viterbo *et al.*, 2007, Leitgeb *et al.*, 2007). Sometimes, a single peptaibol synthetase can be responsible for the production of more than one class of peptaibols (Mukherjee *et al.*, 2011). Among the low molecular weight NRPs of *Trichoderma* origin, the most famous is gliotoxin (Mukherjee *et al.*, 2012b). This is the oldest antibiotic to have been discovered in *Trichoderma*, and the most extensively studied for antibiotic properties against plant pathogens (Weindling and Emerson 1936; Howell 2003). This is produced only by *T. virens*, but not all *T. virens* strains produce gliotoxin. Only the "Q" strains produce gliotoxin and the "P" strains produce another NRP gliovirin (Howell *et al.*, 1993). Gliotoxin is involved in biological control of *Pythium* sp., but not of *R. solani*, even though gliotoxin inhibits the growth of *Rhizoctonia* (Vargas *et al.*, 2014). Gliotoxin was also explored as an anti-tumor agent, but the toxicity on normal cells limits its application in therapy of human diseases. Moreover, gliotoxin is a virulence factor of *Aspergillus fumigatus*, a human pathogen. The siderophores are essential for survival of microbes in natural habitat as it is required to acquire iron and also intracellular storage. Siderophores are also a group of NRPs, and there are a few reports on the role of these compounds in survival and biocontrol properties of *Trichoderma* (Segarra *et al.*, 2010). Polyketides, the products of polyketide synthases, are abundantly produced, but only one of them has been characterized in detail in *Trichoderma*, the pigment polyketide. Using a gene knockout approach, Atanasova *et al.* (2013b) studied the role of the pigment polyketide in *T. reesei*. The pigment polyketide is essential for the conidial green pigmentation, and is involved in cell wall stability and antagonistic properties (mycoparasitism and antibiosis). The deletion of the corresponding gene made the conidia more sensitive to UV light. A hitherto unidentified polyketide/NRP hybrid metabolite is involved in induced resistance (Mukherjee *et al.*, 2012a). Another extensively studied *Trichoderma* secondary metabolite is viridin (and its hydroxyl-derivative viridiol) which is a strong fungistatic compound (Brian and McGowan 1945; Hanson 1995). Viridin is a steroid with anticancer properties too, and has been extensively studied as a therapeutic agent (Anderson *et al.*, 2004; Viswanathan *et al.*, 2012). However, non-target effects on normal cells limited its elevation as a drug in cancer therapy. Nevertheless, interest in this compound is renewed with the more precise drug delivery systems that are being developed. Terpenes are products of the terpenoid biosynthesis machinery and are most abundant of all the metabolites. These are derivatives of isoprene subunits in various combinations and synthesized by the terpene cyclases. After cyclization, the terpenes may undergo several modifications to yield novel compounds. In *T. virens*, a terpene cyclase has been identified that is part of a gene cluster and is responsible for the production of more than 20 volatile sesquiterpenes (Crutcher *et al.*, 2013). Some such compounds have the potential to be developed as biofuels.

## Impact on Human/Animal Health

*Trichoderma* spp. are generally regarded as harmless. However, there are certain species (hence the need for accuracy in identification) that can cause adverse effects on human health (Hatvani *et al.*, 2013). Two species namely *T. longibrachiatum* and *T. citrinoviride* are widely isolated from infected humans, especially immune-compromised patients (Sandoval-Denis 2014). Fortunately, most of the opportunistic human pathogenic *Trichoderma* belong to *Longibrachiatum* group and grow at 37 °C. Hence, the agricultural use of *Trichoderma* that grow at 37 °C should be discouraged. Similarly, there are some species (especially *T. aggressivum* and *T. pleuroticola*) that can infect mushrooms and can cause immense damage (O'Brien *et al.*, 2014; Sobieralski *et al.*, 2010, 2012).

## Conclusion

*Trichoderma* spp. are a group of fungi with immense impact on human welfare, mostly by virtue of its multifarious applications in agriculture and industry. More than 200 species are well defined (Atanosova *et al.*, 2013a). Interestingly, only about 10 per cent of these came from soil, majority being isolated from wild mushrooms/ tree bark. Those isolated from soil have been exploited in agriculture/biotechnology, although most of the remaining 90 per cent have not left the taxonomist's desk. With several hundred more species awaiting discovery, it is very clear that the potential of majority of the species/strains have not been tapped yet. Thus, the potential exploitation of natural *Trichoderma* strains is tremendous, and several effective strains await discovery and applications. This, along with the availability of modern genetic tools makes the utility of these fungi limited only by our imagination. The genome sequences of seven species are publicly available and analyses of some of these genomes have already thrown several surprises. For example, before the genome sequences were published, we knew of only two elicitor-like proteins from *Trichoderma*. An analysis of *Trichoderma* genomes yielded an inventory of a few hundred such elicitor proteins. These newly discovered proteins can be expressed in a suitable host like *E. coli* or *Pichia*, and an effective cocktail of elicitor proteins could be developed for inducing resistance in plants. Similarly, several hydrolases have been identified in *Trichoderma* genomes which could be utilized in biofuels industry. Another outcome of genome sequences is the identification of several novel gene clusters related to secondary metabolism. Secondary metabolism most often is produced by several genes present as a cluster in the genome. Interestingly, many of these gene clusters are not expressed under standard laboratory cultivation conditions. With the development of new molecular tools, it possible to "awaken" such "sleeping" clusters leading to the discovery of novel metabolites (Brakhage and Schroeckh 2011). Natural exploration combined with genetic enhancements is all set to revolutionize the future of *Trichoderma* research and its application in human welfare.

## References

1. Ahluwalia V, Walia S, Sati OP, Kumar J, Kundu A, Shankar J and Paul YS. 2014. Isolation, characterization of major secondary metabolites of *Trichoderma koningii* and their antifungal activity. *Archives of Phytopathology and Plant Protection* 47: 1063–1071.

2. Anderson, EA., Alexanian,EJ and Sorensen EJ. 2004. Synthesis of the furanosteroidal antibiotic viridin. *Angewandte Chemie International Edition* 43: 1998-2001.

3. Atanasova L, Druzhinina IS and Jaklitsch WM. 2013a. Two hundred *Trichoderma* species recognized on the basis of molecular phylogeny. In: Mukherjee PK, Horwitz BA, Singh US, Mukherjee M and Schmoll M. (Eds.). *Trichoderma: Biology and Applications*. CABI, UK. 10-42.

4. Atanasova L, Knox BP, Kubicek CP, Druzhinina IS and Baker SE. 2013b. The polyketide synthase gene Pks4 of *Trichoderma reesei* provides pigmentation and stress resistance. *Eukaryotic Cell* 12: 1499-1508.

5. Bae H, Roberts DP, Lim HS, Strem MD, Park SC, Ryu CM, Melnick RL and Bailey BA. 2011. Entophytic *Trichoderma* isolates from tropical environments delay disease onset and induce resistance against *Phytophthora capsici* in hot pepper using multiple mechanisms. *Molecular Plant-Microbe Interactions* 24: 336-351.

6. Bae H, Sicher RC, Kim MS, Kim SH, Strem MD, Melnick RL and Bailey BA. 2009. The beneficial endophyte *Trichoderma hamatum* isolate DIS 219b promotes growth and delays the onset of the drought response in *Theobroma cacao*. *Journal of Experimental Botany* 60: 3279-3295.

7. Bailey BA, Melnick RL. 2013. The endophytic *Trichoderma*. In: Mukherjee PK, Horwitz BA, Singh US, Mukherjee M and Schmoll M. (Eds.). *Trichoderma: Biology and Applications*. CABI, UK. 152-172.

8. Bailey BA, Bae H, Strem MD, Crozier J, Thomas SE, Samuels GJ, Vinyard BT and Holmes KA.2008. Antibiosis, mycoparasitism, and colonization success for endophytic *Trichoderma* isolates with biological control potential in *Theobroma cacao*. *Biological Control* 46: 24-35.

9. Brakhage AA and Schroeckh V. 2011. Fungal secondary metabolites - strategies to activate silent gene clusters. *Fungal Genetics and Biology* 48: 15-22.

10. Brian PW and McGowan JC. 1945. Viridin: a highly fungistatic substance produced by *Trichoderma viride*. *Nature* 156: 144-145.

11. Buensanteai N, Mukherjee PK, Horwitz BA, Cheng C, Dangott LJ and Kenerley CM. 2010. Expression and purification of biologically active *Trichoderma* virens proteinaceous elicitor Sm1 in *Pichia pastoris*. *Protein Expression and Purification* 72: 131-138

12. Contreras-Cornejo HA, Macias-Rodriguez L, Cortes-Penagos C and Lopez-Bucio J. 2009. *Trichoderma virens*, a plant beneficial fungus, enhances biomass production and promotes lateral root growth through an auxin-dependent mechanism in Arabidopsis. *Plant Physiology* 149: 1579-1592

13. Contreras-Cornejo HA, Ortiz-Castro R, López-Bucio J. 2013. Promotion of plant growth and the induction of systemic defence by *Trichoderma*: physiology, genetics and gene expression. In: Mukherjee PK, Horwitz BA, Singh US, Mukherjee M, Schmoll M. (Eds.) *Trichoderma: Biology and Applications*. CABI, UK.173-194.

14. Contreras-Cornejo HA, Macías-Rodríguez L, Alfaro-Cuevas R, Lopez-Bucio J. 2014. *Trichoderma* spp. improve growth of Arabidopsis seedlings under salt stress through enhanced root development, osmolite production and Na+ elimination through root exudates. *Molecular Plant-Microbe Interactions* 27: 503-514.

15. Crutcher F K., Alexandra P, Schuhmacher R, Mukherjee PK, Zeilinger S and Kenerley CM. 2013. A putative terpene cyclase vir4 is responsible for the biosynthesis of volatile terpene compounds in the biocontrol fungus *Trichoderma virens*. *Fungal Genetics and Biology* 56: 67-77.

16. Dixit P, Mukherjee PK, Ramachandran V and Eapen S. 2011a. Glutathione transferase from *Trichoderma virens* enhances cadmium tolerance without enhancing its accumulation in transgenic *Nicotiana tabacum*. *PLoS One* 6: e16360.

17. Dixit P, Mukherjee PK, Sherkhane PD, Kale SP and Eapen S. 2011b. Enhanced tolerance and remediation of anthracene by transgenic tobacco plants expressing a fungal glutathione transferase gene. *Journal of Hazardous Materials* 192: 270-276.

18. Djonovic S, Pozo MJ, Dangott LJ, Howell CR and Kenerley CM. 2006. Sm1, a proteinaceous elicitor secreted by the biocontrol fungus *Trichoderma virens* induces plant defense responses and systemic resistance. *Molecular Plant-Microbe Interactions* 19: 838-853.

19. Djonovic S, Vargas WA, Kolomiets MV, Horndeski M, Wiest A and Kenerley CM. 2007. A proteinaceous elicitor Sm1 from the beneficial fungus *Trichoderma virens* is required for induced systemic resistance in maize. *Plant Physiology* 145: 875-889.

20. Druzhinina IS, Seidl-Seiboth V, Herrera-Estrella A, Horwitz BA, Kenerley CM,, Monte, E.,Mukherjee PK *et al.*, 2011. *Trichoderma*: the genomics of opportunistic success. *Nature Reviews Microbiology* 9: 749-759.

21. Fadel HHM, Mahmoud MG, Asker MMS and Lotfy SN. 2014. Characterization and evaluation of coconut aroma produced by *Trichoderma viride* EMCC-107 in solid state fermentation on sugarcane bagasse. *Electronic Journal of Biotechnology*: In press.

22. Gorsche R. Mach-Aigner AR, Mach RL. 2013. *Trichoderma* as cell factories. In: Mukherjee PK, Horwitz BA, Singh US, Mukherjee M and Schmoll M (Eds.) *Trichoderma: Biology and Applications*. CABI, UK. 280-291.

23. Hanada, Rogério E, Jorge Souza TD, Pomella AWV, Hebbar KP, O. Pereira JO, Ismaiel A, and Samuels GJ. 2008. *Trichoderma* martial sp. nov., a new endophyte from sapwood of *Theobroma cacao* with a potential for biological control. *Mycological Research*. 1121: 1335-1343.

24. Hanson JR (1995). The viridin family of steroidal antibiotics. *Natural Product Reports* 12: 381–384.

25. Harman GE, Howell CR, Viterbo A, Chet I, Lorito M. 2004a. *Trichoderma* species-opportunistic, avirulent plant symbionts. *Nature Reviews Microbiology* 2: 43-56.

26. Harman GE, Petzoldt R, Comis A and Chen J. 2004b. Interactions between *Trichoderma harzianum* strain T22 and maize inbred line Mo17 and effects of these interactions on diseases caused by *Pythium ultimum* and *Colletotrichum graminicola*. *Phytopathology* 94 (2): 147-153.

27. Harman GE. 2000. Myths and dogmas of biocontrol. Changes in perceptions derived from research on *Trichoderma harzianum* T-22. *Plant Disease* 84: 377-393.

28. Harman GE. 2006. Overview of Mechanisms and Uses of *Trichoderma* spp. *Phytopathology* 96: 190-194.

29. Hatvani L, Manczinger L, Vágvölgyi C, Kredics L. 2013. *Trichoderma* as a human pathogen. In: Mukherjee PK, Horwitz BA, Singh US, Mukherjee M and Schmoll M (Eds.). *Trichoderma: Biology and Applications*. CABI, UK. 292-313.

30. Heraux, Farah MG, Steven GH, and Stephen CW. 2005a. Combining *Trichoderma virens* inoculated compost and a rye cover crop for weed control in transplanted vegetables. *Biological Control* 34: 21-26.

31. Heraux, Farah MG, Steven GH, K. GR, and Stephen C W. 2005b. Composted chicken manure as a medium for the production and delivery of *Trichoderma virens* for weed control. *Horticulture Science* 40: 1394-1397.

32. Hermosa R, Botella L, Keck E, Jimenez JA, Montero-Barrientos M, *et al.*, 2011. The overexpression in *Arabidopsis thaliana* of a *Trichoderma* harzianum gene that modulates glucosidase activity, and enhances tolerance to salt and osmotic stresses. *Journal of Plant Physiology* 168: 1295-1302.

33. Horwitz BA, Kosti I, Glaser F, Mukherjee M. 2013. *Trichoderma* genomes: a vast reservoir of potential elicitor proteins In: Mukherjee PK, Horwitz BA, Singh US, Mukherjee M, Schmoll M. (Eds.) *Trichoderma: Biology and Applications*. CABI, UK.195-208.

34. Howell CR, Stipanovic R and Lumsden R. 1993. Antibiotic production by strains of *Gliocladium virens* and its relation to biocontrol of cotton seedling diseases. *Biocontrol Science and Technology* 3: 435–441.

35. Howell CR. 2003. Mechanisms employed by *Trichoderma* species in the biological control of plant diseases: the history and evolution of current concepts. *Plant Disease* 87 (1): 4-10.

36. Hutchinson CM. 1999 *Trichoderma virens*-inoculated composted chicken manure for biological weed control. *Biological Control* 16: 217-222.

37. Jovanovic B, Mach R and Mach-Aigner A (2013) Characterization of erythrose reductases from filamentous fungi. *AMB Express* 3(1): 43.

38. Kethirabalan C and Annadurai G. 2013. Bioengineered silver nanobowls using *Trichoderma viride* and its antibacterial activity against gram-positive and gram-negative bacteria. *Journal of Nanostructure in Chemistry* 3: 1-7.

39. Kleifeld O and Chet I. 1992.*Trichoderma harzianum*: interaction with plants and effect on growth response. *Plant and Soil* 144(2): 267-272.

40. Kotasthane A, Agrawal T, Kushwah R, Rahatkar OV.2014. In-vitro antagonism of *Trichoderma* spp. against *Sclerotium rolfsii* and *Rhizoctonia solani* and their response towards growth of cucumber, bottle gourd and bitter gourd. *European Journal of Plant Pathology*: In press.

41. Le Crom S, Schackwitz W, Pennacchio L, Magnuson JK, Culley DE, Collett JR, Martin J, Druzhinina IS, Mathis H, Monot F, Seiboth B,Cherry B, Rey M, Berka R, Kubicek CP, Baker SE, Margeot A. 2009. Tracking the roots of cellulase hyperproduction by the fungus *Trichoderma reesei* using massively parallel DNA sequencing. *Proceedings of the National Academy of Sciences, USA* 106: 16151-16156.

42. Leitgeb B, Szekeres A, Manezinger L, Vagvolgyi C and Kredics L. 2007. The history of alamethicin: a review of the most extensively studied peptaibol. *Chemistry and Biodiversity* 4: 1027–1051.

43. Lorito M, Woo SL, Fernandez IG, Colucci G, Harman GE, Pintor-Toro JA, Filippone E, Muccifora S, Lawrence CB, Zonia A, Tuzun S and Scala F. 1998. Genes from mycoparasitic fungi as a source for improving plant resistance to fungal pathogens. *Proceedings of the National Academy of Sciences, USA* 95: 7860-7865.

44. Martinez D, Berka RM, Henrissat B, Saloheimo M, Arvas M, Baker SE, Chapman J, Chertkov O, Coutinho PM, Cullen D, Danchin EG, Grigoriev IV *et al.*, 2008. Genome sequencing and analysis of the biomass-degrading fungus *Trichoderma reesei* (syn. *Hypocrea jecorina*). *Nature Biotechnology* 26: 553-560.

45. Martínez-Medina A, Del Mar Alguacil M, Pascual JA, Van Wees SC. 2014. Phytohormone profiles induced by *Trichoderma* isolates correspond with their biocontrol and plant growth-promoting activity on melon plants. *Journal of Chemical Ecology* 40: 804-815.

46. Mastouri F, Bjorkman T, Harman GE. 2012. *Trichoderma harzianum* enhances antioxidant defense of tomato seedlings and resistance to water deficit. *Molecular Plant Microbe Interactions* 25: 1264-1271.

47. Mastouri, F, Thomas B, and Harman GE. 2010. Seed treatment with *Trichoderma harzianum* alleviates biotic, abiotic and physiological stresses in germinating seeds and seedlings. *Phytopathology* 100: 1213-1221.

48. Masunaka A, Hyakumachi M, Takenaka S. 2011. Plant growth-promoting fungus, *Trichoderma koningii* suppresses isoflavonoid phytoalexin vestitol production for colonization on/in the roots of *Lotus japonicus*. *Microbes and Environments* 26: 128-134.

49. Mishra A, Kumari M, Pandey S, Chaudhry V, Gupta KC and Nautiyal CS. 2014. Biocatalytic and antimicrobial activities of gold nanoparticles synthesized by *Trichoderma* sp. *Bioresource Technology*.166: 235-242

50. Montero-Barrientos M, Hermosa R, Cardoza RE, Gutierrez S, Nicolas C, Monte E. 2010. Transgenic expression of the *Trichoderma harzianum* hsp70 gene increases Arabidopsis resistance to heat and other abiotic stresses. *Journal of Plant Physiology* 167: 659-665.

51. Mukherjee P K, Wiest A, Ruiz N, Keightley A, Moran-Diez ME, McCluskey K, Pouchus YF and Kenerley CM. 2011. Two classes of new peptaibols are synthesized by a single non-ribosomal peptide synthetase of *Trichoderma virens*. *Journal of Biological Chemistry* 286: 4544-4554.

52. Mukherjee P, Roy M, Mandal BP, Dey GK, Mukherjee PK, Ghatak J, Tyagi AK, and Kale SP. 2008. Green synthesis of highly stabilized nanocrystalline silver particles by a non-pathogenic and agriculturally important fungus *T. asperellum*. *Nanotechnology* 19 (7): 075-103.

53. Mukherjee PK, Buensanteai N, Moran-Diez ME, Druzhinina IS, Kenerley CM. 2012b. Functional analysis of non-ribosomal peptide synthetases (NRPSs) in *Trichoderma virens* reveals a polyketide synthase (PKS)/NRPS hybrid enzyme involved in the induced systemic resistance response in maize. *Microbiology* 158: 155-65.

54. Mukherjee PK, Horwitz BA, Singh US, Mukherjee M, Schmoll M. 2013a. *Trichoderma* in agriculture, industry and medicine: an overview. In: Mukherjee PK, Horwitz BA, Singh US, Mukherjee M, Schmoll M. (Eds.) *Trichoderma: Biology and Applications*. CABI, UK.1-9.

55. Mukherjee PK, Horwitz BA, Herrera-Estrella A, Schmoll M and Kenerley CM. 2013b. *Trichoderma* research in the genome era. *Annual Review of Phytopathology* 51: 105-129.

56. Mukherjee PK, Horwitz BA and Kenerley CM. 2012a. Secondary metabolism in *Trichoderma*-a genomic perspective. *Microbiology* 158: 35-45

57. Muvea AM, Meyhöfer R, Subramanian S, Poehling HM, Ekesi S and Maniania NK 2014. Colonization of Onions by Endophytic Fungi and Their Impacts on the Biology of *Thrips tabaci*. *PLoS One* 25;9(9): e108242

58. Nicolas C, Hermosa R, Rubio B, Mukherjee PK and Monte E. 2014. *Trichoderma* genes in plants for stress tolerance- status and prospects. *Plant Science* 228: 71-78.

59. O'Brien M, Grogan H, and Kavanagh K. 2014. Proteomic response of *Trichoderma* aggressivum europaeum Agaricus bisporus tissue and mushroom compost. *Fungal Biology* 118: 785-791.

60. Pazzagli L, Seidl-Seiboth V, Barsottini M, Vargas WA, Scala A and Mukherjee PK. 2014. Cerato-platanins: elicitors and effectors. *Plant Science* 228: 79-87.

61. Qi W and Zhao L. 2013. Study of the siderophore-producing *Trichoderma asperellum* Q1 on cucumber growth promotion under salt stress. *Journal of Basic Microbiology* 53: 355-364.

62. Rawat L, Singh Y, Shukla N and Kumar J. 2012. Seed biopriming with salinity tolerant isolates of *Trichoderma harzianum* alleviates salt stress in rice: growth, physiological and biochemical characteristics. *Journal of Plant Pathology* 94: 353-365.

63. Rawat L, Singh Y, Shukla N and Kumar J. 2013. Salinity tolerant *Trichoderma harzianum* reinforces NaCl tolerance and reduces population dynamics of *Fusarium oxysporum* f. sp. ciceri in chickpea (*Cicer arietinum* L.) under salt stress conditions. *Archives of Phytopathology and Plant Protection* 46: 1442-1467.

64. Rawat L, Singh Y, Shukla N, and Kumar J. 2011. Alleviation of the adverse effects of salinity stress in wheat (*Triticum aestivum L.*) by seed biopriming with salinity tolerant isolates of *Trichoderma harzianum*. *Plant and soil* 347: 387-400.

65. Reglinski T, Rodenburg N, Taylor JT, Northcott GL, Chee AA, Spiers TM, and Hill RA. 2012. *Trichoderma atroviride* promotes growth and enhances systemic resistance to *Diplodia pinea* in radiata pine (*Pinus radiata*) seedlings. *Forest Pathology* 42: 75-78.

66. Reino JL, Guerrero RF, Hernández-Galan R and Collado IG. 2008. Secondary metabolites from species of the biocontrol agent *Trichoderma*. *Phytochemistry Review* 7: 89–123.

67. Ruiz N, Roullier C, Petit K, Sallenave-Namont C, Grovel O and Pouchus YF. 2013. Marine-derived Trichoderma: a source of new bioactive metabolites. In: Mukherjee PK, Horwitz B A, Singh US, Mukherjee M and Schmoll M (Eds.) *Trichoderma: Biology and Applications*. CABI, UK. 247-279.

68. Sandoval-Denis M. 2014. Phylogeny of the clinically relevant species of the emerging fungus *Trichoderma* and their antifungal susceptibilities. *Journal of Clinical Microbiology* 52: 2112-25.

69. Segarra G, Casanova E, Aviles M and Trillas I. 2010. *Trichoderma asperellum* strain T34 controls *Fusarium* wilt disease in tomato plants in soilless culture through competition for iron. *Microbial Ecology* 59: 141-149.

70. Shoresh M, Harman GE and Mastouri F. 2010. Induced systemic resistance and plant responses to fungal biocontrol agents. *Annual Review of Phytopathology* 48: 21-43.

71. Singh HB, Singh BN, Singh SP and Sharma BK. 2012. Exploring different avenues of *Trichoderma* as a potent bio-fungicidal and plant growth promoting candidate-an overview. *Review of Plant Pathology* 2012: 315-426

72. Sivasithamparam K and Ghisalberti E. 1998. Secondary metabolism in *Trichoderma* and *Gliocladium*. In: Kubicek CP and Harman GE. (Eds.) *Trichoderma and Gliocladium: Basic Biology, Taxonomy and Genetics*. London: Taylor and Francis. 139-191.

73. Sobieralski K, Siwulski M, Komon-Zelazowska M and B³aszczyk ML. 2012. Evaluation of the growth of *Trichoderma pleurotum* and *Trichoderma pleuroticola* isolates and their biotic interaction with *Pleurotus* sp. *Journal of Plant Protection Research* 52: 235-239.

74. Sobieralski K, Siwulski M, Gorski R, Fruzynska-Jozwiak D and Nowak-Sowiñska M. (2010). Impact of *Trichoderma aggressivum* f. europaeum isolates on yielding and morphological featurus of *Pleurotus eryngii*. *Phytopathologia* 56: 17–25.

75. Steiger MG, Vitikainen M, Uskonen P, Brunner K, Adam G, Pakula T, Penttilä M, Saloheimo M, Mach RL and Mach-Aigner AR. 2011. Transformation system for *Hypocera jecorina* (*Trichoderma reesei*) that favours homologous integration and employs reusable bidirectionally selectable markers. *Applied and Environmental Microbiology* 77: 114–121.

76. Strieker M, Tanovic A and Marahiel MA. 2010. Nonribosomal peptide synthetases: structures and dynamics. *Current Opinion in Structure Biology* 20: 234-240.

77. Sundaravadivelan C and Padmanabhan MN. 2014. Effect of mycosynthesized silver nanoparticles from filtrate of *Trichoderma harzianum* against larvae and pupa of dengue vector *Aedes aegypti* L. *Environmental Science and Pollution Research International* 21: 4624-4633.

78. Tripathi RM, Gupta RK, Shrivastav A, Singh MP, Shrivastav BR and Singh P. 2013.*Trichoderma koningii* assisted biogenic synthesis of silver nanoparticles and evaluation of their antibacterial activity. *Advances in Natural Sciences: Nanoscience and Nanotechnology* 4: 1-5.

79. Vahabi K. Mansoori GA, and Karimi S. 2011. Biosynthesis of silver nanoparticles by fungus *Trichoderma* reesei (a route for large-scale production of AgNPs). *Insciences Journal* 1(1): 65-79.

80. Vargas WA, Crutcher FK and Kenerley CM. 2011. Functional characterization of a plant-like sucrose transporter from the beneficial fungus *Trichoderma virens*. Regulation of the symbiotic association with plants by sucrose metabolism inside the fungal cells. *New Phytologist* 189: 777-789.

81. Vargas WA, Djonovic S, Sukno SA and Kenerley CM. 2008. Dimerization controls the activity of fungal elicitors that trigger systemic resistance in plants. *Journal of Biological Chemistry* 283: 19804-19815.

82. Vargas WA, Mandawe JC and Kenerley CM. 2009. Plant-derived sucrose is a key element in the symbiotic association between *Trichoderma virens* and maize plants. *Plant Physiology* 151: 792-808.

83. Vargas WA, Mukherjee PK, Laughlin D, Wiest A, Moran-Diez ME and Kenerley CM. 2014. Role of gliotoxin in the symbiotic and pathogenic interactions of *Trichoderma virens*. *Microbiology*, 160: 2319-2330.

84. Verma M, Brar S, Tyagi R, Surampalli R and Valero J. 2007. Antagonistic fungi, *Trichoderma* spp.: panoply of biological control. *Biochemical Engineering Journal* 37: 1-20.

85. Vinale F, Sivasithamparam K, Ghisalberti EL, Ruocco M, Woo SL and Lorito M. 2012. *Trichoderma* secondary metabolites that affect plant metabolism. *Natural Product Communication* 7: 1545-1550.

86. Vinale F, Nigro M, Sivasithamparam K, Flematti G, Ghisalberti EL, Ruocco M, Varlese R, Marra R, Lanzuise S, Eid A, Woo SL and Lorito M. 2013. Harzianic acid: a novel siderophore from *Trichoderma harzianum*. *FEMS Microbiology Letter* 347: 123-129.

87. Viswanathan K, Ononye SN, HCooper HD, Hadden MK, Anderson AC and Wright DL. 2012. Viridin analogs derived from steroidal building blocks. *Bioorganic and Medicinal Chemistry Letters* 22: 6919-6922.

88. Viterbo A, Harel M, Horwitz BA, Chet I and Mukherjee PK. 2005. *Trichoderma* mitogen-activated protein kinase signaling is involved in induction of plant systemic resistance. *Applied and Environmental Microbiology* 71: 6241-6246

89. Viterbo A and Horwitz BA. 2010. Mycoparasitism. In: Borkovich K, Ebbole DJ. (Eds) *Cellular and Molecular Biology of Filamentous Fungi*, Washington DC: ASM Press. 676-693.

90. Viterbo A, Wiest A, Brotman Y, Chet I and Kenerley C. 2007. The 18mer peptaibols from *Trichoderma virens* elicit plant defense responses. *Molecular Plant Pathology* 8: 737-46.

91. Vivek A, Kumar J, Sisodia R, Shakil NA, and Walia S. 2014. Green synthesis of silver nanoparticles by *Trichoderma harzianum* and their bio-efficacy evaluation against *Staphylococcus aureus* and *Klebsiella pneumonia*. *Industrial Crops and Products* 55: 202-206.

92. Weindling R and Emerson H. 1936. The isolation of toxic substance from the culture filtrates of *Trichoderma*. *Phytopathology* 26: 1068-1070.

93. Weindling R. 1932. *Trichoderma lignorum* as a parasite of other soil fungi. *Phytopathology* 22: 837-845.

94. Wells HD, Bell DK and Jaworski CA, 1972. Efficacy of *Trichoderma harzianum* as a biological control for *Sclerotium rolfsii*. *Phytopathology* 62: 442-447

95. Windham MT, Elad Y and Baker R.1986. A *mechanism* for increased plant growth induced by *Trichoderma* spp. Phytopathology 76: 518-521.

96. Yedidia I, Benhamou N and Chet I. 1999. Induction of defense responses in cucumber plants (*Cucumis sativus* L.) by the biocontrol agent *Trichoderma harzianum*. *Applied Environmental Microbiology* 65: 1061-1070

97. Zaidi NW and Singh US. 2013. *Trichoderma* in plant health management. In: Mukherjee PK, Horwitz B A, Singh US, Mukherjee M and Schmoll M (Eds.) *Trichoderma: Biology and Applications*. CABI, UK. 230-246.

98. Zeilinger S and Schuhmacher R. 2013. Volatile organic metabolites of *Trichoderma* spp.: biosynthesis, biology and analytics. In: Mukherjee PK, Horwitz BA, Singh US, Mukherjee M, and Schmoll M.(Eds). *Trichoderma: Biology and Applications*. CABI, UK. 110-127.

*Section IV*

# Microbial Technology

**2016, Biotechnology: Progress and Applications** *Pages* **225–247**
*Editors:* **Saif Hameed and Zeeshan Fatima**
*Published by:* **DAYA PUBLISHING HOUSE, NEW DELHI**

## *Chapter 10*

# Optimization of Bioprocess Parameters for Fermentation Process from Laboratory to Industry

***A.N. Pathak****

*Amity Institute of Biotechnology,*
*Amity University, Jaipur, Rajasthan, India*

## Introduction

Biotechnology is an offspring of the marriage between biological sciences such as biochemistry, microbiology, molecular biology and genetics with engineering sciences such as chemical engineering, public health engineering, environmental engineering and sanitation engineering. Biotechnology is the study of scientific and engineering principles which represents heterogeneous group of subjects to provide opportunity to analyze data for biological systems which help to optimize the production processes for obtaining best production, yield and productivity of the desired primary and or secondary metabolite(s). In order to explain the concept of biotechnology and bioengineering to a non-biologist as also to a non-engineer, is to compare the difference between biochemistry to chemistry, and the difference between biochemical engineering to chemical engineering.

* Corresponding Author: E-mail: anpathak@jpr.amity.edu, anpathak2004@gmail.com

## Biotechnology and Industry

India has identified setting up of industries using modern biology and biotechnology as a promising area for economic growth. The Department of Biotechnology was established in Feb., 1986 by Government of India to promote, inter alia, the development of biotechnology industries in the country.

Among the areas for industrial opportunities, the following thrust areas have been indentified which are gaining attention by Indian researches to develop effective production processes. Much more attention is being paid to the areas mentioned below:

## Thrust Areas in Industrial Biotechnology

**1. Human and Animal Health**

1.1 Diagnostics

1.2 Vaccines

1.3 Antibiotics

1.4 Bioactive Molecules

**2. Animal Husbandry**

2.1 Animal Birth Control Vaccines (TALSUR).

2.2 Embryo Transfer Technology for Cows and Buffaloes.

2.3 Artificial Insemination.

2.4 Preservation and supply of frozen embryos and gametes.

**3. Agriculture**

3.1 Large-scale production of bio-fertilizers.

3.2 Propagation of bamboo, cardamom, eucalyptus etc.

3.3 Hybrid high-yielding seeds.

3.4 Edible oil through oil-palm safflower, sunflower.

3.5 Floriculture by tissue culture techniques.

3.6 Bio-pesticides

**4. Aquaculture**

4.1 Prawn aquaculture.

4.2 Carp aquaculture.

**5. Horticulture**

5.1 Hybrid high-yielding seeds and plants for horticulture products like mango, litchi, grapes, oranges, citrus, jack-fruit, etc.

**6. Use of Biocides in Controlling Mosquitoes**

**7. Microbial Products**

7.1 Solvents-Ethanol, Acetone, Butanol, etc.

7.2 Enzymes-Amylases, Alkaline protease, penicillin-V/Acylase, Penicillin-G/Acylase.

7.3 Organic acids-citric acid and acetic acid.

7.4 Sorbitol.

7.5 Milk Products.

7.6 Microbial leaching of copper from low grade ore.

**8. Processed Fruit Juices and Protein Foods**

**9. Environment**

9.1 Industrial Effluent Treatment and Methane Gas Generation.

9.2 Tannery Waste Treatment and Chrome Recovery.

## Industrial Fermentations

Industrial fermentation deals with all forms of applied microbiology which have an economic aspect. This deals with those areas of applied microbiology on which in some manner, a monetary value can be placed where microbiology involves a fermentation product. In most cases the economic criterion applies to production of useful products through metabolic activities of the microbes. Thus the ability of microorganisms to convert into inexpensive raw material or substrates, to economically valuable and value added organic compounds is considered to be of great industrial importance. The ability of microorganisms to attack, degrade and utilize components of the medium and to accumulate fermentation products through the actions of microbial enzymes such as invertases, amylases, lipases and proteases are of importance. Thus, fermentation products may be components of microbial cells, the cells themselves, intracellular or extracellular, enzymes or biochemicals/chemicals which are produced or altered by the cells.

Biotechnology involves the use in industry of living organisms or their components such as enzymes. It includes the introduction of genetically engineered microorganisms into a variety of industrial processes which are central to the economic success of biotechnology. The fermentation process is essentially a biochemical process of converting raw material into end products which are produced by microorganisms through microbial enzymes.

The scope and efficiency of the fermentation process is widened and refined over the years. The difference between fermentation technology which deals with living organisms where they serve as miniature factories converting raw materials into enzymes, to that of enzyme technology is where by catalysts (enzymes) extracted from those living organisms are used to make value added metabolites known as final product.

Fermentation technologies are useful for production of proteins partly because these are the direct products of genes. But proteins (as enzymes) can also be used in thousands of additional conversions to produce practically most of organic chemicals and in many inorganic ones such as:

1. Carbohydrates, such as fructose sweeteners,
2. Lipids and vitamins such as A, E, and K,

3. Alcohols such as ethanol and butanol,
4. Organic compounds such as acetone.

Fermentation is not the only way to manufacture or isolate these products. Some are traditionally produced using other methods. If a change from one process to another is to occur, both economic and social pressures will help to determine whether an innovative approach will be used to produce a particular product.

The stimuli for change and innovation are as follows:

1. Abundance of potentially useful raw material
2. Scarcity of an established product
3. Discovery of new products
4. Environmental concerns
5. Scarcity of currently used raw material

Each of these factors has tended to accelerate the application of fermentation and are discussed below.

## Abundance of Potentially Useful Raw Materials

The use of raw materials can be the driving force in developing a process. When straight chain hydrocarbons (n-alkanes) were produced on a large scale as petroleum refinery by-products, fermentation processes developed to convert them to single cell proteins for use in animal feed.

## Scarcity of an Established Product

The new found potential for producing human hormones through fermentation technology is a major impetus to the industry today. Similarly many organic compounds obtained by other processes, like citric acid, which was directly extracted from citrus fruits, are now made by fermentation.

## Discovery of a New Product

The discovery that antibiotics were produced by microorganisms sparked searches for an entirely new group of products. Several thousand antibiotics have been produced to date which have clinical applications.

## Environmental Concerns

The problems of sewage treatment and the need for new sources of energy have triggered a search for methods to convert sewage and municipal wastes to methane, the principal compounds of natural gas. Because microorganisms play a major role in the natural cycling of organic compounds, fermentation has been one method used for conversion.

## Scarcity of Currently Used Raw Material

Since the earth's supplies of fossil fuels are rapidly dwindling, there is an intense interest in finding methods for converting other raw materials to fuel. Fermentation offers a major approach to such conversions.

Fermentation technologies can be effective in each of these situations because of their outstanding versatility and relative simplicity. The processes of fermentation are basically identical, no matter what organism is selected, what medium used, what product formed. The apparatus, with minor modifications can be used to produce a drug, an agricultural product, a chemical or an animal feed supplement.

## Biotechnology in Fermentation

Use of biotechnology in fermentation involves either usage of whole living cells in the process or extraction of key enzymes from the living cell and its usage in the process. Both these technologies are discussed in the following paragraphs.

## Living Cell Technology

Originally fermentation used some of the most primitive forms of plant life as cell factories. Bacteria were used to make yoghurt and antibiotics, yeasts to ferment wine, the filamentous fungi to produce organic acids. More recently, fermentation technology has begun to use cells derived from higher plants and animals under growth conditions known as cell or tissue culture. In all cases, large quantities of cell with uniform characteristics are grown under controlled conditions.

In this simplest form, fermentation consists of mixing a microorganism with a liquid broth and allowing the components to react. More sophisticated large scale processes require control of the entire environment so that fermentation proceeds efficiently.

## Types of Processes

Fermentation can be carried out as a batch or continuous process or variations of these processes.

### Batch Process

Batch process may be considered as a closed system containing a limited amount of medium, in which the inoculated culture passes through a number of phases. In batch culture, all the substrate is added at the beginning of the fermentation, whereas in fed batch process, the substrate is added in increments throughout the process. Fed batch cultures may be operated to reduce the viscosity of the medium to reduce the effect of the toxic medium constituents, or to extend the product formation stage of the process for as long as possible.

### Continuous Process

In continuous fermentation, an improvement on the batch process, fermentation goes on without interruption, with a constant input of raw materials and other nutrients. The most recent approaches used microorganisms that have been immobilized in supporting structure. As the solution containing the raw materials passes over the cells, the microorganisms process the material and release the products into the solution flowing out of the Fermenter. The continuous processes have lead to significant improvement in yields as compared to batch processes.

### Surface Process

This process, though ancient, is still practiced an industrial scale even today. In this process the sterile nutrient solution is filled in open trays and inoculated on the surface by being sprayed with suspension of spores.

The whole arrangement is placed in a room which can be kept aseptic. The supply of oxygen and the regulation of the temperature are carried out by aerating the room with sterile conditioned air. In this process, the organisms take up the necessary oxygen via diffusion but this is affected in an uncontrollable manner by the growth of microorganisms on the surface of the nutrient solution. Hence such fermentations are difficult to control and their course remains obscure.

**Submerged Process-Anaerobic**

This is one of the least expensive processes from the point of apparatus required. For this process type, any shape of Fermenter vessel can be used.

In the process, depending on the size of the end stage, it must be preceded by one or more preliminary stages for the multiplication of the microorganisms. The duration and size of such a preliminary cultivation must be determined for each microorganism and this varies between a few hours and several days.

The dissolution or suspension of the nutrients and auxiliary materials in water and the sterlization of the nutrient solution are carried out either in the fermentation vessel or in the preparation vessel. Once the process starts, it is not possible to intervene; but it is possible to carry out control measures and to make additions during the fermentation. The most important measure is the maintenance of optimum process temperature which can be done by passing heat or cooling media through outer jackets or built-in coils. The course of the fermentation in relation to these parameters can readily be followed by taking samples for analysis and by built-in measurement sensors.

**Submerged Process-Aerobic**

In this type of fermentation, several vessels of different sizes are connected in series in order to make it possible to grow the biomass necessary for inoculating the production stage. Because of the aeration with the sterile compressed air or mixture of gases that are always required and the resulting formation of foam during fermentation. Chemical or mechanical foam suppressing arrangement has to be provided. The economy of the process is considerably affected by the energy demand and by mass transfer.

In this process, almost every parameter can be modified, provided the biochemical course of the process is known.

The above methods mentioned belong to the batch process. In contrast to these are the continuous processes in which the microorganisms can be kept for a long time by various provisions in a physiological state which corresponds to optimum production phase. Even though continuous processes are theoretically applicable for many products, they are not much used in practical applications.

## Economics of Production

In general, products obtained by fermentation also can be produced by chemical synthesis. A fermentation process is usually most competitive when the chemical process requires many individual steps to complete the conversion.

The fermentation process employs a variety of raw materials whose availability varies from region to region. The economics of production process varies accordingly. The cost of the raw material can contribute significantly to the cost of production. For large volume, low priced products (such as commodity chemicals), the cost relationship of raw material and the end product is significant. For low volume, high priced products (such as certain pharmaceuticals), the relationship is negligible.

## Enzyme Technology

Fermentation carried out by live cells provided the conceptual basis for designing fermentation processes based on isolated enzymes. A single enzyme situated within a living cell is needed to convert a raw material into a product. But if the actual enzyme responsible for the conversion is identified, it can be extracted from the cell and used in place of a living cell. The purified enzyme carries out the same conversion as the cell, breaking down the raw materials in the absence of any viable microorganism.

### Types of Processes

Both batch and continuous methods are used in enzyme technology. However, in the batch method, the enzymes cannot be recovered economically, and new enzymes, must be added for each production cycle. Furthermore, the enzymes, are difficult to separate from the end product and constitute a potential contaminant. Because enzymes used in the continuous method are reusable and tend not to be found in the product, the continuous method is the method of choice for most processes. Depending on the desired conversion, the immobilized microorganisms can be replaced by an appropriate immobilized enzyme.

#### Enzyme-Immobilization Techniques

The enzyme immobilization techniques can be divided into the following five groups.

1. **Covalent attachment of enzymes to solid supports:** Enzymes are usually immobilized through their amino carboxyl groups in most instances, the immobilization procedure consists of at least two steps: activation of support and enzyme attachment per se.
2. **Adsorption of enzymes into solid supports:** Ion exchangers readily adsorb most proteins and have therefore been used for enzyme immobilization. The immobilization procedure is simple: an enzyme solution is added to the support mixed and surplus enzyme is then removed by washing.
3. **Entrapment of enzymes in polymeric gels:** In this approach, an enzyme is added to a solution of monomers before the gel is formed. Gel formation is then initiated either by altering the temperature or by adding a gel-inducing chemical. As a result, the enzyme becomes trapped in the gel.
4. **Cross-linking of enzymes with bi-functional reagents:** Among the most popular cross-linkers are glutaraldehyde, dimethyl adipimidate and aliphatic diamines. The two directly cross-link enzymes through their amino groups. Diamines cross-link enzymes through carboxyl groups following activation of these groups with carbodiamines.

5. **Encapsulation of enzymes/soluble enzymes in semi-permeable membrane reactors:** In this enzymes are enzymes are enveloped within various forms of membranes (eg.within bollow capsules or fires consisting of semi permeable membranes) that are impermeable to enzymes and other macromolecules, but permeable to substrates and products of low molecular weight.

## Aeration and Agitation

The large scale production of antibiotics and vitamins by submerged aerobic fermentation made necessary a better rapport between biologists and engineers, thus creating biochemical engineering. Aeration and agitation got maximum attention from these biochemical engineers. The preliminary aim of such studies has generally been to improve the operation efficiency of the fermenters with a view to increase yields since much of the operating expense lies in the cost of power used for aeration and agitation.

The purpose of aeration is to provide adequate oxygen to meet the demand of the organisms and to provide adequate means to ensure mixing in order to prevent the settling of solids. While these objectives may be met solely by aeration, this is frequently accompanied by mechanical agitation. Sometimes, aeration helps in the ventilation of inhibitory metabolites such as carbon dioxide.

Dissolved Oxygen (D.O.) concentration plays a decisive role in microbial metabolism. The rate of Oxygen transfer to a fermentation broth is an important parameter in the design of aerobic fomenters as it affects growth rate of microorganisms and product of metabolism. Oxygen is a rather insoluble gas and in a high demand system, microorganism finds a small supply may cause damage to the respiring cells. Thus Oxygen supply must be continuously renewed by aeration and agitation.

The performance of aerobic biological reactors containing high cell densities may be limited by the rate of Oxygen.transfer, since the rate of Oxygen consumption may exceed the rate of Oxygen transfer resulting the zero dissolved Oxygen concentration in the liquid media. The rheological properties of fermentation broth may also have significant influence on the performance of biological reactor. Further, mycelial cells may form pellets for cells growing inside the pellet. The evolution and removal of $CO_2$ and other fermentation products from fermentation media present another mass transfer problem which could adversely affect the performance of biological reactors. Other reactors which can affect the Oxygen transfer are temperature, pressure, rheological properties, surface active agents, ionic strength and also presence of organic substance.

## Bioprocess Control

An industrial fermentation is a fairly sophisticated device with control of temperature, aeration rate, and perhaps pH, concentration. There has been a strong trend to automated data collection and analysis. Ana log control is still very common, but when a computer is available for on-line data collection, it makes sense to use it for control as well. More elaborate measurements are performed with research bioreactors, but each new electrode or assay adds more work, additional costs, and potential headaches. Most of the functional relationships in biotechnology are nonlinear, but this may not hinder control when bioprocess operates over a narrow range of conditions. Furthermore, process control is far advanced beyond the days when the main tools for designing control systems were intended for linear systems.

Many of the sensor problems such as those with steam-sterilizable the most important factor for bioprocessing is the concentration of organisms, but there is no practical method for continuous measurement. Samples of the process fluid must be filtered and dried to get the mass concentration of cells. Numbers can be obtained by direct counting with a microscope or by counting the colonies that form when samples are cultured with nutrient medium in Petri dishes. In lieu of direct measurement, many other ways to estimate cell concentration are tried. Turbidity of the culture fluid can be correlated with cell concentration, but properties and calibration change during the process, and the optical surfaces of the sensors tend to become fouled. Alternatives such as measuring electrical conductivity or capacitance of the fluid sometimes are useful but often are suited only to specific cases. Indirect methods such as measuring protein produced by cells or monitoring nucleic acids are reported, but their proportionality to cell mass may vary during the fermentation. An important advanced was made by developing computer models that can interpret measured variables to calculate cell mass or product concentration that may be difficult or impractical to measure on-line.

Mounting electrodes in a bioreactor is costly and there is an additional contamination risk for sensitive cell cultures. Some other sensors of practical importance are those for dissolved oxygen and for dissolved carbon dioxide. The analysis of gas existing from a bioreactor with an infrared unit that detects carbon dioxide or a paramagnetic unit that detects oxygen (after carbon dioxide removal) has been replaced by mass spectrophotometer. Gas chromatographic procedures coupled with a mass spectrophotometer will detect all the volatile components.

A useful index of process performance is the oxygen uptake rate, OUR, that is calculated from the difference in oxygen concentration of the inlet air and the exiting gas. Also important is the respiration ratio defined as the carbon dioxide evolved by the oxygen consumed.

Although dynamic responses of microbial systems are poorly understood, models with some basic features and some empirical features have been found to correlate with actual data fairly well. Real fermentations take days to run, but many variables can be tried in a few minutes using computer simulation. Optimization of fermentation with models and real-time dynamic control is in its early infancy; however, bases for such work are advancing steadily. The foundations for all such studies are accurate material balances.

The common indices of the physical environment are: temperature, pressure, shaft power input, impeller speed, foam level, gas flow rate, liquid feed rates, broth viscosity, turbidity, pH, oxidation-reduction potential, dissolved oxygen, and exit gas concentrations. A wide variety of chemical assays can be performed; product concentration, nutrient concentration, and product precursor concentration are important. Indices of respiration were mentioned with regard to oxygen transfer and are particularly useful in tracking fermentation behaviour. Computer control schemes for fermentation can focus on high productivity, high product titter, or minimum cost. Computer systems may perform on-line optimization of fermentation. Progress has been slow by empirical methods because there is a multiplicity of variables and because statistical techniques suffer from the relatively poor reproducibility of fermentations. Careful attention to preparation of inoculums, time/temperature factors of sterilization, and the timing of inoculation and feeding can greatly reduce variability of bioprocess performance.

## Process Considerations

Fermentation can be combined with other operations. For example, feedback inhibition of enzymatic hydrolysis of cellulose can be relieved by removal of the product glucose by fermentation as it forms. This is termed simultaneous-saccharification-fermentation (SSF).

Valves and pumps that have a potential path for contaminating organisms are taboo for aseptic operations. Rising stem valves could bring organisms to the sterile side by the in and out motion as the valve operates. Diaphragm valves are still commonly used, but heating, cooling, and the abrasion by solids in the nutrient media are somewhat severe conditions leading to occasional rupture of a diaphragm and contamination of a run. Ball valves or plug valves do not have an absolute seal to the outside, but the direction of motion does not tend to bring organisms in. Contamination is seldom attributed to these valves; they are designed for easy maintenance in place, and there is the very nice human advantage that a glance at the handle tells easily whether the valve is open or closed. Many runs have been spoiled or impaired because a manual valve was left in the wrong position. For plant operations, pumps with diaphragms are satisfactory. In the lab or pilot plant, peristaltic pumps (also known as tubing squeezers) predominate.

Transfer of fluid in a fermentation plant usually makes use of air pressure differences. One or more manifold headers may interconnect many vessels. As transfers may have to be aseptic, headers are pressurized with steam until needed. A typical arrangement of steam seals is shown in Figure 10.1.

Sample lines commonly have steam seals too. A typical layout is shown in Figure 10.2. In the closed position, steam, provides an absolute barrier to contamination. To take a sample, the steam line and the trap line are closed, and fermentation medium is flowed to waste until the pipes are cool to be touch so that sensitive products do not give false assays because of thermal destruction. Cooking takes up to 5 liters of medium if not done carefully and bad practices can waste considerably more. Pilot-sized tanks have less massive fittings that are easier to cool; less medium is waste, but over sampling to the point where the Fermenter volume is low can be a problem. For this reason, alternate sampling methods have been devised. For example, a sterile syringe and needle may be used to sample through a rubber diaphragm in the wall of the tank. Although such methods appear reliable, there is a tendency to scrap all innovations and to return to tried and true steam seals when the factory encounters any period of contamination.

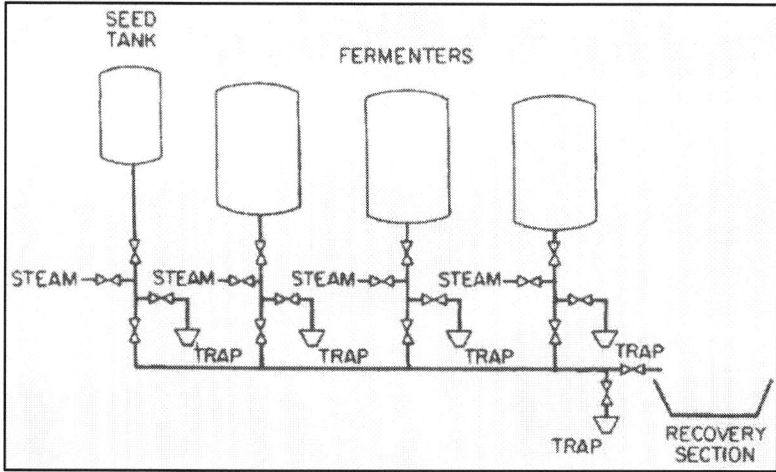


**Figure 10.1: Inoculation and Harvest Header (Perry's hand book of Chemical Engineering).**

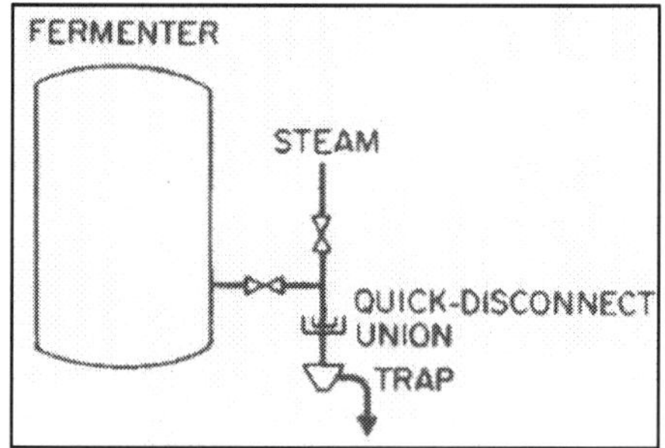


**Figure 10.2: Sample-line Piping.**
**(A valve to the sewer allows bypass of the trap while cooling the time-Perry's hand book of Chemical Engineering).**

All piping to a Fermenter is flushed with steam during the sterilization period. A clever means for sight-glass cleaning uses steam condensate that is naturally sterile in a dead leg. Steam pressure behind the condensate forces this water to the sight glass. Without cleaning, splashing and spray can quickly cover the sight glass with a thick coating of microorganisms and medium.

While it is easy to add materials to a fermentation, removal is difficult. Membrane device have been placed in the Fermenter or in external recycle loops to dialyze away a soluble component. Cells release wastes or metabolites that can be inhibitory; these are sometimes referred to a staling factors. Their removal by dialysis has allowed cell concentrations to reach ten to one hundred times that of control cultures.

Solid substrates such as pulverized wood cannot be stirred when slurry concentration exceeds about 5 per cent. For saccharification prior to ethanol fermentation, keeping sugar concentration high can avoid an evaporation step. In a batch reactor, mixing limitations with the wood results in a dilute sugar solution. This has been circumvented by placing the wood in a column and percolating the solution through. As wood dissolves, more is added. Simultaneous fermentation of the sugars formed in the column is possible.

## Oxygen Transfer

Supplying sufficient oxygen can be a very challenging engineering problem for some aerobic fermentations. Oxygen is sparingly soluble in water; saturation with pressurized air at room temperature provides only 6 to milligrams per liter of oxygen. A vigorous process can deplete the dissolved oxygen to several seconds when aeration is stopped. Mass transfer of gases to liquids is covered in Sec.5. Emphasis is somewhat different for biological systems that commonly have bubble aeration. Because the number and size of bubbles is very difficult to estimate, transfer area is usually lumped with the mass-transfer coefficient as $K_1a$ term. The "1" subscript in $K_1$ signifies that liquid film resistance should greatly predominate for a sparingly soluble gas such as oxygen. The relationship between oxygen concentration and growth is of a Michaelis-Menten type. When a process is rate-limited by oxygen, the specific respiration rate ($Q_{o2}$) also increases steeply with dissolved oxygen concentration until a plateau is reached. The concentration below which respiration is severely limited is termed the critical oxygen concentration, which typically ranges from 0.5 to 2.0 ppm for well0dispersed bacteria, yeast, and fungi growing at 20 to 30°C (68 to 86°F). Above this critical concentration, the specific oxygen uptake increases only slightly with increasing oxygen concentrations.

A plot of specific respiration rate $Q_{o2}$ versus the specific growth rate coefficient μ is linear, with the intercept on the ordinate equal to the oxygen uptake rate for cell maintenance. A formulation of this is:

Uptake rate = uptake for maintenance + uptake for growth

$$Qo_2X = \frac{(Q_{o_2})_M X + \mu X}{Y_g} \tag{1}$$

*where,*

X is organism concentration, the subscript M denotes maintenance, $Y_g$ is yield of cell mass per mass of oxygen, and the Q terms signify oxygen uptake rates in mass $O_2$ per mass of organisms.

This type of correlation applies to almost any substrate involved in cellular energy metabolism and is supported by experimental data and energetic considerations. However, it is based on assumptions true at or near the steady-state equilibrium conditions and may not be valid during transient states. The oxygen-uptake equation should be modified when other cellular activities requiring oxygen can be identified. For example, use of oxygen for product formation would be represented by:

Uptake rate = maintenance uptake + growth uptake + product uptake

$$Qo_2X = (Q_{o_2})_M X + \frac{dX}{dt}\frac{1}{Y_G} + \frac{dP}{dt}\frac{1}{Yp} \tag{2}$$

*where,*

P is the product concentration and Y is the yield of product per unit weight of limiting nutrient. Oxygen uptake is distributed between that for growth and that for cellular activities dependent on cell concentration.

As the oxygen transfer rate under steady-state conditions must equal oxygen uptake, $K_{1a}$ may be calculated:

$$K_{1a} = \frac{\text{Overall oxygen uptake rate}}{(C^o - C)_{mean}} \tag{3}$$

*where,*

$C_o$ – concentration of oxygen in the liquid that would be in equilibrium with the gas-bubble concentration and $K_{1a}$ = the volumetric oxygen transfer rate.

A convenient method for measuring oxygen transfer rates in microbial systems depends on dissolved oxygen electrodes with relatively fast response time. Quite inexpensive organisms electrodes are available for use with open systems, and steam-sterilizable electrodes are available for aseptic systems. There are two basic types: one develops a voltage from an electrochemical cell based on oxygen, and the other is a polar graphic cell whose current depends on the rate at which oxygen arrives. See Figure 10.14. measurement of oxygen transfer properties requires only a brief interruption of oxygen supply. A mass balance for oxygen is:

$$\frac{dO}{dt} = \text{Rate of supply} - \text{Uptake rate} \tag{4}$$

A tracing of the electrode signal during a cycle of turning aeration off and on is shown in Figure 10.15. the rate of supply is zero (after bubbles have escaped) in the first portion of the response curve; thus, the slope equals the uptake rate by the organisms. When aeration is resumed, both the supply rate and uptake rate terms apply. The values for $C_o$ – C can be calculated from the data, the slope of the response curve at a given point is measured to get dC/dt, and the equation can be solved for $K_{1a}$ because all the other values are known.

Measurements of the rate of change in concentration of oxidizable chemicals in aerated vessels have questionable value for assessing rates with biological systems. Not only are flow patterns and bubble sizes different for biological systems, but surface active agents and suspended particles can seriously impair gas transfer. Figure 10.16 shows the effects of various particles. In general, spherical particles have a small effect, elongated particles have more effect, and entangled particles markedly impair transfer. As many mold cultures are intertwined and lipids and proteins are present with strong surface activity, oxygen transfer to a fermentation can be much slower than that to simple aqueous solutions.

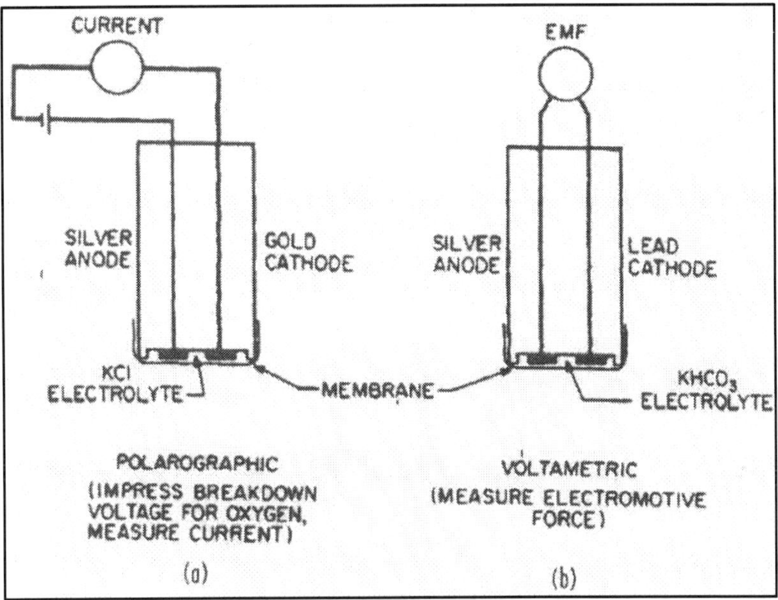


**Figure 10.3: Dissolved Oxygen Electrodes: (a) Palaeographic (impress break- down voltage for oxygen: measure current): (b) volt metric(measure electromotive force) (Perry's hand book of Chemical Engineering).**

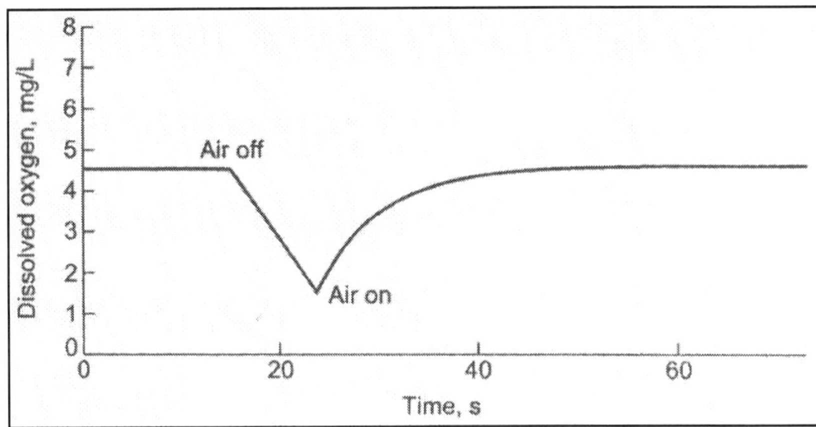


**Figure 10.4: Computer Simulation of Response for Dynamic Measurement of Kla. (Perry's hand book of Chemical Engineering).**

Except as an index of respiration, carbon dioxide is seldom considered in fermentations but plays important roles. Its participation in carbonate equilibria affects pH; removal of carbon dioxide by photosynthesis can force the pH above 10 in dense, well-illuminated algal cultures. Several biochemical reactions involve carbon dioxide, so their kinetics and equilibrium concentrations are dependent on gas concentrations, and metabolic rates of associated reactions may also change. Attempts to increase oxygen transfer rates by elevating pressure to get more driving force sometimes encounter poor process performance that might be attributed to excessive dissolved carbon dioxide.

## Sparger Systems

Gas distributors in tanks are shown in Sec.6. Large openings are desirable for spargers in industrial fermentations to avoid clogging by microbial growth, but the diameter is usually designed for the acoustic velocity that insures small bubbles. Relatively small holes or diffusers are used in activated sludge units for biological waste treatment, but there is commonly a means for swinging a section of the aerator out of the vessel for cleaning. Newer designs for Fermenter were conceived as answers to the problems of oxygen transfer. The air lift Fermenter (Figure 10.4) creates intimate mixing of air and medium while using the buoyancy of the gas to mix the fluid.

Surface active substances also lead to foaming that can be so bad that most of the contents of the fermenter are lost. Mechanical antifoam devices are helpful but cannot function alone except when there is little propensity for foaming. The mechanical foam breakers rupture the large, weak bubbles while allowing tiny, rugged bubbles to the bubbles so that the mechanical shocks are easily transmitted to Figure 10.17. some lipids used a antifoams are metabolized by the culture and must be replaced. The nutrition supplied by these oils may be beneficial, but they are much more expensive than their equivalents in carbohydrate nutrients. Furthermore, it is troublesome to have nutrition coupled to foam control. Several synthetic antifoam agents are not nutrients and tend to persist. Their tendency to be lost by coating solid surfaces in the Fermenter means that more must be added occasionally. These synthetic antifoam agents are toxic to some organisms, but one of the many types is usually satisfactory.

## Scale-Up

Fermentation ranging from about two to over 100 liters (0.07 – 3.5 $ft^3$) have been used for research and development, but the smaller sizes provided too little volume for sampling and are difficult to replicate, while large vessels are expensive and use too much medium. Autoclavable small Fermenter that are placed in a water bath for temperature control are less expensive than vessels with jackets or coils, but much labour is required for handling them. Pressure vessels that are sterilized in place are more convenient, but initial investment is high. Judgement is needed to select the most economical equipment and to plan for cost-effect experimentation.

A suitable means of scale-up for aerobic processes is to measure the dissolved oxygen level that is adequate in small equipment and to adjust conditions in the plant until this level of dissolved oxygen is reached. However, some antibiotic fermentations and the production of fodder yeast from hydrocarbon substrates have very severe requirements, and designers are hard-pressed to supply enough oxygen.

Older method of fermentation scale up insisted on geometric similarity based on proportional physical dimensions. It was thought that applying the same power per unit volume as in the pilot equipment would give an equivalent process performance in large fermenters. Antibiotic fermentations aim for mixer power in the range of 0.2 to 4 Kw/m (0.1 to 2 HP/100 gal). As mixing devices have areas (dimensions squared)

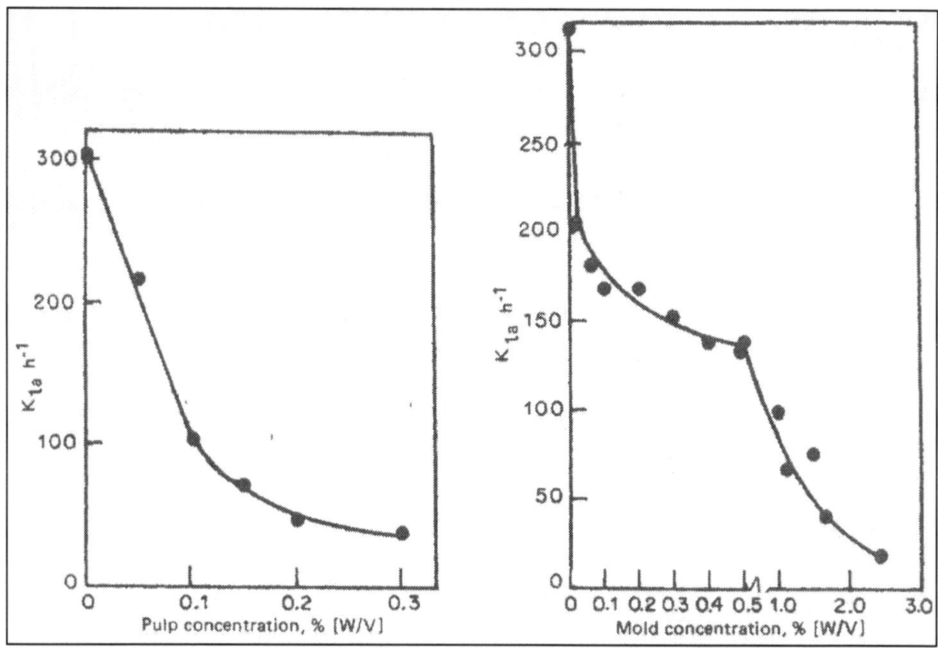


**Figure 10.5: Effect of Solids on Kla. Operating Conditions: Agitator Speed 800r/min; Air Flow, 2.5 min. [M.R. Brierly and R. Steel, Appl. Microbiol,. 7,57(1959). Courtesy of American Society for Microbiology, USA.]**

to supply a volume (dimensions curbed), methods based on dimensional similarity are fundamentally unsound. Scale-up based on equivalent oxygen transfer coefficient $K_{ia}$ has been reasonably successful.

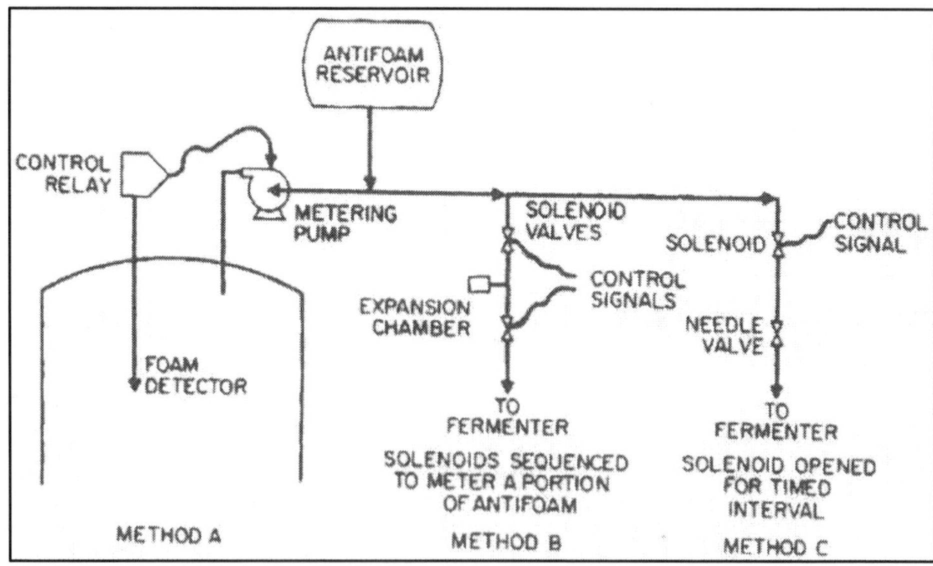


**Figure 10.6: Antifoam Systems (Perry's hand book of Chemical Engineering).**

Impeller Reynolds number and equations for mixing power for particle suspensions are in Sec.5. Dispersion of gasses into liquids is in Sec.14. usually, an increase in mechanical agitation is more effective than is an increase in aeration rate for improving mass transfer.

Other scale-up factors are shear, mixing time, Reynolds number, momentum, and the mixing provided by rising bubbles. Shear is maximum at the tip of the impeller and may be estimated from Eq.(5), where the subscripts *s* and *l* stand for small and large and *Di* is impeller diameter [R, Steel and W.D. Maxon, Biotechnol, Bioengr., 4, 231 (1962)]

$$S_s = S_1 \left[ \frac{Di_s}{Di_1} \right]^{1/3} \tag{5}$$

Some mycelial fermentations exhibit early sporulation, breakup of mycelium, and low yields if the shear is excessive. A tip speed of 250 to 500 cm/s (8 to 16 ft/s) is considered permissible. Mixing time has been proposed as a scale-up consideration, but little can be done to improve it in a large fermenter because gigantic motors would be required to get rapid mixing. Culturing cells from plants or animals is beset by mixing problems because these cells are easily damaged by shear.

Constant Reynolds number is not used for fermentation scale-up; it is only one factor in the aeration task. This is also true for considering the impeller as a pump and attempting scale-up by constant momentum. As mechanical mixing tends to predominate over bubble effects in improving aeration, scale-up equations including bubble effects have had little use.

Fermentation biomass productivities usually range from 2 to 5 g/(l-h). This represents an oxygen demand in the range of 1.5 to 4 g/(l-h). In a 500-m Fermenter, this means achievement of a volumetric oxygen transfer coefficient in the range of 250 to 400 $h^{-1}$. Such oxygen-transfer capabilities can be achieved with aeration rates of the order of 0.5 VVM (volume of air at STP/volume of broth) and mechanical agitation power inputs of 2.4 to 32. Kw/m (1.2 to 1.6 HP/100 gal).

Often heat removal causes design problems for scale-up Mechanical agitation coupled with a metabolic heat from the growing biomass overwhelms cooling capacity of a large Fermenter with only a jacket. External circulation through a heat exchanger or extensive coils inside the Fermenter must be used. In highly viscous fermentations, internal cooling coils are usually not desirable cause of interference with mixing patterns. Numerous schemes exist for heat removal in large Fermenter such as half-coil baffles (plate cols) and draft tubes. Heat removal limits the size of packed-cell bioreactors. There is also a serious problem of gas evocation for bioreactors when the cells are immobilized with a membrane or by retention in a get because the gas can rupture the structure.

Evaporation of medium provides a little cooling. The inlet air to particulate filters must not be near saturation because condensation of moisture on the filter medium mobilizes contaminating microorganisms so that their chances of penetration are greatly increased. Sometimes humidified air is used, and the filter unit is heated to prevent condensation. However, this common only for small equipment where the extra operations are relatively easy to install and maintain.

Following are the main scale up criteria for industrial fermentations

1. Equal power input per unit volume.
2. Geometrical similarity of bioreactors.
3. Equal energy dissipated per unit mass.
4. Volumetric mass transfer coefficient (KLa) to be constant.
5. Impeller tip velocity to be constant.

Once a plant is built, the conditions of agitation, aeration, oxygen transfer, and heat transfer are more less set, and sterilization cycles are defined. Those environmental conditions achievable in plant-scale equipment should be scaled down to the pilot plant and laboratory equipment (shaken flasks) to insure that results can be translated.

## Sterilization

Some old, traditional fermentations such as those for alcohol and pickles are conducted by organisms that are hardy and help their own cause by creating conditions that are unfavourable for competitors. Yeasts, for example, lower pH by producing acids from sugar, and tolerance to alcohol is another powerful advantage that allows their domination. Nevertheless, modern factories use aseptic techniques or extreme care to minimize contamination that can jeopardize product quality by affecting taste, texture, aroma, or appearance. Bioprocesses such as tissue culture to produce vaccines are very easily contaminated because there is an abundance of nutrients and no inherent protection against foreign organisms. Thus bioprocesses range from relatively good self-protection to practically none; the value of the product and the need for quality control determine the extent to which precautions must be taken. Contamination in the practical sense is statistical in that foreign organisms can be present but may not propagate rapidly enough to damage the run. Good defence against contamination is relatively inexpensive, while absolute protection is impossible and attempts to achieve it can be inordinately costly. The production of agents for biological warfare takes extreme pains to keep the organisms away from the workers, yet people operating the fermentations are occasionally killed. The most expensive and best protected industrial fermentations are sometimes contaminated. Sterilization and aseptic techniques to keep bioprocesses uncontaminated can be crucial.

Common ways of sterilization are removal of microorganisms by filtration or killing them with heat or chemicals. Sterilization by filtration follows a standard unit operation that is covered in Sec.18. The differences for biochemical engineering are: (1) the filter medium and the downstream lines are steamed at a pressure where the temperature kills all organisms and spores; (2) the sizes of the particles being removed are in the micrometer range; and (3) the filter medium should be reusable and not degraded by repeated heating.

## Cell Culture

Single plant and animal cells fare much larger than microbial cells and are easily damaged by shear due to intense agitation. Standard stirred tank bioreactor designs tend not to work well for undifferentiated plant and animal cells. Attempts to substitute paddle and spiral agitators have proven successful in some cases, but shear can still cause a morphological change in the structure of the cells. Usually, single cells and cellular aggregates tend to be smaller and produce significantly less secondary metabolites with improper agitation. Airlift reactors may work well, and cultured cells tend not to have severe oxygen requirements. There has been a thrust for new systems that can maintain absolute sterility without depending on antibiotics that can mask the slow development of a low-grade infection in a culture and can affect the metabolism of cultured cells in subtle ways that cannot always be predicted.

Cells that must attach to a surface can be a grown on spongy polymers, arrays of thin tubing or hollow fibers, stacks of thin plates, or microscopically small beads called micro-carriers. Damage from shear can be a serious problem; thus, aeration systems cannot employ the vigorous mixing that aids mass transfer in microbial processes. Airlift bioreactors have been developed that are tapered instead of being cylindrical with a central draft tube. Above a zone of intense aeration are the suspended cells circulating with little more mixing than would result from Brownian motion.

Improvements in fermentation include micro-carriers that not only provide support for anchorage-dependent cells but also aid in harvesting at the end of a run. While micro-carriers based on dextran, polystyrene and polyacrylamide beads have been widely used in the past, new materials for micro-carriers make separating cells from their growth medium easier. Some are made of collagen that is detached from the cells by immersion in dilute scollagenase. However, since some collagen can be left in solution, downstream processing can be made difficult. An alternative is to use plastic beads coated with collagen. The plastic can be easily separated after the cells are released enzymatic ally. Another alternative is glass-coated particles that induce attachment of the cells' long slender filopodia. At the end of a process, a brief incubation in dilute trypsin gently removes the cells from the beads.

Supplying sufficient oxygen can be difficult when dealing with differentiated plant tissues such as root cultures that can reach lengths of several decimetres and can be highly branched and complex structurally. In this case, mechanical agitation is impractical because the mass of roots can occupy approximately 50 per cent of the reactor volume. Alternatives are bubble columns, rotating drums, and trickled bed reactors where the medium is recycled and sprayed over the column of roots.

Because of the differences in primary and secondary metabolism, a reactor may have a dual-stage fed-batch system. In other words, fed-batch operation optimizes growth with little or no product formation. When sufficient biomass has accumulated, a different fed-batch protocol comes into play.

In an industrial fermentation process for biosynthesis of primary and secondary metabolites, the parameters to be optimized are – Temperature, pH, dissolved oxygen (DO), oxygen utilization rate (OUR), RQ value, vent pressure(P0), air supply rate

(VVM), agitation rate (RPM), KLa (Hr.), impeller tip velocity (ĐDN), rpm of agitator, flow rate of oil, anti-foam agent and nitrate addition (Lts/Hr.), rate of precursor addition (kg./Hr.) and soluble phosphorus(ppm) for regulation are some of important parameters to be controlled during fermentation process for optimum productivity, yield and production of the metabolite.

## Preparation for Harvesting and Product Recovery

Methods for the separation and isolation of products are commonly referred to as downstream processing. These methods have long been recognized as technically difficult and expensive. In fact it has now been realized that the downstream processing is a critical limiting factor in the commercial development of biotechnology. Now as plethora of potential techniques is emerging nearly all of them have not yet been tested on commercial scale.

Recovery plants in fermentation unit for downstream processing usually represent a major investment. The isolation costs are a substantial fraction of the manufacturing cost of the product. The unit operations which are normally used in downstream processing of a metabolite produced through fermentation processes are filtration, centrifugation, mechanical dewatering, drying, distillation and heat transfer. The processes which are required for concentration and purification are ultra-filtration, centrifugation and freeze drying, ion exchange chromatography and solvent extraction. The unit operations which are required to achieve sterilization, evaporation, distillation and generation of heat, do not require other non-electric power generation in small plants, but are certainly required to generate power within the plant premises itself. The details of the process which are used in down-stream processing are given below:

**1. Separation**

This can be achieved by the following four operations:

a. Flocculation and floatation
b. Filtration
c. Centrifugation
d. Precipitation by pH changes, solvents such as acetone or salts such as ammonium sulphate

**2. Disintegration**

a. Disintegration of microorganisms
b. Disintegration of animal and plant tissues

**3. Enrichment**

a. Concentration by thermal process
b. Extraction
c. Membrane filtration
d. Freeze concentration
e. Ion exchange processes and adsorption with adsorber resins
f. Precipitation reactions

**4. Purification**

a. Crystallization

b. Chromatography

**5. Drying**

a. Contract drying

b. Conventional drying

c. Freeze drying

All above operations are to be performed for effective down stream processing. Details will be reported subsequently.

## References

1. Asenjo JA, Merchuck JC (1994). *"Bioreactor System Design"*, Dekker, New York.
2. Becker JM, Caldwell GA, Zachgo EA (2007). *" Biotechnology A Laboratory Course"* Second Edition, Academic Press.
3. Bisaria VS. Pathak AN (1973) "Enzyme Engineering and Technology", *Chemical Age of India* 24(7). 419-427.
4. Casida, LE (1996). *" Industrial Microbiology"*, New Age International (P) Limited.
5. Doble M, Gummadi SN (2007). *"Biochemical Engineering"*, Prentice Hall of India Private Limited, New Delhi.
6. Fogler H Scott (2013). *" Elements of Chemical Reaction Engineering"*, PHI Learning Private Limited, New Delhi.
7. Gerald R (2004). *" Prescott and Dunn's Industrial Microbiology"* CBS Publishers and Distributors, New Delhi.
8. Ghose TK and Pathak AN (1973) *"Cellulase-II, Applications" Process Biochemistry (London)* 8(6). 35-36.
9. Ghose TK and Pathak AN (1974) *"Cellulose as a source of future Food"* paper presented in symposium of A.I. Chem. E. Melbourne, Australia.
10. Ghose TK and Pathak AN (1975) *"Cellulose as a source of future food"*, Indian Chemical Engineer, Vol.XVII(4). 3016.
11. Ghose TK and Pathak AN (1975) "Kinetics of cellulase production by Trichoderma Virde" in symposium on Enzymatic Hydrolysis of Cellulose (Ed.Baily. M.) Helsinki, Finland, March 12-14, P-111-126.
12. Harrison RG (ed.) (1994). *Protein Purification Process Engineering Physical Chemistry and Bioanalytical Applications*, Dekker, New York.
13. Inamdar STA (2007). *"Biochemical Engineering Principles and Concepts"*, Prentice Hall of India Private Limited, New Delhi.
14. Inamdar STA (2009). *"Biochemical Engineering Principles and Concepts"*, PHI Learning Private Limited, New Delhi.

15. Mattiasson BN, Holst O (eds.) (1991). *Extractive Bioconversions*, Marcel Dekker, New York.
16. Okotore RO (2008). " *Basic Separation Techniques in Biochemistry*", New Age International (P) Limited.
17. Patel AH (2004) "*Industrial Microbiology*", Macmillan India Ltd.
18. Pathak AN (1974) "Cellulose as a source of Energy", *Chemical Age of India* 25(9). 615-629.
19. Pathak AN (1974). "Rheology in Food Texture" *J. Scientific and Industrial Research*, 33.
20. Pathak AN (2011). *Key Note Address on Nanobiotechnology – Applications and Ethical Issues in International Conference on Green Biotech: Recent Trends and Innovations.*
21. Pathak AN (2012). Mutagenesis of Azetobactor vinelandii strain and production of Poly b – Hydroxybutyrate from Distillers Spent wash published in *Bioprocessing International Journal*, 3(2), pp.45-48.
22. Pathak AN and RaoPattnaik KSK (2007) "*Ethical issues on Nanoscience and Nanotechnology" paper was presented in an International Conference of Nanoscience and Nanotechnology in California (CA). USA.*
23. Pathak AN, Mathur Nidhi (2013). "*Stem Bromelain: In Silico Analysis for Stability and Modification*" Lambert Academic Publishing, Germany. ISBN : 978-3-659-50335-1 International.
24. Pathak AN, Bose D (2012). *Antimicrobial activity of citrus fruit juices.* Lambert Academic Publishing, Germany ISBN : 978-3-659-28108-2
25. Pathak AN, Bose D (2012). *Preservation of commercially important cultures.* Lambert Academic Publishing, Germany. ISBN: 978-3-659-29905-6
26. Pathak AN, Bose D (2012). *Production of Cephalosporin from Acremonium chrysogenum.* Lambert Academic Publishing, Germany. ISBN : 978-3-659-30011-0
27. Pathak AN, Bose Debajyoti (2013). *Fermentative Production of Erythromycin.* Lambert Academic Publishing, Germany. ISBN : 978-3-659-37022-9.
28. Pathak AN, Bose Debajyoti, Parecha Darshan Kumar (2013). "*Recent Developments on Cephalosporin Derivatives.* Lambert Academic Publishing, Germany. ISBN : 978-3-659-44138-7 International.
29. Pathak AN, Lakhawat SS (2014). "Production Technology and Applications of Kojic Acid", *Annual Research and Review in Biology*, 4(21). 3165-3196.
30. Pathak AN, Lakhawat SS (2014). "Production Technology and Applications of Kojic Acid", *Bio Processing Journal* (USA). 13(3) Page No. P 62-79. (ISSN No.- 1538-8786).
31. Pathak AN, Mishra N, Nakhate P (2012). "*Agriculture and Human Values: A n Era Towards a New Dawn.* Lambert Academic Publication ISBN: 978-3-659-37170-7.

32. Pathak AN, Srivastava A(1998) "Erythromycin and its Semi Synthetic Derivatives", *Journal of Scientific and Industrial Research*, Vol.57.

33. Pathak AN, Srivastava A(1998). Modern Technologies in Distillery Effluent treatment, *Journal of Scientific and Industrial Research*, P 388-392.

34. Pathak AN,Lakhawat SS (2012). *Polyhydroxy Butyrate Derived Bioplastics.* Lambert academic publishing Germany ISBN : 978-3-659-25885-5.

35. Pathak, AN (1997) "Membrane Technology with Special Reference to Filtration and Reverse Osmosis" *Journal of Scientific and Industrial Research.* Vol.56, Page 335-345.

36. Pathak, AN (1998) "Antifungal Agents to Manage Mycotic Infections". Paper presented in *National Seminar on "Emerging Trends in the Management of Fungal (Mycotic) Infections"* organized by National Academy of Chemistry and Biology (India) in Lucknow on Nov., 14-16, 1997. Published in Book "Antifungal Agents" Forewarded by Prof. J.G. Cannon, U.S.A.

37. Peppler HJ, Perlaman D (2009). " *Microbial Technology*", Academic Press, Second Edition Volume I.

38. Peppler HJ, Perlaman D (2009). " *Microbial Technology*", Academic Press, Second Edition Volume II.

39. Perry RH, Chilton CH (1973). *"Chemical Engineer's Handbook"* McGraw-Hill Kogakusha, Ltd.

40. Prasad NK (2010). *" Downstream Process Technology-A New Horizon in Biotechnology"*, PHI Learning, New Delhi.

41. Rehm HJ, Reed G. (1988). *Biotechnology, vol. 6b*, VCH Verlagsgesellschaft.

42. Richardson JF, Harker JH, Backhurst JR (2003). *"Coulson and Richardson's Chemical Engineering"* Elsevier.

43. Shuler ML, Kargi F (2008). *" Bioprocess Engineering Basic Concepts"* Second Edition Prentice Hall of India, New Delhi.

44. Srivastava A, Pathak AN (1997) "Status Report on Tannery Wastes with Special Reference to tanneries at Kanpur, Lucknow, Uttar Pradesh" Journal of Scientific and Industrial Research, 55: 435-459.

45. Thakur IS (2010). *" Industrial Biotechnology Problems and Remedies"*, L.K. International Pvt. Ltd., New Delhi.

**2016, Biotechnology: Progress and Applications** *Pages* **249–279**
*Editors:* **Saif Hameed and Zeeshan Fatima**
*Published by:* **DAYA PUBLISHING HOUSE, NEW DELHI**

# *Chapter 11*

# Thermozymes: Biology and Technology at High Temperature

***Sangeeta Kumari***

*Amity Institute of Biotechnology,*
*Amity University Haryana, Gurgaon – 122 413*

*"What's really special about this enzyme is that it loves high temperature and pH,"*

*says* **Vicki Thompson.**

## ABSTRACT

Most of the potential bioprospecting is currently related to the study of the thermozymes and their potential is used in industrial processes. Recently microbial thermozymes has applications in various industries and constitute a major group of industrial enzymes. It discusses the industrial applications of thermophilic enzymes, their cost of production and challenges in thermophilic research especially in the area of improving process economics of enzyme production. Bioprospecting is defined as the exploration of biodiversity for commercially valuable biochemical and genetic resources for achieving economic and conservation goals.In this chapter the state of art and future perspectives of industrial applications of enzymes from hyperthermophilic and extreme thermophilic microorganisms, special attention being paid to the biotechnological methods involved in their industrial exploitation.

***Keywords:*** *Thermozymes, Thermophilic microorganisms, Chaperonins, Biofuel, Starch degradation, Protease, Amylase, Lipase.*

* Corresponding Author: E-mail: sanshnic@gmail.com

## Introduction

Industrial enzymology is an important sector of biotechnology, characterised by a substantial growth over the last decade (Brooke and Freeze,1969). Since a long time the microorganism that grows at high temperature has held a particular fascination for microbiologists and biochemists. As any of their cellular components, their proteins are inherently more stable to heat than those of conventional organisms. This thermal stability is not due to any specific characteristic, however results a consequence of various changes which contributes to the whole stability of the protein in an additive manner. These enzymes are not only more thermostable, also more resistant to chemical agents than their mesophilic homologous, what makes them extremely interesting for industrial processes. Despite this, most of the enzymes used at present in industrial processes have been isolated from mesophiles due to the limited knowledge and difficulties to grow thermophiles in high scale. The objective of this chapter is to consider briefly the importance of the thermostability and the most recent advances in the identification of new thermophilic organisms and enzymes in order to apply enzymes in the industry. We will focus on the potential and applications of hydrolytic enzymes, which are established in industrial scale. To further illustrate the use of thermostable biocatalysts on renewable raw materials in large scale. A survey by Business Communications Company Inc, the global market for industrial enzymes was estimated a total of $2 billion in 2004 (Figure 11.1). Furthermore, the annual growth rate of industrial enzymes is predicted to be between 4 and 5 per cent also with this comes lower prices of enzymes due to an intensified competition on the market. The largest section is technical enzymes where enzymes used for detergents and pulp and paper constitute 52 per cent of the total world market. Leading enzymes in this section are hydrolytic enzymes, classified as proteases and amylases, which comprise 20 or 25 per cent of the total market, respectively. Hydrolases are generally easy to use in bioprocesses, as they normally do not require co-factors or complex substrates.

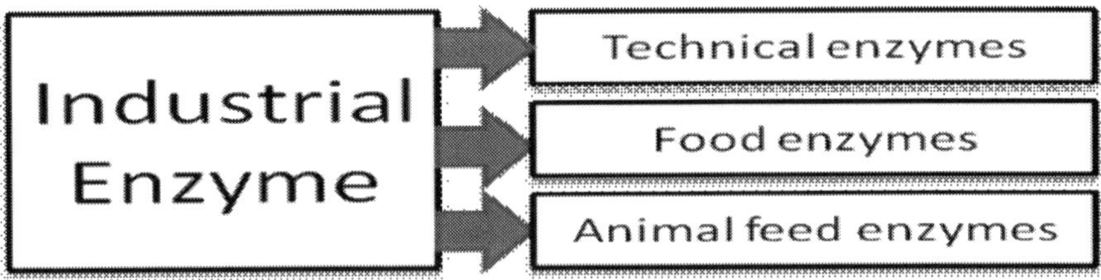


**Figure 11.1: Types of Industrial Enzyme.**

This holds substantial promise for the development of novel compounds for food production and processes, consumer goods, public health, environmental and energy uses. To serve these purposes, existing diversity of microorganisms can act as a resource reservoir from which individual species with special traits can be exploited (Bull *et al*,2000; Egorova and Antranikian, 2005). Due to the environmental related issues and advances in biotechnology, many chemical processes in different Industries such as textile, leather, pulp, paper, fruits and vegetables processing also animal's feed are being replaced with biocatalysts *i.e.* enzymes. Besides playing a crucial role in food production and other consumer goods and services, they have recently become vital in the production of fuel for automobiles. Enzymes can be harvested from a variety of sources *i.e.*, animal's tissue, plants and microbes. Microorganisms represent an attractive source of enzymes because they can be cultured in large quantities in a relatively short time and as such they can produce an abundant, regular supply of desired enzyme products. Moreover, microbial proteins are often more stable than enzymes of similar specificity obtained from plants or animals sources and often may be stored under less than ideal conditions for weeks without significant loss of activity Hence most of the commercial enzymes are derived from the microorganisms (Headon and Walsh, 1994). The recent developments of genetic model systems for moderate and extreme thermophiles are referred. An important source of thermostable enzymes can be found in the bacterial kingdom, where thermophilic prokaryotes are distributed among various groups in the phylogenic tree of Bacterial and Archaeal domains (Kristjansson and Stetter,1992). Compost is a self-heating aerobic medium that results from microbial degradation of organic matter (Bertoldi *et al*,1983) Enzymes synthesized by thermophiles (organisms with optimal growth temperatures > 60°C) and hyperthermophiles (optimal growth temperatures > 80°C) are typically thermostable (resistant to irreversible inactivation at high temperatures) and thermophilic (optimally active at high temperatures, *i.e.*, > 60°C). These enzymes, called thermozymes, share catalytic mechanisms with their mesophilic counterparts. When cloned and expressed in mesophilic hosts, thermozymes usually retain their thermal properties, suggesting that these properties are genetically encoded. Sequence alignments, amino acid content comparisons, and crystal structure comparisons indicate that thermozymes are, indeed, very similar to mesophilic enzymes. No obvious sequence or structural features account for enzyme thermostability and thermophilicity. Thermostability and thermophilicity molecular mechanisms are varied, differing from enzyme to enzyme. Thermostability and thermophilicity are usually caused by the accumulation of numerous subtle sequence differences. In this book we discussed the mechanisms involved in enzyme thermostability and thermophilicity and their relationships with protein rigidity and flexibility and with protein folding and unfolding. Intrinsic stabilizing force (*e.g.*, salt bridges, hydrogen bonds, hydrophobic interactions) and extrinsic stabilizing factors are invoved in the stability of enzyme. There are no major conformational differences with mesophilic enzymes, and a small number of extra salt bridges, hydrophobic interactions, or hydrogen bounds seem to confer the extra degree of stabilization. Currently, over expression of thermozymes in standard *Escherichia coli* allows the production of much larger quantities of enzymes, which are easy to purify by heat treatment. Finally, thermozymes' potential as catalysts for industrial processes.

Thermozymes are often used when the enzymatic process is compatible with existing in process condition at high temperature and replaces the chemicals. The main advantages of performing processes at higher temperatures are reduced risk of microbial contamination, lower viscosity, improved transfer rates, and improved solubility of substrates. However, cofactors, substrates, or products might be unstable or on the other side reactions may occur. Recent studies show that thermophiles are a good source of novel catalysts that are of great industrial interest. Hydrolytic enzymes have many industrial applications in the detergent sector an important role in food, chemical, pharmaceutical, paper, pulp, waste-treatment industries and biofuel industry (Gupta *etal*,2004 and Hasan *et al.*, 2005). Considerable research efforts have been made to understand the stability of thermozymes better. With wider availability and lower cost, thermophilic enzymes will see more applications in industry.

## Mechanisms of Thermostability

Thermostability is optimal activity at high temperatures are inherent properties of the proteins isolated from thermophilic microorganisms. The mechanisms of the thermal stability of thermophilic proteins have attracted the interest of the scientific community ever since they were firstly discovered because most proteins are inactive at such temperatures. On one hand, their amino acid sequence analysis revealed nothing unusual. On the other hand, the sequences from homologous hyperthermophilic and mesophilic proteins resulted to be highly similar, their threedimensional structures are superimposable and they exhibit the same catalytic mechanisms (Vieille and Zeikus,2001). Nevertheless, the fact that such proteins remain thermostable once expressed in mesophilic hosts suggested that thermotolerance resides in the gene sequence. In general, it was observed that thermophilic proteins tend to have hydrophobic cores due to a different folding with respect to mesophilic ones, which probably affords them more stability (Kirino *et al.*, 1994). Therefore, the mere fact of replacing one amino acid by another affecting the folding is enough to render a protein thermoresistant. Although no specific rules for thermostability or generalizations can be derived from a systematic analysis of the sequences of homologous mesophilic and thermophilic proteins, some hydrophobic substitutions in the protein core investigated by site-directed mutagenesis (SDM) revealed an enhanced thermostability (Hough, and Danson, 1999). As mentioned above, a wide variety of genes from hyperthermophiles have been successfully cloned and expressed in mesophiles. Thus, most archaeal genes cloned in *E. coli* have been successfully expressed under the control of strong promoters such as pLac, pTac or the T7 RNA polymerase promoter, owing to the large differences between the archaeal and bacterial transcription systems. A few genes from hyperthermophilic Archaea have also been successfully expressed in yeast systems, which harbour transcription systems are more closely related (D'Auria *et al.*, 1996). Such enzymes might require posttranslational modifications or molecular chaperones in order to attain their stable folded state. Thermophiles (organisms growing between 50 and 80°C) and hyperthermophiles (above 80°C) are reported to have specialized proteins, known as "chaperonins", that are thermostable and resistant to denaturation. Such proteins help others to refold to their native form after denaturation and to restore their functions (Everly and Alberto, 2000). There are several mechanisms used by hyperthermophiles to maintain the stability of their DNA at high temperature. Thus, all hyperthermophiles produce a particular type of DNA topoisomerase, called DNA reverse gyrase, which introduces positive supercoils in the DNA molecule at the expense of ATP, this conferring a greater stability and rendering more resistant to thermal denaturation (Forterre, 1996 and Lopez,1999). Another mechanism, which is not present in all hyperthermophiles, involves some histone-like proteins that help to preserve the duplex structure of DNA. Such proteins increase the fusion temperature of DNA up to 40°C above normal values (Todorova, R. and Atanosov, 2004; Kimura *et al.*, 1984). A third mechanism to protect DNA from denaturation is the presence in the cytoplasm of some hyperthermophilic methanogenes of large amounts of cyclic 2,3-potasium-bi-phospho-glycerate, which avoids chemical damage such as DNA depurinization (Grogan, 2004). Since most hyperthermophiles belong to the Archaea, it is necessary to refer the increased stability of archaeal cytoplasmic membranes with respect to bacterial counterparts, because of their different composition. Thus, whereas the typical fatty acid bilayer structure of the bacterial cytoplasmic membrane would become disrupted at extreme temperature, the archaeal monolayer membrane composed of phytanyl chains connected to glycerol with ether links is much more resistant (Comita *et al.*,,1984). The most hyperthermophilic organism described and classified so far is *Pyrolobus fumarii*, which is able to grow at 113°C(Bloch *et al.*, 1997), although recently an organism called "strain 121" was isolated and shown to be able to grow at 121°C (Kashefi and Lovely,2003).

The upper temperature at which life is possible is still unknown. However, since ATP is spontaneously hydrolyzed in aqueous solution at a temperature of about 140°C, one cannot expect to find any living organism above such temperature. If life above 140°C is found, such finding would contribute entirely new principles to Biochemistry and Biology.

However, there are substantial differences between the conclusions reached with different proteins, and therefore, there is not sufficient evidence for a universal rule for the structural basis of stability. The mechanisms that allow thermophilic enzymes to function appropriately at extremely high temperature are ambiguous and still under research. A primary hypothesis lies in the belief that thermophilic enzymes are more rigid than their mesophilic nature counterparts as a direct result of amino acid sequence and 3-dimensional structure. Initially, however, there are significant differences in certain base pairing frequencies in thermophilic archaeal tRNA and RNA. It has been noted that an increase in extremozyme stability at high temperatures is associated with increased guanine-cytosine base pairs with the maximal three hydrogen bonds (Kaine 1990; Saunders *et al.*, 2003). Though this adaptation indicates improved primary structural stability, the mechanisms by which stability is improved are still under investigation. It has also been hypothesized that thermophilic enzyme amino acid sequences contain much higher concentrations of certain amino acids (Figure 11.2). However, most studies are inconclusive in terms of which specific amino acids provide rigidity (Vogt *et al.*, 1997). In one particular study analyzing the tRNA sequences of several hyperthermophilic Archaea, there was an almost linear trend in the content of Gln, Thr, and Leu as optimal growth temperatures increased (Saunders *et al.*, 2003). Another hypothesis is that thermophilic enzymes have an unusually large number of solvent molecules within their cores. The increase in the aforementioned amino acids leads to a nearly two-fold increase in the solvent-accessible area as opposed to the inaccessible area (Saunders *et al.*, 2003). This indicates that the unusually large number of solvent molecules within thermophilic enzyme cores is a result of the increase of specific amino acids. Surface ion pairs and

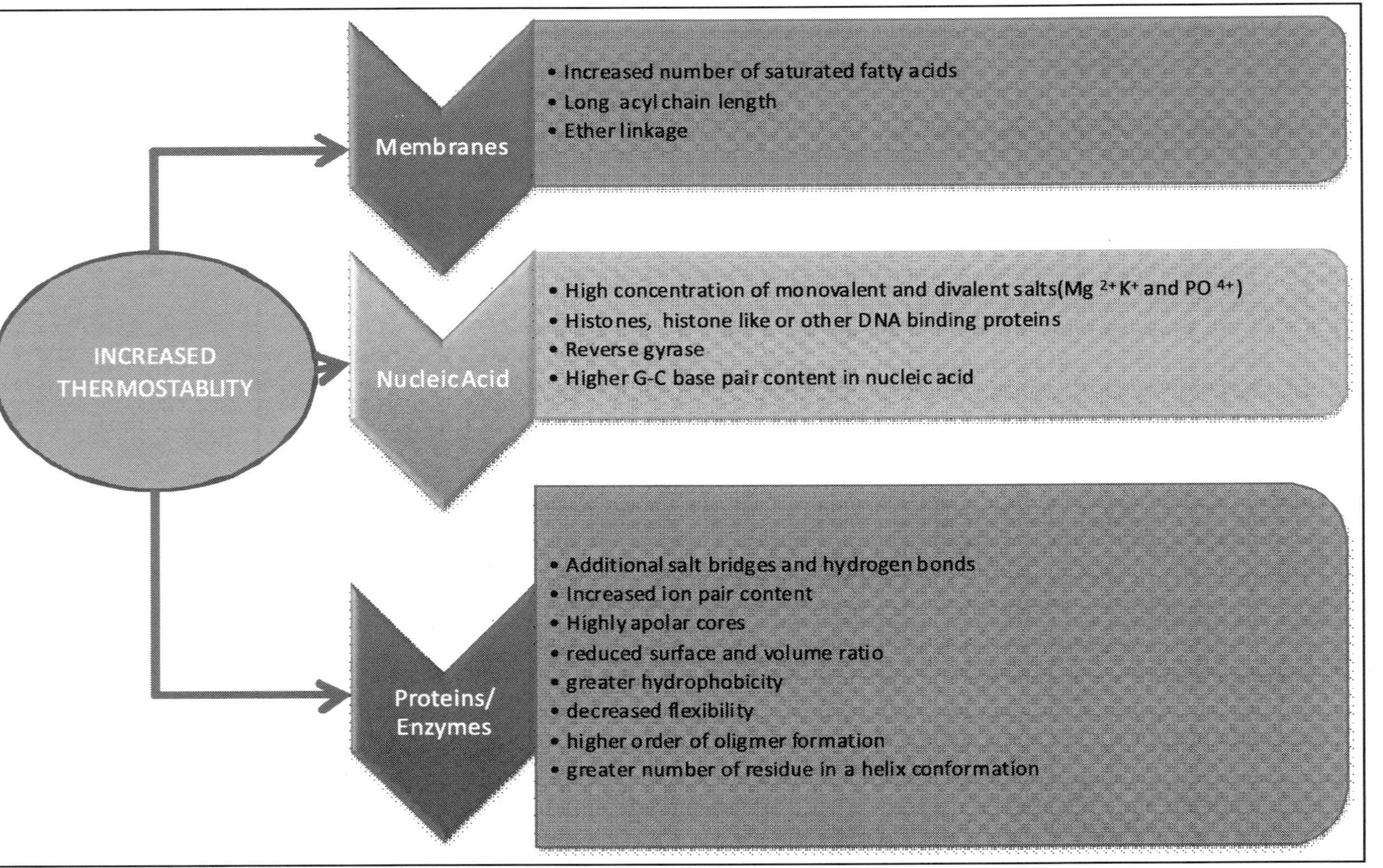


**Figure 11.2: Different Characteristic of Thermophiles to Enhance there Thermostability.**

these solvent filled cavities may provide an extra degree of protection to prevent denaturation under high heat by increasing overall structural stability (Dawson and Hough 1998). Several studies have reported that thermophilic enzymes denature much more rapidly with the disruption of their ion pairs as a result of decreased stability and subsequent unfolding (Cavagnero *et al.*, 1998; Dawson and Hough 1998). Salt links and hydrogen bonding may also be a key component of enzyme stability in thermophiles (Figure 11.2), as they would stabilize both tertiary and quaternary structure in resistance to denaturation (Vieille and Zeikus 2001).

## Advantages of Working with Thermostable Enzymes

Currently thermostable enzymes have engrossed much attention in biotechnology and industry as many common enzymatic industrial reactions are performed at high temperatures. Their increased stability with respect to mesophilic enzymes makes them more suitable for harsh industrial processes. In addition, their thermostability is usually associated with a higher resistance to the chemical denaturants commonly used in many industrial reactions.

Performing enzymatic reactions at high temperature allows higher reaction rates and process yields because of (Figure 11.3):

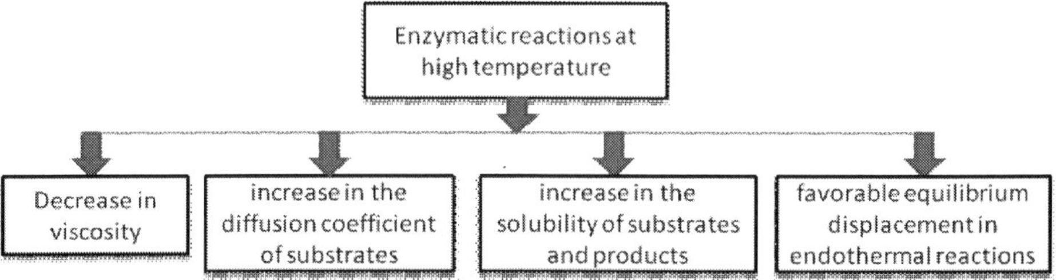


**Figure 11.3: Schematic Representation of Rates of Enzymatic Reaction at High Temperature.**

## Applications in Biotechnology

### As Whole Cell Applications

Thermophilic enzymes are potentially applicable in a wide range of industrial processes mainly due to their extraordinary operational stability at high temperatures and denaturant tolerance. Such enzymes are used in the chemical, food, pharmaceutical, paper, textile and other industries (Zaks A,2001; Gomes and Steiner,2004). Most of these applications utilize recombinant thermostable enzymes that have been expressed in mesophilic hosts. Depending on the type of application, the nature of reactions and product purity, the enzyme preparation can be cell-free (crude, partially purified or homogenous) or cell-associated. For example, the use of cell-free dehydrogenases is hampered by the need for expensive and sensitive co-factors (Stewart,2001) while transaminases suffer from unfavourable reaction equilibria (Taylor *et al.*, 1998) In this regard, whole cell applications can be more attractive. Whole cell applications have also been reported in food processing, making use of recombinant thermophilic glucosidase expressed in *Lactococcus lactis* (Giuliano *et al.*, 2004)The usage of whole cells is of special interest for transformation of lignocellulosics. The bioconversion involves two major steps; saccharification and fermentation. Saccharification is the hydrolysis of carbohydrate polymers into sugar, and this hydrolysate is then utilized as substrate in the fermentation step by microorganisms that transform it into metabolic products. Whole-cell microbial bioconversion offers an attractive possibility of a single step transformation, in which the microorganisms produce saccharolytic enzymes that degrade the lignocellulose and ferment the liberated sugars, which could lead to higher efficiency than in the common multistep lignocellulosic conversions (Lynd *et al.*, 2002).

Among pre-treatment methods, high temperature pre-treatment using liquid hot water is shown to make the biomass cellulose more accessible to enzymatic attack. Development of fermentation systems for thermophiles is here appealing, as it allows energy savings by reducing the cooling cost after steam pre-treatment, lowering the risk of contamination, and improving saccharification and fermentation rates. Moreover, in production of ethanol, thermophilic conditions result in continuous ethanol evaporation allowing harvest during fermentation. Simultaneous fermentation and product recovery can decrease product inhibition of the fermentation process, reduce the volume of water consumed for distillery cooling, and the time required for distillation, leading to a more efficient process. A problem associated with lignocellulose pretreatment procedures is, however, liberation of degradation products that can inhibit microbial growth, but some thermophilic bacteria have shown promising results in fermenting lignocellulosic hydrolysates to ethanol, like the xylanolytic anaerobic thermophilic bacterium, *Thermoanaerobacter mathranii*, shown to ferment the xylose in the hemicellulose fraction from alkaline wet oxidized wheat straw to ethanol with no prior detoxification (Kinke *et al.*, 2004). Moreover, the insolubility of lignocellulosics also create problems in maintaining homogeneity in reactors making monitoring and control of process parameters difficult. Therefore, for their mesophilic counterparts, efficient utilization of thermophiles in integrated bioprocesses needs thorough investigation. In the last few years, reports have been made on solid state cultivation of thermophiles on lignocellulosics (Patel *et al.*, 2006). In some cases, compared to the more traditional submerged liquid fermentation, better conversion has been reached under solid sate cultivation (Chinn, 2006).

Use of naturally occurring microorganisms is, however, generally not efficient enough in transforming the substrate into higher value products. Thus, it is imperative to enhance the robustness of the microbes towards increased substrate hydrolysis through metabolic engineering ethanol and higher product yields (Tao *et al.*, 2001). However, such metabolic engineering reports have been very rare for thermophiles but may increase with the availability/development of genetic tools (Desai *et al.*, 2003). Several thermophilic organisms such as *Thermoanaerobium brockii* (Ben-Bassat *et al.*, 1981) *Clostridium thermohydrosulfuricum* (Lovitt *et al.*, 2005) and *Moorella sp.* HUC22-1 (Sakai *et al.*, 2008), have been studied for ethanol production. Metabolic engineering of such thermophiles to improve ethanol productivity and efficiency of utilizing different substrates like cellulose, hemicellulose and pectin can be very interesting.

## Proteases

Proteolytic enzymes produced by thermophiles are of substantial interest because they are stable and active at elevated temperatures. Likewise they are resistant to organic solvents, detergents, lowand high pH and other denaturants, these properties allow many technological processes. Commercial exploitation of thermostable proteases are available are very limited. One of them is alcalase isolated from *Bacillus licheniformis*. The major ingredient of this preparation is subtilisin, which is an endoprotease of serine type, exhibiting highest activity at 60°C and pH of 8.3. Alcalase found many applications in the food industry,*e.g.*, by reason of their low specificity towards different proteins from plant and animal sources. For example, this enzyme is important in the processing of soy meal which results in soluble, non-bitter hydrolyzate, used as component of protein-fortified soft drinks and dietetic food (Synowiecki, 2008). Alcalase has great importance for recovery of proteins from by-products of the meat and fish industry and from crustacean shell waste during chitin production. In addition, thermostable proteases that are resistant to anionic or non-ionic surfactants and are active at temperatures above 60°C found application as component of dishwashing detergents (Banerjee *et al.*, 1999; Niehaus *et al.*, 1999). Such enzymes can be also used for cleaning ultrafiltration membranes at high temperatures, increasing the efficiency of this process (Bruins *et al.*, 2001). The other potential application of heat-resistant proteases is meat tenderizing. It is due to the great difference of enzymatic activity at moderate and high temperatures. The mesophilic proteases injected into the tissue show a residual activity during the whole period of post-slaughter storage of the meat cuts leading to an excessive fragmentation of the protein molecules. The use of enzymes with an essential activity only during cooking allows the stoppage of proteolysis just by cooling. The stability of heat-resistant proteases in aqueous/organic and nonaqueous media leads to the modification of the reaction equilibria, creating new peptide bonds. Such reverse reactions may be used to improve nutritional quality and functionality of protein hydrolyzates and to reduce their bitterness. It is achieved by increased molecular weight of peptides as well as by introduction of desirable amino acid residues. Among the thermostable proteases currently used on an industrial scale, immobilized thermolysin from *Bacillus thermoproteolyticus* is involved in the synthesis of aspartylphenylalanine- 1-methyl ester, known as aspartame (De Martin *et al.*, 2001). This product is commonly used as sweetener in many low-caloric food and beverages. Enzymatic synthesis eliminates contamination of the product by non-sweet and bitter isomers. Poultry industry generates a large amount of feathers. Traditionally, this by-product is degraded by alkali hydrolysis and steam pressure cooking. Such way of processing destroy some essential amino acids and lead to formation of non-nutritive lysinoalanine and lantionine. biotechnological utilization of keratin-containing wastes. This insoluble protein can be broken down by some microorganisms secreting keratinases and obtained hydrolyzates have been used as fertilizers and dietary protein supplement for animal feed (Grazziotin *et al.*, 2006). In addition, keratinases have potential use as de-hairing agent in leather and cosmetic industry, and as component of detergents and edible films (Cortezi *et al.*, 2008). Furthermore, thermophiles belonging to *Thermoanaerobacter* and *Thermococcus* were found to hydrolyze the thermally denatured amyloid form of prion and could be used for decontamination of animal wastes (Suzuki *et al.*, 2006). The activity of the membrane-bound proteases from thermophilic *Geobacillus colagenovorans* MO-1 towards collagen shows that their substrate binding-domains could be useful in enhancing drug delivery in some tissues. Currently, thermostable protease from *Thermus* sp. Is produced for cleanup of DNA before polymerase chain reaction (PCR). The application of thermostable enzyme for this purpose is caused by compatibility with an existing high-temperature process (Bruins *et al.*, 2001).

## Amylases

Starch from cultivated plants is one of the most abundant and accessible energy sources in the world. It consists of amylose and amylopectin. Corn is the mostly used crop in starch processing in industries, however wheat, potato and tapioca are also important crops while rice, sorghum, sweet potato, arrowroot, sago and mung beans are used in a lesser extent. Hydrolases acting on starch are members of the ?-amylase superfamily, which consists of a large number of primary sequence-related enzymes with a retaining catalytic mechanism (Sinnott ML,1990). Specific consensus sequences, and a varying number of domains, are believed to be responsible for the specificity variations, leading to hydrolysis or transferase activity, as well as differing substrate specificity.

Processed starch is mainly used for glucose, maltose, and oligosaccharide production, but a number of products intermediates can also be produced via cyclodextrins. Glucose can be further converted to high-fructose syrups, crystalline dextrose and dextrose syrups, which are used in food applications. Glucose can of course also be fermented to produce ethanol, amino acids or organic acids. Conversion to high-fructose syrup by glucose isomerase usually runs at 55–60°C and pH 7.0–8.5, requiring a thermostable enzyme. Fructose is a popular sweetener, partly because of the availability of bulk quantities of corn starch at low cost. Starch processing is usually performed in a two-step hydrolysis process of liquefaction and saccharification.Liquefaction is the conversion of granular starch into soluble, shorter-chain-length dextrins [DE (dextrose equivalents) 9–14]. In liquefaction, starch of corn is gelatinized by thermal treatment requiring at a temperature of around 70– 90°C, but to assure the removal of all lipid-amylose complexes, a preferred process temperature is above 100°C (Crabb and Mitchinson,1997).When the starch-slurry is cooled down it forms a thermo-irreversible gel, by a process known as retrogradation, in which the amylose chains interact by hydrogen bonding (Kaper *et al.*, 2004). The crystalline order is then lost and the starch granules swell as the amylose and amylopectin chains are hydrated (Tester and Debon,2000). The starch-slurry is then flash-cooled to 95°C and kept at that temperature for 60–120 mins to complete the enzymatic liquefaction, (Van der Veen *et al.*,2006). Consequently, a highly thermostable enzyme is required which will be active during the whole procedure. Nowadays there are, in addition to the originally used enzymes from *Bacillus stearothermophilus* or *B. licheniformis*, numerous examples available and marketed *e.g.* the Valley "Ultra-thin™" from Valley Research/Diversa, The water content in the starch-slurry is generally quite high (35 per cent), as a high viscosity increases the melting temperature of starch (Souza and Andrade, 2002). Saccharification involves hydrolysis of remaining oligosaccharides (8–12 glucose units) into either maltose syrup by α-amylase or glucose/glucose syrups by glucoamylase (Pandey,1995). The process runs at pH 4.2–4.5 and 60°C, at which temperature the currently used *Aspergillus niger* glucoamylase is stable. Still, the temperature has to be cooled down after liquefaction and the pH has to be adjusted, in order for the glucoamylase to act. More economically feasible would be to utilize an enzyme active in the same pH and temperature range as the liquefaction enzymes. Kim *et al.*, have recently reported on a glucoamylase from *Sulfolobus solfataricus*, which is optimally active at 90°C and pH 5.5–6.0. This enzyme also formed less isomaltose, a common side reaction, than the commercially available fungal glucoamylase(Kim *et al.*,,2004).To increase the efficiency in saccharification, a debranching enzyme, such as pullulanase, can be added to the process. Thermostable enzyme mixes are today available on the market containing both glucoamylase and pullulanase, *e.g.* OPTIMAX® from Genencor.

Several industrially relevant thermostable and thermoactive amylomaltases are known to date (*Thermus* species, *Thermococcus* species, and *Aquifex aeolicus* with optimal temperatures between 75 and 90°C. Amylomaltase catalysis results in conversion into a therthermoreversible starch gel that consists of amylopectin with shortened and elongated side-chains, but free of amylase (Kaper *et al.*, 2004). The obtained gel behaves similar to gelatin and has many uses in the food industry. Applications of amylomaltases on starch also include formation of cycloamyloses (Tereda,1999) and production of isomalto-oligosaccharides (Lee,2002). Cyclodextrins (CDs) are other starch-derived products with a range of possible applications, due to the apolar interior that can host "guest molecules" and solubilize and stabilize them (Szejtli,1998). There are CDs of different sizes, suitable for different applications. Examples of applications of CDs and derivatives there of are: carriers for therapeutically important peptides, proteins and oligonucleotides (Irie and Uekama K,1999),solubilization and stabilization of a range of pharmaceutical molecules (Davis and Brewster,2004), analytical separations (Muderawan and Ong, 2004) and various applications in foods,cosmetics, textiles, and adhesives (Szejtli,1997). There are also large cyclic dextrins, commonly known as cycloamyloses (Strater *et al.*,,2002). Cycloamyloses can be used as a coating material, in adhesives, for biodegradable plastics, as a high energy additive to soft drinks, as a retrogradation retardant for bread improvement, for freeze resistant jellies and for production of non-sticky rice as described by Larsen, (Larsen 2002). Cycloamyloses have also been proposed to aid in protein refolding by acting as an artificial chaperone(Machida *et al.*,2002).

## Lipases

Microbial lipases have an enormous biotechnological potential owing to their unique characteristics, including stability in organic solvents, broad substrate specificity, and high enantioselectivity. Although reports on lipases from hyperthermophilic microorganisms are scant, some such enzymes have been isolated from moderately thermophilic microorganisms and exhibit activity at temperatures of 90°C (Li and Zhang,2005; Rathi *et al.*,,2005).Thermostable lipases from hyperthermophilic archaea have been isolated from *Pyrobaculum calidifontis, Pyrococcus furiosus* and *Pyrococcus horikoshii* and cloned in *E. coli* (Hotta el al,2002; Ikeda and Clark, 1998; Ando *et al.*, 2002].

Lipases are also most widely used group of biocatalysts for biotechnology. They catalyse both the hydrolysis of longchain acylglycerols into glycerol and fatty acids and the reverse reaction. Thus, these enzymes can be applied industrially as hydrolases and as synthetases (Jaeger and Eggert, 2002). Among the many applications of lipases as hydrolases is their use as additives in the detergent industry. In the food industry, lipases are also used to modify the structure of some triglycerides trying to enhance their physical, nutritional or sensory properties. Additionally, these enzymes are also used to obtain poly-unsaturated fatty acids (PUFAs) from plant and animal lipids, which can be subsequently used to produce pharmaceuticals. Another application of lipases as hydrolases is their use in removing the hydrophobic components of the wood ("pitch") in the pulp and paper industry (Jaeger and Reetz, 1998)

Lipases and esterases are also being used for the synthesis of new biopolymeric materials, such as polyphenols, polysaccharides and polyesters. These compounds are of special interest from the environmental point of view because they are biodegradable and can be produced from renewable natural resources. Recently, microbial lipases have been used to obtain biodiesel by transesterification of plant fats. Biodiesel is an alternative source of energy that reduces sulphur oxide production. It is currently being used in public transport, although its long-scale production is still hampered by the high costs of the biocatalyst. Another main characteristic of microbial lipases is their enantioselectivity, which makes them suitable for the production of a number of fine chemicals. This is of great importance in the case of certain pharmaceuticals and agrochemicals.

## Biorefining

The initial step in crop biorefining is fractionation. This is achieved by physical, chemical and biological processes. After starting the physical step, often milling, the biological process employs different hydrolases, depending on what kind of crop is fractionated. Fractionation is often accompanied by elevated temperature, which demands thermostable and thermoactive enzymes. Chemical processes may be used for some applications, but may generate toxic and unwanted side products, and we will not focus on those methods here. Instead enzymatic degradation of starch from grains and utilization of products gained from this will serve as an example of the potential of thermostable enzymes in this type of processing. The straw may also be processed to utilize the carbohydrates present in the lignocellulosic fraction.

## Biodegradation of Lignocellulose

Lignocellulose is an important example of an abundant raw material, produced in large quantities for the production of forest products, often leaving a significant fraction of unutilized waste products. Agricultural waste, such as straw, also has significant lignocellulose content. Enzymes that hydrolyze the polymeric lignocelluloses into shorter metabolizable intermediates, or that reduce viscosity of non-starch polysaccharide in feed cereals can be used to improve utilization of the lignocellulosic carbohydrate fraction. Feed enzymes have been in the market for the last 15 years and the estimated value of this market is nearly $US360 million (Brufau *et al.*, 2006). Feed processing is normally performed at high temperature (Pasamontes *et al.*, 1997) so use and development of stable and robust enzymes has been imperative. Lignocelluloses of plant cell walls are composed of cellulose, hemicellulose, pectin, and lignin. Cellulose is the major constituent of all plant material and the most abundant organic molecule on Earth,while hemicelluloses and pectins are the matrix polysaccharides of the plant cell wall. Many enzymes are involved in the degradation of this biomass resource (Ward and Moo young, 1989)and they are often built up by discrete modules, which is of great environmental interest due to the possibility to decrease chemical bleaching consumption in subsequent steps(Suurnakki *et al.*, 1997). Due to process conditions, enzymes functioning at high temperature and high pH-values are desirable in the following bleaching process. Enzymes from thermophiles meet the temperature demand, as they display intrinsic thermostability, and maximum activity at high temperature, and *e.g.* the xylanase Xyn10A from *R. marinus* has been shown to improve brightness in bleaching sequences of hardwood and softwood kraft pulps prepared by Kraft processing, when introducing the enzyme treatment step at 80°C(Pfabigan, 2002). Several patents have been filed on thermostable xylanases in relation to use in pulping (Bergquist *et al.*, 1997;Sandal, *et al.*, 1997; Fagerstromet al,1999), Xylanases are also produced in industrial scale as additives in feed for poultry (Niehaus *et al.*, 1999)and as additives to wheat flour for improving the quality of baked products [Kaper et al, 2004]. Mannanases have potential in pulp bleaching, especially in combination with xylanase(Clarke,2000) and applications in food and feed include viscosity decreasing action in coffee extracts for instant coffee production (Sachslehner *et al.*, 2005).

## Conversion of Pectins

Pectins is the third main structural polysaccharide group of plant cell walls, abundant in sugar beet pulp and fruit, *e.g.* in citrus fruit and apple, where it can form up to half of the polymeric content of the cell wall (Brummell,2006). The pectin backbone, which consists of homo galacturonic acid regions, and regions of both rhamnose and galacturonic acid, has neutral sugar sidechains made up from L-rhamnose, arabinose, galactose and xylose(Kashyap,2001).Pectin has found widespread commercial use, especially in the textile industry (Henriksson,2001)and in the food industry as thickener, texturizer, emulsifier, stabilizer, filler in confections, dairy products, bakery products,etc (Liu *et al.*, 2006). It is also studied for its potential in drug delivery and in the pharmaceutical industry (Champin *et al.,*,2006) and is interesting as a dietary supplementation to human due to its possible cholesterol-lowering effect (Fernandez *et al.*, 2006). Pectin also has a potential in making biodegradable films (Hoagland and Parris, 1996). In some applications, it can be more proficient to use thermostable enzymes, particularly when using substrates that are poorly soluble at ambient temperatures, such as naringin and rutin, present in fruits. Thermostable pectinases are not so frequently described, but reports show a few thermostable α-L-rhamnosidases, *e.g.* from *Clostridium stercorarium* (Zverlov *et al.,*,2000) and from a strain closely related to *Thermomicrobium* (Birgisson,2004). A thermostable polygalacturonase from a thermophilic mould, Sporotrichum thermophile, optimally active at 55°C has also been reported and may be relevant for the fruit juice industry (Kaur *et al.*, 2004).

## Cellulose Conversion by Cellulases

Cellulose is a homopolysaccharide composed of β-D-glucopyranose units, linked by β-(1→4)-glycosidic bonds. The smallest repetitive unit is cellobiose, as the successive glucose residues are rotated 180° relative to each other. The cellulose hydrolysing enzymes (*i.e.* cellulases) is divided into three major groups: endoglucanases, cellobiohydrolases (and exoglucanases), and glucosidases, all three attacking β-1,4-glycosidic bonds.

A considerable industrial importance for cellulases reached high during the ninties, mainly within textile, detergent paper and pulp industry. Several thermostable enzymes have been characterized, and there have been many trials in these areas as thermostability is highly relevant for the performance of the enzymes. Degradation of cellulose into fermentable sugars for commodity product production is a biorefining area that has invested enormous research efforts as it is a prerequisite for the subsequent production of energy. It is likely to be performed at least partly at high temperature to facilitate the degradation, thus making making thermostable enzymes desirable. Although cellulases cleave a single type of bond, the crystalline substrates with their extensive bonding pattern necessitate the action of a consortium of free enzymes or alternatively multi-component complexes called cellulosomes (Bayer *et al.*, 1998). Carbohydrate-binding modules connected by linkers to the catalytic modules can also give significant contribution to the action of the enzymes, and improve the degradation efficiency, especially on complex lignocellulosic substrates (Bolam *et al.*, 1998;Kittur *et al.*, 2003; Zeverlov *et al.*, 2001). Further improvements in the efficiency level in cellulose degradation, would create both environmental and economic benefits, motivating trials using enzyme blends, as well as engineered cells, and is still a key challenge open for research (Himmel *et al.*, 1999).

## Biofuel

An exemplar shift in industry moving from fossils towards renewable resource utilization, the increased need of microbial thermostable selective biocatalysts.During the world oil crisis, the interest in the use of cellulases to produce fermentable sugars from cellulosic wastes was awakened both in the United States and in Europe. Bioethanol is the most common renewable fuel today, and *e.g.* the "Biofuels Initiative" in the U.S. (US Department of Energy), strives to make cellulosic ethanol cost effective and to become less dependent on oil and reduce the oil imports. Ethanol fermentation from starch can be improved by utilizing better enzymes and strains and preferably hydrolyze the starch from whole grains without a chemical pre-treatment and with simultaneous liquefaction, saccharification and fermentation (Bhargava, 2005).

Agricultural and forest biomass are available in large enough quantities to be considered for large-scale production of alcohol-based fuels. Urban wastes are an additional source of biomass; it is estimated that cellulose accounts for 40 per cent of municipal solid waste (Tomme *et al.*, 1995). Cellulose-based products can be competitive with products derived from fossil resources provided processing costs are reduced (Whyman,2003). Unfortunately, because of the complex and crystalline structure of lignocellulose, this material is much more difficult to hydrolyze than starch. Efficient conversion of lignocellulosic material to fermentable sugars is necessary, but requires better strains or enzyme systems which are able to convert both pentoses and hexoses and tolerate stress conditions (Zaldiver *et al.*, 2005). Thermostable cellulases, hemicellulases, of thermophilic microorganisms in the degradation of the lignocellulosic material offers an advantage by minimizing the risk of contamination and could enable a single-step process of enzymatic hydrolysis, fermentation, and distillation of formed ethanol (Schiraldi and De Rosa 2002), large scale commercial cultivation of thermophiles for enzyme production remains an economical challenge.

The high cost of large-scale fermentation processes to produce enzymes by thermophiles and hyperthermophiles is justifiable only for very few specific applications. Recombinant enzyme production in mesophilic and thermophilic hosts Reduction of the production cost of thermophilic enzymes is fundamental for their breakthrough in large scale. One alternative to reduce production costs and increase the yield of these processes is to use recombinant technology. An extensive variety of thermostable enzymes have been cloned and successfully expressed in mesophilic organisms, such as *Escherichia coli* (Ramachuran *et al.*, 2002), *Bacillus subtilis* (Soutschek and Staudenbauer, 1987), *Saccharomyces cervisae* (Moracci et al., 1992), *Kluyveromyces lactis* (Walsh *et al.*, 1998) and *Trichoderma reesei* (Bergquist, 2002).

Today, the hydrolysis and fermentation steps are separate.The fermentation step is usually performed by *Saccharomyces cerevisiae* or *Zymomonas mobilis*, but this can be a disadvantage,since the temperature has to be reduced from the hydrolysis step, which is better performed at higher temperature, at least 50°C (Stenberg *et al.*, 2000). Thermoactive yeast, *Kluyveromyces marxianus*, active up to 50°C, performed equally well as S. cerevisiae (Singh *et al.*, 1998), but even higher temperature are desired. The fermentation can also be done by thermo-active anaerobic bacteria. For example, some thermophiles isolated from Icelandic hot springs performed quite well in ethanol production from lignocellulolytic hydrolysates (Sommer *et al.*, 2004).

Enzymatic cellulose hydrolysis to glucose is today predominantly carried out by fungi, *e.g. Trichoderma, Penicillium* and *Aspergillus* (Galbe and Zacchi,2002), but to compete with results from acid hydrolysis, more efficient degradation, presumably at higher temperature is needed, and some relevant enzymes have been described from thermophiles and hyperthermophiles.

An endoglucanase from *Acidothermus cellulolyticus*, which was fused to *T. reesei* cellobiohydrolase and expressed in *T. reesei* was for example enhancing saccharification yields (Bower *et al.*, 2005). Endoglucanase and cellobiohydrolase activity is however not sufficient, as the degradation product (cellobiose) inhibits the former enzymes and blocks further depolymerization of the cellulose. To solve this product inhibition, β-glucosidases have to be added, or engineered into production strains that are able to ferment cellobiose and cellotriose to ethanol (Wood and Ingram, 1992). Thermophiles have not yet played any major role in metabolic engineering, due to the limited amount of vectors and tools available for their modification. Instead, well-known mesophiles like *S. cerevisiae* are used, and has recently been modified with genes from a fungal xylose pathway and from a bacterial arabinose pathway, which resulted in a strain able to grow on both pentose and hexose sugar with improved ethanol yields (Karhumaa *et al.*, 2006). Better technologies for biomass pretreatment are also needed. Mechanical, chemical, biological or thermal pre-treatments enhance the cellulase accessibility by removing lignin and hemicelluloses and by partially disrupting the fiber structure. Production possibilities of the biocatalysts An important consideration when selecting a biocatalyst is the prospect of producing it in sufficient amount. These considerations include the choice of either producing by the native host, or if the gene encoding an enzyme of interest should be transferred to a selected host for recombinant production. Generally, gene expression is not a problem related to the thermophilicity of the target protein and those originating from thermophilic resources meet the same production bottlenecks as their counterparts from mesophiles. Another important consideration, crucial for the implementation of biocatalysts, is the production cost, and a few years ago *e.g.* Genencor International was working under a subcontract from the office of Biomass Program(USA), to reduce the cellulase costs in order to make degradation into fermentable sugar more cost-effective (Mitchinson, 2003).

Cellulose degradation by cellulases in large scale is usually carried out by fungal but to introduce more thermoactive enzymes there is a possibility for heterologous production in bacterial hosts, which generally have higher growth rates than fungi. The difficulty using bacterial cellulases is that they are larger, more complex enzymes and often part of a cellulosome with many different activities. Research has also been aimed towards improving presently used fermentation strains by metabolic engineering.

Enzyme production by thermophiles Cultivation of thermophiles at high temperature is technically and economically interesting as it reduces the risk of contamination reduces viscosity, thus making mixing easier, and leads to a high degree of substrate solubility.The low cell yield poses problems for both large and small scale production, which makes extensive studies of their enzymes very difficult. This has triggered considerable research aiming to improve thermophilic cell yield. Several reports on media compositions and culture optimization of different thermophiles are available (Krahe, 1996). Special equipments and specific processes have been developed to improve fermentation processes of thermophiles and hyperthermophiles.

## Other Synthetic Utility

### Synthetic Glucose

Though a vast number of extremozymes are investigated, a very high number of glycosyl hydrolases, which hydrolyze bonds between two or more carbohydrates and between carbohydrates and other entities. Carbohydrates are of particular interest because of their energetic importance in numerous cellular functions such as cell regulation and differentiation. Synthetic production of oligosaccharides is difficult with conventional chemical methods, and new techniques using archaeal enzymes may provide enhanced control of the stereochemistry, or positioning, of bond

**Table 11.1: Growth Conditions, Habitat and Metabolic Properties of some Available Thermophilic Variants**

| *Organisms* | *Growth Conditions* | *Isolation/Habitat* | *Metabolic Properties* |
|---|---|---|---|
| *Thermofilum pendens* (*376*) | 85–90°C, pH 5.0–6.0 | Solfataric fields, Iceland | Heterotrophic anaerobe, mildly acidophile; grows by S0 respiration on complex peptide substrates; requires S0, $H_2S$, and a polar lipid fraction from ***T. tenax*** |
| **Desulfurococcales** | | | |
| *Desulfurococcus mobilis* | 85°C, pH 6.0 | Solfataric fields, Iceland | Strict heterotrophic anaerobe; grows on peptide substrates; S0 respiration or fermentation |
| *D. amylolyticus* | 90–92°C, pH 6.4 | Thermal springs, Kamchatkapeninsula | Strict heterotrophic anaerobe; grows on peptide substrates and polysaccharides; S0 stimulates growth |
| *Staphylothermus marinus* | 92°C, pH 4.5–8.5, 1–3.5 per cent NaCl | Heated sea floor, Vulcano, Italy | Strict anaerobe; S0 dependent; heterotrophic growth on complex organic substrates; produces $CO_2$, acetate, isovalerate, and $H_2S$ |
| *Thermosphaera aggregans* | 85°C, pH 6.5, 0 per cent NaCl | Yellowstone, Obsidian pool | Heterotrophic anaerobe (YE, AA mix, glucose); S0 inhibits growth |
| **Pyrodictiales** | | | |
| *Pyrodictium occultum* | 105°C, pH 5.5, 1.5 per cent NaCl | Marine solfataric fields, Vulcano, Italy | Strict anaerobe; autotrophic growth on $H_2$ 1 $CO_2$ 1 S0 (produces $H_2S$); in the presence of YE, can grow by reduction of $S_2O_3$ 22 |
| *P. abyssi* | 97°C, pH 5.5, 0.7–4.2 per cent NaCl | Deep-sea MHTV, Guaymas, Mexico; shallow MHTV, Kolbeinsey Ridge, north offlce land | Anaerobe, strict heterotroph; grows by fermenting carbohydrates, cell extracts, proteins, and acetate; produces $CO_2$, isovalerate, isobutyrate, and butanol, reduces S0 and $S_2O_3$ 22 in the presence of $H_2$ |
| *P. brockii* | 105°C, pH 5.5, 1.5 per cent NaCl | Marine solfataric fields, Vulcano, Italy | Strict anaerobe; autotrophic growth on $H_2$ 1 $CO_2$ 1 S0 (produces $H_2S$); YE stimulates growth yield; Reduces $SO_3$ 22, not $S_2O_3$ 22 |

*Contd...*

**Table 11.1**–*Contd...*

| *Organisms* | *Growth Conditions* | *Isolation/Habitat* | *Metabolic Properties* |
|---|---|---|---|
| *Hyperthermus butylicus* | 95–106°C, pH 7.0,1.7 per cent NaCl | Marine solfataric field, Azores | Heterotrophic anaerobe; uses peptide mixtures as carbon and energy sources; forms $H_2S$ from S0 1 $H_2$ as accessory energy source; produces $CO_2$, L-butanol, acetate |
| *Thermodiscus maritimus* | 85°C, pH 6.5 | Hot sea water, Vulcano, Italy | Obligate autotroph |
| *Pyrolobus fumarii* | 106°C, pH 5.5, 1.7 per cent NaCl | Deep-sea MHTV (3,650 m), Mid-Atlantic Ridge | Obligate $H_2$-dependent chemolithoautotroph, grows by $NO_3$ 2, $S_3O_3$ 22, or O2 (0.3 per cent) reduction; S0 and several organic nutrients inhibit growth; no growth at 85°C and below |
| **Unclassified** | | | |
| *Aeropyrum pernixa* | 90–95°C, pH 7.0, 3.5 per cent salt | Coastal solfataric MHTV, Japan | Strict aerobe, heterotroph; grows on complex peptide substrates; no $H_2S$ production |
| *Caldococcus litoralis* | 88°C, pH 6.4, 2.5 per cent NaCl | Shallow MHTV, KurileIslands | Strict anaerobic chemoorganotroph; grows on complex peptide substrates and amino acids; S0 stimulates growth (reduced to $H_2S$) |
| **Archaea: Euryarchaeota** | | | |
| **Thermococcales** | | | |
| *Palaeococcus ferrophilus* | 83°C, pH 6.0, 4.7 per cent sea salt | Deep-sea MHTV, Ogasawa-Bonin Arc, Japan | Strict anaerobic chemoorganotroph; grows on proteinaceous substrates in the presence of S0 or Fe21 |
| *Thermococcus aggregans* | 88°C, pH 7.0 | Guaymas basin, Mexico | Chemoorganotrophic strict anaerobe |
| *T. barophilus* | 85°C, pH 7.0, 2–3 per cent NaCl | MHTV (3,550 m), Mid-Atlantic Ridge | Obligate heterotroph; S0 stimulates growth; obligate barophile at 95–100°C |
| *T. guaymasensis* | 88°C, pH 7.2 | Guaymas basin, Mexico | Chemoorganotrophic anaerobe |
| *T. celer* | 88°C, pH 5.8, 4 per cent NaCl | Shallow marine solfataricfield, Vulcano, Italy | Obligate heterotrophic anaerobe; grows on peptide substrates by S0 espiration or by fermentation; NaCl required |

*Contd...*

**Table 11.1**–*Contd...*

| *Organisms* | *Growth Conditions* | *Isolation/Habitat* | *Metabolic Properties* |
|---|---|---|---|
| *T. acidaminovorans* | 85°C, pH 9 0, 1–4 per cent NaCl | Shallow MHTV, Italy | Obligate heterotroph; grows on amino acids as sole carbon and energy source; S0 stimulates growth |
| *T. chitonophagus* | 85°C, pH 6.7, 2 per cent NaCl | Deep-sea MHTV, Guaymas, Mexico | Obligate heterotrophic anaerobe; grows on chitin, YE, and meat extract; produces H2 (H2S in the presence of S0), $CO_2$, $NH_3$, acetate, and format |
| *T. barossii* | 82.5°C, pH 6.5–7.5, 1–4 per cent NaCl | Juan de Fuca Ridge | Obligate heterotrophic anaerobe, grows on peptides; S0 required for growth |
| *T. litoralis* | 85°C, pH 6.0, 1.8–6.5 per cent NaCl | Marine solfataras, Vulcanoand Naples, Italy | Obligate heterotrophic anaerobe; grows in complex peptide substrates; S0 stimulates growth |
| *T. profundus* | 80°C, pH 7.5, 2–4 per cent NaCl | MHTV (1,400 m), Mid-Okinawa Trough, Japan | Obligate heterotrophic anaerobe; S0 dependent; uses complex peptide substrates, starch, pyruvate and maltose |
| *T. stetteri* | 75°C, pH 6.5, 2.5 per cent NaCl | Marine solfararic fields, Northern Kurils | Strict anaerobe, S0 dependent; uses complex peptide substrates, starch, and pectin; production of $CO_2$, acetate, isobutyrate, isovalerate, and $H_2S$ |
| *T. hydrothermalisb* | 85°C, pH 6.0 2–4 per cent NaCl | Deep-sea MHTV, East Pacific Rise | Obligate heterotrophic anaerobe; grows on proteolysis products, AA mix, and maltose in the presence of S0 |
| *Pyrococcus furiosusb* | 100°C, pH 7.0, 2 per cent NaCl | Marine solfataric fields, Vulcano, Italy | Obligate heterotrophic anaerobe; grows on peptide substrates and carbohydrates; S0 stimulates growth, probably by detoxifying $H_2$ (forming $H_2S$) |
| *P. woesei* | 100–103°C, pH 6.0–6.5, 3 per cent NaCl | Marine solfataras, Vulcano, Italy | Obligate heterotrophic anaerobe (YE, peptides, PS); S0 respiration, no fermentation |
| *P. abyssi* | 96°C, pH 6.8, 3 per cent NaCl | Deep-sea MHTV, North Fiji Basin | Obligate chemoorganotroph, fermenting peptide substrates; Produces $CO_2$, $H_2$, acetate, propionate, isovalerate, and isobutyrate; produces $H_2S$ in the presence of S0; facultative barophilic; NaCl required |
| *P. horikoshiia* | 98°C, pH 7.0, 2.4 per cent NaCl | Okinawa Trough, western Pacific | Obligate heterotrophic anaerobe; Trp auxotroph |

*Contd...*

**Table 11.1**–*Contd...*

| *Organisms* | *Growth Conditions* | *Isolation/Habitat* | *Metabolic Properties* |
|---|---|---|---|
| **Archaeoglobales** | | | |
| *Archaeoglobus fulgidusa* | 83°C, pH 5.5–7.5 | Heated sea floor, Vulcano, Italy | Strict anaerobe; chemolithoautotroph in the presence of $H_2$, $CO_2$, and $S_2O_3$ 22; heterotrophic growth on formate, formamide, lactate, glucose, starch, and peptide substrates; produces traces of methane |
| *A. profundus* | 82°C, pH 4.5–7.5, 0.9–3.6 per cent NaCl | Deep-sea MHTV, Guaymas, Mexico | Strict anaerobe, mixotroph, requires $H_2$ for growth; uses organic acids, YE, peptide substrates as carbon sources; electron acceptors include sulfate, $S_2O_3$ 22, and sulfite |
| **Methanococales** | | | |
| *Methanococcus jannaschiia* | 85°C, pH 6.0, 2–3 per cent NaCl | Deep-sea MHTV (2,600 m), East Pacific Rise | Autotrophic anaerobe, methanogen; NaCl and sulfide required for growth |
| *M. vulcanius* | 80°C, pH 6.5, 2.5 per cent NaCl | East Pacific Rise | Anaerobe, methanogen; growth stimulated by YE, selenate, and tungstate; reduces S0 in the presence of $CO_2$ and $H_2$ |
| *M. fervens* | 85°C, pH 6.5, 3 per cent NaCl | Guaymas Basin, Mexico | Anaerobe, methanogen; growth stimulated by YE, selenate, and tungstate, Casamino Acids, and trypticase |
| *M. igneus* | 88°C, pH 5.7, 1.8 per cent NaCl | Shallow MHTV, Mid-Atlantic Ridge, north off Iceland | Anaerobe, methanogen, obligate chemolithoautotroph; S0 inhibits growth |
| *M. infernus* | 85°C, pH 6.5, 2.5 per cent salt | Deep-sea MHTV, Mid-Atlantic Ridge | Chemolithotroph, obligate anaerobe, methanogen, reduces S0; YE stimulates growth |
| **Methanobacteriales** | | | |
| *Methanothermus fervidus* | 83°C, pH 6.5 | Icelandic hot spring | Anaerobe, methanogen; requires YE to grow in artificial medium |
| *M. sociabilis* | 88°C, pH 6.5 | Continental solfatara fields, Iceland | Anaerobic S-independent autotroph; methanogen |
| **Methanopyrales** | | | |
| *Methanopyrus kandleri* | 98°C, pH 6.5, 1.5 per cent NaCl | Deep-sea MHTV, Guaymas, Mexico | Strict anaerobe chemolithoautotroph; methanogen |

formation (Moracci *et al.*, 2001). α-Amylase, a starch-degrading enzyme, randomly hydrolyzes linkages yielding glucose and oligosaccharides (Eichler 2001). Several hyperthermophilic Archaea, such as *Pyrococcus furiosus*, *Pyrococcus woesei* and *Thermococcus litoralis* (Table 11.1) have produced effective amyolytic enzymes (Bergquist *et al.*, 2004; Lee *et al.*, 1996). Two different types of α-amylase are also successfully produced by the archaeon *Thermococcus profundus*, (Table 11.1) one of which was able to be cloned and expressed in *E. coli* (Lee *et al.*, 1996). The ability of an isolated enzyme to be expressed in a mesophilic host at the same temperature with the same molecular weight suggests that the enzyme can be produced in a host and also retain its functional properties at extremely high temperatures. In this instance, the activity of the recombinant α-amylase is 155 times higher than the enzyme produced by the parent strain when observed under what would be considered its natural conditions. High temperatures are required to liquefy starch to make it suitable for enzymatic processing and allow its eventual conversion to more biochemically valuable products such as glucose and dextrin (Bergquist *et al.*, 2004). Therefore, these types of enzymes could prove valuable for future use in synthetic oligosaccharide and glucose production (Lee *et al.*, 1996).

**Synthetic Hydrocarbon**

Alcohol dehydrogenases play an important role in catalyzing the reversible oxidation of alcohols to the correct aldehyde or ketone. Though many alcohol dehydrogenases have been found to be important building blocks in drug manufacturing, they are difficult to work with because of their overall low structural stability. Alcohol dehydrogenases from thermophilic Archaea have been cloned and expressed in *E. coli*, showing correct substrate specificity and rate of substrate hydrolysis (Griffith and Ewart 1995). This makes alcohol dehydrogenases a potential candidate for use as biocatalysts for industrial chiral aliphatic alcohol, or artificial alcohol hydrocarbon, syntheses (Hirakawa *et al.*, 2004).

On the other hand, due to factors such as requirement of complex and expensive media, low solubility of gas at high temperature, and low specific growth rates and product inhibition.Large scale commercial cultivation of thermophiles for enzyme production remains an economical challenge. The high cost of large-scale fermentation processes to produce enzymes by thermophiles and hyperthermophiles is justifiable only for very few specific applications.

## Conclusion

Enzymes from thermophilic microorganisms offer a potential solution for industrial processes carried out at high temperatures.The applications discussed in this chapter also present far-reaching implications and limitless potential in improving the area of biotechnological applications by manipulating and understanding the archaeal domain. Discoveries of new archaeal microorganisms continue to excite the scientific community. Thermophilic enzymes are potentially applicable in a wide range of industrial processes mainly due to their extraordinary operational stability at a high temperatures and denaturant tolerance. Their unique adaptations that cater to hyperthermic, and hypothermic circumstances has driven the research to manipulate those attributes for use in every aspect of life. Adaptations in membrane, enzymes, and protein structures and components have potential applications in areas including electronics, agriculture, aquaculture, medicine, pharmaceuticals, food science, and nutrition other industries. Thermophiles and especially thermophilic enzymes have to date gained a great of interest both as analytical tools, and as biocatalysts for application in large scale.Utilization of these enzymes are however still today, despite many efforts, often limited by the cost of the enzymes. With an increasing market for the enzymes, leading to production in higher volumes, the cost is however predicted to decrease. One of the main problems in most enzymatic industrial processes is how to achieve an appropriate compromise between harsh working conditions and enzyme stability. Since only a very few species from this group of microorganisms have been isolated to date, there seems to be a large number of hyperthermophilic catalysts with unique properties are awaiting discovery. The development of molecular tools together with biotechnological research should lead to the future availability of new hyperthermophilic enzymes to meet all types of industrial demands.

## References


1. Ando, S., Ishida, H., Kosugi, Y. and Ishikawa, K. (2002). *Appl. Environ. Microbiol. 68,* 430-433.
2. Banerjee V, Saani K, Azmi W, Soni R (1999). Thermostable alkaline protease from *Bacillus brevis* and its characterization as a laundry additive. *Proc. Biochem.* 35: 213-219.
3. Bayer EA, Chanzy H, Lamed R, Shoham Y (1998). Cellulose, cellulases and cellulosomes. *Current Opinion in Structural Biology,* 8: 548-557.
4. Ben-Bassat A, Lamed R, Zeikus JG (1981). Ethanol production by thermophilic bacteria: metabolic control of end product formation in Thermoanaerobium Brockii. *Journal of Bacteriology* 146: 192-199.
5. Bergquist PL, Gibbs MD, Morris D (1997). Xylanase from *Dictyoglomus thermophilum* and its use in bleaching of cellulose products. *Patent.* WO 9736995.
6. Bergquist, P.L. *et al.* (2004). Recombinant enzymes from thermophilic micro-organisms expressed in fungal hosts. *Biochemical Society Transactions* 32, 293-297.
7. Bertoldi M, Vallini G, Pera A (1983). The biology of composting: a review. *Waste Manage Res* 1: 157–176
8. Bhargava S, Frisner H, Bisgard-Frantzen H, Tams JW (2005). A process of producing a fermentation product. *Patent.* WO2005113785.
9. Birgisson H, Hreggvidsson GO, Fridjonsson OH, Mort A, Kristjansson JK, Mattiasson B (2004). Two new thermostable α-L-rhamnosidases from a novel *Thermophilic bacterium. Enzyme and Microbial Technology* 34: 561-571.
10. Bloch, E., Rachel, R., Burgraff, S., Hafenbradl, D., Jannasch, H.W. and Stetter, K.O. (1997). *Extremophiles* 1, 14-21.

11. Bolam DN, Ciruela A, McQueen-Mason S, Simpson P, Williamson MP, Rixon JE, Boraston A, Hazlewood GP, Gilbert HJ (1998). Pseudomonas cellulose- binding domains mediate their effects by increasing enzyme substrate proximity. *Biochemical Journal*, 331: 775-781.

12. Bower B, Larenas E, Mitchinson C (2005). Exo-endo cellulase fusion protein. *Patent*. WO2005093073.

13. Brock TD, Freeze H: Thermus aquaticus gen. N. And sp. N., a nonsporulating extreme thermophile. *Journal of Bacteriology* 1969, 98: 289-297.

14. Brufau J, Francesch M, Pérez-Vendrell AM (2006). The use of enzymes to improve cereal diets for animal feeding. *Journal of the Science of Food and Agriculture*, 86: 1705-1713.

15. Bruins ME, Janssen AEM, Boom RM (2001). Thermozymes and their applications. *Appl. Biochem. Biotechnol.* 90: 155-186.

16. Brummell DA (2006). : Cell wall disassembly in ripening fruit. *Functional Plant Biology*, 33: 103-119.

17. Bull, A.T.; Ward A.C.; Goodfellow, M. (2000). Search and Discovery. Strategies for Biotechnology: the Paradigm Shift. *Microbiol. Mol. Biol. Rev.*, 46(3), 573-606.

18. Cavagnero, S. D. *et al.* (1998). Kinetic role of electrostatic interactions in the unfolding of hyperthermophile and mesophilic rubredoxins. *Biochemistry* 27, 3369-3376.

19. Chambin O, Dupuis G, Champion D, Voilley A, Pourcelot Y (2006). Colonspecific drug delivery: Influence of solution reticulation properties upon pectin beads performance. *International Journal of Pharmaceutics*, 321: 86-93.

20. Chinn MS, Nokes SE, Strobel HJ (2006). Screening of thermophilic anaerobic bacteria for solid substrate cultivation on lignocellulosic substrates. *Biotechnology Progress*, 22: 53-59.

21. Clarke JH, Davidson K, Rixon JE, Halstead JR, Fransen MP, Gilbert HJ, Hazlewood GP (2000). A comparison of enzyme-aided bleaching of softwood paper pulp using combinations of xylanase, mannanase and α-galactosidase. *Applied Microbiology and Biotechnology*, 53: 661-667.

22. Comita, P.B., Gagosian, R.B., Pang, H. and Costello, C.E. (1984). *J.Biol. Chem.* 259, 15234-15241.

23. Cortezi M, Contiero J, De Lima CJB, Lovaglio RB, Monti R (2008). Characterization of a feather degrading by *Bacillus amyloliquefaciens* protease: A new strain. *World J. Agric. Sci.* 4: 648-656.

24. Crabb WD, Mitchinson C (1997). : Enzymes involved in the processing of starch to sugars. *Trends in Biotechnology*, 15: 349-352.

25. D'Auria, S., Morana, A., Febbraio, F., Carlo, V., De Rosa, M. and Nucci, R. (1996). *Protein Expr Purif.* 7, 299-308.

26. Davis ME, Brewster ME (2004). Cyclodextrin-based pharmaceutics: Past, present and future. *Nature Reviews Drug Discovery*, 3: 1023-1035.

27. Dawson, M. J. and D. W. Hough (1998). Structure, function and stability of enzymes from the Archaea. *Trends in Microbiology* 6, 307-314.

28. De Martin L, Ebert C, Gardossi L, Linda P (2001). High isolate yields in thermolysin catalyzed synthesis of Z-L-aspartyl-L-phenylalanine methyl ester in toluene at controlled water activity. *Tetrahedron Lett.* 42: 3395-3397.

29. Desai SG, Guerinot ML, Lynd LR (2004). Cloning of L-lactate dehydrogenase and elimination of lactic acid production via gene knockout in *Thermoanaerobacterium saccharolyticum* JW/SLYS485. *Applied Microbiology and Biotechnology*, 65: 600-605.

30. Doran JB, Cripe J, Sutton M, Foster B (2000). Fermentations of pectinrich biomass with recombinant bacteria to produce fuel ethanol. *Applied Biochemistry and Biotechnology*, 84–6: 141-152.

31. Egorova, K.; Antranikian, G. (2005). Industrial relevance of thermophilic Archaea. *Curr Opin. Microbiol.*, 8, 649-655.

32. Eichler, J. (2001). Biological uses of archaeal extremozymes. *Biotechnology Advances* 19, 261 278.

33. Everly, C. and Alberto, J. (2000). *Front. Biosci.* 5, 780-786.

34. Fagerstrom RB, Paloheimo M, Lantto R, Lahtinen T, Suominen P(1999). Xylanases and their use. *Patent. US* 5922579.

35. Fernandez ML, Sun DM, Tosca MA, McNamara DJ (2006). : Citrus pectin and cholesterol interact to regulate hepatic cholesterol homeostasis and lipoprotein metabolism: a dose-response study in guinea pigs. *American Journal of Clinical Nutrition*, 59: 869-878.

36. Firn, R.D. (2003). Bioprospecting-Why is it so unrewarding? *Biodiversity and Conservation*, 12, 207-216.

37. Forterre, P., Bergerat, A. and López-García, P. (1996). *FEMS Microbiol. Rev.* 18, 237-248.

38. Galbe M, Zacchi G (2002). : A review of the production of ethanol from softwood. *Applied Microbiology and Biotechnology*, 59: 618-628.

39. Giuliano G, Schiraldi C, Marotta MR, Hugenholtz J, De Rosa M (2004). Expression of *Sulfolobus solfataricus* α-glucosidase in *Lactococcus lactis*. *Applied Microbiology and Biotechnology*, 64: 829-832.

40. Gomes J, Steiner W (2004). The biocatalytic potential of extremophiles and extremozymes. *Food Technology and Biotechnology*, 42: 223-235.

41. Grazziotin A, Pimentel FA, De Jong EV, Brandelli A (2006). Nutritional improvement of feather protein by treatment by microbial keratinase. *Anim. Feed Sci. Technol.* 126: 135-144.

42. Griffith, M. and K. V. Ewart (1995). Antifreeze proteins and their potential use in frozen foods. *Biotechnology Advances* 13(3), 375-402.

43. Grogan, D.W. (2004). *Curr. Issues. Mol. Biol.* 6, 137-144.

44. Gupta R, Gupta N, Rathi P (2004). Bacterial lipases: an overview of production, purification and biochemical properties. *Appl Microbiol Biotechnol* 64: 763–781

45. Haki, G.D. and Rakshit, S.K. (2003). *Biores. Technol.* 89, 17-34

46. Haney, P.J., Stees, M. and Konisky, J. (1999). *J. Biol. Chem.* 274, 841-848.

47. Hasan F, Shah AA, Hameed A (2005). Industrial applications of microbial lipases. *Enzyme Microb Technol* 39: 235–251

48. Hayashi T. and K. Hayashi (1996). Calcium spirulan, an inhibitor of enveloped virus replication, from a bluegreen alga Spirulina platensis. *Journal of Natural Products* 59, 83–87.

49. Headon, D.R.; Walsh, G. (1994). The industrial production of enzymes. *Biotechnol. Adv.*, 12(4), 635-646.

50. Henriksson G, Akin DE, Slomczynski D, Eriksson KEL (1999). Production of highly efficient enzymes for flax retting by *Rhizomucor pusillus*. *Journal of Biotechnology* 1999, 68: 115-123.

51. Himmel ME, Ruth MF, Wyman CE(1999). : Cellulase for commodity products from cellulosic biomass. *Current Opinion in Biotechnology*, 10: 358-364.

52. Hirakawa, H. *et al.* (2004). Properties of an alcohol dehydrogenase from the hyperthermophilic archaeon Aeropyrum pernix K1. *Journal of Bioscience and Bioengineering* 97, 202-206.

53. Hoagland PD, Parris N (1996). Chitosan/pectin laminated films. *Journal of Agricultural and Food Chemistry*, 44: 1915-1919.

54. Hotta, Y., Ezaki, S., Atomi, H. and Imanaka, T. (2002). *Appl. Environ. Microbiol.* 68, 3925-3931.

55. Hough, D.W. and Danson, M.J. (1999). *Curr. Opin. Chem. Biol.* 3, 39-46.

56. Hutnan M, Drtil M, Mrafkova L (2000). Anaerobic biodegradation of sugar beet pulp. *Biodegradation* 2000, 11: 203-211.

57. Ikeda, M. and Clark, D. (1998). *Biotechnol. Bioeng. 57*, 624-629.

58. Ilmen, M.; Saloheimo, A.; Onnela, M.A.; Penttila, M.E. (1997). Regulation of cellulase gene expression in the filamentous fungus *Trichoderma reesei*. *Appl. Environ. Micobiol.*, 63, 1298-1306.

59. Irie T, Uekama K (1999). Cyclodextrins in peptide and protein delivery. *Advanced Drug Delivery Reviews*, 36: 101-123.

60. Jaeger, K.E. and Reetz, M.T. (1998). *Trends Biotechnol. 16*, 396- 403

61. Kaine, B.P. (1990). Structure of the archaebacterial 7S RNA molecule. *Molecular and General Genetics* 221(3), 315-321.

62. Kaper T, Talik B, Ettema TJ, Bos H, van der Maarel M, Dijkhuizen L (2005). Amylomaltase of *Pyrobaculum aerophilum* IM2 produces thermoreversible starch gels. *Applied and Environmental Microbiology*, 71: 5098-5106.

63. Kaper T, van der Maarel MJEC, Euverink G-JW, Dijkhuizen L (2004). Exploring and exploiting starch-modifying amylomaltases from thermophiles. *Biochemical Society Transactions*, 32: 279-282.

64. Karhumaa K, Wiedemann B, Hahn-Hagerdal B, Boles E, Gorwa- Grauslund MF (2006). Co-utilization of L-arabinose and D-xylose by laboratory and industrial *Saccharomyces cerevisiae* strains. *Microbial Cell Factories*, 5: 1-23.

65. Kashefi, K. and Lovley, D.R. (2003). *Science* 15 (301).

66. Kashyap DR, Vohra PK, Chopra S, Tewari R (2001). : Applications of pectinases in the commercial sector: a review. *Bioresource Technology*, 77: 215-227.

67. Kaur G, Kumar S, Satyanarayana T(2004). Production, characterization and application of a thermostable polygalacturonase of a thermophilic mould *Sporotrichum thermophile Apinis*. *Bioresource Technology*, 94: 239-243.

68. Kim M-S, Park J-T, Kim Y-W, Lee H-S, Nyawira R, Shin H-S, Park CS, Yoo S-H, Kim Y-R, Moon T-W (2004). Properties of a novel thermostable glucoamylase from the hyperthermophilic archaeon Sulfolobus solfataricus in relation to starch processing. *Applied and Environmental Microbiology*, 70: 3933-3940.

69. Kimura, M., Kimura, J., Davie, P., Reinhardt, R. and Dijk, J. (1984). *FEBS Lett.* 176, 176-178.

70. Kirino, H., Aoki, M., Aoshima, M., Hayashi, Y., Ohba, M., Yamagishi, T., Wakagi, T. and Oshima, T. (1994). *Eur. J. Biochem.* 220, 275-281.

71. Kittur FS, Mangala SL, Abu Rus'd A, Kitaoka M, Tsujibo H, Hayashi K (2003). Fusion of family 2b carbohydrate-binding module increases the catalytic activity of a xylanase from Thermotoga maritime to soluble xylan. *FEBS Letters*, 549: 147-151.

72. Klinke HB, Thomsen AB, Ahring BK (2004). Inhibition of ethanol-producing yeast and bacteria by degradation products produced during pre-treatment of biomass. *Applied Microbiology and Biotechnology*, 66: 10-26.

73. Krahe M, Antranikian G, Mäirkl H (1996). Fermentation of extremophilic microorganisms. *FEMS Microbiology Reviews*,18: 271-285.

74. Kristjansson JK, Stetter KO (1992). Thermophilic bacteria. In: Kristjansson JK (ed). *Thermophilic Bacteria*. CRC press, Boca Raton, pp. 1–18

75. Kuhad RC, Singh A, Eriksson K-EL: Microorganisms and enzymes involved in the degradation of plant fiber cell walls. In *Biotechnology in the Pulp and Paper Industry* Edited by: Eriksson K-EL. Berlin: Springer-Verlag; 1997: 45-125.

76. Larsen KL(2002). Large cyclodextrins. *Journal of Inclusion Phenomena*, 43: 1-13.

77. Lee H-S, Auh J-H, Yoon H-G, Kim M-J, Park J-H, Hong S-S, Kang MH, Kim T-J, Moon T-W, Kim J-W, (2002). Cooperative action of alpha-glucanotransferase and maltogenic amylase for an improved process of isomaltooligosaccharide (IMO). production. *Journal of Agricultural and Food Chemistry*, 50: 2812-2817.

78. Lee, J.T. *et al.* (1996). Cloning, nucleotide sequence, and hypertension of α–Amylase gene from an archaeon, *Thermococcus profundus*. *Journal of Fermentation and Bioengineering* 82, 432-438.

79. Li, H. and Zhang, X. (2005). *Protein Expr. Purif.* 42, 153-159.

80. Liu Y, Shi J, Langrish TAG (2006). : Water-based extraction of pectin from flavedo and albedo of orange peels. *Chemical Engineering Journal*, 120: 203-209.

81. Lopez, G. (1999). *J. Mol. Evol.* 46, 439-452.

82. Lovitt RW, Longin R, Zeikus JG (1998). Ethanol production by thermophilic bacteria: physiological comparison of solvent effects on parent and alcohol-tolerant strains of *Clostridium thermohydrosulfuricum*. *Applied and Environmental Microbiology*, 48: 171-177.

83. Lynd LR, Van Zyl WH, McBride JE, Laser M(2005). Consolidated bioprocessing of cellulosic biomass: an update. *Current Opinion in Biotechnology*, 16: 577-583.

84. Machida S, Ogawa S, Xiaohua S, Takaha T, Fujii K, Hayashi K (2000). : Cycloamylose as an efficient artificial chaperone for protein refolding. *FEBS Letters*, 486: 131-135.

85. Marchessault RH, Sundararajan PR: Cellulose. In *The Polysaccharides* Volume 2. Edited by: Aspinall GO. London: Academic Press Inc; 1983: 11-95.

86. Mitchinson C(2003). Improved cellulose products for biomass conversion. In *Abstract Book Groningen Netherlands: 5th Carbohydrate Bioengineering Meeting*.

87. Moracci M, La Volpe A, Pulitzer JF, Rossi M, Ciaramella M (1992). Expression of the thermostable β-galactosidase gene from the archaebacterium *Sulfolobus solfataricus* in *Saccharomyces cerevisiae* and characterization of a new inducible promoter for heterologous expression. *Journal of Bacteriology*, 174: 873-882.

88. Moracci, M. *et al.* (2001). Enzymatic synthesis of oligosaccharides by two glycosyl hydrolases of Sulfolobus solfactaricus. *Extremophiles* 5, 145-152.

89. Muderawan IW, Ong TT, Ng SC(2006). Urea bonded cyclodextrin derivatives onto silica for chiral HPLC. *Journal of Separation Science* 2006, 29: 1849-1871.

90. Niehaus F, Bertoldo C, Kahler M, Antranikian G (1999). Extremophiles as a source of novel enzymes for industrial applications. *Appl. Microbiol. Biotechnol.* 51: 711-729.

91. Niehaus F, Bertoldo C, Kähler M, Antranikian G (1999). Extremophiles as a source of novel enzymes for industrial application. *Applied Microbiology and Biotechnology*, 51: 711-729.

92. Pandey A (1995). Glucoamylase research – an overview. *Starch*, 47: 439-445.

93. Pasamontes L, Haiker M, Wyss M, Tessier M, van Loon APGM (1997). Gene cloning, purification, and characterization of a heat-stable phytase from the fungus Aspergillus fumigatus. *Applied and Environmental Microbiology*, 63: 1696-1700.

94. Patel MA, Ou MS, Harbrucker R, Aldrich HC, Buszko ML, Ingram LO, Shanmugam KT (2006). Isolation and characterization of acid-tolerant, thermophilic bacteria for effective fermentation of biomass- derived sugars to lactic acid. *Applied and Environmental Microbiology*, 72: 3228 3235.

95. Pfabigan N, Nordberg Karlsson E, Ditzelmueller G, Holst O (2002). Prebleaching of kraft pulp with full-length and functional domains of a thermostable xylanase from Rhodothermus marinus. *Biotechnology Letters*, 24: 1191-1197.

96. Ramchuran SO, Nordberg Karlsson E, Velut S, Maré de L, Hagander P, Holst O (2002). : Production of heterologous thermostable glycoside hydrolases and the presence of host-cell proteases in substrate limited fed-batch cultures of *Escherichia coli* BL21(DE3). *Applied Microbiology and Biotechnology* 60: 408-416.

97. Rathi, P., Sapna, B., Sexena, R. and Gupta, R. (2000). *Biotechnol.Lett.* 22, 495-498.

98. Sachslehner A, Foidl G, Foidl N, Gubitz G, Haltrich D (2000). Hydrolysis of isolated coffee mannan and coffee extract by mannanases of *Sclerotium rolfsii*. *Journal of Biotechnology*, 80: 127-134.

99. Saha BC: α-L-Arabinofuranosidases – biochemistry, molecular biology and application in biotechnology. *Biotechnology Advances* 2000, 18: 403-423.

100. Sakai S, Nakashimada Y, Yoshimoto H, Watanabe S, Okada H, Nishio N(2004). Ethanol production from H2 and CO2 by a newly isolated thermophilic bacterium, Moorella sp.HUC22-1. *Biotechnology Letters*, 26: 1607-1612.

101. Sandal T, Kofod LV, Kauppinen MS, Andersen LN, Dybdal L (1997). Enzyme with xylanase activity. *Patent*. WO 9727292.

102. Saunders, N.F.W. *et al.* (2003). Mechanisms of thermal adaptation revealed from the genomes of the Antarctic Archaea Methanogenium frigidum and Methanococcoides burtonii. *Genome Research* 13, 1580-1588.

103. Schiraldi C, De Rosa M (2002). The production of biocatalysts and biomolecules from extremophiles. *Trends in Biotechnology* 2002, 20: 515-521.

104. Sheehan J, Himmel M(1999). Enzymes, energy, and the environment: A strategic perspective on the U.S. Department of Energy's research and development activities for bioethanol. *Biotechnology Progress*, 15: 817-827.

105. Singh D, Nigam P, Banat IM, Marchant R, McHale AP (1998). Ethanol production at elevated temperatures and alcohol concentrations: Part II – Use of *Kluyveromyces marxianus* IMB3. *World Journal of Microbiology and Biotechnology*, 14: 823-834.

106. Singh SA, Ramakrishna M, Appu Rao AG (1999). Optimisation of downstream processing parameters for the recovery of pectinase from the fermented bran of *Aspergillus carbonarius*. *Process Biochemistry*, 35: 411-417.

107. Sinnott ML (1990). Catalytic mechanisms of enzymic glycosyl transfer. *Chemical Reviews*, 90: 1171-1202.

108. Sjöström E. *Wood Chemistry: Fundamentals and Applications*. London: Academic Press Inc; 1993.

109. Sommer P, Georgieva T, Ahring BK (2004). Potential for using thermophilic anaerobic bacteria for bioethanol production from hemicellulose. Bioch*emical Society Transactions*, 32: 283-289.

110. Soutschek-Bauer E, Staudenbauer WL (1987). Synthesis and secretion of a heat-stable carboxymethylcellulase from *Clostridium thermocellum* in *Bacillus subtilis* and *Bacillus stearothermophilus*. *Molecular and General Genetics*, 208: 537-541.

111. Souza RCR, Andrade CT (2002). Investigation of the gelatinization andextrusion processes of corn starch. *Advances in Polymer Technology*, 21: 17-24.

112. Spagnuolo M, Crecchio C, Pizzigallo MDR, Ruggiero P (1999). Fractionation of sugar beet pulp into pectin, cellulose, and arabinose by arabinases combined with ultrafiltration. *Biotechnology and Bioengineering*, 64: 685-691.

113. Stenberg K, Galbe M, Zacchi G (2000). The influence of lactic acid formation on the simultaneous saccharification and fermentation (SSF). of softwood to ethanol. *Enzyme and Microbial Technology*, 26: 71-79.

114. Stewart JD (2001). Dehydrogenases and transaminases in asymmetric synthesis. *Current Opinion in Chemical Biology* 2001, 5: 120-129.

115. Sträter N, Przylas I, Saenger W, Terada Y, Fujii K, Takaha T(2002). Structural basis of the synthesis of large cycloamyloses by amylomaltase. *Biologia, Bratislava*, 57(Suppl 11). 93-99.

116. Sung WL, Yaguchi M, Ischikawa K: Modification of xylanases to improve thermophilicity, alkalophilicity, and thermostability for pulp bleaching. Patent. EP 828002 1998.

117. Suurnäkki A, Tenkanen M, Buchert J, Viikari L: Hemicellulases in the bleaching of chemical pulps. In: *Biotechnology in the Pulp and Paper Industry* Volume 57. Edited by: Eriksson K-EL. Berlin: Springer- Verlag; 1997: 261-287.

118. Suzuki Y, Tsujimoto Y, Matsui H, Watanabe K (2006). Decomposition of extremely hard-to-degrade animal proteins by thermophilic bacteria. *J. Biosci. Bioeng*. 102: 73-81.

119. Synowiecki J (2008). Thermostable enzymes in food processing, in Recent Research Developments in Food Biotechnology. Enzymes as Additives or Processing Aids, Research Signpost, Kerala.

120. Szejtli J (1997). Utilization of cyclodextrins in industrial products and processes. *Journal of Materials Chemistry* 1997, 7: 575-587.

121. Szejtli J (1998). : Introduction and general overview of cyclodextrin chemistry. *Chemical Reviews*, 98: 1743-1754.

122. Tao H, Gonzalez R, Martinez A, Rodriguez M, Ingram LO, Preston JF, Shanmugam T (2001). Engineering a homo-ethanol pathway in *Escherichia coli*: Increased glycolytic flux and levels of expression of glycolytic genes during xylose fermentation. *Journal of Bacteriology*, 183: 2979-2988.

123. Taylor PP, Pantaleone DP, Senkpeil RF, Fotheringham IG(1998). Novel biosynthetic approaches to the production of unnatural amino acids using transaminases. *Trends in Biotechnology*,16: 412-418.

124. Teeri TT: Crystalline cellulose degradation: new insight into the function of cellobiohydrolases. *Trends in Biotechnology* 1997, 15: 160-167.

125. Terada Y, Fujii K, Takaha T, Okada S (1999). *Thermus aquaticus* ATCC 33923 amylomaltase gene cloning and expression and enzyme characterization: Production of cycloamylose. *Applied and Environmental Microbiology*, 65: 910-915.

126. Tester RF, Debon SJJ 2000: Annealing of starch – a review. *International Journal of Biological Macromolecules*, 27: 1-12.

127. Todorova, R. and Atanosov, B. (2004). *Int. J. Biol. Macromol.* 34, 135-147.

128. Tomme P, Warren RAJ, Gilkes NR(1995). Cellulose hydrolysis by bacteria and fungi. In: *Advances in Microbial Physiology* Volume *37*. Edited by: Poole RK. London: Academic Press.

129. Van der Veen ME, Veelaert S, Van der Goot AJ, Boom RM: Starch hydrolysis under low water conditions: A conceptual process design. *Journal of Food Engineering* 2006, 75: 178-186.

130. Vieille, C. and G.J. Zeikus (2001). Hyperthermophilic Enzymes: Sources, Uses, and Molecular Mechanisms for Thermostability. *Microbiology and Molecular Biology Reviews* 65, 1-43.

131. Vieille, C. and Zeikus, G.J. (2001). *Microbiol. Mol. Biol. Rev.* 65, 1-43

132. Vogt, G. *et al.* (1997). Protein thermal stability, hydrogen bonds, and ion pairs. Journal of Molecular Biology 269, 631-643. Vieille, C. and Zeikus, G.J. (2001). *Microbiol. Mol. Biol. Rev.* 65, 1-43.

133. Walsh DJ, Gibbs MD, Bergquist PL (1998). Expression and secretion of a xylanase from the extreme thermophile *Thermotoga* strain FjSS3-B1 in *Kluveromyces lactis*. Extremophiles, 2: 2-Bergquist P, Te'o V, Gibbs M, Cziferszky A, de Faria FP, Azevedo M,Nevalainen H (2002). Expression of xylanase enzymes from thermophilic microorganisms in fungal hosts. *Extremophiles*, 6.

134. Walsh, G.; Headon, D. (1994). *Protein Biotechnology*. Chicester, Wiley, UK.

135. Ward OP, Moo young M (1989). Enzymatic degradation of cell-wall and related plant polysaccharides. *Critical Reviews in Biotechnology*, 8: 237-274.

136. Wood BE, Ingram LO (1992). : Ethanol production from cellobiose, amorphous cellulose, and crystalline cellulose by recombinant *Klebsiella oxytoca* containing integrated *Zymomonasmobilis* genes for ethanol production and plasmids expressing thermostable cellulase genes from *Clostridium thermocellum*. *Applied and Environmental Microbiology*, 58: 2103-2110.

137. Wyman CE(2003). Potential synergies and challenges in refining cellulosic biomass to fuels, chemicals, and power. *Biotechnology Progress*, 19: 254-262.

138. Wyman CE, Dale BE, Elander RT (2005). Coordinated development of leading biomass pretreatment technologies. *Bioresource Technology*, 96: 1959-1967.

139. Zaks A (2001). Industrial biocatalysis. *Current Opinion in Chemical Biology*, 5: 130-136.

140. Zaldivar J, Nielsen J, Olsson L(2001). Fuel ethanol production from lignocellulose: a challenge for metabolic engineering and process integration. *Applied Microbiology and Biotechnology*, 56: 17-34.

141. Zverlov VV, Hertel C, Bronnenmeier K, Hroch A, Kellermann J, Schwarz WH (2000). The thermostable α-L-rhamnosidase RamA of *Clostridium stercorarium*: biochemical characterization and primary structure of a bacterial α-L-rhamnoside hydrolase, a new type of inverting glycoside hydrolase. *Molecular Microbiology*, 35: 173-179.

142. Zverlov VV, Volkov IY, Velikodvorskaya GA, Schwarz WH (2001). The binding pattern of two carbohydrate-binding modules of laminarinase Lam16A from Thermotoga neapolitana: differences in beta-glucan binding within family CBM4. *Microbiology*, 147: 621-629.

*Section V*

# Environmental Biotechnology

**2016, Biotechnology: Progress and Applications** *Pages* **283–302**
*Editors:* **Saif Hameed and Zeeshan Fatima**
*Published by:* **DAYA PUBLISHING HOUSE, NEW DELHI**

# *Chapter 12*

# Detoxification Mechanism and Remediation of Heavy Metals by Use of Plant and Microorganisms

***Shaili Srivastava[1]* and Madan Kumar [2]***

[1]*Amity School of Earth and Environmental sciences,*
*Amity University Gurgaon, Gurgaon – 122 413, Haryana, India*
[2]*School of Environmental Sciences,*
*Jawaharlal Nehru University, New Delhi – 110 067, India*

## ABSTRACT

Environmental pollution by heavy metals is among the most serious contamination to soil, water, plant and air is of great concern due to potential impact on human and health. A number of physicochemical methods have been developed for removal of heavy metals from soil and wastewater. However these methods have disadvantage such as unpredictable metal ion removal, high reagent requirement, generation of toxic sludge etc. Cheaper and effective technologies are needed to protect the precious natural resources and biological lives. The use of plant species for cleaning polluted soils and waters named as phytoremediation has gained increasing attention since last decade, as an emerging cheaper technology. Biosorption is a process which represents biotechnological innovation as well as cost effective tool for removing heavy metals from contaminated waste water. Numerous plant species have been identified and tested for their traits in the uptake and accumulation of different heavy metals.

* Corresponding Author: E-mail: ssnigam@ggn.amity.edu

There are genetic variations among plant species and even among the cultivar of the same species. The mechanisms of metal uptake, accumulation, exclusion, translocation, osmoregulation and copartmentation vary with each plant species and determine its specific role in phytoremediation. Mechanisms of metal uptake at whole plant and cellular levels have been investigated. The use of microorganism for removal and uptake mechanism of heavy metals have been successful reviewed for remediation of toxic heavy metals. This chapter provides selective overview and mechanism of heavy metal tolerance and removal by plant as well as microorganism have been reviewed and reported.

## Introduction

The pollution of heavy metals is becoming more serious with increasing industrialization and disturbance of natural biogeochemical cycles. Heavy metals are essentially non biodegradable and therefore accumulate in the environment. The accumulation of heavy metals in soils and waters poses a serious health hazard and risk to the environment.

Some heavy metals are carcinogenic, mutagenic, while others cause neurological and behavioral changes. The use of physico-chemical methods for remediation of heavy metals has limitations due to high cost and alteration in soil microflora.While phytoremediation is the use of plants and associated soil microbes to reduce the concentrations or toxic effects of contaminants in the environment. This chapter provides an overview of the variety of potential mechanisms and environment friendly remediation methods for decontamination of heavy metal-polluted soils and potential mechanism and reclamation may be involved in the detoxification and tolerance to heavy metals at the cellular level in plants and microorganisms.

## Heavy Metal

"Heavy metals" are chemical elements with a specific gravity that is at least 5 times the specific gravity of water. The specific gravity of water is 1 at 4°C (39°F). Simply stated, specific gravity is a measure of density of a given amount of a solid substance when it is compared to an equal amount of water. The term heavy metal refers to all metals with atomic number greater than five (Albert and Geoffery, 1970) The term 'heavy metal' is somewhat imprecise, but includes most metals with an atomic number greater than 20, and excludes alkali metals, alkaline earths, lanthanides and actinides.

Heavy metals are the pillars of all major civilization, past or present. Metals are not biodegradable and it cannot be destroyed. Heavy metals are a group of 65 elements. Which have many similar physical and chemical properties. Among these Pd, Cd, Hg does not have any biological significance and are known to be extremely toxic. Other toxic metals are Cr, Cu, Mn, Ni, Sn and Zn once dispersed in the biosphere, these metals cannot be recovered or degraded. Hence environment effect of metal pollution are said to be Permanent (Sharma, 2005).

Regarding their role in biological systems, heavy metals are classified as essential and non-essential. Essential heavy metals are those, which are needed by living organisms in minute quantities for vital physiological and biochemical functions. Some of essential heavy metals are Fe, Mn, Cu, Zn, Co, Mg, Mo and Ni.

Heavy metals have also adverse effects on human health. Many heavy metals and metalloids are toxic and can cause serious and adverse effect on living being. The most toxic heavy metals are Hg, Cd, Pb, As, Cu, Zn, Sn, and Cr. The toxicity of heavy metal may vary on valence state or oxidation states. Some metals are more toxic in their one speciation form they may be least toxic or nontoxic in other oxidation states.

## National Scenario of Heavy Metal Users

Data of CPCB shows that Gujrat, Maharastra and Andhra Pradesh contribute 80 per cent of heavy metal emission from point and non point sources in to the atmosphere. Aluminum pollution is associated with bauxite mining. The demand of aluminium for industrial and domestic uses increasing day by day. India ranks sixth in bauxite mining and 8th in aluminum production. The state of Orissa is worst affected.

## International Scenario of Heavy Metal Users

Three countries the United States, Germany, and Russia with only 8 per cent of the world's population consume about 75 per cent of the world's most widely used metals. The United States, with 4.5 per cent of the world's population, uses about 20 per cent of the world's metal population and 25 per cent of the fossil fuels produced each year (Sharma, 2005).

## Sources of Heavy Metals

There are number of heavy metals in the environment. Some of these are toxic and the rest non-toxic. Heavy metals are either required as catalyst or ingredients in manufacturing processes in the various industries. They could be also part of the manufactured products. The industries that are mainly responsible for the emission of heavy metals in the atmosphere are mining industry, chemicals and leather industries. Thermal power plants also contribute heavily to the heavy metal load in the atmosphere.

Metals are introduced in aquatic systems as a result of the weathering of soils and rocks, from volcanic eruptions, and from a variety of human activities involving the mining, processing, or use of metals and/or substances that contain metal pollutants. The most common heavy metal pollutants are arsenic, cadmium, chromium, copper, nickel, lead and mercury. There are different types of sources of pollutants: point sources (localized pollution), where pollutants come from single, identifiable sources. The second type of pollutant sources are nonpoint sources, where pollutants come from dispersed sources. There are only a few examples of localized metal pollution, like the natural weathering of ore bodies and the little metal particles coming from coal-burning power plants via smokestacks in air, water and soils around the factory. The most common metal pollution in freshwater comes from mining companies. They usually use an acid mine drainage system to release heavy metals from ores, because metals are very soluble in an acid solution. After the drainage process, they disperse the acid solution in the groundwater, containing high levels of metals.

Many metal ions are essential as trace elements but at higher concentrations they become toxic. Heavy metals are difficult to remove from the environment and unlike many other pollutants cannot be chemically or biologically degraded and are ultimately indestructible. Today, many heavy metals constitute a global environmental hazard. Remediation of the heavy metal pollution is only by as their immobilization in a non-bioavailable form, or their re-speciation into less toxic forms. While these approaches do not solve the problem altogether.

**Table 12.1**

| *Sl.No.* | *Metal* | *Industries* |
|---|---|---|
| 1 | Chromium (Cr) | Mining, industrial coolants, chromium salts manufacturing, leather tanning |
| 2 | Lead (Pd) | lead acid batteries, paints, E-waste, Smelting operations, coal- based thermalpower plants, ceramics, bangle industry |
| 3 | Mercury (Hg) | Chlor-alkali plants, thermal power plants, fluorescent lamps, hospital waste (damaged thermometers, barometers, phygmomanometers), electrical appliances etc |
| 4. | Arsenic (As) | Geogenic/natural processes, smelting operations, thermal power plants, fuelburning |
| 5. | Copper (Cu) | Mining, electroplating, smelting operations |
| 6. | Nickel (Ni) | Smelting operations, thermal power plants, battery industry |
| 7. | Cadmium (Cd) | Zinc smelting, waste batteries,e-waste, paint sludge, incinerations and fuelcombustion |
| 8. | Zinc (Zn) | Smelting, electroplating |
| 9. | Pd, Cu, Sb | Defence works and defence establishments, including ordinance storage and training areas where live firing is carried out |

*Source*: CPCB Report 2011.

## Toxicity of Heavy Metal

Heavy metals become toxic when they are not metabolized by the body and accumulate in the soft tissues. Heavy metals may enter the human body through food, water, air, or absorption through the skin when they come in contact with humans in agriculture and in manufacturing, pharmaceutical, industrial, or residential settings. Industrial exposure accounts for a common route of exposure for adults. Ingestion is the most common route of exposure in children (Hu¨ttermann *et al.*, 1999).

The Agency for Toxic Substances and Disease Registry (ASTDR) in Atlanta, Georgia, (a part of the U.S. Department of Health and Human Services) was established by congressional mandate to perform specific functions concerning adverse human health effects and diminished quality of life associated with exposure to hazardous substances. The ASTDR is responsible for assessment of waste sites and providing health information concerning hazardous substances, response to emergency release situations, and education and training concerning hazardous substances. ASTDR was listed top 20 hazardous and toxic heavy metal including arsenic (no. 1), lead (no. 2), mercury (no. 3), cadmium (no.7), aluminium and iron.

Arsenic is number 1 on the ASTDR's "Top 20 List." Arsenic is released into the environment by the smelting process of copper, zinc, and lead, as well as by the manufacturing of chemicals and glasses. Target organs are the blood, kidneys, and central nervous, digestive, and skin systems.

Lead is number 2 on the ASTDR's "Top 20 List." Every year, industry produces about 2.5 million tons of lead throughout the world. Most of this lead is used for batteries. Lead accounts for most of the cases of pediatric heavy metal poisoning. Target organs are the bones, brain, blood, kidneys, and thyroid gland.

Number 3 on ASTDR's "Top 20 List" is mercury. Mercury is generated naturally in the environment from the degassing of the earth's crust, from volcanic emissions. It exists in three forms: elemental mercury and organic and inorganic mercury. Mining operations, chloralkali plants, and paper industries are significant producers of mercury. Medicines, such as mercurochrome and merthiolate, are still available. Algaecides and childhood vaccines are also potential sources. Inhalation is the most frequent cause of exposure to mercury. The organic form is readily absorbed in the gastrointestinal tract (90-100 per cent); lesser but still significant amounts of inorganic mercury are absorbed in the gastrointestinal tract (7-15 per cent). Target organs are the brain and kidneys.

Cadmium is a byproduct of the mining and smelting of lead and zinc and is number 7 on ASTDR's "Top 20 list." It is used in nickel-cadmium batteries, PVC plastics, and paint pigments. It can be found in soils because insecticides, fungicides, sludge, and commercial fertilizers that use cadmium are used in agriculture. Inhalation of cadmium accounts for 15-50 per cent of absorption through the respiratory system; 2-7 per cent of ingested cadmium is absorbed in the gastrointestinal system. Target organs are the liver, placenta, kidneys, lungs, brain, and bones.

Iron does not appear on the ASTDR's "Top 20 List," but it is a heavy metal of concern, particularly because ingesting dietary iron supplements may acutely poison young children. Target organs are the liver, cardiovascular system, and kidneys.

Aluminium is readily available for human ingestion through the use of food additives, antacids, buffered aspirin, astringents, nasal sprays, and antiperspirants; from drinking water; automobile exhaust and tobacco smoke; and using aluminum foil, aluminum cookware, cans, ceramics, and fireworks. Aluminum might have a possible connection with developing Alzheimer's disease as researchers found what they considered to be significant amounts of aluminum in the brain tissue of Alzheimer's patients. Although aluminum was also found in the brain tissue of people who did not have Alzheimer's disease, recommendations to avoid sources of aluminum received widespread public attention (Sharma, 2005).

## Phytoremediation of Heavy Metals

Heavy metals are non biodegradable, released into the environment by anthropogenic activities and they accumulate in the environment, as a result contaminating food chain. The term 'Phytoremediation' consists of Greek word *Phyto* means plant and Latin word *remedium* means to correct or remove. Phytoremediation is an emerging technology and can be defined as the use of plant and its associated microbes for the removal, sequestration and stabilization of contaminants from the environment. This is a cost-effective, efficient, in situ applicable, eco-friendly approach to remove heavy metals, radionuclides and organic contaminants. Plants have the ability to uptake accumulate, tolerate and detoxify the heavy metals through various mechanisms. The technique or mechanism to remove the heavy metals is phytoextraction, phytostabilisation, phytofiltration, and phytovolatilization (Ali *et al.*, 2013).

## Techniques of Phytoremediation

### Phytoextraction

Phytoextraction (also known as phytoaccumulation or phytoabsorption) is the uptake or absorption of heavy metals by plant roots from soil or water and their translocation and accumulation into the aerial parts of plant (Shoots). Metal accumulated in stems and leaves are harvested then burned or incinerated to gain energy and recovery of metals.

### Phytostabilisation

Phytostabilisation (also known as phytoimmobilization or phytorestoration) is a plant base technique for the stabilization or immobilization of contaminants present in the soil and water, reducing their mobility, bioavailability and preventing their entry into the groundwater or food chain. The immobilization of contaminants takes place through accumulation in plant tissues, adsorption by roots, precipitation and complexion within the root zone. Actually phytostabilisation is a management strategy and the objective is to stabilize and reduce the risk of the toxic metal contaminants to health and environment and not to remove them from site (Hall, 2002).

### *Phytofiltration*

Phytofiltration is the removal of contaminants from water and aqueous waste streams (agriculture run-off, industrial discharge and nuclear material processing waste) by the process of adsorption and absorption by plants. The types of phytofiltration are rhizofiltration (roots used), blastofiltration (seedlings used) and caulofiltration (excised plant shoots used) and minimizing their movement.

### Phytovolatilization

Phytovolatilization is the uptake of contaminants by plants in elemental form and their biological conversion or modification to volatile form followed by their release into the atmosphere through the stomata. Metal contaminants such as As, Hg and Se can be remediated by plants through this technique. It does not remove the contaminants completely but only transferred from soil to atmosphere.

Among the above mentioned techniques phytoextraction is the most promising technique for the commercial application to remove heavy metal contaminants present in soil, sediment and water. The effectiveness of phytoremediation depends upon plant species to be used, soil properties, avoiding food chain contamination and bioavailability and distribution of heavy metals in the soil. Plants having higher growth rate, high biomass, adaptability to metal stress are preferred for phytoextraction for *e.g.* grasses are preferred over shrubs and trees. Phytoextraction potential of plant depends on two factors i.e shoot metal concentration and shoot biomass. On the basis of these two factors the two methods can be applied i.e use of hyper accumulator plants as they accumulate very high concentration of heavy metal in shoot portion and plants that accumulate heavy metal in fewer amounts but produce high above-

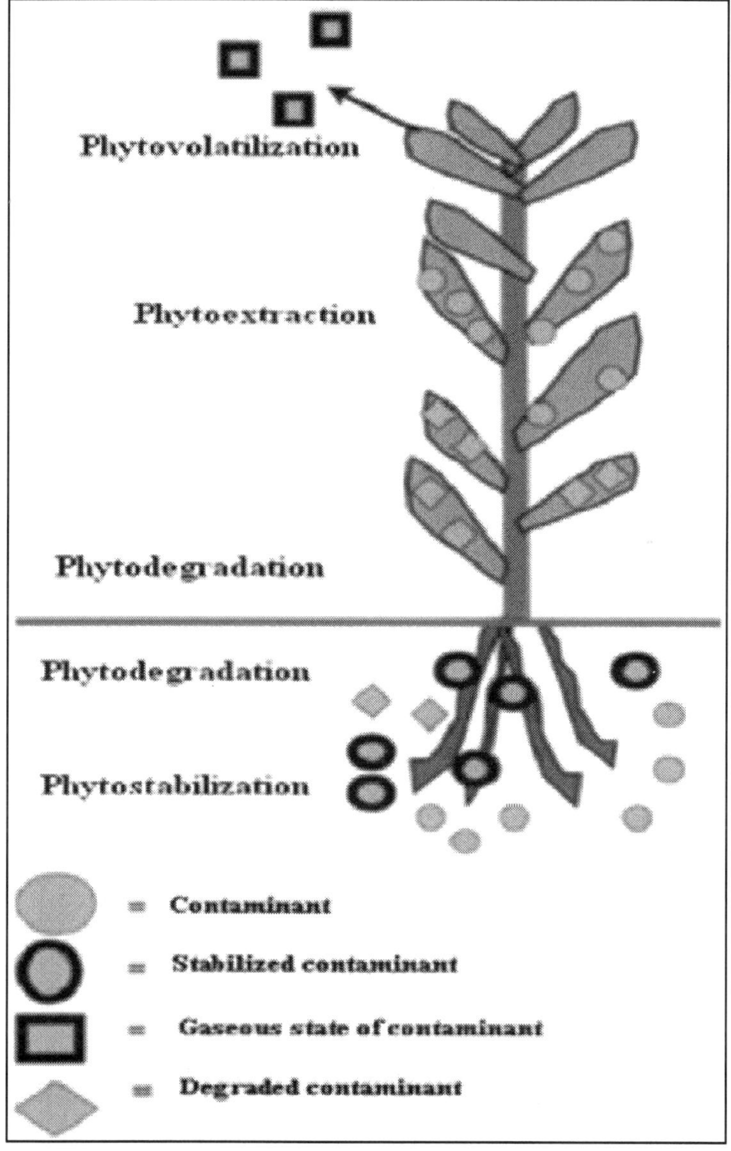


**Figure 12.1: Schematic Model of different Phytoremediation Technologies Involving Removal and Containment of Contaminants (*Source*: Nature Education, 2011).**

ground biomass giving overall accumulation comparable to hyper accumulators. The phytoextraction efficiency can be quantified by calculating bioconcentration or accumulation factor and translocation factor. These are the main factors considered for selection of plants for phytoextraction. Bioconcentration factor is defined as efficiency to accumulation heavy metals into the tissue by plants from the surrounding environment. Translocation factor is defined as the efficiency of plants to translocate the accumulated heavy metal from its roots to shoots (Hall, 2002).

## Factors Affecting Heavy Metal Uptake

There are several factors that can affect the process of uptake of heavy metal by plants and knowledge on effect of these factors on uptake mechanism can help to greatly improve the process.

### Plant Species and Properties of Medium

Plant uptake mechanism varies from species to species and the plant species having higher potential of accumulation or tolerance to heavy metal are of great significance. The process itself depends on the selection of plant species that can accumulate large amount of contaminants and biomass under already established conditions. The efficiency of accumulation can be further enhanced by change in media properties (pH, organics, and phosphorus). Pb uptake is affected by pH and phosphorus content. Pb uptake is reduced by lowering the pH.

### Bioavailability of Metals in soil

Bioavailability of metal can be categorized into readily bioavailable (Zn, As, Se, Cu, Cd, Ni,), moderately bioavailable (Mn, Co, Fe) and least bioavailable (Cr, Pb, U) (Prasad, 1999). Availability of heavy metal in the plant- obtainable form from the soil is critical factor affecting the phytoremediation of metals. The redox potential, organic matter and pH of the metal bound soil will determine whether the metal will be present in the plant-available form or not. Plants also devised the mechanism to increase the bioavailability of metal by root exudates secretion for lowering the pH as a result increase metal desorption, secreting metal-mobilizing substances and rhizospheric microbes improve metal availability to plants (Ali *et al.*, 2013)

### Addition of Chelators

Phytoextraction can be enriched by addition of chelating agents that will increase the bioavailability of heavy metals in soil for plant uptake. The chelating agents form water soluble complex with the heavy metals and assist desorption of metal from soil particles. The chelating agents used to increase the bioavailability of heavy metal for the plant uptake are EDTA, ammonium sulphate, citric acid and elemental sulphur. The synthetic chelating agent can be toxic to the plant and its associated rhizospheric microbes for *e.g.* EDTA is non-biodegradable and can leach into the ground water causing another environmental risk so caution must be taken before trying synthetic agents for induction of phytoremediation.

## Advantages of Phytoremediation

The phytoremediation offers advantage such as:

- Public acceptance and aesthetically pleasing
- Less disruptive than physical and chemical processes.
- Inexpensive technology
- Wide range of contaminants can be removed
- Effective in reduction of contaminant even if present in very low concentration
- Eco-friendly technology

## The Disadvantages of Phytoremediation

- Time consuming process
- Growth rate and productivity of biomass
- Root contact area and age of plants
- Climatic conditions
- Soil chemistry and level of contaminants
- The concentration of contaminants present
- Effect of contaminated vegetation

# Mechanisms of Heavy Metal Accumulation and Tolerance in Plants

## The Root Exudates and Cell Wall

The uptake of heavy metals by plants can take place either through symplast (intracellular) or apoplast (extracellular) pathway depending on the type of heavy metal. Plant can tolerate the toxic effects of heavy by preventing its excess entry into the plant. Precipitation or complex formation mechanism of heavy metal with root exudates and plant induced change in pH and redox reaction guides the entry of heavy metal into the plants from soil. Root exudates has diverse role as it can retards or enhance the uptake of heavy metal depending upon the type of type of compound exudated. Ni uptake was reduced by Ni-chelating histidine and citrate accumulated in root exudate in Ni non-hyperaccumulating plant as compared to Ni hyperaccumulating plant. Secretion of oxalic acid from roots upon Al stress accumulated non-toxic Al-oxalate in the leaves of Buckwheat. Secretion of malate in sorghum and citrate in maize and oxalate from tomato in the form of root exudate in response to cadmium stress have also been reported (Hossain *et al.*, 2012).

The cell wall of root remains in direct contact to heavy metal so can accumulate range heavy metals in the interface of cell wall and plasma membrane. This mechanism needs further clarification as very few convincing reports are available. Heavy metal tolerant *Silenevulgaris*ssp.*humilis* and accumulated a range of heavy metals in cell wall by binding to protein or as silicates. *Loliummultiflorum* cell wall bound 60 per cent Cu in the cell wall. These findings clearly suggest the importance of these mechanisms in heavy metal tolerance by plants.

## Plasma Membrane and Vacuolar Compartmentalization

Plasma membrane is the target for heavy metal toxicity and can be rapidly affected by presence of high concentration of heavy metal. The various mechanism responsible for damage to membrane are inhibition of vital membrane proteins, oxidation and cross linking of thiols, change in membrane lipid composition and fluidity (Mehrag, 1993). The plant tolerance mechanism included prevention of increased leakage of solute from cells, protect the integrity of the membrane and enhance membrane repair after damage. Cell membrane maintains heavy metal homeostasis either through efflux mechanism or by reducing or preventing entry into the cell. Some of these metals are essential hence selective efflux takes place. The effect of metal on the membrane is metal specific such as Cu causes membrane leakage by increased $K^+$effluxs from *Agrostiscapillaris* roots in contrast Zn protects membrane against leakage and oxidation. Reduced uptake mechanism is responsible for tolerance of arsenic toxicity. Arsenic tolerant genotype of *Holcuslanatus* showed lower uptake of phosphate and arsenate as compared to non tolerant genotype as arsenate and phosphate are taken up by the same systems. After the entrance of heavy metal into the cells either it will be transported out of the cell or accumulated into the vacuoles to maintain cell homeostasis. Vacuole is the site of accumulation of heavy metals such as Zn, Cd and Mo.Tthe vacuolar uptake is facilitated by two vacuolar proton pumps i.e vacuolar protonpyrophosphatase (V-Ppase) and vacuolar proton-ATPase (V-ATPase). The plant owns different types of transporters and channels that are involved in the uptake of heavy metal maintainance of homeostasis. With the advancement in molecular techniques, gene families responsible for heavy metal uptake, vacuolar accumulation and remobilization, xylem mediated transportation have been discovered. The Nt CBP4 transporter, a calmodulin-binding protein associated with reduced Ni accumulation and higher Pb accumulation. The Nramps). There are other class of metal transporters responsible for heavy metal transportation such as CPx-ATPases (Cu, Zn, Cd, Pb transportation), the Nramps (uptake of Fe and Cd), the CDF (cation diffusion facilitator, Co, Cd and Zn transportation) family, the ZIP family (Zn, Fe, Cd and Mn), ABC transporters (ATP-binding Cassette), COPT (Cu transporter), CAX ($Ca^{2+}$cation antiporter), etc. The localization of Cd and Zn into the vacuole was shown in *Nicotianarustica*var Pavonii seedlings and Festucarubra roots (Hall, 2002).

## Phytochelatins (PCs)

Chelation of heavy metals by high-affinity ligands in the cytoplasm plays important role in tolerance and detoxification of heavy metal. Amino acids, organic acids, Phytochelatins and metallothioneins act as high- affinity ligands.The phytochelatins (PCs) are widely studied in plants and are induced rapidly in plants after treatment with heavy metal. They are family of metalcomplexing cysteine rich polypeptides having a general structure ($\gamma$-Glu-Cys) n-Gly where n = 2–11and $\gamma$- Ser, Ala, Gln, or Glu but the most common have n= 2–4 peptides.PCs are synthesized non-translationally using glutathione as a substrate by $\gamma$-glutamylcysteinyldi-peptidyltranspeptidase (PC synthase) by activation of PC synthase in the presence of metal ions such as, Ag, Cu, Ni, Au, Pb, As, and Zn, Cd, Hg but Cd is the strongest inducer. The thiol group of Cys forms complex with the Cd ion and the complex gets accumulated into the vacuole mediated by ABC transporters in the *Brassica juncea*. Binding of Cd with PCs and its accumulation in vacuole was shown in water hyacinth (*Eichhorniacrassipes*). The metal such as Zn, Ni, Cu, Hg formed complex with PCs and maintained the homeostasis in cytoplasm in diffent plants. Since Pcs is secreted by both metal tolerant and sensitive plants and other findings also concluded that PCs may not be primarily responsible for accumulation of Zn, Ni or Pb. Further study in this direction will help to understand the role and mechanistic details of PCs in metal tolerance (Salt *et al.*, 1998).

## Metallothioneins (MTs)

MTs are gene encoded polypeptides having low molecular weight (4-8 KDa),Cys rich and binds to the metal via thiol group of Cys residue in plants. The role of MTs is not fully explained but suggested that it might help in heavy metal accumulation, maintaining cytoplasmic homeostasis and protect against oxidative stress. MT gene have been discovered in plants and divided into three classes on the basis of arrangement of Cys residue such as Cys-Cys, Cys-X-Cys, and Cys-X-X-Cys motifs where X-any amino acid. Tissue specific pattern of expression was shown for different isoform of MT having potential to interact with different metal ions. The expression of Mt gene is induced by several factors such as hormones, heavy metals and cytotoxins. In *Arabidopsis*, MT2 was strongly induced by Cu but and slightly by Zn and Cd in seedlings. Different isoform of MT genes was regulated differentially under different heavy metal (Cu, Zn) stress in *Brassica rapa*. Further study in this area will elucidated the molecular mechanism responsible for transportation and maintenance of homeostasis in plants (Prasad, 1999).

## Organic Acids, Amino Acids, and Phosphate Derivatives

The organic acids, amino acids and phosphate derivatives are the potential ligands for the chelation of heavy metal that change the mobility and bioavailability of metal ion and help in tolerance and detoxification of heavy metal in plants. Heavy metal tolerance mechanism can be of external exclusion and internal tolerance type. Some of the molecules responsible for external exclusion have been discussed above. The presence of carboxylic acid, phosphate derivatives (Phytate), amino acid such as histidine, citrate, malate, oxalate and Nicotianamine (NA) are the prospective ligands for heavy metals complexion and play important role in heavy metal tolerance and detoxification. In the case of internal tolerance the potential ligands can chelate with heavy metal in the cytoplasm and can be converted or transformed into nontoxic form. In a Ni-hyperaccumulating plant *Alyssum lesbiacum* 36 fold increase in histidine content of xylem sap was reported on Ni exposure (Kramer *et al.*, 1996) but the absence of histidine response mechanism in other hyperaccumulating plant clearly indicated its plant specific role. Citrate can chelate with strong affinity to Fe, Zn, Cd, Ni and Co, Zn-phytate and Al-phytate has been reported in *Deschampsia caespitosa* roots and *Zea mays* vavuoles respectively but the detailed mechanism of tolerance or detoxification is still unclear. The synthesis of NA takes place by condensation of three molecules of Sadenosyl-Met (SAM) by NA Synthase enzyme. NA is mobile and transport heavy metal in different part of plants, can chelate with Zn, Cu, Ni and Fe.

NA have been reported in maintaining the Zn homeostasis and NA Synthase in *Arabidopsis* (Hossain *et al.*, 2002).

## Mechanisms of Hyper Accumulation in Plants

The hyperaccumulator plant has the capability to grow on metalliferous soils and accumulate unusually high concentration of heavy metal in their above ground (aerial) parts without developing any toxic symptoms. The hypothesis proposed to explain the hyper accumulation function in plants is "elemental defense" and "joint effect" hypothesis and the available evidence supported elemental defense hypothesis (Rascio and Navari-Izzo, 2011). According to elemental defense hypothesis the plant accumulate heavy metal to provide defense against natural enemies and the joint effect hypothesis states that heavy metal accumulation acts jointly along with organic compounds to provide enhanced defense to plants. The shoot: root ratio of metal concentrations is unusually high as compared to metal tolerant plant. These plants can hyperaccumulate metal even after low concentration of metal in the surroundings. The mechanism for the hyperaccumulation is coordinated by the constitutive expression of efficient metal tranporters, chelators, antioxidants in response to metal hperaccumulation, efficient intracellular compartmentalization of metals (Figure 12.2). The processes responsible for hyper accumulation are (1) loading of metal from symplast into apoplast xylem in root for transportation with the action of transpiration pull to the shoots. (2) Storage of metals into the vacuoles of leaves effectively. (3) Enhanced cellular uptake rate and effective cell to cell metal distribution. (4) Protection of root metal toxicity by efficient mechanism of metal transfer from root to shoot.

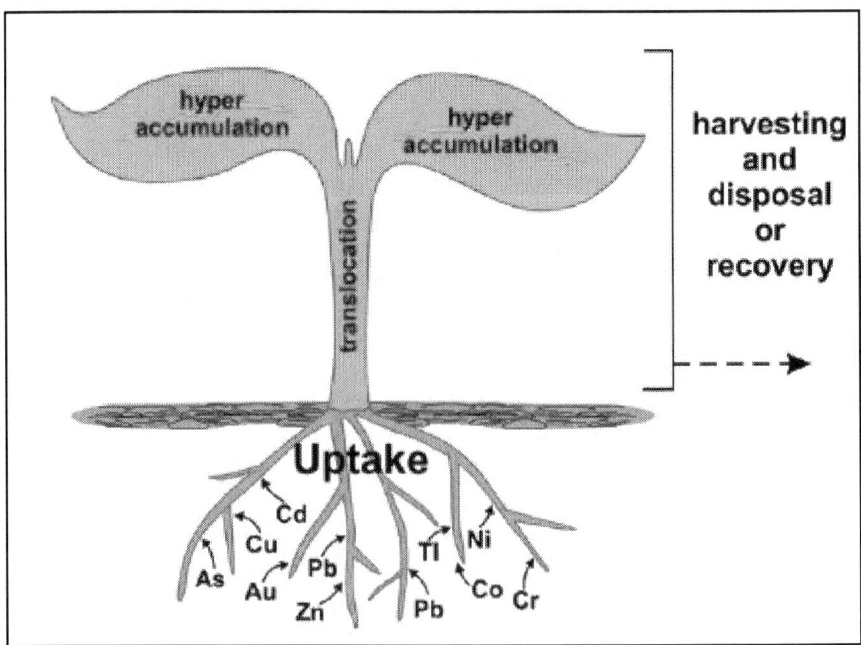


**Figure 12.2: Phytoremediation and Phytomining of Heavy Metal Rich Soil by Using Plant (*Source*: Rascio and Navari-Izzo, 2011).**

## Chemical Modification

The presence of metal modifying enzymes helps to assimilation of metal into organic molecules (selanate to dimethylselenide) and causes change in oxidation state of metal (toxic to less toxic form). The Fe and Cu are reduced by the action of reductases thus helping the plants in the mechanism of detoxification. The accumulation of Cr in reduced state (Cr(III)) has been observed in *Eichhornia crassipes* after supplying Cr(VI).

## Transcription Factors (TFs)

TFs plays important role in regulation of genes under different conditions and decides the response of the cell upon induction. The TFs have been discovered in various plant species conferring tolerance to heavy metals. TFs induced in response to metal induction such as metal responsive transcriptional factor 1(MTF-1) activates the set of genes responsible for the metal uptake, transportation, detoxification and tolerance. They induce the cascade of signal transduction pathway in response to stress. TFs responsible for Cd tolerance belongs to different families such as basic leucine zipper (bZIP), ethylene-responsive factor (ERF), myeloblastosis protein (MYB) and WRKY induce a signal transduction pathway genes in response to Cd stress.

## Antioxidant Defense and Glyoxalase in Metal Tolerance

The production of reactive oxygen species (ROS) such as hydrogen peroxide ($H_2O_2$), hydroxyl radical (HO), singlet oxygen ($^1O_2$) and superoxide radical ($O_2^-$) takes place in higher plants during different metabolic processes in the cell organelles i.e mitochondria, chloroplast and Peroxisomes. The antioxidants scavenge the ROS immediately from the site of production as soon as they are produced. During stress conditions the rate of production of ROS is dramatically elevated surpassing the capacity of locally present antioxidants to scavenge leads to leakage of ROS into the cytoplasm and other compartments. Methylglyoxal (MG) is αβ-dicarbonyl aldehyde and is highly reactive molecule that is proposed to be formed from metabolic intermediates of carbohydrates, lipids and proteins metabolism. In plants they are produced enzymatically or non-enzymatically from glycolytic and photosynthetic intermediates such as glyceraldehyde-3-phosphate (G3P), and dihydroxyacetone phosphate (DHAP). The production of MG is spontaneous and unavoidable as the rate of glycolysis increases and the unstable triose phosphates leads to the formation of MG.

The increase in ROS and MG level during stress has been proposed to act as signals for the activation of signaling pathway genes in response to stress. They can react with nucleic acids, proteins, amino acids and lipids and in the absence of proper defense mechanism they can damage cell structure and function, irreparable metabolic dysfunction. Plant protects them from oxidative stress by activating antioxidant defense mechanism by producing various antioxidant molecules and enzymes. The antioxidant are superoxide dismutase (SOD) and catalase (CAT), monodehydroascorbatereductase (MDHAR), dehydroascorbate reductase (DHAR), glutathione reductase (GR), glutathione peroxidase (GPX), glutathione *S*-transferase (GST), glyoxalase I (lactoylglutathionelyase), glyoxalase II (hydroxyacylglutathione hydrolase), and water-soluble compounds such as ascorbate (AsA), cysteine, *O*-acetylserine and glutathione (GSH). Both AsA and GSH act as cofactor for the antioxidant (ascorbate-glutathione cycle) and glyoxylase pathways and also regulate expression of genes related to stress. GSH prevent plant from oxidative damage as shown in *N. goesingense* form heavy metal stress such as Ni, Co, Zn and Cd(30, 32 annual reviews.). GSH can acts as chelator and substrate for synthesis of PCs. Both glyoxalase I and glyoxalase II act in coordination and catalyze the irreversible reaction to convert MG and other 2-oxoaldehydes into 2-hydroxyacids using GSH. Since GSH is the mediator for the defense response its biosynthesis, utilization and regenerating enzymes level has been found elevated in the heavy metal tolerant and hyper accumulating plants (Hall, 2002).

## Role and Mechanisms of Plant-Associated Microbes in Uptake of Heavy Metals

Plants accumulate and tolerate higher concentration of metals and their associated microbes plays important role in metal or exclusion, mobilization, immobilization, transformation, detoxification and bioaccumulation of metals from contaminated site. Microbes can affect plant in several ways either they inhibit the plant growth or promote the plant growth through various mechanisms by releasing chelators, altering soil pH, mediating redox reactions. The plant associated microbes act synergistically to enhance the metal uptake potential of plant and also promotes plant growth. This plant-microbe association can be exploited as an approach for the enhancement of plant biomass and accumulation of heavy metal. Microbial action that helps in metal mobilization or immobilization are

### Siderophores

Siderophores are iron chelators having high association constant and low molecular mass (400-1000 Da) that are produced by bacteria and fungi in response to low iron level in the zoot zone of plant. The main function of siderophore is to form complex with the unavailable metal bound to soil and solubilize it as a result increasing avalibility for plant uptake.They can form stable complexes with other metals also such as Pb, Al, Cu, and Zn. On the bais of their chemical nature of donating oxygen ligands to $Fe^{3+}$binding they are broadly classified into catecholates (enterobactin), hydroxamates (desferrioxamines) and (α-hydroxy-)carboxylates (aerobactin)

Siderophore produced by microbes form complex with metal and the plant uptake metal from siderophore-metal complex through root mediated processes such as direct uptake of metal-siderophore complex, chelate degradation and metal uptake and by ligand exchange process. Siderophore such as pyoverdin and pyochelin has been reported in *pseudomonas aeruginosa* helping Cr and Pb uptake by maize plant. *Streptomyces* sp. has been reported for producing siderophores and help plant to enhance uptake of metal such as Al, Cu, Fe, Mn, Ni, U, and Pb. Uptake of metal is retarded by the production of siderophores as reported in some strains of *pseudomonas*. Cd uptake reduction by *Pseudomonas aeruginosa* strain KUCd1 in *Brassica juncea* and Ni uptake in chickpea by *pseudomonas* sp. This different response may be due to soil factors, bioavailability of metal, type of plant species,etc. Ectomycorrhizal fungi such as *Rhizopogon luteolus, Suillus luteus* and *Scleroderma verrucosum* were reported to produce catecholates and hydroxamates siderophores under iron deficient condition in pinus (Rajkumar *et al.*, 2012).

## Polymeric Substances and Glycoprotein

Extracellular polymeric substance (EPS) and glycoproteins produced by rhizospheric microbes plays important role in heavy metal mobilization in the polluted site. They form complex with the metals and as a result decrease their mobility in the soil. EPS produced by *Azotobacter* sp. reduced the uptake of Cd and Cr by *Triticum aestivum*.Arbuscular mycorrhizal fungi (AMF) produce insoluble glycoprotein, glomalin that form complex with Cu, Pb and Cd as prevent their mobilization. The structure and mechanistic detail of glomalin influencing plant metal uptake are not well defined. There is correlation between amount of glomalin and metal binding; hence efficient AMF producing glomalin will offer great application in phytostabilisation of toxic compounds.

## Organic Acids

Organic acids are carboxylic group containing molecules, secreted by plant associated microbes and forms complex with the metal and plays important role in their solubility and mobilization. The stability of metal-organic acid complex depends on the number of carboxylic group present, pH of soil and metal present. The stability of complex containing less number of carboxylic groups is less stable and vice-versa. Bacteria and fungi secrete various types of organic acids to improve the heavy metal bioavailability such as citric acid (*Aspergillus niger, Penicilliumbilaiae, pseudomonas,*Solubilise Ni, Cu, Fe, Zn, and Pb), gluconic acid (*Pseudomonas aeruginosa* CMG 823, Solubilize ZnO and $Zn_3$ $(PO_4)_2$, 5-ketogluconic acid (*Gluconacetobacter,* Solubilize ZnO, $Zn_3(PO_4)_2$ and $ZnCO_3$), oxalic acid (*Beauveria caledonica, burkholderia cepacia,* Solubilize Pb and Zn), formic acid, tartaric acid and acetic acid (*Burkholderia cepacia,* Solubilize ZnO, $ZnCO_3$ and $CdCO_3$)(Ali *et al.*, 2012).

## Biosurfactants

Biosurfactants are amphiphilic molecules consists of nonpolar tail made up of long chain of saturated and unsaturated hydroxylated fatty acids and polar head group consists of polysaccharides and proteins. Biosurfactant produced by plant associated microbes increases the mobilization of heavy metal by forming complex with metal at soil interface and increases their solubility and bioavailability in contaminated site thus facilitates metal uptake by plants.There are reports available clearly indicated the role of biosurfactants produced by microbes in mobilization of metals. Biosurfactants produced by bacteria are di-rhamnolipid produced by *P. aeruginosa* BS2 solubilizes Cd and Pb. Lipopeptide produced by *Bacillus* sp. J119 mobilize Cd in tomato, maize and Rhamnolipids by *Pseudomonas aeruginosa* mobilize Cu. Further study in the direction biosurfactant producing microbe and their mechanism of action on metal mobilization and its uptake by plants will significantly improve our understanding.

## Biosorption

Biosorption can be defined as the adsorption of metals by the plant-associated microbes by active or passive process. Biosorption mechanism plays important role in plant metal uptake and generally reduced the uptake of metals by plants. Plant-associated microbes can bind to the heavy metal and reduce their bioavailability to plants for *e.g. Burkholderia* sp. and *Magnaportheoryzae* reduced the uptake of Ni and Cd in tomato and *Brevibacillus* sp B-I reduced the Zn uptake in *Trifoliumrepens.* Fungi act as a biofilter and prevent the heavy metal translocation from root to shoot of plant. Ectomycorrhizal fungi (EMF) have large surface area that helps to strongly adsorb heavy metal from soil. The EMF, Scleroderma *citrinum, Amanita muscaria* and *Lactariusrufus* reduced Pb, Zn and Cd uptake from root to shoot in pinus through biosorption of metal by inner and outer mycelia component. This clearly indicated that inoculation of plant with these microbes can provide protection to plant against heavy metal and phytostabilisation of toxic metals present in contaminated sites. The survival and colonization potential of plant-associated microbes depends on various factors such as pH, tolerance to toxic metals, indigenous microflora and nutrient deficiency greatly influence the process of phytoremediation. The application of microbe for remediation of contaminated site depends on types of metal present, its concentration, plant type, growth conditions, etc (Rajkumar *et al.*, 2012).

## Metal Reduction and Oxidization

The plant-associated microbes can influence the mobility of heavy metal in the polluted site through redox reactions. The metal oxidizing bacteria reduces the soil pH and increases the mobility and uptake of metals by plants such as sulfur oxidizing bacteria enhances Cu mobilization by reducing soil pH through conversion of reduced sulfur to sulfates. The rhizospheric microbes can immobilize the heavy metals by reduction reaction such as *Cellulosimicrobium cellulansa* Cr resistant bacterium reduced the uptake of toxic Cr (VI) in green chilli plant by converting into nontoxic Cr(III). *Stenotrophomonasmaltophilia* bacteria reduce the uptake by converting toxic Se (IV) to insoluble Se (0) in the rhizosphere of *Astragalus bisulcatus.* Application of both metals oxidizing (Fe or S) and reducing (Fe) microbes for metal mobilization has been studied for Cu, Cd, Hg and Zn. These findings clearly indicates role of redox processes carried out by plant-associated microbes and its potential application in phytoremediation technology (Rajkumar *et al.*, 2012, Ali *et al.*, 2013).

## Biological Control

Plant-associated microbes protect the plant from bacterial, viral, fungal infection, pest like nematode and insects through several mechanisms by the production of antibiotic metabolites, cell wall lytic enzymes, siderophores and induction of systemic resistance in plants. Plants are mainly infected by fungus and the rhizospheric microbes protects the plant from infection by production of enzymes such as chitinase, beta 1,3 glucanase, protease, lipase that causes lysis of fungal cell wall. Siderophores produced by rhizospheric microbes reduces the availability of iron to the pathogen in the soil as the pathogen depends on availability of iron. The interaction of plant with the rhizospheric microbes and pathogenic microbes in metal contaminated soil are poorly understood and further study in this direction will enhance our understanding of the processes responsible for resistance to plants.

## Plant Growth Promotion by Plant-Associated Microbes

### 1-aminocyclopropane-1-carboxylic Acid (ACC)

ACC act as an intermediate in the biosynthetic pathway of plant hormone ethylene. Ethylene plays important role in regulating the growth and cellular metabolism of plant such as stress tolerance, disease resistance, plant-microbe interaction and nutrient cycle. The overproduction of ethylene causes inhibition of root elongation, root hair formation and lateral root growth. The plant-associated bacteria improves the stress response of plants by hydrolyzing ACC secreted from root exudates into α-ketobutyrate and ammonia by the help of ACC deaminase enzyme before being oxidized by plant ACC oxidase to form ethylene. The ammonia was utilized by microbes as sole nitrogen source and thereby decreasing ACC in the plants as a result maintain low ethylene level in plant. The low level of ethylene helps plant in response to metal stress and enhances plant biomass production. This clearly indicated that plant-associated bacteria can promote the plant growth by regulating the synthesis of ethylene.

### Production of Phytohormones by Plant-Associated Microbes

Phytohormones regulates the physiology and metabolism in plants related to growth, germination, reproduction and protection against biotic and abiotic stress. The phytohormones produced by plant-associated microbes are indole-3-acetic acid (IAA), cytokinins, and gibberellins. IAA production pathway depends on the plant-microbe interactions, the beneficial rhizospheric bacteria follows indole-3-pathway and pathogenic bacteria follows indole-3-acetamide pathway for IAA synthesis. IAA level modulates plant microbe interaction, promotes root elongation and contributes in plants defense system development. *Pseudomonas putida* GR12-2 produced low level of auxin (2 μg $ml^{-1}$) and caused root elongation 2-3 fold but its mutant form produced high level of auxin (8.2 μg $ml^{-1}$) and inhibits elongation of canola root. The inhibition of root elongation was due to the synthesis of ACC by the interaction of high IAA concentration and ACC synthase acts as signal molecule and regulates the transcription of ACC synthase gene in plant initiating cross-talk between IAA and ethylene and the IAA signal transduction pathway was inhibited by ethylene. This suggests that bacteria facilitates the plant growth and regulates hormonal balance directly or indirectly through the synthesis of ethylene. Cytokinins and gibberellin are also produced by plant-associated microbes that are shown to regulate the growth and morphology of plants under stress and non-stress conditions (Rajkumar *et al.*, 2012).

### Enhancement of Antioxidant Defense of Plants by Plant-Associated Microbes

Plant has acquired several defense mechanisms to protect them from oxidative damage by producing various types of antioxidant molecules and enzymes. Plant rhizospheric microbe has been shown to elevate the antioxidant enzyme production (catalase and peroxidases) in bacterial inoculated *R. communis* and *H. annuus* plant under Ni stress. *Agrobacterium radiobacter* increased superoxide dismutase and catalase production in *Populous deltoides* LH05-17 under As stress and *Rhizobium* sp. RP5 increased glutathione reductase activity in pea plant under Ni and Zn stress.

## Plant-Associated Microbes as Nutrient Provider

Phosphorus is an essential nutrient required for proper plants growth and development and its uptake by plant is affected by the presence of heavy metal in soil and its deficiency in plants leads to growth retardation. The plant-associated bacteria convert the unavailable phosphates into bioavailable form through chelation, acidification, exchange reaction and mineralization of organic phosphates by secreting extracellular phosphatase. Inoculation of plants with phosphate solubilizing bacteria was shown to increase the uptake of phosphorus. Nitrogen is essential to plants growth and development and the fixation of atmospheric nitrogen is the ability of endophytes and rhizobacteria and their association benefits the plants. The rhizobacteria genera are *Mesorhizobium, Rhizobium, Bradyrhizobium,* etc provide nitrogen to the plant through nitrogen fixation and also shown to help plant growth under heavy metal stress (Rajkumar *et al.,* 2012).

## Fate of Plant Used for Phytoextraction

Plant biomass containing accumulated heavy metals can be combusted to get energy and the remaining ash is considered as 'bio-ore'. This bio-ore can be processed for the recovery or extraction of the heavy metals. An advantage of phytomining is the sale of energy from combustion of the biomass cultivation of energy maize in the Campine region in Belgium and the Netherlands could result in the generation of 30000–42000 kWhel+th of renewable energy per hectare. Thus phytomining is an environment and eco-friendly option as compared to the conventional extraction methods. However, the commercial viability of phytomining depends on many factors like the efficiency of phytoextraction and current market value of the processed metals. Phytomining has been commercially used for Ni and it is believed that it is less expensive than the conventional extraction methods (Anderson *et al.,* 1999; Meers *et al.,* 2010).The schematic diagram (Figure 12.3) showing plants after burning, can be either disposed as hazardous waste safely in specialized dumps or if economically feasible, processed for biorecovery of precious and semiprecious metals.

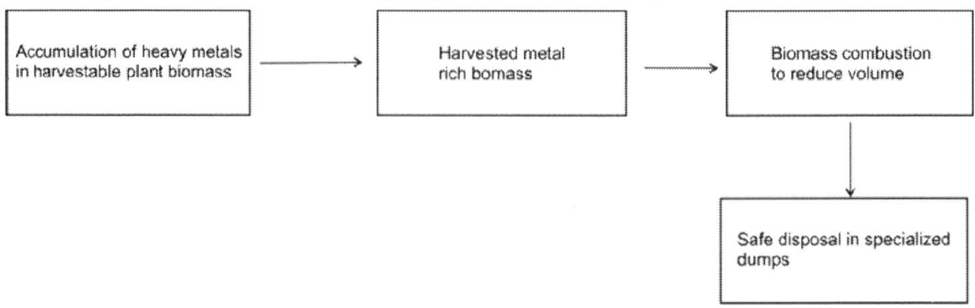


**Figure 12.3: Main Route of Post-harvest Treatment of Phytoremediator Plants (Source: Ali *et al.,* 2013).**

## Conclusion

Since contamination of soils and waters by toxic heavy metals is a serious environmental problem, therefore effective remediation methods are necessary. Physical and chemical methods for cleanup and restoration of heavy metal-contaminated soils have serious limitations like high cost, irreversible changes in soil properties, Destruction of native soil microflora and creation of secondary. Heavy metals uptake, by plants using phytoremediation technology, seems to be a prosperous way to remediate heavy metals-contaminated environment pollution problems. In contrast, phytoremediation is a better solution to the problem. Phytoremediation is environment-friendly and ecologically responsible solar-driven technology with good public acceptance. It is a relatively recent technology and is mostly in research stage. Its research is highly interdisciplinary in nature and requires background knowledge in soil chemistry, plant biology,ecology and soil microbiology as well as environmental engineering. Fortunately, interdisciplinary studies and research are appreciated and highly encouraged in broad-minded scientific communities across the globe and it is fully hoped that the integration of scientific disciplines will be highly fruitful. Research is in progress to screen native plants for phytoremediation of target heavy metals and to evaluate the effect of different parameters on phytoremediation efficiency. Phytoremediation has some limitations as compared to other conventional technologies. The most important factor is use of suitable plant and microorganisms for remediation of heavy metals. Furthermore, research is being conducted to genetically modify some suitable plants for better phytoremediation of heavy metals and other xenobiotics. Studies are also being done to identify and characterize the different proteins involved in cross-membrane transport and vacuolar sequestration of heavy metals. Advancement and achievements in such molecular studies will greatly help in understanding the mechanism and enhancing the efficiency of phytoremediation. However it is time to consider the commercialization of phytoextraction using high-biomass hyperaccumulator plants becomes widespread, considering that not only will it remediate contaminated sites but will generate income from agricultural lands. An improved understanding of heavy metal uptake by plants from soil will also help in promoting phytomining a plant-based eco-friendly mining of metals, which can be used for extraction of metals even from low-grade ores. Phytoextraction of heavy metals is expected to be a commercially viable technology for phytoremediation and phytomining of heavy metals in near future.

## References

1. Albert, C.F Geoffery, 1970. *Advanced Inorganic Chemistry*. Wiley Eastern Pvt. Ltd. New Delhi. p. 141.

2. Ali, H., Khan, E., Sajad, M.A. 2013. Phytoremediation of heavy metals: Concepts and applications. *Chemosphere*, 91: 869–881.

3. Ampiah-Bonney, R. J., Tyson, J. F., Lanza, G. R. 2007. Phytoextraction of arsenic from soil by Leersia oryzoides. *International Journal of Phytoremediation* 9, 31–40.

4. Anderson, C.W.N., Brooks, R.R., Stewart, R.B., Simcock, R., 1999. Gold uptake by plants. *Gold Bull.* 32, 48–52.

5. Greipsson, S. 2011. *Phytoremediation*. Nature Education Knowledge 3(10): 7
6. Hall, J.L.2002. Cellular mechanisms for heavy metal detoxification and tolerance. *Journal of Experimental Botany* 53, 1–11.
7. Hossain, M.A., Piyatida, P., Jaime A., Silva,T.D., Fujita,M. 2012. Molecular Mechanism of Heavy Metal Toxicity and Tolerance in Plants: Central Role of Glutathione in Detoxification of Reactive Oxygen Species and Methylglyoxal and in Heavy Metal Chelation. *Journal of Botany* 37, 2011-17
8. Hu¨ttermann, A., Arduini I., Godbold, D.L. 1999. Metal pollution and forest decline. In: Prasad NMV, Hagemeyer J, eds. *Heavy metal stress in plants: from molecules to ecosystems*. Berlin: Springer-Verlag, 253–272.
9. Kra¨mer, U., Cotter-Howells J.D., Charnock, J.M., Baker, A.J.M., Smith, J.A.C. 1996. Free histidine as a metal chelator in plants that accumulate nickel. *Nature* 379, 635–638.
10. Meers, E., Slycken, S.V., Adriaensen, K., Ruttens, A., Vangronsveld, J., Laing, G.D., Witters, N., Thewys, T., Tack, F.M.G. 2010. The use of bio-energy crops (Zea mays) for 'phytoremediation' of heavy metals on moderately contaminated soils: a field experiment. *Chemosphere* 78, 35–41.
11. Meharg, A.A. 1993. The role of the plasma lemma in metal tolerance in angiosperms. *Physiologia Plantarum* 88, 191–198.
12. Prasad, M.N.V. 1999. Metallothioneins and metal binding complexes in plants. In: Prasad MNV, Hagemeyer J, eds. *Heavy metal stress in plants: from molecules to ecosystems*. Berlin: Springer-Verlag, 51–72.
13. Rajkumar, M. Sandhya, S. Prasad, M.N.V., Freitas, H. 2012. Perspectives of plant-associated microbes in heavy metal phytoremediation. *Biotechnology Advances* 30, 1562–1574.
14. Rascioa, N., Navari-Izzo, F. 2011. Heavy metal hyperaccumulating plants: How and why do they do it? what makes them so interesting? *Plant Science* 180, 169–181
15. Salt D.E, Smith R.D, Raskin I. 1998. Phytoremediation. *Annual Review of Plant Physiology and Plant Molecular Biology* 49, 643–668.
16. Sharma, B. K. 2005. *Environmental Chemistry*. Krishna Prakashan Media Pvt Ltd. Meerut.

**2016, Biotechnology: Progress and Applications** *Pages* **303–318**
*Editors:* **Saif Hameed and Zeeshan Fatima**
*Published by:* **DAYA PUBLISHING HOUSE, NEW DELHI**

# *Chapter 13*

# Bioenergy–A Future Fuel: Prospects and Constraints

***Rajesh Singh Tomar and Shuchi Kaushik***

*Amity Institute of Biotechnology,*
*Amity University Madhya Pradesh, Gwalior – 474 005, M.P.*

## ABSTRACT

The current patterns of energy use, and associated links to emissions of greenhouse gases and climate change, are of major global concern. Fossil fuel substitution is a key role of renewable energy, leading to emissions being reduced or further increases being avoided. Waste to energy has the potential to make a significant contribution to sustainable development by helping meet the Kyoto obligations. Conservation of natural resources is a common objective. There are several policy issues evolving under the umbrella of sustainable development, including energy self-reliance, security of supplies, diversity of sources, sustainable development, clean air and climate change initiatives. Bioenergy has emerged as an alternative to fossil fuels and is being promoted as a cleaner source of energy. Bioenergy refers to renewable energy produced from biomass, which is organic material such as trees, plants (including crops), and waste materials (*e.g.*, wood waste from mills, municipal wastes, manure, landfill gas (LFG), and methane from wastewater treatment facilities). Bioenergy systems include reducing the use of non-renewable resources, maintaining the quality of resources, or recycling and reusing resources. Sustainable approaches to producing bioenergy on existing cropland include using waste, by-products and dedicated energy crops. Growth of bioenergy depends on how much affordable, sustainably-sourced biomass is

* Corresponding Author: E-mail: rstomar33@hotmail.com

available. Many of the technical barriers to bioenergy use have been overcome; the remaining hurdles are those of a less tangible nature, *i.e.*, financial and institutional. Given a supportive policy environment, bioenergy can provide a sustainable solution to future energy demands.

## Introduction

Global energy demand is growing rapidly. It is the major concern for most countries, mainly in terms of security of supply and costs. The relationship between fossil fuel consumption and increasing atmospheric concentrations of greenhouse gases is causing concern, resulting in actions to militate against or adapt to an indecisive future climate. Escalating fuel prices and the comprehensive uproar about rapidly-declining reserves of fossil fuels have impelled power producers to examine other, commercially viable and environment-friendly alternatives. Bioenergy is, one of the strongest contenders for the top spot as the next-best substitute for petrol and diesel.

## Bioenergy and Biomass

The Earth is warming due to the accumulation of greenhouse gases (GHGs) from the intensive use of fossil fuels and other natural resources. GHGs include carbon dioxide ($CO_2$), methane, nitrogen oxides, water vapor, ozone and others, but $CO_2$ is responsible for more than half of the anthropogenic greenhouse effects. These GHGs are trapping heat over the Earth's surface, resulting in changes in temperature and other climatic processes.

Bioenergy has emerged as an alternative to fossil fuels and is being promoted as a cleaner source of energy. Rise in oil prices have also increased the level of interest in bioenergy. Bioenergy refers to products of biomass that have been converted into liquid, solid or gas form, depending on the raw material base and the technology employed, for energy generation. Biomass encompasses a wide spectrum of plant materials ranging from agricultural, forestry and municipal wastes to crops grown specifically to make biofuels, such as bioethanol and biodiesel. Over the past decades, the modern use of biomass has increased rapidly in many parts of the world. In the light of the Kyoto GHG reduction targets, many countries have ambitious targets for further biomass utilisation.

### Resources of Biomass

The resource base includes a wide range of primary resources, and secondary and tertiary residues.

#### 1. Forest Resources

*a. Primary*

- ☆ Logging residues from conventional harvest operations and residues from forest management and land clearing operations
- ☆ Removal of excess biomass (fuel treatments) from timberlands and other forestlands
- ☆ Fuelwood extracted from forestlands

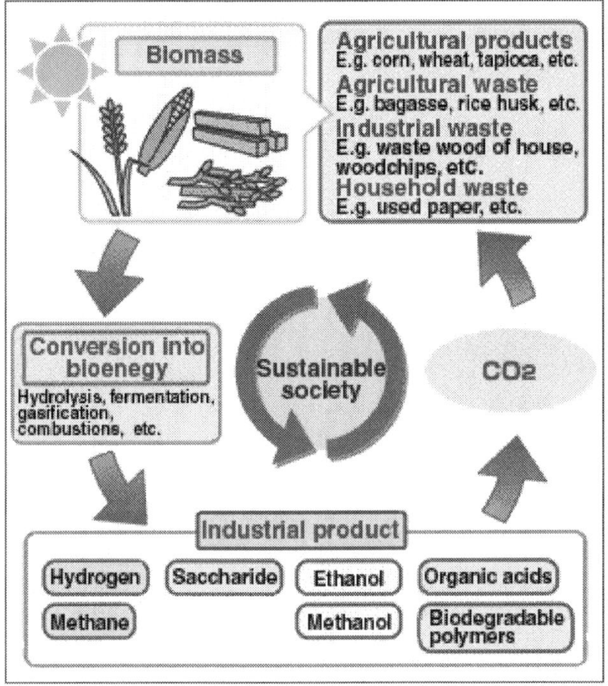


**Figure 13.1: Summarizing the Concept of Bioenergy.**

*b. Secondary*

- ☆ Primary wood processing mill residues
- ☆ Secondary wood processing mill residues
- ☆ Pulping liquors (black liquor)

*c. Tertiary*

- ☆ Urban wood residues — construction and demolition debris, tree trimmings, packaging wastes and consumer durables

**2. Agricultural Resources**

*a. Primary*

- ☆ Crop residues from major crops — corn stover, small grain straw, and others
- ☆ ·Grains (corn and soybeans) used for ethanol, biodiesel, and bioproducts
- ☆ Perennial grasses
- ☆ Perennial woody crops

*b. Secondary*

- ☆ Animal manures
- ☆ Food/feed processing residues

*c. Tertiary*

- ☆ MSW and post-consumer residues and landfill gases

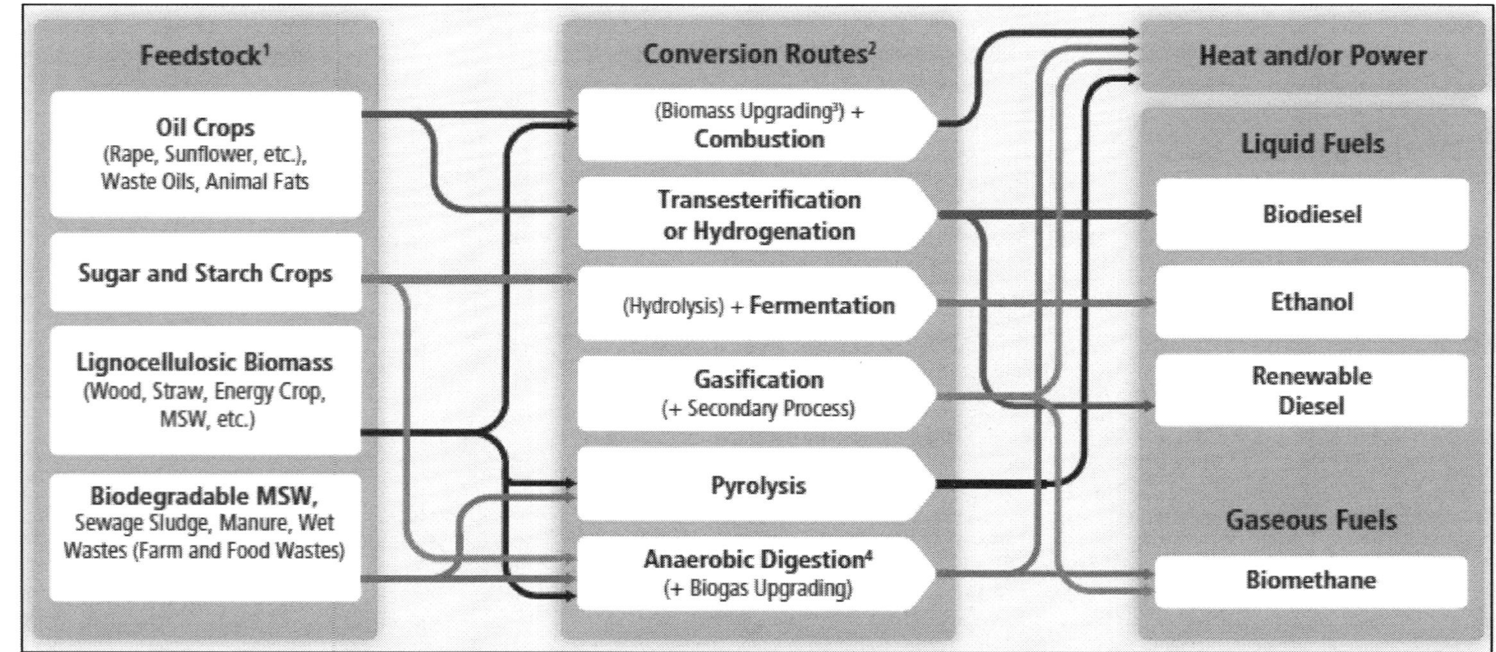


**Figure 13.2: Schematic View of Commercial Bioenergy Routes**
**(Modified from IEA, Bioenergy, 2009 and taken from chapter by Chum *et al.*, 2011).**

**Notes: 1. Parts of each feedstock, for example, crop residues, could also be used in other routes. 2. Each route also gives co-products. 3. Biomass upgrading includes any one of the densification processes (pelletization, pyrolysis, etc.). 4. Anaerobic digestion processes release methane and $CO_2$ and removal of $CO_2$ provides essentially methane, the main component of natural gas; the upgraded gas is called biomethane.**

## Current Scenario of Bioenergy/Biofuels across the Globe with Special Reference to India

Biomass is a versatile raw material that can be used for production of heat, power, transport fuels, and bioproducts. When produced and used on a sustainable basis, it is a carbon-neutral carrier and can make a large contribution to reducing greenhouse gas emissions. Currently, biomass-driven combined heat and power, co-firing, and combustion plants provide reliable, efficient, and clean power and heat. Production and use of biofuels are growing at a very rapid pace. Sugar cane-based ethanol is already a competitive biofuel in tropical regions.

A mandatory 5-10 per cent blending of biofuels with petrol and diesel stipulated by some countries in the last 10-20 years triggered a rapid growth in the biofuel sector in the last decade. By 2007-08, world biofuel production had touched 62.2 billion tons (t), of which around 88 per cent was in the form of ethanol. The two largest ethanol producers, Brazil and the United States of America (USA), account for almost 87 per cent of its total production. Biodiesel production that accounts for a smaller proportion of liquid biofuels, increased from 0.01 million ton in 1991 to 9.0 million ton by 2008. The European Union (EU) produces over 60 per cent of the global share with a significantly smaller contribution coming from the USA (17 per cent). The Bureau of Indian Standards (BIS) has already evolved a standard (IS-15607) for Bio-diesel (B 100), which is the Indian adaptation of the American Standard ASTM D-6751 and European Standard EN-14214. BIS has also published IS: 2796: 2008 which covers specification for motor gasoline blended with 5 per cent ethanol and motor gasoline blended with 10 per cent ethanol.In India, the Sugarcane Breeding Institute has developed varieties of energy cane, including a high fiber and low sugar variety that can be used by existing sugarcane mills suitable for first generation ethanol and co-generation power, and an even higher fiber, high biomass variety suitable for cellulosic ethanol and co-gen. (Biofuel digest). China and India – the world's two most populous countries – dominate Asia's alternative fuels landscape, according to Lux Research. However, even strong growth will leave these nations short of their self-imposed targets.

Ethanol is booming in China, thanks to the nation's 10 per cent blend mandate, equivalent to 3.3 billion gallons per year for 2020, but as total fuel demand explodes as well, ethanol will creep up to only 4 per cent of gasoline in 2017, placing that target in doubt. India is betting on risky jatropha for biodiesel as it races to meet an aggressive 20 per cent biofuels mandate for 2017, but biofuels will account for under 0.6 per cent of its diesel fuel and 0.3 per cent of its gasoline in 2017 (Biofuel digest).

Unlike other countries that are still struggling with the problem of devoting land that could potentially be used for agricultural purposes to the exclusive cultivation of fuel-producing crop, India is uniquely poised to tap into this eco-friendly fossil-fuel alternative. In addition to helping the country to dramatically reduce its carbon footprint, biofuels can also earn valuable foreign income for the country through exports, while also providing the agricultural community with a means to generate more revenue.

The rapid pace of expansion within the domestic automobile industry will only increase the already overwhelming demand on the nation's meagre fossil-fuel reserves-an eventuality that can be easily circumvented with the help of biofuel. According to Indian chemical Council, projected demand and supply of alcohol in India in 2014-15 is 1900 million liters (potable sector), 1280 million liters (industry sector), 1260 million liters (5 per cent blending), 4440 million liters (total alcohol required), Highest expected alcohol availability could be 2400 million liters and deficit would be 2040 million liters. All these values based on past trends, the growth rates are assumed to be 5 per cent for the industrial sector, 7 per cent for the potable sector and 5 per cent for blending.

In 2008, the Government of India announced its National Biofuel Policy, which envisions meeting close to 20 percent of the domestic diesel demand from biofuels. This means that 14, 00, 000 sq. km of land will have to be set aside for the cultivation of fuel-producing plants, up from today's 5,000 sq. km. Of this, the government has already identified 4, 00, 000 sq. km land for Jatropha crop.

The Jatropha plant can easily be cultivated in wastelands. Hence, unlike other sources of biodiesel, growing this crop will not encroach upon valuable farmland, and will also serve as an additional means of income for Indian farmers. According to former Indian President, Dr. Abdul Kalam, a strong believer in Jatropha as a biofuel, nearly half of the country's 6, 00, 000 sq. km of wasteland is suitable for Jatropha cultivation.

Jatropha oil is cost effective, since it can be used directly after extraction, without the need for any additional refining. The cultivation of Jatropha crop is relatively easier, as the leaves of these plants are unpalatable to livestock. This ensures that the plants remain intact during the sapling stage, unlike many other types of tree saplings. The prolific use of this oil will eventually reduce the country's dependence on fossil fuel imports. The money thus saved can be diverted to other, more productive uses such as for building infrastructure.

In a bid to encourage Indian farmers to take up Jatropha cultivation, the government has signed a Memorandum of Understanding with bio-oils specialist D1 Mohan Bio Oils Ltd to hand out loans worth Rs 1.3 billion to farmers; D1 Mohan will sponsor the purchase of the Jatropha seeds. Several Indian states are also looking into Jatropha cultivation with renewed interest, with the state governments teaming up with private players to finance and promote Jatropha-planting projects. For example, the Hindustan Petroleum Corporation Ltd (HPCL) and the Maharashtra State Farming Corporation Ltd have jointly set up a Jatropha seed-based biodiesel venture, while Chattisgarh is well on its way to becoming a biofuel self-reliant state by the year 2015.

## Conversion Technologies for Bioenergy and Biofuels

Bioenergy and biofuels are important components within a country's green energy portfolio. While there are numerous renewable energy options for heat and electricity generation, biofuels are currently the only means of displacing liquid fossil fuels such as gasoline, diesel, and aviation fuels.

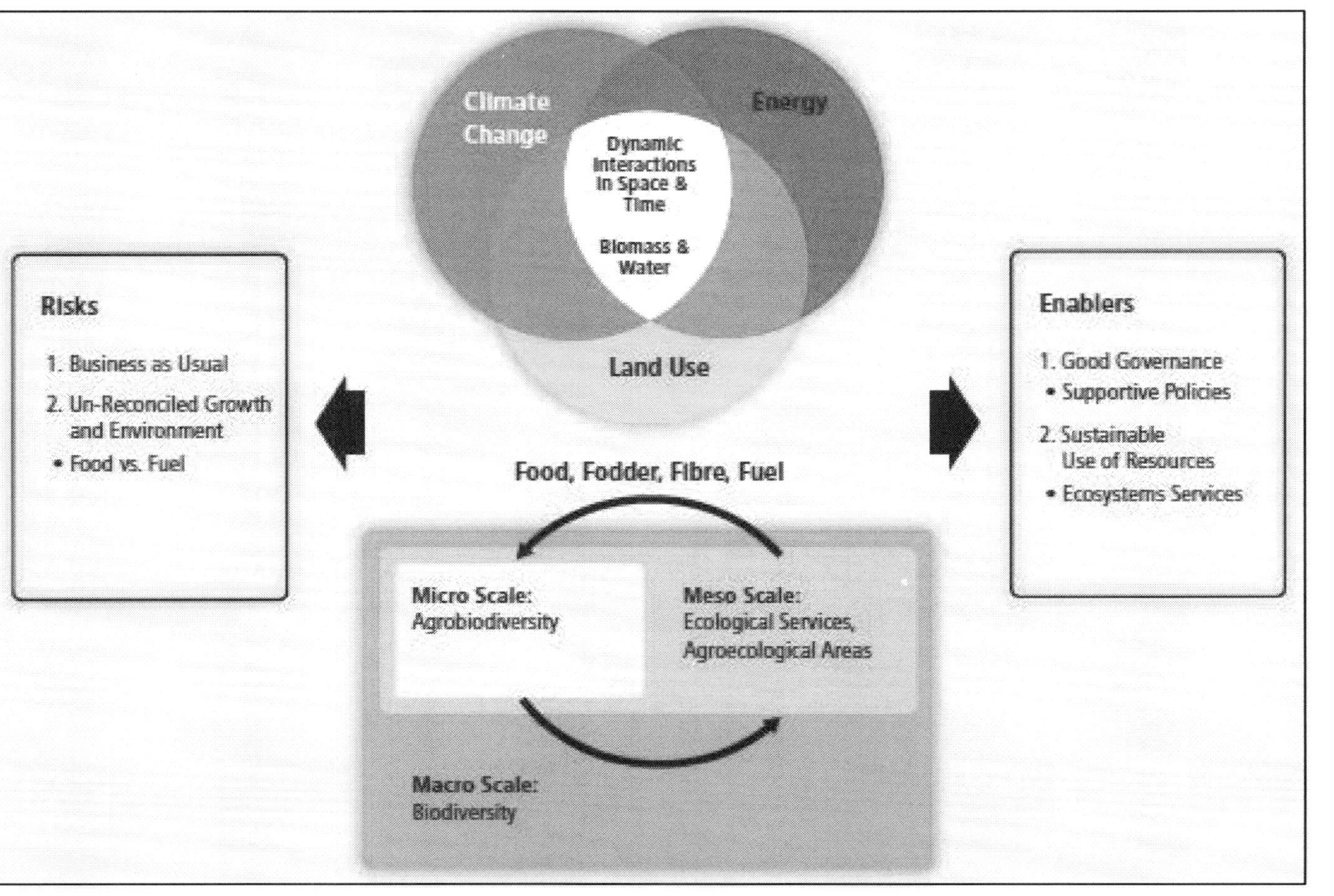


**Figure 13.3: Summarizing the Issues Related to Bioenergy.**

**Table 13.1: Environmental and Socioeconomic Impacts of Bioenergy: Example Areas of Concern with Selected Impact Categories (Synthesized from the literature review by van Dam *et al.*, 2010)**

| *Example Areas of Concern* | *Examples of Impact Categories* |
|---|---|
| Global, regional, off-site environmental effects | GHGs; albedo; acidification; eutrophication; water availability and quality; regional air quality |
| Local/onsite environmental effects | Soil quality; local air quality; water availability and quality; biodiversity and habitat loss |
| Technology | Hazards; emissions; congestion; safety; genetically modified organisms/plants |
| Human rights and working conditions | Freedom of association; access to social security; job creation and average wages; freedom from discrimination; no child labour and minimumage of workers; freedom of labour (no forced labour); rights of indigenous people; acknowledgment of gender issues |
| Health and safety | Impacts on workers and users; safety conditions at work |
| Food security | Replacement of staple crops; safeguarding local food security |
| Land and property rights | Acknowledgment of customary and legal rights of land owners; proof of ownership; compensation systems available; agreements by consent |
| Participation and well-being of local communities | Cultural and religious values; contribution to local economy and activities; compensation for use of traditional knowledge; support to local education; local procurement of services and inputs; special measures to target vulnerable groups |

A central problem for sustainable development is striking a balance between resource use and conservation so that high levels of development can be ensured. Therefore, conservation and wise use of natural resources has been an important global issue to be addressed from individual, community, national as well as international levels. In this concern, various biotechnological tools hold great promise in different areas of medicine, agriculture, forestry and environment to enhance food productivity as well as in the meantime maintaining human and animal health, environmental integrity and biological diversity at ecosystem, species and intra-specific levels.

Several traditional crops that are grown for food and other uses can also be used to produce bioenergy, primarily as biofuels. Crops currently used as biomass feed stocks include: Corn, Rapeseed, Sorghum, Soybeans and Sugarcane.

Other crops that are planted and harvested specifically for use as biomass feed stocks in the production of bio-energy are referred to as "energy crops." Energy crops are fast-growing and grown for the specific purpose of producing energy (electricity or liquid fuels) from all or part of the resulting plant. The advantages of using crops specifically grown for energy production include consistency in moisture content, heat content, and processing characteristics, which makes them more cost-effective to process efficiently (U.S. EPA, 2007a). Emerging energy crops include:

### Microalgae

The oil in microalgae can be converted into jet fuel or diesel fuel (National Renewable Energy Laboratory (NREL), 2006). Microalgae with high lipid content are best suited to production of liquid fuel.

Microalgae are highly productive, do not use agricul-tural land or products, and are carbon-neutral (Ali *et al.*, 2014). More than 50 companies are researching microalgal oil production, including development of new bioreactors and use of biotechnologies to influence microalgal growth (NREL, 2008).

### Switch Grass, Poplar and Willow Trees

These energy crops are not yet being grown commercially in the United States for bioenergy, but may have the great-est potential for dedicated bioenergy use over a wide geographic range. The U.S. Department of Energy (U.S. DOE) estimates that about 190 million acres of land in the United States could be used to produce energy crops such as switch grass and poplar and willow trees (U.S. EPA, 2007a). Several states in the Midwest and South could produce significant biopower using switch grass, which is currently grown on some Conservation Reserve Program acres and on hay acres as a forage crop (U.S. EPA, 2007a; Ugarte *et al.*, 2010).

## Processing of Biomass for Bioenergy/Biofuel Production

Conversion routes for biomass are generally thermochemical or biochemical, but can also include chemical and physical. In thermochemical conversion, electricity and heat are generated from biomass by direct combustion, co-firing, and gasification systems. An important smaller scale biochemical conversion technology is anaerobic digestion.

### Thermochemical Conversion

Thermochemical conversion uses heat and chemicals to break down cel-lulosic feedstock into syngas. Depending upon the pro-cess being used, the gas can be converted to liquid fuels such as ethanol, bio-butanol, methanol, mixed alcohols, or bio-oil (through pyrolysis). Thermochemical con-version is particularly useful for lignin, which cannot be easily converted to ethanol using the biochemical process described above; up to one-third of cellulosic feedstock can be composed of lignin. Forest and mill residue feedstocks generally have high lignin contents, and thus would be more suitable for thermochemical ethanol conversion than biochemical conversion.

The thermochemical conversion process involves:

Drying the cellulosic feedstock

↓

Gasification (using heat to convert the feedstock to a syngas) or pyrolysis (using heat and pressure to produce oil)

↓

Contaminant removal

Conversion of the syngas to ethanol, bio-oil, or other products

↓

Distillation to separate ethanol from water (if producing ethanol)

**Biochemical Conversion**

Biochemical conversion for ethanol production from cellulosic feedstocks involves

Pretreatment of the feedstock using high-tempera-ture, high-pressure acid; enzymes; or other methods to break down the lignin and hemicellulose that surround the cellulose.

Hydrolysis using enzymes and acids to break down the cellulose into sugars

Fermentation to convert the sugars into ethanol (as in conventional production)

Distillation to produce purer ethanol (as in conventional production)

## Types of Biofuels

Solid biofuels are plant matter such as wood chips, and other solid or woody biomass, that can be directly used as fuel, mainly in traditional cook stoves. Two of the most widely used forms are wood chips and bagasse. Bagasse has been used for decades for electricity generation at sugar mills. Millions of people in developing countries depend on traditional biofuels for their most basic cooking and heating needs; this dependence is most prevalent in sub-Saharan Africa and South Asia. In some countries, the share of biomass in energy use reaches 90 per cent.

- ☆ Liquid biofuels are used for heating, cooking, lighting, transport and power generation. Bioethanol, biodiesel and pure plant oils are the most common forms of liquid biofuels.
- ☆ Gaseous biofuels include biogas, which is produced by digesting organic waste and is generally used for cooking, lighting and power generation at the village level.

## Opportunities for Bioenergy

Biomass is a versatile energy source that can be used for production of heat, power, and transport fuels, as well as biomaterials and, when produced and used on a sustainable basis, can make a large contribution to reducing GHG emissions. Biomass

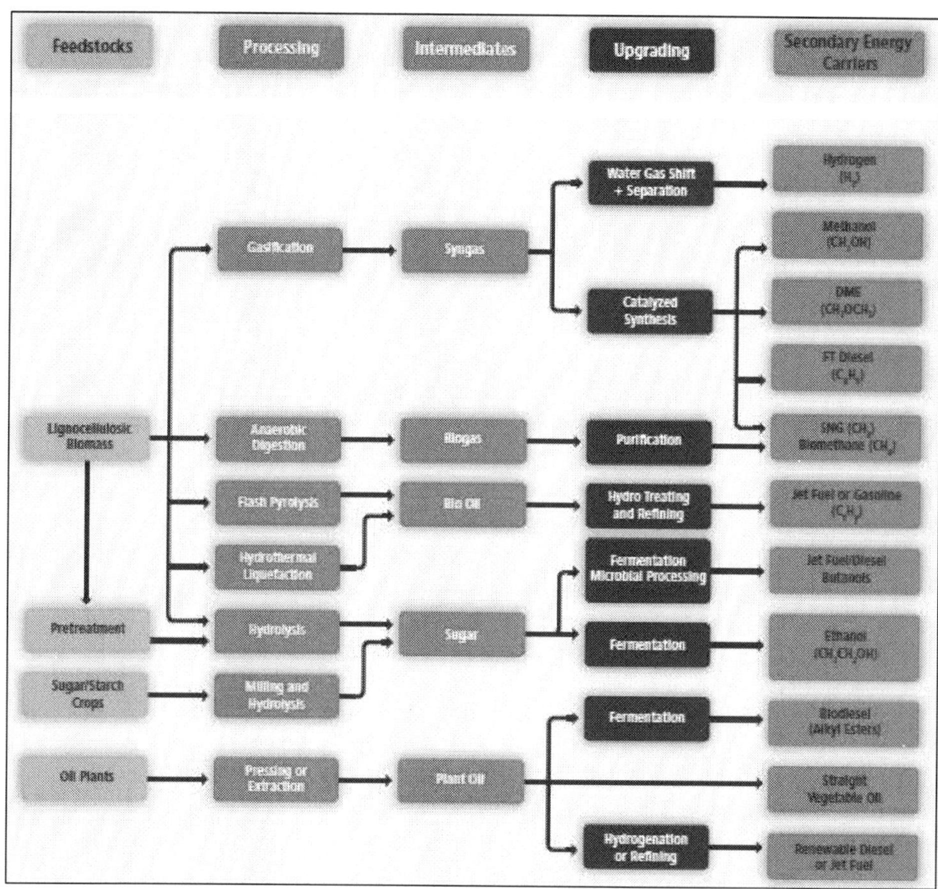


**Figure 13.4: Overview of Lignocellulosic Biomass, Sugar/Starch Crops and Oil Plants (Feedstocks) and the Processing Routes to Key Intermediates, which can be Upgraded through Various Routes to Secondary Energy Carriers, such as: liquid and Gaseous Biofuels. Fuel product examples are (1) oxygenated biofuels to blend with current gasoline and diesel fuels or to use in pure form, such as ethanol, butanols, methanol, liquid ethers, biodiesel, and gaseous DME (dimethyl ether); (2) hydrocarbon biofuels such as Fischer Tropsch (FT) liquids, renewable diesel and some microbial fuels (which are compatible with the current infrastructure of liquid fuels because their chemical composition is similar to that of gasoline, diesel, and jet fuels), or the simplest hydrocarbon methane for natural gas replacement (SNG) from gasifi cation or biomethane from anaerobic digestion; and (3) $H_2$ for future transportation (adapted from Hamelinck and Faaij, 2006)**

is the most important renewable energy option at present and is expected to maintain that position during the first half of this century and likely beyond that [IPCC, 2007 a-d; IEA, 2006]. Currently, combined heat and power (CHP), co-firing and various combustion concepts provide reliable, efficient, and clean conversion routes for converting solid biomass to power and heat. Production and use of biofuels are growing at a very rapid pace.

Although the future role of bioenergy will depend on its competitiveness with fossil fuels and on agricultural policies worldwide, it seems realistic to expect that the current contribution of bioenergy of 40-55 EJ per year will increase considerably. A range from 200 to 400 EJ may be expected during this century, making biomass a more important energy supply option than mineral oil today–large enough to supply one-third of the world's total energy needs.

Bioenergy markets provide major business opportunities, environmental benefits, and rural development on a global scale. If indeed the global bioenergy market is to develop to a size of 300 EJ over this century (which is quite possible given the findings of recent global potential assessments) the value of that market at E4-8/GJ (considering pre-treated biomass such as pellets up to liquid fuels such as ethanol or synfuels) amounts to some E1.2-2.4 trillion per year.

Feedstocks can be provided from residues from agriculture, forestry, and the wood industry, from biomass produced from degraded and marginal lands, and from biomass produced on good quality agricultural and pasture lands without jeopardising the world's food and feed supply, forests, and biodiversity. The pre-condition to achieve such a situation is that agricultural land-use efficiency is increased, especially in developing regions.

## 2$^{nd}$ Generation Biofuels

These biofuels are cellulosic ethanol, waste based on a complex carbohydrate material, derived from non-grain parts of plants including switch grass, prairie grasses, trees and forestry. Biofuels are produced through hydrolysis/fermentation, gasification or pyrolysis technologies. Future R and D needs to focus more on making second generation sources economically viable.

Research and Development focus on plantations, biofuel processing and production technologies, as well as on maximizing efficiencies of different end-use applications and utilization of by-products. High priority should be accorded to indigenous R and D and technology development based on local feedstocks and needs, which would be benchmarked with international efforts and patents would be registered, wherever possible. Multi-institutional, time-bound research programmes with clearly defined goals and milestones should be developed and supported.

Emphasis should be given to advanced conversion technologies for first generation biofuels and emerging technologies for second generation biofuels including conversion of ligno-cellulosic materials to ethanol such as crop residues, forest wastes and algae, biomass-to-liquid (BTL) fuels, bio-refineries, etc and utilization of by-products of bio-diesel and bio-ethanol production processes such as oil cake, glycerin, bagasse, etc.

## The Future of Biofuels

In light of the fact that commercial biofuel production and use is a relatively new concept, all stakeholders – including the Indian government, small farmers, investors, researchers have an array of possible actions and stances regarding biofuels.

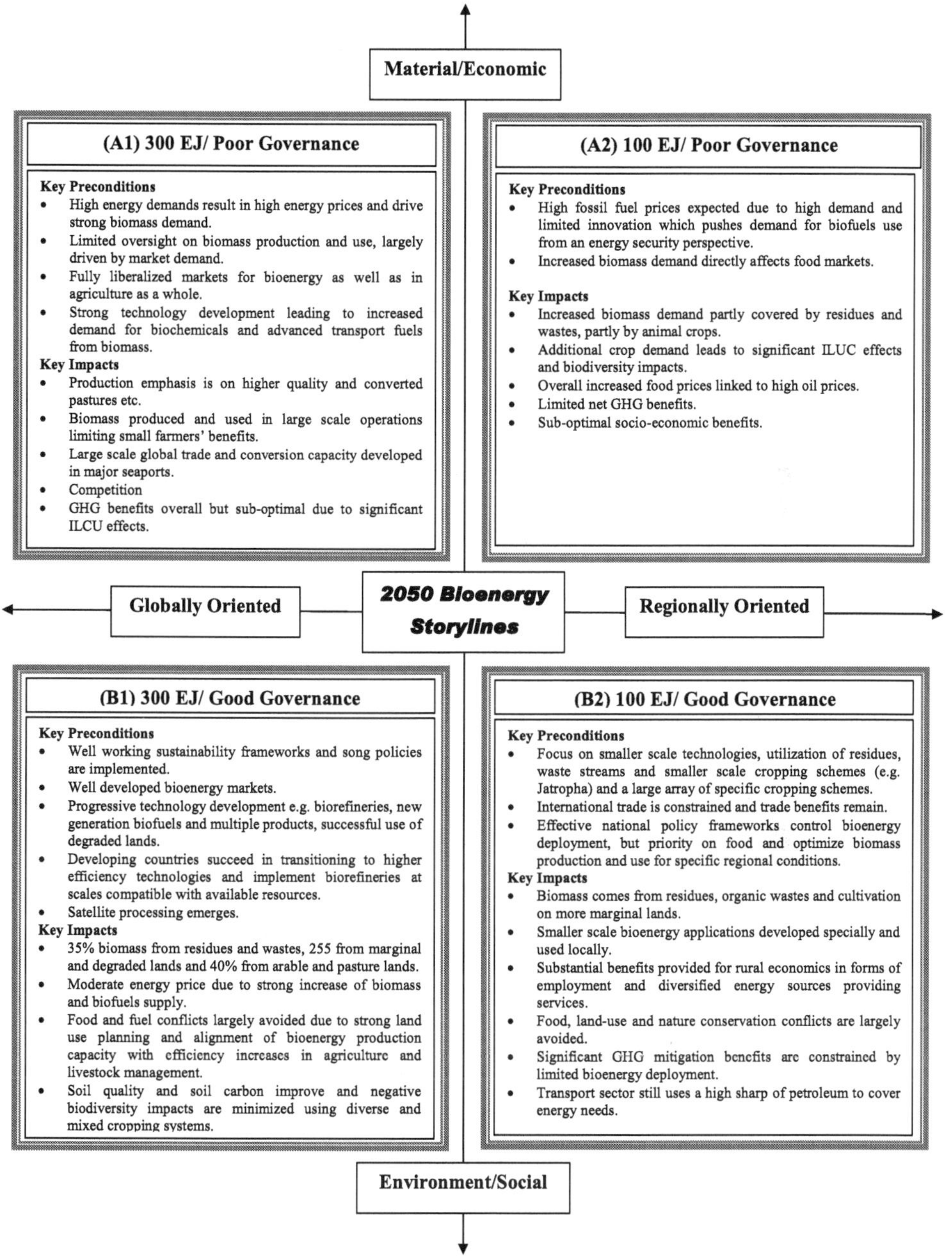

Material/Economic

**(A1) 300 EJ/ Poor Governance**

**Key Preconditions**
- High energy demands result in high energy prices and drive strong biomass demand.
- Limited oversight on biomass production and use, largely driven by market demand.
- Fully liberalized markets for bioenergy as well as in agriculture as a whole.
- Strong technology development leading to increased demand for biochemicals and advanced transport fuels from biomass.

**Key Impacts**
- Production emphasis is on higher quality and converted pastures etc.
- Biomass produced and used in large scale operations limiting small farmers' benefits.
- Large scale global trade and conversion capacity developed in major seaports.
- Competition
- GHG benefits overall but sub-optimal due to significant ILCU effects.

**(A2) 100 EJ/ Poor Governance**

**Key Preconditions**
- High fossil fuel prices expected due to high demand and limited innovation which pushes demand for biofuels use from an energy security perspective.
- Increased biomass demand directly affects food markets.

**Key Impacts**
- Increased biomass demand partly covered by residues and wastes, partly by animal crops.
- Additional crop demand leads to significant ILUC effects and biodiversity impacts.
- Overall increased food prices linked to high oil prices.
- Limited net GHG benefits.
- Sub-optimal socio-economic benefits.

Globally Oriented — ***2050 Bioenergy Storylines*** — Regionally Oriented

**(B1) 300 EJ/ Good Governance**

**Key Preconditions**
- Well working sustainability frameworks and song policies are implemented.
- Well developed bioenergy markets.
- Progressive technology development e.g. biorefineries, new generation biofuels and multiple products, successful use of degraded lands.
- Developing countries succeed in transitioning to higher efficiency technologies and implement biorefineries at scales compatible with available resources.
- Satellite processing emerges.

**Key Impacts**
- 35% biomass from residues and wastes, 255 from marginal and degraded lands and 40% from arable and pasture lands.
- Moderate energy price due to strong increase of biomass and biofuels supply.
- Food and fuel conflicts largely avoided due to strong land use planning and alignment of bioenergy production capacity with efficiency increases in agriculture and livestock management.
- Soil quality and soil carbon improve and negative biodiversity impacts are minimized using diverse and mixed cropping systems.

**(B2) 100 EJ/ Good Governance**

**Key Preconditions**
- Focus on smaller scale technologies, utilization of residues, waste streams and smaller scale cropping schemes (e.g. Jatropha) and a large array of specific cropping schemes.
- International trade is constrained and trade benefits remain.
- Effective national policy frameworks control bioenergy deployment, but priority on food and optimize biomass production and use for specific regional conditions.

**Key Impacts**
- Biomass comes from residues, organic wastes and cultivation on more marginal lands.
- Smaller scale bioenergy applications developed specially and used locally.
- Substantial benefits provided for rural economics in forms of employment and diversified energy sources providing services.
- Food, land-use and nature conservation conflicts are largely avoided.
- Significant GHG mitigation benefits are constrained by limited bioenergy deployment.
- Transport sector still uses a high sharp of petroleum to cover energy needs.

Environment/Social

**Figure 13.5: Possible Futures for 2050 Biomass Deployment for Energy: Four illustrative contrasting sketches describing key preconditions and impacts following world conditions typical of the IPCC SRES storylines (IPCC, 2000a).**

There are numerous different opinions about what the logical next steps are. The following are a few potential options for India:

1. Much research is undoubtedly necessary to better understand the capabilities and downfalls of biofuels. A huge number of questions remain and more are still surfacing. For example, whether jatropha will be economically viable compared to oil, whether indirect and environmental costs of biofuels will outweigh the direct benefits, or whether impoverished farmers will significantly benefit, all are still unanswered. Some believe that research needs to come before the Indian government invests major sums into biofuel development policy particularly those policies based around jatropha. Others have indicated that India must work to improve its agricultural practices before it moves forward with the rest of its biofuel agenda.
2. There are improvements the government could start making in order to give the biofuel industry a solid support system from which it can grow. Advances in agriculture through water saving methods, intercropping, seasonal planting, and crop rotation could create more efficiency in the agricultural sector and therefore increase yields. These improvements could create a variety of positive effects downstream in the biofuel industry and the rest of Indian society.
3. Investing time and money studying and domesticating jatropha also has potential benefits. The Indian government has indicated through its policies and public statements that it hopes to make jatropha a central part of the biofuel industry. Unfortunately, however, there still is not enough information about jatropha to ensure that the Indian government would get a substantial return on its investment.
4. Sweet sorghum stalk has been found to be a potential source of raw material for commercial ethanol production. Sweet sorghum does not compromise on food, feed or fodder production when used for energy production, thereby meeting the biofuel program's vision of not compromising on food security. Sweet sorghum cultivation involves the judicious use of scarce resources like water and other inputs. The crop uses less than a third of the inputs used by sugarcane, such as water, electricity and fertilizers, making it a promising alternative feedstock (Reddy *et al.*, 2008; Rao *et al.*, 2004). Sweet sorghum scores favorably on all parameters compared to alternative feedstocks. As per a pilot study conducted by Vasantdada Sugar Institute (VSI), Pune, India sweet sorghum-based ethanol production has 25 per cent of the biological oxygen dissolved (BOD), i. e., 19,500 mg/l and lower chemical oxygen dissolved (COD), i.e, 38,640 mg liter-1 compared to molasses-based ethanol production. Hence, besides molasses, there is a need for clear guidelines to promote alternative feedstocks like sweet sorghum.

## References

1. Ali N, Zhang T, Yasir HK, Ammar AM, Vakil A, Idrees AM. 2014. Making biofuels from microalgae - A review of technologies. *Journal of Food Science and Technology*: 1 (2): 007-014.

2. Rao DB, Ratnavathi CV, Karthikeyan K, Biswas PK, Rao SS, Viyay Kumar BS and Seetharama N (2004). Sweet sorghum cane for biofuel production: A SWOT analysis in Indian context. Hyderabad, Andhra Pradesh, India: National Research Center for Sorghum: 20.

3. Chum H, Faaij A, Moreira J, Berndes G, Dhamija P, Dong H, Gabrielle B, Goss Eng A, Lucht W, Mapako M, Masera Cerutti O, McIntyre T, Minowa T, Pingoud K (2011). Bioenergy: In IPCC Special Report on Renewable Energy Sources and Climate Change Mitigation [Edenhofer O, Pichs-Madruga R, Sokona Y, Seyboth K, Matschoss P, Kadner S, Zwickel T, Eickemeier P, Hansen G, Schlomer S, von Stechow C (eds)], Cambridge University Press, Cambridge, United Kingdom and New York, NY, USA.

4. Hamelinck CN and Faaij APC (2006). Outlook for advanced biofuels. *Energy Policy*, 34 (17): 3268-3283.

5. IEA. 2006a. World Energy Outlook, 2006. OECD/IEA, Paris, France: 601.

6. IEA. 2006b. Energy Technology perspectives – Scenario's and strategies to 2050, OECD/IEA, Paris, France.

7 IPCC (2000a). *Special Report on Emissions Scenarios.* Nakicenovic N and Swart R (eds.), Cambridge University Press: 570.

8. IPCC (2007a). *Climate Change 2007: Impacts, Adaptation and Vulnerability. Contribution of Working Group II to the Fourth Assessment Report of the Intergovernmental Panel on Climate Change.* Parry ML, Canziani OF, Palutikof JP, van der Linden PJ, and Hanson CE (eds.), Cambridge University Press: 979.

9. IPCC (2007b). *Climate Change 2007: Synthesis Report. Contributions of Working Groups I, II and II to the Fourth Assessment Report of the Intergovernmental Panel on Climate Change.* Core Writing Team, Pachauri RK and Reisinger A (eds.), Cambridge University Press: 104.

10. IPCC (2007c). *Climate Change 2007: The Physical Science Basis. Contribution of Working Group I to the Fourth Assessment Report of the Intergovernmental Panel on Climate Change.* S. Solomon S, Qin D, Manning M, Chen Z, Marquis M, Averyt AB, Tignor M and Miller HL (eds.), Cambridge University Press: 996.

11. IPCC (2007d). *Climate Change 2007: Mitigation of Climate Change. Contribution of Working Group III to the Fourth Assessment Report of the Intergovernmental Panel on Climate Change.* Metz B, Davidson OR, Bosch PR, Dave R and Meyer LA (eds.), Cambridge University Press: 851.

12. Reddy BVS, Ramesh S, Ashok KA, Wani SP, Ortiz R, Ceballos H and Sreedevi TK (2008). Biofuel crops research for energy security and rural development in developing countries. *Bioenergy Research*, 1: 248-258.

13. Ugarte DLT, DGL He, Jensen KL and English BC (2010). Expanded ethanol production: Implications for agriculture, water demand, and water quality. *Biomass and Bioenergy*, 34 (11): 1586-1596.

14. Van Dam J, Junginger M and Faaij APC (2010). From the global efforts on certification of bioenergy towards an integrated approach based on sustainable land use planning. *Renewable and Sustainable Energy Reviews,* 14 (9): 2445-2472.

## Web Resources for Detailed Information

1. www.nrel.gov/biomass/pdfs/39436.pdf
2. www.nrel.gov/biomass/pdfs/40742.pdf
3. www.epa.gov/chp/documents/biomass_chp_catalog.pdf

**2016, Biotechnology: Progress and Applications** *Pages* **319–330**
*Editors:* **Saif Hameed and Zeeshan Fatima**
*Published by:* **DAYA PUBLISHING HOUSE, NEW DELHI**

*Chapter 14*

# Scope of Biotechnology and Bioinformatics towards Rare and Endangered *Flacourtia jangomas* (Lour.) Raeusch.

***Ashutosh Pathak, Rajesh Kumar, Shashi Kant Shukla and Anupam Dikshit****

*Biological Product Lab, Department of Botany, University of Allahabad, Allahabad – 211 002, U.P., India*

## ABSTRACT

*Flacourtia jangomas* (Lour.) Raeusch member of Flacourtiaceae commonly known as Paniyala in hindi - belongs to the ethnic group of Gorakhpur, Uttar Pradesh, India with unique fruit which when eaten as such become sour in taste but when consumed after softening pulp by exerting pressure turned sweat in taste. Besides it serve the purpose of food, ornamentation of gardens and medicine. Ethno-medicinally folks used the leaf decoction to halt diarrhoea, powdered dried leaves are employed to relieve bronchitis and cough. *F. jangomas* has therapeutic uses such as antidiabetic, antimicrobial, antimalarial and hepatoprotective activitiy. So, far it is not declared or considered noxious by any government authorities. The cultivation of *F. jangomas* is needed to be promoted with biotechnological approaches, which enhances the socio-economic status of farmers and health of human beings. The potential of biotechnology has open a new horizon for the

* Corresponding Author: E-mail: anupambplau@rediffmail.com

scientist or researchers worldwide to exploit the biological resources. Since, *F. jangomas* is a rare and endangered plant and very little work has been done regarding propagation and needs to be focused. Apart from biotechnology, bioinformatics provides evidence for the exploitation of other substitutes such as *Xylosoma benthamii, Xylosoma congesta* and *Scolopia spinosa.*

## Introduction

Since last fifty years biotechnology has been an emerging and continuously growing field to fulfill the needs or requirements of human race. Although it is difficult to define biotechnology but it can be defined as exploitation of biological resources for the betterment of human race which includes several sub disciples such as genetic engineering, tissue culture etc. The potential of biotechnology has open a new horizon for the scientist or researchers worldwide to exploit the biological resources. In this current chapter we discuss the requirement and importance of biotechnological approach to solve a problem regarding the cultivation of a rare, endangered plant *Flacourtia jangomas* (Lour.) Raeusch. *F. jangomas* (Flacourtiaceae) (hindi - Paniyala) belongs to the ethnic group of Gorakhpur, Uttar Pradesh, India with unique fruit which when eaten as such become sour in taste but when consumed after softening pulp by exerting pressure turned sweat in taste. Fruits, are stewed as dessert, made into juice, syrup, jam, marmalade and pickles and also used in chutneys. The fruits are eaten to overcome biliousness, nausea and diarrhoea [1]. It also serve the purpose of food, ornamentation of gardens, medicine. Ethno-medicinally folks used the leaf decoction to halt diarrhoea, powdered dried leaves are employed to relieve bronchitis and cough [1]. So, far it is not declared or considered noxious by any state government authorities [2]. In acute toxicity study, no toxic symptoms were observed for methanol extract *F. jangomas* of up to dose of 2 g/kg body weight *i.e.* for a normal person of 60 kg it is safe to eat 160 g of extract [3]. The cultivation of *F. jangomas* is needed to be promoted with biotechnological approaches, which enhances the socio-economic status of farmers and health of human beings [1].

## An Overview about *Flacourtia jangomas* (Lour.) Raeusch

### Taxonomic Hierarchy according to Bentham and Hooker (1862) [4]

- ☆ **Kingdom-** Plantae
- ☆ **Division-** Phanerogamia
- ☆ **Class-** Dicotyledons
- ☆ **Subclass-** Polypetalae
- ☆ **Series-** Thalamiflorae
- ☆ **Cohors-** Parietales
- ☆ **Order-** Bixinae
- ☆ **Tribe-** Flacourtieae
- ☆ **Subtribe-** Euflacourtieae
- ☆ **Family-** Flacourtiaceae

- ☆ **Genus-** *Flacourtia* Comm. ex L'Hér.
- ☆ **Species-** *F. jangomas* (Lour.) Raeusch. Nomencl. Bot. ed. 3,280. 1797.
- ☆ **Synonyms-** *Flacourtia cataphracta* Roxb. ex Willd, *Stigmarota jangomas* Lour.
- ☆ **Vernacular name-** Pachnala, Paniyala, Jamnua, Indian plum.

## Description

Large shrubs or small trees, 5-10 m tall, deciduous; trunk and older branches usually unarmed, young branches with simple or divaricate spines; bark yellow-brown, reddish brown, or light brown, flaky; young branches smooth, glabrous or sparsely pubescent, lenticellate. Petiole 4-8 mm, pubescent or glabrescent; leaf blade dark green abaxially, shiny adaxially, in fresh state pinkish to reddish or orange-brown when young, narrowly ovate, ovate-elliptic, or ovate-oblong, rarely oblong-lanceolate or (slightly) obovate-lanceolate, 7-14 × 2-5 cm, thinly leathery to papery, both surfaces practically glabrous, any hairs present very short, midvein slightly raised on both surfaces, lateral veins 3-6 pairs, conspicuous adaxially, base acute, obtuse, or rounded, margin entire or serrate to crenate, apex obtuse or gradually tapering to narrowly acuminate, rarely more abruptly acuminate. Inflorescences axillary, racemose; rachis 0.5-2 cm, puberulous. Pedicels 5-10(-15) mm, very slender, minutely and sparsely puberulous or glabrous; bracts ovate, 0.5-1 mm, outside glabrous or sparsely hairy, inside pubescent, margin entire, ciliate. Flowers appearing with or before young leaves, white to greenish, honey-scented. Sepals 4 or 5, ca. 2 mm, ovate-triangular, apex obtuse, outside practically glabrous, inside pubescent, margin ciliate, hairs very short, often barely visible in female flowers. Staminate flowers: stamen filaments 2-3 mm, glabrous. Pistillate flowers: ovary bottle-shaped to globose, 2-3 mm; styles 4-6, united into a distinct column ca. 1 mm, not or slightly free at their apices; stigmas slightly reniform, dilated, recurved. Fruit brownish red or purple, finally blackish, subglobose, fleshy, 1.5-2.5 cm in diam., in dried material sometimes constricted at equator, style column persistent. Seeds 4 or 5(-10); Flowering: Apr-May; Fruiting: May-Oct (6) (Figures 14.1 and 14.2).

## Distribution

*F. jangomas* is very important fruit crop of Gorakhpur region and indigenous to North-Eastern region Uttar Pradesh India and having restricted distribution [1]. Sub himalayan foot hill zone in India extending to South East Asia, China [6].

## Biochemical Properties

The ripe fruits of *F. jangomas* contains highly antioxidant compounds such as alkaloids, flavonoids, phenolic compounds, tannins and contains beta-carotene followed by lutein, zeaxanthin, retinol and phylloquinone (vit.K) [1]. The *F. jangomas* posses antioxidant as well as analgesic properties [8].

## Therapeutic Studies Conducted on *Flacourtia*

### Hepatoprotective Activity

Several studies conducted on the other species which is endemic to India *i.e. Flacourtia indica*. The hepatoprotective activity had been seen against hepatotoxicity

**Figure 14.1: Tree (A), Fruit (B), Seeds(C) [5].**

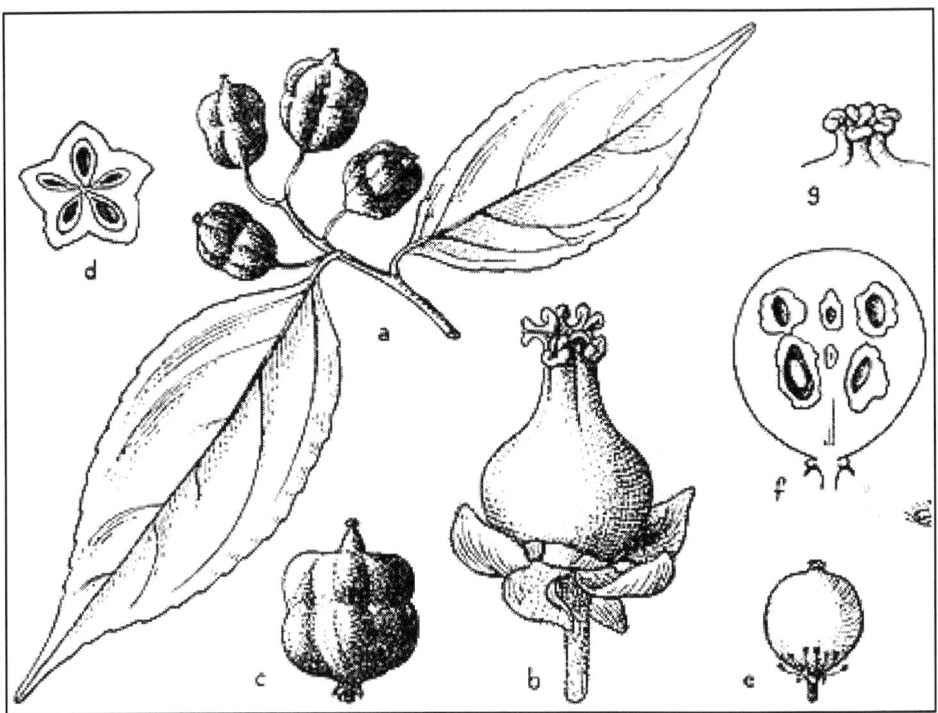


**Figure 14.2: Illustration of *Flacourtia jangomas* (Lour.) Raeusch. (a) Twig with fruits (b) Female flower (c) Fruit (d) Fruit cross section (e) Friut (f) Fruit longitudinal section (g) Stigma (adopted from Flora Malesiana series I, Vol. 5 page 72)[7].**

caused by three different inducing agents first methanotrexate, second paracetamol and third carbon tetrachloride. Petroleum ether extract of aerial parts was effective against methanotrexate and paracetamol induced hepatotoxicity whereas aqueous extraxt of aerial parts is found to be effective against carbon tetrachloride induced hepatotoxicity. It lowers the total bilirubin, lipid peroxides, serum levels of SGOT and SGPT.[9-12]

### Antimalarial and Antibacterial Activity

Three compounds *viz.*, poliothrysoside, purocatechol and homaloside D, isolated from *F. indica* decoction have antiplasmodial activity on chloroquine-resistant strain of *Plasmodium falciparum*; causal organism of malaria.[13]. *F. jangomas* chloroform extract is found to be effective against *E. coli* and *Bacillus cereus* by disk diffusion technique [14].

## Miscellaneous Use

The *F. jangomas* extract can be used as a steel inhibitor against mild steel corrosion. Study conducted showed that *F. jangomas* leaf extract is effective in 1M HCl and 0.5M $H_2SO_4$ was found to inhibit the mild steel corrosion up to 98 per cent and 95 per cent respectively [15].

## Role of Biotechnology

Biotechnology comprised of old and new biotechnology. Old biotechnology comprised of simple processes like fermentation where as new biotechnology comprised of genetic engineering, recombinant DNA technology, karyotyping, tissue culture, antibiotic production etc.

## Karyomorphological Study

*F. jangomas* chromosomes are classified into four groups *i.e.* one pair nearly metacentric long chromosome with nearly subterminal secondary constriction; two pairs of nearly metacentric to sub- metacentric long chromosomes; three pairs of very small chromosomes in which no centromere could be detected. Nine bivalents were found *i.e.* total number of chromosomes is eighteen (2n = 18) [16].

## Propagation of *F. jangomas*

Vegetative propagation is a traditional and less expensive method generally used for the cultivation of many crops. Propagation of *F. jangomas* has been tested by two methods so far first, propagation by seeds and secondly by clonal propagation by stem cutting methods [5].

### Propagation by Seeds

Seeds have been extracted by depulping of fully ripe fruits which are uniform in size. Uniformly sized fruits were collected to reduced the non-treatment variations at the germination as the seedling strength is directly proportional to the seed size[18].

### Clonal Propagation by Stem Cutting

Two year old shoots of *F. jangomas* has been used for the clonal propagation. The two nodes cutting with two leaves trimmed to half has been used and treated with fungicide (Diathane M45) to prevent from any fungal infection[19]. (A detailed procedure can be downloaded from the reference).

The above mentioned procedures are useful where *F. jangomas* has already been cultivated and for garden purpose and it is time consuming as well. But, for the generation of large amount of stock from very little amount of explants we need to look forward to the modern techniques available to us in the form of biotechnology.

### Tissue-Culture

Tissue culture techniques provides us an opportunity for high quality uniform plantlet production or planting material anywhere throughout the globe. The aseptic and controlled conditions provided *in vitro* made the whole year production possible irrespective of seasonal variations. However, it is costly and needs more effort, labour, electricity and clean water supply. Although, manual labour is available in many third world country but the trained human resource is not available. The conditions required for each species of plant differs and depends upon temperature, light period, humidity and nutritional requirements.

There are many advantage of tissue culture technique over conventional approaches. Tissue culture eliminates the disease present in stock plant which may results into the increased growth, branching, flowering and higher yield. The method may succeed to propagate plants where seeds or vegetative propagation is not possible or difficult or undesirable as in the case of *F. jangomas*. As the money investment on mother plants is reduced to almost negligible, there is greater plant uniformity and the availability in the mass at any time and a better programming of the production is possible. Two possible techniques are discussed below:

*Oraganogenesis*

Organogenesis is a technique used in biotechnology for the regeneration of plant organs which exploits the totipotent property of a plant meristematic cell to generate whole plant. So far, *in-vitro* studies conducted on organogenesis plantlets regenerated from adventitious meristems may originate either from shoot organogenesis or somatic embryogenesis. In case of shoot organogenesis, first shoots formation takes place, then roots are induced from the base of shoots, followed by well-developed vascular connection between shoots and maternal explants. In somatic embryogenesis, somatic embryo (a bipolar structure) gives rise to the shoots and roots from embryogenic shoots and roots and there is no vascular connection between somatic embryos and parental explants [20, 21].

*Callus*

Callus, an unorganized tissue mass growing *in vitro* on a solid substrate or naturally on plants in response to wounding, infestations, or at graft unions. In *in vitro* conditions elevated levels of harmones such as cytokinins and auxin induces the callus formation and it is primarily a central procedure in tissue culture technique. A single callus can be multiplied over and over again to produce large number of uniform TCPs. In another approach one can insert desired gene into cells and obtain TCPs of superior quality, disease resistant, drought resistant etc from transformed callus.

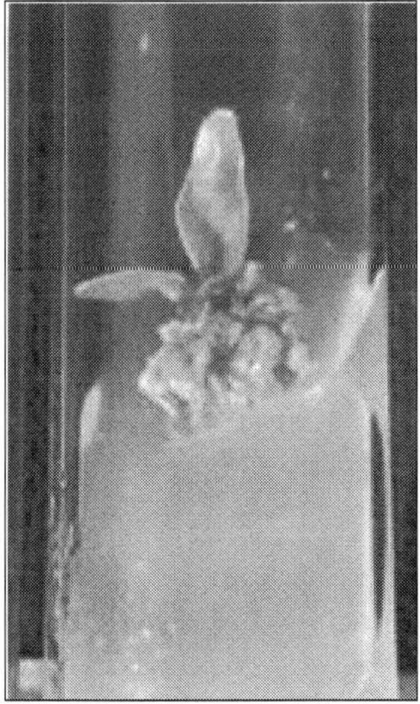

**Figure 14.3: Explant Derived Callus Tissue *F. jangomas* [21].**

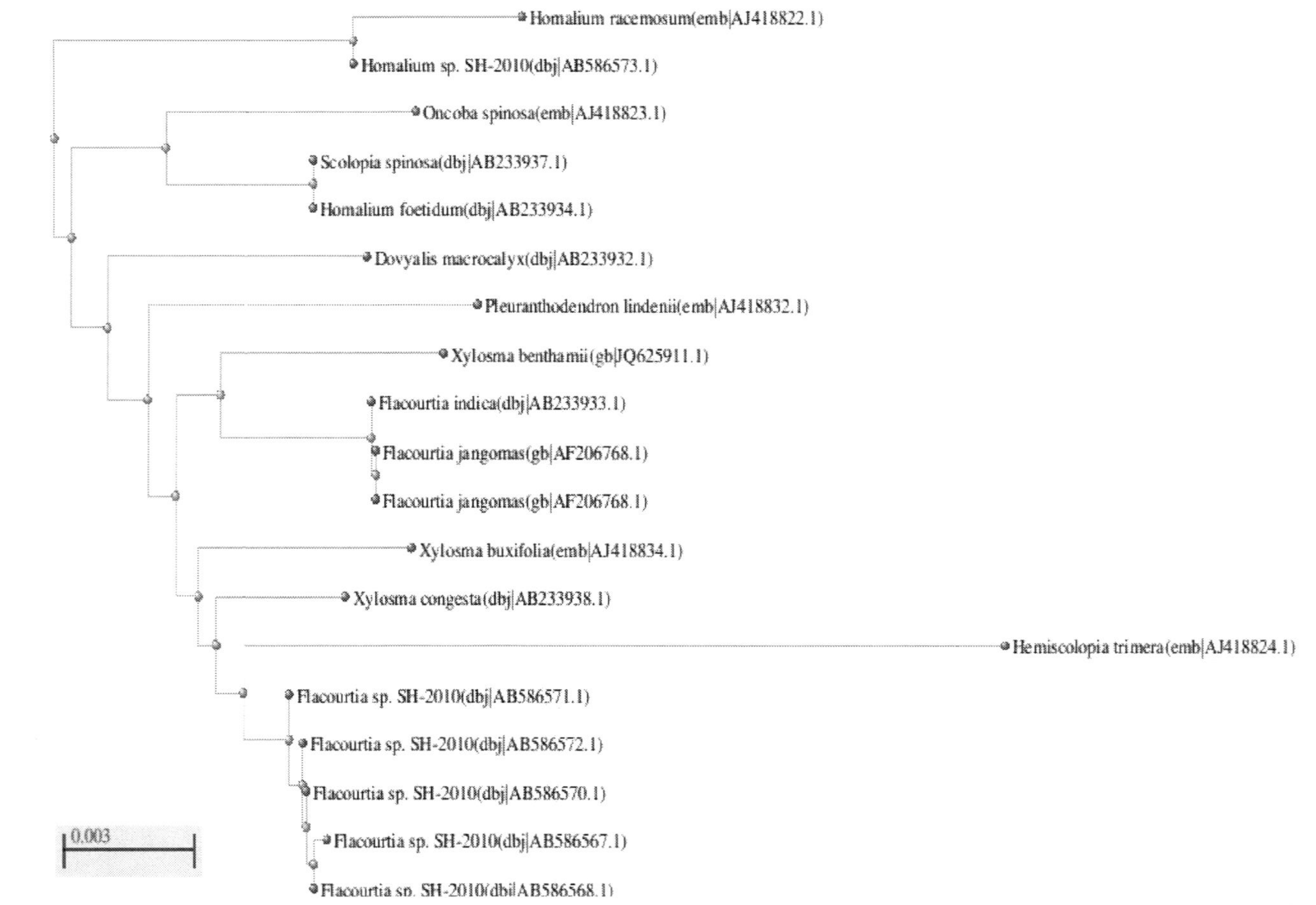


**Figure 14.4: Phylogenetic Tree on the Basis of rbcl Gene of *F. jangomas* (Acc No. AF206768.1) and Closely Related Species.**

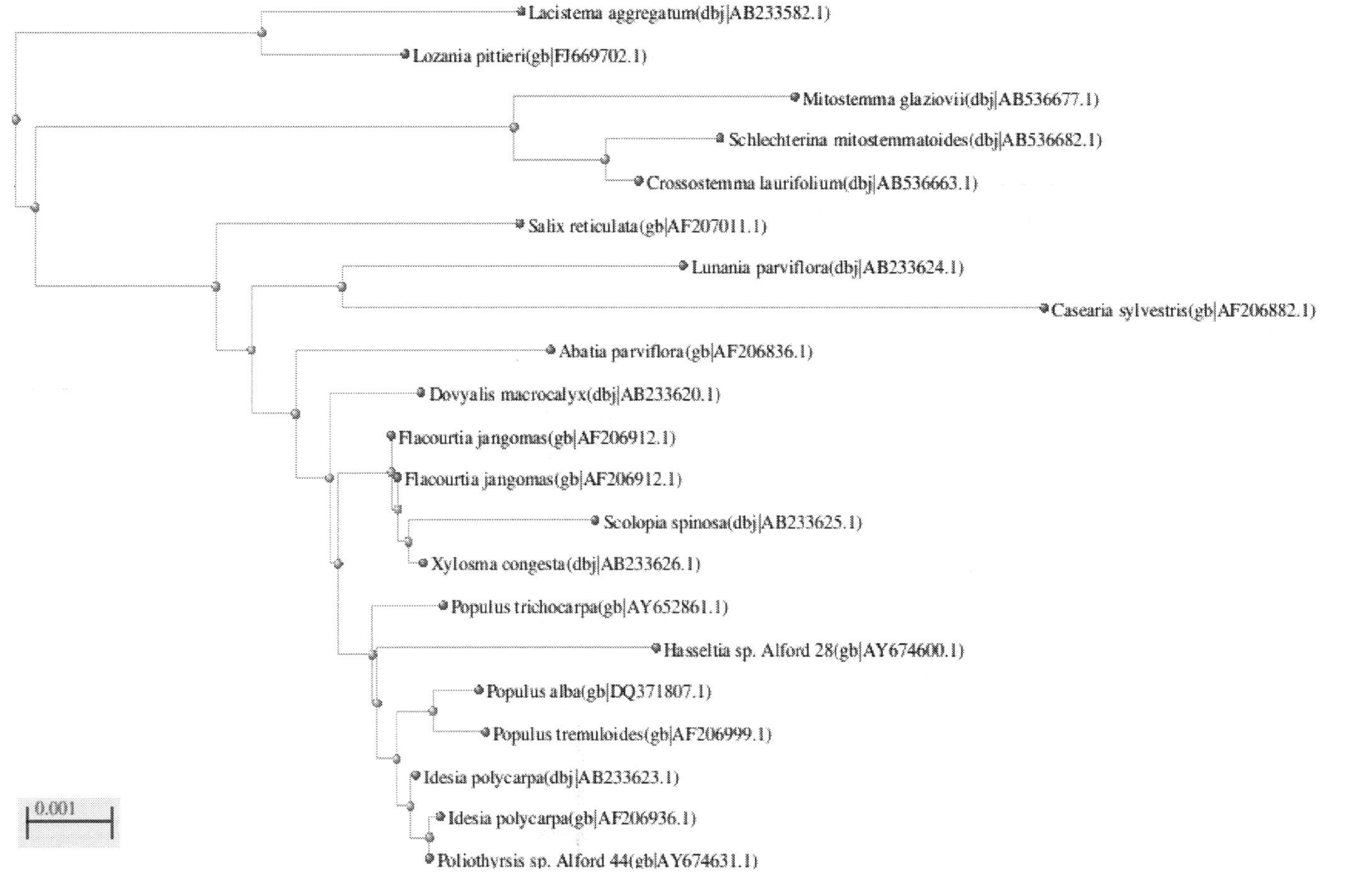


**Figure 14.5: Phylogenetic Tree on the Basis of 18S rDNA of *F. jangomas* (Acc No. AF206912.1) and Closely Related Species.**

An attempt of *in vitro* regeneration of *F. jangomas* was attempted by Chandra and Bhanja (2002) through nodal segment derived from callus tissue (Figure 14.3). Promising results were obtained in their study but the hardening process haven't been done or published and more efforts have to be put. Hardening or field transfer of cultured plantlets must be done for the conservation and betterment of humans [21].

## Role of Bioinformatics

Bioinformatics provides us an option of data mining through which we can deduce the phylogeny of a particular organism or its closely related organisms. As due course of evolution every organisms has been evolved individually but the closely related ones showed similarities. The data gathered so far by the eminent scientist and researchers has been used to deduce the closely related species of *F. jangomas* through National Center for Biotechnology Information (NCBI) Blast program. Two dendrograms were plotted using different regions of plant genome; first, rbcl gene present in the chloroplast genome of a plant, which codes for the long chain of RUBISCO enzyme (Figure 14.4); second, 18S ribosomal DNA (Internal Transcribed Spacer region- ITS), ITS is a non-coding region present in between the r-DNA of plant genome (Figure 14.5). The phylogenetic tree constructed clearly showed that the closely related species according to molecular data available is *Xylosoma benthamii* in case of rbcl gene and for the ITS region it is *Xylosoma congesta* and *Scolopia spinosa*.

The aforementioned species can be studied for their biochemical, therapeutic, antioxidant and other properties like toxicity and can be utilized as a substitute for the *F. jangomas*.

## Conclusion

*F. jangomas* (ver. Hindi Name *Paniyala*) of Flacourtiaceae a rare, endangered plant, and have a tremendous potential. The scientific explanation to the mystery of the fruit of the plant which when eaten as such showing sourness but when consumed after softening the pulp by exerting pressure become sweet, is very much required. Besides, to overcome the cultivation problem of *F. jangomas*, application of biotechnology is the only available tool. The alternative sources which can be utilized instead of *F. jangomas* are *Xylosoma benthamii*, *Xylosoma congesta* and *Scolopia spinosa* as demonstrated through aforementioned phylogenetic tree.

## Acknowledgement

Thanks are due to University Grants Commission and University of Allahabad for providing financial assistance and library facilities, respectively.

## References

1. Dubey N and Pandey VN (2013). Ethnobiological Importance of Flacourtia jangomas (Lour.). Raeusch. *Trends in Biosciences*. 6: 532-534.
2. http: //keyserver.lucidcentral.org/weeds/data/03030800-0b07-490a-8d04-0605030c0f01/media/Html/Flacourtia_jangomas.htm.

3. Singh AK and Singh J (2010). Evaluation of anti-diabetic potential of leaves and stem of *Flucourtia jangomas* in streptozotocin-induced diabetic rats. *Indian J. Pharmacol.* 42(5): 301-305.

4. Bentham G and Hooker JD (1862). *Genera Plantarum.* Londini-Typis JE Taylor et Soc.; Little Queen Street, Holborn. Reprint- International Book Distributers, Dehradun, India. 1: 122.

5. Hossain A, Sen M, Jewel MIU and Kabir MA (2011). Propagation of *Flacourtia jangomas*: an approach towards the domestication of a wild fruit species in Bangladesh. *Dendrobiology.* 65: 63-71.

6. Flora of China. 13: 119 Online issue. http: //www.efloras.org/florataxon.aspx?flora_id=2 and taxon_id=242422473.

7. http: //www.plantillustrations.org/illustration.php?id_illustration=232197.

8. Talukader C, Saha S, Adhikari S, Mondal HK, Islam K and Anisuzzman Md. (2012). Evaluation of antioxidant, analgesic and antidiarrhoeal activity of *F. jangomas* (Lour.). Raeusch leaves. *Pharmacologyonline.* 3: 20-28.

9. Joyamma V and Thomas J (2011). Protective effect of *F, indica* (Burm.*f.*). Merr. In Methotrexate induced hepatotoxicity. *An International Journal of Advances in Pharmaceutical Sciences.* 2(2-3): 115-123.

10. Jollow DJ, Thorgeirsson SS, Potler WZ, Hashimoto M and Mitchel Jr. (1974). Acetaminophen-induced hepatic necrosis. *Pharmacology.* 12: 251-271.

11. Ganaprakash K (2010). Aqueous extract of *F. indica* prevents carbon tetrachloride induced hepatotoxicity in rat. *International Journal of Biological and Life Sciences.* 6(1): 51-55.

12. Kota GC, Karthikeyan M, Kannan M ad Rajasekar (2012). *Flacourtia indica* (Burm. f. Merr.- A Phytopharmacological review. *International Journal of Research in Pharmaceutical and Biomedical Sciences.* 3(1): 78-81.

13. Kaou AM, Mahiou-eddet V, Canlet C, Debrauwer L and Sebastien H (2010). Antimalarial compounds from the aerial parts of *Flacourtia indica* (Flacourtiaceae). *Journal of Ethnopharmacology.* 130(2): 272-274.

14. Sarker GC, Zahan R, Badrul Alam M, Saiful Islam MD Mosaddik MA and Kramul Haque ME (2011). Antibacterial activity of *Flacourtia jangomas* and *Flacourtia sepiaria. Int. J. of Pharm and Life Sci.* 2(7): 878-883.

15. Hasan SK and Sisodia P (2011). Paniala (*Flacourtia jangomas*). plant extract as eco-friendly inhibitor on the corrosion of mild steel in acidic media. *Rasayan J. Chem.* 4(3): 548-553.

16. Chandra I and Bhanja P (2012). Karyomorphological Study of *Flacourtia jangomas* (Lour.). Raeusch. *International Journal of Research in Plant Science.* 2(4): 74-75.

17. Indira EP and Basha SC (1999). Effects of seeds from different sources on germination and growth in Teak (*Tectona grandis* L. f.). in the nursery. *Annals of Forestry*. 7: 39–44.

18. Puigderrajols P, Celestino C, Mònica S, Mariano T and Marisa M (2000). Histology of organogenic and embryogenic responses in cotyledons of somatic embryos of *Quercus suber* l. *International Journal of Plant Sciences*. 161, 353- 362.

19. Chengalrayan K., Hazra, S and GalloMeagher M (2001). Histological analysis of somatic embryogenesis and organogenesis induced from mature zygotic embryo-derived leaflets of peanut (*Arachis hypogaea* L.). *Plant Science*. 161: 415–421.

20. Chandra I and Bhanja P (2002). Study of organogenesis *in vitro* from callus tissue of *Flacourtia jangomas* (Lour.). Raeusch through scanning electron microscopy. *Current Science*. Indian Academy of Sciences, Bangalore. 83(4): 476-479.

*Section VI*

# Infectious Diseases

**2016, Biotechnology: Progress and Applications** *Pages* **333–343**
*Editors:* **Saif Hameed and Zeeshan Fatima**
*Published by:* **DAYA PUBLISHING HOUSE, NEW DELHI**

# *Chapter 15*

# Glyoxylate Cycle: A Promising Antimicrobial Drug Target

***Saif Hameed, Moiz A. Ansari and Zeeshan Fatima****

*Amity Institute of Biotechnology, Amity University Haryana, Gurgaon – 122 413, Manesar, Haryana, India*

## ABSTRACT

During last decades, the ever increasing burden of antimicrobial drug resistance is on the mount against the present drug regimes. Moreover with the rise in immunocompromised conditions such as AIDS, organ transplantation and chemotherapy, the problem of microbial infections has compounded. Thus there is an urgent need to find newer drugs and efficient drug targets. Again discovery of newer drugs is impeded by their cost effectiveness, toxicity. Therefore utilizing specific drug targets still remains the central theme of the antimicrobial chemotherapy. One strategy that has been used as potential antimicrobial drug targets in great extent is to target specific metabolic pathways which provide the advantage to these microorganisms in the hostile niche. One such metabolic pathway is glyoxylate cycle (GC) which is absent in humans and functional in most of the microorganisms. This chapter focuses on the significance of GC in context of antimicrobial chemotherapy and how GC can be used as an efficient drug target.

***Keywords:*** *Glyoxylate cycle, Drug target, Isocitrate lyase, Malate synthase.*

* Corresponding Author: E-mail: drzeeshanfatima@gmail.com

## Introduction

Human beings are one of the most conducive environments for the growth of microorganisms residing in as microflora or infecting them from outside as pathogens. Many of the commensal organisms could also turns to their virulent form during some immune-compromised conditions *viz.*, HIV- infection, diabetic patients, individual under gone organ transplantation; severe burn patients etc. which makes the body easy target for diseases and thereby leading to extensively high mortality rates. Moreover, the pathogens infecting from outside can also cause severe diseases and could be hazardous to human health. There are many drugs that are customarily used as the therapeutic agents against the various infecting microorganisms for their effective antimicrobial activity. But the major shortcoming towards the usage of current therapeutic regime includes toxicity and cost effectiveness. The search for novel drugs with no significant side effects on patients continues, and research has been focused mainly on specific cell components and pathways. Furthermore, tremendous practice of antimicrobial drugs has led to the emergence of the resistance of microbes against these remedies by a known as Multi Drug Resistance (MDR). The appearance of MDR constrained us to explore the new salutary treatments having newer targets than the drug that are already in use.

Apart from focusing on specific cell components like cell wall or ergosterol synthesis, recently, specific metabolic pathways that can be used as potential antimicrobial drug targets have been explored in great extent. These pathways contribute to the metabolic flexibility and allow microbial cells to survive in nutrient-limited host niches during infection. Metabolic flexibility is particularly crucial for assimilation of diverse carbon sources thereby affecting not only the adaptability of microorganisms but also their pathogenicity and virulence. Among the various metabolic pathways, one such pathway that has gained staunch interest of researchers and the focus of this chapter is glyoxylate cycle (GC). The following sections will deal with significance and biochemistry of GC and how it can be useful as drug targets among various pathogenic microorganisms such as fungi, bacteria and protozoa.

## Biochemistry of GC

More than half a century ago, Kornberg and his coworkers discovered a modified form of tricarboxylic acid (TCA) cycle referring as GC (Kornberg and Kreb, 1957; Kornberg and Madison 1957). The occurrence of GC is widespread and well documented in bacteria, fungi and protozoans. TCA cycle usually starts when Acetyl-CoA reacts with oxaloacetate and converting into CoA to form citrate, where eight enzymatic steps occurs forming different intermediates that noshes various pathways. Thus it becomes difficult for any organism being human or microorganism to survive without TCA cycle. The GC pathway on other hands starts with two initial steps of TCA cycle where citrate synthase and aconitase catalyses the formation of citrate and isocitrate. Post this point GC bypasses the decarboxylation steps directly to form glyoxylate and succinate. There are two critical and specific steps in GC. Firstly, isocitrate (C6) is hydrolyzed to succinate (C4) and glyoxylate (C2) by isocitrate lyase (ICL). In the next step, acetyl-CoA (C2) condenses with glyoxylate to produce malate

(C4) by malate synthase (MS). From this step onwards again uptill isocitrate productions all the steps are shared between glyoxylate and TCA cycle. Malate, from TCA cycle, is converted to form oxaloacetate and then to citrate by adding another molecule of acetyl-CoA, following with the production of isocitrate (Figure 15.1). Thus each round of GC utilizes two molecules of acetyl-CoA producing one molecule of succinate and malate which is further utilized for the biosynthetic rationale.

Despite the fact that GC is an alternate form of TCA cycle it occurs in distinct cell organelles in different cell types. For instance, GC in some plants takes place in glyoxysomes, a specialized peroxisomes and TCA cycle in mitochondria (Graham 2008). However, both cycles collaborate with each other to utilize the succinate formed during GC and the oxaloacetate formed during TCA cycle. The oxaloacetate from TCA cycle (in mitochondria) is transferred to glyoxysome in the form of aspartate. This aspartate is converted to oxaloacetate to enter in GC resulting to forms succinate which again enters the mitochondria and take part in TCA cycle.

Pathogens such as MTB expresses two types of ICL *i.e.* ICL1 and ICL2 encoded by genes *icl1* and *icl2* [Muñoz, 2005]. MTB ICL is homo-tetramer when it is biologically active form having of four identical subunits [Nakazawa 2005]. A conserved sequence "KKCGH" is common to all ICL for bacterium, prokaryotic or eukaryotic. The conserved catalytic residues contain cysteine present on a supple loop which experience large conformational change post binding of substrate. This results in absolute closing of the active site from the bulk solvent [Smith 2004]. It was also seen that the divalent cations like $Cd^{2+}$ and $Zn^{2+}$ functions as inhibitors that results in fractional unfolding of the enzyme and an inclusive thrashing of enzyme activity, although, the quaternary structure *viz.*, tetrameric structure of the enzyme is not distraughted [Kumar 2008].

## Significance of GC

Drugs targeting GC could be a better alternative to work on because of its two fold advantage. Firstly, the GC is not absent in human as there are no reported orthologs of GC cycle enzymes hence can be specifically targeted by drugs or compounds inhibiting GC enzymes. Secondly, GC being a modified form of TCA Cycle and a metabolic pathway enables organisms to synthesize carbohydrates from two carbon compounds *viz.*, ethanol and acetate (Kornberg and Krebs, 1957). The important carbon supply in most organisms is glucose that is rehabilitated into five-carbon sugars (such as xylulose, ribose and deoxyribose) through pentose phosphate pathway and catabolized to acetyl-CoA via glycolysis. Since, glucose is often not available to microorganisms in the nutrient limited host niches, thus simple carbon compounds are the only available supply of carbon like acetate, ethanol, fatty acids which can act as fuel and basis of phosphoenolpyruvate to synthesize carbohydrate. TCA cycle having two decarboxylation steps does not allow the incorporation of carbon and thus it does not permit to synthesize macromolecules from C2 compounds. The GC on other hand bypasses these decarboxylation steps thereby preventing the two carbon loss and enabling even C2 compound to act as a carbon source for gluconeogenesis and successive incorporation into amino acids, DNA and RNA. Moreover, GC also acts as an anaplerotic pathway for TCA cycle where the pool of TCA cycle intermediates necessary for gluconeogenesis and other biosynthetic

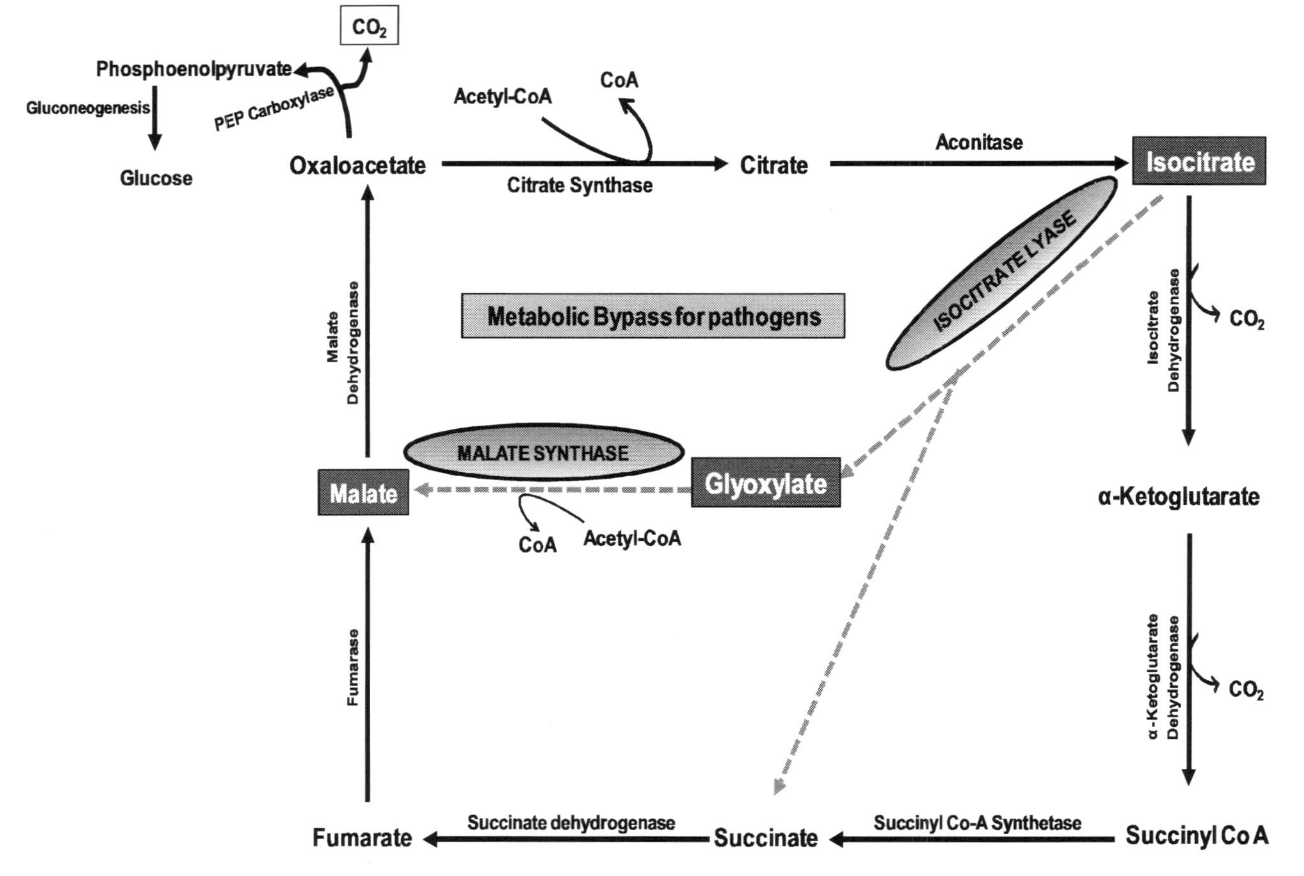


**Figure 15.1: Schematic Representation of the Specific and Common Steps in GC and TCA Cycles. GC specific steps are shown in red and GC specific enzymes are highlighted in boxes with GC specific intermediates.**

processes is replenished thereby aiding the survival of pathogens. For instance, microorganisms like *Mycobacterium tuberculosis* (MTB), the bacterium responsible to cause tuberculosis, because of GC is able to reside for extended periods within the macrophage, in which it can thrive and consequently spread all over the body. Similarly, despite differences in the evasion mechanisms in macrophages in the opportunistic fungal pathogen *Candida albicans*, survival within the macrophages is contributed by GC as well (Lorenz and Fink 2002). Therefore, ICL and MS as described above becomes the most specific and vital enzymes of this pathway to be targeted for antimicrobial rehabilitation. The ICL and MS family proteins have been found in a wide range of organisms, such as bacteria (Kornberg 1966), fungi (Lorenz and Fink 2001), archaea (Serrano, Camacho, and Bonete 1998) and higher plants (Eastmond and Graham 2001).

## GC and Fungal Infections

Over the last few decades the systemic fungal infections causing severe diseases has became the predominant predicament around the globe, explicitly with the neutropenic patients facing the immunocompromised conditions such as AIDS, cancer chemotherapy, organ transplantations, burn, diabetes and pregnancy (Sims 2005). The prevalent usage of antifungal agents turns out to be the foundation of MDR and encumbers the appropriate treatment to the fungal infections. Currently many innovative antifungal drugs along with the novel drug targets are being explored by many researchers. One of the metabolic targets that have gained significant weightage is GC due to the reasons described earlier.

The genome-wide expression profiling to mimic the events that take place in *Saccharomyces cerevisiae* and *C. albicans* on intake by a mammalian macrophage has not only shown the induction of GC genes but also reduced virulence in the knock out mutants lacking ICL (Lorenze and Fink, 2001). Recently, many workers have also worked on natural compounds to demonstrate ICL and MS as the impending antifungal targets. For instance several natural compounds targeting ICL such as, caffeic acid, apigenin and rosmarinic acid have been studied to efficiently act as the antifungal drug target (Figure 15.2) by performing ICL enzyme assay (Chean *et al.*, 2014). Similarly, Suvanine sesterterpenes which were isolated from a tropical sponge *Coscinoderma* sp. were checked for their inhibitory actions toward recombinant ICL from *C. albicans* where a potent ICL inhibitor in the form of suvanine salt, having a sodium counterion was identified (Lee *et al.*). *Cryptococcus* is another pathogenic fungus that causes meningitis in immunocompromised patients. It has been reported that the capsule formation which is the major virulence factor of *C. neoformans* is linked with cell cycle regulation and upregulation of GC pathway. Proteomic assays hints that in the cln1 mutant strain that is affected in cell cycle regulations, upregulation in the GC cycle occurs (García-Rodas *et al.*, 2014).

*Aspergillus* a causative agent of another commonly occurring fungal disease aspergillosis is also one of the major rationales for broadening of fungal infections and could not be ignored. Many workers have worked to find the novel targets against the MDR developed by it by exploring some novel targets for its eradication and GC is one of them. It is known that acuD and acuE are genes that encode the two GC

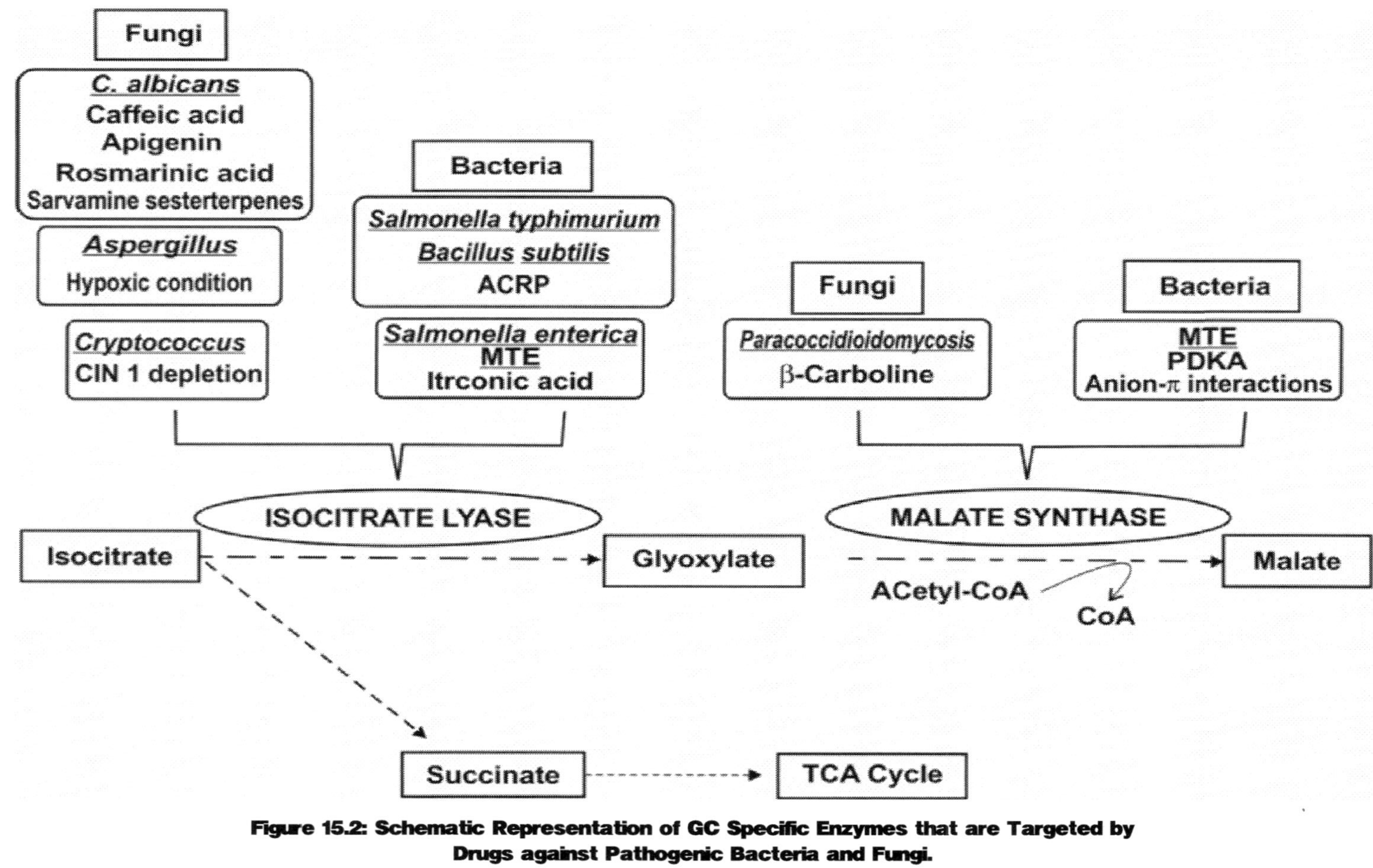


**Figure 15.2: Schematic Representation of GC Specific Enzymes that are Targeted by Drugs against Pathogenic Bacteria and Fungi.**

specific enzymes, ICL and MS in *Aspergillus nidulans* but a study have also suggested that AcuL could also links GC with gluconeogenesis by bartering cytoplasmic succinate for mitochondrial fumarate (Flippi 2014). Similarly, the response in the genetic profiling of *Aspergillus oryzea* and *A. nidulans* under hypoxic condition shows that both fungi abolish NADH accretion under hypoxic environment as both fungi up-regulates genes for alcoholic fermentation and the γ-aminobutyrate shunt of the TCA cycle, whereas *A. oryzae* upregulates only the GC (Terabayashi 2012). GC has been targeted against another pathogenic fungus *Paracoccidioides* a causative agent of paracoccidioidomycosis. The effect of β-Carboline Alkaloids from Galianthe ramose on GC have also been reported against this fungal foe (Figure 15.2) (de Freitas 2014).

## GC and Bacterial Infections

Bacterial infections are on the mount and one of the most common problems faced by us in our day to day life. For instance, MTB causative agent of Tuberculosis (TB) makes human lives to suffer each year more than any other bacterial pathogen. Instead of eradicating the disease completely through the therapeutics, this bacterium is developing MDR against them. Despite of urgency in the need for new therapies targeting remorseless bacteria, the knowledge of bacterial metabolism and its mode of infection still remain rudimentary. Among different bacteria many studies have been conducted against *Mycobacterium* sp. and similar to fungal therapeutics; it is evident that the assimilation and activity of the ICL in *M. avium* and MTB is enhanced when grown under minimal growth conditions (Honer 1999). GC has also paved way in acting as efficient target for drug in pathogenic bacteria. It has been shown more than a decade back that perseverance and virulence of MTB in mice is assisted by ICL (McKinney *et al.*, 2000). It has also been estimated that isoforms of the ICL *i.e.* ICL1 and ICL2 in prokaryotic and eukaryotic are mutually requisite for fatty acid catabolism and virulence in MTB by studying on the icl isoforms (icl1 and icl2) knockout mutants (Muñoz-Elías 2005). The effect of Long-range in anion-ð interactions was studied on MTB and the finding estimate that it inhibits the MS enzyme that is required for GC (Bauzá 2014). Similarly in MTB phenyl-diketo acid (PDKA) was shown as the inhibitor of MS enzymein GC cycle (Krieger 2012) (Figure 15.2). Apart from studies on *Mycobacterium* the studies on some other pathogenic bacteria have also been conducted. In recent study natural compounds have also been explored which acts on GC in bacteria. For instance, in *Achatina fulica* the *Achatina* C-reactive protein (ACRP) was found to inhibit some key enzymes such as isocitrate dehydrogenase in TCA cycle, ICL in GC in *Salmonella typhimurium* and *Bacillus subtilis* (Mukherjee 2014) (Figure 15.2). Similarly, itraconic acid, an organic compound, act as antibacterial on *Salmonella enterica* and MTB to inhibit ICL, the key enzyme of the glyoxylate shunt (Michelucci 2013). In a study with sensitive and resistance strains of *Helicobacter pylori* against metronidazole (Mtr), no ICL activity was detected in metronidazole sensitive (Mtrs) strains of *H. pylori*. It was estimated that ICL activity was expressed by sensitive strains that is transformed into the resistant strain when carrying the DNA from Metronidazole resistant (Mtrr) strain. This study indicates that resistance to Metronidazole could be a recessive trait which could possibly triggers inactivation of a regulatory gene, which constitutively expresses the gene for ICL synthesis (Hoffman *et al.*, 1996). In another study with *Rhodobacter sphaeroides*, a purple photosynthetic bacteria which is anaerobic phototropic and aerobically chemotropic in the absence of light, it was revealed that it is having any alternate pathway for the assimilation of acetate considered as alternate GC (Alber *et al.*, 2006). Thus the fact that GC enzymes could be an efficient antibacterial drug target is evident from wide ranges of studies.

## GC and Protozoa Infections

Protozoans are another class of microorganisms residing in the living environment including humans. Protozoa are assorted faction of habitually motile unicellular eukaryotic organisms. They are also the source of many diseases occurring in humans and affecting the daily life such as *Leishmania*, which is a main causative source of leishmaniasis and *Entamoeba* causing amoebiasis. Many studies have been conducted to study their GC and relate to its antimicrobial efficiency. In a study it was confirmed the GC and/or the consequent biosynthetic pathway from fructose-6-phosphate to mannan regulates with each other by incorporating [2-14C]acetate and [1-14C]laurate into mannan. They have found that Glucose does not affect label inserted into mannan, but glutamate increases four-fold of the incorporation from [2-14C] acetate and a 2.4-fold augmentation from [1-14C]laurate. Staurosporine, a protein kinase inhibitor which inhibits glutamate and alanine oxidation, was also not inhibiting the label incorporation from [2-14C] acetate into mannan which suggest that GC must be working efficiently during these stresses (Keegan 1993). In earlier studies the presence of GC specific enzymes was shown in *Leishmaniae* by Mukkada on four *Leishmaniae* species *viz., L. brasiliensis, L. donovani, L. mexicana and L. tropica* (Mukkada 1977). Similar work was proposed by Simon *et al.*, where they had also evidently proved the presence of GC enzymes ICL and MS in five species of *Leishmaniae viz., L. brasiliensis, L. donovani, L. mexicana, L. tarentolae, and L. tropica* (Simon 1978). However, despite confirmation of functional GC within protozoan, fewer studies have been performed to check the antiprotozoan drug targeting this efficient pathway.

## Conclusion

Since in the current scenario, increment in the resistance of the microorganisms against the current therapeutic regimes is on mount, there comes the need to find the new therapies along with the novel drug targets. GC is the effective target to work on as the absence of the crucial enzymes of this cascade *viz.*, ICL and MS in humans makes it easy target to work against the microbes as antimicrobial therapy against bacteria, fungi as well as protozoa.

## Acknowledgement

Financial assistances in the form of Young Scientist awards to S.H (SR/FT/LS-12/2012) and Z.F (SR/FT/LS-173/2010) from Science and Engineering Research Board (SERB), New Delhi and (2013/37B/45/BRNS/1903) to Z.F and S.H. from Board of Research in Nuclear Sciences (BRNS), Mumbai is deeply acknowledged. We thank Prof. S. M. Paul Khurana, Dean, Faculty of Science, Engineering and Technology for encouragement.

## References


1. Alber BE, Spanheimer R, Ebenau-Jehle C, Fuchs G. Study of an alternate glyoxylate cycle for acetate assimilation by *Rhodobacter sphaeroides*. *Mol Microbiol*. 2006; 61(2): 297-309.
2. Bauzá A, Quiñonero D, Deyà PM, Frontera A. Long-range effects in anion-α interactions: their crucial role in the inhibition mechanism of *Mycobacterium tuberculosis* malate synthase. *Chemistry*. 2014; 20(23): 6985-90.
3. Cheah HL, Lim V, Sandai D. Inhibitors of the glyoxylate cycle enzymes ICL1 in *Candida albicans* for potential use as antifungal agents. *PLoS One*. 2014; 9(4): e95951.
4. de Freitas CS, Kato L, de Oliveira CM, Queiroz LH Jr, Santana MJ, Schuquel IT, Delprete PG, da Silva RA, Quintino GO, da Silva Neto BR, Soares CM, Pereira M. β-Carboline Alkaloids from *Galianthe ramosa* Inhibit Malate Synthase from *Paracoccidioides* spp. *Planta Med*. 2014; 80(18): 1746-52.
5. Dunn MF, Ramýrez-Trujillo JA, Hernandez-Lucas I. Major roles of isocitrate lyase and malate synthase in bacterial and fungal pathogenesis. *Microbiology* 2009; 155: 3166-75.
6. Eastmond, P. J. and Graham, I. A. Re-examining the role of the glyoxylate cycle in oilseeds. *Trends Plant Sci*. 2001; 6: 72–78.
7. Flipphi M, Oestreicher N, Nicolas V, Guitton A, Vélot C. The *Aspergillus nidulans* acuL gene encodes a mitochondrial carrier required for the utilization of carbon sources that are metabolized via the TCA cycle. *Fungal Genet Biol*. 2014; 68: 9-22.
8. García-Rodas R, Cordero RJ, Trevijano-Contador N, Janbon G, Moyrand F, Casadevall A, Zaragoza O. MBio. Capsule growth in *Cryptococcos neoformans* is coordinated with cell cycle progression. 2014; 5(3): e00945-14.
9. Graham IA. Seed storage oil mobilization. Annu Rev Plant Biol. 2008; 59: 115-42.
10. Hoffman PS, Goodwin A, Johnsen J, Magee K, Veldhuyzen van Zanten SJ. Metabolic activities of metronidazole-sensitive and resistant strains of *Helicobacter pylori*: repression of pyruvate oxidoreductase and expression of isocitrate lyase activity correlate with resistance. *J Bacteriol*. 1996; 178(16): 4822-9.
11. Höner Zu Bentrup K, Miczak A, Swenson DL, Russell DG. Characterization of activity and expression of isocitrate lyase in *Mycobacterium avium* and *Mycobacterium tuberculosis*. *J Bacteriol*. 1999; 181(23): 7161-7.
12. Keegan FP, Blum JJ. Incorporation of label from acetate and laurate into the mannan of *Leishmania donovani* via the glyoxylate cycle. *J Eukaryot Microbiol*. 1993; 40(6): 730-2.
13. Kornberg, H. L. and Krebs, H. A. 1957. Synthesis of cell constituents from C2-units by a modified tricarboxylic acid cycle. *Nature*. 1957; 179: 988–991.
14. Kornberg, H. L. The role and control of the glyoxylate cycle in *Escherichia coli*. *Biochem. J*. 1966; 99: 1–11.

15. Krieger IV, Freundlich JS, Gawandi VB, Roberts JP, Gawandi VB, Sun Q, Owen JL, Fraile MT, Huss SI, Lavandera JL, Ioerger TR, Sacchettini JC. Structure-guided discovery of phenyl-diketo acids as potent inhibitors of *M. tuberculosis* malate synthase. *Chem Biol.* 2012; 19(12): 1556-67.
16. Kumar R, Bhakuni V. *Mycobacterium tuberculosis* isocitrate lyase (MtbIcl): Role of divalent cations in modulation of functional and structural properties. *Proteins.* 2008; 72(3): 892-900.
17. Lee SH, Won TH, Kim H, Ahn CH, Shin J, Oh KB. Suvanine sesterterpenes from a tropical sponge *Coscinoderma* sp. inhibit isocitrate lyase in the glyoxylate cycle. *Mar Drugs.* 2014; 12(10): 5148-59.
18. Lorenz MC, Fink GR. Life and death in a macrophage: Role of the glyoxylate cycle in virulence. *Eukaryot Cell.* 2002; 1: 657-62.
19. Lorenz, M. C. and Fink, G. R. The glyoxylate cycle is required for fungal virulence. *Nature.* 2001; 412: 83–86.
20. Michelucci A, Cordes T, Ghelfi J, Pailot A, Reiling N, Goldmann O, Binz T, Wegner A, Tallam A, Rausell A, Buttini M, Linster CL, Medina E, Balling R, Hiller K. Immune-responsive gene 1 protein links metabolism to immunity by catalyzing itaconic acid production. *Proc Natl Acad Sci USA.* 2013; 110(19): 7820-5.
21. Mukherjee S, Barman S, Mandal NC, Bhattacharya S. Anti-bacterial activity of *Achatina* CRP and its mechanism of action. *Indian J Exp Biol.* 2014; 52(7): 692-704.
22. Mukkada AJ. Tricarboxylic acid and glyoxylate cycles in the Leishmaniae. *Acta Trop.* 1977; 34(2): 167-75.
23. Muñoz-Elías EJ, McKinney JD. *Mycobacterium tuberculosis* isocitrate lyases 1 and 2 are jointly required for in vivo growth and virulence. *Nat Med.* 2005; 11(6): 638-44.
24. Nakazawa M, Minami T, Teramura K, Kumamoto S, Hanato S, Takenaka S, *et al.*, Molecular characterization of a bifunctional glyoxylate cycle enzyme, malate synthase/isocitrate lyase, in *Euglena gracilis*. *Comp Biochem Physiol B Biochem Mol Biol.* 2005; 141: 445-52.
25. Serrano, J. A., Camacho, M. and Bonete, M. J. Operation of glyoxylate cycle in halophilic archaea: presence of malate synthase and isocitrate lyase in *Haloferax volcanii*. *FEBS Lett.* 1998; 434: 13–16.
26. Sharma R, Das O, Damle SG, Sharma AK. Isocitrate lyase: a potential target for anti-tubercular drugs. *Recent Pat Inflamm Allergy Drug Discov.* 2013; 7(2): 114-23.
27. Simon MW, Martin E, Mukkada AJ. Evidence for a functional glyoxylate cycle in the leishmaniae. *J Bacteriol.* 1978; 135(3): 895-9.
28. Sims CR, Ostrosky-Zeichner L, Rex JH. Invasive candidiasis in immunocompromised hospitalized patients. *Arch Med Res.* 2005; 36(6): 660-71.
29. Smith CV, Sharma V, Sacchettini JC. TB drug discovery: Addressing issues of persistence and resistance. *Tuberculosis.* 2004; 84: 45-55.

30. Terabayashi Y, Shimizu M, Kitazume T, Masuo S, Fujii T, Takaya N. Conserved and specific responses to hypoxia in *Aspergillus oryzae* and *Aspergillus nidulans* determined by comparative transcriptomics. *Appl Microbiol Biotechnol.* 2012; 93(1): 305-17.

31. Vanni P, Giachetti E, Pinzauti G, McFadden B. Comparative structure, function and regulation of isocitrate lyase, an important assimilatory enzyme. *Comp Biochem Physiol B.* 1990; 95(3): 431-58.

2016, Biotechnology: Progress and Applications *Pages* 345–372
*Editors:* Saif Hameed and Zeeshan Fatima
*Published by:* DAYA PUBLISHING HOUSE, NEW DELHI

*Chapter 16*

# Infectious Microbes and their Biotechnological Applications

*Narendra Kumar**

*Amity Institute of Biotechnology, Amity University, Gurgaon – 122 413, Manasar, Haryana, India*

## ABSTRACT

Infections are caused by infectious agents such as viruses, viroids, and prions, microorganisms such as bacteria, nematodes such as round worms and pinworms, arthropods such as ticks, mites, fleas, and lice, fungi such as ringworm, and other macroparasites such as tapeworms Infectious diseases are illnesses caused by germs (microbes). It is important to realize that not all germs (bacteria, viruses, fungi, and parasites) cause disease. In fact, a host of bacteria normally live on the skin, eyelids, nose, and mouth and in the gut. These bacteria are called *normal flora*and are considered normal inhabitants. Factors in bacteria, viruses, and fungi include genes that determine how harmful (virulent) the microbe can be. Some germs make toxins that cause illness by themselves or contribute to infections caused by the germ. Examples include enterotoxins, which cause diarrhea; tetanus toxin, which causes lock jaw; and toxic shock toxin, which leads to low blood pressure and collapse (shock).Sometimes infections takes the form of a pandemic (or global epidemic) disease that affects people over an extensive geographical area.Disease can arise if the host's protective immune mechanisms are compromised and the organism inflicts damage on the host. Microorganisms can cause tissue damage by releasing a variety of toxins or destructive enzymes.For infecting organisms to survive and repeat the infection cycle in other hosts, they

* Author: E-mail: narendra.microbiology@rediffmail.com

(or their progeny) must leave an existing reservoir and cause infection elsewhere. Infection transmission can take place via many potential routes.

Although infectious microbes are infecting agents but they have wider applications also. *Listeria monocytogenes* is a Gram-positive bacterium, the use of attenuated strains of *monocytogenes* L. as vaccines has gained considerable interest because they can be used to express heterologous antigens as well as to somatically deliver recombinant DNA to eukaryotic cells. A novel vaccine concept, the use of non-viable but metabolically active bacteria to induced immunoprotective responses, has been developed with *monocytogenes* L. These implications has arised a new era of Patho-biotechnology that describes the application of pathogen derived (*ex vivo* and *in vivo*) stress survival strategies for the design of more technologically robust and effective probiotic cultures with improved biotechnological and clinical applications as well as the development of novel vaccine and drug delivery platforms.Infectious viruses have given rise to virotherapy. Virotherapy is a treatment using biotechnology to convert viruses into therapeutic agents by reprogramming viruses to treat diseases

## Terminology Related with Infectious Microbes

**Bacteria**-two types free living, normal inhabitants (normal flora); pathogens that produces disease

**Carrier**-a child who is colonized but not sick

**Colonization**-presence of a germ in or on you without disease

**Commensalism-** is a symbiotic relation in which one organism benefits and the other is not harmed. Our bodies are populated by an extremely large number of bacteria that can, but don't always, benefit us. The bacteria always benefit. The *E. coli* in our colon are examples.

**Contagious**-able to spread the illness

**Fungi**-molds and yeasts fungi colonize and are pathogens

**Incubation**-time between infection and illness

**Infection**-a germ causing an illness your body will react by making antibodies

**Intoxication**-illness due to toxin made by a germ

**Latent infection**-a germ(most often a virus) in a resting stage

**Normal flora**-bacteria that live on or in a child

**Parasites**-forms range from single cells (amoeba,protozoa) to worms

**Parasitism** -occurs when the infecting organism benefits and the host is harmed. If the host sustains injury or pathological changes in response to the parasite, the process is an infectious disease.

**Pathogen**-a germ that causes a disease

**Prions**-infectious proteins, the smallest known infectious agents

**Reactivation**-the latent germs wakes up and reproduces

**Viruses**-very small viruses take over your cells to reproduce themselves

## Introduction

**Infection** is the invasion of a host organism's bodily tissues by disease-causing organisms, their multiplication, and the reaction of host tissues to these organisms and the toxins they produce. **Infectious diseases**, also known as transmissible diseases or communicable diseases, comprise clinically evident illness (*i.e.*, characteristic medical signs and/or symptoms of disease) resulting from the **infection**, presence and growth of pathogenic biological agents in an individual host organism.

Infections are caused by infectious agents such as viruses, viroids, and prions, microorganisms such as bacteria, nematodes such as round worms and pinworms, arthropods such as ticks, mites, fleas, and lice, fungi such as ringworm, and other macroparasites such as tapeworms.

Hosts can fight infections using their immune system. Mammalian hosts react to infections with an innate response, often involving inflamation, followed by an adaptive response (Signore, 2013). Pharmaceuticals can also help fight infections.

The branch of medicine that focuses on infections and pathogens is infectious disease medicine.

## History of Pandemic Microbial Infections

A pandemic (or global epidemic) is a disease that affects people over an extensive geographical area.

- Plague of Justinian, from 541 to 750, killed between 50 per cent and 60 per cent of Europe's population.
- The Black Death of 1347 to 1352 killed 25 million in Europe over 5 years. The plague reduced the world population from an estimated 450 million to between 350 and 375 million in the 14$^{th}$ century.
- The introduction of smallpox, measles, and typhus to the areas of Central and South America by European explorers during the 15th and 16th centuries caused pandemics among the native inhabitants. Between 1518 and 1568 disease pandemics are said to have caused the population of Mexico to fall from 20 million to 3 million.
- The first European influenza epidemic occurred between 1556 and 1560, with an estimated mortality rate of 20 per cent (Dobson *et al.*, 1996).
- Smallpox killed an estimated 60 million Europeans during the 18th century[3] (approximately 400,000 per year).[4] Up to 30 per cent of those infected, including 80 per cent of the children under 5 years of age, died from the disease, and one-third of the survivors went blind.[5]
- In the 19th century, tuberculosis killed an estimated one-quarter of the adult population of Europe;[6] by 1918 one in six deaths in France were still caused by TB.
- The Influenza Pandemic of 1918 (or the Spanish Flu) killed 25-50 million people (about 2 per cent of world population of 1.7 billion).[7] Today Influenza kills about 250,000 to 500,000 worldwide each year.

## History of Infectious Microbes

Ideas of contagion became more popular in Europe during the Renaissance, particularly through the writing of the Italian physician Girolamo Fracastoro.[8]

Anton van Leeuwenhoek (1632–1723) advanced the science of microscopy by being the first to observe microorganisms, allowing for easy visualization of bacteria.

In the mid-19th century John Snow and William Budd did important work demonstrating the contagiousness of typhoid and cholera through contaminated water. Both are credited with decreasing epidemics of cholera in their towns by implementing measures to prevent contamination of water.[9]

Louis Pasteur proved beyond doubt that certain diseases are caused by infectious agents, and developed a vaccine for rabies.

Robert Koch, provided the study of infectious diseases with a scientific basis known as Koch's postulates.

Edward Jenner, Jonas Salk and Albert Sabin developed effective vaccines for smallpox and polio, which would later result in the eradication and near-eradication of these diseases, respectively.

Alexander Fleming discovered the world's first antibiotic Penicillin which Florey and Chain then developed.

Gerhard Domagk developed sulphonamides, the first broad spectrum synthetic antibacterial drugs.

## Fossil Data

Evidence of infection in fossil remains is a subject of interest for paleopathologists, scientists who study occurrences of injuries and illness in extinct life forms. Signs of infection have been discovered in the bones of carnivorous dinosaurs. When present, however, these infections seem to tend to be confined to only small regions of the body. A skull attributed to the early carnivorous dinosaur *Herrerasaurus ischigualastensis* exhibits pit-like wounds surrounded by swollen and porous bone. The unusual texture of the bone around the wounds suggests they were afflicted by a short-lived, non-lethal infection. Scientists who studied the skull speculated that the bite marks were received in a fight with another *Herrerasaurus*. Other carnivorous dinosaurs with documented evidence of infection include *Acrocanthosaurus*, *Allosaurus*, *Tyrannosaurus* and a tyrannosaur from theKirtland Formation. The infections from both tyrannosaurs were received by being bitten during a fight, like the *Herrerasaurus* specimen. [10]

## Emerging Diseases

In most cases, microorganisms live in harmony with their hosts via mutual or commensal interactions. Diseases can emerge when existing parasites become pathogenic or when new pathogenic parasites enter a new host.

1. Coevolution between parasite and host can lead to hosts becoming resistant to the parasites or the parasites may evolve greater virulence, leading to immunopathological disease.

2. Human activity is involved with many emerging infectious diseases, such as environmental change enabling a parasite to occupy new niches. When that happens, apathogen that had been confined to a remote habitat has a wider distribution and possibly a new host organism. Parasites jumping from nonhuman to human hosts are known as zoonoses. Under disease invasion, when a parasite invades a new host species, it may become pathogenic in the new host.[11]

Several human activities have led to the emergence of zoonotic human pathogens, including viruses, bacteria, protozoa, and rickettsia,[12] and spread of vector-borne diseases,[39].

- *Encroachment on wildlife habitats*: The construction of new villages and housing developments in rural areas force animals to live in dense populations, creating opportunities for microbes to mutate and emerge.[13]
- *Changes in agriculture*: The introduction of new crops attracts new crop pests and the microbes they carry to farming communities, exposing people to unfamiliar diseases.
- *The destruction of rain forests*: As countries make use of their rain forests, by building roads through forests and clearing areas for settlement or commercial ventures, people encounter insects and other animals harboring previously unknown microorganisms.
- *Uncontrolled urbanization*: The rapid growth of cities in many developing countries tends to concentrate large numbers of people into crowded areas with poor sanitation. These conditions foster transmission of contagious diseases.

*Modern transport*: Ships and other cargo carriers often harbor unintended "passengers", that can spread diseases to faraway destinations. While with international jet-airplane travel, people infected with a disease can carry it to distant lands, or home to their families, before their first symptoms appear.

## (i) Microbial Infection Details

An infection is the detrimental colonization of a host organism by a foreign species.

In infection, the infecting organism seeks to utilize the host's resources in order to multiply (usually at the expense of the host).The infecting organism, or pathogen, interferes with the normal functioning of the host and can lead to chronic wounds, gangrene, loss of an infected limb, and even death. The host's response to infection is inflammation. A symbiosis between parasite and host, whereby the relationship is beneficial for the former but detrimental to the latter, is characterized as parasitism.Wound colonization refers to non replicating microorganisms within the wound. Replicating organisms exist in infected wounds and tissue is injured. All multicellular organisms are colonized to some degree by extrinsic organisms, and the vast majority of these exist in either a mutualistic or commensal relationship with the host.An example of the former would be the anaerobic bacteria species which colonize the mammalian colon, and an example of the latter would be the various species of staphylococcus which exist on human skin. Neither of these colonizations would be considered infections.

The difference between an infection and a colonization is often only a matter of circumstance. Organisms which are normally non-pathogenic can become pathogenic under the right conditions, and even the most virulent organism requires certain circumstances to cause a compromising infection. Some colonizing bacteria, such as *Corynebacteria* sp. and viridans streptococci, prevent the adhesion and colonization of pathogenic bacteria and thus have a symbiotic relationship with the host, preventing infection and speeding wound healing.

The variables involved in the outcome of a host becoming inoculated by a pathogen and the ultimate outcome include:

- ☆ Route of entry of the pathogen and the access to host regions that it gains
- ☆ Intrinsic virulence of the particular organism
- ☆ Quantity or load of the initial inoculant
- ☆ Immune status of the host being colonized

As an example, the staphylococcus species present on skin remain harmless on the skin, but, when present in a normally sterile space, such as in the capsule of a joint, will multiply without resistance and create a huge burden on the host.

### (ii) Infectious Disease

Disease can arise if the host's protective immune mechanisms are compromised and the organism inflicts damage on the host. Microorganisms can cause tissue damage by releasing a variety of toxins or destructive enzymes. For example, *Clostridium tetani* releases a toxin that paralyzes muscles, and s*taphylococcus* releases toxins that produce shock and sepsis. Not all infectious agents cause disease in all hosts. For example less than 5 per cent of individuals infected with polio develop disease. On the other hand, some infectious agents are highly virulent. The prion causing mad cow disease and Creutzfeldt–Jakob disease kills almost all animals and people that are infected.

Persistent infections occur because the body is unable to clear the organism after the initial infection. Persistent infections are characterized by the continual presence of the infectious organism, often as latent infection with occasional recurrent relapses of active infection. There are some viruses that can maintain a persistent infection by infecting different cells of the body. Some viruses once acquired never leave the body. A typical example is the herpes virus, which tends to hide in nerves and become reactivated when specific circumstances arise.

Persistent infections cause millions of deaths globally each year. Chronic infections by parasites account for a high morbidity and mortality in many underdeveloped countries.An infectious disease or communicable disease is caused by a biological agent such as by a virus, bacterium or parasite. Infectious diseases are the invasion of a host organism by a foreign replicator, generally microorganisms, often called **microbes**, that are invisible to the naked eye.Microbes that cause illness are also known as pathogens. The most common pathogens are various **bacteria** and **viruses**, though a number of other microorganisms, including some kinds of **fungi** and **protozoa**, also cause disease. An infectious disease is termed *contagious* if it is easily transmitted from one person to another.

An organism that a microbe infects is known as the host for that microbe. In the human host, a microorganism causes disease by either disrupting a vital body process or stimulating the immune system to mount a defensive reaction. An immune response against a pathogen, which can include a high fever, inflammation, and other damaging symptoms, can be more devastating than the direct damage caused by the microbe.

### (iii) Diagnosis of Infectious Diseases

Diagnosis is initially by medical history and physical examination, and imaging (such as X-rays), but the principal tool in infectious disease is the microbiological culture.

**Microbial Culture**

A **microbiological culture** is a method of multiplying microbial organisms by letting them reproduce in predetermined culture media under controlled laboratory conditions. Microbial cultures are used to determine the type of organism, its abundance in the sample being tested, or both. It is one of the primary diagnostic methods of microbiology and used as a tool to determine the cause of infectious disease by letting the agent multiply in a predetermined medium. For example, a throat culture is taken by scraping the lining of tissue in the back of the throat and blotting the sample into a medium to be able to screen for harmful microorganisms, such as *Streptococcus pyogenes*, the causative agent of strep throat. Furthermore, the term culture is more generally used informally to refer to "selectively growing" a specific kind of microorganism in the lab.

Microbial cultures are foundational and basic diagnostic methods used extensively as a research tool in molecular biology. It is often essential to isolate a pure culture of microorganisms. A pure (or *axenic*) culture is a population of cells or multicellular organisms growing in the absence of other species or types. A pure culture may originate from a single cell or single organism, in which case the cells are genetic clones of one another.

For the purpose of gelling the microbial culture, the medium of agarose gel (agar) is used. Agar is a gelatinous substance derived from sea weed. A cheap substitute for agar is guar gum, which can be used for the isolation and maintenance of thermophiles. Microbiological cultures can be grown in petri dishes of differing sizes that have a thin layer of agar-based growth medium. Once the growth medium in the petri dish is inoculated with the desired bacteria, the plates are incubated at the best temperature for the growing of the selected bacteria (for example, usually at 37 degrees Celsius for cultures from humans or animals, or lower for environmental cultures).

Another method of bacterial culture is liquid culture, in which the desired bacteria are suspended in liquid broth, a nutrient medium. These are ideal for preparation of an antimicrobial assay. The experimenter would inoculate liquid broth with bacteria and let it grow overnight (they may use a shaker for uniform growth). Then they would take aliquots of the sample to test for the antimicrobial activity of a specific drug or protein (antimicrobial peptides).

As an alternative, the microbiologist may decide to use static liquid cultures. These cultures are not shaken and they provide the microbes with an oxygen gradient.

**Microscopy**

Another principal tool in the diagnosis of infectious disease is microscopy. Virtually all of the culture techniques rely, at some point, on microscopic examination for definitive identification of the infectious agent. Microscopy may be carried out with simple instruments, such as the compound light microscope, or with instruments as complex as an electron microscope. Samples obtained from patients may be viewed directly under the light microscope, and can often rapidly lead to identification. Microscopy is often also used in conjunction with biochemical staining techniques, and can be made exquisitely specific when used in combination with antibody based techniques. For example, the use of antibodies made artificially fluorescent (fluorescently labeled antibodies) can be directed to bind to and identify a specific antigens present on a pathogen. A fluorescence microscope is then used to detect fluorescently labeled antibodies bound to internalized antigens within clinical samples or cultured cells. This technique is especially useful in the diagnosis of viral diseases, where the light microscope is incapable of identifying a virus directly.

Other microscopic procedures may also aid in identifying infectious agents. Almost all cells readily stain with a number of basic dyes due to the electrostatic attraction between negatively charged cellular molecules and the positive charge on the dye. A cell is normally transparent under a microscope, and using a stain increases the contrast of a cell with its background. Staining a cell with a dye such as Giemsa stain or crystal violet allows a microscopist to describe its size, shape, internal and external components and its associations with other cells. The response of bacteria to different staining procedures is used in the taxonomic classification of microbes as well. Two methods, the Gram stain and the acid-fast stain, are the standard approaches used to classify bacteria and to diagnosis of disease. The Gram stain identifies the bacterial groups Firmicutes and Actinobacteria, both of which contain many significant human pathogens. The acid-fast staining procedure identifies the Actinobacterial genera *Mycobacterium* and *Nocardia*.

## (iv) Kinds of Infections

Most infections caused by microbes fall into three major groups:

1. Acute infections
2. Chronic infections
3. Latent infections

Infectious disease requires an agent and a mode of transmission (or **vector**). A good example is malaria, caused by Plasmodial parasites - *Plasmodium falciparum* but does not affect humans unless the vector, the Anopheles mosquito, is around to introduce the parasite into the human bloodstream.

But in some cases the vector does not have to be biological. Many infectious diseases are transmitted by droplets which enter the airway (*e.g.* common cold and tuberculosis).

## Disease Transmissions

For infecting organisms to survive and repeat the infection cycle in other hosts, they (or their progeny) must leave an existing reservoir and cause infection elsewhere. Infection transmission can take place via many potential routes:

- **Droplet contact**, also known as the *respiratory route*, and the resultant infection can be termed airborne disease. If an infected person coughs or sneezes on another person the microorganisms, suspended in warm, moist droplets, may enter the body through the nose, mouth or eye surfaces.
- **Fecal-oral transmission**, wherein foodstuffs or water become contaminated (by people not washing their hands before preparing food, or untreated sewage being released into a drinking water supply) and the people who eat and drink them become infected. Common fecal-oral transmitted pathogens include *Vibrio cholerae, Giardia* species,rotaviruses, *Entameba histolytica, Escherichia coli*, and tape worms.[14] Most of these pathogens cause gastroenteritis.
- **Sexual transmission**, with the resulting disease being called sexually transmitted disease
- **Oral transmission**; Diseases that are transmitted primarily by oral means may be caught through direct oral contact such as kissing, or by indirect contact such as by sharing a drinking glass or a cigarette.
- **Transmission by direct contact**; Some diseases that are transmissible by direct contact include athlete's foot, impetigo and warts
- **Vertical transmission'**; directly from the mother to an embryo, fetus or baby during pregnancy or childbirth. It can occur when the mother gets an infection as an intercurrent disease in pregnancy.
- **Iatrogenic transmission**, due to medical procedures such as injection or transplantation of infected material.
- **Vector-borne transmission**, transmitted by a vector, which is an organism that does not cause disease itself but that transmits **infection** by conveying pathogens from one host to another.[15] Culex mosquitos (*Culex quinquefasciatus*) are biological vectors that transmit West Nile Virus.

The relationship between virulence versus transmissibility is complex; if a disease is rapidly fatal, the host may die before the microbe can get passed along to another host.

## Scale of Sizes and some Details of Infectious Microbes

There are many agents of infectious diseases, ranging in size from microscopically small to macroscopically large. Since most microbes are really, really small, we need to get a feeling for the units in which they are measured, nanometers. One nanometer is one-billionth of a meter.

Viruses can range in size from 28–200 nm and t hey have from 5K–500K bases. Bacteria run from 1000–2000 nm in size and contain 1000K–9000K base pairs.

## Prions

During the 1950s, anthropologists studying in the Fore people in Papua, New Guinea recorded the presence of a strange fatal disease, called kuru, which attacked the brain. Victims lost coordination, fine motor control, and the ability to swallow. They developed tremors and showed signs of dementia.

Of a population of 35,000, fully 3700 cases of this disease were recorded among this tribe. It seemed to center almost exclusively on adult females, although it was present at lower levels equally in boys and girls. Some villages suffered so many deaths, they had triple the number of living males as females. Records indicated the disease took between four and forty years to manifest itself. After extensive epidemiological study, it became apparent that there was a very strong association between kuru and the practice of ritual cannibalism, whereby the relatives "honored" the recently deceased member of the family by consuming her/his body in its entirety. Almost exclusively adult women practiced the details of the ritual—men rarely participated. Women and children would bake the internal organs, brain, and bone, grind the results into a powder, and then they would mix the powder with vegetables and eat the results.

Scrapie is a disease of goats and sheep that has been present in Europe for at least two centuries. The disease migrated from England to Canada and made its American debut in 1947. Transmission is thought to be from the ewe to her offspring or to other lambs due to contact with placental materials and fluids. Symptoms do not appear for two to five years. Infected sheep seem to suffer from an interminable itch. They frequently "scrape" themselves against a tree or fence post until they wear off the outer layers of tissue. This is accompanied by a loss of ccorresponding difficulties in walking.

The disease is widespread in Europe (200–600 individual cases per year in the UK, endemic in other countries), the US (900 flocks infected to date), and the Middle East. There do not seem to be any infections in either Australia or New Zealand. There are recorded transmissions from sheep to goats. Prions are extremely robust with respect to sterilization. They are known to resist radiation, boiling, microwaves, and chemical agents. Surgical instruments used on Creutzfeldt-Jakob disease patients that had been carefully sterilized have shown traces of PrPsc. Iatrogenic CJD is estimated to comprise 5 per cent of all cases. There are documented transmissions due to corneal transplants, liver transplants, dura mater graft transplants, and contaminated instruments and electrodes. Disease has been induced by an injection of small doses of PrPsc and feeding meat of infected animals to animals of the same and different species. Epidemiological studies have shown a very strong link between the ingestion of brain and spinal cord tissue from infected animals, mostly cattle, and variant Creutzfeldt-Jakob disease. Ordinarily, CJD occurs in about one in a million people. It has no standard single course of progression of symptoms. Its first signs of incidence are rare in those under forty years-of-age, more common in those aged 50 to 75, and rare again for those over 75. The average age at death is about 60. Most brain damage occurs in the cerebellum. There is an indication that the disease may have an incubation period at least as long as thirty or as many as fifty years, but once the protein refolding starts, it proceeds at an accelerating rate and patients rarely live more than two years. Nevertheless, some people infected in the 1990s showed symptoms (and died) much faster.

## Viruses

A virus is an obligate intracellular parasite (meaning that it must exist within the cells of its host in order to replicate). A virus is metabolically inert outside a cell. Viruses are not living cells. They cannot provide their own nutrition, nor can they replicate on their own. Viruses have no organized cellular structures but simply a protein coat, called the capsid, surrounding a nucleic acid core, called a genome, of either RNA or DNA, but never both. The capsid together with the genome is called the nucleocapsid. The nucleocapsid may be surrounded by an envelope that is composed of a lipid bilayer containing protein spikes. An entire virus particle is called a virion. Viruses are Classified by the categories: DNA or RNA; single strand or double strand; enveloped or non-enveloped; by their symmetry—helical, icosahedral, or complex; family; and species.

The bacteriophage is a virus that uses a bacterium as its host. There are poxviruses that infect mammals, snakes, and insects. In fact, some experts attribute the control of the exponential growth of insects to the presence of such viruses. As viruses go, smallpox, with its 187,000 bases and about 200 genes, is well armed to confront its hosts. Its 11–14 day incubation period gives it ample opportunity to spread itself far and wide in this time of rapid travel. The influenza virus, which is transmitted on exhaled aerosol droplets, is no slouch either.

Much in the news is the norovirus orNorwalk-like virus, which has been responsible for gastrointestinal disorders frequently seen aboard cruise ships. Worldwide, the virus causes 23,000,000 infections per year and has an infectious dose of between 5 and 100 virions—far fewer than almost any other virus (except smallpox which can infect a person with around a dozen virions). So far, no one has been able to grow a norovirus for the length of time needed to research possible vaccines.

Another particularly frightening microbc is thc Ebola virus which causes a particularly lethal hemorrhagic (causing internal bleeding) disease. It is known to infect humans, gorillas, chimpanzees, and small antelopes called duikers. In outbreaks among humans it has had a high mortality rate (>60 per cent) but killed only around a thousand

Human Papillomavirus (HPV) DNA has been identified in 99.7 per cent of cervical cancers. HPV 16 accounts for 50–60 per cent of cervical cancers, HPV 18 accounts for 10–15 per cent, and HPV 31, 33, 45, 52, and 58 account for 18 per cent of such cancers. HPV 6 and 11 are noncancer-causing, but are associated with external anogenital warts, low-grade genital dysplasias, and recurrent respiratory papillomatosis. The vaccine Gardasil targets HPV 6, 11, 16, and 18, while Cervarix targets HPV 16 and 18. On the other hand, it has attacked western lowland gorillas and killed at least 5500 and perhaps as many 30,000 of them.

## Influenza

is caused by an RNA virus whose genome is constructed of eight separate fragments (which code for 11 proteins). The latest version making the rounds is characterized as an A:H1N1 virus. As virions break, the eight fragments can intermingle with those of other forms of influenza, thus contributing to antigenic drift and antigenic shift. Once a virus has infiltrated a cell and been transformed to a provirus, that cell is fully infected. As an indication of the ramifications of this, in May 2004 an organ donor diagnosed with a brain hemorrhage died and five people received the harvested organs. It turns out the donor was infected with the rabies virus and each of the transplanted organs were likewise infected. One recipient died of surgical complications, but the other four died of rabies in June of 2004.

Viruses and cells of other species of microbe can mutate. The RNA-based influenza virus has a 10 gene, segmented genome, meaning that instead of a single piece of nucleic acid, it is composed of several pieces. If two different varieties of influenza infect the same cell, these pieces can get intermixed resulting in an entirely different form of the virus. This is precisely how antigenic shift occurs among strains of influenza viruses.

## Bacteria

Bacteria (singular bacterium) are autonomously replicating unicellular organisms lacking both an organized nucleus (which defines the class of cells called prokaryotes) and organized intracellular organelles. They have only a single circular chromosome of double-stranded DNA (dsDNA), some extrachromosomal DNA, and most have a cell wall containing the polymer peptidoglycan. Bacterial DNA has no introns, the DNA sequence consists only of codons.

Many bacteria possess a quorum-sensing signaling system, whereby chemical signaling occurs when the colony reaches a certain size. In some species this is the marker to begin producing toxins that injure the host. One such species is *Bacillus anthracis*,the causative agent for anthrax.

It is known that infection by the bacterium *Corynebacterium diphtheriae* results in the disease diphtheria only when a specific bacteriophage infects the microbe. The toxin causing the disease is produced as a result of a phage gene inserted into the microbe's DNA. Similarly, infection by *Vibrio cholerae* results in the disease cholera due to phage genes that carry instructions for manufacture of a certain protein toxin. Also *Escherichia coli* produces Shiga toxin when a phage begins to reproduce inside the bacterium. The toxin is not released until the bacterium lyses.

The most current estimates are that there are 1030 bacteria living on our planet and each of us has a mere 1014 of them (that's a factor of 10,000 trillion fewer) living in oBacteria are responsible for most food-borne illnesses. The cold-loving bacterium *Listeria monocytes* tends to colonize improperly processed meats, deli foods, hot dogs, milk, and soft cheeses leading to about 2500 cases of illness, most of which require hospitalization, and 500 deaths each year. They are also present in so-called biofilms that coat many ordinary surfaces.

Vibrios (which cause cholera) are about 1000 by 3000 nm long and 500 to 800 nm wide. When ingested without food an infectious dose (with probability 50 per cent) is anywhere from 100 to 10,000 bacteria, whereas when ingested with food a range of 1000 to 1,000,000 are needed. *Yersinia pestis* (which cause Plague) are between 1000 and 200 nm long and 500 nm wide. Each infected flea bite will inject from 25,000 to 100,000 Y. pestis and an infectious dose There are many special types of bacteria that deserve separate mention.

Spirochetes are bacteria that are helical-shaped and have filaments wound about the cell wall along the entire length of the cell. The best-known disease caused by spirochetes is syphilis.

Mycoplasmas are unicellular prokaryotes without rigid cell walls that are capable of independent replication.

Because they have no cell walls, they cannot be identified by Gram-staining. They are extremely simple structures, about one third the size of the typical bacteria. The smallest are around 200 nm and have as few as 580 kilobase pairs. This is very nearly the lower limit for sustaining life. In humans most mycoplasmas are commensals, although some species of mycoplasmas can be responsible for some atypical pneumonias, genital infections, and vertically transmitted respiratory infections—mostly to low birth weight infants.

Mycobacteria comprise a class of Gram-positive bacteria that pass through a phase, called acid-fastness, in their life cycle when they can be stained, and thus identified. Tuberculosis is caused by a species of Mycobacteria, *Mycobacterium tuberculosis*. This organism can infect and attack almost any organ or tissue of the body. In common usage, the term refers to an infection of the lungs. The lesions associated with the disease are called tubercles and their innermost spheroidal volumes consist of dead or dying tissue. Hansen's disease, leprosy, is caused by *Mycobacterium leprae*. It can be treated with multiple drug therapy and has not yet been cultured in the laboratory. One reason may be its doubling time, which is the longest of all known bacteria. Although one associates it with biblical times, it is very much alive and thriving to this day.

*Rickettsiae* and *chlamydiae* are obligate intracellular pathogens with a rigid peptidoglycan cell wall, which reproduce asexually by binary fission, and can contain both DNA and RNA. *Rickettsiae* depend on the host cell for essential nutrients and they can infect arthropods without causing disease in them. They cause the potentially deadly diseases typhus and Rocky Mountain spotted fever. *Chlamydiae* scavenge intracellular leftovers for nutrition. They are transmitted directly and are responsible for some bronchitis, sinusitis, some pneumonias, the common sexually transmitted infection (STI) referred to as chlamydia urethritis, the relatively rare STI lymphogranuloma venereum, and trachoma. In fact, trachoma is the leading cause of preventable blindness, and chlamydia is the most common STI in the world today.

Chlamydia pneumonia has been strongly implicated as a causative agent atherosclerosis (hardening of the arteries).

## Fungi

Fungi are free-living eukaryotic saprophytes (organisms that live on dead things)consisting of yeasts and molds (also called filamentous fungi). Fungi can cause two broad classes of disease: superficial and deep. The superficial diseases are either cutaneous, growing on the skin, or subcutaneous, growing just below the skin. The deep diseases are either systemic, they colonize most of the body, or opportunistic (fungi that normally live in the body but become pathogenic only when the immune system is suppressed).

*Candida albicans* is an example of an opportunistic infection, since most of us have *C. albicans* as a standard part of the microbiota of our mouths. They can also infect internal organs.

Coccidioidomycosis, cryptococcosis, and histoplasmosis are three other fungal infections common to AIDS patients.

## Other Eukaryotic Parasites

Protozoa are unicellular eukaryotes that lack cell walls. Their life cycles can be quite complex. There are four classes of protozoa: (1) sporozoa, which are intracellular parasites, (classes of protozoa: (1) sporozoa, which are intracellular parasites, (2) flagellates, (3) amoebas, and (4) ciliates.

Although most protozoa are saprophytes, some have adapted to the human host, where they can cause many diseases. They can be spread host-to-host, from a common source—like amoebic dysentery and giardiasis, or via an arthropod vector—like malaria, one of the top killers in the world today.

An article circulated by the Associated Press on September 29, 2007 described a species of amoeba Naegleria fowleri, that is found in warm bodies of water and soil. The amoeba can enter the nose and can attach itself to the olfactory nerve. It then destroys tissue and makes its way to the brain. Those infected suffer from a stiff neck, headaches, and fevers. As the disease damages the brain, the patient has hallucinations and behavioral changes.

There is no known cure and death can follow in as little as two weeks after infection. Between 1995 and 2004, there have been 23 known deaths in the US. During 2007, there were six cases—three in Florida, two in Texas, and one in Arizona. Using noseclips while swimming is an effective preventative. Helminths are worms. There are three forms: roundworms, tapeworms, and flukes. The majority of helminths can reproduce within the host or within an intermediate host. Transmission can occur by ingestion of their fertilized eggs, penetration of the skin by infectious larval stages of their development, or by an arthropod vector. They can cause lymphatic filariasis, hookworm, pinworm, schistosomiasis, tapeworm, trichinosis, etc. A typical lymphatic filariasis infection is a chronic infestation with threadlike filaria nematodes. They congregate in the lymph nodes and surrounding tissue and eventually block the flow of lymph. This results in pain and significant swelling in the area involved.

Helminths come in all sizes, ranging from a few millimeters long to as many as twenty meters in length for a full-grown tapeworm. In fact, the human species can be inhabited by as many as 54 species of tapeworm, which is little more than an inside-out intestine that absorbs nutrients from our gut and reproduces at an astronomical rate. They can shed millions of eggs at a time, none of which will survive unless they leave the host.

Arthropods—mosquitoes, ticks, fleas, mites, chiggers, and lice—are blood feeders and the ultimate vectors. They either insert their proboscis into your skin to withdraw blood or they chew off a section of skin, attach themselves to you, and suck out the blood. As they withdraw blood, they either inject a substance that prevents coagulation or they regurgitate or excrete into the wound they caused. In either case, any pathogens living in their salivary glands are inserted nearly directly into your blood stream. Ticks usually need to feed on you for anywhere from 24 to 48 hours before they are sufficiently engorged and able to transmit anything to you.

The resettlement of refugees and rapid urbanization of many countries in the developing world has lead to crowding, inadequate sewer systems, inadequate water supplies, and a rapid increase in the number and size of breeding grounds for mosquitoes. Infection of a single person bitten by these insects rapidly leads to infections of many people in the surrounding areas. It is not uncommon that newcomers to an urban area are more susceptible to many of these arthropod-borne diseases. Some creatures coexist with humans with no ill infects for us. One such is the 0.3 mm long Demodex mite which lives in the sebaceous glands of the eyelash. Its presence is more common among older people because children produce relatively little sebum. The mites usually live in pairs and can live for several weeks. In the dark they are able to leave the eyelash and walk about the skin. Even mosquito-carried microbes as large as filaria are extraordinarily damaging. When injected into the mosquito's body, they take one to three weeks to develop. After that, the mosquito is essentially a flying syringe ready to inject these threadlike worms into a human victim. As the filaria move to the lymphatic system, disease symptoms may take years to develop, if they ever do. Even an asymptomatic infection can result in damage to the lymph system and the kidneys.

Estimates are that over 120 million people have been infected and fully one-third of them are either incapacitated or disfigured by the disease. Two-thirds of the victims are equally distributed between Africa and India. The remaining third is spread over Southeast Asia, Micronesia, and the Americas.

## Involvement of Microbe in Human Disease

### Bacteria

Bacteria are the most common cause of infections in people with cancer. Types of bacteria that cause infection include

- *Staphylococcus*
- *Streptococcus*
- *Pseudomonas*
- *Clostridium difficile*

- *Escherichia coli*
- *Listeria monocytogenes*

*Staphylococcal* infections (often shortened to Staph) mainly affect the skin. Two common types are *Staphylococcus epidermis* and *Staphylococcus aureus*. They usually cause mild infections, but these can be more serious in people with cancer. MRSA is a type of Staph aureus infection that standard antibiotics can't treat. MRSA stands for Methicillin Resistant *Staphylococcus aureus*.

*Streptococci* (often shortened to Strep) are common bacteria that can cause tonsillitis and skin infection (cellulitis). They are usually treated with antibiotics like penicillin. Pneumococcal pneumonia is a chest infection caused by a type of strep, especially in people who have had their spleen removed.

Another type of bacteria that can affect people with cancer are Enterococci. These can cause bladder infections and blood poisoning (septicaemia).

*Pseudomonas* infections are more rare, but can be a problem for people who have weak immune systems or who are in hospital for some time. The germs live in soil, water and on the skin.

*Clostridium difficile* infection is most common in people who are already on antibiotics. Certain bacteria can live normally in the bowel in just the right numbers to keep it healthy. But being on antibiotics can upset the balance and allow some bacteria to multiply and become harmful. This is what causes *Clostridium difficile* or *C. diff*, as it is often called. Symptoms include diarrhoea, abdominal pain and fever.

*Escherichia coli* (*E. coli*) infections also affect the gut causing diarrhoea and fever.

*Listeria monocytogenes* is commonly known as Listeria. The infection is caused by eating contaminated food. It is rare, but can be serious if caught by pregnant women, unborn or newborn babies, elderly people, or those with a weak immune system. A recent study showed that people with cancer may have an increased risk of developing the illness. You can reduce your risk of listeria infection by avoiding certain foods, including

- Pre packed and sliced delicatessen meats
- Soft cheeses
- Smoked fish
- Paté
- Pre prepared cooked chilled meals
- Pre prepared sandwiches
- Un pasteurised milk

## Viruses

Viruses are tiny particles that can't live on their own. To reproduce, they need to infect a living cell, such as a cell in a human body.

It is a virus that causes flu. The most common types of viruses that cause infections in people with cancer include:

- ☆ Common cold viruses
- ☆ Herpes simplex
- ☆ Varicella zoster
- ☆ Cytomegalovirus (CMV)
- ☆ Flu (influenza)

Common cold viruses occur often in healthy people and do not usually cause major problems. But these viruses can cause serious infections, such as pneumonia, in people whose immune systems are weak. If you are having cancer treatment that weakens your immune system, you should try to stay away from people with colds. This may seem impossible if you are having treatment during winter in the UK, when everyone you meet seems to have a cold! You can only do your best with this one.

Herpes simplex is a virus that causes cold sores and genital herpes. In people with healthy immune systems, these infections are usually mild. But the virus can cause serious infections in those with weakened immune systems, including some people with cancer.

Varicella zoster is related to the herpes simplex virus. Varicella zoster is the same virus that causes chicken pox. It can cause very serious and sometimes fatal infections, such as pneumonia, in people with cancer. The virus also causes a painful condition called shingles. You can only get shingles if you have had chicken pox in the past. This is because after you've had chicken pox the virus does not fully go away. It lies inactive in the body, but can become active again if the immune system is weak. This is why people can get shingles after chemotherapy treatment.

Cytomegalovirus (CMV) is common and can cause several types of infections. Cytomegalo just means large cell, which is how the infected cells look under a microscope (swollen and big). Most people will have had a CMV infection by the time they are adults. It isn't generally serious. But the virus can remain inactive in your body for many years and cause recurrent infections. This usually only becomes a problem in people who have very weakened immune systems. For example, it can cause serious chest infections after a stem cell or bone marrow transplant.

Flu is an infection caused by viruses. It is very infectious. You mainly catch it from coughs and sneezes of people who have a flu virus. When you have flu you feel unwell more quickly than with a cold. Flu often causes

- ☆ High temperatures (fever)
- ☆ Aching muscles
- ☆ Coughing
- ☆ Headaches

Many cancer treatments can affect the immune system, so you are less able to fight infections. It will take you longer to recover from flu if you do get it. And you are more likely to develop complications such as chest infections. Your doctor may suggest you have the flu jab and will tell you how and when to get it.

## Fungal Infections

Fungi can live in our bodies without causing any problems. But if a fungal infection occurs when your immune system is weak, it can cause problems. The most common fungal infection in people with cancer is *Candida*, often called thrush. If you have a sore mouth from chemotherapy or radiotherapy treatment you are more at risk from thrush. Chlorhexidine mouthwashes can help to prevent thrush in the mouth. You may notice white patches with red and sore skin underneath.

Thrush can be treated with an anti fungal drug called nystatin. This is given in liquid form which coats the inside of the mouth and throat. Or a gel such as miconazole may be applied to the affected areas. If the infection does not respond to either of these treatments, doctors may uses a drug such as fluconozole. Women can also get vaginal thrush, which is treated with anti fungal creams or pessaries.

Less common fungal infections are aspergillosis and pneumocystis. These infections may be a serious problem for people who have very weak immune systems, for example after a stem cell or bone marrow transplant. Aspergillus can cause a serious chest infection that needs treatment with anti-fungal drugs through a drip. And the pneumocystis bug can cause a serious form of pneumonia called PCP. This stands for pneumocystis carinii pneumonia. Doctors use the drug co-trimoxazole (Septrin) to help prevent or treat it.

## Protozoal Infections

Protozoa are the smallest animals known to man. Toxoplasmosis is an infection caused by protozoa. It causes a mild illness in healthy people, but those with severely weakened immune systems can get serious infections that spread to the brain.

Remember many of these infections are only a problem if your resistance to infection is very low. If you are worried about your risk of getting infections do talk with your doctor or specialist nurse.

## Short Listing of some Disease

**Prions:** Kuru, Creutzfeld-Jakob disease, Gerstmann-Sträussler-Scheinker syndrome, fatal familial insomnia

**Viruses:** Chicken pox, common cold, diarrhea, encephalitis, Epstein-Barr Syndrome,gingivitis, hepatitis, herpes, hoof and mouth disease, influenza, HIV disease, HTLV disease, Kaposi's sarcoma, measles, meningitis, molluscum contagiosum, monkey pox, mumps, pneumonia, polio, rabies, rubella, severe acute respiratory syndrome (SARS), T cell leukemia, various hemorrhagic fevers, warts, yellow fever, etc.

**Bacteria:** Anthrax, bacillary dysentery, boils, brucellosis, cat scratch fever, chancroid, cholera, diarrhea, diphtheria, endocarditis, food poisoning, gangrene, gonorrhea, haemophilus influenza, Legionnaire's disease, Listerosis, meningitis, necrotizing fasciitis, pneumonia, septicemia, scarlet fever, tetanus, toxic shock syndrome, typhoid fever, ulcers, urethritis, whooping cough, etc.

**Spirochetes:** Leptospirosis, Lyme disease, relapsing fever, syphilis, yaws

**Mycobacteria:** Hansen's disease (leprosy), pneumonia, tuberculosis

**Rickettsiae and Chlamydiae:** Chlamydia, lymphogranuloma venereum, pneumonia, Q fever, Rocky Mountain spotted fever, typhus, urethritis

**Fungi:** Aspregillosis, candidiasis, coccidioidomycosis, cryptococcosis, histoplasmosis, Pneumoncystis jerovecii (carinii) pneumonia, thrush, tinea pedis (athlete's foot), tinea cruex (jock itch)

**Protozoa:** Amoebic dysentery, cryptosporidiosis, diarrhea, giardiasis, leishmaniasis, liver abscess, malaria, onchocerciasis (river blindness), toxoplasmosis, trichomoniasis, trypanosomiasis (sleeping sickness/Chagas' disease)

## Prevention

### Medical Specialists

The medical treatment of infectious diseases falls into the medical field of **Infectiology** and in some cases the study of propagation pertains to the field of Epidemiology. Generally, **infections** are initially diagnosed by primary care physicians or internal medicinespecialists. For example, an "uncomplicated" pneumonia will generally be treated by the internist or the pulmonologist (lung physician).The work of the infectiologist therefore entails working with both patients and general practitioners, as well as laboratory scientists,immunologists, bacteriologists and other specialists.

An infectious disease team may be alerted when:

- The disease has not been definitively diagnosed after an initial workup
- The patient is immunocompromised (for example, in AIDS or after chemotherapy);
- The infectious agent is of an uncommon nature (*e.g.* tropical diseases);
- The disease has not responded to first line antibiotics;
- The disease might be dangerous to other patients, and the patient might have to be isolated

Washing one's hands, a form ofhygiene, is an effective way to prevent the spread of infectious disease.

Techniques like hand washing, wearing gowns, and wearing face masks can help prevent infections from being passed from the surgeon to the patient or vice versa. Frequent hand washing remains the most important defense against the spread of unwanted organisms.[15]. Cooking foods well and avoiding foods that have been left outside for a long time is also important.

Antimicrobial substances used to prevent transmission of infections include:

- Antiseptics, which are applied to living tissue/skin
- Disinfectants, which destroy microorganisms found on non-living objects.

The long term use of antibiotics leads to resistance and chances of developing opportunistic infections like clostridium difficile colitis.[16] Thus, avoiding using antibiotics longer than necessary helps preventing such infectious diseases. When infection attacks the body, *anti-infective* drugs can suppress the infection. Four types of *anti-infective* or drugs exist: antibacterial (antibiotic), antiviral, antitubercular, andantifungal.[17] Depending on the severity and the type of infection, the antibiotic may be given by mouth, injection or may be applied topically. Severe infections of the brain are usually treated with intravenous antibiotics. Sometimes, multiple antibiotics are used to decrease the risk of resistance and increase efficacy. Antibiotics only work for bacteria and do not affect viruses. Antibiotics work by slowing down the multiplication of bacteria or killing the bacteria.

The most common classes of antibiotics used in medicine include penicillin, cephalosporins, aminoglycosides, macrolides, quinolones and tetracyclines.

One of the ways to prevent or slow down the transmission of infectious diseases is to recognize the different characteristics of various diseases.[18] Some critical disease characteristics that should be evaluated include virulence, distance traveled by victims, and level of contagiousness. The human strains of Ebola virus, for example, incapacitate their victims extremely quickly and kill them soon after. As a result, the victims of this disease do not have the opportunity to travel very far from the initial infection zone.[19] Also, this virus must spread through skin lesions or permeable membranes such as the eye. Thus, the initial stage of Ebola is not very contagious since its victims experience only internal hemorrhaging. As a result of the above features, the spread of Ebola is very rapid and usually stays within a relatively confined geographical area. In contrast, the Human Immunodeficiency Virus (HIV) kills its victims very slowly by attacking their immune system.[20] As a result, many of its victims transmit the virus to other individuals before even realizing that they are carrying the disease. Also, the relatively low virulence allows its victims to travel long distances, increasing the likelihood of an epidemic.

Another effective way to decrease the transmission rate of infectious diseases is to recognize the effects of small-world networks.[21] In epidemics, there are often extensive interactions within hubs or groups of infected individuals and other interactions within discrete hubs of susceptible individuals. Despite the low interaction between discrete hubs, the disease can jump to and spread in a susceptible hub via a single or few interactions with an infected hub. Thus, infection rates in small-world networks can be reduced somewhat if interactions between individuals within infected hubs are eliminated. However, infection rates can be drastically reduced if the main focus is on the prevention of transmission jumps between hubs. The use of needle exchange programs in areas with a high density of drug users with HIV is an example of the successful implementation of this treatment method. [22] Another example is the use of ring culling or vaccination of potentially susceptible livestock in adjacent farms to prevent the spread of the foot-and-mouth virus in 2001.[23]

A general method to prevent transmission of vector-borne pathogens is pest control.

## Immunity

Mary Mallon (a.k.a. Typhoid Mary) was an asymptomatic carrier of typhoid fever. Over the course of her career as a cook, she infected 53 people, three of whom died.

Infection with most pathogens does not result in death of the host and the offending organism is ultimately cleared after the symptoms of the disease have waned. This process requires immune mechanisms to kill or inactivate the inoculum of the pathogen. Specific acquired immunity against infectious diseases may be mediated by antibodies and/or T lymphocytes. Immunity mediated by these two factors may be manifested by:

- A direct effect upon a pathogen, such as antibody-initiated complement-dependent bacteriolysis, opsonoization, phagocytosis and killing, as occurs for some bacteria,
- Neutralization of viruses so that these organisms cannot enter cells,
- or by T lymphocytes which will kill a cell parasitized by a microorganism.

The immune system response to a microorganism often causes symptoms such as a high fever and inflammation, and has the potential to be more devastating than direct damage caused by a microbe.[24]

Resistance to infection (immunity) may be acquired following a disease, by asymptomatic carriage of the pathogen, by harboring an organism with a similar structure (crossreacting), or by vaccination. Knowledge of the protective antigens and specific acquired host immune factors is more complete for primary pathogens than for opportunistic pathogens.

Immune resistance to an infectious disease requires a critical level of either antigen-specific antibodies and/or T cells when the host encounters the pathogen. Some individuals develop natural serum antibodies to the surface polysaccharides of some agents although they have had little or no contact with the agent, these natural antibodies confer specific protection to adults and are passively transmitted to new borns

## Infectious or Not

One way of proving that a given disease is "infectious", is to satisfy Koch's postulates (first proposed by Robert Koch), which demands that the infectious agent be identified only in patients and not in healthy controls, and that patients who contract the agent also develop the disease. These postulates were first used in the discovery that Mycobacteriaspecies cause tuberculosis. Koch's postulates can not be applied ethically for many human diseases because they require experimental infection of a healthy individual with a pathogen produced as a pure culture. Often, even diseases that are quite clearly infectious do not meet the infectious criteria. For example, *Treponema pallidum*, the causatives pirochete of syphilis, cannot be cultured *in vitro* - however the organism can be cultured in rabbit testes. It is less clear that a pure culture comes from an animal source serving as host than it is when derived from microbes derived from plate culture. Epidemiology is another important tool used to study disease in a population. For infectious diseases it helps to determine if a disease outbreak is sporadic (occasional occurrence), endemic (regular cases often occurring in a region), epidemic (an unusually high number of cases in a region), or pandemic (a global epidemic).

## Contagiousness

Infectious diseases are sometimes called "contagious disease" when they are easily transmitted by contact with an ill person or their secretions (*e.g.*, influenza). Thus, acontagious disease is a subset of infectious disease that is especially infective or easily transmitted. Other types of infectious/transmissible/communicable diseases with more specialized routes of infection, such as vector transmission or sexual transmission, are usually not regarded as "contagious," and often do not require medical isolation (sometimes loosely called quarantine) of victims. However, this specialized connotation of the word "contagious" and "contagious disease" (easy transmissibility) is not always respected in popular use.

## Biotechnological Applications

Listeriosis, a serious infection usually caused by eating food contaminated with the **bacterium.** Listeria **monocytogenes**, is an important public health problem in the United States. The disease primarily affects older adults, **pregnant women**, newborns, and adults with weakened immune systems. *Listeria monocytogenes* has become a paradigm for the study of host–pathogen interactions and bacterial adaptation to mammalian hosts[25]. *Listeria monocytogenes* infection has been a useful model for evaluation of the cellular interactions that are crucial for the initiation of the host Tcell response. Indeed, the ability of *monocytogenesL.* to invade and replicate in different cell types has been extensively studied and has revealed the sophisticated relationship between the bacterium and its host. As a facultative intracellular bacterium, *monocytogenes* L. has adapted to live within the cytosol of the host cell. Cytosolic delivery of an antigen facilitates the generation of a productive CD8+ Tcell response to the particular antigen. A new concept for vaccines was recently propagated using *monocytogenes* L. whereby, recombinant psoraleninactivated *monocytogenes* L. lacking *uvr*AB genes vaccines induced potent CD4(+) and CD8(+) Tcell responses and protected mice against virus challenge in an infectious disease model as well as provided therapeutic benefit in a mouse cancer model [26] (Brockstedt *al.et*, 2005). Microbially inert but metabolically active vaccines used either as a recombinant vaccine platform or as a modified form of the pathogen itself may have broad use for the treatment of infectious disease and cancer.

As *Listeria* survives within antigenpresenting cells after phagocytosis, it is an ideal vector for the delivery of antigens to be processed and presented through both the class I and II antigenprocessing pathways [27]

Listeriolysin Odeficient *monocytogenes* L. impart protective immunity and that this highly attenuated strain may be a useful platform for vaccine delivery [28]. The function of LLO as a pHsensitive phagosomal lysin has been exploited for vaccine development to intracellular pathogens and cancer cells either for delivery of pHsensitive liposomes containing nucleic acidbased compounds or protein antigens or expression of LLO by attenuated bacterial vaccine strains expressing the desired antigens or carrying DNA vaccine vectors[29] [30] Indeed, an LLO expressing tuberculosis vaccine strain *Mycobacterium bovis* bacille CalmetteeGuérin (BCG) is currently being tested in clinical trials for vaccine efficacy.[31]

A major obstacle to achieving effective DNAbased therapeutics is efficient delivery of the DNA to its site of action in the cell. *Listeria* has been used for DNA delivery in two ways. Intracellular bacteria have been shown to transfer eukaryotic expression plasmids to mammalian host cells *in vitro* and *in vivo*. This can be used to induce immune responses towards protein antigens encoded by the plasmid, to complement genetic defects or even to direct the production of proteins in appropriate organs. The ease of generating such vehicles makes this a highly attractive area for further research [32]. Functional transfer of GFPCFTR (cystic fibrosis transmembrane conductance regulator) to eukaryotic cells using attenuated *monocytogenes* L. mediated gene transfer (bacteriofection) has been demonstrated[33] [34]

Second, LLO, the haemolytic toxin of *monocytogenes* L. is capable of facilitating transfection using an anionic liposomeentrapped polycationcondensed DNA delivery system (LPDII). This anionic delivery vehicle represents the successful combination of the LPDII system for condensation of the DNA with the unique endosomolytic properties of LLO for improved transfection using plasmid DNA [35]*Drosophila melanogaster* has been used as a model for many genetic and immunological studies and it has been successfully used to test *monocytogenes* L. virulence [36]. Many new host factors important for entry into the host cell, escape from the vacuole and intracellular growth of *monocytogenesL.*, have been revealed using genomewide RNAinterference screens in *melanogasterD.* S2 cells (which are macrophagelike cells). [37] [38]

An emerging area is the use of *monocytogenes* L. as a model to screen for efficacy of antibiotics during intracellular growth. Discrepancies between resistance *in vitro* and therapeutic efficacy *in vivo* can result from differential *in vitroin vivo* expression of bacterial determinants of antibiotic susceptibility. Thus Hpt, an adaptive factor promoting listerial virulence is induced *in vivo*, and render cells *monocytogenes* L. very sensitive to fosfomycin treatment during intracytoplasmic growth but not during growth in broth cultures. [39]

*Listeria monocytogenes* have evolved sophisticated strategies to overcome host defences, to interact with the immune system and to interfere with essential host systems. Exploring the knowledge of these strategies can help in the design of more technologically robust and effective biotechnological and clinical applications. The term 'pathobiotechnology' has been coined to describe the exploitation of these valuable traits in biotechnology, medicine and food [40]. The approach shows promise for the development of novel vaccine and drug delivery systems. The genetic tractability of *monocytogenes* L., the availability of the complete genome sequence of this intracellular pathogen, its ability to cope with stress and its ability to traverse the gastrointestinal tract and induce a strong cellular immune response make *monocytogenesL.* an ideal model organism for demonstrating the pathobiotechnology concept.

## Patho-Biotechnology

Pathogenic bacteria have evolved sophisticated strategies to overcome host defences, interact with the immune system and interfere with essential host systems. We coin the term 'patho-biotechnology' to describe the exploitation of these valuable traits in biotechnology and biomedicine. This approach shows promise for the design of more technologically robust and effective probiotic cultures with improved biotechnological and clinical applications as well as the development of novel vaccine and drug delivery platforms. [41].

Virotherapy is a treatment using biotechnology to convert viruses into therapeutic agents by reprogramming viruses to treat diseases.

The immune modulating virotherapy known as RIGVIR developed at the Institute of Microbiology in Latvia is the first virotherapy to pass all stages of clinical development and was approved in 2009. [42].

Since 2004 Rigvir is approved in Latvia (national registration), but since 2008 Rigvir is available in pharmacies of Latvia. Right now virotherapy with Rigvir is successfully used in Latvia and 25 other countries around the world.

In 2004, researchers from University of Texas genetically programmed a type of common cold virus Adenovirus Delta-24-RGD to attack glioblastoma multiforme. Later other researchers [43] have tried tests on mice where 9 out of 10 mice have shown degeneration of tumours and prolonged survival. A drug grade virus was approved for clinical trials on humans in 2009.[44]

In 2006 researchers from the Hebrew University succeeded in isolating a variant of the Newcastle disease Virus (NDV-HUJ), which usually affects birds, in order to specifically target cancer cells.[45] The researchers tested the new virotherapy on patients with glioblastoma multiforme and achieved promising results for the first time.

Vaccinia virus, a virus credited for the eradication of smallpox, is being developed as an oncolytic virus, *e.g.* GL-ONC1 and JX-594. [46].

Promising research results warrant its clinical development in human patients.

The experimental virotherapy that has progressed the farthest in clinical trials (as of 2013) is Talimogene laherparepvec. It is based on an engineered version of herpes simplex virus which has also been engineered to express GM-CSF. This virus is being developed by Amgen who reported that a pivotal phase 3 study in melanoma had met its primary endpoint (durable response rate) with a very high degree of significance in March 2013, the first positive phase 3 study with an oncolytic virus in the western world. [47]. [48].

ProSavin is one of a number of therapies in the Lentivector platform under development by Oxford BioMedica. It delivers to the brain the genes for three enzymes important in the production of dopamine, a deficiency of which causes Parkinson's disease.TNFerade (a non replicating TNF gene therapy virus) failed a phase III trial for pancreatic cancer.Trovax is an immunotherapy that uses a pox-virus bearing the tumour antigen 5T4, to induce an immune response against a variety of cancer types.

The therapy was developed by Oxford BioMedica and failed to improve overall survival in a phase 3 trial in renal cell carcinoma. New phase II trials have since begun at Cardiff University (UK) withcolorectal cancer and at the Velindre Cancer Centre (Cardiff, UK) with malignant pleural mesothelioma.

## Conclusion

Infectious microbes are unlimited in number and size so a very large diversity occurs.Pathogenic bacteria have evolved sophisticated strategies to overcome host defences, interact with the immune system and interfere with essential host systems. These microbes gave rise the development of novel vaccine and new drug delivery platforms.Thus development of a new field pathobiotechnology.

Engineered probiotics thus have the potential to alleviate the symptoms of chronic gastrointestinal disorders and associated sequelae, to fight infection, modulate the immune system, and act as delivery vehicles for bioactive molecules

However, the increasing emergence of antibiotic resistance, coupled with a significant decline in production of new antibacterials, means that probiotics are finally coming of age, representing a real alternative to traditional drug based therapies.Due to very burning problems of infectious viruses have also gave rise to a new branch *i.e.*, virotherapy. Virotherapy is a treatment using biotechnology to convert viruses into therapeutic agents by reprogramming viruses to treat diseases.

## References


1. Alberto Signore (2013). "About inflammation and infection". *EJNMMI Research*, **8** (3).
2. Dobson, Andrew P. and E. Robin Carter (1996). Infectious Diseases and Human Population History. *Bioscience*; 46 2.
3. Smallpox. North Carolina Digital History.
4. Smallpox and Vaccinia. National Center for Biotechnology Information.
5. Smallpox: The Triumph over the Most Terrible of the Ministers of Death.
6. Multidrug-Resistant Tuberculosis. Centers for Disease Control and Prevention.
7. Influenza of 1918 (Spanish Flu) and the US Navy.
8. Beretta M (2003). "The revival of Lucretian atomism and contagious diseases during the renaissance". *Medicina nei secoli* **15** (2): 129–54.
9. Robert Moorhead, "William Budd and typhoid fever". (2010). *J R Soc Med.*, **95(11):** 561–564.
10. Molnar, R. E., 2001, Theropod paleopathology: a literature survey: In: *Mesozoic Vertebrate Life*, edited by Tanke, D. H., and Carpenter, K., Indiana University Press, p. 337-363.
11. Krauss H, Weber A, Appel M (2003). *Zoonoses: Infectious Diseases Transmissible from Animals to Humans* (3rd ed.). Washington, D.C.: ASM Press. ISBN 1-55581-236-8.

12. Potter P (2013). "Summer buzz". *Emerg Infect Dis* **19 (3):** 1184.
13. Daszak *et al.* (2000). *Emerging Infectious Diseases of Wildlife: Threats to Biodiversity and Human Health.*, USA.
14. Intestinal Parasites and Infection fungusfocus.com.
15. Pathogens and vectors. MetaPathogen.com.
16. Medicine Health. *"Bacterial and Viral Infections"* 2010-02-08.
17. *Nursing Pharmacology Made Incredibly Easy*. Lippincott Williams and Wilkins. 2009.p.483
18. Watts, Duncan (2003). *Six degrees: the science of a connected age*. London: William Heinemann. ISBN 0-393-04142-5
19. Preston, Richard (1995). The hot zone. Garden City, N.Y.: Anchor Books. ISBN 0-385-49522-6.
20. Ryan KJ; Ray CG (editors) (2004). *Sherris Medical Microbiology* (4th ed.). McGraw Hill. ISBN 0-8385-8529-9
21. Watts, Duncan (2003). Six degrees: the science of a connected age. London: William Heinemann. ISBN 0-393-04142-5
22. Bacterial vs. Viral Infections - Do You Know the Difference? National Information Program on Antibiotics.
23. Ferguson NM, Donnelly CA, Anderson RM (2001). "The foot-and- mouth epidemic in Great Britain: pattern of spread and impact of interventions". Science,**292** : 1155–60.
24. Ryan KJ; Ray CG (editors) (2004). Sherris Medical Microbiology (4th ed.). McGraw Hill. ISBN 0-8385-8529-9.
25. Hamon M., Bierne H., Cossart P. *Listeria monocytogenes*: a multifaceted model. Nat Rev Microbiol. 2006;4: 423–434.
26. Brockstedt D.G., Bahjat K.S., Giedlin M.A., Liu W., Leong M., Luckett W. Killed but metabolically active microbes: a new vaccine paradigm for eliciting effector Tcell responses and protective immunity. *Nat Med*. 2005;11: 853–860.
27. Singh R., Paterson Y. *Listeria monocytogenes* as a vector for tumorassociated antigens for cancer immunotherapy. *Expert Rev Vaccines*. 2006;5: 541–552
28. Hamilton S.E., Badovinac V.P., Khanolkar A., Harty J.T. Listeriolysin Odeficient *Listeria monocytogenes* as a vaccine delivery vehicle: antigenspecific CD8 T cell priming and protective immunity. *J Immunol*. 2006;177: 4012–4020.
29. Provoda C.J., Lee K.D. Bacterial poreforming hemolysins and their use in the cytosolic delivery of macromolecules. *Adv Drug Deliv Rev*. 2000;41: 209–221.
30. Dietrich G., Spreng S., Favre D., Viret J.F., Guzman C.A. Live attenuated bacteria as vectors to deliver plasmid DNA vaccines. *Curr Opin Mol Ther*. 2003;5: 10–19.
31. Grode L., Seiler P., Baumann S., Hess J., Brinkmann V., Nasser E.A. Increased vaccine efficacy against tuberculosis of recombinant *Mycobacterium bovis* bacille

CalmetteGuerin mutants that secrete listeriolysin. *J Clin Invest*. 2005;115: 2472–2479.

32. Weiss S., Chakraborty T. Transfer of eukaryotic expression plasmids to mammalian host cells by bacterial carriers. *Curr Opin Biotechnol*. 2001;12: 467–472.
33. Krusch S., Domann E., Frings M., Zelmer A., Diener M., Chakraborty T., Weiss S. *Listeria monocytogenes* mediated CFTR transgene transfer to mammalian cells. *J Gene Med*. 2002;4: 655–667.
34. Helmer A., Krusch S., Koschinski A., Rohde M., Repp H., Chakraborty T., Weiss S. Functional transfer of eukaryotic expression plasmids to mammalian cells by *Listeria monocytogenes*: a mechanistic approach. *J Gene Med*. 2005;7: 1097–1112.
35. Lorenzi G.L., Lee K.D. Enhanced plasmid DNA delivery using anionic LPDII by listeriolysin O incorporation. *J Gene Med*. 2005;7: 1077–1085.
36. Mansfield B.E., Dionne M.S., Schneider D.S., Freitag N.E. Exploration of hostpathogen interactions using *Listeria monocytogenes* and *Drosophila melanogaster*. *Cell Microbiol*.2003;5: 901–911.
37. Agaisse H., Burrack L.S., Philips J.A., Rubin E.J., Perrimon N., Higgins D.E. Genomewide RNAi screen for host factors required for intracellular bacterial infection. *Science*. 2005;309: 1248–1251.
38. Cheng L.W., Viala J.P., Stuurman N., Wiedemann U., Vale R.D., Portnoy D.A. Use of RNA interference in *Drosophila* S2 cells to identify host pathways controlling compartmentalization of an intracellular pathogen. *Proc Natl Acad Sci USA*. 2005;102: 13646–13651.
39. Scortti M., LacharmeLora L., Wagner M., ChicoCalero I., Losito P., VázquezBoland J.A. Coexpression of virulence and fosfomycin susceptibility in *Listeria*: molecular basis of an antimicrobial *in vitroin vivo* paradox. *Nat Med*. 2006;12: 515–517.
40. Sleator R.D., Hill C. Pathobiotechnology; using bad bugs to make good bugs better. *Sci Prog*. 2007; 90: 1–14
41. Sleator RD,Hill C.Patho-biotechnology; using bad bugs to make good bugs better. *Sci Prog*. 2007; 90(Pt 1): 1-14.
42. "Register of Human Medicines". State Agency of Medicines of Latvia.
43. Witlox AM, Van Beusechem VW, Molenaar B, Bras H, Schaap GR, Alemany R, Curiel DT, Pinedo HM, Wuisman PI, Gerritsen WR., Conditionally replicative adenovirus with tropism expanded towards integrins inhibits osteosarcoma tumor growth *in vitro* and *in vivo*. *Clin Cancer Res*. 2004 Jan 1;10(1 Pt 1): 61-7
44. Clinical Trial for Delta-24-RGD for Recurrent Malignant Gliomas
45. Isracast news article on virotherapy March 2006
46. "Welcome to Genelux - intro". Genelux.com

47. Zhang, Q; Yu, YA; Wang, E; Chen, N; Danner, RL; Munson, PJ; Marincola, FM; Szalay, AA (2007). "Eradication of solid human breast tumors in nude mice with an intravenously injected light-emitting oncolytic vaccinia virus". *Cancer Research* **67** (20): 10038–46.

48. Kelly, KJ; Woo, Y; Brader, P; Yu, Z; Riedl, C; Lin, SF; Chen, N; Yu, YA; Rusch, VW; Szalay, Aladar A.; Fong, Yuman (2008). "Novel oncolytic agent GLV-1h68 is effective against malignant pleural mesothelioma". *Human gene therapy* **19** (8): 774–82.

*Section VII*

# Pharmaceutical Biotechnology

**2016, Biotechnology: Progress and Applications** *Pages* **375–410**
*Editors:* **Saif Hameed and Zeeshan Fatima**
*Published by:* **DAYA PUBLISHING HOUSE, NEW DELHI**

# *Chapter 17*

# Biological Significance of Coumarin Derivatives

***Simpi Mehta, Rakesh Kumar, Monica Vats, Anurag Sharma and Seema R. Pathak***

*Department of Chemistry, Amity School of Applied Sciences, Amity University Haryana, Gurgaon College of Engineering, India*

## ABSTRACT

Coumarins belong to the family of lactones. It is also known as -2H-chromen-2-one or (2H-1-benzopyran-2-one which consists of a benzene ring joined to a pyrone nucleus. Coumarin and its derivatives are of wide occurrence in nature that can be isolated from plants as well as can be synthesized in the laboratory.

Benzopyrones have been mentioned in ancient literature and herbal remedies. They are naturally found in vegetables, fruits, seeds, nuts, coffee, tea, and wine. In view of their medicinal values, relative cheapness and low toxicity benzopyrones have drawn attention of many medicinal chemist to explore its various pharmacological potentials like anti coagulant activity, anti-microbial, anti-viral, anti-diabetic, anti-cancer activity, anti-oxidant, anti-parasitic, anti-helmintic, anti-proliferative, anti-convulsant, anti-inflammatory and anti-hypertensive activities neuroprotective properties etc.

The coumarins are of great interest due to their pharmacological properties. In particular, their physiological, bacteriostatic and anti-tumor activity makes these compounds attractive backbone for derivatisation and screening as novel therapeutic agents. There are reports on anti breast cancer activity of coumarin which selectively induce apoptosis in both primary and metastatic breast cancer cell lines.

Owing to the importance of coumarin derivatives, present chapter aims to highlight research reported on the biological potential of coumarin-linked heterocycles during past few decades. Literature study of various research papers and other publications which provide detailed work and recent study of its derivatives is taken into consideration.

***Keywords:*** *Coumarins, Benzopyrone, Anti-inflammatory, Antimicrobial, Antioxidant, Anticancer, Anticonvulsant, Anti-hepatitis.*

## Introduction

The prototypical coumarin is known as 1, 2-benzopyrone or, less commonly, as 5-hydroxycinnamic acid and lactone. Coumarins (2*H*-1-benzopyran-2-one) (**I**) are made of fused benzene and α-pyrone rings [1]. Coumarins were initially found in tonka bean (*Dipteryx odorata* Wild) and are reported in about 150 different species distributed over nearly 30 different families, of which a few important ones are Rutaceae, Umbelliferae, Clusiaceae, Guttiferae, Caprifoliaceae, Oleaceae, Nyctaginaceae, and Api- aceae. More than 1300 types of coumarins derivatives have been identified as secondary metabolites from plants, bacteria, and fungi [2].

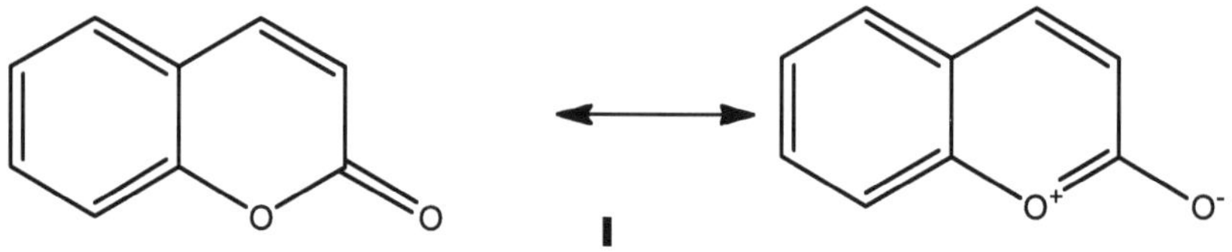

O O ⟷ O+ O-

I

Mostly coumarins are distributed in all parts of the plant, but at the highest levels in the fruits (Bael fruits (Aegle marmelos) [3], Tetrapleura tetraptera TAUB (Mimosaceae) [4], bilberry, and cloudberry), seeds (tonka beans) (Calophyllum cerasiferum Vesque and Calo- phyllum inophyllum Linn) [5],some essential oils such as cassia oil [6] Occurance of coumarin in different plant parts depends on environmental conditions and seasonal changes.

The laboratory syntheses of coumarins and their derivatives has attracted considerable attention of organic and medicinal chemists for many years as the large number of natural products contains this heterocyclic nucleus Coumarins were first synthesized via the Perkin reaction in 1868, and many simple coumarins are still prepared through this method. In the early 1900s, the Knoevenagel reaction emerged as an important synthetic method to synthesize coumarin derivatives with carboxylic acid at the 3-position. To date, many other synthetic methods for coumarins have been reported, including the Pechmann, Reformatsky and Wittig reactions [7].

The pharmacological properties as well as therapeutic applications of coumarins depend upon the pattern of substitution and recently they are reported to possess many pharmacological activities namely anticoagulation, antibiotic, antifungal, antipsoriasis, cytotoxic, anti-HIV, anti-inflammatory. Especially 7-hydroxycoumarin has antioxidant [8,9, 10] properties and cytostatic,[11,12] antibacterial,[9] antiviral,[9] xanthine oxidase inhibitor,[9] antihyperglycemic, [13] casein kinase 2 inhibitor[14] activities. Coumarins are widely used as additives in food, perfumes, cosmetics, pharmaceuticals, optical brighteners, dispersed fluorescent and laser dyes. Thus the synthesis of this heterocyclic nucleus is of much interest [15].

When a coumarin ring fused with other rings, a synergistic effect of both the rings in their biological activities are obtained, such compounds are exploited in development of various important molecule which provides pharmacophore for drug development. Various moieties when combined with coumarin can produces same or different effects but with different potencies. Details are presented here for such recent derivatives of coumarins with different activities. Various activities which are specifically studied here are:

## Biological Activities

### ACE Inhibitory Activity

6, 7- Dimethoxycoumarin derivatives carrying a protonable benzylamino group, linked to position 3 by suitable linkers inhibit acetylcholinesterase (AchEIs). Ensaculin (1) induce a modest improvement in memory and cognitive functions [16].

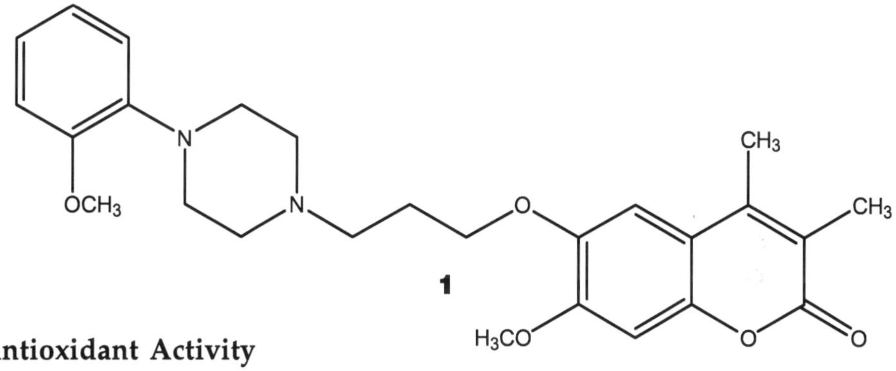

### Antioxidant Activity

Natural coumarin (2), being significant in reducing tissue oedema [17]. In comparison to quercetin and trolox, (3) have significant radical scavenging activity. Facile incorporation of the coumarin motif Lead to development of new and improved synthetic antioxidants [18].

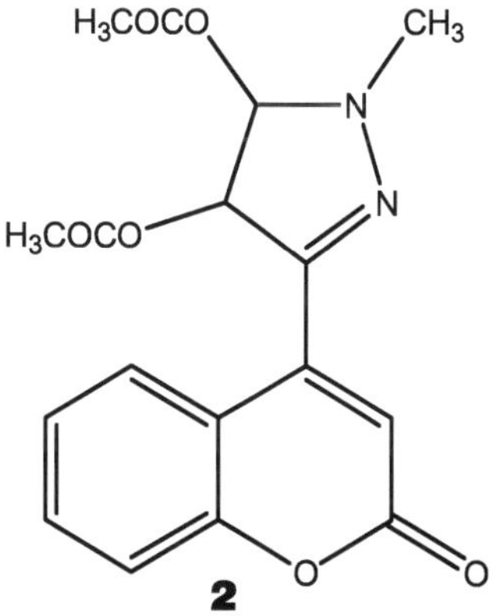

A series of 3-alkanoyl/aroyl/heteroaroyl-2H-chromene-2-thiones (4) were synthesized and these compounds were found to scavenge DPPH free radicals efficiently. The newly synthesized compounds exhibited profound antioxidant activity.

Coumarin analogues (5) bearing a substituted phenyl ring on position 3 show antioxidant activity. A series of novel coumarin-3-carboxamides (6) also show antioxidant activity. Therefore structure activity relationships (SAR) were developed in order to define the structural features required for activity.[19]

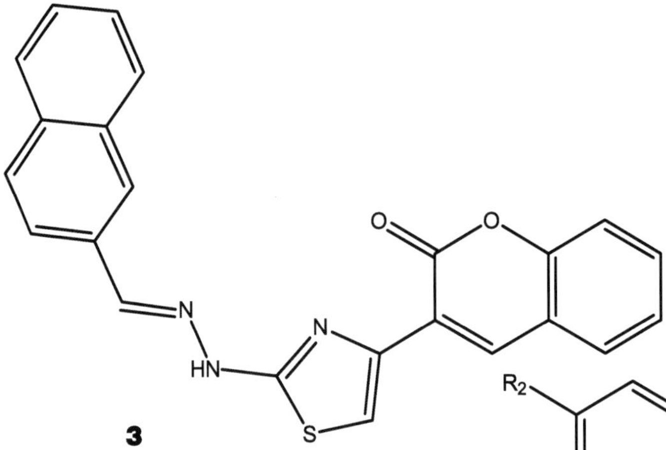

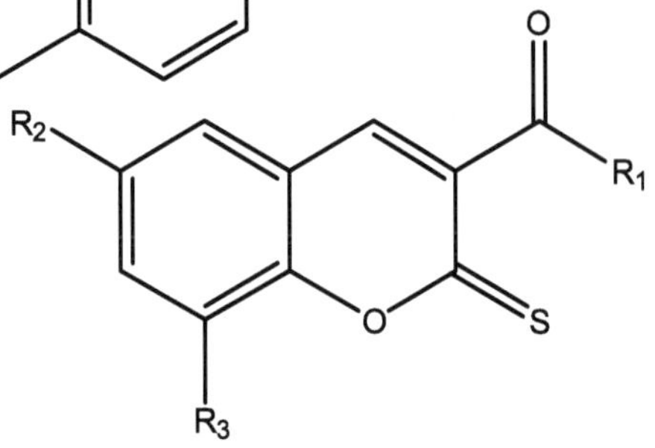

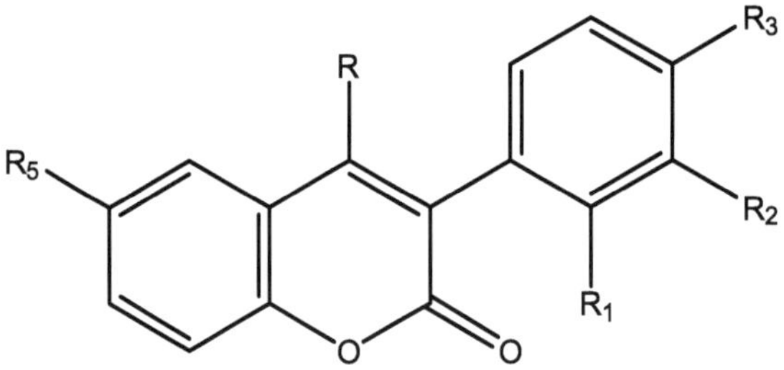

In compound (7) the presence of electron-withdrawing groups on the phenyl ring of position 3 increases antioxidant activity [20]. A series of 7- or 8-hydroxybenzo[f]coumarins were prepared by the treatment of phenols and naphthalenediols, respectively, with malic acid and $H_2SO_4$ under microwave irradiation. Compound (8) shows higher anti-oxidant activity among various hydroxybenzocoumarins being synthesized. [21]

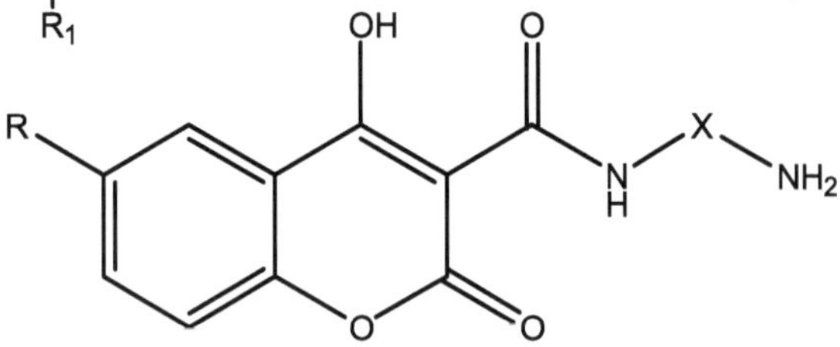

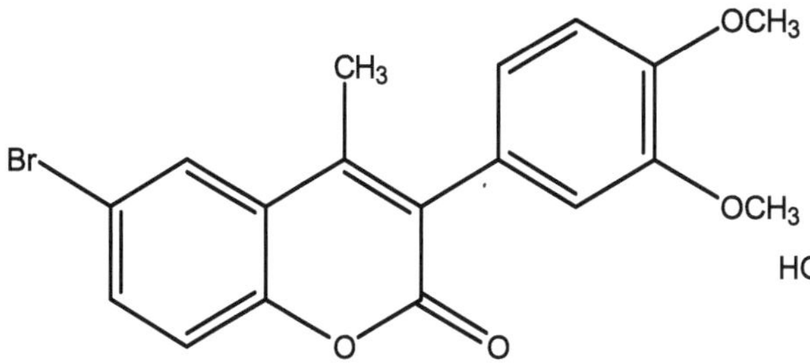

Br CH3 OCH3 OCH3 O O

**7**

HO O O

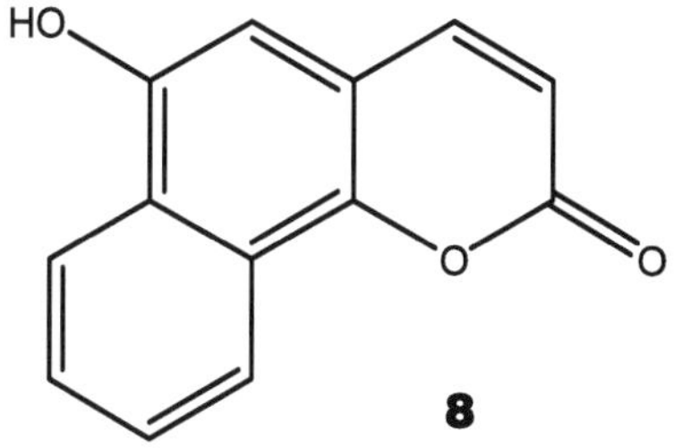

**8**

A series of substituted 3, 3'-arylidenebis-4-hydroxycoumarins and tetrakis-4-hydroxycoumarin derivative (9) were synthesized by 4-hydroxycoumarin and aromatic aldehydes containing different groups in ortho, meta or para positions and condensing them in boiling ethanol or acetic acid. The formation of hydrogen bonds may be an important factor in assisting the synthesized dicoumarols to attain the correct configuration for antioxidant activities. [22]

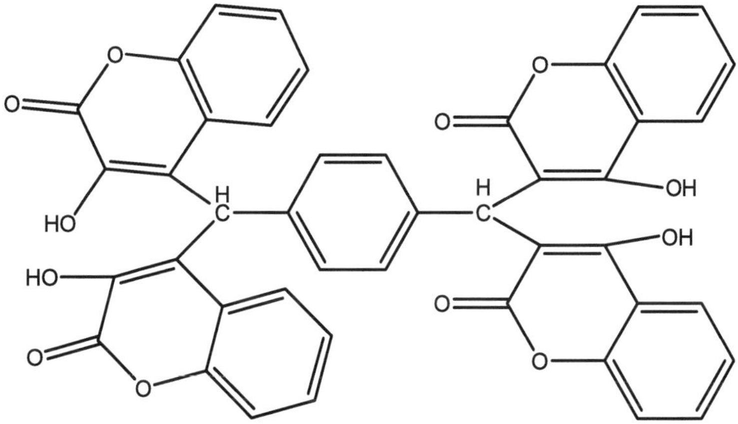

**9**

When checked at high concentration (0.5mM) one of the components Fraxin (10) showed free radical scavenging effect [23]. When the effects on human whole blood leukocytes was evaluated few compounds such as grandivittin (11), agasyllin (12), aegelinol benzoate (13), and osthol (14) show antioxidant activity.

## Anticancer Activity

Coumarin-linked 3,4- dihydropyrazole (15) have shown cytotoxic activity against gastric SGC-7901, PC-3 (human prostate cancer), and A431 (human epidermoid carcinoma) cell lines[24]. Various 7- mercapto-chromone (16) analogs were designed, synthesized and evaluated in vitro exhibited potent cytotoxic activity against four

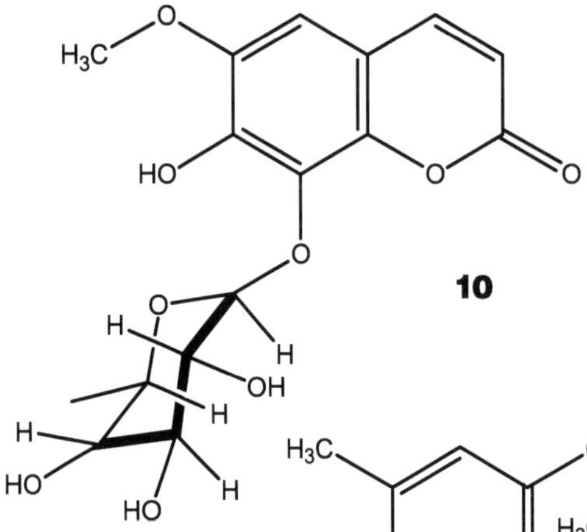

**10**

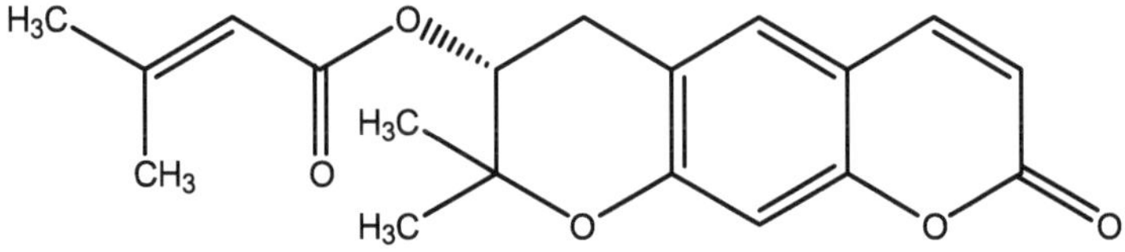

**11**

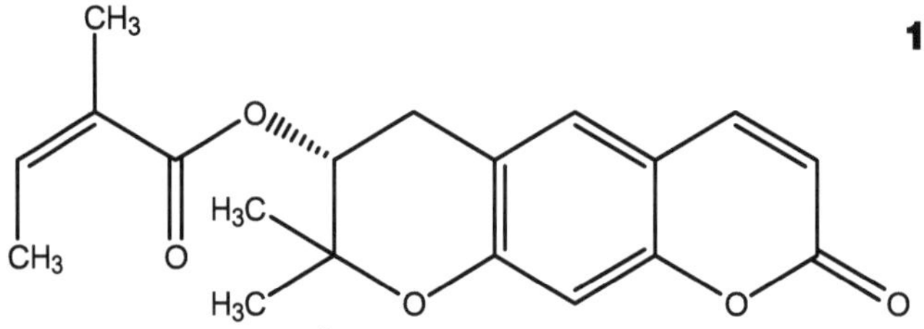

**12**

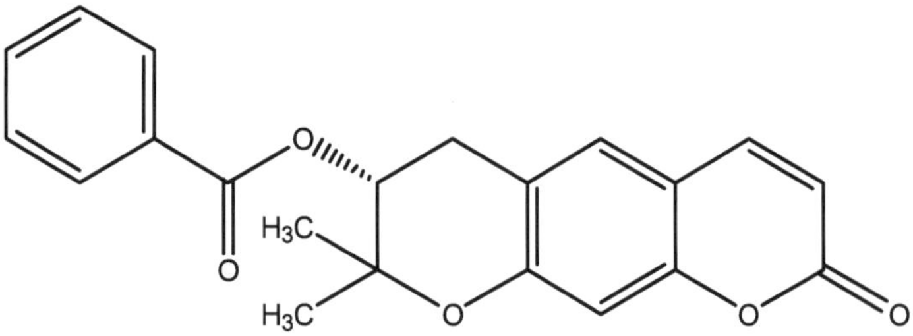

**13**

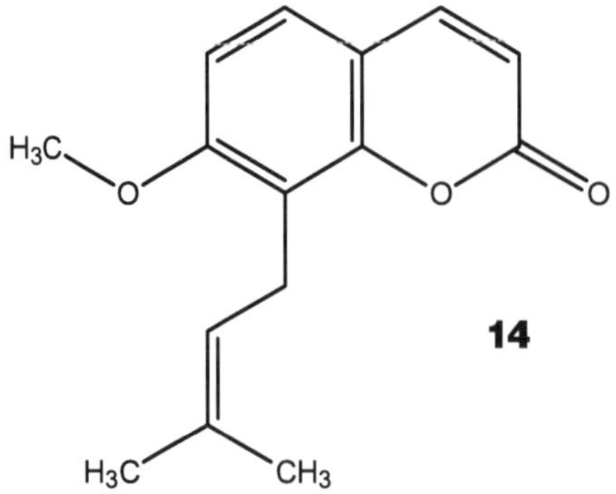

**14**

human tumor cell lines [KB(nasopharyngeal), KB-vin (vincristine-resistant subline), A549 (lung) and DU145 (prostate)].[25]

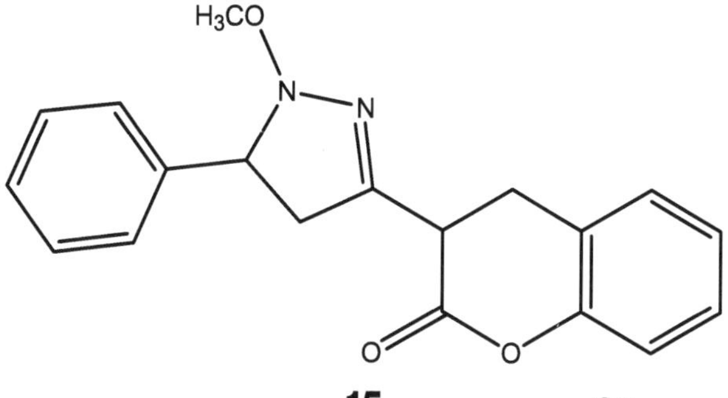

$H_3CO$ N N O O

**15**

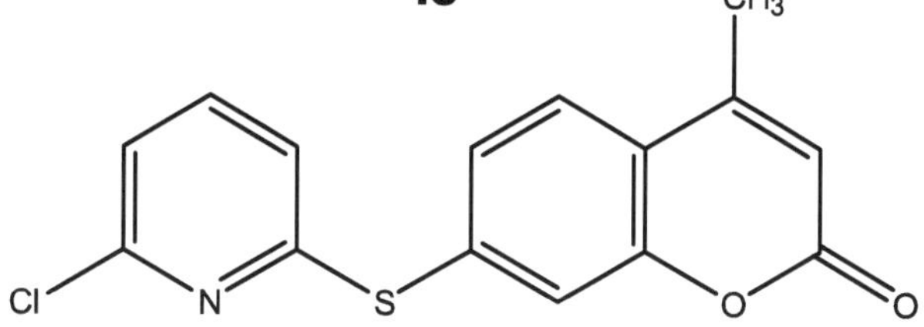

$CH_3$ Cl N S O O

**16**

The cytotoxic effects (17, 18, 1 9) were determined on two cell lines NALM-6 cells and HL-60 cells. IC50 values for NALM-6 are about nine times lower than HL-60 cell lines showed that these compounds were highly toxic to NALM-6 cells. By an increase of the hydrophobic parameters in the region of the substituents at the 2-, 3- and 4-positions of the benzopyrone skeleton of these compounds increased their cytotoxic effect. [26]

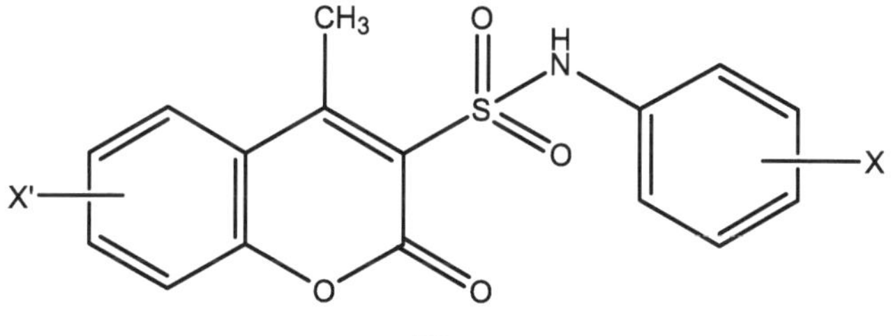

$CH_3$ O H N S O X X' O O

**17**

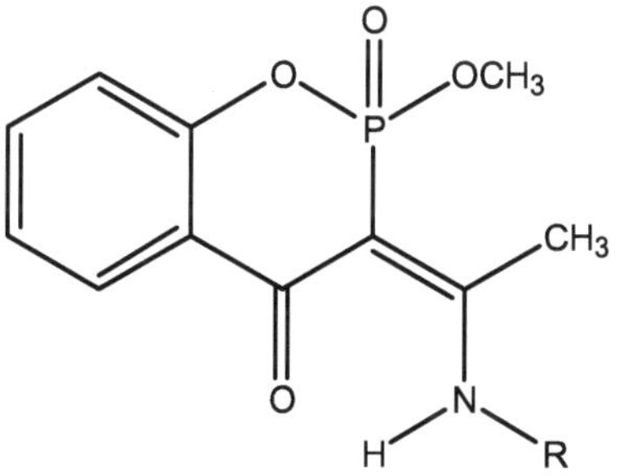

O O P $OCH_3$ $CH_3$ O N H R

**18**

O O O P $R_2$ $H_3CO$ $R_2$ $OCH_3$

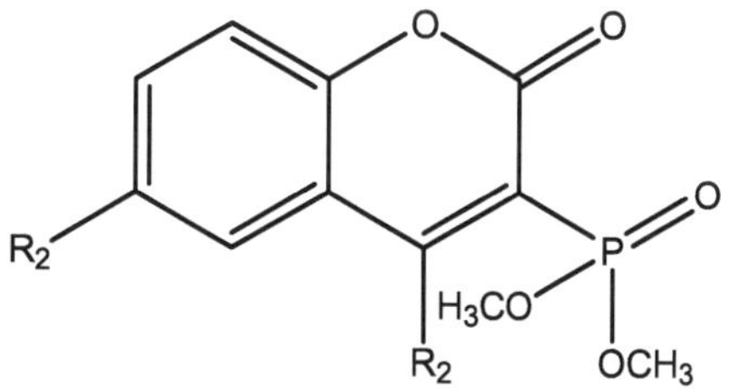

**19**

It has been found that (20), significantly inhibited the proliferation of MCF-7 (IC50~2.4µM), T47D (IC50~3.1µM) and MDA-MB-231 (IC50~3.9µM) breast cancer cell lines at all concentrations in a time dependent manner and induced apoptosis specifically in breast cancer cells irrespective of ER status and did not induce apoptosis in normal cells. To check the fact that (20) treatment induces apoptosis in both primary as well as metastatic breast cancer cells MCF-7 and MDA-MB-231, caspase-3 activation has been measured.

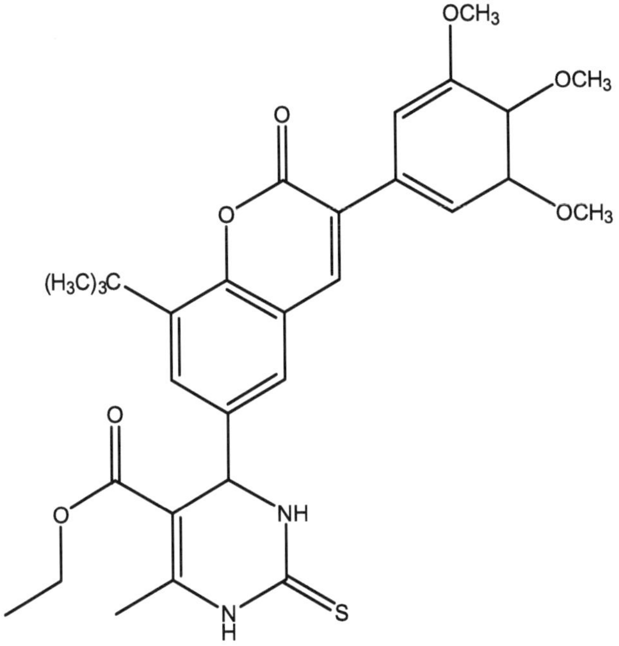

$OCH_3$ $OCH_3$ O $OCH_3$ O $(H_3C)_3C$ O NH N H S

**20**

A series of 7-diethylaminocoumarin compounds were synthesized and the cytotoxicities were tested against human umbilical vein endothelial cell (HUVEC) and some cancer cells. The inhibition activity of 21a against tested cancer cell lines was comparable to a positive control, paclitaxel, and compounds 21b and 21c showed

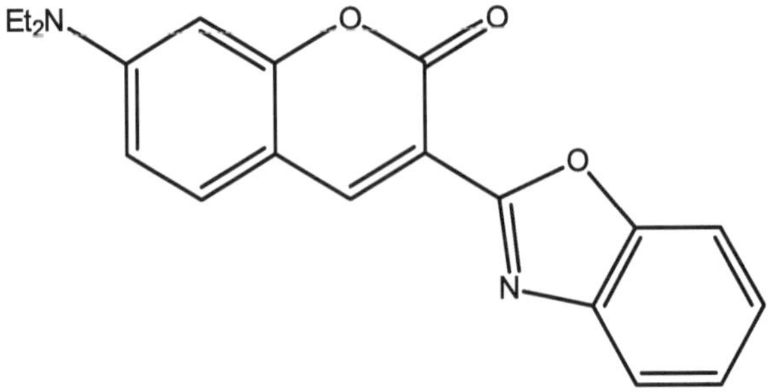

$Et_2N$ O O O N

**21a**

a high selectivity for HUVEC, which in turn can be potentially utilized as lead compounds to develop nontoxic angiogenesis inhibitors. [27]

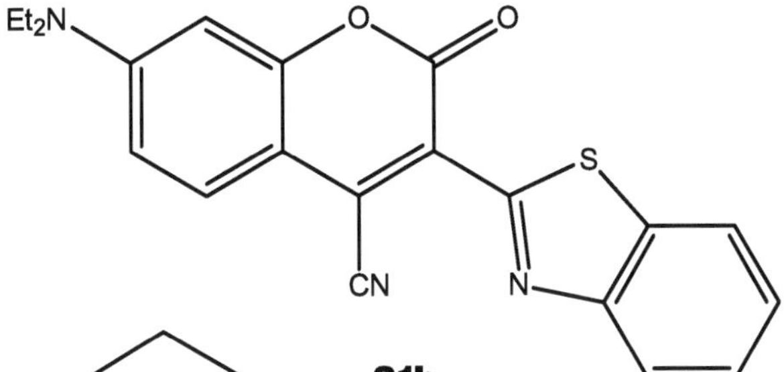

**21b**

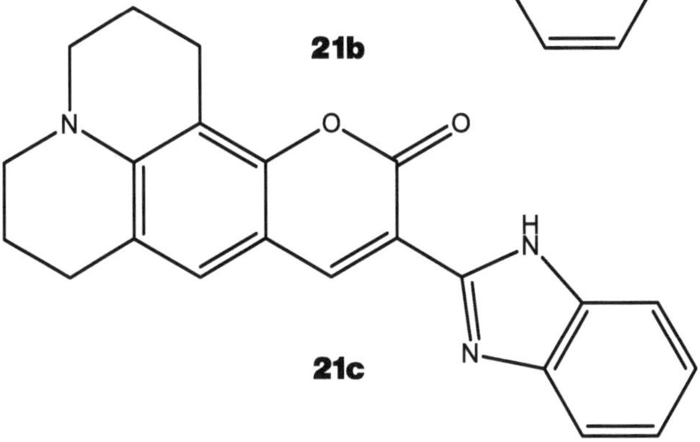

**21c**

Imperatorin (22) and Esculetin (23) exhibited anti-tumor activities. Osthole (14) grandivittin (11), agasyllin (12), aegelinol benzoate (13) from Ferulago campestris plant exhibited marginally cytotoxic activity against the A549 lung cancer cell line.

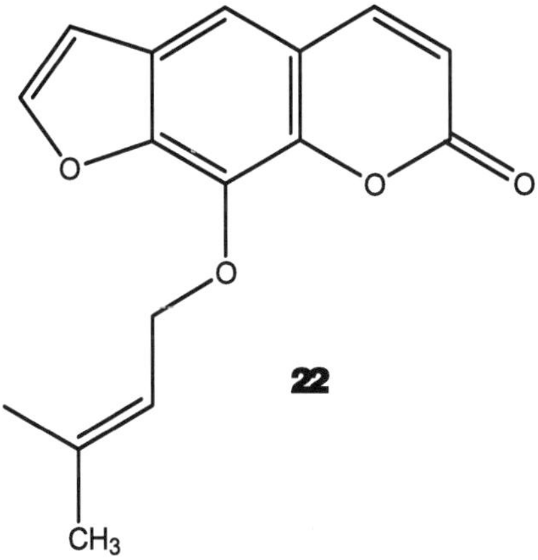

**22** 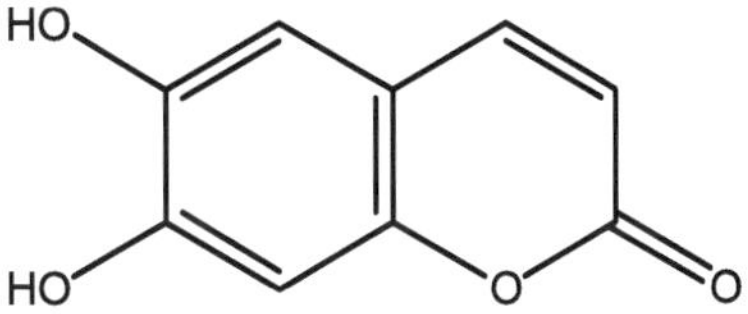 **23**

## Antimicrobial Activity

Compounds (24a-24b) from series of 4-[(5-mercapto-4-phenyl-4H-1,2,4-triazol-3-yl)-methoxy]-2Hchromen-2-one showed good activity as antifungals against fluconazole standard drug[28].

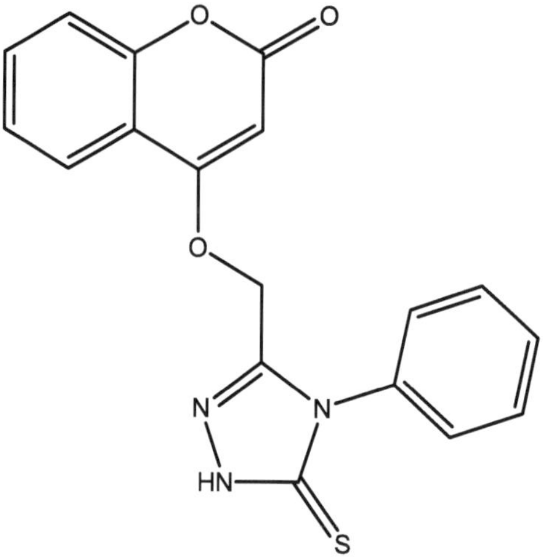

24a

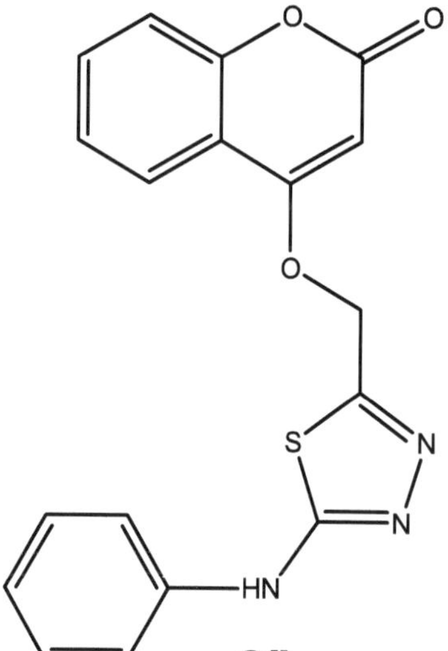

24b

Compound (26) possess maximum antibacterial activity against Gram positive bacteria *i.e. Staphylococcus aureus* and Gram negative bacteria *i.e. Escherichia coli* as compared to standard drug amoxicillin at 0.1ml concentration.[29].

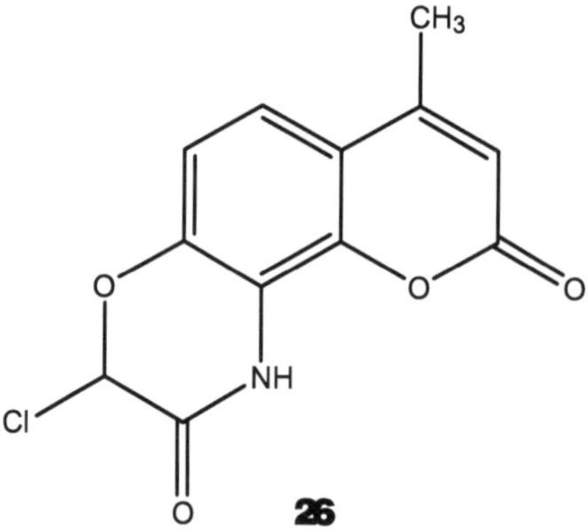

26

Compound (27) from series of novel 3-[{(3-(2′-Nitrophenyl)}-prop2-enoyl]-4-hydroxy-6-methyl- 2H-chromene-2-ones were synthesized and and their in-vitro antimicrobial activity screened against four strains of bacteria such as *Staphylococcus aureus, Bacillus megaterium, Escherichia coli,* and *Proteus vulgaris* and one fungi *Aspergillus niger.* Zone of inhibition of highly active compound was 25mm as antibacterial agent against *Escherichia coli* compared with standards ampicillin (16mm), amoxicillin (17mm), ciprofloxacinin (26mm), erythromycin (22mm) and was 23mm as antifungal agent against *Aspergillus niger* compared with standard drug griseofulvin (21mm)[30].

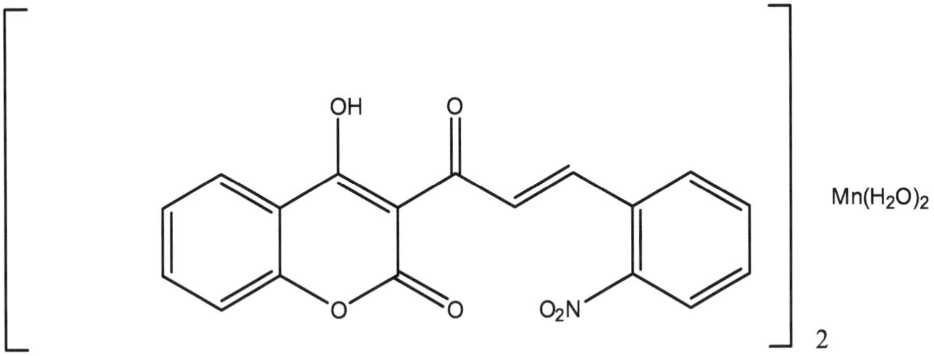

Mn$(H_2O)_2$

2

**27**

A series of 3-(4-(4-(substituted phenyl) prop-1-ene-3-one) phenylimino) methyl)-4-chloro-2H-chromen-2-ones compounds were investigated in-vitro against gram positive bacteria, *Staphylococcus aureus* (ATCC 9144), *Bacillus subtilis* (ATCC 6633) and *Staphylococcus epidermis* (ATCC 12228) and gram negative bacteria, *Escherichia coli* (ATCC 25922), *Salmonella typhi* and *Psuedomonas aeruginosa* (ATCC 9027) and the antifungal activity was evaluated against *Aspergillus niger* (ATCC 10594) and *Clostridium albicans* (ATCC 10231) using amoxicillin and fluconazole as standard drugs for antibacterial and antifungal activities respectively. One compound (28) was found to be most active with an MIC of 20μgml–1against all the tested organisms [31].

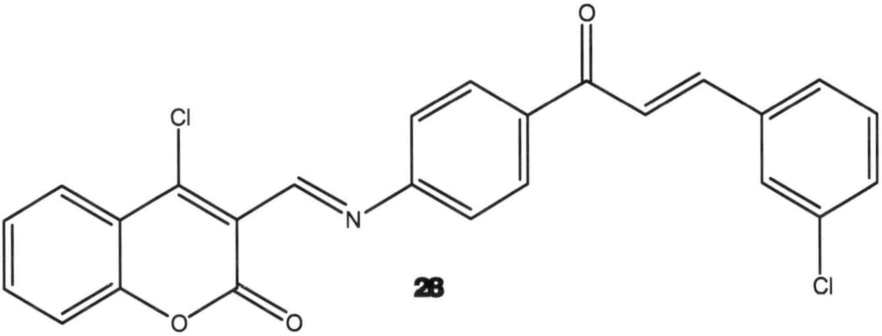

**28**

A series of thiazolidin-4-one ring were screened for their antibacterial activity against Gram positive bacteria *Staphylococcus aureus, Bacillus subtilis* and m negative bacteria *Klebsiella pneumonia,* and *Escherichia coli* at the concentration of 0.001mol ml$^{-1}$ compared with standard drug Ciprofloxacin. Zone of inhibition of highly active compound (29) was 20 mm against *Staphylococcus aureus* and *Bacillus subtilis* [32].

All compounds from a series of 4-methyl-3-phenyl-6-[4-(3-aryl-1-phenyl-1H-pyrazol-4-yl)-6-arylpyridin- 2-yl] coumarin derivatives (30)were found to be lesser active as compared to standard drug when screened for anti-bacterial activity against *Escherichia coli, Bacillus subtillis* and anti-fungal activity against *Candida albican* [33].

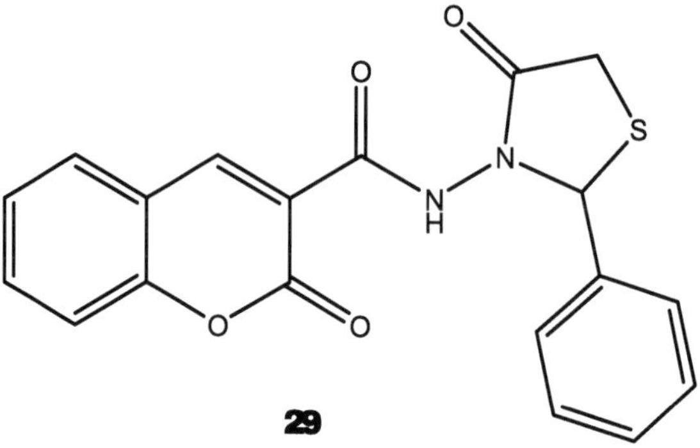

**29**

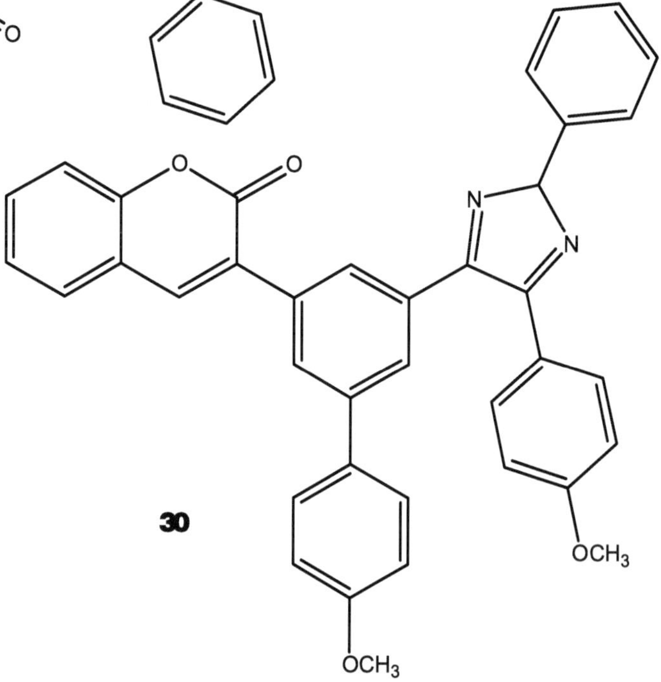

**30**

A series of novel 4-hydroxy-chromene-2-one derivatives were synthesized and screened for their antibacterial activity against Gram positive bacteria *Staphylococcus aureus, Bacillus subtili* and Gram negative bacteria *Klebsiella pneumonia, Escherichia coli* and their antifungal activity against *M. mucedo, C. albicans.* Streptomycin was used as standard antibacterial drug and ketoconazole as standard antifungal drug. One compound (31) had activity equal to that of standard drug ketoconazole (31.25$\mu gml^{-1}$) against *M. mucedo* [34].

A series of of 3-[(2'-Substituted benzylidene amino thiazol-4'-yl)amino] coumarins (32a-32b) were evaluated for antibacterial activity against various bacteria, *Staphylococcus aureus* 209 P, *E. coli* ESS2231, *Proteus vulgaris, K. pneumoniae* were used and antifungal activity was performed against *Candida albicans* ATCC10231 and results were compared with gattifloxacin and ciprofloxacin for antibacterial and fluconazole for antifungal activities respectively and propylene glycol treated group served as control.

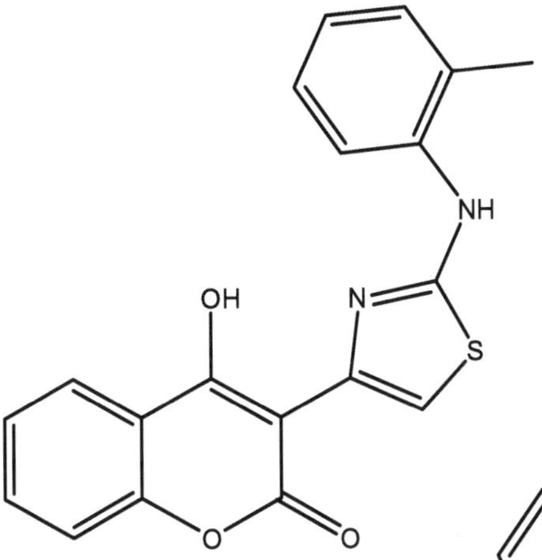

**31**

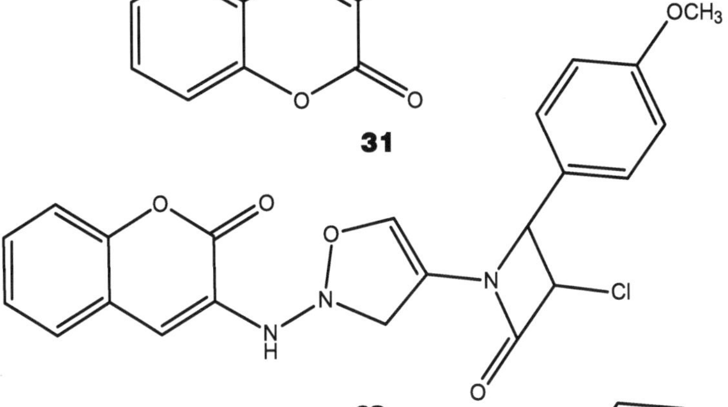

**32a**

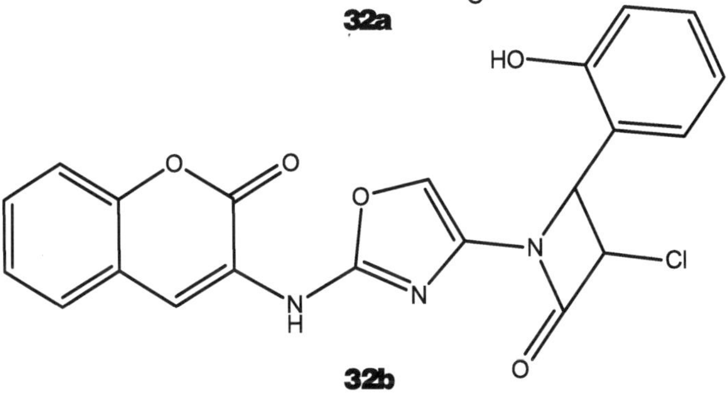

**32b**

A series of 3-coumrinoyl pyridinium bromides (33) were tested for antimicrobial activity and compared with that of gentamycin and amoxycillin. Coumarinoyl pyridinium salts having R = -H and R′ = 4-$COOC_2H_5$, R = -Cl and R′ = 4-$COOC_2H_5$, R = -H and R′ = 3-$COOC_2H_5$ and R = -Cl and R′ = 4-$COOCH_3$ were found to be more active than that of other test compounds [35].

A new series of 7–methoxy-4-methyl-8-[5-(substituted aryl)isoxazol-3-yl]–2H–benzopyran-2-ones were tested for their antimicrobial activity against 24 hr old cultures of *Escherichia coli, Pseudomonas aeruginosa, Staphylococcus aureus* and *Bacillus subtilis*. The fungi used were *Aspergillus niger, Aspergillus flavus* and *Candida albicans*

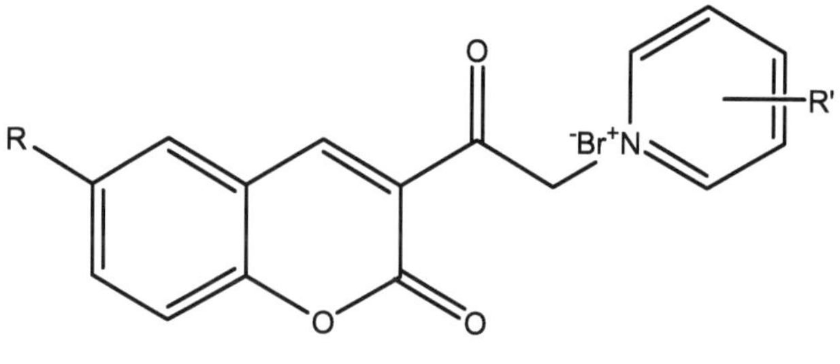

**33**

at concentrations of 25µg/ml in dimethylformamide against all the organisms. Ciprofloxacin (25µgml–1) and fluconazole (25µgml–1) were used as standard drugs for antibacterial and antifungal activities respectively. Among the compounds tested for antibacterial activity, one compound (34) showed highest zone of inhibition against *S. aureus* and *B. subtilis* and minimum inhibition against *E. coli* and *P. aeruginosa*.

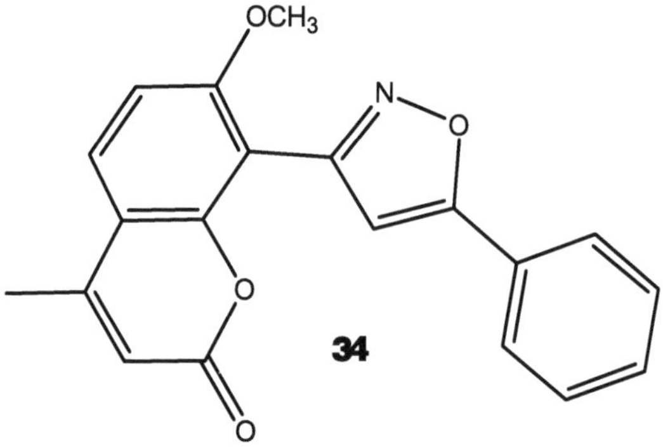

**34**

A series of 4-aryloxmethylcoumarins (35) were synthesized. Their antibacterial activity was carried out against two Gram positive bacteria, *viz., Staphylococcus aureus,* and *Streptococcus faecalis* and three Gram negative bacteria, *viz., Escherichia coli, Psuedomonus aeruginosa, Klebsiella pneumonia.* Antifungal activity was carried out against five fungi, *viz., Aspergillus flavus, Aspergillus fumigatus, Candida albicans, Penicillium notatum* and *Rhizopus.* Ciprofloxacin and Fluconazole were used as standard antibacterial and antifungal drug at different concentrations of 500, 250, 100 and 50µgml$^{-1}$respectively. The compounds possessing methoxy,chloro, bromo substituents at C-6 position of coumarin showed higher activity[36].

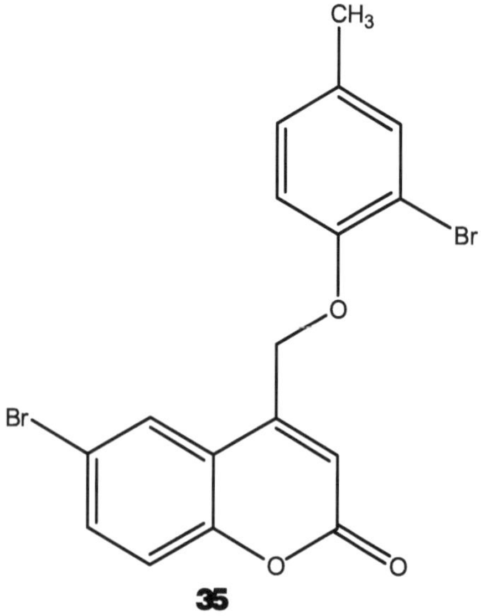

**35**

A series of 2-(substitutedphenyl)-3-[3-(2-oxo-2H-coumarin-3-yl)-5- thioxo-1,5-dihydro-4H-1,2,4-triazol-4-yl]-1,3-thiazolidin-4-ones screened for their antimicrobial activity against Gram positive *Staphylococcus aureus* and Gram negative *Escherichia coli* stains and antifungal activity against *Clostridium albicans.* Ciprofloxacin and ketoconazole were used as the standard antibacterial and antifungal drugs respectively. The test compounds and standards drugs were evaluated at concentration of 100µg/mL. DM F (N,Ndimethylformamide) was used as solvent and control. One compound (36) showed 92 per cent, 80 per cent and 90 per cent growth inhibition against *Staphylococcus aureus, Escherichia coli* and *Clostridium albicans* respectively [37].

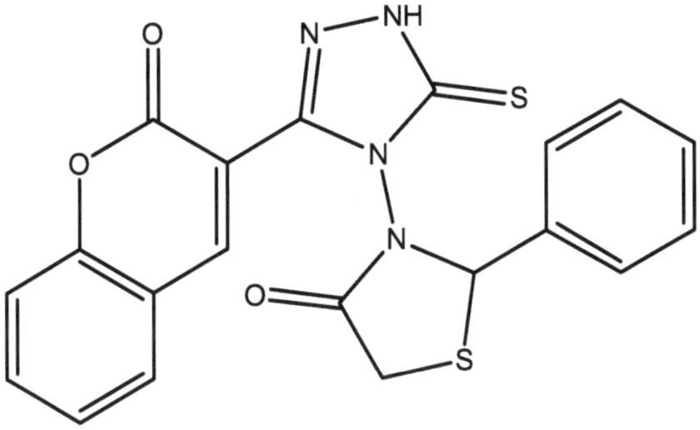

O N NH S O N N O S

**36**

A series of Schiff bases and Cu(II) complexes (37a-37b) were synthesized. All of the free ligands and their metal complexes were tested for their antifungal activity compared with ketoconazole and amphotericin B. The ligands showed no antimicrobial activity whereas a number of the metal complexes exhibited potent antimicrobial activity when compared with their respective standard drugs[38].

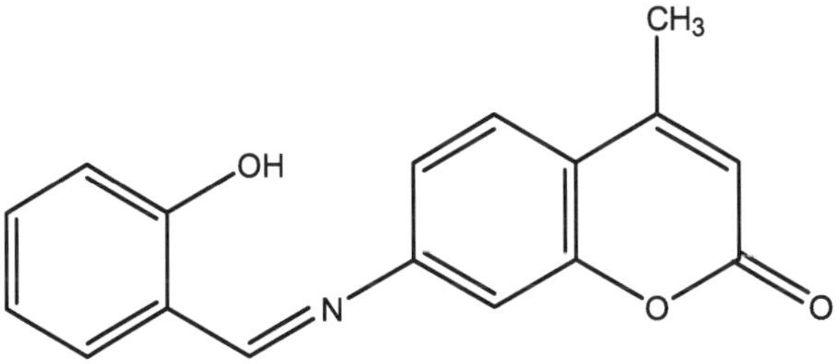

$CH_3$ OH N O O

**37a**

Compound (38) from the series of 4-chloro-3-((substituted-phenylimino) methyl)-2H-chromen-2-ones was found to be most active against all the tested organisms with an MIC of 15µgml$^{-1}$. Their antimicrobial activity tested in-vitro against Gram positive *Staphylococcus aureus* (ATCC 29737) and *Bacillus subtilis* (ATCC 2063) and Gram negative bacteria *Escherichia coli* (ATCC 20931) and fungi Aspergillus niger (ATCC 16404) and *Candida albicans* (ATCC 10231). Amoxicillin was standard for antibacterial activity and fluconazole for antifungal activity [39].

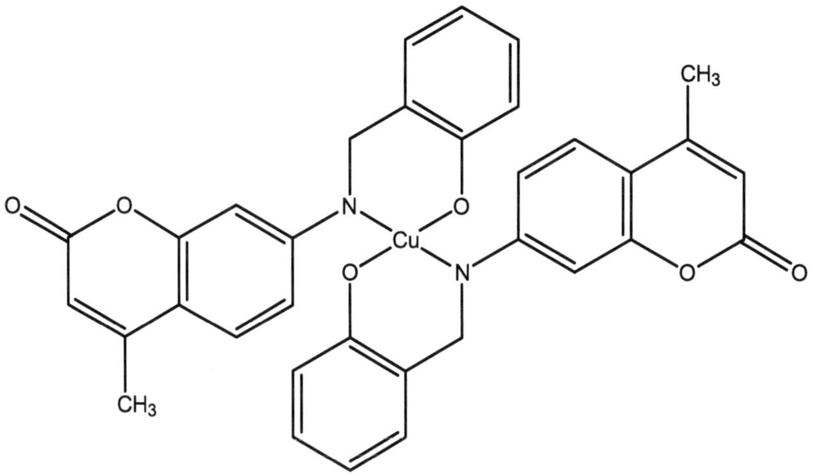

**37b**

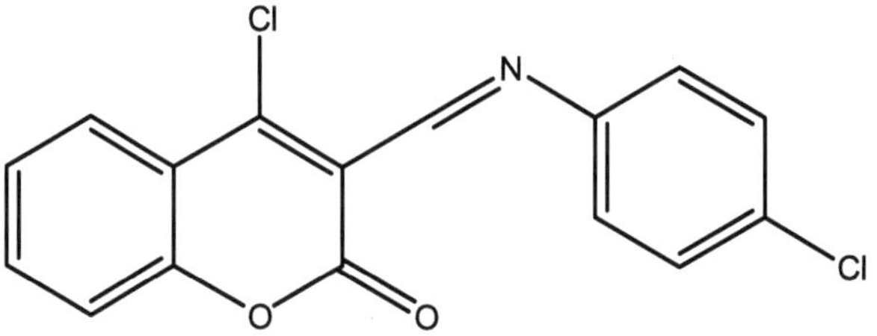

**38**

A new series of coumarin inhibitors of DNA gyrase B bearing a N-propargyloxycarbamate at C-3′ of various 5′,5′-di-alkylnoviose including RU79115 (39) were synthesized and their antibacterial activities against *Escherichia faecium* and *Staphylococcus aureus* was time dependent and similar to that of standard drug vancomycin in the case of *Staphylococcus aureus*[40].

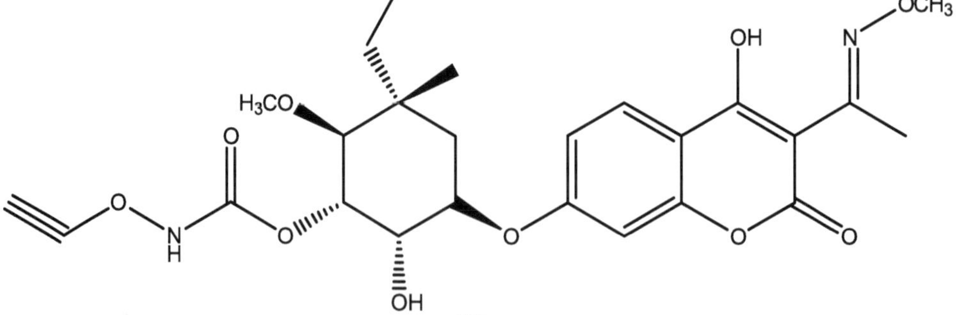

**39**

## Monoamine Oxidase Inhibitor Activity

Coumarins linked with dihydrofuranyl (40) have been studied for monoamine oxidase (MAO) inhibitors have been proved to be the most important drugs for the clinical management of depression and Alzheimer disease. MAO inhibitory activity was found to be a function of drug concentration showing a specific IC50 value[41].

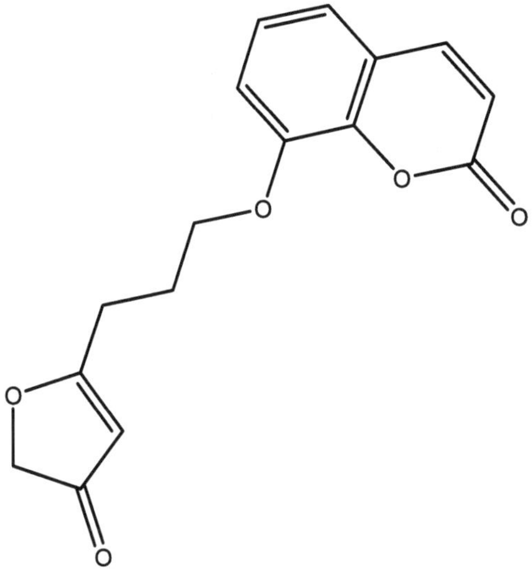

**40** 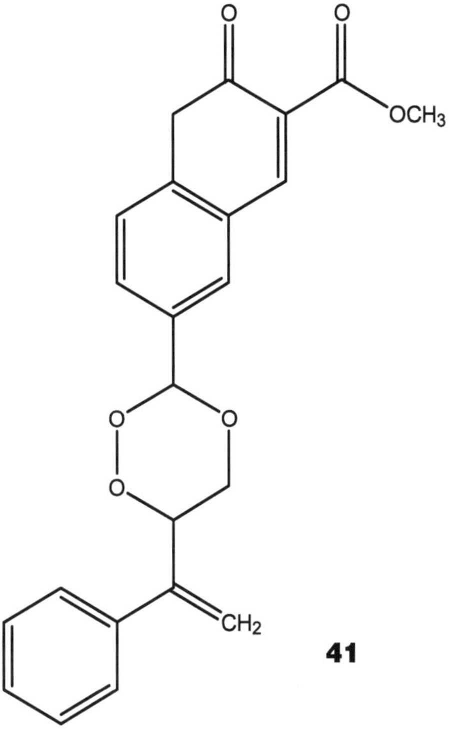 **41**

## Anti Malarial Activity

Increase in deaths as a result of malaria is attributed to the spread of *Plasmodium falciparum* strains resistant to the antimalarial chloroquine. For tackling of drug resistance, the concept of hybrid antimalarials molecules act as multiple therapeutic targets due to two different, covalently fused pharmacophores. This concept led a new series of coumarin-linked trioxane (41) and been evaluated for their antimalarial activity against a sensitive strain of P. falciparum[42].

## Antifungal Activity

The antifungal activity has been studied for 4- aminomethyl coumarin crown ether derivatives. Compound (42) has been explained in terms of the size-fit concept which states that crown ether forms a more stable complex with the cation which is more suitable in size for the crown ether cavityCoumarins fused with thiazoles (43) have been assessed against a fungal strain of *Candida albicans* [87].

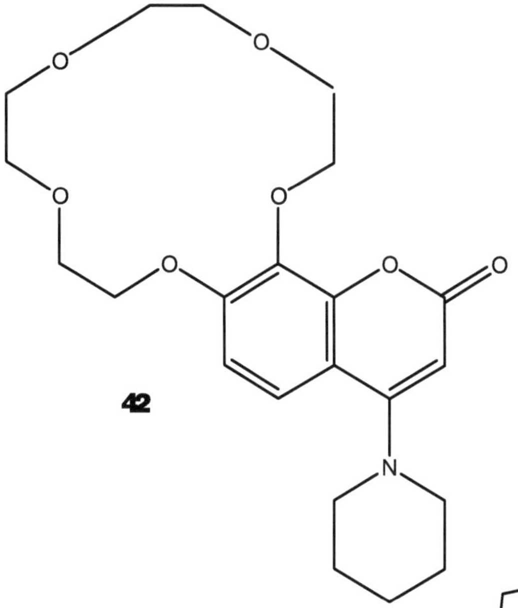

Osthole (14) exhibited wide spectrum of antifungal activity against important plant pathogens such as *Rhizoctonia solani, Phytophtora capsici, Botrytis cinerea, Sclerotinia sclerotiorum,* and *Fusarium graminearum* [43]

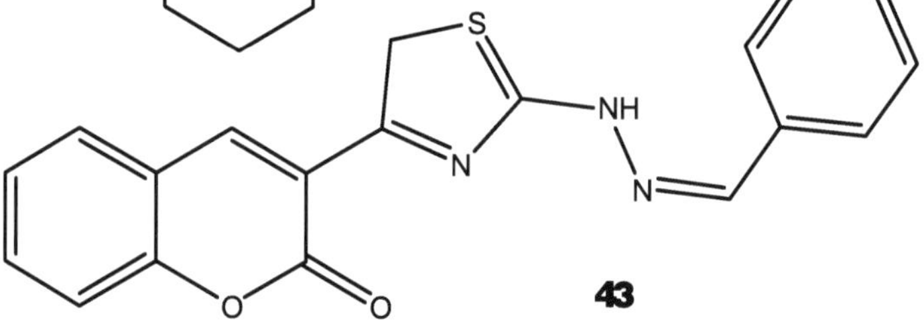

## Anti-Inflammatory Activity

The activity of coumarinyl ethers such as coumarin-linked to benzofuran (44) and coumarin-linked triheterocyclic thiazoles (46) was studied using carrageenan-induced paw edema in rats. The inhibition of edema was evaluated by the comparison of paw volume measured with plethysmometer coupled to a peristaltic pump, immediately before and 3 hrs after the injection of carrageenan [44] (46) found to be slow acting anti-inflammatory agents [45]. Styryl carbonyl derivatives (45) possess appreciable anti-inflammatory activity. Coumarin and related derivatives have been used as inhibitors of lipoxygenase (LOX) and cyclooxygenase (COX) in arachidonic acid pathway [46].

A novel series of 5-(substituted)aryl-3-(3 coumarinyl)-1-phenyl-2- pyrazolines (47) were synthesized compounds having 4-Cl-$C_6H_4$ and 2,4-$(Cl)_2$-$C_6H_3$ showed considerable activity in chronic inflammation such as adjuvant induced arthritis and were compared with diclofenac (13.5 mgkg–1) as a standard drug.

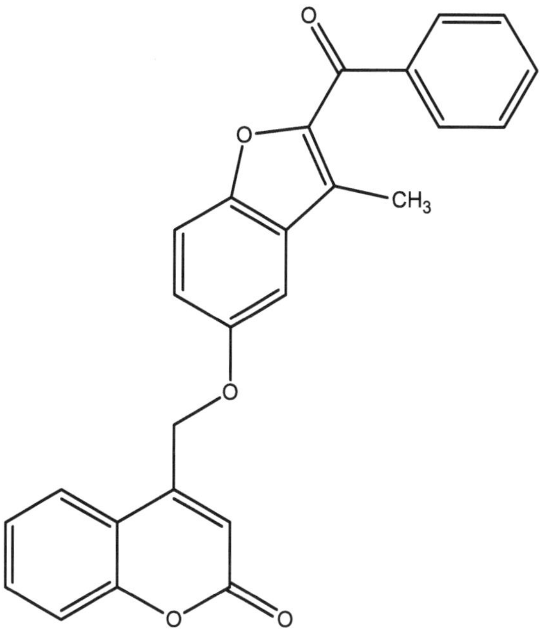

O
O
$CH_3$
O
O
O
**44**

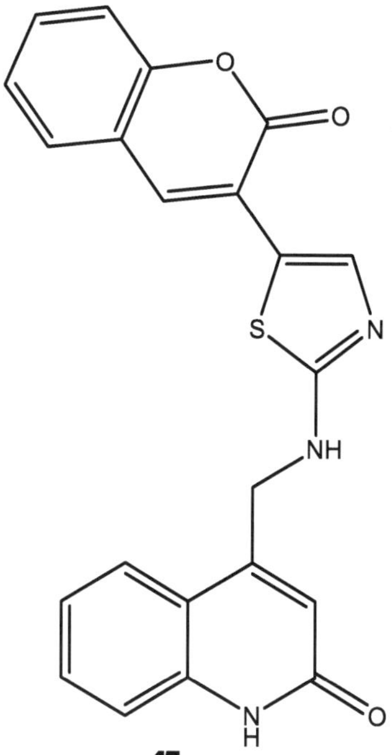

O
O
S
N
NH
N
H
O
**45**

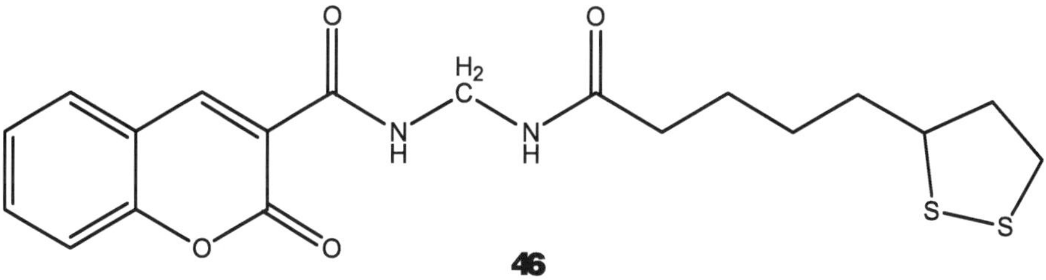

O
$H_2$
C
O
N
H
N
H
O
O
S
S
**46**

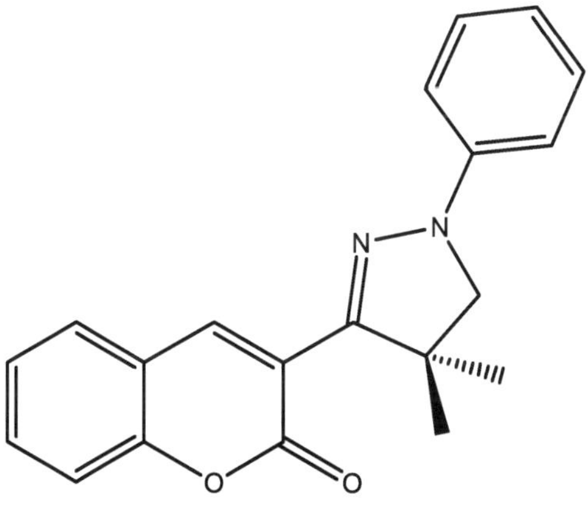

N
N
O
O
**47**

Coumarin-based carbamates (48) exhibited potent inhibitory activity of TNF-α (Tumor necrosis factor) which is a pro inflammatory cytokine secreted by a variety of cells, including monocytes and macrophages, in response to many inflammatory stimuli or external cellular stress[48]. Substitution at C-6 position of the coumarin ring system was found to have the most influence on TNF-α inhibitory activity. This observation led to the discovery of 6-halo derivatives with influential activity[49].

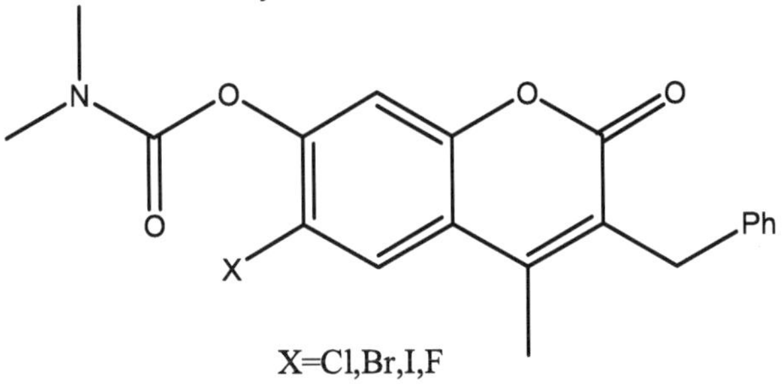

X=Cl,Br,I,F

**48**

## Analgesic and Ulcerogenic Activity

A novel series of 3-(2-amino-6-pyrimidin-4-yl)-6-bromo-2H-chromen-2-one (49) from 3-acetyl-6-bromo-2H-chromen-2-one were screened for in-vivo analgesic activity at a dose of 20 mg/kg body weight. Among them, compounds having o-chloro, m-chloro and m-bromo phenyl exhibited significant analgesic activity and compounds having 2,4-dichloro and 2,6-dichloro phenyl exhibited highly significant activity comparable with standard drug-Diclofenac sodium using acetic acid induced writhing model. Compounds having o-chloro phenyl, 2,4-dichloro and 2,6-dichloro phenyl were further evaluated for acute-ulcerogenic activity. Among them, compound having 2,6-dichloro phenyl was found to be most promising analgesic agent devoid of ulcerogenic effects[50].

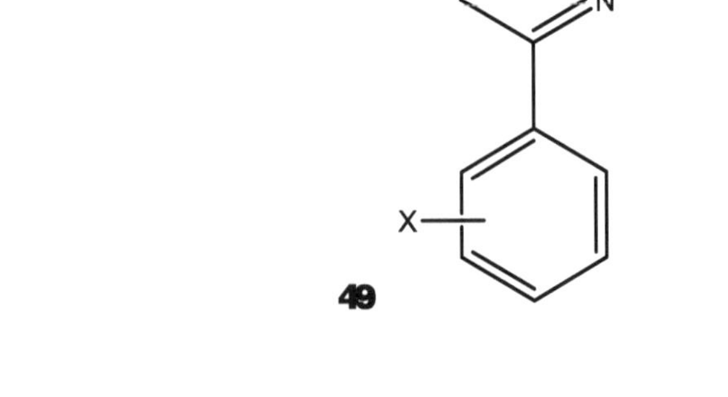

**49**

## Anticonvulsant Activity

A series of hetero aryl semicarbazones (50) were tested for anticonvulsant activity utilizing pentylenetetrazole induced seizure (PTZ) and maximal electroshock seizure (MES) tests at 30, 100 and 300 mg/kg dose levels. Neurotoxicity of the compounds was also assessed at the same dose levels. Three compounds having 3,4-Cl.$C_6H_3$, 2-$OCH_3$.$C_6H_4$ and 4-Br.$C_6H_4$ exhibited significant anticonvulsant activity at 30 mg/kg dose level comparable to the standard drug-phenytoin[51].

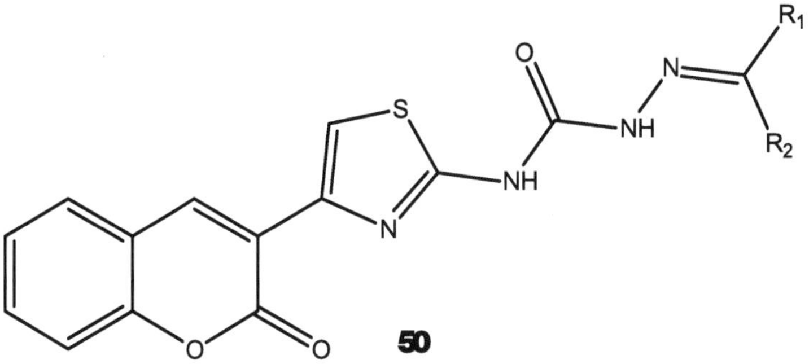

50

Imperatorin (22) showed anticonvulsant action in mice and the ED50 values ranged between 167 and 290mg/kg. Osthole (14) exhibited anticonvulsant action in mice and the ED50 values ranged between 253 and 639mg/kg and the acute eurotoxic effects with the TD50 values ranged between 531 and 648 mg/kg [52].

## Antihyperlipidemic Activity

A series of novel coumarin bisindole heterocycles (51) evaluated for antihyperlipidemic activity that have both coumarin and indole entities in one molecule. In these compounds, as far as coumarin pharmacophore is considered, it revealed that the substitution at position 3 play a pivotal role and the presence of ethyl ester over methyl is preferred for pronounced activity. On the other hand, cursory look at the lower indole pharmacophore highlighted that the unsubstituted indoles have good activity profile compared to substituted indoles. Among twelve compounds tested, one compound having R = -$C_2H_5$ and $R_1$ = $R_2$ = -H showed potent activity and was found to decrease the plasma triglyceride levels (TG) by 55 per cent, total cholesterol (TC) by 20 per cent, accompanied by an increase in HDL-C/

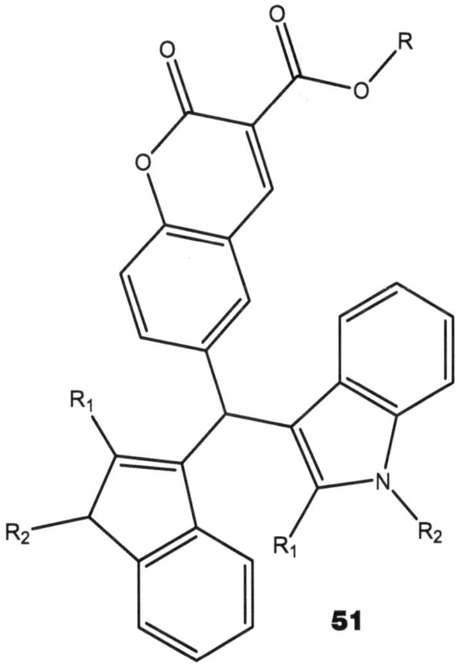

51

TC ratio by 42 per cent in hyperlipidemic rats to a greater degree than some of the reference statins[53].

## Anti-Parkinsonism Activity

A new series of 8-bromo-6-methyl-3-phenylcoumarin derivatives(52) were evaluated as MAO-A and MAO-B inhibitors using R-(-)-deprenyl (selegiline) and Iproniazide as reference inhibitors, most of them showing MAO-B inhibitory activity in the nanomolar range. Compound with 3-methoxy group, loses activity and selectivity in respect to the mono and dimethoxy derivatives. The substitution of methoxy group at 4′ position (IC50 = 3.23±0.49nM) found to be most potent molecule. These compounds did not showed MAO-A inhibitory activity for the highest concentration tested (100 μM)[54].

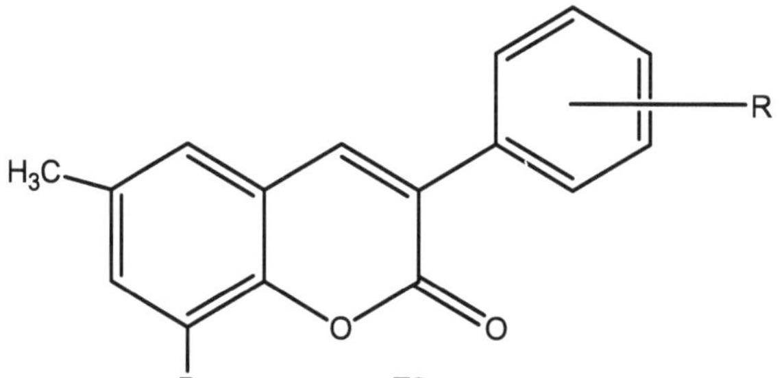

$H_3C$ R O O Br

**52**

## Anti Pyretic Activity

A novel series of 4-[4-(6-phenyl-pyrimidin-4-yl)-phenoxymethyl]-chromen-2-ones were Synthesized by linking two active pharmacophores pyrimidines and coumarins which are likely to exhibit interesting biological activity of anti pyretics activity. Compounds 53a, 53b and 53c exhibited significant anti-pyretic activities comparable with standard drug aspirin. Amongst the tested compounds, those containing $-NH_2$ functional group at 2-position of pyrimidine ring increases their analgesic and anti-pyretic activities [55].

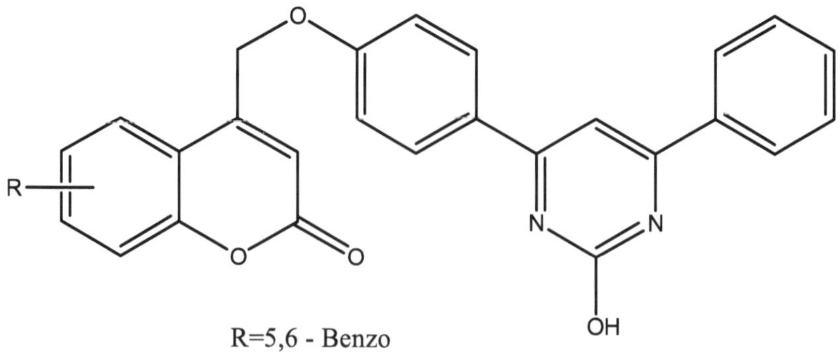

O R O O N N OH

R=5,6 - Benzo

**53a**

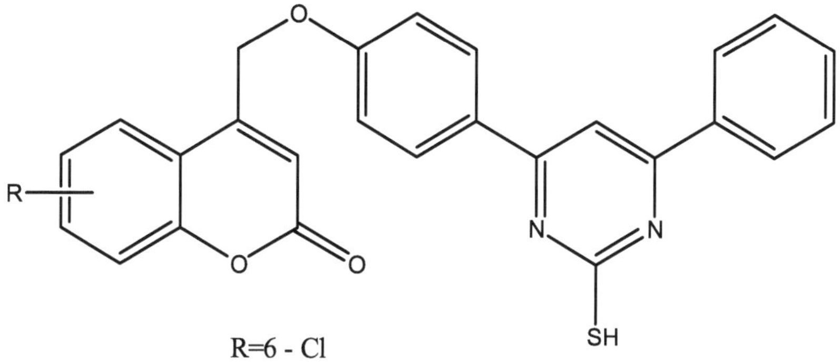

R=6 - Cl

**53b**

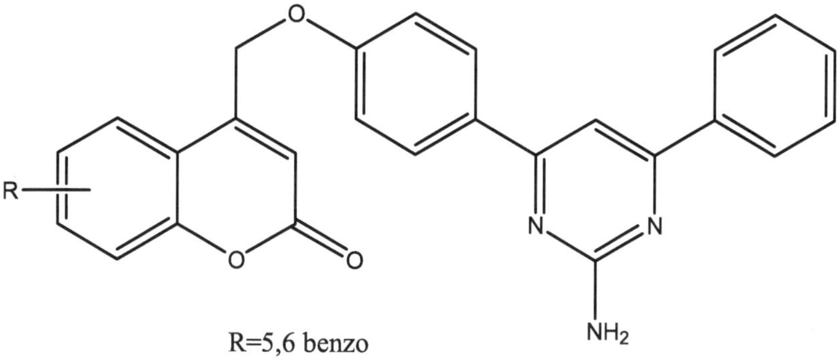

R=5,6 benzo

**53c**

## Anti Hepatitis and Hepatoprotective Activity

A series of substituted benzimidazole– coumarin conjugated compounds were synthesized. Among the various synthesized derivatives of benzimidazole–coumarin conjugates 72a and glucoside 72b inhibited HCV replication with $EC_{50}$ values of 3.4 μM and 4.1 μM, respectively. It was found that enhancement of the selectivity from substituents in the coumarin nucleus was found of the order OMe<H<Br; attachment of the $-CH_2-$coumarin moiety onto the benzimidazole-2-thione nucleus connected with a pyranose enhanced the HCV inhibitory activity by >9.4-fold and selectivity by 6.7-fold; incorporation of a β-d-glucose peracetate moiety into the benzimidazole–

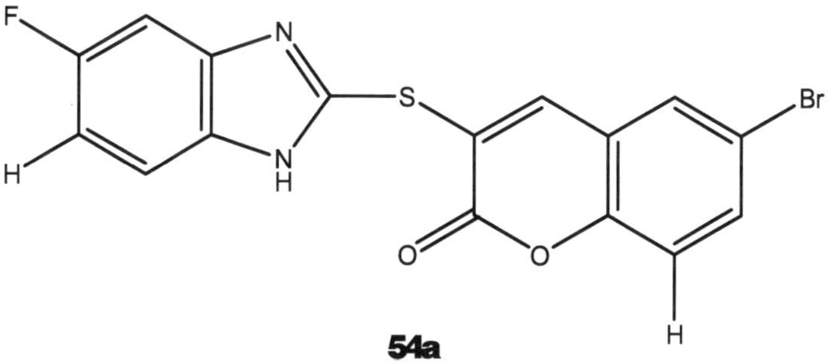

**54a**

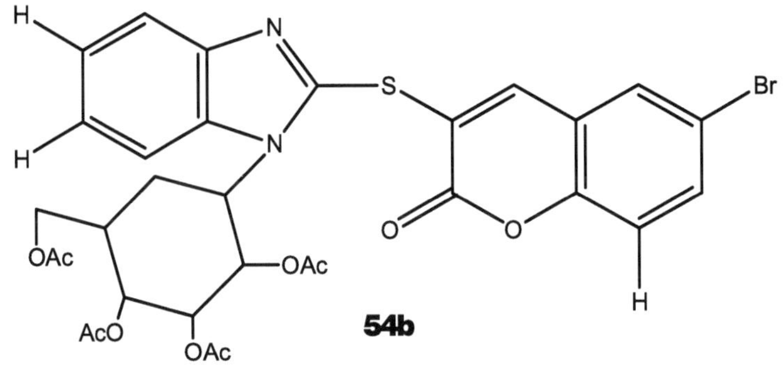

H
N
S
Br
H
N
O
O
OAc
OAc
H
AcO
OAc
**54b**

coumarin conjugate resulted in a 4.8-fold increase in anti-HCV activity. Thus compounds 54a and 54b may be considered as potential lead anti-HCV compounds for further selectivity optimization against hepatitis-C virus. [56]

It has been found that coumarin-linked benzoxazole-5-carboxylic acids (55) showed Inhibitory activity against hepatitis virus RNA polymerase. The thio methylene linker has been used to connect a heterobicycle with various aromatic rings by synthetic methods to form hybrid compounds for antiviral bioassays. The coumarin can be linked with heterobicycles like benzimidazole, imidazopyridine, purine, benzoxazole, and benzothiazole to evaluate the activity[57].

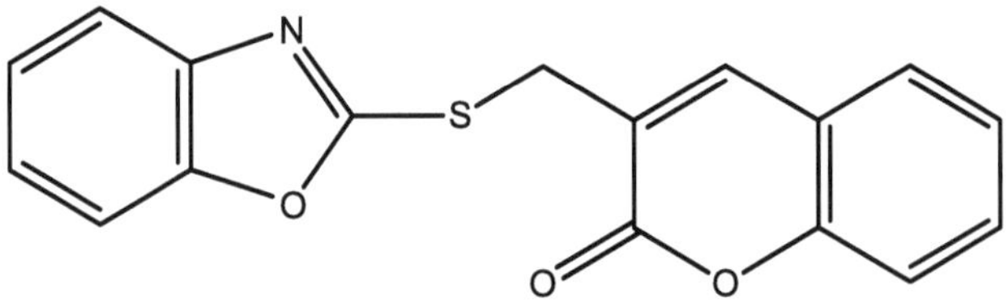

N
S
O
O
O

**55**

## Coumarins for Anticoagulant Activity

Dicoumarol (56) was found in sweet clover and exhibited anticoagulant activity [58].Coumarins are vitamin K antagonists that produce their anticoagulant effect by interfering with the cyclic interconversion of vitamin K and its 2,3 epoxide (vitamin K epoxide) [59]. Vitamin K is a cofactor for the posttranslational carboxylation of glutamate residues to 5-carboxyglutamates on the N-terminal regions of vitamin K-dependent proteins (Figure 17.1) [60–65].

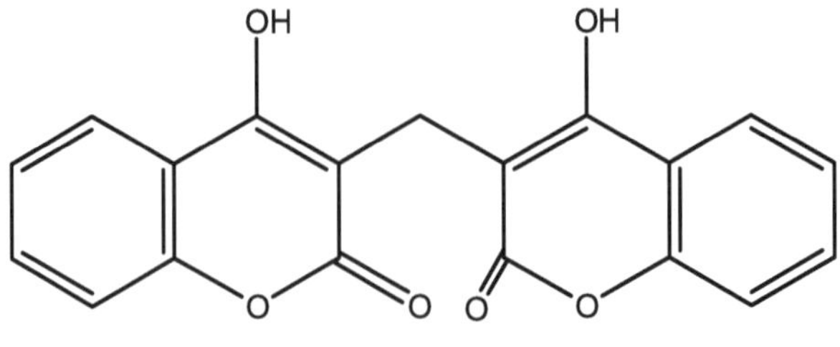

OH
OH
O
O
O
O

**56**

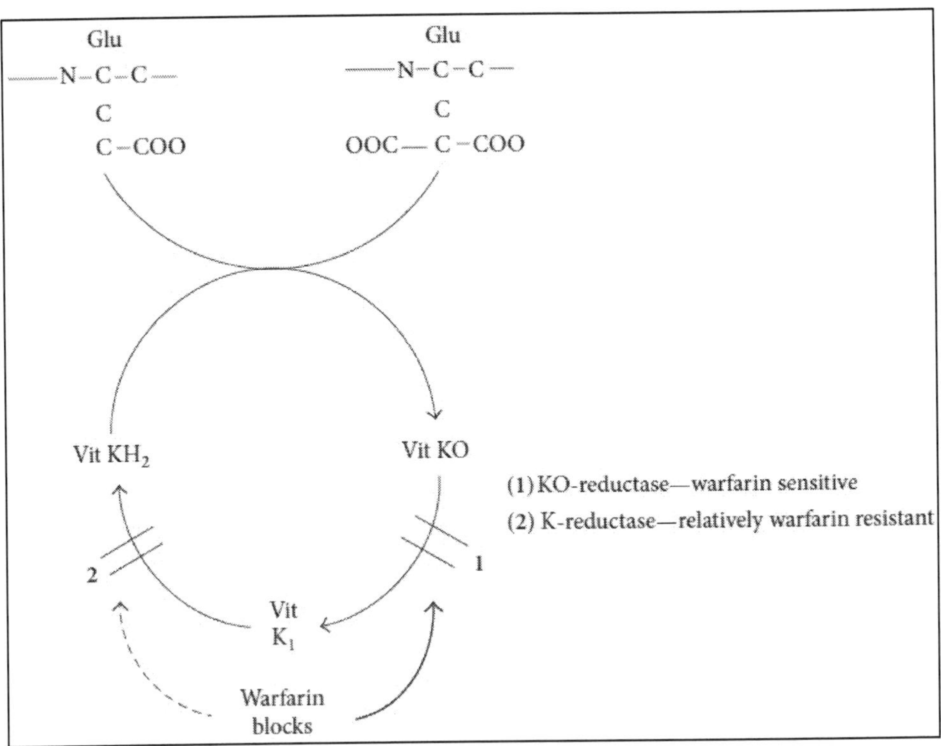


**Figure 17.1**

## Coumarins as Antihypertensive Activity

Antihypertensive effect of dihydromammea C/OB that has been isolated from the seeds of the West African tree Mammea.Scopoletin (57) was isolated form the fruits of Tetrapleura tetraptera TAUB (Mimosaceae) and it produces hypotension in laboratory animals in vitro and in vivo through its smooth muscle relaxant activity [66]. Visnadine (**58**), an active ingredient extracted from the fruit of Ammi visnaga,

OH (CH2)4CH3 HO O O O CH3 H3C

**57**

H3CO HO O O

**58**

exhibited peripheral and coronary vasodilator activities. Khellactone (59) was isolated from Phlojodicarpus sibiricus and it exhibited vasodilatory action.

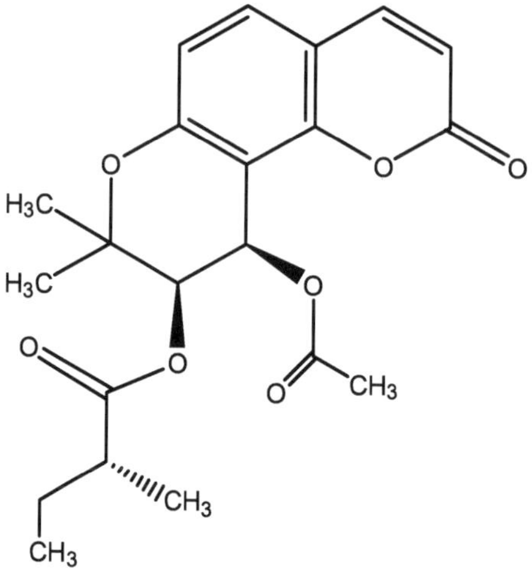

59

## Anti-Tubercular Activity

The compounds scopoletin (57) and umbelliferone (60) are found to be active against Mycobacterium tuberculosis $H_{37}$ Rv with MIC values of 42 and 58.3 5g/mL, respectively [67]. Compounds phellodenol A(61), (+)-(S)-marmesin (62) and xanthyletin (65) exhibited activity at 60 5g/mL and the remaining compounds exhibited activity at more than 119 5g/mL. Umbelliferone (60), phellodenol A (61), psoralen (64) and scopoletin (57), bergapten (63), (+)-(S)-marmesin (62), (+)-(S)-rutaretin (66), and

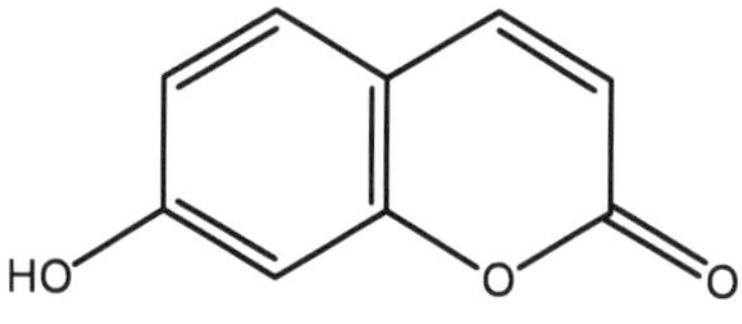

60

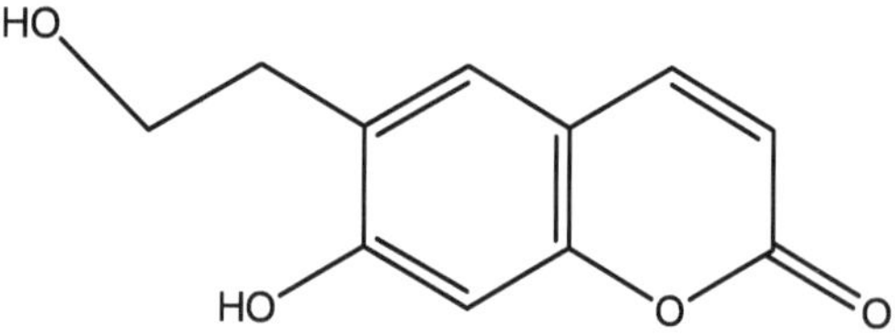

61

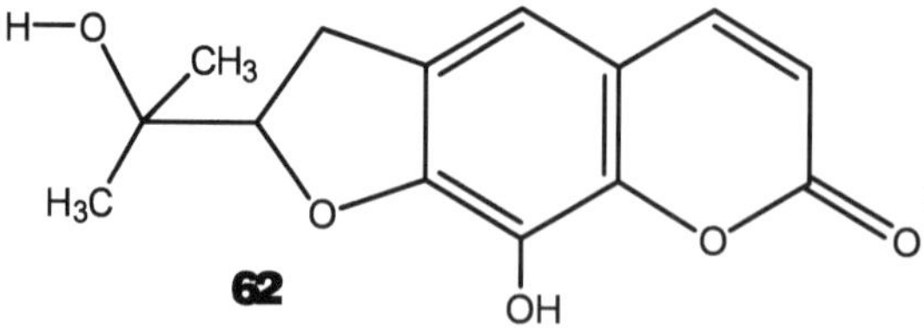

62

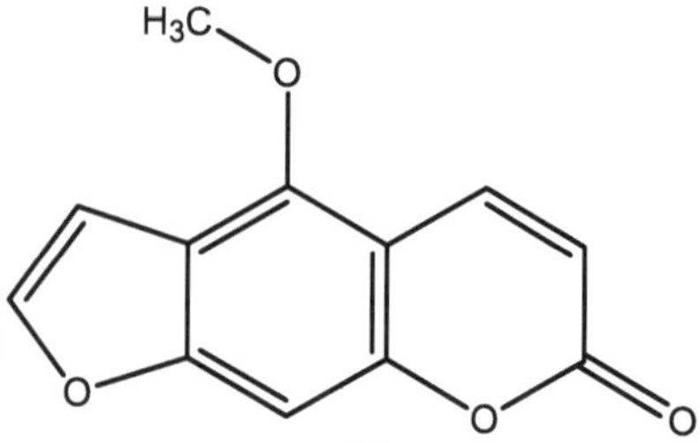

63

xanthyletin (63) were isolated from the whole plants of Fatoua pilosa. Phellodenol A(61) was also isolated from the leaves of *Phellodendron amurense* var. *wilsonii* [68].

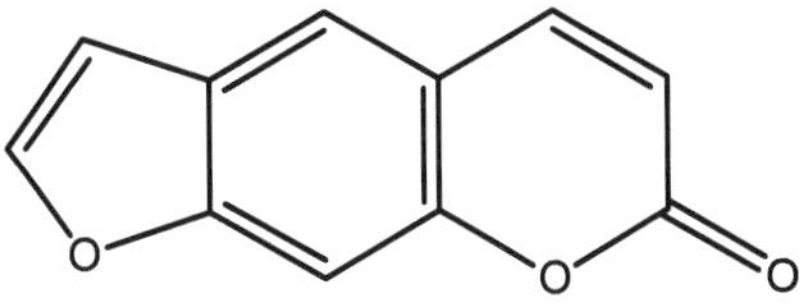

64

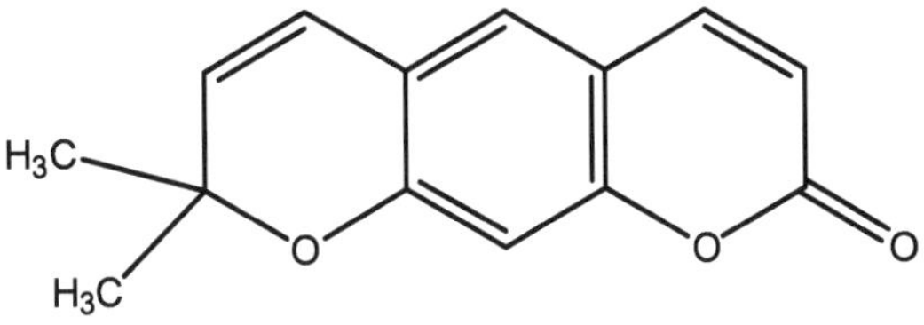

65

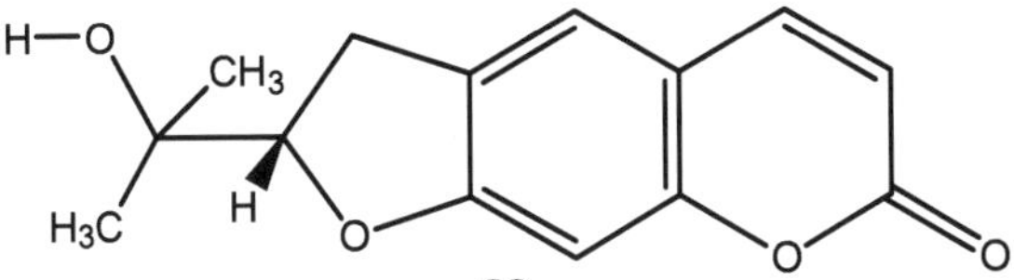

66

## Coumarins for Multiple Sclerosis

Osthole (14) is a bioactive coumarin derivative extracted from medicinal plants such as *Angelica pubescens, Cnidium monnieri,* and *Peucedanum ostruthium*. Osthole (14) could be a potential therapeutic agent for the treatment of multiple Sclerosis [69].

## Coumarins for Antihyperglycemic Activity

Fraxidin (42) inhibited the formation of inducible nitric oxide synthase [70] and showed antihyperglycemic activity [71].

## Coumarins for Neuroprotective Activity

Esculetin (23) also exhibited neuroprotective effects on cerebral ischemia/ reperfusion injury in a middle cerebral artery occlusion model in mice at 20 5g/mL and was administered intracerebroventricularly at 30 min before ischemia [72].

## Coumarins as Phytoalexins

Phytoalexins are produced in plants in response to fungal infection, physical damage, chemical injury, or a pathogenic process and inhibit or destroy the invading agents such as bacteria, insects, and viruses. Phytoalexins are oxygenated coumarin derivatives. Ayapin (67) is one among them.

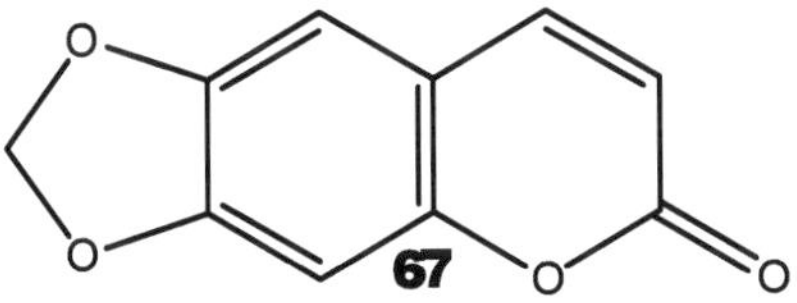

67

## Platelet Aggregation Inhibitor Activity

Coumarin-linked piperazine moiety (68) show notably high activity towards all the platelet aggregation inducers used, and the most active one, 8-methyl-4-(1-

piperazinyl)-7-(3- pyridylmethoxy)coumarin, proved to be a potent in-vitro antiplatelet agent. The screening of this compound can be carried out by conducting test in which human blood from healthy volunteers was added to a 130 mM trisodium citrate aqueous solution, then centrifuged to give platelet-rich plasma (PRP) which was diluted with platelet-poor plasma (PPP) and pre incubated at 37°C with drug solution before the addition of the platelet aggregation agent. PRP aggregation was induced by inducers mentioned. The compound 8-methyl-4-(1-piperazinyl)-7-(3 pyridylmethoxy)-coumarin proved to be a potent *in vitro* inhibitor of human platelet aggregation towards all the platelet aggregation inducers used, suggesting proper heteroarylmethoxy groups as novel useful 7- alkoxy substituents[73].

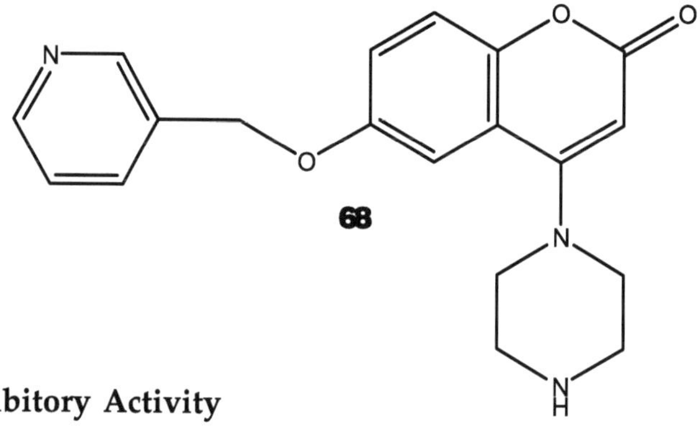

68

## TNF-α Inhibitory Activity

A coumarin-linked with piperidinylmethyl (69), dimethyl-carbamic acid 3-benzyl-4-methyl-2-oxo-2H-chromen-7-yl ester has been found to be a moderate inhibitor of TNF-α. Which is secreted by a variety of cells, including monocytes and macrophages, in response to many inflammatory stimuli or external cellular stress. It is a key cytokine in the inflammation cascade, causing the production and/or release of other cytokines and agents. TNF-a inhibitory activity was assessed by in vivo inhibition of serum TNF-production in the rats[74].

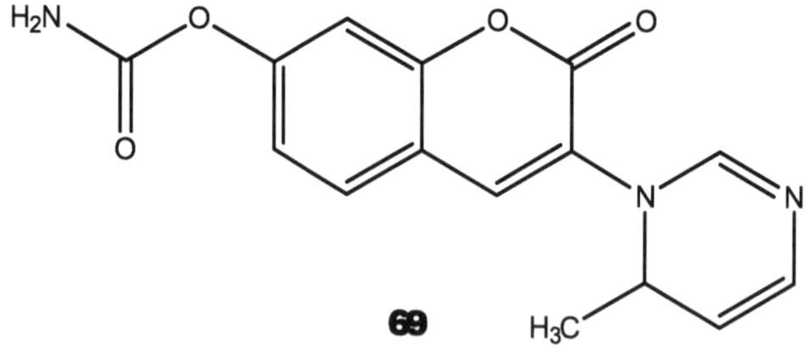

69

## BASE-1 Inhibitory Activity

Coumarin-linked with substituted phenyl-piperazine (70) have good β- secretase inhibitory activity like new nonpeptidomimetic derivatives which incorporate in their structure various heterocyclic moieties: coumarin, quinoline or chromene. These moieties have been linked through an amide bond to form phenyl-piperazine scaffold which have shown modified BACE inhibitory properties[75].

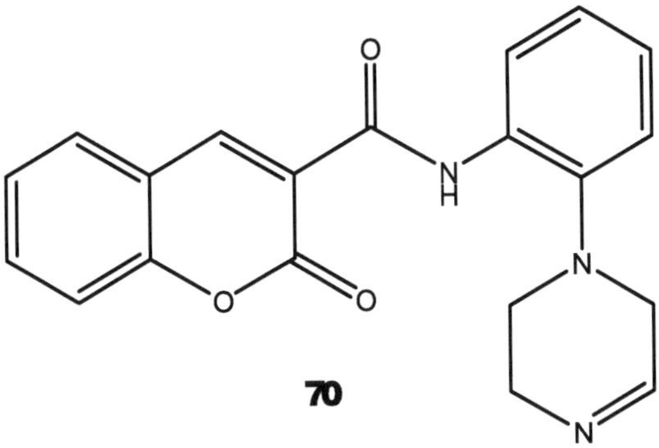

**70**

## Lipid Lowering Activity

Atherosclerotic cardiovascular disease is the leading cause of death in developed countries. Epidemiological studies have identified a variety of independent risk factors that contribute to coronary artery disease (CAD), most notable are high low density lipoproteins (LDL)-cholesterol, triglyceride (TG) and low high density lipoproteins (HDL)-cholesterol Current therapies mostly focus on lowering LDL-cholesterol and statins used for this purpose are very effective and also safe. Yet they are not effective in lowering TG and HDL. Plasma triglyceride lowering activity of coumarin-linked to heterocycle was determined.The compound (71) has shown maximum (45 per cent) plasma triglycerides lowering activity and exhibits significant reduction in TG at 3 mg/kg/day dose[76].

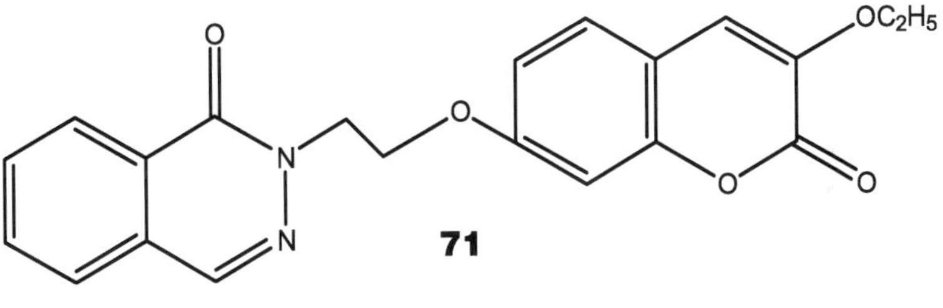

**71**

## Casein Kinase-2 Inhibitory Activity

Casein kinase-2 (CK2) is probably the most pleiotropic protein kinase known, with more than 300 protein substrates already recognized, active CK2 is ubiquitous, essential, and implicated in a wide variety of important cell functions, evidence has been accumulating that its catalytic subunits may behave as oncogenes, consistent with the observation that they display an anti-apoptotic effect in prostate cancer cell

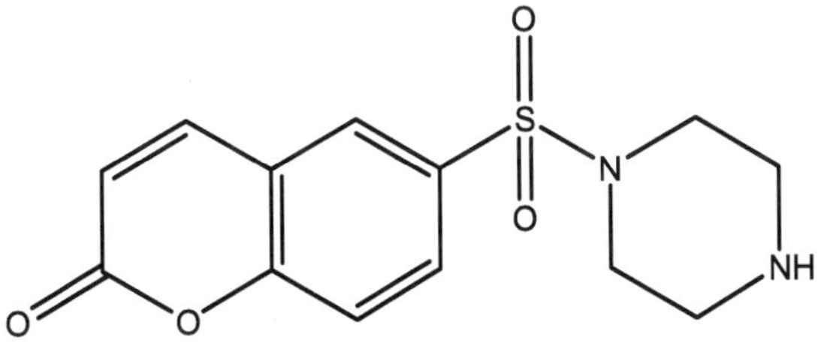

**72**

lines. The most promising inhibitor, coumarin linked piperazine (72) has been also crystallized in complex with CK2, and the experimental binding mode has been used to derive a linear interaction energy (LIE) model [77].

## Cytochrome P450 Enzyme Inhibitory Activity

Cytochrome P450 (CYP) is a large family of hemoproteins present in most forms of life (plants, bacteria, and mammals). They are concerned with metabolism in vital processes and many of them with the ability to activate carcinogens have been implicated as risk factors for cancers Coumarin-linked imidazoles (73) have been proved to be most promising compounds as CYP450 inhibitors. It has been shown to display stronger CYP19 inhibitory activity (IC50 = 47 nM) than the standard drug fadrozole (IC50=52 nM) with excellent selectivity over CYP11B1 and CYP11B2[78].

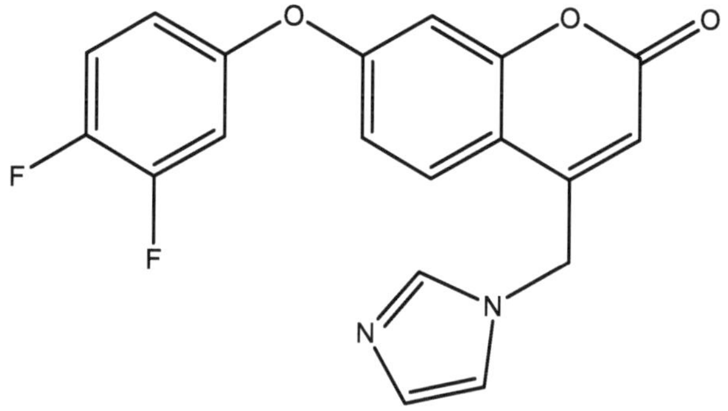

**73**

## Tyrosinase Inhibitor Activity

A series of coumarin-resveratrol hybrids (74) were synthesized.

Tyrosinase activity assays were performed with L-DOPA as substrate with slight modifications and activity of mushroom tyrosinase was determined by spectrophotometric technique. IC50 values of these compounds were measured. The results showed that these compounds exhibited tyrosinase inhibitory activity. 3-(3′,4′,5′-trihydroxyphenyl)-6,8-dihydroxycoumarin was found to be the most potent

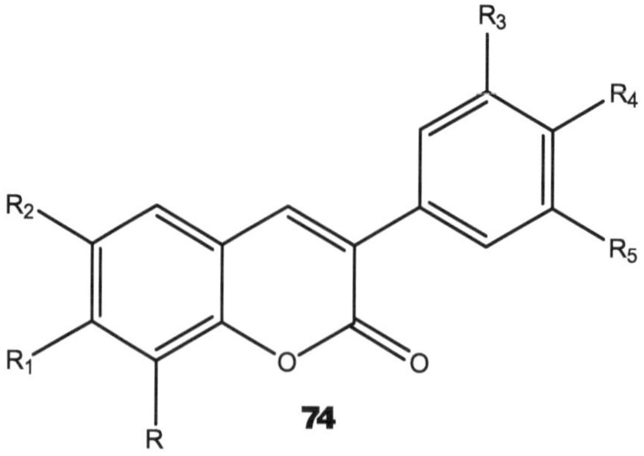

**74**

compound (0.27 mM) more than umbelliferone (0.42 mM) used as reference compound. The kinetic studies revealed that this compound caused non-competitive tyrosinase inhibition and the number and the position of free hydroxyl groups play an important role in determining the activity[79].

## References


1. Y. Aoyama, T. Katayama,M. Yamamoto,H. Tanaka, and K. Kon,"A new antitumor antibiotic product, demethylchartreusin.Isolation and biological activities," *The Journal of Antibiotics*,vol. 45, no. 6, pp. 875–878, 1992.
2. M. Iranshahi, M. Askari, A. Sahebkar, and D. Hadjipavlou-Litina, "Evaluation of antioxidant, anti-inflammatory and lipoxygenase inhibitory activities of the prenylated coumarin umbelliprenin," *DARU*, vol. 17, no. 2, pp. 99–103, 2009.
3. W. C. Evans, *Trease and Evans Pharmacognosy*, Elsevier Ltd., 16th edition, 2009.
4. J. A.Mead, J. N. Smith, and R. T.Williams, "Studies in detoxication.72. The metabolism of coumarin and of o-coumaric acid,"*The Biochemical Journal*, vol. 68, no. 1, pp. 67–74, 1958.
5. C. Spino, M. Dodier, and S. Sotheeswaran, "Anti-HIV coumarins from calophyllum seed oil," *Bioorganic and Medicinal Chemistry Letters*, vol. 8, no. 24, pp. 3475–3478, 1998.
6. J. Choi, K. T. Lee, H. Ka,W. T. Jung, H. J. Jung, and H. J. Park, "Constituents of the essential oil of the *Cinnamomum cassia* stem bark and the biological properties," *Archives of Pharmacal Research*, vol. 24, no. 5, pp. 418–423, 2001.
7. Shaabani A, Ghadari R, Rahmati A, Rezayan AH. Coumarin Synthesis via Knoevenagel Condensation Reaction in 1,1,3,3-N,N,N',N'-*Tetramethyl guanidinium* Trifluoroacetate Ionic Liquid. *J Iran Chem Soc* 2009; 6(4): 710-714
8. K. C.Fylaktakidou; D. J.Hadjipavlou-Litina; K. E. Litinas; D. N. Nicolaides; *Curr. Pharm. Des*. 2004, 10, 3813.
9. F.Borges; F. Roleira; N. Milhazes;, L. Santana; E. Uriarte; *Curr. Med. Chem*.2005, 12,887.
10. C. Kontogiorgis; D. Hadjipavlou-Litina, *J. Med. Chem*. 2005, 48, 6400.
11. A.Lacy; R. O'Kennedy, *Curr. Pharm. Des*. 2004, 10, 3797.
12. I. Kostova; *Curr. Med. Chem*, 2005, 5, 29.
13. B. Ramesh.; K. V. Pugalendi; *J. Med. Food*, 2006, 9, 562.
14. A. Chilin; R. Battistutta; A. Bortolato; G. Cozza; S. Zanatta; G. Poletto; M. Mazzorana; G. Zagotto;, E. Uriarte; A. Guiotto; L. A. Pinna; F. Meggio; S. Moro; *J. Med. Chem*. 2008, 51, 752.
15. Rajithaa B, Kumara VN, Someshwara P, Madhava JV, Reddy PN, Reddy YT. Dipyridine copper chloride catalyzed coumarin synthesis via Pechmann condensation under conventional heating and microwave irradiation. *ARKIVOC* 2006; xii: 23-27.

16. S.F.Razavi, M.Khoobi, H. Nadri, A.Sakhteman, A.Moradi, S.Emami, A.Foroumadi, A.Shafiee. *Eur. J. of Med. Chem.*, 64 (2013) 252-259.

17. A.A.Emmanuel-Giota, D.N.Nicolaides, K.C.Fylaktakidou. *J. Heterocyclic Chem.*, 38 (2001) 717-721.

18. H. Osman, A. Arshad, C.Kit-Lam, M.Bagley. *Chemistry Central Journal*, (2012) 1-10.

19. Melagraki G, Afantitis A, Markopoulou OI, Detsi A, Koufaki M, Kontogiorgis C, Synthesis and evaluation of the antioxidant and anti-inflammatory activity of novel coumarin-3-aminoamides and their alpha-lipoic acid adducts, *Eur J Med Chem*, 2009, 44, 3020-3026.

20. Marina Roussaki; Christos A. Kontogiorgis; Dimitra Hadjipavlou-Litina; Stylianos Hamilakis;Anastasia Detsi; Bioorganic and Medicinal Chemistry Letters; 2010, xxx, xxx–xxx.

21. Theodoros Symeonidis; Michael Chamilos; Dimitra J. Hadjipavlou-Litina,; Michael Kallitsakis; Konstantinos E. Litinas; *Bioorganic and Medicinal Chemistry Letters*; 2009, 19, 1139–1142.

22. Naceur Hamdi; M. Carmen Puerta; Pedro Valerga; *European Journal of Medicinal Chemistry*; 2008, 43,2541-2548.

23. W. K. Whang, H. S. Park, I. Ham *et al.*, "Natural compounds, fraxin and chemicals structurally related to fraxin protect cells from oxidative stress," *Experimental and Molecular Medicine*, vol. 37, no. 5, pp. 436–446, 2005.

24. X.H.Liu, H.F.Liu, H.L.Zhu. *Bioorg. Med. Chem. Letters*, 20 (2010) 5705-5708.

25. Y.Chen, H.R.Liu, K.H.Lee, P.Xia, *Eur. J. of Med. Chem.*, 2012, 49, 74-85.

26. Budzisz E, Brzezinska E, Krajewska U, Rozalski M, Cytotoxic effects, alkylating properties and molecular modelling of coumarin derivatives and their phosphonic analogues, *Eur J Med Chem*,2003, 38, 597-603.

27. Seokjoon Lee; Krishnamoorthy Sivakumar; Woon-Seob Shin; Fang Xie; Qian Wang; *Bioorganic and Medicinal Chemistry Letters*; 2006, 16, 4596–4599.

28. Al-AmieryAA, Kadhum AAH, Mohamad AB,Antifungal activities of new coumarins, *Molecules*, 17, 2012, 5713- 5723.

29. Sahoo SS, Shukla S, Nandy S, Sahoo HB, Synthesis of novel coumarin derivatives and its biological evaluations, *Eur Journal of Experimental Bio*, 2(4), 2012, 899-908.

30. Vyas KB, Nimavat KS, Joshi KM, Jani GR, Synt hesis of 3-{(3-(2′-nitrophenyl)}-prop-2-enoyl]-4-hydroxy-6-methyl-2Hchromene-2-one andits metalcomplexes asanantimicrobial agent,*Jrnl of Chemical and Pharm Res*, 4(5), 2012, 2720-2723.

31. Kudale SD, Deodhar MN, Synthesis and evaluation of some coumarin containing potential antimicrobial agents, *E-Journal of Chemistry*, 9(4), 2012, 2493-2500.

32. RamaGanesh CK, Yadav DB, Venkatesh KB, Synthesis and biological evaluation of some innovative coumarin derivatives containing thiazolidin-4-one ring, Ind *Jrnl of Chem*, 49B, 2010,1151-1154.

33. Brahmbhatt DI, Kaneria AR, Patel AK, Patel NH, Synthesis and antimicrobial screening of some 3-4-(3-aryl-1-phenyl-1H-pyrazol-4-yl)-6-aryl-pyridin-2-yl. and 4-methyl-3- phenyl-6-4-(3- aryl-1-phenyl-1H-pyrazol-4-yl)-6-aryl-pyridin-2-yl]coumarins, *Ind Jrnl of Chem, 49B*, 2010, 971-977.

34. M ladenovic M, Vukovic N, Sukdolak S, Solujic S, Design of novel 4-hydroxy-chromene-2-one derivatives as antimicrobial agents, *Molecules*, 15, 2010, 4294-4308.

35. Porwal B, Jayashree BS, Attimarad M. Synthesis of some new 3-coumarinoylpyridinium and quinolinium bromides for their antimicrobial activity. *Jrnl Basic Clin Pharm*, 2009; 1(1), 29-32.

36. Basanagouda M, Kulkarni MV, Sharma D, Gupta VK, Pranesha, Sandhyarani P, Rasal VP, Synthesis of some new 4-aryloxmethylcoumarins and examination of their antibacterial and antifungal activities, *J Chem Sci*, 121(4), 2009, 485–495.

37. Bhat MA, Siddiqui N, Khan SA, Mohamed MI, Synthesis of triazolothiazolidi none derivatives of coumarin with antimicrobial activity, *Acta Poloniae Pharmaceutican Drug Res*, 66(6), 2009, 625-632

38. Creaven BS, Devereux M, Karcz D, Kellett A, McCann M, Noble A, Walsh M,Copper(II) complexes of coumarin-derived Schiff bases and their anti-Candida activity, *J of Inorg Biochem*,103(9), 2009, 1196-1203.

39. Bairagi S, Bhosale A, Deodhar M N, Design, synthesis and evaluation of schiff's bases of 4-chloro-3-coumarin aldehyde as antimicrobial agents, *E-Journal of Chemistry*, 6(3),2009, 759-762.

40. Musicki B, Periers AM, Laurin P, Ferroud D, Benedetti Y, Lachaud S, Chatreaux F, Haesslein JL, Iltis A, Pierre C, Khider J, Tessot N, Airault M, Demassey J, Dupuis-Hamelin C, Lassaigne P,Bonnefoy A, Vicatc P, Klicha M, Improved antibacterial activities of coumarin antibiotics bearing 50,50-dialkylnoviose: biological activity of RU79115, *Bioorg and Med Chem Lett*, 10, 2000, 1695-1699.

41. P.O.Patil, S.B.Bari, S.D.Firke, P.K.Deshmukh. *Bioorg. and Med. Chem.*, 21 (2013) 2434-2450.

42. K.V.Shashidhara, A.Kumar, R.P.Dodda, N.N.Krishna. Bioorg. *and Med. Chem. Letters*, 22 (2012) 3926-3930.

43. A.Arshad, H.Osman, M.C.Bagley. *Eur. J. of Med. Chem.*, 46 (2011) 3788-3794.

44. C.M.Wang,W. Zhou,C.X. Li, H. Chen,Z.Q. Shi, andY. J. Fan,"Efficacy of osthol, a potent coumarin compound, in controlling powdery mildew caused by Sphaerotheca fuliginea," *Journal of Asian Natural Products Research*, vol. 11, no. 9, pp. 783–791, 2009.

45. M.Ghate, R.A.Kusanur, M.V.Kulkarni. *Eur. J. of Med. Chem.*, 40 (2005) 882-887.

46. R.G.Kalkhambkar, G.M.Kulkarni, H.Shivkumar, R.N.Rao. *Eur. J. of Med. Chem.*, 42 (2007) 1272-1276.

47. G.Melagraki, A.Afantitis, O.Markopoulou, D.J.Hadjipavlou-litina. *Eur. J. of Med. Chem.*, 44 (2009) 3020-3026.

48. R. M.Locksley; N.Killeen; M. Lenardo; *J. Cell*; 2001,104, 487.

49. Jie-Fei Cheng, Mi Chen; David Wallace; Sovouthy Tith,; Thomas Arrhenius; Hirotaka Kashiwagi; Yoshiyuki Ono; Akira Ishikawa,; Haruhiko Sato; Toshiro Kozono; Hediki Sato; Alex M. Nadzan; *Bioorganic and Medicinal Chemistry Letters*; 2004, 14, 2411–2415.

50. Gupta JK, Sharma PK, Dudhe R, Chaudhary A, Verma PK, Synthesis, analgesic and ulcerogenic activity of novel pyrimidine derivative of coumarin moiety, Analele UniversitãÑii din Bucuresti- Chimie (serie nouã), 2010, 19(2), 09-21.

51. Siddiqui N, Arshad MF, Khan SA. Synthesis of some new coumarin incorporated thiazolyl semicarbazones as anticonvulsants. *Acta Poloniae Pharmaceutica-Drug Research*, 2009;66(2): 161-167.

52. J. J. Luszczki, E.Wojda, M. Andres-Mach *et al.*, "Anticonvulsant and acute neurotoxic effects of imperatorin, osthole and valproate in the maximal electroshock seizure and chimney tests in mice: a comparative study," *Epilepsy Research*, vol. 85, no. 2-3, pp. 293–299, 2009.

53. Sashidhara KV, Kumar A, Kumar M, Srivastava A, Puri A, Synthesis and antihyperlipidemic activity of novel coumarin bisindole derivatives, *Bioorg Medicinal Chem Lett*, 2010, 20, 6504-6507.

54. Matos MJ, Vina D, Janeiro P, Orallo F, Uriarte E and Santana L, Synthesis and pharmacological evaluation of coumarins as new scaffold on the Parkinson´s disease, *13th International Electronic Conference on Synthetic Organic Chemistry* (ECSOC-13), 1-30,November 2009.

55. S. Keri Rangappa; M. Hosamani Kallappa, V. Shingalapur Ramya; H. Hugar Mallinath; *European Journal of Medicinal Chemistry*; 2010,45,2597-2605.

56. Jih Ru Hwua; Raghunath Singha; Shih Ching Hong; Yung Hsiung Chang; Asish R. Das; Inge Vliegen; Erik De Clercq; Johan Neyts; *Antiviral Research*; 2008, 77, 157–162.

57. J.Neyts, J.R.Hwu, E.D.Clercq, R.Singha. J. of Med. Chem., 52 (2009) 1486-1490.

58. S. K. Poole and C. F. Poole, "Thin-layer chromatographic method for the determination of the principal polar aromatic flavour compounds of the cinnamons of commerce," *The Analyst*, vol. 119, no. 1, pp. 113–120, 1994.

59. J. Hirsh, J. E. Dalen, D. R. Anderson *et al.*, "Oral anticoagulants: mechanism of action, clinical effectiveness, and optimal therapeutic range," *Chest*, vol. 119, no. 1, supplement, pp. 8S–21S, 2001.

60. G. L. Nelsestuen, T. H. Zytkovicz, and J. B. Howard, "The mode of action of vitamin K. Identification of 5 carboxyglutamic acid as a component of

prothrombin," *The Journal of Biological Chemistry*, vol. 249, no. 19, pp. 6347–6350, 1974.

61. J. Stenflo, P. Fernlund, W. Egan, and P. Roepstorff, "Vitamin K dependent modifications of glutamic acid residues in prothrombin," *Proceedings of the National Academy of Sciences of the United States of America*, vol. 71, no. 7, pp. 2730–2733, 1974.

62. D. S. Whitlon, J. A. Sadowski, and J. W. Suttie, "Mechanism of coumarin action: significance of vitamin K epoxide reductase inhibition," *Biochemistry*, vol. 17, no. 8, pp. 1371–1377, 1978.

63. L. S. Trivedi, M. Rhee, J. H. Galivan, and M. J. Fasco, "Normal and warfarin-resistant rat hepatocyte metabolism of vitamin K 2,3-epoxide: evidence formultiple pathways of hydroxyvitamin K formation," *Archives of Biochemistry and Biophysics*, vol. 264, no. 1, pp. 67–73, 1988.

64. M. J. Fasco, E. F. Hildebrandt, and J. W. Suttie, "Evidence that warfarin anticoagulant action involves two distinct reductase activities," *The Journal of Biological Chemistry*, vol. 257, no. 19, pp. 11210–11212, 1982.

65. I.A.Choonara, R. G.Malia, B. P. Haynes *et al.*, "The relationship between inhibition of vitamin K1 2,3-epoxide reductase and reduction of clotting factor activity with warfarin," *British Journal of Clinical Pharmacology*, vol. 25, no. 1, pp. 1–7, 1988.

66. J. A.Mead, J. N. Smith, and R. T.Williams, "Studies in detoxication. 72. The metabolism of coumarin and of o-coumaric acid," *The Biochemical Journal*, vol. 68, no. 1, pp. 67–74, 1958.

67. C. C. Chiang, M. J. Cheng, C. F. Peng, H. Y. Huang, and I. S. Chen, "Anovel dimeric coumarin analog and antimycobacterial constituents from Fatoua pilosa," *Chemistry and Biodiversity*, vol. 7, no. 7, pp. 1728–1736, 2010.

68. A. J. Cohen, "Critical review of the toxicology of coumarin with special reference to interspecies differences in metabolism and hepatotoxic response and their significance to man," *Food and Cosmetics Toxicology*, vol. 17, no. 3, pp. 277–289, 1979.

69. X. Chen, R. Pi, Y. Zou *et al.*, "Attenuation of experimental autoimmune encephalomyelitis in C57 BL/6 mice by osthole, a natural coumarin," *European Journal of Pharmacology*, vol. 629, no. 1–3, pp. 40–46, 2010

70. N. Y. Kim, H. O. Pae, Y. S. Ko *et al.*, "In vitro inducible nitric oxide synthesis inhibitory active constituents from *Fraxinus rhynchophylla*," *Planta Medica*, vol. 65, no. 7, pp. 656–658, 1999.

71. D. M. Fort, K. Rao, S. D. Jolad, J. Luo, T. J. Carlson, and S. R. King, "Antihyperglycemic activity of *Teramnus labialis* (Fabaceae)," *Phytomedicine*, vol. 6, no. 6, pp. 465–467, 2000.

72. C. Wang, A. Pei, J. Chen *et al.*, "A natural coumarin derivative esculetin offers neuroprotection on cerebral ischemia/reperfusion injury in mice," *Journal of Neurochemistry*, vol. 121, no. 6, pp. 1007–1013, 2012.

73. M.Braccio, G.Grossi, G.Roma. *Eur. J. of Med. Chem.*, 39 (2004) 397-409.
74. J.Cheng, M.Chen, D.Wallace, S.Tith. *Bioorg. and Med. Chem. Letters*, 14 (2004) 2411-2415.
75. C.Garino, N.Pietrancosta, Y.Laras, J.L.Kraus. *Bioorg. and Med. Chem. Letters*, 16 (2006) 1995-1999.
76. G.R.Madhavan, V.Balraju, B.Mallesham. *Bioorg. and Med. Chem. Letters*, 13 (2003) 2547-2551.
77. A.Chilin, S.Moro, R.Battistutta, A.Bortolato. *J. of Med. Chem.*, 51 (2008) 752-759.
78. L.Zhang, X.Peng, L.V.Damu, R.Geng. *Med. Res.Rev.* (2013) 1-98.
79. Fais A, Corda M, Era B, Fadda MB, Matos MJ, Quezada E *et al.*, Tyrosinase Inhibitor Activity of Coumarin-Resveratrol Hybrids. *Molecules* 2009; 14: 2514-2520.

**2016, Biotechnology: Progress and Applications** *Pages* **411–448**
*Editors:* **Saif Hameed and Zeeshan Fatima**
*Published by:* **DAYA PUBLISHING HOUSE, NEW DELHI**

# *Chapter 18*

# Biotechnlogical Interventions for the Production of Bioactive Compounds in Medicinal Plants

*Pravej Alam*[1]*, *Malik Zainul Abdin*[2], *Jawaid A. Khan*[1] *and Zahid Ali*[2]*

[1]*Department of Biosciences, Jamia Millia Islamia, New Delhi – 110 025, India*
[2]*Department of Biotechnology, Jamia Hamdard, New Delhi – 110 062, India*

## ABSTRACT

Medicinal plants are rich source of bioactive compounds which are important for human healthcare. The production of bioactive compounds such as artemisinin, terpene indole alkaloids and esculin are limited in plants. Due its low yield it could not be commercialized worldwide. The economic limitations of chemical synthesis are not feasible due it complexity and cost effectiveness. The development of cell culture and molecular engineering of plants for the production of these bioactive compounds have been studied by different groups. In *in vitro* conditions, plant cell tissue and hairy root culture offer promising alternatives for the production of these compounds because totipotency enables plant cells and organs to produce useful secondary metabolites. The genetic modifications together with the establishment of *in vitro* plant regeneration and *Agrobacteium* mediated transformation systems facilitates efforts to engineer secondary product in

* Corresponding Authors: E-mail: alamprez@gmail.com; zahidnoble@gmail.com,

medicinal plants for enhanced production of these bioactive compounds. The role and importance of today's used host plant is critically described, and the approaches discussed to give an outlook for future trends and possibilities to enhance the level of bioactive compounds like artemisinin, Indole alkaloids and esculin in medicinal plants. In this chapter, an attempt has been made to summarize the biotechnological method (transgenic and non-transgenic) to boost the production of bioactive compounds level in plants systems in a precise way to help the researchers for enhanced production of metabolites.

***Keywords:*** *Artemisia annua, Artemisinin, C. roseus, TIAs, Chicorium intybus, Esculin, Tissue culture, Genetic engineering.*

## Introduction

The average life expectancy in industrialized countries has approximately doubled during the last century. Resistance can be caused by genetic modification of antibiotic targets, enzymatic degradation of drugs by the parasite and active export of drugs from parasite cells. The spreading of resistance is a direct consequence of the medical and non-medical applications of anti-infective agents. The WHO considers phytotherapy in its health programs and suggests the basic procedures for the validation of plant derived drugs in developing countries (Vulto and Smet, 1988; OMS, 1991). Currently the global demand of plant derived drugs is increasing and it is estimated that current global herbal drug market has reached a level of US $62 billion, which is expected to grow to US $5 trillion by the year 2050 (Joshi *et al.*, 2004). Natural products (secondary metabolites) have been the most successful source of potential life saving drugs obtained from plants (Haefner, 2003; Cragg and Newman, 2005; Mishra and Tiwari, 2011). Higher plants synthesize an array of bioactive compounds, either as an off shoot of normal growth and development or in response to pathogen attack or stress. Since the ability to synthesize particular class of secondary metabolites is restricted to certain plant groups, these compounds are clearly not essential for survival. In addition to their natural role, plant metabolites also represent a vast resource of complex molecules that are valued and exploited by man for their pharmacological and other properties. Plant natural products have thus been grouped under categories depending upon the presence of different chemical groups which include: alkaloids (hyoscymine, scopolamine, codeine, morphine, vincristine, vinblastine, atropine, cocaine, nicotine), terpenoids (artemisinin menthol, taxol), glycosides (digitoxin), phenols (vanillin), coumarins (esculin, psoralen), etc.

In recent years, the demand for bioactive compounds have increased dramatically. The supply of the source plants, however, is often limited because of overexploitation, disease, changes in climate, economical development and shrinking habitat. Plant cell and tissue culture offers an alternative source for controlled production of these products. In terms of cellular economy, secondary metabolites are generally metabolically expensive to produce and therefore frequently present in plants in much smaller quantities (Bajaj *et al.*, 1988). The relative low yield of these bioactive compounds are serious problem to the commercialization of the drugs due to its low concentration (Laughlin, 1995). The chemical synthesis of bioactive compounds are also possible but it is complicated and economically unviable due to poor yields and complexity (Ravindranathan *et al.*, 1990) Since the total synthesis of bioactive compounds are difficult and expensive, a biotechnological approach has been consider as a feasible alternative for its production. At present, the supply of artemisinin (0.01 per cent -0.60 per cent) dry weight (DW), vinblastine and vincretine (0.0005 per cent) as well as esculin (0.5-1.5 per cent; Alam *et al.*, 2014, Alam and abdin, 2012, Mujeeb *et al.*, 2012; Ali *et al.*, 2013) are far from enough in the international market because of the relative low content of present in plants. Considerable efferent has been focused on enhancing the production of these compunds for some long time. However, chemosynthesis conventional breeding of plants showing high yields of metabolites, and the manipulation of culture conditions, growth media and hormone leaves to increase the yield of bioactive compounds in tissue and cell cultures have not been successful (Abdin *et al.*, 2003). Alternative method such as transgenic technology may also the another useful approach for production of metabolites via transgenic plants or engineering the biosynthetic pathway in host cells. In this chapter we have discuss the production of bioactive compounds from three different plants such as *Artemisia annua* L., *Catharanthus roseus* and *Chicorium intybus*.

## I. *Artemisia annua* L.

### Bioactive Compound: The Classical Anti-malarial Drug-Artemisinin

Artemisia annua L.

Artemisinin has been reported to accumulate in the green stems, buds, flowers and seed of *A. annua*. Its content varies with the plant development stages. There are mainly two suggestions on the highest content of artemisinin in plant development (Abdin *et al.*, 2003) one is that the highest content of artemisinin is reached before plant flowering, while the other is that it is reached in the full flowering period (Wang *et al.*, 2004).

## Pharmacology

### Malaria

Malaria is characterized by fever, headache, nausea, and muscle pain and caused by the parasite *Plasmodium* (van Agtmael *et al.*, 1999). It is a very serious and even fatal infectious disease and transmitted mainly through female *Anopheles* mosquito. There are four species of *Plasmodium* resulting in malaria including *P. falciparum*, *P. ovale*, *P. vivax*, and *P. malariae* (WHO, 1998). *P. falciparum* accounts for the most serious version of the disease and death due to its high levels of death rate and spread of antimalarial drug- resistance (Mendis *et al.*, 2001). Every year, about 2 to 3 million deaths (Moorthy *et al.*, 2004) are caused by malaria, of which at least a million deaths are from *P. falciparum* infection (WHO, 2005a); and 40 per cent of the world population is at risk of malaria. *P. falciparum* infects as many as 400 million people a year (WHO, 2005a). To treat malaria, quinine and quinoline-based drugs such as chloroquine, mefloquine, and primaquine were widely used before the early sixties (Woerdenbag *et al.*, 1990). Some *P. falciparum* strains, however, became resistant to those antimalarial drugs and multiple side effects also exhibited (Abdin *et al.*, 2003). Artemisinin and its derivatives (Figure 18.1) have attracted more and more attention and in 2001, WHO recommended that Artemisinin-based Combination Therapies (ACTs) should be adopted to treat malaria (Mandelbaum-Schmid, 2005) due to little or no cross-resistance with other antimalarial drugs, rapid reduction of the parasite, and efficacious activity against *P. falciparum* strains (Meshnick *et al.*, 1996).

## Artemisinin: Uses other than Malaria

Besides malaria, artemisinin and its derivatives have been shown to be effective against a number of viruses, including human cytomegaloviruses, herpes simplex, and hepatitis B and C (Efferth *et al.*, 2002; Romero *et al.*, 2005). The drug is also effective against other prokaryotic and eukaryotic agents such as *Toxoplasma gondii*, *Schistosoma* (Utzinger *et al.*, 2001), and *Pneumocystis carinii* (Merali *et al.*, 1991). Artemisinin has further been shown to be useful in the treatment of some cancers. Various artemisinin derivatives have been tested for their ability to induce apoptosis in cultured human cancer cell lines (Singh *et al.*, 2004). Efficacy apparently stems from the anti-angiogenesis activities of artemisinin along with its possible role in protein alkylation in high iron environments (Chen *et al.*, 2003). Cancer cells maintain a higher iron content than do non-cancerous cells, with the exception of erythrocytes, making this a viable working hypothesis for the specific activity of artemisinin. In addition, dihydroartemisinin has been shown to induce cell cycle arrest and apoptosis in certain cancer cell lines (Efferth, 2007). Limited clinical trials are in progress using artesunate in conjunction with standard chemotherapies as a treatment for uveal melanoma with some positive results (Berger *et al.*, 2005). Artemisinin and its derivatives therefore, may have wide applicability in the treatment of various cancers.

## Biosynthesis of Artemisinin

Artemisinin is an endoperoxide sesquiterpene lactone which belongs secondary plant metabolites isoprenoid groups of compound. Mevalonate is a six-carbon intermediate in the pathway, arising from the sequential condensation of three acetyl-CoA units to generate HMG-CoA, which is converted to mevalonate in an irreversible reaction catalyzed by HMG-CoA reductase (HMGR). Due to the irreversible nature of this reaction, early workers correctly surmised that this step was a likely regulatory point of sterol biosynthesis in mammalian systems and eventually correlated the absolute rate of cholesterol biosynthesis with the level of this enzyme activity (Goldstein and Brown, 1990). Whether this enzyme plays a similar rate-limiting role in controlling plant isoprenoid biosynthesis remains unresolved (Bach, 1986; Narita and Gruissem, 1989; Choi *et al.*, 1992). The six-carbon mevalonate is sequentially phosphorylated and decarboxylated to generate IPP, which along with dimethylallyl diphosphate, an interconvertible isomer of IPP, represent the "activated" monomer building blocks for all other isoprenoids. Isopentenyl diphosphate (IPP) and its

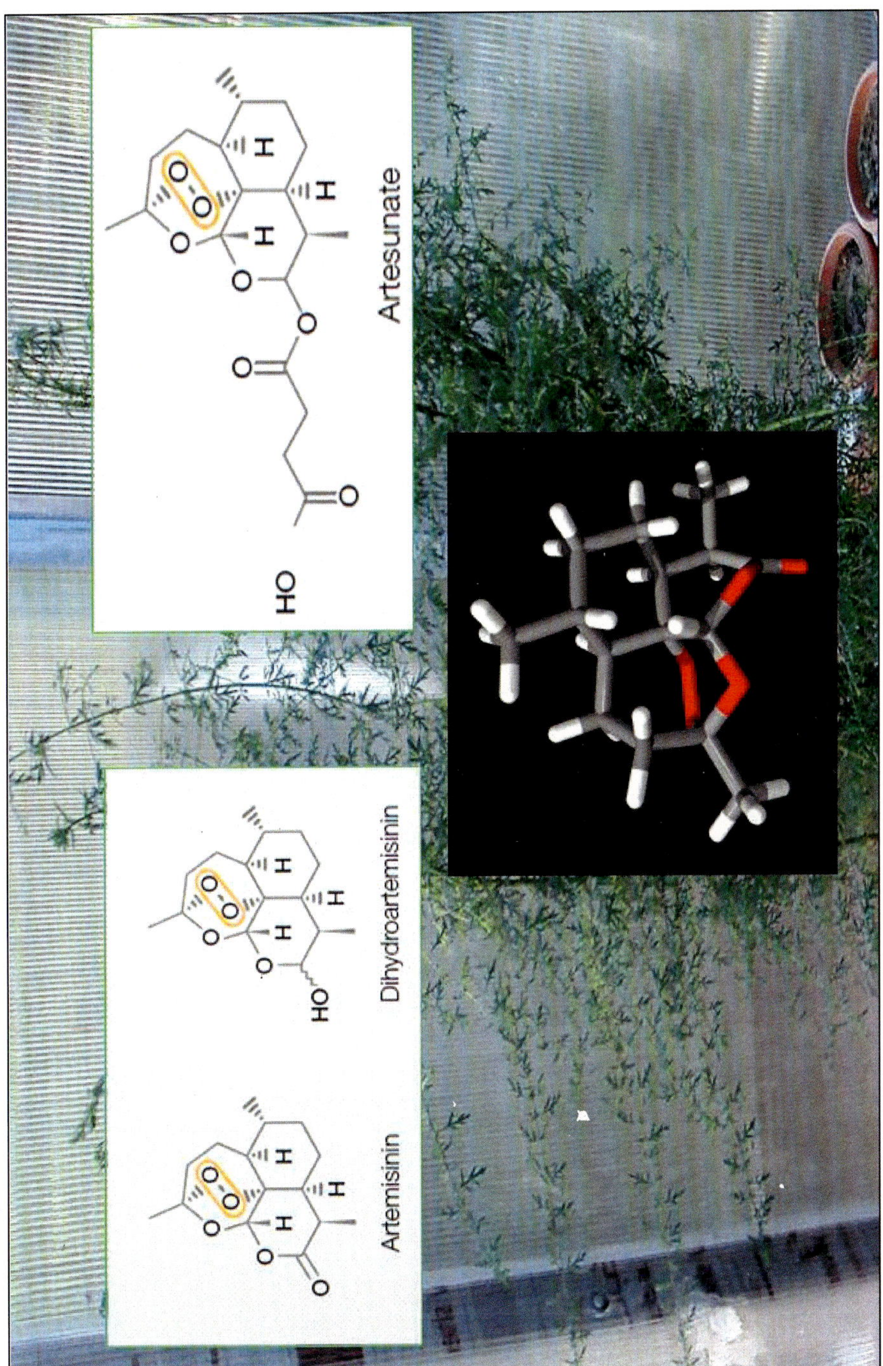


**Figure 18.1: Structures of 3-D form of Artemisinin and its Derivatives.**

isomer, dimethyllallyl diphosphate (DMAPP) are the primary precursors for the biosynthesis of terpenoids. Two different biosynthetic pathways lead to the production of IPP, namely the mevalonic acid dependent pathway (MVA pathway) present in the cytosol and mevalonic acid independent pathway (MEP pathway) present in plastids (Arsenault *et al.*, 2008; Nguyen *et al.*, 2011). The condensation of C5 substrates, DMAPP and IPP derived from MVA pathway, leads to the biosynthesis of C15 product, farnesyl diphosphate (FPP) *via* farnesyl diphosphate synthase (FPS) through enzymatic catalysis. Then FPP is converted into various isoprenoids, such as artemisinin, phytoalexins, sterols etc. in *A. annua* L. plants. The first committed step in artemisinin biosynthesis from FPP is catalyzed by amorpha-4,11-diene synthase (ADS) a sesquiterpene cyclase, that leads to the synthesis of amorpha-4,11-diene (AD), a cyclic sesquiterpene (Figure 18.2). It is followed by the oxidation of AD into artemisinic alcohol and then to artemisinic aldehyde/dihydroartemisinic aldehyde. The study of kinetic parameters has shown that *Aldh1* converts dihydroartemisinic aldehyde (DHA) to dihydroartemisinic acid (DHAA) more efficiently than RED1 that converts DHA into dihydroartemisinic alcohol (Ryden *et al.*, 2010). So, most of the DHA is converted to DHAA rather than dihydro artemisinic alcohol. In similar manner, Dihydroartemisinic acid is known as the direct precursor of artemisinin. However, no enzyme has been found to catalyze the conversion of dihydroartemisinic acid to artemisinin. Finally, the formation of artemisinin from DHAA occurs through non-enzymatic photo-oxidative step (Brown and Sy, 2004; Bertea *et al.*, 2005).

## Current Trends in Artemisinin Production

### (i) Non-transgenic Efforts

Much of the non-transgenic work to increase artemisinin content in *A. annua* L. has focused primarily on selective breeding and *in vitro* cell culture approaches including the formation and testing of calli, suspension cells, hairy root and shoot cultures (Arsenault *et al.*, 2008). Compared to suspension cultures, hairy roots are more stable, grow faster, and may be easier to scale-up. Many different culture conditions including light, elicitors, and culture in bioreactors have been investigated (Towler *et al.*, 2006; Weathers *et al.*, 2006). However, artemisinin yields in hairy roots are not yet high enough to be economically attractive. Alternatively, shoot cultures of *A. annua* L. are also being studied. Different culture conditions such as carbon sources, sugar concentration, $NH_4^+/NO^{3-}$ ratio, phosphate concentration, phytohormones (Basile *et al.*, 1993; Woerdenbag *et al.*, 1993; Liu *et al.*, 1998), addition of precursors, such as mevalonic acid (Abdin *et al.*, 2003), elicitors, or addition of metabolic inhibitors have been studied (Ahlawat *et al.*, 2014). Shoots cultured in bioreactors (Liu *et al.*, 1998) have also been studied. Unfortunately, artemisinin production in shoot cultures, although greater than in hairy roots, is also still much less than in whole plants. In view of the low concentrations of artemisinin detected in the plants, tissue culture systems have also been used with great interest as alternative method for the production of this drug. However, the results so far are not very encouraging. The artemisinin content that has been detected so far in shoot culture of *A. annua* L. is lower than in intact plants. The first trial of using tissue culture technique for artemisinin production came from He *et al.* (1983). The yield was 0.008 per cent in

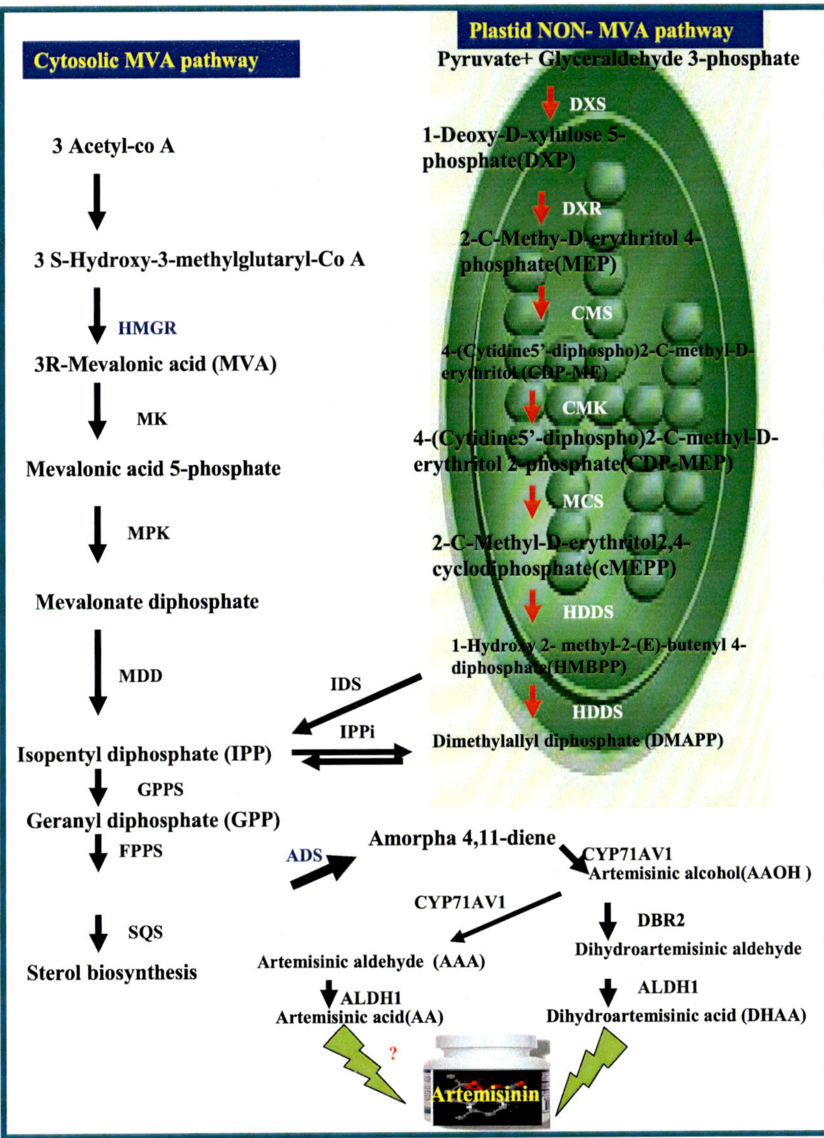


**Figure 18.2: Artemisinin Biosynthetic Pathway in *A. annua* L. (Adapted from Abdin and Alam, 2015).**

***Abbreviations*: HMGR: 3-hydroxyl-3-methyglutaryl CoA reductase; MK: mevalonate kinase; MPK: mevalonate-5-phosphate kinase; MPD: mevalonate pyrophosphate decarboxylase; DXS: 1-deoxy-D-xylulose-5-phosphate synthase;: DXR: 1-deoxy-D-xylulose-5-phosphate reductoisomerase; CMS: 4-diphosphocytidyl-2C-methyl-D-erythritol 4-phosphate synthase; CMK: 4-diphosphocytidyl-2-C-methyl-D-erythritol kinase; MCS: 2-C-methyl-D- erythritol 2:4-cyclodiphosphate synthase; HDDS: 1-hydroxy-2-methyl-2-(E)-butenyl 4-diphosphate synthase; IDS: IPP/DMAPP synthase; FPPS: farnesyl diphosphate synthase; SQS: Squalene synthase; ADS: Amorpha-4:11: diene synthase; CYP71AV1: cytochrome P450 monooxygenase;DBR2: artemisinic aldehyde double bond reductase; ALDH1: Aldehyde dehydogenase.**

shoots from the callus of *A. annua* plants of Chinese origin. Martinez and Staba (1988) did extensive monitoring with shoot cultures and were able to isolate high (0.3mg/g dw) and low (0.03mg/g dw) artemisinin synthesizing lines. In the present investigation the artemisnin content of *A. annua* shoot cultures was found to be 0.3 mg/g dw on MS basal medium supplemented with 1.0 mg/l Kinetin. Similar results were reported by Simon *et al.* (1990) where concentrations of artemisnin ranged from 0.03 per cent to 0.05 per cent on dw basis. In our study the artemisinin content in *in vitro* grown non-transgenic and transgenic shoots was found to be 0.03 and 0.04 per cent, respectively. The yields as high as 0.4 to 0.7 mg/g dw however, have been reported by Whipkey *et al.* (1992). Gulati *et al.* (1996) obtained *in vitro* flowering from various explants by supplementing the medium with $GA_{3,}$ where artemisinin content reached upto 0.1 per cent. Woerdenbag *et al.* (1993) reported a high percentage of artemisinin content (0.16 per cent of dry weight) from shoot cultures using MS medium supplemented with 0.05 mg/l NAA, 0.2mg/l BAP and 1 per cent sucrose. Most workers (Martinez and Staba, 1988; Tawfiq *et al.*, 1989; Kim and Kim, 1992; Gulati *et al.*, 1996; Ferreira and Janick, 1996) did not detect artemisinin in roots. Roots do not contain artemisinin, but evidently enhance its production in cultured shoots. When root initiation occurred from shoot cultures, active synthesis of arteannuin B and artemisinin was observed, suggesting that a certain degree of differentiation is required for the production of artemisinin (Martinez and Staba, 1988; Fulzele *et al.*, 1991). Removal of roots from shoots cultured in hormone free liquid medium reduced artemisinin content in shoot by 53 per cent (Ferreira and Janick, 1996). This might be the reason for high artemisnin content reported in some of the earlier studies.

The accumulation of artemisinin by plant tissue cultures *in vitro* occurs in calli, cell suspension, shoot and hairy root cultures (Weathers *et al.*, 1994; Cai *et al.*, 1995). Hairy root cultures of *A. annua* L. with rapid growth and high productivity of artemisinin, therefore, represent a potential source of commercial artemisinin Variation among hairy root lines developed from the same plant, but different plant parts due to differences in their biosynthetic capabilities is well reported for production of secondary metabolites (Kim *et al.*, 2002). Selection of high yielding root line is one of the most important steps to develop a commercial process for production of secondary metabolites by hairy root cultures (Baldi *et al.*, 2008). The mannopine or agropine-type strains of *A. rhizogenes* used to transport auxin-sensitive characteristics to the resulting transformants. Physiological character is manifested either via a heightened sensitivity to auxin, or by a hypersensitivity to auxin along with elevated levels of IAA inside the tissues (Giri *et al.*, 2001; Arroo *et al.*, 1995). Hairy root auxin sensitivity has received great attention. By and large, researchers investigating the physiological role of exogenously applied auxin in root growth and secondary metabolite production have established that these signal molecules can great impact plant tissue stability and secondary product accumulation either individually or through phytohormone-phytohormone interactions. Again, these reports show species specificity in auxin's effect on secondary production in hairy roots. In *A. annua*, addition of cytokinin (BAP) to the medium of hairy roots resulted in the disorganization of the root matrix with little effect on production of artemisinic compounds (Bunk, 1997). The effect of different environmental factors like temperature, pH, cultivation media and carbon source on growth and artemisinin production were studied in shake flask cultures. Detailed batch growth and production kinetics with sugar consumption profile were also established. Maximum volumetric productivity of 390 µg $L^{-1}$ $day^{-1}$ was obtained in hairy root cultures (Ahlawat *et al.*, 2012). Fulzele *et al.* (1995) concluded that exogenous addition of ethylene to the culture medium stimulated terpenoid synthesis in *A. annua* L. plantlet cultures grown in a bioreactor. In our lab Bunk (1997) observed clone specific results in *A. annua* hairy roots supplied with ethylene (as ethephon). *A. annua* clone YUM-290 grew unaffected in the presence of ethylene. When *A. annua* clone YUT-16 was provided with 5.0 or 15.0 mg/l ethylene (as ethephon), roots often grew poorly usually developing dense callused tissue or suspension cells. Smith *et al.* (1997) observed that $GA_3$ concentrations of 1-10 ng per liter provided the most signifi.cant increase in biomass during exponential growth in *A. annua* hairy roots. Conversely, Liu *et al.* (1997) and Bais *et al.* (2001) observed that the addition of GA to the growth medium increased artemisinin content in the hairy roots of *A. annua* L. Since GA is known to induce flowering in some plants (Crozier *et al.*, 2000), and maximum artemisinin accumulation in field-grown crops occurs at or near the onset of flowering in most strains (Woerdenbag *et al.*, 1994; Morales *et al.*, 1993; Ferreira and Janick, 1995; Laughlin, 1995), this may explain the increase in artemisinin.

Most attempts at stimulating artemisinin production, however, have not been encouraging. There are clearly numerous points of control for artemisinin. One control point occurs after IPP condensation at the branch point to synthesize sterols or sesquiterpenes. The isoprenoid biosynthesis pathways are under coordinate control as demonstrated in both yeast and plants through either genetic engineering or inhibition of specific enzymes (Arsenault *et al.*, 2008). For example, when miconazole, an inhibitor of squalene synthase, was fed to *A.annua* L. shoots (Kudakasseril *et al.*, 1987; Woerdenbag *et al.*, 1993) and seedlings (Towler *et al.*, 2006) artemisinin concentrations rose significantly compared to uninhibited controls. These studies indicate that inducible shunting of carbon may be a reasonable appoach for engineering increased levels of artemisinin in plants.

**(ii) Transgenic Efforts**

Early attempts at using transgenic approaches for increasing artemisinin levels in *A. annua* have included the insertion of genes affecting flowering (Wang *et al.*, 2004; 2007). Since miemisinin levels have been reported as highest during flowering, genes stimulating flowering were transferred into *A. annua*. The flowering promoting factor 1 gene (*fpfl*) and CONSTANS (CO), an early flowering gene, both from Arabidopsis, were transferred into *A. annua* using *Agrobacterium tumefaciens*. Although *fpfl* plants flowered 20 d earlier than nontransgenic plants, no significant difference in artemisinin level was observed upon flowering (Wang *et al.*, 2004). Similar results were observed for CO transformed plants; flowering occurred earlier than in the controls with no change in artemisinin concentration leading Wang *et al.* (2007) to conclude that there appears to be no direct link between artemisinin production and flowering. Vergauwe *et al.* (1996) developed an *Agrobacterium tumefaciens*-mediated transformation system for *A. annua* L. plants with high transformation rates (75 per cent regenerants harboring foreign gene). Artemisinin content in the leaves of regenerated plants was 0.17 per cent DW, a little bit higher than that present in the leaves of normally cultured plants (0.11 per cent DW).The factors influencing *A. tumefaciens*-mediated transformation of *A. annua* L., which included the age of the explant, *A. tumefaciens* strain, and plant genotype (Vergauwe *et al.,* 1998). Ghosh *et al.* (1997) transformed organ cultures of *A. annua* L. with two wild type nopaline strains of *A.tumefaciens*. The artemisinin content in shoot teratomas was 0.083±0.002g per100g dw. Non-transformed shoots, however, were found to synthesize only 0.0179±0.002 g per100g dw of the shoot. Chen *et al.* (1998) developed an efficient transformation system. They transformed leaf explants of *A. annua* L. using *A. rhizogenes* with a cotton cadinene synthase cDNA, cad-14. In another study, seven hairy root lines with the properties of fast growth and high artemisinin contents were selected from 747 hairy roots induced by transformation of *A. annua* L. with *A. rhizogenes,* ATCC 15834. The difference of growth rates and artemisinin contents among the seven selected hairy root lines were extremely significant. Among these lines, HR-9 gave the highest yield of artemisinin (Liu *et al.,* 1998). Chen *et al.* (1999) reported transferring green fluorescent protein (GFP) reporter gene into *A. annua* L. with *A. tumefaciens* and regenerating transgenic plants. They induced fascicled shoot from the leaves, buds, and flowers. They found that artemisinin production with the shoot cultures induced from flower organ explants was two times of that from the leaves. Chen *et al.* (2000) expressed a chimeric farnesyl diphosphate synthase gene in *A. annua* L. The resulting transformed plants have 2-3 times higher artemisinin content than that of the control. The effect of *isopentenyl transferase* gene expression on the physiological and biochemical characteristics of *A. annua* L. was studied by Sa *et al.* (2001). It was found, in the *ipt* gene expression, that cytokinins, chlorophyll, and artemisinin contents increased by different degrees. Content of cytokinins was elevated two- to threefold, chlorophyll 20–60 per cent, and artemisinin 30–70 per cent compared with the control plants. A direct correlation was found between the content of cytokinins, chlorophyll, and artemisinin. Han *et al.* (2005) have optimized the transformation system for *A. annua* L. using *A. tumefaciens* strains LBA 4404 and EHA105 harbouring the binary vector pBI carrying GFP and have successfully increased the frequency of transgenic fascicled shoot to approximately 4 per cent to 10 per cent. Because one cluster of fascicled shoots consists of 10-40 shoots in their case, the transformation efficiency was quite high.Farnesyl diphosphate synthase (FPS) was overexpressed in high-yielding strain of *A. annua* L. to increase the artemisinin content. The FPS activity in transgenic *A. annua* L. was two- to threefold greater than that in non-transgenic plants and the highest artemisinin content in transgenic plants was approximately 0.9 per cent (dry weight), which was 34.4 per cent higher than that in non-transgenic *A. annua* L. (Han *et al.,* 2006).

In our studies, we have developed transgenic lines over expressing exogenous *hmgr via A. tumefaciens* mediated genetic transformation of low and high artemisinin yielding chemotypes of *A. annua* L. plants and achieved 22 per cent and 38 per cent enhancements in artemisinin contents, respectively (Nafis *et al.,* 2011; Aquil *et al.,* 2009). Ma *et al.* (2009a) showed 100, 65 and 59 per cent increments in artemisinin, artemisinic acid and dihydroartemisinin contents, respectively in transgenic *A. annua*

L. plants by over-expressing *ads* gene. Recent past, we have over-expressed both *hmgr* and *ads* genes simultaneously in *A. annua* L. plants and found 7.68-fold enhancement in artemisinin content in the transgenic lines as compared to the non-transgenic plants (Alam and Abdin, 2011). The results of our study and others, hence, indicate that artemisinin biosynthesis can be up-regulated and artemisinin content can be increased in transgenic *A. annua* L. plants by over-expressing the genes encoding rate limiting enzymes of mevalonate and artemisinin biosynthetic pathways (Alam *et al.*, 2014). Zhang *et al.* (2009) developed transgenic *A. annua* L. by suppressing the expression of SQS (squalene synthase), a key enzyme of sterol pathway (a pathway competitive with that of artemisinin biosynthesis) by means of a hairpin-RNA-mediated RNAi (RNA interference) technnology. Artemisinin content was significantly increased, with the highest values reaching 31.4 mg/g dry weight, which was about 3.14-fold the content observed in untransformed control plants. Also the expression of *sqs* was suppressed and sterol was efficiently decreased in the transgenic plants. Their study demonstrated that genetic-engineering strategy of RNAi is an effective means of increasing artemisinin content in plants.

## II. *Catharanthus roseus*

### Bioactive Compounds: Vinblastine, Vincrestine, Vindolin, Catha ranthine

**Catharanthus roseus**

Plant-derived secondary metabolites are a most exclusive source of life-saving drugs and constitute an important resource for the discovery of active principles besides its other use as biopesticides (Cordell, 2002; Oksman-Caldenty and Inze, 2004; Sosa and Tonn, 2008). Cancer is a dreaded disease posing challenge long since to the medical science. The most successful plant-derived chemicals used in cancer chemotherapy are the alkaloids of *Catharanthus roseus*. A group of 20 dimeric alkaloids in this plant is found to contain antineoplastic activity (Noble, 1990). Two of these dimeric alkaloids, namely, vinblastine and vincristine, are extracted commercially from *C. roseus* and are used either alone or in combination with other forms of chemotherapy for cancer treatment. Vinblastine and vincristine are present as minor constituents (0.001 per cent) in *C. roseus*, and the commercial extraction of plant material is costly and tedious thus making the cost of these life-saving drugs very expensive (Giri, and Narasu, 2009; Hisiger and Jolicoeur, 2007; Hong *et al.*, 2006; Peebles *et al.*, 2006).

Plant cell and tissue cultures can be engineered rather than whole plants to procude these therapeutic compounds in higher quantities. *C. roseus* (L.) G. Don has been extensively investigated in order to utilize cell culture system for the production of important anticancer compounds, vinblastine and vincristine (Giri *et al.*, 2009; Van Der Heijden *et al.*, 2004). Metabolic and enzyme studies have suggested that dimeric alkaloids are biosynthesised in specifically differentiated tissue, their biosynthesis is under strong developmental control and modulated by tissue-specific and light-dependent factors (Aerts and De Luca, 1992; Hirata *et al.*, 1993; Miyamoto and Miura, 1988; Moreno-Valenzuela *et al.*, 1998).

## Pharmacology

India possess a rich biodiversity of the medicinal plants that were still not explored completely to produce the life saving drugs worldwide. The need for the novel pharmaceutical products out from the plant has attained a great interest in the present research world due to the cost and the higher side effects that are associated with the chemically manufactured drugs. *Catharanthus roseus*, which is a potent medicinal plant many of the pharmacological actions such as antimicrobial, antioxidant, anthelmintic, antifeedant, antisterility, antidiarrheal, antidiabetic effect etc (Table 18.1). That is used to treat many of the fatal diseases. Moreover, more than 130 alkaloids have been isolated from different parts; amongst which two important alkaloids (Vinblastine and Vincristine used in cancer treatment) present in very low concentrations (Junaid *et al.*, 2010).

**Table 18.1: Medicinal Uses of *Catharanthus roseus world wide***

| *Plant Parts* | *Disease* | *Country* |
|---|---|---|
| Leaves | Menorrhagia, diabetes | Australia |
| Entire Plants | Diabetes mellitus | Brazil |
| Aerial parts | Menstrual regulators | China |
| Entire plants | Diabetes | England |
| Leaves | Diabetes | Europe |
| Entire Plants | Antigalactagogue | France |
| Entire Plant | Cancer | India |
| Dried ovule | Diabetes | Pakistan |
| Leaves | Diabetes | South Africa |

The pharmacology of the plant was found to be associated mostly especially with the alkaloids that occupies almost most of the parts of the plant.

## Biosynthesis of TIAs

Terpenoid indole alkaloids biosynthetic pathway in *C. roseus* is complex with multiple steps and is under strict molecular regulation. The broad class of TIAs in *C. roseus* is formed by the coupling of an indole moiety of tryptamine and a monoterpenoid component of secologanin. The tryptamine derives from the Shikimate pathway (Indole pathway), while the secologanin from the terpenoid pathway. The Shikimate pathway links carbohydrate metabolism to the synthesis of aromatic amino acids (phenylalanine, tyrosine and tryptophan), which can in turn act as precursors for various primary and secondary metabolites. In *C. roseus*, anthranilate synthase (AS) catalyzes the conversion of chorismate to anthranilate and is feedback-inhibited by tryptophan, which is the first committed step of tryptophan synthesis. The

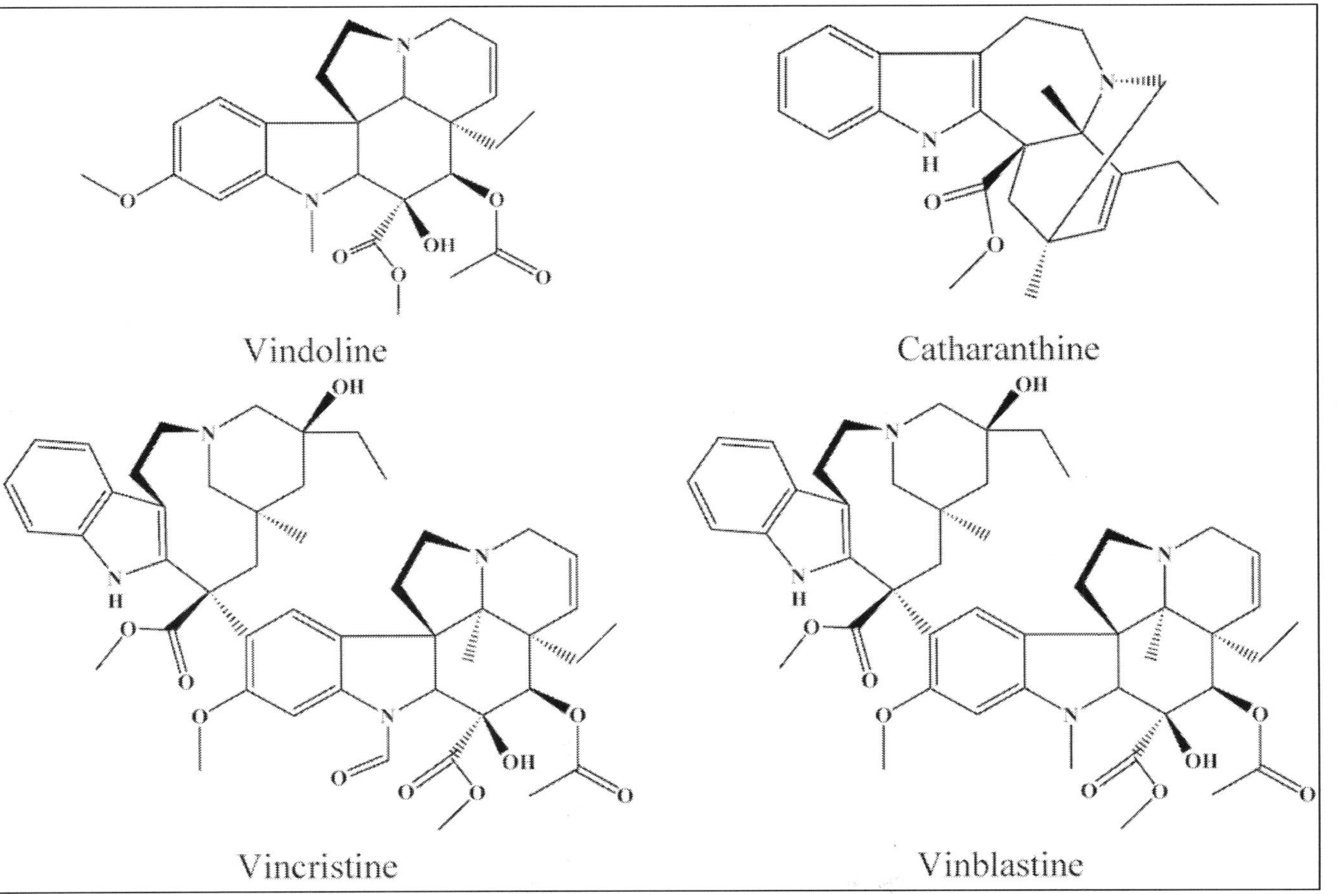


**Figure 18.3: Structure of Vindolin, Cathranthine, Vincrestine and Vinblastine.**

enzymatic reactions from anthranilate to tryptophan remain unclear in *C. roseus* so far. At the last step in the Shikimate pathway, tryptophan is converted to tryptamine catalyzed by tryptophan decarboxylase (TDC; EC 4.1.1.28). There are two independent pathways in plants to synthesize isopentenyl diphosphate (IPP) and its isomer dimethyally diphosphate (DMAPP). IPP and DMAPP are universal precursors of terpene biosynthesis. The first pathway is the cytosolic mevalonate pathway (MVA pathway). The other pathway, an alternative non-mevalonoid 1-deoxy-Dxylulise-5-phosphate pathway (non-MVA pathway or DXP pathway), exists in plastid (Lichtentheler *et al.*, 1997; Eisenreich *et al.*, 2001). Because MEP is the first committed intermediate of the non-mevalonate pathway, it is also called the MEP pathway. In the past, the MVA pathway was considered to be the sole biosynthetic route leading to IPP. Recent studies have established that this novel DXP pathway is involved in the biosynthesis of geraniol in bacteria and plants. It is realized now that the DXP pathway appears to be largely responsible for the biosynthesis of monoterpenes, diterpenes, tetraterpenes (carotenoids), and polyprenols. More importantly, the DXP pathway was recently confirmed to be the primary source of IPP for secologanin synthesis in *C. roseus* cell culture (Contin *et al.*, 1998; Veau *et al.*, 2000; Hong *et al.*, 2003). The first step of the DXP pathway in *C. roseus* is the condensation of pyruvate and D-glyceraldehyde 3-phosphate (GA-3-P) into 1-deoxy-D-xylulose 5-phosphate (DXP; Veau *et al.*, 2000). Finally, the MEP is converted to IPP and its isomer DMAPP catalyzed consecutively by hydroxymethylbutenyl 4-diphosphate synthase (HDS) and IPP/DMAPP synthease (IDS) (Rodr´ýguez-Concepci´on and Boronat, 2002; Dubey *et al.*, 2003). After the formation of IPP and DMAPP, different prenyltransferases catalyze the sequential chain elongation of DMAPP with IPP units to geranyl diphosphate (GPP), farnesyl diphosphate (FPP), and geranylgeranyl diphosphate (GGPP), which ultimately serve as precursors of themonoterpenes (C10), sesqui- and triterpenes/sterols (C15/30) and di- and tetraterpenes (C20/40).

The key precursor of monoterpene biosynthesis is GPP produced by GPP synthase (GPPS, EC 2.5.1.1) which catalyzes the condensation of IPP and DMAPP to the C10 product (McGarvey and Croteau, 1995). Although it is currently unknown how GPP is converted to geraniol, some progress has been made recently on the reactions from geraniol to secologanin in *C. roseus*. Further, the enzyme NADPH:cytochrome P450 reductase (CPR) is essential for the activity of G10H and other cytochrome P450 mono-oxygenases, since it functions in electron transfer to cytochrome P450 proteins (Meijer *et al.*, 1993). At the end of the pathway, 7-deoxyloganin is converted to secologanin by two consecutive reactions (Irmler *et al.*, 2000). Deoxyloganin 7-hydroxylase (DL7H) first catalyzes 7-deoxyloganin into loganin, then the loganin is synthesized into secologanin by secolaganin synthase (SLS). Irmler *et al.* (2000) identified that cytochrome P450 CYP72A1 acted as secologanin synthase. the first step is the condensation of tryptamine and secologanin by strictosidine synthase (STR; EC 4.3.3.2) to form strictosidine, a key intermediate and the first committed step in the biosynthesis of TIAs (McKnight *et al.*, 1990) (Liu *et al.*, 2007; Figure 18.4). All TIAs in *C. roseus* are derived from the central precursor strictosidine, which is a fusion product of the shikimate pathway derived tryptamine moiety and the plastidic nonmevalonate pathway-derived secologanin moiety. Starting from the amino acid tryptophan and the monoterpenoid geraniol, the biosynthesis of bisindole alkaloids in *C. roseus* involves at least 35 intermediates and 30 enzymes (Zhou *et al.*, 2009; vander Heijden *et al.*, 2004). Tryptamine and secologanin are condensed by strictosidine synthase (STR) to form strictosidine, the common precursor to all TIAs. The first of six steps involved in the conversion of tabersonine to vindoline consists of hydroxylation at the C-16 position by tabersonine 16-hydroxylase (T16H) was characterized as a P450-dependent mono-oxygenase. Subsequent to the 16-hydroxylation of tabersonine, the next three steps in vindoline biosynthesis has been now accepted as 16-*O*-methylation, an *S*-adenosyl-L-methionine (SAM)-dependent*O*-methyltransferase (OMT) that methylates 16-*O*-demethyl-4-*O*-deacetylvindoline. The final step in vindoline biosynthesis is catalyzed by acetylcoenzyme A deacetylvindoline 4-*O*-acetyltransferase (DAT). Further, vindoline and cathranthine converted in to vinblastine and vincretine respectively through PRX.

## Current Trends in TIAs Production

**(i) Non-transgenic Efforts**

The production of TIAs in plant cell and tissue cultures have been studied by different groups (Mujeeb *et al.*, 2012, Moreno-Valenzuela *et al.*, 1998). In this prospects, there are three main culture systems introduced: cell culture, hairy root culture and somatic embryogenesis to enhance the level of metabolites. To establish cell suspension culture of *C. roseus*, firstly, the choice of explants, medium (MS, B5, N6, White etc) composition supplemented with carbon source and different combination and concentration of PGRs should be optimized. The role of plant growth regulators (PGRs) in the regulation of TIAs has been studied by different group (Mujeeb *et al.*, 2012; Zhou *et al.*, 2009; Liu *et al.*, 2007; El-Sayed and Verpoorte, 2007a, Roytrakul and Verpoorte, 2007) in *C. roseus*. Recentely, methyl jasmonate (MeJA), abscisic acid (ABA), salicylic acid (SA), ethylene (ethephon) and gibberellic acid (GA3) play a important role on TIAs production by different investigators in *C. roseus* cell suspensions cultures and hairy roots (Ruiz-May *et al.*, 2008; El-Sayed and Verpoorte, 2004; El-Sayed and Verpoorte, 2005; Lee-Parsons *et al.*, 2004). The commercial production of secondary metabolites by plant cells in suspension culture is mainly affected by the low productivity of the cultures.

The secondary-metabolite productivity of a cell culture is defined by three factors, i) specific growth rate, ii) product formation rate and iii) biomass concentration. A higher product formation rate might be achieved by the screening or selection of cell lines and optimization of the culture conditions (Hooper *et al.*, 1994). Pati *et al.* (2011) analyzed of total alkaloids and other pharmaceutical important constituents such as vincristine and vinblastine in *C. roseus* plants tissue. There was a significant increase in the total alkaloid content in tissue culture-raised plants (0.888 per cent w/w) compared to the mother plant (0.526 per cent w/w) in liquid culture system for shoot proliferation. However, the concentration of vincristine (0.00028 per cent w/w) and vinblastine (0.00032 per cent w/w) did not change much. But, It was interesting that the total alkaloid content of multiple shoots increased 1.5 times in agar gelled medium than multiple shoots grown in liquid medium. Compact callus clusters were developed from *C. roseus* stem explants in a modified liquid medium containing NAA (α-

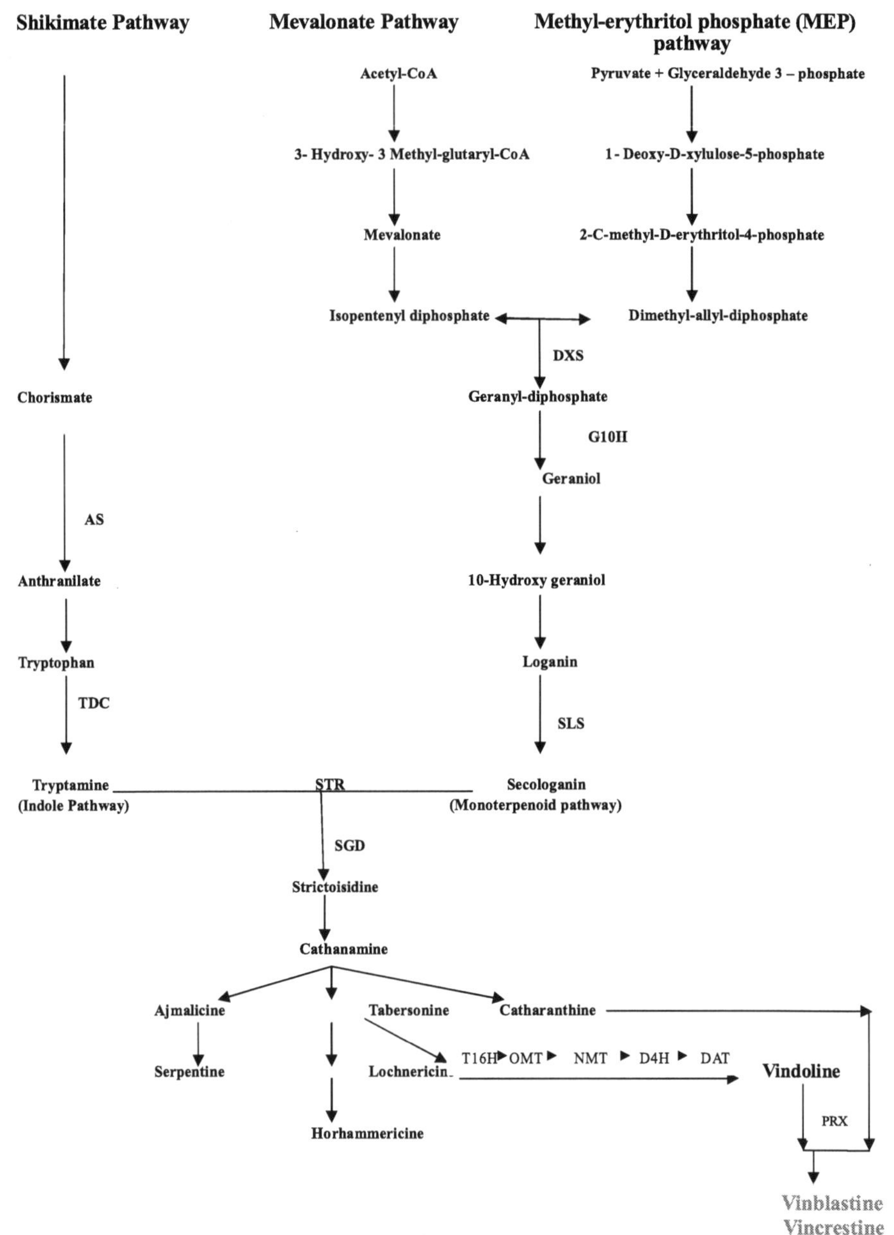


**Figure 18.4: Vinblastine and Vincrestine Biosynthetic Pathway in *C. roseus*.**

**AS: Anthranilate synthase; DXS: 1-deoxy-D-xylulose 5-phosphate synthase; G10H: Geraniol 10-hydroxylase; STR: Strictosidine synthase; T16H:Tabersonine 16-hydroxylase; TDC: Tryptophan decarboxylase; SGD: Strictosidine b-D-glucosidase; SLS Secologanin synthase;T16H: Tabersonine 16-hydroxylase; D4H: desacetoxyvindoline 4-hydroxylase; DAT: deacetylvindoline 4-O-acetyltransferase; OMT: 16-hydroxytabersonine O-methyltransferase; NMT: 2:3-dihydro-3-hydroxytabersonine N-methyltransferase; PRX: peroxidise (Adapted from Mujeeb *et al.*, 2012).**

naphthaleneacetic acid; 5.37 μM) and kinetin (4.65 μM).The production of ajmalicine and catharanthine was enhanced more than 2.8 fold in cell suspension cultures of *C. roseus*. Azra Ataei-Azimi *et al.* (2008) reported that the use of different concentration of NAA and Kin play important role to enhance the level of indole alkaloid levels in callus and new roots where petiole used as a explants. In this study, they achieved 10-fold catharanthine, 125-fold serpentine, 0.5-fold vindoline and 0.34-fold ajmalicine in newly developed root callus and roots from petioles in the presence of 0.1-20 mg/l Kin and NAA. In shoot cultures of *C. roseus* by using Murashige and Skoog medium supplemented with IAA (11.42 μM) and a low BA (2.22 μM), Satdive *et al.*, 2003 achieved high accumulation of ajmalicine concentrations. Vindoline synthesis in multiple shoot culture of *C. roseus* was investigated under continuous light conditions (16 h) photoperiod and the content of vindoline reached values similar to those found in leaves from field-grown plants(Campos-Tamayo *et al.*, 2008). The agitated liquid medium have also seen productive for multiple-shoot regeneration, growth processes and high incidences of root formation (Mujib *et al.*, 1995,2012).

*In vitro* cultures of hairy root derived from *C. roseus* exhibited more genetic stability as compared with cell suspensions and accumulate higher levels of TIAs. The influence of buffered media upon the growth and alkaloid productivity of *C. roseus* hairy-root culture was examined through the changes in pH, use of elicitors, carbon and nitrogen sources. An optimized hairy root culture protocol was established on MS with sucrose as carbon and lactalbumin as nitrogen source, in which, enhanced level of total alkaloids was noticed compared to other explants and calli (Sun and Zeng 2005). Treated with antagonists (verapamil or EGTA and Cdcl2) that block the $Ca2^{++}$ led to 25 per cent enhancement in the concentration of in the *C. roseus* hairy-root system (Moreno-Valenzuela *et al.* (2003). Various bioreactors and complex fermenters were also tried to improve the biomass of hairy-roots that accumulated better production of secondary metabolites (Nuutila, 1994). Efforts to engineer *C. roseus* hairy roots to produce commercially significant amounts of valuable compounds like ajmalicine vinblastine and vincristine may require the development of tools to study the effects of over expressing key metabolic and regulatory genes. In this direction, a glucocorticoidinducible and an alcohol-inducible promoter system were used to study the effects of gene overexpression within the terpenoid indole alkaloid pathway on metabolite production (Peebles *et al.*, 2007). The effect of the engineered indole pathway on accumulation of phenolic compounds in response to ectopic overexpression of tryptophan feedback resistant anthranilate synthase holoenzyme (AS alpha and beta) was analyzed in *C. roseus* hairy roots and it was noted that the un induced and induced AS alpha and beta hairy roots accumulated up to 1.2 and 4.5 mg/g DW respectively over a period of 72 h (Chung *et al.*, 2007).

**(ii) Transgenic Efforts**

TIAs production in *C. roseus* is developmentally regulated, with the highest levels occurring in younger, faster dividing regions of the plant (De Luca and St-Pierre, 2000). Furthermore, some steps are depend on environmental conditions. For example, light proves to be necessary for the synthesis of vindoline, an important precursor of vincristine and vinblastine in *C. roseus* (De Luca *et al.*, 1989; De Luca and St-Pierre, 2000). The lack of differentiation in cell cultures has also been a barrier to successful alkaloid production. *C. roseus* hairy roots, a fast growing, differentiated tissue culture generated by *Agrobacterium rhizogenes* infection, are known to accumulate far higher levels of alkaloids than undifferentiated cell and callus cultures (Moreno-Valenzuela *et al.*, 1998). The genetically increasing alkaloid flux in *C. roseus* remains limited, a few notable metabolic engineering attempts have been made that further elucidate the network. Regarding pathways relevant to terpenoid indole alkaloid metabolism, previous attempts have included constitutive overexpression of tryptophan decarboxylase (TDC) in cell culture and crown gall calluses (Canel *et al.*, 1998; Goddijn *et al.*, 1995), HMG-CoA reductase in hairy roots (Ayora-Talavera *et al.*, 2002), the transcriptional regulator ORCA3 in cell cultures (van der Fits and Memelink, 2000), and strictosidine synthase in cell culture (Canel *et al.*, 1998). Gains in particular indole alkaloids were made in the ORCA3, STR, and TDC cell culture studies under certain precursor feeding conditions (van der Fits and Memelink, 2000; Whitmer *et al.*, 2002a; Whitmer *et al.*, 2002b). In addition to an increasing knowledge base, the development of new tools for plant metabolic engineering is encouraging (Zuo and Chua, 2000).

More recently, the production of pharmaceutically important terpenoid indole alkaloids (TIAs) from *Catharanthus roseus* is partly regulated at the transcriptional level was studded by Goklany *et al.* (2009). In this study, limitations in TIA biosynthesis from *C. roseus* hairy root cultures were assessed through gene expression profiling and precursor feeding. The transcript levels of key TIA pathway genes (G10h, Tdc, Str, and Sgd) and metabolite levels associated with the TIA pathway (tryptamine, loganin, secologanin, strictosidine, ajmalicine, serpentine, and tabersonine) were monitored using quantitative RT-PCR and HPLC, respectively. In cultures elicited with methyl jasmonate (250 microM MeJA on day 21), G10h, Tdc, Str, and Sgd expression increased by 9.1, 3.1, 6.7, and 8.3-fold, respectively, after 24 h. Up-regulation of gene expression was followed by a 160, 440, and 420 per cent increase in strictosidine, ajmalicine, and tabersonine levels, respectively, after 5 days. In order to improve the production of the anticancer dimeric indole alkaloids in *Catharanthus roseus*, much research has been dedicated to culturing cell lines, hairy roots, and efforts to elucidate the regulation of the monoterpenoid indole alkaloid (MIA) biosynthesis. In this study, the ORCA3 (Octadecanoid-derivative Responsive *Catharanthus* AP2-domain) gene alone or integrated with the G10H (geraniol 10-hydroxylase) gene were first introduced into *C. roseus* plants. Transgenic *C. roseus* plants overexpressing ORCA3 alone (OR lines), or co-overexpressing G10H and ORCA3 (GO lines) were obtained by genetic modification. ORCA3 overexpression induced an increase of AS, TDC, STR and D4H transcripts but did not affect CRMYC2 and G10H transcription. All OR and GO lines showed higher levels of monoindole alkaloids vindoline, catharanthine and ajmalicine than the control. The vindoline accumulation ranged from 0.78–2.83 mg $g^{-1}$ DW in OR lines and from 1.25–3.00 mg $g^{-1}$ DW in GO lines when the average in control lines was 0.70 mg $g^{-1}$ DW. The catharanthine accumulation was variable with the range of 3.65–5.79 mg $g^{-1}$ DW in OR lines and 3.78–5.36 mg $g^{-1}$ DW in GO lines when the average in control lines was 1.99 mg $g^{-1}$ DW. Ajmalicine was accumulated with the range of 0.07–0.10 mg $g^{-1}$ DW in OR lines and 0.14–0.49 mg $g^{-1}$ DW in GO lines, while the average in control lines was 0.05 mg $g^{-1}$ DW. However, only two OR lines and two GO lines showed an increase of bisindole alkaloids anhydrovinblastine and vinblastine. The change of vinblastine accumulation was inconsistent with that of anhydrovinblastine accumlation. Line OR43 had the highest level of anhydrovinblastine and vinblastine, at 10.25 mg $g^{-1}$ DW with 1.08-fold up and at 0.27 mg $g^{-1}$DW with 10.19-fold up compared with the controls, respectively. The results implied that ORCA3 and G10H overexpression had a stronger influence on the biosynthesis of monoindole alkaloids than that of bisindole alkaloids. The accumulation of vinblastine was more predictable of the level of anhydrovinblastine than that of their precursors vindoline and catharanthine (Pan *et al.*, 2012). The AP2/ERFdomain transcription factor ORCA3 of *Catharanthus roseus* has been recognized. Metabolite analysis using HPLC established that the average content of catharanthine and vindoline in the transgenic hairy root extracts was increased 2.49 fold and 4.2 fold in comparison to the control hairy root lines, respectively. However, vinblastine content could not been detected by HPLC method in transgenic and control hairy root cultures.The AP2/ERF-domain transcription factor ORCA2 from *Catharanthus roseus* was identified and expressed earlier to regulate the expressions of Str gene, an important gene involved in the terpenoid indole alkaloids biosynthetic pathway in *C. roseus* cells (Tang *et al.*, 2011).Metabolite analysis using high performance liquid chromatography (HPLC) analysis established that the average content of catharanthine and vindoline in the transgenic hairy root extracts was increased up to 2.03 and 3.67-fold in comparison to the control lines, respectively (Liu *et al.*, 2012).

## III. *Cichorium intybus*

*C. intybus*, a perennial herb of the Asteraceae, with blue flowers, is also known as blue sailors, endive succory, coffee weed and kashen,s or kasini. It is native to the temperate part of the Old World. It is cultivated in Bihar, Gujarat, Himachal Pradesh and Tamil Nadu. Historically, chicory was grown by the ancient Egyptians, as a medicinal plant, coffee substitute, vegetable crop, and occasionally for animal forage. For its special chemical composition and physiological bioactivities, natural products extracted from chicory have been used for developing herbal medicine (Wang *et al.*, 2009). The plant is used as antihepatotoxic, antiulcerogenic, anti-inflammatory, appetizer, digestive stomachic, depurative and diuretic. Pharmacological investigations also reveals anticancer properties. Esculin is an important component used in pharmaceutical industry. It is a glycosidic coumarin (monosaccharide of esculetin) belonging to a group of phenolic compounds distributed in natural plants (Ali *et al.*, 2012). Esculin is a well known UV-B protective agent and one of the phytomedicines in practice for the treatment of various peripheral vascular disorders (Martindale *et al.*, 1989). It is an elasticity giving and sealing material for blood vascular system, vitamin P

*Cichorium intybus*

activity, protects cells against DNA damage and inhibits carcinogenesis (Kaneko *et al.*, 2004).

## Bioactive Compound: Esculin

Esculin (6, 7-dihydroxy coumarin) belongs to coumarins an important group of compounds found in plants. It is among the most important constituents used in pharmaceutical industry (Buszewski *et al.*, 1993). It serves as elasticity giving and sealing material for blood vascular systems (Pietrogrande *et al.*, 1992). Esculin finds application as anti-cancerous, cytotoxic, cytostatic, antiflogistic and anti-mutagenic agent (Kostovan 2005). It is also used as a marker in microbiological media. It is a well-known natural UV-B protective agent and one of the phytomedicines used for the treatment of various peripheral vascular disorders (Stanic *et al.*, 1999) and for vitamin P activity (George *et al.*, 1999; Rehman *et al.*, 2003). It has earlier been isolated using solid phase extraction in *Flos hippocastani* and drugs Venescin and Venacron (Buszewski *et al.*, 1993). Williams *et al.* (1996) have characterized esculin in the leaf samples of *Taraxacum officinale* along with cichorin. Content variability of esculin in *Aesculus hippocastanum* has been established by analyzing bark samples of the tree during the four seasons and esculin content was highest through summer and from oldest bark samples (Stanic *et al.*, 1999).

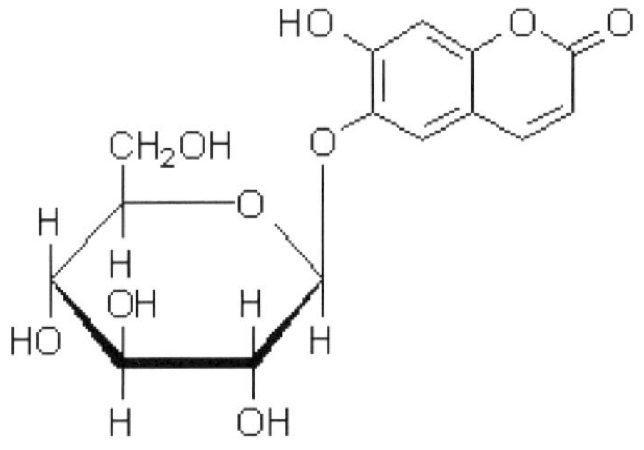

**Figyre 18.5: Chemical Structure of Esculin.**

## Biosynthesis of Esculin

Indeed, the biosynthesis of coumarins in plants is not well understood, although these metabolic pathways are often found in the plant kingdom. Esculin is belongs coumarins group of plant natural products derived from the phenylpropanoid pathway. At least 1000 natural occurring coumarins, among them about 300 simple coumarins such as esculetin, fraxetin, isofraxidin, isoscopoletin, daphnetin, have been found in many families of higher plants. (Berenbaum *et al.*, 1991). Coumarins are derived from 1,2-benzopyrones. These molecules are found in higher plants where they originate from the general phenylpropanoid pathway (Harborne, 1999) and are subject to numerous modifications. Coumarins continue to receive attention for their diverse bioactivities (Bourgaud *et al.*, 2006).

In plants coumarins are derived from L-phenylalanine pathway. In this pathway, 4-coumarate is formed from L-phenylalanine in two steps, catalyzed by phenylalanine/tyrosine ammonia-lyase and *trans*-cinnamate 4-monooxygenase, which is a cytochorme P450 hydroxylase (Figure 18.6). This pathway starts with the enzyme tyrosine ammonia-lyase, which converts L-tyrosine to 4-coumarate in a single step. In addition of the plant enzymes 4-coumarate—CoA ligase and 4-coumaroyl 2'-hydroxylase further synthesized of umbelliferone. The formation of umbelliferone proceeds from 4-coumaric acid or its ester derivatives. The conversion of cinnamic acid to 4-coumaric acid is catalyzed by cinnamate 4-hydroxylase, a cytochrome P450 monooxygenase from the CYP73A family (Teutsch *et al.*, 1993). The formation of esculetin (6,7-dihydroxy coumarin) was examined in *Cichorium intybus* (Brown, 1985). These studies revealed that umbelliferone was an efficient precursor but not suggesting 6-hydroxylation of umbelliferone, probably by the action of a P450 monooxygenase. This deserves mentioning, because the conversion of caffeic acid to esculetin is readily accomplished in vitro with various plant extracts containing phenoloxidase activity (Kneusel 1987; Sato 1967).

## Current Trends in Esculin Production

### (i) Non-transgenic

The primary goal of tissue culture is the mass scale production of genetically identical, and pathogen free plantlets which can be acclimatized in a reduced time period and at a lower cost. Plant cell/tissue culture is a rapidly developing technology which holds promise of restructuring agricultural, horticultural and forestry practices. Cultured explants undergo frequent genetic changes which are expressed at biochemical or molecular level. The genetic variability expressed in regenerated plants can be transmitted to the progeny through sexual or vegetative propagation (Razdan 2000). *C. intybus* has been regenerated from the primary explants such as storage roots (Margara and Rancillac 1966), green leaves (Toponi 1963), etiolated leaves (Vasseur 1979), floral stem (Bouriquet and Vasseur 1973), suspension culture (Yasseen and Splittstoesser 1995), leaf and petiole (Rahman *et al.*, 2003; Velayutham *et al.*, 2006; Kumari *et al.*, 2007), pollen (Castano and De Proft 2000) and protoplasts (Sasksi *et al.*, 1986; Varotto *et al.*, 1997). Plants have also been regenerated from tissue cultures of mature tap–roots (Heirwiegh *et al.*, 1985). A regeneration protocol from the leaf explants has earlier been described by Vermeulen *et al.* (1992) and Genga *et al.* (1994). Most investigations on *in vitro* culture of C. *intybus* have involved root and leaf segment cultures (Grazia *et al.*, 1990). Apical and axillary meristems have often been used to establish vegetative multiplication. Paulet and Nitsch (1964) and Nitsch *et al.* (1966) obtained normal flowers from plants regenerated from root tissues cultured using Knop (1966) and Murashige and Skoog (MS) (1962) modified media. Moreover, they found that the addition of Indole-3-Acetic acid (IAA) to the culture medium prevented the formation of flower buds even at concentrations that did not inhibit the initiation and development of vegetative buds, and that gibberellic acid ($GA_3$) ($10^{-7}$-$10^{-6}$ M) promoted flowering. Phenolic compounds such as p-coumaric acid ($10^{-5}$ M) and coumarin ($10^{-6}$ M) increased the percentage of flower buds. Senet *et al.* (1983) obtained root production on explants in Heller nutrient agar medium supplemented with Naphthalene acetic acid (NAA) ($10^{-6}$ M), kinetin (kn) ($10^{-7}$ M) and glucose (10g $l^{-1}$).

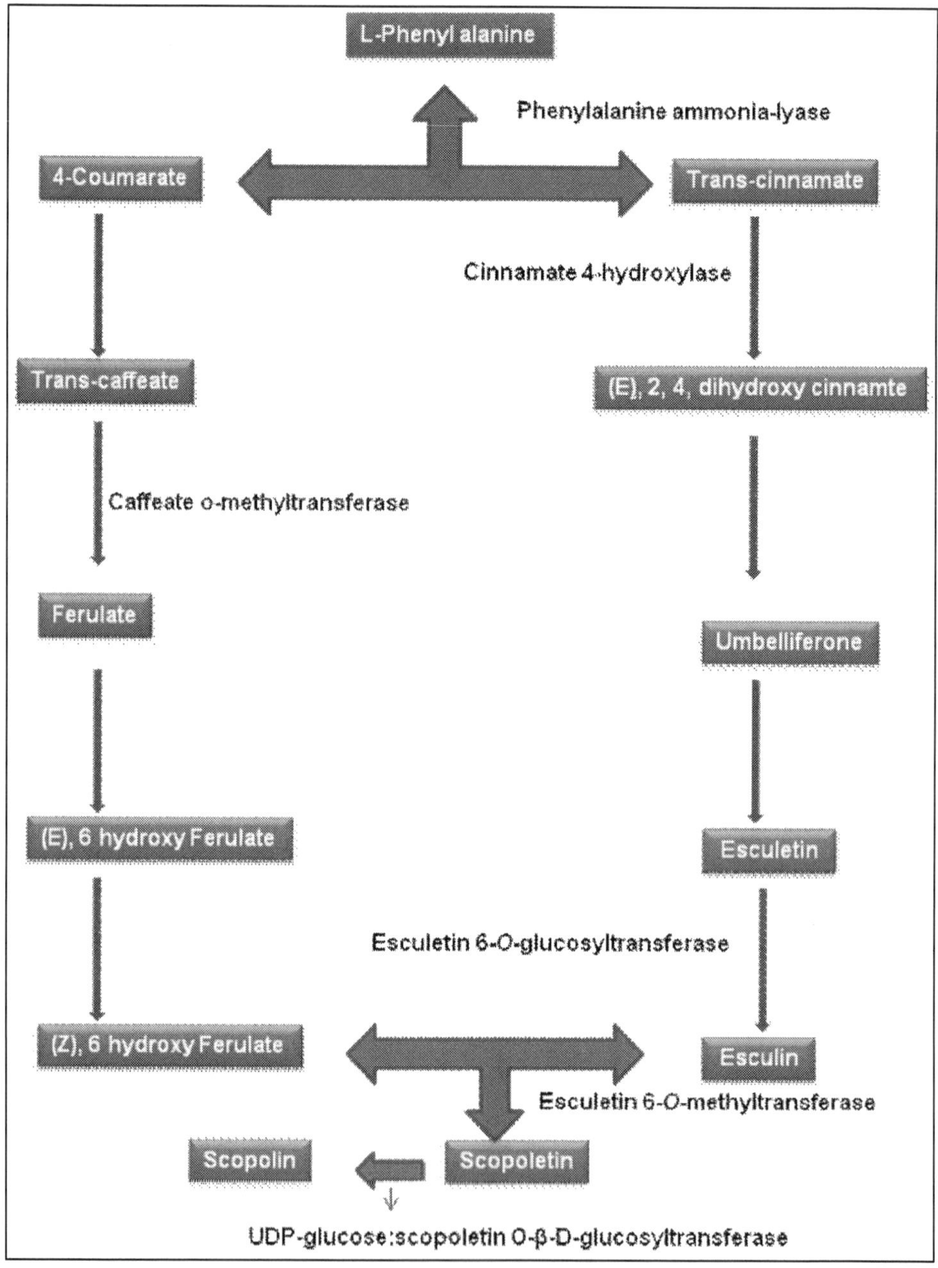


**Figure 18.6: Esculin Biosynthetic Pathway Adapted from MetaCyc Pathway Database.**

In root explants of chicory supplemented with $GA_3$ ($10^{-3}$ M) induced flowering stem under long days, and vegetative growth under short day (Demeulemeester *et al.*, 1995). An efficient and rapid procedure for plantlet regeneration from chicory mesophyll protoplasts has been developed by Nenz *et al.* (2000). Putrescine (Put) and $AgNO_3$ influences shoot multiplication, *in vitro* flowering and endogenous titers of polyamines in C. *intybus* L. (Bais *et al.*, 2000). Exogenous administration of Put at a concentration of 40 mM resulted in maximum tissue response in terms of shoot numbers and shoot lengths on MS media supplemented with 2-iP (2.0 mg $l^{-1}$) and $GA_3$ (0.5 mg $l^{-1}$). Similar results were obtained when of 40 μ M $AgNO_3$ was applied externally on the same medium. Endogenous titers of conjugated spermidine decreased sharply from day 7–21, whereas endogenous conjugated spermine levels peaked on day 28, after treatment with 40 mM Put. $AgNO_3$ fed samples resulted in higher titers of endogenous conjugated spermine on day 14. All other treatments showed decreasing endogenous levels during the whole culture period. Both Put (40 mM) and $AgNO_3$ (40 μM) resulted in floral initiation and floral development on day 28 and 14, respectively. Rehman *et al.* (2003) developed a protocol for the regeneration of plantlets from leaf explants of witloof chicory. They obtained regeneration via callus on modified MS semisolid medium containing 2 μM IAA in combination with 5 μM Kn and 1000 mg $l^{-1}$ casein hydrolysate, whereas Lim *et al.* (2004) obtained regenerated shoots from leaf explants on solid MS medium supplemented with different concentrations of cytokinins and auxins. The maximum numbers of shoots were obtained on MS medium supplemented with BAP and NAA (0.1 mg $l^{-1}$) and maximum numbers of roots were formed in the medium supplemented with IAA. Velayutham *et al.* (2006) reported maximum number of shoots and roots on MS medium supplemented with BAP (4 μM) + IAA (1μM) followed by 12.5 μM IBA. Nandagopal and Kumari (2006) used young leaf as explants and obtained callus induction, shoot multiplication and flowering on MS and Gamborg's ($B_5$) media supplemented with different concentration and combination of IAA, BAP, 6-furfurylaminopurine and Adenine sulphate. The highest percentage of callus induction and multiple shoot proliferation (93.6 per cent) was observed on MS+B5 medium containing BAP (6.66 μM) + IAA (2.852 μM) + adenine sulphate (1.360 μM). Maximum root induction (100 per cent) in regenerated shoots was observed on MS+B5 medium containing IBA (4.92 μM)). Maximum number of flowers was observed on MS+B5 medium supplemented with 6.66 μM BAP, 2.852μM IAA and 1.360 μM Adenine sulphate. Nandagopal and Kumari (2007) developed a protocol for the root culture of *C.intybus* by using leaf and hypocotyl explants. On half strength MS medium supplemented with different conc. of auxins IAA, NAA and IBA. Maximum rooting (94 per cent) was obtained on 0.5 mg$l^{-1}$ NAA and 0.1 mg$l^{-1}$ IBA supplemented medium.

Hui *et al.* (2011) established a high frequency regeneration of C. *intybus* using leaf segments of aseptic seedlings. Callus and adventitious buds of chicory were induced by inoculating explants on MS medium supplemented with BAP and NAA at different concentrations. When lower part of leaves derived from 20-day-old seedlings was used as explant and inoculated on MS medium containing 2.0 mg$l^{-1}$ 6-BA, 0.5 mg$l^{-1}$ NAA and 40 g$l^{-1}$ sucrose, the frequency of adventitious bud formation was 90.0 per cent. When the regenerated shoots were cultured in half MS medium containing 0.1 mg$l^{-1}$ NAA, the frequency of root formation was maximum (88.3 per cent). Gibberellic acids (GA) are widespread and so far ubiquitous in angiosperms and gymnosperms as well as pteridophytes (Crozier *et al.*, 2000). GAs are products of the terpenoid pathway (Crozier *et al.*, 2000). Following are some of the responses known for gibberellins. In studies of the effects of GA (Gibberellic acids) on hairy root morphology, Bais *et al.* (2001) reported that low concentrations (0.5 mg/l) of GA increased the biomass accumulation in the hairy roots of *Cichorum intybus*, due to high levels of root elongation, lateral root branching, and primary root growth almost 2-fold over the hormone-free controls.

**(ii) Transgenic Efforts**

Limited understanding of the regulation of compounds has increased speculation on the application of genetic engineering techniques for large-scale production of secondary metabolites (Banthorpe 1994). Transgenic cultures have been developed in plant systems to produce secondary metabolites from medicinal plants. Various strategies to enhance metabolites of interest include metabolic engineering of secondary metabolites (Verpoorte *et al.*, 1998, Verpoorte *et al.*, 1999; Alam and Abdin 2011; Singh *et al.*, 2011). The most efficient way of natural genetic transformation was studied by Tepfer (1984) using *Agrobacterium rhizogenes*. Natural genetic transformation by exogenous DNA has been documented in bacteria (Romanowoski 1993) and DNA transfer from bacteria to higher plants has also been documented previously (Tepfer 1984; Romanowoski 1993). There is interest in genetically engineered chicory to obtain higher yields and create new potentials. Work related to the development of transgenic chicory has been in progress in recent years. The conversion of Belgian endive from biennial to annual flowering was achieved by using *A. rhizogenesis* transformation (Sun *et al.*, 1991).

Vermeulen *et al.*, 1992 developed an efficient transformation method for introducing acetolactate synthase gene that confers resistance to chlorsulfuron (herbicide) into chicory through *Agrobacterium tumifaciens*. Frulleux *et al.* (1997) carried out transformation experiment on shoot buds of *Cichorium intybus* L. cv. *sativum* (root chicory) using *A. tumefaciens* carrying a reporter gene namely *uidA* coding for β-glucuronidase and a selective marker *nptII*, coding for neomycin phosphotransferase conferring resistance to kanamycin. No toxicity or allergenicity and successful degradation in the digestive tract were found (Zoden 1998). Residue assessment shows that only 50 per cent of plants of crops derived from transgenic chicory will be tolerant to glufosinate ammonium (Zoden 1998). This herbicide is used only for selection purposes in breeding and is not authorised for use in chicory crop production (Zoden 1998). Male sterile lines of chicory have been obtained by incorporating the barnase gene (Zoden 1998). Limami *et al.* (1998) showed that transformation by Ri-TDNA from *A. rhizogenes* could lead to a beneficial or deleterious phenotype of potential adaptive significance in *C. intybus*. It was observed that the transgenic plants obtained through *A rhizogenes* mediation had altered phenotypic character in terms of metabolic level, inulin content and amino acid profile as compared to normal chicory plants (Limami 1998). It is also reported that a non-symbiotic haemoglobin gene can be cloned and expressed in chicory during the early stages of somatic embryogenesis (Hendriks 1998). Similarly, the sucrose-fructan-6-fructosyltransferase (6-SFT) gene was cloned and expressed in chicory, which resulted in overproduction of trisaccharide kestose and fructan of phlein type. (Spranger 1997). Transformation of chicory by *A. rhizogenes* mediation resulted in altered phenotypes, associated with an altered morphogenetic behaviour with regard to shoot multiplication and *in vitro* flowering, and also altered polyamine and ethylene biosynthesis (Bais 2000).

The transgenic plants of chicory showed lower levels of endogenous polyamines. It was observed that exogenous administration of putrescine and $AgNO_3$ influences the morphogenesis in transgenic roots (Bais 2000). A gene (*lol1*) from over-wintering rye grass (*Lolium perenne* L.) that encodes a protein that inhibits the re-crystallisation of ice was introduced via *Agrobacterium*-mediated transformation into cotyledons excised from 17d-old seedlings of chicory. Extracts from leaves of different ages excised from 140 day old transgenic plants inhibited ice recrystallisation; inhibition was greater with extracts from younger leaves compared with extracts from older leaves. Leaf extracts from non-transformed plants failed to inhibit ice re-crystallisation. Chicory has been transformed with *Mycobacterium tuberculosis* antigen ESAT6 through *A. tumifaciens* (Matvieieva *et al.*, 2011a), that could be used for edible vaccine synthesis. In another experiment, Matvieieva (2011b) regenerated transgenic chicory plants after transformation by a wild type *A. rhizogenes* A4 strain containing either *ifn-α2b* gene encoding human leukocytic interferon or *esxA*::fbpB$\Delta^{TMD}$ genes encoding *M. tuberculosis* antigens ESAT6 and Ag85B.

## Conclusions

The commercial importance of bioactive compounds such as artemisinin, esculin, vinblastine and vincristine had prompted many attempts to produce them using plant tissue culture methodology in past decades. Most of the efforts, however, have not focused on genetically engineered cell cultures, tissue cultures, or plants but rather on non-engineered cell lines studied under elicitation, with precursor feeding, or in optimized media. While these classical methods have hastened characterization efforts, metabolic engineering offers the most promising method for improved product composition and increased alkaloid yield of plants and cultured cell systems (Hughes and Shanks 2002). The pathways of interest often involve multiple organellar compartments, resulting in transport limitations and sequestered pools of metabolites. This compartmentation is particularly important to consider as incorrect targeting of transgenes will leave them without the necessary substrates (Herminghaus *et al.*, 1996).Owing to problems of instability, the production of secondary metabolites using de-differentiated cells (callus) is not always feasible on an industrial scale. The cloning of strong tissue specific promoters and their use in either over-expressing gene(s) encoding rate limiting steps in artemisinin, TIAs and esculin biosynthetic pathways, or silencing the genes encoding rate limiting steps of competing pathways, could be a better strategy to enhance the metabolites production in plant systems. The over-expression of transcription factors, *viz.*, ORC WRKY1, AaERF1 and AaERF2 may also resulted into increased biosynthesis and accumulation of bioactive compounds in those plants. It is due to the fact that the extra carbon and energy in this case will only be used in those cells / tissues where those are biosynthesized. Hence, the plant resources will be used more effectively and economically in the process.

As the supply of plant-derived bioactive compounds such as artemisinin, esculin, vinblastine and vincristine are subjected to the seasonal variations and stability, a common problem with many other plants-based commodities, this may lead to the fluctuation in supply and price of artemisinin that is used in the manufacture of ACTs. The commercial scale production of semi-synthetic production of bioactive compounds could be another better platform to maintain the supply over existing plant-derived bioactive compounds to meet the global demand for plant derived drugs. Both semi-synthetic and direct use of dried leaves will also help to check the price of life saving drugs such as antimalarial, anticancer as well as antidiabetic.

## References

1. Abdin MZ, Israr M, Rehman RU, Jain SK (2003). Artemisinin, a Novel Antimalarial Drug: Biochmical and Molecular Approaches for Enhanced Production. *Planta Med.* 69: 289-299.

2. Abdin MZ, Alam, P (2015). Genetic engineering of artemisinin biosynthesis: Prospects to improve its production. *Acta Physiol Plant,* 37: 33.

3. Ahlawat S, Saxena P, Alam P, Mauji Ram, Nafis T, Mohd A, Abdin MZ (2012). Influence of *Agrobacterium rhizogenes* on induction of hairy roots for enhanced production of artemisinin in *Artemisia annua* L. plants. *Afr J Biotecnol.* 1: 8684–8691.

4. Alam P, Abdin MZ (2011). Overexpression of HMGCoA reductase and amorpha4,11diene synthase genes in *Artemisia annua* L. and its influence on artemisinin content. *Plant Cell Rep* 10: 1919–1928.

5. Alam, P, Kamaluddin, Khan MA, Mohammad A, Khan R, Abdin MZ (2014). Enhanced artemisinin accumulation and metabolic profiling of transgenic *Artemisia annua* L. plants overexpressing by rate limiting enzymes from isoprenoid pathway, *J Plant Inter.* doi: 10.1080/17429145.2014.893030.

6. Ali Z, Showkat HG, Alka N, Abdin MZ, Srivastava PS (2013). Efficient regeneration and high esculin content in *Cichorium intybus* L., a potent medicinal plant. *Int J Pharm Bio Sci* 4(3): 1333 - 1340

7. Aquil S, Husaini AM, Abdin MZ, Rather GM (2009). Overexpression of the HMGCoA Reductase Gene Leads to Enhanced Artemisinin Biosynthesis in transgenic *Artemisia annua* Plants. *Planta Med* 75: 1-6.

8. Arroo RR, Develi J, Meijers A, Van de Westerlo E, Kemp AK, Croes AF, Wullens GJ (1995). Effects of exogenous auxin on root morphology and secondary metabolism in *Tagetes patula* hairy roots cultures. *Physiol. Plant.* 93: 233–240.

9. Arsenault PR, Wobbe KK, Weathers PJ (2008). Recent advances in artemisinin production through heterologous expression. *Curr Med Chem* 15: 2886.

10. Ataei-Azimi A, Delnavaz HB, Ebrahimzadeh H, Ahmad M (2008). High in vitro production of anticancer indole alkaloids from periwinkle (*Catharanthus roseus*). tissue culture. *AJB* 7 (16): 2834-2839.

11. Ayora-Talavera T, Chappell J, Lozoya-Gloria E, Loyola-Vargas VM (2002). Overexpression in *Catharanthus roseus* hairy roots of a truncated hamster 3-hydroxy-3-methylglutaryl-CoA reductase gene. *Appl Biochem Biotechnol.* 97(2): 135-45.

12. Bach TJ (1986). Hydroxymet hylglutaryl – CoA reductase, a key enzyme in phytosterol synthesis. *Lipid* 21: 121–125.

13. Bais HP, Sudha G, George J, Ravishankar GA (2001). Influence of exogenous hormones on growth and secondary metabolite production in hairy root cultures of Cichorium intybus L. cv. Lucknow Local. *In Vitro Cell Dev Biol Plant*. 37: 293–299.

14. Bais HP, Sudha G, Ravishankar GA (2000). Putrescine and silver nitrate influences shoot multiplication, *in vitro* flowering and endogenous titers of polyamines in *Cichorium intybus* L. cv. Lucknow Local. *J Plant Growth Regul* 19: 238-248.

15. Baldi A, Dixit VK (2008). Yield enhancement strategies for artemisinin production by suspension cultures of *Artemisia annua*. *Bioresource Technol*. 99: 4609–4614.

16. Basile DV, Akhtari N, Durand Y, Nair MSR (1993). Toward the production of artemisinin through tissue culture: determining nutrient- hormone combinations suitable for cell suspension cultures. *In Vitro Cell. Dev. Biol. Plant* 29: 143-147.

17. Berenbaum MR (1991). "Coumarins." In: Herbivores: Their interactions with secondary metabolites. Vol 1, Rosenthal GA, Berenbaum MR (eds.), Academic Press, San Diego, 221-249.

18. Berger TG, Dieckmann D, Efierth T, Schultz ES, Funk JO. Baur A, Schuler G (2005). *Oncol. Rep*. 14: 1599-1603.

19. Bertea CM, Freije JR, Van der Woude H, Verstappen F W, Perk L, Marquez V, De Kraker J W, Posthumus MA, Jansen BJM, De Groot AE, Franssen MCR, Bouwmeester H J. (2005). Identification of intermediates and enzymes involved in the early steps of artemisinin biosynthesis in *Artemisia annua*. *Planta Med*. 71: 40–47.

20. Bourgaud, F., Hehn, A., Larbat, R., Doerper, S., Gontier, E., Kellner, S., *et al*. (2006). Biosynthesis of coumarins in plants: a major pathway still to be unravelled for cytochrome P450 enzymes. *Phytochem. Rev*. 5, 293–308.

21. Bouriquet R, Vasseur J (1973). Croissance et bourgeommement des tissus de feuilles d'endive en function de l' age et du lieu de prelevement des explantats. *Bull soc Bot Fr* 120: 27-32.

22. Brown GD, Sy LK (2004). In vivo transformations of dihydroartemisinic acid in *Artemisia annua* plants. *Tetrahedron* 60: 1139–1159.

23. Brown SA (1985). Biosynthesis of 6,7-dihydroxy-coumarin in *Cichorium intybus*. *Can. J Biochem. Cell Biol*. 63: 292–295.

24. Bunk JG (1997). Hormone Influences on Growth and Secondary Metabolite Production in Transformed Root Cultures of *Artemesia annua* L. *M.Sc. Thesis*. Worcester Polytechnic Institute.

25. Buszewski H, Kawka S, Suprynowior Z, Wolski T (1993). Simultaneous isolation of rutin and esculin from plant material and drugs using solid-phase extraction. *J Pharm Biomed Anal* 11: 211-215.

26. Cai G, Li G, Ye H, Li G (1995). Hairy root culture of *Artemisia annua* L. by Ri plasmid transformation and biosynthesis of artemisinin. *Chinese Journal of Biotechnology*. 11: 227-235.

27. Canel C, Lopes-Cardoso MI, Whitmer S, Van-Der-Fits L, Pasquali G, Van-Der-Heijden R, Hoge JH, and Verpoorte R (1998). Effects of over-expression of strictosidine synthase and tryptophan decarboxylase on alkaloid production by cell cultures of *Catharanthus roseus*. *Planta*, 205(3): 414-419.

28. Castano CI, De PM P (2000). *In vitro* pollination of isolated ovules of *Cichorium intybus* L. *Plant cell Rep* 19: 616-621.

29. Chen DH, Ye HC, Li GF (1999). *Artemisia annua* mRNA for HMG-CoA reductase. Submitted Department of Cell and Genetic Engineering, Institute of Botany, the Chinese Academy of Sciences, No.20, Nanxincun, Xiangshan, Beijing 100093, P. R. China.

30. Chen HH, Zhou HJ, Fang X (2003). Inhibition of human cancer cell line growth and human umbilical vein endothelial cell angiogenesis by artemisinin derivatives *in vitro*. *Pharmacal. Res.*, 48: 231- 236.

31. Chen-DaHua, Meng Y, Ye HC, Li GF, Chen XY, Chen DH (1998). Cultures of transgenic *Artemisia annua* hairy root with cadinene synthase gene. *Acta Bot Sin* 40: 711-4.

32. Choi D, Ward BL, Bostock RM (1992). Differential induction and suppression of potato 3-methylglutaryl coenzyme A reductase genes in response to *Phytophthora infestans* and to its elieitor arachidonic acid. *Plant Cell* 4; 1333-1344.

33. Chung IM, Hong SB, Peebles CA, Kim JA, San KY (2007). Effect of the engineered indole pathway on accumulation of phenolic compounds in *Catharanthus roseus* hairy roots. *Biotechnol Prog*. 23: 327–332.

34. Contin A, Van Der Heijden R, Lefeber WM, Verpoorte R (1998). The iridoid glucoside secologanin is derived from the novel triose phosphate/pyruvate pathway in a *Catharanthus roseus* cell culture. *FEBS Lett*. 434, 413–416.

35. Cordell G. (2002). Natural products in drug discovery: creating a new vision. *Phytochemistry Reviews*, 1 (3), 261-273.

36. Crozier A, Kamiya Y, Bishop G, Yokota T (2000). Biosynthesis of hormones and elicitor molecules. In: Buchanan BB, Gruissem W, Jones RL (eds). *Biochemistry and molecular biology of plants*. American Society of Plant Physiology, Rockville, pp. 850–929.

37. Crozier A, Turnbull CGN (1984). Gibberellins: Biochemistry and action in extension growth. *What's New Plant Physiol*. 15: 912.

38. De Luca V, St. Pierre B (2000). The cell and developmental biology of alkaloid biosynthesis. *Trends in Plant Sci*. **5:** 349-364.

39. De Luca, VC Marineau, Brisson N (1989). Molecular cloning and analysis of cDNA encoding a plant tryptophan decarboxylase: comparision with animal dopa decarboxylases. *Proc. Natl. Acad. Sci. USA* 86: 2582-2586.

40. Demeulemeester MAC, Voet A, Proft MPD (1995). Induction of stem elongation on *in vitro* chicory root explants under unfavourable photoperiodic conditions by gibberellin A3. *Plant Growth Regul* 16: 239-241.

41. Dubey VS, Bhalla R, Luthra R (2003). An overview of the non-mevalonate pathway for terpenoid biosynthesis in plants. *J. Biosci.* 28, 637–646.

42. Efferth T, Li PCH, Konkimalla VSB, Kaina B (2007). From traditional Chinese medicine to rational cancer therapy. *Trends Mol Med* 13: 353–361.

43. Efferth T, Olbrich A, Bauer R (2002). mRNA expression profiles for the response of human tumor cell lines to the anti-malarial drugs artesunate, arteether, and artemether. Biochem. Pharmacol. 64: 617–623.Romero MR, Efferth T, Serrano MA, Castano B, Macias RI, Briz O, Marin JJ (2005). Effect of artemisinin/artesunate as inhibitors of hepatitis B virus production in an *'in vitro'* system. *Antiviral Res* 68: 75–83.

44. Eisenreich W, Rohdich F, Backer A (2001). Deoxylulose phosphate pathway to terpenoids. *Trends Plant Sci.* 6, 78–84.

45. El-Sayed M, Verpoorte R (2004). Growth, metabolic profiling and enzymes activities of *Catharanthus roseus* seedlings treated with plant growth regulators. *Plant Growth Regul.*, 44: 53-58.

46. El-Sayed M, Verpoorte R (2005). Methyljasmonate accelerates catabolism of monoterpenoid indole alkaloids in *Catharanthus roseus* during leaf processing. *Fitoterapia.* 76(1): 83-90.

47. El-Sayed M, Verpoorte R (2007). *Catharanthus* terpenoid indole akaloids: biosythesis and regulation. *Phytochem Rev* 6: 277 – 305.

48. Ferreira JFS and Janick J (1996). Distribution of artemisinin in *Artemisia annua.* J. Janick (ed.), *Progress in new crops.* ASHS Press, Arlington, VA. P. 579-584.

49. Ferreira JFS, Simon JE, Janick J (1995). Relationship o artemisinin content of tissue cultured, green house and field grown plants of *Artemisia Anuua. Planta Med.* 61: 351-355.

50. Frulleux F, Weyens G, Jacobs M (1997). *Agrobacterium tumefaciens*-mediated transformation of shoot-buds of chicory. *Plant Cell Tiss Org Cult* 50: 107-112.

51. Fulzele DP, Sipahimalani AT, Heble MR (1991). Tissue cultures of *Artemisia annua*: organogenesis and artemisinin production *Phytother Res* 5: 149-53.

52. Genga A, Giansante L, Bernacchia G, Allavena A (1994). Plant regeneration from *Cichorium intybus* L. Leaf explants transformed by *Agrobacterium tumifaciens. J Genet Breed* 48: 25-32.

53. George J, Bais H and Ravishankar GA (1999). Production of esculin by hairy root cultures of chicory (*Cichorium intybus* L. cv. Lucknow local). *Indian Journal of Exp Biol* 37: 269 -273.

54. Ghosh B, Mukherjee S, Jha S (1997). Genetic transformation of *Artemisia annua* by *Agrobacterium tumefaciens* and artemisinin synthesis in transformed cultures. *Plant Sci* 122: 193-9.

55. Giri A, Ravindra ST, Dhingra V, Narasu ML (2001). Influence of different strains of Agrobacterium rhizogenes on induction of hairy roots and artemisinin production in *Artemisia annua*. *Curr Sci.* 81: 378–382.

56. Giri CC, Giri A, Narasu ML (2009). *Catharanthus roseus* (Periwinckle). a potential drug source for cancer chemotherapy and biotechnological interventions: An overview. *Medicinal Plants: Utilization and Conservation* (2nd ed., pp. 475–504). Jaipur, India: Aviskar Publishers, Distributors.

57. Goddijn OJM, Pennings EJM, Helm PV, Schilperoort RA, Verpoorte R, Hoge JHC, Vander- Helm P (1995). Overexpression of a tryptophan decarboxylase cDNA in *Catharanthus roseus* crown gall calluses results in increased tryptamide levels but not in increased terpenoid indole alkaloid production. *Trans Res* 4: 315–323.

58. Goklany S, Loring RH, Glick J, Lee-Parsons CW (2009). Assessing the limitations to terpenoid indole alkaloid biosynthesis in *Catharanthus roseus* hairy root cultures through gene expression profiling and precursor feeding. *Biotechnol Prog* 25: 1289-1296.

59. Goldstein JL, Brown MS (1990). Regulation of the mevalonate pathway. *Nature* 343, 425-430.

60. Grazia S, Rochetta G, Pieragostini N (1990). Somaclonal variation in chicory, in *Biotechnology in Agriculture and Forestry*, Vol II. Springer, Berlin.

61. Gulati A, Bharel S, Jain SK, Abdin MZ, Srivastava PS (1996). *In vitro* micro propagation and flowering in *Artemisia annua*. *J Plant Biochem Biotech* 5: 31-35.

62. Han JL, Liu BY, Ye HC, Wang H, Li ZQ, Li GF (2006). Effects of Overexpression of the Endogenous Farnesyl Diphosphate Synthase on the Artemisinin Content in *Artemisia annua* L. *Journal of Integrative Plant Biology* 48 (4): 482–487.

63. Han JL, Wang H, Ye HC, Liu Y, Li ZQ, Zhang Y, Zhang YS, Yan F, Li GF (2005). High efficiency of genetic transformation and regeneration of *Artemisia annua* L. via *Agrobacterium tumefaciens*-mediated procedure. *Plant Sci* 168: 73–80.

64. Harborne JB (1999). Classes and functions of secondary products from plants. In Walton NJ, Brown DE (eds). *Chemicals from plants*. Imperial College Press, London, pp. 1–25

65. He XC, Zeng MY, Li GF, Liang Z (1983). Callus induction and regeneration of plantlets from *Artemisia annua* and changes in quinghaosu contents. *Acta Bot Sin* 25: 87-90.

66. Heirwiegh KMG, Banerjee N, Neram K van, De Langhe E (1985). Somatic embryogenesis and plant regeneration in *Chicorium intybus* L. (witloof Compositae). *Plant Cell Rep* 4: 108-111.

67. Hendriks T, Scheer I, Quillet MC, Randoux B, Delbreil B, Vasseur JA (1998). Nonsymbiotic hemoglobin gene is expressed during somatic embryogenesis in Cichorium. *Biochim Biophys Acta* 1443: 193-197.

68. Hirata K, Yatini E, Asada M, Miyamoto K, Miura Y (1993). Effect of near ultra-violet light on alkaloid production in *Catharanthus roseus* plants. *Planta Medica*,59, 46–50.

69. Hisiger S, Jolicoeur M (2007). Analysis of *Catharanthus roseus* alkaloids by HPLC. *Phytochemistry Reviews*, 6, 207–234.

70. Hong SB, Hughes EH, Shanks JV, San KY, Gibson SI (2003). Role of the non-mevalonate pathway in indole alkaloid production by *Catharanthus roseus* hairy roots. *Biotechnol. Prog.* 19, 1105–1108.

71. Hong, SB, Peebles CAM, Shanks JV, San KY, Gibson SI (2006). Terpenoid indole alkaloid production by *Catharanthus roseus* hairy roots induced by *Agrobacterium tumefaciens* harboring rol ABC genes. *Biotechnology and Bioengineering*, 93, 386–390.

72. Hoopen HJG, Gulik T, Van WM, Schlatmann JE, Moreno PRH, Vinki JL, Heijnen JJ, Heijnen R, Verpoorte R, Ten Hoopen HJG, Gulik V (1994). Ajmalicine production by cell cultures of *Catharanthus roseus*: from shake flask to bioreactor. *Plant Cell Tissue Organ Cult* 38: 85–91.

73. Hui S, Lina C, Beijiu C, Xiaoning C (2011). Tissue culture and plant regeneration of chicory. *Animal Husbandry Feed Sci* 3 (2). 38-41.

74. Irmler S, Schroder G, St-Pierre B, Crouch NP, Hotze M, Schmidt J *et al.* (2000). Indole alkaloid biosynthesis in *Catharanthus roseus*: New enzyme activities and identification of cytochrome P450 CYP72A1 as secologanin synthase. *Plant J.* 24, 797–804.

75. Junaid A, Sheba HK, Zahid HS, Zohra F, Mehpara M, *et al.* (2010). *Catharanthus roseus* (L.). G. Don. An Important Drug: It's Applications and Production. *Pharmacie Globale (IJCP)*, 1: 4 (12).

76. Kai K, Mizutani M, Kawamura N, Yamamoto R, Tamai M, Yamaguchi H, Sakata K, Shimizu B (2008). "Scopoletin is biosynthesized via ortho-hydroxylation of feruloyl CoA by a 2-oxoglutarate-dependent dioxygenase in Arabidopsis thaliana." *Plant J* 55(6);989-99.

77. Kaneko T., S. Tahara, F. Takabayashi and N. Harada (2004). Inhibitory effect of esculin on oxidative DNA damage and carcinogenesis induced by N-nitrosobis (2-oxopropyl). amine in hamster pancreas. *Biofactors*. 22: 83-85.

78. Kim JT, Park JY, Seo HS, Oh HF, Noh JW, Kim JH, Kim DY, Youn, HJ (2002). In vitro antiprotozoal effects of artemisinin on Neospora caninum. *Vet. Parasitol.* 103: 53–63.

79. Kim NC, Kim SU (1992). Biosynthesis artemisinin from 11,12-dihydroarteannuic acids. *J. Kor. Agric. Chem. Soc. Rev.* 35: 106-109.

80. Kneusel RE (1987). Phenolische Verbindungen in der pflanzlichen Abwehr. Eine 4-Cumaroyl-CoA 3- Hydroxylase und eine S-Adenosyl-L-methionin: Kaffeoyl-CoA 3-O-Methyltransferase in Zellsuspensionskulturen von Petersilie (Petroselinum crispum). Diploma Thesis, Universita¨t Freiburg.

81. Kostovan I (2005). Synthetic and natural coumaris as cytotoxic agents. *Curr Med Chem Anti-Cancer Agents* 5: 29-46.

82. Kudakasseril GJ, Lam L, Staba EJ (1987). Effect of sterol inhibitors on the incorporation of $^{14}$C-isopentylpyrophosphate into artemisinin by a cell-free system from *Artemisia annua* tissue cultures and plants. *Planta Med.* 53: 280-284.

83. Kumari BDR, Velayutham P, Anitha S (2007). A comparative study on inulin and esculin content of in vitro and in vivo plants of Chicory (*Cichorium intybus* L. Cv. Lucknow Local). *Adv Biol Res* 1: 22-25.

84. Laughlin JC (1995). The influence of distribution of anti-malarial constituents in *Artemisia annua* L. on time and method of harvest. *Acta Hort.* 390: 67–73.

85. Lee-Parsons CWT, Ertürk S, Tengtrakool J (2004). Enhancement of ajmalicine production in *Catharanthus roseus* cell cultures with methyl jasmonate is dependent on timing and dosage of elicitation. *Biotechnol. Lett.*, 26: 1595-1599.

86. Lichtenthaler HK, Schwender J, Disch A, Rohmer M (1997). Biosynthesis of isoprenoids in higher plant chloroplasts proceeds via mevalonate independent pathway. *FEBS Lett.* 400, 271–274.

87. Lim Jung-Dae, Yang Deok-chun, Lee Hyeon-Yong, Kim Jong- Dai Lee Jin- Ha, Sung Eun-soo, Yu Chang-yeon (2004). Improved Micropropagation of root Chicory, *Cichorium intybus* L. var. *sativus*. *Korean J. Plant* Res 7 (1): 60-64.

88. Limami MA, Sun YL, Douat C, Hegleson J, Tepfer D (1998). Natural genetic transformation by *Agrobacterium rhizogenes*; annual flowering in two biennials, Belgian endive and carrot (witloof chicory). *J Plant Growth Regul* 18: 33-37.

89. Liu B, Ye H, Li G, Chen D, Geng S, Zhang Y (1998). Studies on dynamics of growth and biosynthesis of artemisinin in hairy roots of *Artemisia annua* L. *Biotehchnol.* 14: 248-254.

90. Liu DH, Jin HB, Chen YH, Cui LJ, Ren WW, Gong YF, Tang KX (2007). Terpenoid indole alkaloids biosynthesis and metabolic engineering in *Catharanthus roseus*. *J Integr Plant Biol.* 49: 961–974.

91. Loyola-Vargas VM, Mende-Zeel M, Monforte-Gonzales M, Miranda-Ham ML (1992). Serpentine accumulation during greening in normal and tumor tissues of *Catharanthus roseus*. *J Plant Physiol* 140: 213–217.

92. Ma C, Wang HH, Lu X, Wang H, Xu G, Liu B (2009a). Terpenoid metabolic profiling analysis of transgenic *Artemisia annua* L. by comprehensive twodimensional gas chromatography timeofflight mass spectrometry. *Metabolomics* 5: 497–506.

93. Mandelbaum-Schmid J (2005). World Health Organization efforts to boost antimalarial drug supply. News release 6 June 2005. Communications Officer, Roll Back Malaria WHO/Geneva http: //www.who.int/mediacentre/news/releases/2005/pr24/en/index.htmltin.

94. Margara J, Rancillac M (1966). Recherches experimentales sur la neoformation de bourgeons inflorescentiels ou vegetatifs *in vitro* a partir d' explantats d'endive *Cichorium intybus* L.II. Observations sur la vernalisation prealable de la racine. *Ann Physiol Veg* 8: 39-37.

95. Martinez BC, Staba J (1988). The production of artemisinin in *Artemisia annua* L.tissue cultures. *Adv. Cell. Cult.* 6: 69-87.

96. Martindale (1989). *The Extra Pharmacopoeia, 29th Ed.*, The Pharmaceutical Press, London. p. 1539.

97. Matvieieva NA, Kishchenko OM, Potrochov AO, Shakhovsk AMY, Kuchuk MV (2011a). Regeneration of transgenic plants from hairy roots of *Cichorium intybus* L. var. *Foliosum Hegi. Cytol Genet* 45 (5): 277-281.

98. Matvieieva NA, Vasylenko MY, Shahovsky AM, Bannykova MO, Kvasko OY, Kuchuk NV (2011b). Effective *Agrobacterium*-mediated transformation of Chicory (*Cichorium intybus* L.). with *Mycobacterium Tuberculosis* Antigene ESAT6. *Cyto Genet* 45: 7-12.

99. McGarvey DJ, Croteau R (1995). Terpenoid metabolism. Plant Cell 7, 1015–1026.

100. McKnight TD, Roessner CA, Devagupta R, Scott AI, Nessler CL (1990). Nucleotide sequence of a cDNA encoding the vacuolar protein strictosidine synthase from *Catharanthus roseus*. *Nucleic Acids Res.* 18, 4939.

101. Meijer AH, Cardoso MIL, Voskuilen JT, de Waal A, Verpoorte R, Hoge HC (1993). Isolation and characterization of a cDNA clone from *Catharanthus roseus* encoding NADPH: Cytochrome P450 reductase, an enzyme essential for reactions catalyzed by cytichrome P450 mono-oxygenases in plants. *Plant J.* 4, 47–60.

102. Mendis K, Sina BJ, Marchesini P, Carter R (2001). The neglected burden of *Plasmodium vivax* malaria. *American Journal of Tropical Medicine and Hygiene* 64 (Suppl. 1-2): 97-105.

103. Merali S and Meshnick SR (1991). Susceptibility of *Pneumocystis carinii* to artemisinin *in vitro* antimicrobial agents. *Chemother.* 35: 1225-1227.

104. Meshnick SR, Taylor TE, Kamchonwongpaisan S (1996). Artemisinin and the antimalarial endoperoxides: from herbal remedy to targeted chemotherapy. *Microbiological Reviews* 60(2): 301–315.

105. Miura Y, Hirata K, Kurano N, Miyamo K, Kazuhisa M, Keiichi (1988). Formation of Vinblastine in Multiple Shoot Culture. *Planta Med.* 54(1): 18-20.

106. Moorthy VS, Good MF, Hil AVS (2004). Malaria vaccine developments. *The Lancet* 363: 150-156.

107. Morales MR, Charles DJ, Simon JE (1993). Seasonal accumulation of artemisinin in *Artemisia annua* L. *Acta Hort.* 344: 416-420.

108. Moreno-Valenzuela OA, Galaz-Avalos RM, Minero-Garcia Y, Loyola-Vargas VM, (1998). Effect of differentiation on the regulation of indole alkaloid production in *Catharanthus roseus* hairy roots. *Plant Cell Rep.* 18, 99–104.

109. Moreno-Valenzuela OA, Minero-Garcý´a Y, Chan W, Mayer-Geraldo E, Carbajal E, Loyola-Vargas VM (2003). Increase in the indole alkaloid production and its excretion into the culture medium by calcium antagonists in *Catharanthus roseus* hairy roots. *Biotechnol Lett.* 25: 1345–1349.

110. Mujib A, Das S, Dey S, Bhattacharya B (1995). Influence of agitation in in vitro cultivation of *Catharanthus roseus* (L). G. Don multiple shoot. *Phytomorphology* 45: 239–245.

111. Mujib A, Ilah A, Aslam J, Fatima S, Siddiqui ZH, Maqsood M (2012). *Catharanthus roseus* alkaloids: application of biotechnology for improving yield. *Plant Growth Reg*. 68(2): 111–127.

112. Murashige T, Skoog F (1962). A revised medium for growth and bioassay with tobacco tissue culture. *Physiol Plant* 15: 473-497.

113. Nafis T, Akmal M, Mauji Ram, Alam P, Ahlawat S, Mohammad A, Abdin MZ (2011). Enhancement of artemisinin content by constitutive expression of HMG CoA Reductase gene in high yielding strain of *Artemisia annua* L. *Plant Biotechnol* Rep. 5: 53–60.

114. Nandagopal S, Kumari BDR (2006). Adenine sulphate induced high frequency shoot organogenesis in callus and *in vitro* flowering of *Cichorium intybus* L. cv. Focus - a potent medicinal plant. *Acta agric Slovenica* 87: 415-425.

115. Nandagopal S, Kumari BDR (2007). Effectiveness of auxin induced *in vitro* root culture in chicory. *J Central European Agric* 8: 73-80.

116. Narita JO, Gruissem W (1989). Tomato hydroxymethylglutaryl-CoA reductase is required early in fruit development but not during ripening. *Plant cell* 1, 181-190.

117. Nenz E, Varotto S, Lucchin M, Parrini P (2000). An efficient and rapid procedure for plantlet regeneration from Chicory mesophyll protoplasts. *Plant Cell Tissue Org Cult* 62: 85-88.

118. Nguyen KT, Arsenault PR, Weathers PJ (2011). Trichomes +roots + ROS = artemisinin: regulating artemisinin biosynthesis in *Artemisia annua* L. *in vitro. Cell Dev Biol Plant* 47: 329–338.

119. Nitsch JP, Bui Dang Ha, Nitsch C (1966). Extraction d'un facteur de bourgeonnement de l'endive (*Cichorium intybus* L). *Bull Soc Bot Fr* 113: 425-429.

120. Noble RL (1990). The discovery of the *Vinca* alkaloids-chemotherapeutic agents against cancer. *Biochem Cell Biol* 68: 1344–1351.

121. Nuutila AM (1994). Bioreactor studies on hairy root cultures of *Catharanthus roseus*. Composition of three bioreactor types. *Biotechnol. Tech.* 8: 61–66.

122. Pati PK, Kaur J, Singh P. 2011.A liquid culture system for shoot proliferation and analysis of pharmaceutically active constituents of *Catharanthus roseus* (L.). G. Don. *PCTOC*. 105, 299-307.

123. Paulet P, Nitsch JP (1964). Neoformation de eurs in vitro sur des cultures de tissus de racines de *C. intybus*. *CR Acad Sci Paris* 258: 5952-5955.

124. Peebles CA, Gibson SI, Shanks JV, San KY (2007). Characterization of an ethanol-inducible promoter system in *Catharanthus roseus* hairy roots. *Biotechnol. Prog.* 23: 1258–1260.

125. Peebles CA, Hong SB, Gibson SI, Shanks JV, San KY (2005). Transient effects of overexpressing anthranilate synthase alpha and beta subunits in *Catharanthus roseus* hairy roots. *Biotechnol Prog*. 5: 1572-6.

126. Pietrogrand M Ch, Reschiglian P, Domi F, Kahie YD, Bartolasi V (1992). Correlations between high-performance liquid chromatographic retention, X-ray structural and 13- C NMR spectro- scopic data of flavonoid compounds. *J. Chromat.* 592: 65-73.

127. Ravindranathan T, Anil Kumar M, Menon RB, Hiremath SV(1990). Stereo selective synthesis of artemisinin. *Tetrahedron Lett*. 31: 755-758.

128. Razdan MK (2000). *An introduction to plant tissue culture.* Oxford and IBH Publishing Co. Pvt. Ltd. New Delhi 398.

129. Rehman RU, Israr M, Srivastava PS, Bansal KC and Abdin MZ (2003). *In vitro* regenenation of witloof chicory (*Cichorium intybus* L.). from leaf explants and accumnlation of esculin. *In vitro Cell Dev Biol Plant* 39: 142-146.

130. Rodr´ýguez-Concepci ´on M, Boronat A (2002). Elucidation of the methylerythritol phosphate pathway for isoprenoid biosynthesis in bacteria and plastids. A metabolic milestone achieved through genomics. *Plant Physiol.* 130, 1079–1089.

131. Romanowoski G, Lorenz MG, Wackernagel W (1993). Plasmid DNA in a ground water aquifer microcosm adsorption, DNAase resistance and natural genetic transformation of *Bacillus subtilis*. *Mol Ecol* 2: 171-181.

132. Roytrakul S, Verpoorte R (2007). Role of vacuolar transporter proteins in plant secondary metabolism: *Catharanthus roseus* cell culture. *Phytochem. Rev.* 6: 383-396.

133. Rydén AM, Ruyter-Spira C, Quax WJ, Osada H, Muranaka T, Kayser O, Bouwmeester H. (2010). The molecular cloning of dihydroartemisinic aldehyde reductase and its implication in artemisinin biosynthesis in *Artemisia annua*. *Planta Med.* 76(15): 1778-83.

134. Sa G, Mi M, He-chun Y, Ben-ye L, Guo-feng L, Kang C (2001). Effect of ipt gene expression on the physiological and chemical characteristices of *Artemisia annua* L. *Plant Sci.* 160: 691-698.

135. Sasksi N, Dubois J, Millicamps JL, Vasseur J (1986). Regeneration de plantes de chicoree witloof cv. Zoom a partir de protoplastes: influence de la nutrition glucidique et azotee. *CR Acad Sci Paris* 302: 165-170.

136. Sato M (1967). Metabolism of phenolic substances by the chloroplasts–III : Phenolase as an enzyme concerning the formation of esculetin. *Phytochemistry* 6: 1363– 1373.

137. Shanks JV, Bhadra R, Morgan J, Rijhwan R, Vani S (1998). Quantification of metabolites in the indole alkaloid pathways of *Catharanthus roseus*: implications for metabolic engineering. *Biotechnol. Bioeng.* 58, 333–338.

138. Simon JE, Charles D, Cebert E, Grant L, Janick J, Whipkey A. (1990). *Artemisia annua* L., a promising aromatic and medicinal. In: *Advances in New Crops*. Janick J, Simon JE, editors. West Lafayette: Timber Press, pp.522-526.

139. Singh A, Nirala NK, Das S, Narula A, Rajam MV, Srivastava PS (2011). Overexpression of odc (ornithine decarboxylase). in *Datura innoxia* enhances the yield of scopolamine. *Acta Physiol Plant* 33: 2453-2459.

140. Singh NP, Lai HC (2004). Artemisinin induces apoptosis in human cancer cells. *Anticancer Res* 24(4): 2277-2280.

141. Smith TC, Weathers PJ, Cheetham RC (1997). Effect of gibberellic acid on hair root cultures of *Artemisia annua* growth and artemisinin production. *In vitro Cell Dev. Biol. Plant* 33: 75-79.

142. Sosa ME, Tonn CE (2008). Plant Secondary Metabolites from Argentinean Semiarid Lands: Bioactivity against Insects. *Phytochemistry Reviews*, 7: 3-24.

143. Spranger N, Schellenbaum L, van Dun K, Boller T and Weiken A (1997). Fructan synthesis in transgenic tobacco and chicory plants expressing barley sucrose: fructan 6G-fructosyltransferase. *FEBS Lett* 400: 355-358.

144. Stanic G, Jurisic B, Brkic D (1999). HPLC analysis of esulin and fraxin in horse chestnut bark (*Aesculus hippocastanum* L.). *Croatica Chemica Acta* 72: 827-834.

145. Tang ZH, Liu YJ, Guo XR, Zu YG. 2011. The combined effects of salinity and nitrogen forms on *Catharanthus roseus*: the role of internal ammonium and free amino acids during salt stress. *J Plant Nutrition Soil Sci*. 174: 135–144.

146. Tawfiq NK, Anderzon LA, Roberts MF, Phillipson JD, Bray DH, Warhurst DC (1989). Antiplasmodial activity of *Artemsia annua* plant cell culture. *Plant Cell Rep* 8: 425-428.

147. Tepfer D. (1984). Transformation of several species of higher plants by *Agrobacterium rhizogenes*: sexual transmission of the genotype and phenotype. *Cell* 37: 959-967.

148. Toponi M (1963). Action combine de la kinetine et de l'acide indolylacetique sur la neoformation d'organes par des fragments de feuilles d'endive (*Cichorium intybus* L.). cultives *in vitro*. *CR Acad Sci Paris* 257: 3030-3033.

149. Towler MJ, Kim Y, Wyslouzil BE, Correll MJ, Weathers PJ (2006). Design, development and applications of mist bioreactors for micropropagation and hairy root culture. In: Dutta Gupta S, Ibaraki Y (eds). *Plant tissue culture engineering, series: focus on biotechnology*, vol 6. Springer, Dordrecht, pp. 119–134.

150. Utzinger J, Xiao S, N'Goran EK, Berquist R, Tanner M (2001)The potential of artemether for the control of schistosomiasis. *Int J Parasitol* 31: 1549–1562.

151. Van Agtmael MA, Eggelte TA, van Boxtel CJ (1999). Artemisinin drugs in the treatment of malaria: from medicinal herb to registered medication. *Trends Pharmacol Sci* 20, 199–205.

152. Van der Fits L, Memelink J (2000). ORCA3, a jasmonate-responsive transcriptional regulator of plant primary and secondary metabolism. *Science* 289: 295–297.

153. van der Heijden R, Jacobs DJ, Snoeijer W, Hallard D, Verpoorte R (2004). The *Catharanthus* alkaloids: pharmacognosy and biotechnology. *Curr Med Chem* 11: 1241–1253.

154. Varotto S, Lucchin M, Parrini P (1997). Plant regeneration from protoplasts of Italian red chicory red chicory (*Chicorium intybus.* L.*). J Genet Breed* 51: 17-22.

155. Vasseur J (1979). Etude du bourgeonnement de fragments de feuilles etiolees d' endive. II. Formation des bourgeons en function de caracteristiques biochimiques. *Rev Gen Bot* 86: 113-190.

156. Veau B, Courtois M, Oudin A, Chenieux JC, Rideau M, Clastre M (2000). Cloning and expression of cDNAs encoding two enzymes of the MEP pathway in *Catharanthus roseus. Biochim. Biophys. Acta* 1517, 159–163.

157. Velayutham P, Ranjithakumari BD, Bhaskaran P (2006). An efficient in vitro plant regeneration system for *Cichorium intybus* L.: An important medicinal plant. *J Agric Technol* 2 (2): 287-298.

158. Vergauwe A, Cammaert R, Vanberghe D, Genetello C Van Montagu DM, Van den Eeckhout E (1996). *Agrobacterium tumefaciens*-mediated transformation of *Artemisia annua* L. and regeneration of transgenic plants, *Plant Cell Rep* 15: 929–933.

159. Vergauwe A, Van Geldre E, Inze D, Van Montagu M, Van den Eeckhout E (1998). Factors influencing *Agrobacterium tumefaciens*-mediated transformation of *Artemisia annua. Plant Cell Rep* 18: 105–110.

160. Vermeulen A, Vaucheret H, Pautot V, Chupeau Y (1992). Agrobacterium mediated transfer of a mutant Arabidopsis acetolactate synthase gene confers resistance to chlorsulfuron in chicory (*Cichorium intybus* L.). *Plant Cell Rep* 11: 243-247.

161. Verpoorte R, van der Heijden R, Memelink J (1998). In: *The Alkaloids,* Ed. Cordell GA, Academic Press, San Diego, CA 50: 453-508.

162. Verpoorte R, van der Heijden R, ten Hoopen HJG, Memelink J (1999). Metabolic engineering of plant secondary metabolite pathways for the production of fine chemicals. *Biotechnol Lett* 21: 467-479.

163. Wang H, Ge L, Ye HC, Chong K, Lie BY, Li GF (2004). Studies on the effects of fpf1 gene on *Artemisia annua* flowering time and on the linkage between flowering and artemisinin biosynthesis. *Planta Med.* 70: 347–352.

164. Wang H, Lill Y, Ching K, Liu B, Ye HC, Li Z, Yan F, Li GF (2007). Earlier flowering induced by over-expression of co gene does not accompany increase of artemisinin biosynthesis in *Artemisia annua Plant Bial. (stuttg).* 9: 442-446.

165. Weathers PJ, Cheethan RD, Follansbee E, Theoharides K (1994). Artemisinin production by transformed roots of *Artemisia annua. Biotechnol. lett.* 16: 1281-1286.

166. Weathers Pl, Elkholy S, Wobbe, KK (2006). Artemisinin: the biosynthetic pathway and its regulation in *Artemisia annua*, a terpenoid-rich species. *In vitro Cell Dev. Biol. Plant* 42: 309-317.

167. Whipkey A, Simon JE, Charles DJ, Janick J (1992). In vitro production of artemesinin from *Artemisia annua* L. *Phytother. Res.* 1: 15-25.

168. Whitmer S, van der Heijden R, Verpoorte R (2002a). Effect of precursor feeding on alkaloid accumulation by a strictosidine synthase over-expressing transgenic cell line S1 of *Catharanthus roseus*. *Plant Cell Tiss Org Cult* 69(1): 85–93.

169. Whitmer S, van der Heijden R, Verpoorte R (2002b). Effect of precursor feeding on alkaloid accumulation by a tryptophan decarboxylase over-expressing transgenic cell line T22 of *Catharanthus roseus*. *J Biotechnol* 96(2): 193–203.

170. WHO (1998). Roll Back Malaria. A Global Partnership. Geneva: World Health Organization.

171. WHO Roll Back Malaria Department (2005a). World Health Organization drives efforts to boost antimalarial drug supply.

172. Williams CA, Goldastone F, Greenham J (1996). Flavonoid, cinnamic acids and coumarins from the different tissues and medicinal preparations of *Taraxacum officinale*. *Phytochemistry* 42: 121-127.

173. Woerdenbag HJ, Luers JFJ, van Uden W, pras N, Malingre TM, Alfermenn AW (1993). Production of the new antimalarial drug artemisinin in shoot cultures of *Artemisisa annua*. *Plant cell Tiss. Org. Cult.* 32: 247-257.

174. Woerdenbag HJ, Lugt CB, Pras N (1990). *Artemisia annua* L.: a source of novel antimalarial drugs. Pharmaceutisch Weekblad. Scientific edition 12(5): 169-81.

175. Yasseen MY, Splittstoesser WE (1995). Somatic embryogenesis from leaf of witloof chicory through leaf of suspension culture. *Plant Cell Rep* 14: 804-806.

176. Zhang L, Jing F, Li F, Li M, Wang Y, Wang G, Sun X, Tang K (2009). Development of transgenic *Artemisia annua* (Chinese wormwood). plants with an enhanced content of artemisinin, an effective antimalarial drug, by hairpinRNA mediated gene silencing. *Biotechnol Appl Biochem* 52: 199–207.

177. Zhou ML, Shao JR, Tang YX (2009). Production and metabolic engineering of terpenoid indole alkaloids in cell cultures of the medicinal plant *Catharanthus roseus* (L.). G. Don (Madagascar periwinkle). *Biotechnol Appl Biochem.* 52: 313-23.

178. Zoden B (1998). Opinion of the scientific committee on plants regarding chicory derived from genetically modified male sterile, glufosinate tolerant parental lines (RM-3, RM3-4, RM 3-6). Notification Number C/NL/94/25-A.

179. Zuo and Chua (2000). Chemical-inducible systems for regulated expression of plant genes. *Curr. Opin. Biotechnol.*, 11: 146–151.

*Section VIII*

# Cancer

**2016, Biotechnology: Progress and Applications** *Pages* **451–466**
*Editors:* **Saif Hameed and Zeeshan Fatima**
*Published by:* **DAYA PUBLISHING HOUSE, NEW DELHI**

# *Chapter 19*

# Role of FOXO Gene Family in Breast Cancer

***Maria Habib* and Syed Akhtar Husain***

*Department of Biosciences, Jamia Millia Islamia, New Delhi – 110 025, India*

## ABSTRACT

Breast cancer is a tumor that starts from cells of the breast tissue, either in cells that line the ducts that carry milk to the nipples (ductal cancer) and/or in cells that line the lobules, which are glands involved in milk production. Owing to the lack of awareness of this disease and in absence of a breast cancer screening program, the majority of breast cancers are diagnosed at a relatively advanced stage. Although there are some centers of excellence providing multimodality protocol-based treatment at par with the best anywhere in the world, the vast majority of breast cancer patients undergo inadequate and inappropriate treatment due to lack of high-quality infrastructure and sometimes skills, and above all financial resources. Given that cancer and related disorders affect a wide spectrum of the world's population, and in most cases are progressive in nature, it is essential that future care must overcome the present limitations of existing therapies in the absence of toxic side effects. Mammalian forkhead transcription factors of the O class (FOXOs) may fill this niche since these proteins are increasingly considered to represent unique cellular targets directed against human cancer in light of their pro-apoptotic effects and ability to lead to cell cycle arrest. FOXOs can also significantly affect normal cell survival and longevity, requiring new treatments

* Corresponding Author: E-mail: habibmaria@hotmail.com

for neoplastic growth to modulate novel pathways that integrate cell proliferation, metabolism, inflammation and survival. In this respect, members of the FOXOs family are extremely compelling to consider since these transcription factors have emerged as versatile proteins that can control angiogenesis, stem cell proliferation, cell adhesion and autoimmune disease.

## Introduction

Breast cancer is the second most common cancer in the world and also the most frequent cancer among women. According to GLOBOCAN 2011, 1.67 million new breast cancer cases were diagnosed in 2012 (25 per cent of all cancers) making it the most common cancer in women both in more and less developed regions with slightly more cases in less developed (883,000 cases) than in more developed (794,000) regions. Incidence rates vary nearly four-fold across the world regions, with rates ranging from 27 per 100,000 in Middle Africa and Eastern Asia to 96 in Western Europe. Breast cancer ranks as the fifth cause of death from cancer overall (522,000 deaths) and while it is the most frequent cause of cancer death in women in under developed countries (324,000 deaths, 14.3 per cent of total), it is now the second cause of cancer

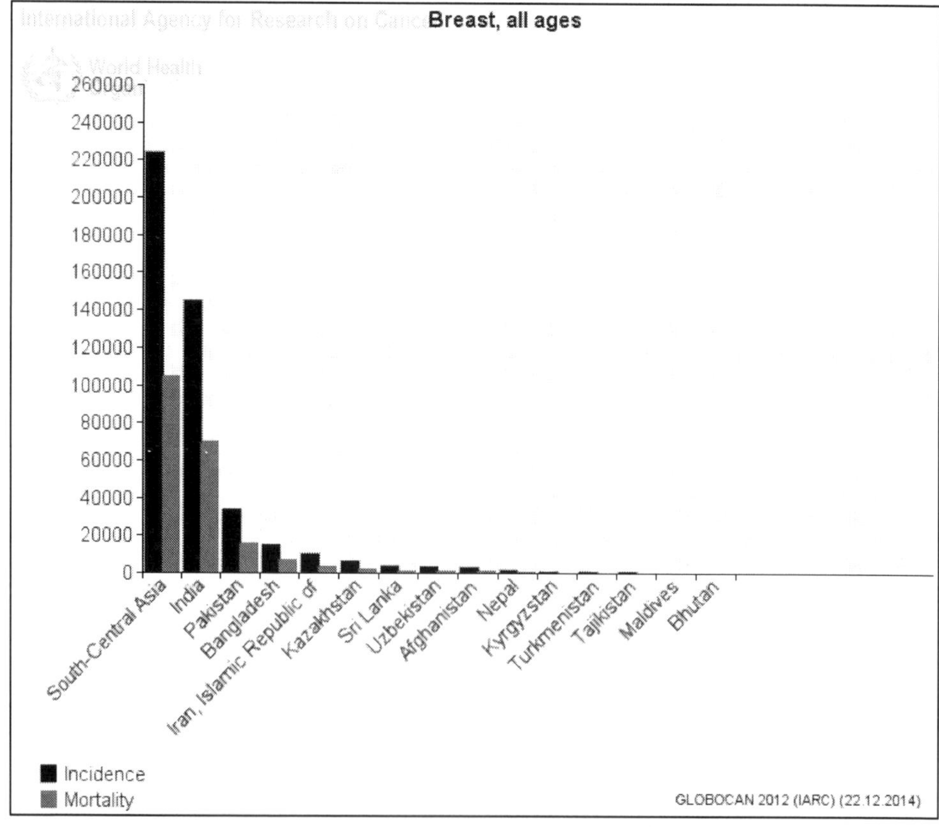


**Figure 19.1: Comparison between Incidence and Mortality for Breast Cancer in different Asian Countries.**

death in developed nations (198,000 deaths, 15.4 per cent). The range in mortality rates between world regions is less than that for incidence because of the more favourable survival of breast cancer in (high-incidence) developed regions, with rates ranging from 6 per 100,000 in Eastern Asia to 20 per 100,000 in Western Africa.

Breast cancer is tumour that originates from cells of breast tissue and can be benign or malignant. About 1.3 lakh fresh cases of breast cancer are reported in India every year. Breast cancer is the common most cancer of urban Indian women and second most common among rural women. There are an estimated 100000-125000 new breast cancer cases reported in India every year. This number is estimated to increase 2 folds by 2025. In the metropolitan cities, breast cancer is the leading cancer diagnosis in women, with rates nearly twice as common as cervical cancer. It was estimated that in 2008 there were 115,251 new cases of breast cancer with an age standardised incidence rate of 22.9 per 100,000. It is now estimated that by 2030 the number of new cases of breast cancer in India will reach just under 200,000 per year. Moreover, the average age of developing a breast cancer has undergone a significant shift over last few decades. More than 80 per cent of Indian patients are younger than 60 years of age and a significant proportion of Indian breast cancer patients are younger than 35 years of age Fig. 19.2.

Maximum inherited cases of breast cancer are associated with two abnormal genes: BRCA1 (BReast CAncer gene one) and BRCA2 (BReast CAncer gene two). Changes in other genes are also associated with breast cancer such as, *CDH1*, *STK11*, and *TP53* genes. Those cases of breast cancer which are not inherited are associated with genetic changes that occur only in breast cancer cells (somatic mutations) and occur during a person's lifetime.

## PI3K Pathway

The PI3K/AKT/mTOR pathway is very commonly deregulated in a wide range of human cancers including breast cancer (Andreopoulou, 2011). Its activation has a number of downstream effects which either promote protein synthesis or constrain protein breakdown and so it has become a major focus of attention owing to its critical regulatory role in diverse cellular processes, including cancer progression. Numerous components of PI3K/AKT pathway are targeted by amplification, mutation and translocation more recurrently than any other pathway in cancer patients (Altomare *et al.*, 2005). This warrants exploiting the PI3K/AKT pathway for cancer drug discovery. Many established nodal points in the PI3K pathway (*i.e.*PIK3CA, AKT, PTEN) have been linked to cancer. Therefore, most treatment strategies are focused on these entrenched altered regulators, though with restricted success.

Modifications succeeding in activation of PI3K/Akt/mTOR signalling are perhaps the most recurrent events observed in solid tumors, including breast cancer, and contribute to neoplastic transformation. The PI3K/mTOR pathway can be activated by overproduction of growth factors or chemokines, loss of phosphatase and tensin homolog (PTEN) expression, or by mutations in growth factor receptors Ras, PTEN, or PI3K itself (Andreopoulou 2011). Activation of this pathway endorses cell cycle proliferation, growth, cell cycle entry, survival, cell motility, protein synthesis, and glucose metabolism, all of which are important aspects of tumorigenesis (Yuan

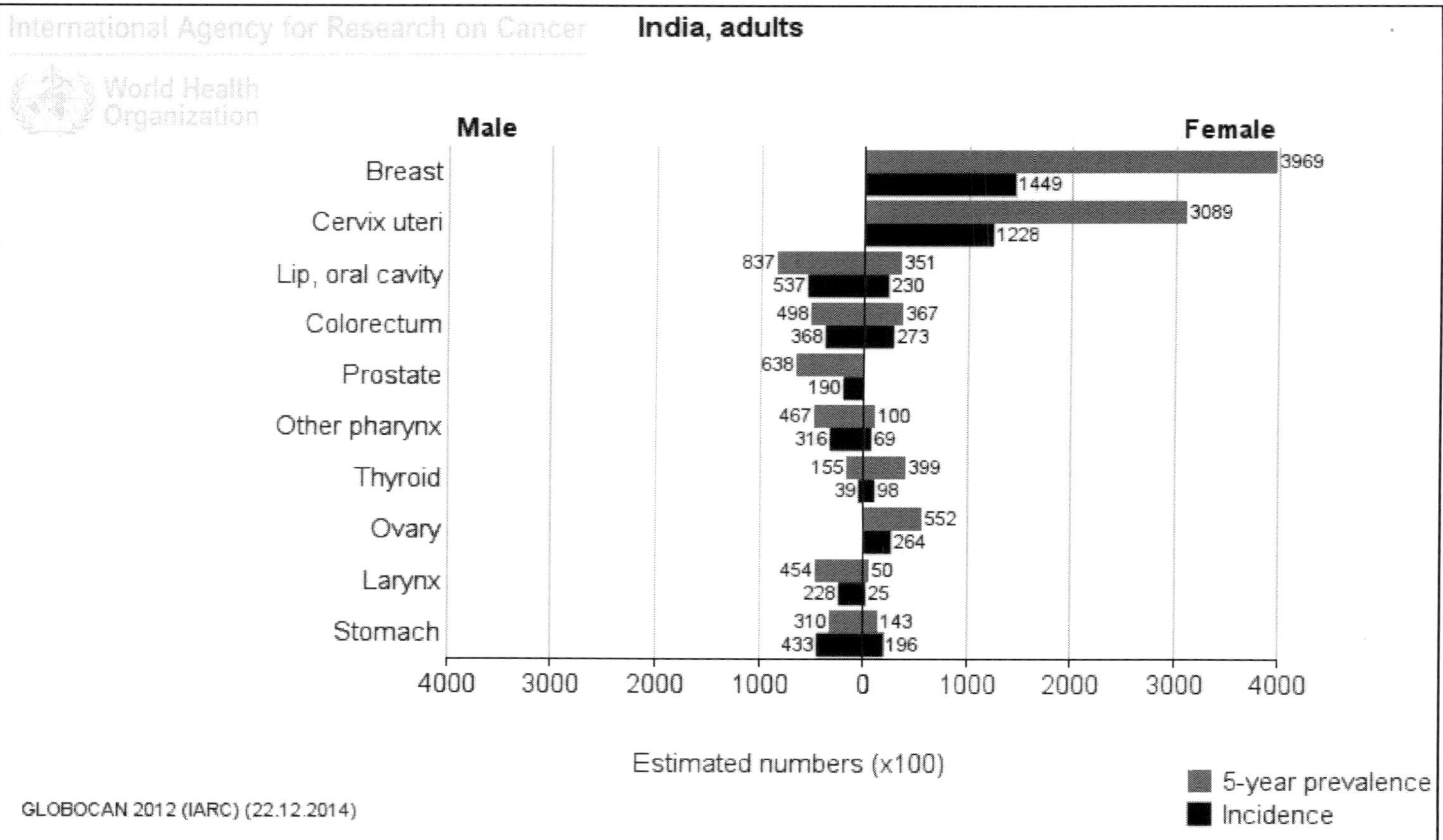


**Figure 19.2: A Comparison if Incidence and 5 Year Prevalence of different Cancers in India.**

*et al.*, 2008). Upon activation, the PI3K/AKT/mTOR pathway can be propagated to various substances, including mTOR which is a master regulator of protein translation. Initial activation of the pathway occurs at the cell membrane, where the signal for pathway activation is circulated through class IA PI3K. Activation of PI3K results in increased formation of the lipid PIP3 from PIP2, which in turn results in recruitment of AKT to the plasma membrane and its subsequent phosphorylation and activation. PIK3CA gene mutations represent one of the most common genetic aberrations in human breast cancer and have been reported in 20 per cent to 40 per cent of cases, although the complex biology of downstream effects remains unclear (Benedetti *et al.*, 2004). In experimental models, these mutations have been shown to activate AKT and induce oncogenic transformation, rendering tumors highly sensitive to therapeutic PI3K/mTOR inhibition (Soni *et al.*, 2008).

However, exploring the possibility that other PI3K pathway members which are frequently mutated can be effective therapeutic targets, may prove advantageous. The PIK3CA gene, is located on chromosome 3q26.3 and consists of 20 exons (Gene Card). Mutations in the p110_ subunit of *PI3K*, called *PIK3CA*, are often responsible for activation of the phosphatidylinositol 3-kinase (PI3K)/AKT/mammalian target of rapamycin (mTOR) pathway and have been reported in various human cancers. *PIK3CA* mutations can trigger neoplastic transformation and promote cancer

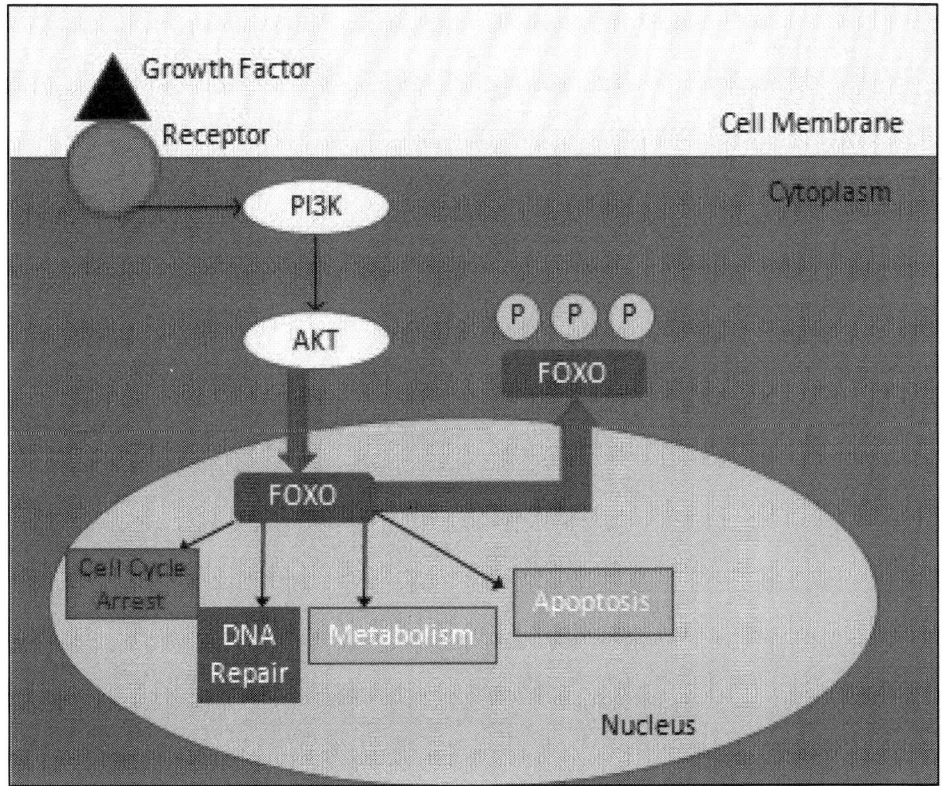


**Figure 19.3: Diagrammatic Representation of PI3K/AKT/FOXO Pathway.**

progression. The PI3K/AKT/mTOR pathway is often deregulated in gynecologic and breast cancers, and *PIK3CA* mutations have been reported in approximately 18 per cent of breast 17 per cent to 33 per cent of cervical, 39 per cent of endometrial and 12 per cent of ovarian cancers (Castenada *et al.*, 2010). Preclinical studies proposed that *PIK3CA* mutations could predict response to PI3K and mTOR inhibitors, although mutations in the mitogen-activated protein kinase (MAPK) pathway (*KRAS, NRAS, BRAF*) might mediate resistance (Yuan and Cantley, 2008).

## FOXO Gene Family

FOX (Forkhead box) proteins are a family of transcription factors which are evolutionarily conserved and play important roles in regulating the expression of genes involved in cell growth, proliferation, differentiation, and longevity (Zhang *et al.*, 2011). Transcription factors are shown to have important roles in cancer, if they impact the activity of genes involved in the cell cycle (Archer, 2011). In addition, transcription factors can be the products of oncogenes or tumor suppressor genes. They function in the nucleus in the presence of genes and their nuclear transport can influence the activity of genes present in nucleus. Moreover, the post translational modifications such as phosphorylation also control the activity of transcription factors. FOXO transcription factors are at the interface of crucial cellular processes, orchestrating programs of gene expression (production of proteins) that regulate apoptosis (cellular programmed death), cell-cycle progression, and oxidative- stress resistance.

The FOXO sub family of Forkhead transcription factors is a direct downstream target of PI3K/AKT pathway (Kops *et al.*, 1999). These transcription factors share the ability to be inhibited and translocated out of the nucleus upon phosphorylation by proteins such as PKB/AKT in PI3K signalling pathway. Mammalian FOXO proteins, homologous to *Caenorhabditis elegans* protein DAF-16, belongs to O (other) class of FOX superfamily (Barthel *et al.*, 2005). As transcription factors in the nucleus, their primary function is to bind to their associated DNA targeting sequences. They seem to be involved in numerous signalling pathways and control a wide range of biochemical processes including tumour suppression (Paik *et al.*, 2007). Due to certain post translational modifications, they interact with other transcription factors to regulate transcription of target genes. The FOXO proteins encoded by FOXO gene family are regulated at multiple levels which are not just confined to phosphorylation, acetylation and ubiquitylation but are also ideal to regulate tumour growth through pro apoptotic pathways and blocking of cell cycle progression. This gene family functions at the interface of tumour suppression, energy metabolism and organismal longevity.

Growth factors have been shown to down regulate the expression and activity of FOXOs (Arden, 2006). They are regulated by a broad variety of external stimuli such as epidermal growth factor receptors (EGFR), insulin, insulin like growth factor, neurotrophins, nutrients, cytokines and oxidative stress. These stimuli control FOXO expression, sub cellular localization or DNA binding and transcriptional activity (Yang *et al.*, 2009).

Transcriptionally, FOXOs either activate or inhibit downstream target genes. Their functions are regulated by multiple signalling pathways. Many such pathways are deregulated in cancer suggesting that FOXO factors may play a tumour suppressor role in variety of cancers. It was initially postulated that 3 out of 4 known FOXO genes are found at chromosomal breakpoints in certain tumours. Contribution of FOXO inactivation in cellular transformation can also be inferred from the fact that FOXOs negatively regulate cell survival and cell cycle progression in mammalian cells. It has been hypothesized that FOXO factors are liberalized in several tumours including Breast cancer (Neilson, 2004).

On certain rare occurrences, FOXOs can also associate themselves with oncogenes. Forkhead subfamily member FOXM1 functions like a classic oncogene. Elevated levels of this gene are documented in cancers of liver, prostate, lung, colon and breast (Myatt and Lam 2007; Pelarsky *et al.*, 2014). FOXOs act as effective transcriptional activators, but gene array analyses indicate that they can also act as transcriptional repressors (Ramaswamy *et al.*, 2002)

Also the deletion of FOXOs have given an insight into their functions. Global deletion of FOXO1 leads to embryonic cell death due too incomplete vascular development (Hosaka 2004), deletion of FOXO3 causes organ inflammation, lymph proliferation (Lin, 2004), age dependent infertility (Hosaka, 2004) and decline in neural stem cell pool (Renault, 2009). FOXO4 deletion triggers colitis and FOXO6 deletion displays impaired memory consolidation (Salih, 2012).

In humans, FOXOs have 4 members; FOXO1, FOXO3, FOXO4 and FOXO6. FOXO1 and FOXO3 are expressed in practically all tissues. FOXO4 is greatly expressed in muscles, kidney and colorectal tissue, while FOXO6 is chiefly expressed in brain (Wang *et al.*, 2014).

The biological characteristics of a tumour are used to estimate prognosis and select appropriate therapy for patients. Immunohistochemical (IHC) markers are widely used to guide treatment decisions. Results of IHC studies in FOXO gene family have provided significant results for prognosis of different cancers.

## FOXO1

Genomic location of FOXO1 is chromosome 13 (13q14.1) (GeneCards). It is by far the best studied member of the FOXO sub family. At least 100 Forkhead genes and 19 human subgroups ranging from FOXA to FOXS are known till date. As reported by Labied *et al.*, 2006, FOXO1 expression in human endometrium increases during secretory phase of the cycle. Goto *et al.*, 2008, demonstrated that FOXO1 functions as a tumour suppressor in endometrial cancer cells and is involved in normal growth control and maintenance of genomic stability. Similar observations were conveyed by Myatt *et al.*, in 2009 with respect to endometrial cancer. Goto *et al.*, 2008 also reported down regulation of FOXO1 in ovarian cancer. In another study by Dong *et al.*, 2006, the outcomes reflected that not only hemizygous deletion of FOXO1A is frequent in human prostate cancer, but also this gene is also down regulated in prostate cancer. Several reports also indicate that down regulation of FOXO1 is observed in cervical cancer (Narayan *et al.*, 2014), lung cancer (Zhao *et al.*, 2014).

As reported by Lo *et al.*, 2010, FOXF1 transcription factor has a novel role to play in cell cycle regulation and its downstream effectors may also serve as novel molecular targets for detection and treatment of breast cancer and other human cancers. Gutilla *et al.*, 2009, acknowledged the novel mechanism for FOXO1 down regulation in breast cancer upsetting the transformation of cells and maintenance of an oncogenic state. FOXO1 is also observed to be down regulated in ovarian cancer cells (Goto 2008). Recently, FOXO1 down regulation is reported in cervical cancer (Narayan *et al.*, 2014), small cell lung cancer (Zhao *et al.*, 2014) and Hodgkins Lymphoma (Xie *et al.*, 2014).

In case of Non-small cell lung carcinoma (NSCLC), IHC staining of clinical samples revealed that FOXO1 was present in both cytoplasm and nucleus. Therefore it is postulated that FOXO1 localization in NSCLC may not be a strict measure of activity (Maekawa, 2009). In case of Hodgkins Lymphoma, IHC confirmed high localization of FOXO1 in B-cell Lineage (Xie *et al.*, 2014). In case of Primary mediastinal B-cell lymphoma. IHC depicted low/negative levels of FOXO1 (Ushmora *et al.*, 2014). In normal endometrial cells, FOXO1 staining was observed to be strong in both nucleus and cytoplasm, whereas in case of cancerous cells the localization was weak in nucleus and was confined to epithelial cells (Myatt, 2009). For breast cancer cell line, the localization was strong in nucleus (Figure 19.4).

## FOXO3

Genomic location for FOXO3 (FOXO3a) is chromosome 6 (6q21) (GeneCards). As seen for FOXO1, FOXO3 also overpowers tumour growth and refurbishment of its activity promotes tumour cell death. It is seen in chronic leukaemia as well as breast cancer that use of anticancer drugs (STI571 and Paclitaxel) inhibit tumour growth by increasing levels of Bim expression via up regulation of FOXO3 (Essafi *et al.*, 2005; Sunters *et al.*, 2010). Hui *et al.*, 2008 published a study highlighting the role of FOXO3a in sensing cellular stress induced Doxorubicin, implicating the effect of this transcription factor in affecting cell cycle arrest in Leukemic cells. Jiang *et al.*, 2013 specified in his study that FOXO3a expression is a favourable prognostic marker in breast cancer. In 2011, a report by Habashy *et al.*, also postulated that FOXO3a nuclear localization is associated with good prognosis in luminal like breast cancer.

In a study by Hu *et al.*, 2004, FOXO3a localized in cytoplasm of breast cancer cells and it was postulated that this predominant cytoplasmic FOXO3a localization is inversely associated with patient survival. Contradictory results were obtained by Chen *et al.*, 2010 who on the basis of IHC staining, proposed that predominant nuclear localization of FOXO3a co relates itself with enhanced PI3K/AKT activity. Habashy *et al.*, 2011 suggested for breast cancer that nuclear localization is associated for good prognosis. In case of colorectal cancer, FOXO3a was strongly expressed in nucleus of normal tissue but the localization was cytoplasmic for cancerous tissues (Gong *et al.*, 2013).

Some recent reports project that, besides its function as a tumour suppressor gene, FOXO3 might also expedite cancer cell survival under certain circumstances. In certain tumour types, FOXO3 persuades detoxification and stress resistance which

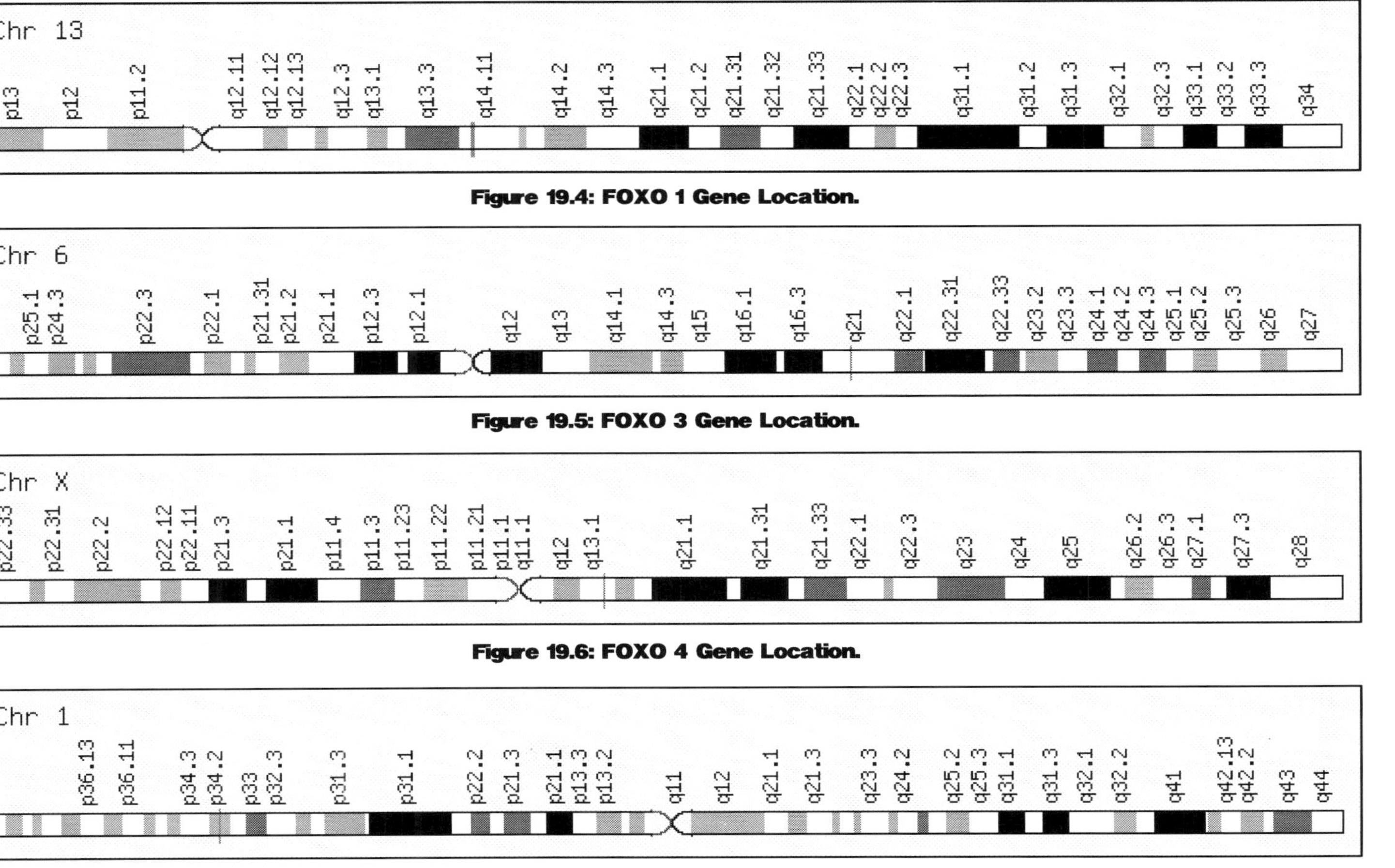


**Figure 19.4: FOXO 1 Gene Location.**

**Figure 19.5: FOXO 3 Gene Location.**

**Figure 19.6: FOXO 4 Gene Location.**

**Figure 19.7: FOXO 6 Gene Location.**

eventually leads to tumour stem cell renewal and protection of cancer cells from eradication during chemotherapy (Figure 19.5).

### FOXO4

Genomic location for FOXO4 is chromosome X (Xq13.1) (GeneCards). It is also recognized as a tumour suppressor gene. Its expression is significantly decreased or deleted in colorectal cancer tissue as reported by Xiang-quang, 2011. Also it was postulated by Yang *et al.*, 2005 that active FOXO4 can reduce tumour onset, size and progression in HER2 positive breast cancer cells. It can also inhibit growth factor mediated tumour metastasis in cholangiosarcoma (Lee *et al.*, 2009). An expression study was conducted by Su *et al.*, 2014 on gastric cancer cells concluded that FOXO4 expression reduced significantly in not only gastric cancer tissues but also in various gastric cancer cell lines. They also proposed that up regulation of FOXO4 induces significant G1 arrest, S phase reduction and down regulation of Vimentin and can thus serve as a potential therapeutic target for gastric cancer.

Localization of FOXO4 in case of NSCLC is studied by Lui *et al.*, 2014. They reported the localization of FOXO4 in nucleus for cancerous tissues. For gastric cancer, non-tumorous tissues showed nuclear localization (Su *et al.*, 2014) (Figure 19.6).

### FOXO6

This transcriptional activator is located on chromosome 1 (1p34.2) (GeneCards). Due to limited no of known biological functions, it is the least studied member of FOXO gene family. In a pilot study by Qinyu 2013, FOXO6 expression was found to be up regulated in gastric cancer tissues. It was also seen that over expression of FOXO6 promoted proliferation.

Clinical significance of FOXO6 in cancer biology is still to be recognized as little is known about its role in different cancers. It can thus be expected that FOXO6 may have a significant impact in breast cancer too (Figure 19.7).

## Future Perspectives

Biomarkers are important tools for cancer detection and monitoring. They serve as trademarks for the physiological status of a cell at a given time and change during the disease process. Gene mutations, alterations in gene transcription and translation, and alterations in their protein products can all potentially serve as specific biomarkers for any disease. Owing to the critical role played by FOXO gene family in tumour suppression, a better understanding of the mechanisms which regulate FOXO gene family is required. Their contribution in cancer initiation, progression and drug resistance may render these transcription factors as crucial prognostic markers and therapeutic targets for breast cancer and other malignancies. Although there has been a substantial progress in deciphering the functions and regulation of FOXOs a lot still remains to be revealed. Since inactivation of FOXOs is a crucial step in tumorigenesis, restoring activity of these factors promises a potential effective therapeutic strategy.

Given that cancer and related disorders affect a wide spectrum of the world's population, and in most cases are progressive in nature, it is essential that future care must overcome the present limitations of existing therapies in the absence of toxic side effects. Mammalian forkhead transcription factors of the O class (FOXOs) may fill this niche since these proteins are increasingly considered to represent unique cellular targets directed against human cancer in light of their pro-apoptotic effects and ability to lead to cell cycle arrest. Yet, FOXOs also can significantly affect normal cell survival and longevity, requiring new treatments for neoplastic growth to modulate novel pathways that integrate cell proliferation, metabolism, inflammation and survival. In this respect, members of the FOXO family are extremely compelling to consider since these transcription factors have emerged as versatile proteins that can control angiogenesis, stem cell proliferation, cell adhesion and autoimmune disease. Further elucidation of FOXO protein function during neoplastic growth should continue to lay the foundation for the successful translation of these transcription factors into novel and robust clinical therapies for cancer. The development of automated sequencing technology and the completion of the human genome project has allowed high throughput mutational analysis of cancer genomes. This approach has been used to evaluate gene families involved in signal transduction and chromosomal instability for genetic alterations.

## Conclusions

The members of FOX gene family encode transcription factors that share a common, structurally related, DNA binding domain called the forkhead domain. Many members of this family have been found to play important roles in embryonic development as well as in the control of cell cycle, cell growth, cellular metabolism, life span, and immune responses. More FOXO target genes remain to be discovered which may act as regulators of FOXO function. An exciting area of future exploration will be to determine how FOXO factors mediate non-autonomous cellular processes in the entire organism. The genetic alteration and deregulation of members of several FOX subfamilies has been linked to the progression of certain types of cancers. A comprehensive understanding of expression profiles, genetic alterations, and epigenetic changes of FOX family genes as well as their binding proteins and target genes is important to develop novel therapeutics and preventives for human diseases, especially for cancer. Genes linked to apoptosis may not necessarily lead to cell death since current studies have suggested additional roles for genes normally associated with apoptosis that can involve cellular replication and transcription. More importantly, regulation of apoptotic pathways appears to serve a critical juncture for the control of tumor growth and unregulated cell proliferation. Interestingly, the induction of apoptosis in cells through FOXO proteins may require pathways aligned with oxidative stress. Since FOXO proteins were originally identified in human tumors, and because they play an important role in cell-cycle arrest, DNA repair, and apoptosis (cell functions that are altered in cancer), the FOXO family is thought to coordinate the balance between longevity and tumor suppression. Oncogenic mutations are attractive drug targets in many malignancies. From a clinical perspective, dysregulation of FOXO transcription factors has been described in several other cancers, prostate cancer and leukaemia. If the observations in breast cancer are recapitulated in vivo then inhibiting FOXO mRNA turnover may be necessary to enhance the efficacy of such compounds in the treatment. Detection, treatment, and prediction of consequence for cancer patients progressively rests on molecular understanding of tumor development and behavior and this molecular understanding of cancer development and progression is an important step toward the identification of additional biomarkers with increased specificity for invasive cancer and of new tumor-specific therapeutic targets. Achieving this will lead to better and long term survival of cancer patients as well as a better quality of life.

## References

1. Arden KC (2006). Multiple roles of FOXO transcription factors in mammalian cells point to multiple roles in cancer. *Experimental Gerontology* 41 709–717.
2. Bardelli A, Parsons DW, Silliman N, Ptak J, Szabo S, Saha S, Markowitz S, Willson JK, Parmigiani G, Kinzler KW, Vogelstein B, and Velculescu VE.(2003). Mutational analysis of the tyrosine kinome in colorectal cancers. *Science.* May 9; 300(5621): 949.
3. BrunetA, Bonni A, Zigmond MJ, Lin MZ, Juo P, Hu LS, Anderson MJ, Arden KC, Blenis J and Greenberg, M. E. (1999). Akt promotes cell survival by phosphorylating and inhibiting a Forkhead transcription factor. *Cell* 96, 857–868.
4. Brunet A, Kanai F, Stehn J, Xu J, Sarbassova D, Frangioni JV, Dalal SN, DeCaprio JA, Greenberg ME, Yaffe MB.(2002). 14-3-3 transits to the nucleus and participates in dynamic nucleocytoplasmic transport. *J Cell Biol.,* 156: 817–828
5. Byun DS, Cho K, Ryu BK, Lee MG, Park JI, Chae KS, Kim HJ and Chi SG (2003). Frequent monoallelic deletion of PTEN and its reciprocal associatioin with PIK3CA amplification in gastric carcinoma. *International Journal of Cancer* 104 318–327.
6. Carlsson P, Mahlapuu M. (2002)Forkhead transcription factors: key players in development and metabolism. *Dev Biol.,* 250: 1–23.
7. Carroll JS, Liu XS, Brodsky AS, (2005)Chromosome-wide mapping of estrogen receptor binding reveals long-range regulation requiring the forkhead protein FoxA1. *Cell,* 122: 33–43.
8. Dijkers PF, Medema R H, Pals C, Banerji L, Thomas NS, Lam EW, Burgering BM, Raaijmakers JA, Lammers JW and Koenderman L. (2000). Forkhead transcription factor FKHR-L1 modulates cytokine-dependent transcriptional regulation of p27(KIP1). *Mol. Cell. Biol.* 20, 9138-9148.
9. Gilley J, Coffer PJ, Ham J. (2003). FOXO transcription factors directly activate bim gene expression and promote apoptosis in sympathetic neurons. *J Cell Biol.,* 162: 613–622.
10. Hayes MP, Wang H, Espinal-Witter R, Douglas W, Solomon GJ, Baker SJ and Ellenson LH (2006). PIK3CA and PTEN mutations in uterine endometrioid carcinoma and complex atypical hyperplasia. *Clinical Cancer Research* 12: 5932–5935.

11. Hribal ML, Nakae J, Kitamura T, Shutter JR, Accili D.(2003). Regulation of insulin-like growth factor-dependent myoblast differentiation by Foxo forkhead transcription factors. *J Cell Biol.*, 162: 535–541.

12. Hui AB, Lo KW, Yin XL, Poon WS and Ng HK (2001). Detection of multiple gene amplifications in glioblastoma multiforme using array-based comparative genomic hybridization. *Laboratory Investigation* 81 717–723.

13. Jemal A, Siegel R, Ward E, Hao Y, Xu J, Murray T, Thun MJ (2008). Cancer statistics. *CA Cancer J Clin.*, 58: 71–96.

14. Kalinichenko VV, Major ML, Wang X (2004). Foxm1b transcription factor is essential for development of hepatocellular carcinomas and is negatively regulated by the p19ARF tumor suppressor. *Genes Dev.*, 18: 830–50.

15. Kang S, Bader AG and Vogt PK (2005). Phosphatidylinositol 3-kinase mutations identified in human cancer are oncogenic. *PNAS* 102 802–807.

16. Karakas B, Bachman KE and Park BH (2006). Mutation of the PIK3CA oncogene in human cancers. *British Journal of Cancer* 94 455–459.

17. Katoh M, Katoh M. (2004). Human FOX gene family (review). *Int J Oncol.*, 25: 1495–500.

18. Kau TR, Schroeder F, Ramaswamy S, Wojciechowski CL, Zhao JJ, Roberts TM, (2003). A chemical genetic screen identifies inhibitors of regulated nuclear export of a Forkhead transcription factor in PTEN-deficient tumor-cells. *Cancer Cell.*, 4: 463–476.

19. Kitamura T, Kitamura YI, Funahashi Y, Shawber CJ, Castrillon DH, Kollipara R, DePinho RA, Kitajewski J, Accili D. (2007).A Foxo/Notch pathway controls myogenic differentiation and fiber type specification. *J Clin Invest.*, 117: 2477–2485.

20. Kops GJ, and Burgering BM (1999). Forkhead transcription factors: New insights into protein kinase B (c-akt). signaling. *J. Mol. Med.*, 77: 656–665.

21. Kops GJ, de Ruiter ND, De Vries-Smits AM, Powell DR, Bos JL and Burgering BM. (1999). Direct control of the Forkhead transcription factor AFX by protein kinase B. *Nature,* 398: 630–634.

22. Kops GJ, Dansen TB, Polderman PE, Saarloos I, Wirtz KW, Coffer PJ, Huang TT, Bos JL, Medema RH. and Burgering BM. (2002). Forkhead transcription factor FOXO3a protects quiescent cells from oxidative stress. *Nature* 419: 316-321.

23. Kozaki K, Imoto I, Pimkhaokham A, Hasegawa S, Tsuda H, Omura K and Inazawa J (2006). PIK3CA mutation is an oncogenic aberration at advanced stages of oral squamous cell carcinoma. *Cancer Science,* 97: 1351–1358.

24. Kumaraguruparan R, Subapriya R, Kabalimoorthy J, and Nagini S.(2002). "Antioxidant profile in the circulation of patients with fibroadenoma and adenocarcinoma of the breast,"*Clinical Biochemistry,* 35(4): 275–279.

25. Kumaraguruparan R, Kabalimoorthy J, and Nagini S. (2005). "Correlation of tissue lipid peroxidation and antioxidants with clinical stage and menopausal

status in patients with adenocarcinoma of the breast," *Clinical Biochemistry*, vol. 38, no. 2, pp. 154– 58.

26. Lai E, Clark KL, Burley SK, Darnell JE, Jr. (1993). Hepatocyte nuclear factor 3/ fork head or "winged helix" proteins: a family of transcription factors of diverse biologic function. *Proc Natl Acad Sci U S A;* 90: 10421–3.
27. Liotta L, Petricoin E. Molecular profiling of human cancer (2000). *Nat Rev Genet.,* 1: 48–56.
28. Liu ZP, Wang Z, Yanagisawa H, Olson EN (2005). Phenotypic modulation of smooth muscle cells through interaction of Foxo4 and myocardin. *Dev Cell.,* 9: 261–270.
29. Ma YY,Wei SJ, Lin YC, Lung JC, Chang TC, Whang-Peng J, Liu JM, YangDM, YangWK and Shen CY (2000). PIK3CA as an oncogene in cervical cancer. *Oncogene* 19 2739–2744.
30. Mani SA, Yang J, Brooks M, *et al.* (2007). Mesenchyme Forkhead 1 (FOXC2). plays a key role in metastasis and is associated with aggressive basallike breast cancers. *Proc Natl Acad Sci U S A,* 104: 10069–74.
31. Massion PP, Kuo WL, Stokoe D, Olshen AB, Treseler PA, Chin K, Chen C, Polikoff D, Jain AN, Pinkel D (2002). Genomic copy number analysis of non-small cell lung cancer using array comparative genomic hybridization: implications of the phosphatidylinositol 3-kinase pathway. *Cancer Research* 62 3636–3640.
32. Medema RH, Kops GJ, Bos JL, Burgering BM (2000). AFX-like Forkhead transcription factors mediate cell-cycle regulation by Ras and PKB through p27kip1. *Nature,* 404: 782–787.
33. Miller CT, Moy JR, Lin L, Schipper M, Normolle D, Brenner DE, Iannettoni MD, Orringer MB and Beer DG (2003). Gene amplification in esophageal adenocarcinomas and Barrett's with high-grade dysplasia. *Clinical Cancer Research* 9 4819–4825.
34. Myatt SS, Lam EW (2007). The emerging roles of forkhead box (Fox). proteins in cancer. *Nat Rev.,* 7: 847–59.
35. Nair AK, Sugunan D, Kumar H and Anilkumar G (2012). Association analysis of common variants inFOXO3 with type 2 diabetes in a South Indian Dravidian population Gene, Volume 491, Issue 2, 10 January 2012, Pages 182–186 http: // dx.doi.org/10.1016/j.gene.2011.09.032.
36. Nakae J, Kitamura T, Silver DL, Accili D. (2001). The forkhead transcription factor Foxol (Fkhr). confers insulin sensitivity onto glucose-6-phosphatase expression. *J Clin Invest.,* 108: 1359–1367.
37. Nakae J, Kitamura T, Kitamura Y, Biggs WH, 3rd, Arden KC, Accili D.(2003). The forkhead transcription factor Foxol regulates adipocyte differentiation. *Dev Cell.,* 4: 119–129.
38. Nemoto S, Fergusson MM, Finkel T. (2004). Nutrient availability regulates SIRT1 through a forkhead-dependent pathway. *Science* 306: 2105–2108.

39. Pal I and Mandal M (2012). PI3K and Akt as molecular targets for cancer therapy: current clinical outcomes, *Acta Pharmacologica Sinica* 33: 1441–1458.
40. Patel H, Polanco-Echeverry G, Segditsas S, Volikos E, McCart A, Lai C, Guenther T, Zaitoun A, Sieber O, Ilyas M (2007). Activation of AKT and nuclear accumulation of wild type TP53 and MDM2 in anal squamous cell carcinoma. *International Journal of Cancer* 121 2668–2673.
41. Phillips WA, Russell SE, Ciavarella ML, Choong DY, Montgomery KG, Smith K, Pearson RB, Thomas RJ and Campbell IG (2006). Mutation analysis of PIK3CA and PIK3CB in esophageal cancer and Barrett's esophagus. *International Journal of Cancer* 118 2644–2646.
42. Pongtheerat T, Pakdeethai S, Purisa W, Chariyalertsak S, Petmitr S. (2011). Promoter methylation and genetic polymorphism of glutathione S-transferase P1 gene (GSTP1). in Thai breast-cancer patients. *Asian Pac J Cancer Prev.*, 12: 2731–2734.
43. Qazi AK, Hussain A (2013). Recent Development in Targeting PI3K-Akt-Mtor Signaling for Anticancer Therapeutic Strategies. Anticancer Agents Med Chem.
44. Qiu W, Scho¨nleben F, Li X, Ho DJ, Close LG, Manolidis S, Bennett BP and Su GH (2006). PIK3CA mutations in head and neck squamous cell carcinoma. *Clinical Cancer Research* 12 1441–1446.
45. Qiu W, Tong GX, Manolidis S, Close LG, Assaad AM and Su GH (2008). Novel mutant-enriched sequencing identified high frequency PIK3CA mutations in pharyngeal cancer. *International Journal of Cancer* 122 1189–1194.
46. Ray G, Batra S., Shukla N. K. (2000.). "Lipid peroxidation, free radical production and antioxidant status in breast cancer," *Breast Cancer Research and Treatment*, vol. 59, no. 2, pp. 163– 170.
47. Samuels Y and Ericson K (2006). Oncogenic PI3K and its role in cancer. *Current Opinion in Oncology* 18 77–82.
48. Samuels Y, Wang Z, Bardelli A, Silliman N, Ptak J, Szabo S, Yan H, Gazdar A, Powell SM, Riggins GJ (2004). High frequency of mutations of the PIK3CA gene in human cancers. *Science* 304 554.
49. Santarpia L, El-Naggar AK, Cote GJ, Myers JN and Sherman SI (2008). Phosphatidylinositol 3- kinase/AKT and ras/raf-mitogen- activated protein kinase pathway mutations in anaplastic thyroid cancer. *Journal of Clinical Endocrinology and Metabolism* 93 278–284.
50. Scho¨nleben F, Qiu W, Ciau NT, Ho DJ, Li X, Allendorf JD, Remotti HE and Su GH (2006). PIK3CA mutations in intraductal papillary mucinous neoplasm/ carcinoma of the pancreas. *Clinical Cancer Research* 12 3851–3855.
51. Shayesteh L, Lu Y, Kuo WL, Baldocchi R, Godfrey T, Collins C, Pinkel D, Powell B, Mills GB and Gray JW (1999). PIK3CA is implicated as an oncogene in ovarian cancer. *Nature Genetics* 21 99–102.

52. Snehal M Gaikwad, Pritha Ray (2012). Non-invasive imaging of PI3K/Akt/mTOR signalling in cancer. *Am J Nucl Med Mol Imaging*, 2(4): 418-431.

53. Srinivas PR, Kramer BS, Srivastava S (2001). Trends in biomarker research for cancer detection. *Lancet Oncol* 2: 698–704.

54. Sunters A, Fernandez de Mattos S, Stahl M, Brosens JJ, Zoumpoulidou G, Saunders CA (2003). FoxO3a transcriptional regulation of Bim controls apoptosis in paclitaxel-treated breast cancer cell lines. *J Biol Chem* 278: 49795–49805.

55. Tran H, Brunet A, Grenier JM, Datta SR, Fornace AJ, Jr, DiStefano PS, Chiang LW, Greenberg ME (2002). DNA repair pathway stimulated by the forkhead transcription factor FOXO3a through the Gadd45 protein. *Science*, 296: 530–534.

56. Wang Z, Cummins JM, Shen D, Cahill DP, Jallepalli PV, Wang TL, Parsons DW, Traverso G, Awad M, Silliman N, Ptak J, Szabo S, Willson JK, Markowitz SD, Goldberg ML, Karess R, Kinzler KW, Vogelstein B, Velculescu VE, Lengauer C. (2004). Three classes of genes mutated in colorectal cancers with chromosomal instability. *Cancer Res*. May 1;64(9): 2998-3001.

57. Wang Z, Shen D, Parsons DW, Bardelli A, Sager J, Szabo S, Ptak J, Silliman N, Peters BA, van der Heijden MS, Parmigiani G, Yan H, Wang TL, Riggins G, Powell SM, Willson JK, Markowitz S, Kinzler KW, Vogelstein B, Velculescu VE.(2004b). Mutational analysis of the tyrosine phosphatome in colorectal cancers, *Science*. May 21;304(5674): 1164-6.

58. Widschwendter M, Jones PA. (2002)DNA methylation and breast carcinogenesis. *Oncogene*, 21: 5462–5482.

59. Woenckhaus J, Steger K, Werner E, Fenic I, Gamerdinger U, Dreyer T and Stahl U (2002). Genomic gain of PIK3CA and increased expression of p110alpha are associated with progression of dysplasia into invasive squamous cell carcinoma. *Journal of Pathology* 198 335–342.

60. Wu G, Mambo E, Guo Z, Hu S, Huang X, Gollin SM, Trink B, Ladenson PW, Sidransky D and Xing M (2005). Uncommon mutation but common amplifications of the PIK3CA gene in thyroid tumors. *Journal of Clinical Endocrinology and Metabolism* 90 4688–4693.

**2016, Biotechnology: Progress and Applications** *Pages* **467–501**
*Editors:* **Saif Hameed and Zeeshan Fatima**
*Published by:* **DAYA PUBLISHING HOUSE, NEW DELHI**

*Chapter 20*

# Genetic Polymorphisms in Xenobiotic Metabolizing Enzymes: Clinical and Pharmacogenetic Aspects in Head and Neck Squamous Cell Carcinoma (HNSCC)

***Munindra Ruwali***

*Amity Institute of Biotechnology, Amity University Haryana, Gurgaon (Manesar) – 122 413, Haryana, India*

## ABSTRACT

Head and neck squamous cell carcinoma (HNSCC) is amongst the most common malignancy worldwide and is one of the leading cancers amongst the males in India. HNSCC occurs through a complex multistage process and tobacco and alcohol are shown to have independent and synergistic roles in the development of HNSCC. Besides tobacco and alcohol, genetic factors such as single nucleotide polymorphisms (SNPs) also play an important role in determining risk to HNSCC. Many of the commonly used drugs in chemotherapy of HNSCC are metabolized by drug metabolizing CYPs and SNPs in these drug

* Corresponding Author: E-mail: mruwali@ggn.amity.edu

metabolizing CYPs account for inter-individual variability in the metabolism and disposition of several therapeutic agents and could be responsible for significant differences that are observed in the clinical response to these drugs. Genetic polymorphisms in xenobiotic metabolizing enzymes can therefore not only lead to therapeutic failures but also to the possible increase in an individual's susceptibility to HNSCC. SNP analysis of these susceptible genes, particularly those involved in increased risk to HNSCC when exposed to environmental triggers, would eventually lead to the development of biomarkers that can be of help to track initiation and progression of HNSCC as well as prove useful in management of chemotherapeutic interventions.

## Single Nucleotide Polymorphisms (SNPS): Origin, Genotyping and Applications

With the completion of the human genome sequence, the study of human genome has shifted its focus to the identification and characterization of variations. This variation in DNA sequence influences individual characteristics such as physical appearance, susceptibility to disease and response to medical treatments. The single nucleotide polymorphism (SNP) is the most frequent variation in the human genome (Wang *et al.*, 1998). A SNP is a single base difference at a given position in the genome with a frequency of at least 1 per cent in at least one population. It is estimated that SNPs account for more than 90 per cent of total variation in the human genome. However, other types of DNA variations also exist which include variable nucleotide repeat, microsatellites, and complete copies of genes or even regions of a chromosome. In 2001, 1.42 million variants were identified in the analysis of human genome (Sachidanandam *et al.*, 2001), which dramatically increased to 15 million (Genomes Project Consortium *et al.*, 2010) and the latest figures from dbSNP release documents are 60 million uniquely identified SNPs (NCBI dbSNP Build 138, 2013). A SNP may be categorized as either 'major' or 'minor' allele based on their observed frequency in the general population.

At a given SNP, a person can have one of several genotypes: homozygous for the major allele, heterozygous, or homozygous for the minor allele due to the diploid number of human chromosomes. Typically, SNPs are bi-allelic, although very rarely triallelic forms can also be found (Hodgkinson and Eyre-Walker, 2010). SNPs, however, are not evenly distributed across the genome, and SNP frequencies can vary by several hundred-fold between two regions. On the basis of nucleotide change, SNPs are of two types; Transitions which involve replacement of pyrimidine base by pyrimidine base (or purine by purine), and Transversions that involve replacement of a pyrimidine base by a purine base or vice versa. Out of these two types, Transitions (A/G or T/C) are the most commonly occuring substitutions (Brookes, 1999). In terms of locations, the SNPs can occur in coding or non-coding regions with most SNPs being located in non-coding regions. Though SNPs in non-coding regions do not alter encoded proteins, they serve as important genetic or physical markers for comparative or evolutionary genomics studies. Studies show that such changes in non-coding regions involving sites participating in gene regulation or mRNA processing which may impact the genome, mRNA transcripts and functional/

structural proteins (Chamary *et al.*, 2006). SNPs may also be present in regulatory sites of a gene affecting the rates of transcription leading to changes in the levels of the expressed protein. SNPs in the coding regions can affect protein structure and hence function, resulting in an altered response to a drug or environmental toxin. Such SNPs can serve as molecular markers in several diseases and pharmacogenomic studies.

A high prevalence of SNPs, despite their simplicity and rather limited polymorphic content, is thus the main reason for the current enormous interest amongst the researchers in the medical field. SNPs can be used as markers to identify the genes that underlie complex diseases such as cancer and to realize the full potential of pharmacogenomics in analyzing variable response to drugs. The increasing numbers of SNPs further support the use of SNPs in disease association studies which will help in identifying the populations at a greater risk for a particular disease and could emerge as an important tool of predictive epidemiology. Such analyses can help understand the basis for the phenotype and support the development of tests to identify high risk individuals and aid in prediction of prognosis in affected individuals in case of common disorders. In addition, the data could provide clues to identify targets for the development of effective intervention strategies.

## Origin of SNPs

SNPs most often result from base substitutions occurring through a non repaired error that occurs during DNA replication (Fredman *et al.*, 2004). The mutation rate at most positions of the human genome is relatively low due to which the vast majority of allelic differences between individuals are inherited rather than new mutations. Two individuals which share the same allele at one position are most likely identical by descent rather than through two independent mutations. The types of mutations as well as their distribution across the genome are not random since transitions are more likely to occur than transversions. This bias is due to differential repair mechanisms (Strachan and Read, 1999) and due to the importance of CpGs which have a mutation rate 3–10 times higher than the average mutation rate with cytosines often methylated and more likely to mutate into thymines (Shen *et al.*, 1994). The interaction of two evolutionary forces, random genetic drift and natural selection, determine the fate of a mutant allele. Drift affects the distribution of a SNP in the population by random sampling of different alleles at each generation while natural selection affects the probability that a particular variant is passed on to the next generation. The principles of natural selection are considered to favour prevalence of alleles that are beneficial for the subjects though deleterious alleles can also survive and spread, a phenomenon that can be explained on the basis of three mechanisms primarily proposed for survival of a new SNP (Hurst, 2009). Firstly, the inheritance of DNA segments as blocks may result in linkage disequilibrium (LD) that may not allow the alleles to segregate independently. Secondly, the non-Mendelian inheritance patterns involving favoured transmission of an allele to more than 50 per cent of the progeny and finally, a complex phenomenon of kin selection proposes that deleterious alleles can survive if they promote fitness of other members of the community (Hurst, 2009). A substitution would therefore have a greater chance of survival if it is in LD with a conserved and functionally critical allele and also if it becomes homozygous in some individuals.

The selection mechanisms can also be classified on the basis of their influence on resultant allele frequencies (outcome) as either positive selection or directional selection or negative or purifying selection. Positive selection leads to spread and fixation of a beneficial allele in a given population or a species while negative or purifying selection leads to elimination of a deleterious or weaker allele. There is another category of selection called as balancing selection in which heterozygosity derived survival advantage leads to maintenance of both the alleles in the population. The ratio of non-synonymous changes to synonymous changes (NS:S) is used to determine whether a particular locus is under positive or purifying selection based on the assumption that non-synonymous changes that do not affect protein function would be fixed at the same rate as synonymous changes leading to the ratio of NS and S as 1. Thus, a ratio of NS: S > 1 would favor the possibility of positive selection while a ratio of NS: S < 1 favors the possibility of purifying selection (Massingham and Goldman, 2005). The genes having the variants with a ratio of NS: S < 1 under purifying selection include deleterious variants which are relevant for health and disease associations. Interestingly, a higher frequency of non-synonymous substitutions compared to synonymous substitutions could be interpreted as a basis of adaptive evolution (Bielawski *et al.*, 2000). Thus, the effect of selection can lead to a lower or higher frequency of individuals homozygous for a allele and some tests such as Ewens-Watterson homozygosity test of neutrality allows discrimination between balancing or directional selection using haplotype data under the assumption of Hardy Weinberg Equilibrium (HWE) (Kaur *et al.*, 2007).

## SNP Genotyping Methods

The relative simplicity of SNP genotyping technologies coupled with the abundance of SNPs in human genome have made them very popular in recent years. SNP genotyping methods are very diverse and suitability of a genotyping method for a particular application is directly related to the sensitivity of the assay and the pool size of target sequence availability. Genotyping typically involves the generation of allele-specific products for SNPs of interest followed by their detection for genotype determination. Most of the current genotyping technologies with only a few exceptions require the PCR amplification step. In most technologies, PCR amplification of a desired SNP-containing region is performed initially to introduce specificity and increase the number of molecules for detection following allelic discrimination (Kim and Misra, 2007). Though there is an availability of broad range of methods for SNP genotyping, some important ones will be discussed.

### Allele Discrimination Strategies

Allele discrimination is done with allele-specific biochemical reactions, and there are four popular methods: primer extension (nucleotide incorporation), hybridization, ligation, and enzymatic cleavage.

*(a)* ***Primer Extension***: This method involves allele-specific incorporation of nucleotides in primer extension reaction with a DNA template, utilizing enzyme specificity to achieve allelic discrimination (Sokolov, 1990).

(b) ***Hybridization***: This method uses differences in thermal stability of double-stranded DNA so as to distinguish between perfectly matched and mismatched target-probe pairs for achieving allelic discrimination (Syvanen, 2001).

(c) ***Ligation***: Ligation methods use the property of specificity of ligase enzymes to perform allelic discrimination by joining the two oligonucleotides which hybridize to single-stranded template DNA with perfect complementarity and are adjacent to each other. Out of three probes which are used in this assay, two probes are allele specific and bind to the template at the SNP site while the third probe is common and binds to the template adjacent to the SNP immediately next to the allele-specific probe (Landegren *et al.*, 1988). Binding of the allele-specific probe at the SNP site with perfect complementarity results in DNA ligase joining it with the common probe and the ligation products are then detected to reveal base identity at the SNP position.

(d) ***Enzymatic Cleavage***: Enzymatic cleavage uses restriction enzymes to cleave DNA by recognition of specific sequences and structures and thus discriminating between alleles when SNP sites are located in an enzyme recognition sequence and allelic differences affect recognition of these sequences.

### Mass Based Allele Detection Method

This method makes use of a small organic molecules called matrix that absorb energy from a laser source of certain wavelength for ionization. Analytes are mixed and cocrystallized with matrix and they are ionized in the form of intact molecules owing to transfer of energy from the matrix molecules (Tanaka *et al.*, 1988). The resulting ions are accelerated by a voltage gradient through a field-free region and separated by the time of flight of each ion which depends on the mass and charge of the ion.

### Fluorescence Signal-Based Detection

This method uses dideoxynucleotides labeled with different fluorescent dyes in a Sanger sequencing reaction to produce a ladder of fluorescently tagged extension products (Kwok, 2002). The primer extension products are then separated by electrophoresis and specific fluorescence signal from each extension product reveals the identity of the base.

## Applications of SNPs

SNPs accounts for most of the genetic differences between individuals and has been helpful in the search for identifying genetic contributions in individual drug response, disease susceptibility and individual identification. The major applications of SNPs are as follows:

### SNPs and Human Diversity

Mutations are generated essentially randomly in DNA molecules and the process of random genetic drift affects the chromosome as a whole but the selection process acts differently on distinct regions of the genome. This results in different levels of polymorphism in different portions of the genome. This is exemplified in the regions of noncoding DNA (introns and intergenic regions) which typically harbour a much higher diversity than coding DNA (Graur and Li, 2000), regulatory elements (such as promoters or splicing sites) or conserved nongenic elements (Drake *et al.*, 2006). Similarly, within exons, synonymous and non-synonymous positions differ in their diversity (Graur and Li, 2000). Diversity also varies along the genome over several megabases, even after correcting for gene or Guanine-Cytosine (GC) content which represents differences in recombination rates that affect the frequency of SNPs (Hellmann *et al.*, 2005).

Analyses of the genetic diversity among humans reveal very little differentiation among populations. Grouping individuals according to their geographical origin is feasible but requires large data sets of highly informative genetic markers (Rosenberg *et al.*, 2005). Differences between individuals from very distant populations only account for roughly 10 per cent of the total variability at most loci, whereas differences between individuals within a particular population represent 90 per cent (Barbujani and Goldstein *et al.*, 2004; Ramachandran *et al.*, 2005). Additionally, the geographic distribution of the diversity seems to be best explained by large gradients of allele frequency (Serre and Paabo, 2005; Rosenberg *et al.*, 2005). In this perspective, the data produced by Perlegen (Hinds *et al.*, 2005) and the International HapMap Consortium (Altshuler *et al.*, 2005) should not be considered a description of the entire worldwide human diversity. Nevertheless, these projects were particularly useful in the investigation of other populations because they offer a detailed description of the allele frequencies at the extremes of some of the worldwide genetic diversity.

**SNP and Pharmacogenomics**

Pharmacogenomics deals with the study of how an individual's genetic make-up affects the body's response to drugs. The term was coined using the words pharmacology and genomics and thus represents the intersection of pharmaceuticals and genetics. Pharmacogenomics might give way for drugs that might be tailor-made for individuals and adapted to each person's own genetic makeup. Other parameters such as environment, diet, age, lifestyle, and state of health can influence a person's response to medicines, but understanding an individual's genetic makeup is thought to be the key to creating personalized drugs with greater efficacy and safety (Ingelman-Sundberg and Rodrquez-Antona, 2005). SNPs in genes that code for drug metabolizing enzymes can influence their ability to metabolize certain drugs as less active or inactive forms of these enzymes would be unable to break down and efficiently eliminate drugs from the body leading to drug overdose in patients (Weinshilboum R, 2003). Clinical trials researchers are now using genetic tests for studying variations in drug metabolizing enzymes to screen and monitor patients. In addition, many pharmaceutical companies screen their chemical compounds to see how well they are broken down by variant forms of CYP enzymes (Hodgson and Marshall, 1998). An example is an enzyme called TPMT (thiopurine methyltransferase) which plays an important role in the chemotherapy treatment of common childhood leukemia by breaking down a class of therapeutic compounds called thiopurines. A small percentage of Caucasians have genetic variants that prevent them from producing an active form of this protein due to which thiopurines elevate to toxic levels in the patient as the inactive form of TMPT is unable to break down the drug.

**SNPs and Disease Susceptibility**

SNPs provide the opportunities for understanding the genetic basis for interindividual differences in the susceptibility to various common and complex diseases such as diabetes, asthma, and cancer (Suh and Vijg, 2005). A SNP that is associated with a disease can provide information on the mechanistic basis for disease while a SNP that is in LD with a genetic allele that confers disease predisposition may be used to identify susceptible individuals (Brookes AJ, 1999). Mostly common diseases are associated with multiple genetic and environmental factors that contribute to individual susceptibility to such diseases. There are two possible approaches for identifying the SNPs in genes and their association with polygenic diseases. The first approach involves examining inheritance patterns for genetic polymorphisms in causal genes in family studies and the second uses case-control studies which compares genotype frequency for candidate disease genes in unrelated individuals with the disease and in healthy controls. Out of these two approaches, use of family studies is generally the preferred approach in scenarios where genetic component of the disease is relatively strong, DNA samples are available from other family members and the disease is relatively easy to diagnose and is not stigmatized. For population based case-control studies several factors need to be taken into account for successful analysis. The factors such as study design, methods for recruitment of cases and controls, functional significance of polymorphisms chosen for study and statistical analysis of data require close attention to ensure that only genuine associations are detected and no false positive associations are observed. In genetic association studies, the frequencies of the alleles or genotypes of the target gene are compared in population of disease affected and non-affected healthy subjects. A higher frequency of minor allele or causative allele in affected subjects provides evidence that the allele or genotype is associated with increased risk of disease. Such studies help us to conclude that the SNP being studied either affects risk of disease directly or is a marker for some nearby genetic variants that affects the disease.

**SNPs and Forensic Science**

Single nucleotide polymorphisms (SNPs) are increasingly having a role in forensics in both human identification and description (Pakstis *et al.*, 2007). The importance of SNPs in forensic science is continuously increasing for many reasons which makes them very suitable DNA marker for forensic studies. Amongst the main reasons, SNPs have very low mutation rate which makes them an ideal candidate for paternity testing, have highly stable inheritance, are very suitable for analysis using high throughout technologies and they do not require comparative allelic sizing. SNPs confer the practical advantage over other genetic markers in terms of the length of the studied DNA region as the DNA sequence needed is much shorter. This is particularly significant for forensic medicine and anthropology studies where often only small or degraded samples are available for DNA analysis (Pakstis *et al.*, 2007).

## Genetic Variations in Tobacco Metabolizing Cytochrome P450s and their Association with Susceptibility to HNSCC

The Cytochrome P450s (CYPs) are the most important super family of phase I xenobiotic metabolizing enzymes which are widely distributed from bacteria to humans (Nelson *et al.*, 1996). Humans have 18 families and 44 subfamilies of CYP genes (Nelson, 2006). The CYP1 to 3 families are responsible for 70-80 per cent of all phase I dependent metabolisms of xenobiotics and others are involved in the metabolism of endogenous substances. Table 20.1 lists cytochrome P450 families and their main functions.

### Genetic Variations in *CYP1* Gene Family

*CYP1* family members include three genes, *CYP1A1*, *CYP1A2* and *CYP1B1* (Nelson *et al.*, 1996; Nebert and Russel, 2002). *CYP1A1* gene is located on the long arm of chromosome 15 (q22–24) and contains seven exons of which the first is non-coding (Hildebrand *et al.*, 1985). The *CYP1A1 m1* polymorphism (*CYP1A1*2A*), consists of a T-C substitution in the 3′ noncoding region of the gene that creates an *MspI* restriction enzyme cleavage site. The *CYP1A1, m2* polymorphism (*CYP1A1*2C*) is due to an A to G substitution at nucleotide 4889 in exon 7, that results in substitution of isoleucine by valine at the amino acid position 462 of the protein, a region that encodes a heme-binding domain of *CYP1A1* (Hayashi *et al.*, 1991a; Kawajiri *et al.*, 1996). The *CYP1A1 m3* polymorphism, creating a *Bsa1* site, is found only in African-Americans (Hayashi *et al.*, 1991a; Crofts *et al.*, 1993; Kawajiri *et al.*, 1996). The *CYP1A1 m4* polymorphism (*CYP1A1*4*) which leads to C to A substitution results in substitution of threonine by asparagine at 461 amino acid of the protein and is only two base pairs away from the m2 polymorphism and consequently has been less studied phenotypically (Cascorbi *et al.*, 1996). While, both *CYP1A1*2A* and *CYP1A1*2C* has been reported in Asians (Song *et al.*, 2002), the frequency of *CYP1A1*2C* is rare in the Caucasians (Hung *et al.*, 2003). Among control population, the frequency of the *CYP1A1 MspI* variant allele was reported as 37 per cent in Japanese, 9 per cent in Caucasians, and 42 per cent in Hawaiians. The reported *CYP1A1 "Val"* allele frequency has been observed to be 24 per cent in Japanese, 3 per cent in Caucasians, and 16 per cent in Hawaiians.

Cytochrome P4501A2 (*CYP1A2*) is involved in the bioactivation of Polycyclic Aromatic Hydrocarbons (PAHs), aromatic amines, many food-derived promutagenic heterocyclic amines (McManus *et al.*, 1990) and tobacco-specific nitrosamines (Smith *et al.*, 1996). The *CYP1A2* gene is located on chromosome 15 and spans 7.8 kb, comprising 7 exons and 6 introns. Functionally important polymorphisms in the *CYP1A2* are identified in upstream sequence and intron 1 and some of these may affect *CYP1A2* protein expression. *CYP1A2*1C* results from a single nucleotide change from G to A at position -3858 in 5′ flanking region (Nakajima *et al.*, 1999). The frequency of variant allele of *CYP1A2*1C* is estimated to be around 0.02-0.03 per cent in Caucasians and higher frequency is found in Japanese (0.10) and Chinese population (0.11). The second variant allele is a C to A transversion (*CYP1A2*1F*) in intron 1 at position 734 downstream of the first transcribed nucleotide. A significant distribution of this polymorphism in different populations has been reported. African populations

**Table 20.1: Cytochrome P450 Families and their Main Functions**

| *Family* | *Members* | *Names* | *Function* |
|---|---|---|---|
| *CYP1* | 3 subfamilies, 3 genes, 1 pseudogene | ***CYP1A1, CYP1A2, CYP1B1*** | PAHs metabolism |
| *CYP2* | 13 subfamilies, 16 genes, 16 pseudogenes | ***CYP2A6, CYP2A7, CYP2A13, CYP2B6, CYP2C8, CYP2C9, CYP2C18, CYP2C19, CYP2D6, CYP2E1, CYP2F1, CYP2J2, CYP2R1, CYP2S1, CYP2U1, CYP2W1*** | Drug metabolism |
| *CYP3* | 1 subfamily, 4 genes, 2 pseudogenes | ***CYP3A4, CYP3A5, CYP3A7, CYP3A43*** | Arachidonic acid metabolism |
| *CYP4* | 6 subfamilies, 11 genes, 10 pseudogenes | ***CYP4A11, CYP4A22, CYP4B1, CYP4F2, CYP4F3, CYP4F8, CYP4F11, CYP4F12, CYP4F22, CYP4V2, CYP4X1, CYP4Z1*** | Xenobiotic and steroid metabolism |
| *CYP5* | 1 subfamily, 1 gene | ***CYP5A1*** | Fatty acid hydroxylation |
| *CYP7* | 2 subfamilies, 2 genes | ***CYP7A1, CYP7B1*** | Throboxane synthesis |
| *CYP8* | 2 subfamilies, 2 genes | ***CYP8A1*** (prostacyclin synthase), ***CYP8B1*** (bile acid biosynthesis) | Cholesterol 7 α-hydroxylation |
| *CYP11* | 2 subfamilies, 3 genes | ***CYP11A1, CYP11B1, CYP11B2*** | Prostaglandin synthesis, Cholesterol side-chain cleavage, Steroid 11 β-hydroxylation, Aldosterone synthesis |
| *CYP17* | 1 subfamily, 1 gene | ***CYP17A1*** | Steroid -17α hydroxylation |
| *CYP19* | 1 subfamily, 1 gene | ***CYP19A1*** | Androgen aromatization |
| *CYP20* | 1 subfamily, 1 gene | ***CYP20A1*** | Steroid 21-hydroxilation |
| *CYP21* | 2 subfamilies, 2 genes, 1 pseudogene | ***CYP21A2*** | Steroid 24-hydroxilation |
| *CYP24* | 1 subfamily, 1 gene | ***CYP24A1*** | Retinoic acid hydroxylation |
| *CYP26* | 3 subfamilies, 3 genes | ***CYP26A1, CYP26B1, CYP26C1*** | |
| *CYP27* | 3 subfamilies, 3 genes | ***CYP27A1*** (bile acid biosynthesis), ***CYP27B1*** (vitamin D3 1-alpha hydroxylase, activates vitamin D3), ***CYP27C1*** (unknown function) | Steroid 27-hydroxilation |
| *CYP39* | 1 subfamily, 1 gene | ***CYP39A1*** | Unknown |
| *CYP46* | 1 subfamily, 1 gene | ***CYP46A1*** | Cholesterol 24- hydroxylation |
| *CYP51* | 1 subfamily, 1 gene3 pseudogenes | ***CYP51A1*** (lanosterol 14-alpha demethylase) | Steroid biosynthesis |

(0.43) exhibit high frequency of variant allele of *CYP1A2*1F* than European (0.31), Chinese (0.33) and Japanese (0.37) populations. The *CYP1A2*1D* (-2467delT), located in the regulatory region, is a relatively common polymorphism and occurs in high frequency in Asians and Caucasians respectively (Nordmark *et al.*, 2002).

The human *CYP1B1* gene is located at chromosome 2p21–22 and consists of three exons. The open reading frame starts in the second exon and is 1629 bp in length, encoding a protein of 543 amino acids. Four functionally important SNPs, C142G (R48G), G355T (A119S), C4326G (L432V), and A4390G (N453S), are known to result in amino acid substitutions. The C142G and G355T polymorphisms (–*CYP1B1**2) are tightly linked and result in amino acid substitutions and are located near the hinge region of *CYP1B1*. The C4326G transition (*CYP1B1**3) leading to the corresponding amino acid transition (L432V) is associated with increased catalytic activity of the CYP1B1 enzyme. The A4390G polymorphism (*CYP1B1**4) leads to the corresponding amino acid transition (N453S) (Bandiera *et al.*, 2005). The variant alleles of *CYP1B1* exhibit different frequency in different populations such as variant allele of Arg48Gly (*CYP1B1**2) polymorphism of *CYP1B1* exhibit high frequency in Japanese (0.33) and Europeans (0.31) than in Chinese (0.19) populations (Sasaki *et al.*, 2003, Long *et al.*, 2006). Similarly high frequency of variant allele of Ala119Ser (*CYP1B1**2) is found in Europeans (0.31) than in Japanese (0.14) and Chinese (0.19) populations (Sasaki *et al.*, 2003, Long *et al.*, 2006). LD studies of two polymorphisms (Arg48Gly and Ala119Ser) of *CYP1B1* also showed that these SNPs exhibit complete LD in Caucasian and Chinese populations but not in Japanese population (Sasaki *et al.*, 2003; Long *et al.*, 2006). Worldwide population studies data indicate that European population exhibit high frequency (0.23) of variant allele of *CYP1B1*3* (Leu432Val) than the Chinese (0.13) and Japanese (0.15) populations (Sasaki *et al.*, 2003; Long *et al.*, 2006). Likewise *CYP1B1*4* (Asn453Ser) was not found to exist in Oriental populations but was found in European population (Sasaki *et al.*, 2003; Long *et al.*, 2006).

## Genetic Variations in *CYP2* Gene Family

*CYP2* is the largest CYP family amongst the cytochrome P450s, consisting of 13 subfamilies, involved in drug metabolism in mammals (Nebert and Russel, 2002). The *CYP2A6* gene spans a region of approximately 6 kilobase pairs, contains 9 exons and has been mapped to the long arm of chromosome 19 (between 19q12 and 19q13.2) (Miles *et al.*, 1989). The location of *CYP2A6* is within a 350-kilobase pair gene cluster together with the *CYP2A7* and *CYP2A13* genes, two *CYP2A7* pseudogenes, as well as genes in the *CYP2B* and *CYP2F* subfamilies (Hoffman *et al.*, 1995). In the *CYP2A6* gene, several mutated alleles have been reported. The *CYP2A6*1A* is a wild type of the *CYP2A6* gene while *CYP2A6*1B* allele has a gene conversion with *CYP2A7* in the 3′-untranslated region (Oscarson *et al.*, 1999). The *CYP2A6*3* allele has gene conversions with *CYP2A7* in exons 3, 6, and 8 while *CYP2A6*2*, *CYP2A6*5*, *CYP2A6*6*, *CYP2A6*7*, *CYP2A6*8*, and *CYP2A6*11* alleles have single amino acid substitutions of L160H, G479V, R128Q, I471T, R485L, and S224P, respectively. The *CYP2A6*9* allele has a point mutation in the TATA box (T-48G), *CYP2A6*10* allele has two simultaneous amino acid substitutions of *CYP2A6*7* and *CYP2A6*8*. There are also polymorphisms that lead to gene deletion and duplication. The *CYP2A6*4* allele causes the deletion of whole *CYP2A6* gene while the *CYP2A6*1X2* allele has a duplication of the *CYP2A6* gene (Rao *et al.*, 2000). *CYP2A6* variant alleles which result in reduced enzyme activity are more common among Asian populations (Chinese, Japanese, and Korean) compared with Caucasian and African North American populations. Whole-gene deletion alleles (*CYP2A6*4B* and *CYP2A6*4C*) exist in the Japanese population, resulting in defective phenotypes. *CYP2A6*2* and *CYP2A6*3* are found in the European population, but the allele is not detected among Asian populations, including Japanese (Oscarson *et al.*, 1999). On the contrary, *CYP2A6*4* is detectable in as high as 10–20 per cent in Asian populations, but is low (about 1 per cent) in the European population (Oscarson *et al.*, 1999). *CYP2A6*7* is observed at a high frequency (15.7 per cent) in the Japanese population.

The other important member of *CYP2* gene family is *CYP2E1* which is involved in the metabolism of ethanol, low molecular weight solvents and tobacco specific nitrosamines. It is located on the 10th chromosome, consisting of 9 exons and 8 introns and contains a typical TATA-box and occupies 11413 bp of genomic DNA. A number of genetic polymorphisms in the *CYP2E1* gene have been reported, most of which are located in introns and appear to have no functional significance (Hu *et al.*, 1999). Currently, more than ten different polymorphic sites of *CYP2E1* are identified in 5′-regulatory, intron and transcribed regions and out of these polymorphisms only few are commonly studied using Restriction Fragment Length Polymorphism (RFLP). *PstI* polymorphism consist of G to C substitution at -1293bp in the 5′noncoding region of the *CYP2E1* gene that creates *PstI* restriction enzyme cleavage site whereas *RsaI* polymorphism constitute C to T substitution at -1053 in 5′noncoding region of the *CYP2E1* gene that deletes *RsaI* restriction enzyme cleavage site. These two polymorphism (*RsaI* and *PstI*) were found to be in linkage disequilibrium and together called as *CYP2E1*5B*, giving rise to the c1 and c2 allele. Hayashi *et al.* (1991b) classified the genotypes as type A (homozygous "normal" alleles; cl,cl), type C (homozygous alleles with the nucleotide exchanges; c2,c2) and type B (heterozygous, having a normal and an altered allele; cl,c2). The *CYP2E1* c2 (mutated) allele that lacks the *RsaI* restriction site is associated with about 10-fold higher gene transcription, protein level, and enzyme activity than the wild type c1 allele (Watanabe *et al.*, 1994). Another polymorphism present at 5′ flanking region at position from -2270 to -1672bp, a two-allele insertion, was identified using DNA restriction endonuclease *XbaI*. Allele containing 6 repeats with length of 42–60 bp is designated as *CYP2E1*1C*, and a rare allele with 8 repeats as *CYP2E1*1D* (Hu *et al.*, 1999). Several polymorphic sites were also identified in *CYP2E1* intron region. One of the important polymorphism identified in intron region is *DraI* which constitutes a T to A substitution at 7668bp in intron 6 that delete *DraI* restriction enzyme cleavage site and is represented as *CYP2E1*6* allele. It has three genotypes, two homozygous designated as CC and DD and one heterozygous genotype, CD. However, *CYP2E1*6* does not influence the enzyme expression and catalytic activity. In contrast, Uematsu *et al.* (1991) reported altered CYP2E1 catalytic activity with *CYP2E1*6* allele but did not find impact of this polymorphism on gene transcription. Another polymorphism present in intron 6 region of *CYP2E1* gene is the *MspI* polymorphism which constitutes A to G replacement at 6827 bp causing deletion of *MspI* restriction enzyme cleavage site. Little information is available about this polymorphism and it was not found to be associated with increased expression or enzyme activity.

There are several polymorphisms of *CYP2E1* which are present on coding region and lead to amino acid replacements. Ethnic variations have been reported in the distribution of major *CYP2E1* genotypes. Most of the polymorphisms of *CYP2E1* have been reported in Caucasians as well as in the Oriental population (Hu *et al.*, 1999). However, only two polymorphisms (*RsaI* and *DraI*) are reported in Indian populations (Mittal *et al.*, 2005; Soya *et al.*, 2005). The frequency of the *RsaI* c2 allele varies in different population. The minor allele (c2) frequency of *RsaI* polymorphism was found to be 1 per cent in Indian population (Mittal *et al.*, 2005; Soya *et al.*, 2005) which is similar to that reported in the Caucasians (1 per cent - 4 per cent). The Orientals (Chinese and Japanese) carry much higher frequency (20 per cent - 30 per cent) of the minor allele. Similarly, the minor allele frequency of *DraI* polymorphism (*CYP2E1*6*) was also reported to be relatively high (19 per cent) in the Indian population (Mittal *et al.*, 2005).This polymorphism is also reported in the Caucasians, though its frequency (7 per cent - 12 per cent) was relatively less when compared to Indian or Oriental (20 per cent - 30 per cent) populations (Hildesheim *et al.*, 1997). Some of the functionally important polymorphisms in carcinogen metabolizing CYPs are listed in Table 20.2.

## Association of Genetic Variations in Tobacco Metabolizing Cytochrome P450s with Susceptibility to HNSCC

Conflicting reports are available regarding the association of *CYP1A1* polymorphism with HNSCC risk. In the Japanese case control studies, the *Msp1* polymorphism was found to be associated with an increased risk of oral cavity carcinoma. However, no statistically significant difference in genotype distribution was observed between the cases and controls in the European populations. Presence of *CYP1A1*2A* polymorphism was found to be closely associated with increased risk for oral squamous cell carcinoma (OSCC) in Korean population regardless of smoking behavior. Likewise, a meta-analysis of 12 case control studies revealed no significant difference between the Ile/Val polymorphism of *CYP1A1* and HNSCC risk (Hashibe *et al.*, 2003). The Val/Val genotype was, however, found to be associated with an increased risk of HNSCC in Japanese and North American case control studies. Haplotype analysis revealed that the frequency of the haplotypes T-A-A and T-G-C was significantly increased in the cases when compared to the controls. Though no such association of *CYP1A1* haplotypes has been reported with HNSCC in either Caucasians or the Asians, both of these haplotypes carrying variant genotypes could be associated with increased enzyme activity. The rare allelic variants of *CYP1A1*2C* and *4 were reported to exhibit significant increase in the production of the ultimate carcinogenic species from benzo[a]pyrene (BP). *CYP1A1*2A* and *2C have been found to be associated with a higher inducible form of the enzyme. Chang *et al.* (2003) have also earlier shown that C-A-C haplotype which carries variant *CYP1A1*2A* allele is associated with higher enzyme activity and prostate cancer susceptibility. Another study has shown that C-G-C haplotype of *CYP1A1* is significantly associated with smoking-induced lung cancer (Shah *et al.*, 2008). That the combination of polymorphisms may enhance the susceptibility to HNSCC was further shown in a study indicating significantly increased risk in the cases carrying the combination of variant genotypes of *CYP1A1* (*CYP1A1*2A* or *CYP1A1*2C*) with *GSTM1* (-) (Singh *et al.*, 2009). A statistically significant increase in the frequency of combined genotypes of *CYP1A1*2A* and *GSTM1*(-) was also observed in cases suffering from OSCC in the Japanese case control studies. It has been hypothesized that individually variants in carcinogen metabolizing genes are modest to moderate risk factors, explaining the inconsistent results seen in epidemiological studies. However, combinations of these genotypes may add up to the risk large enough to be observed in epidemiological studies. Anantharaman *et al.* (2007) also observed that exposure to smokeless tobacco increased the risk to oral cancer in Indian cases carrying the variant genotype of *CYP1A1* and null genotype of *GSTM1*.

**Table 20.2: Functionally Important Polymorphisms in Carcinogen Metabolizing CYPs**

| *Allele* | *Protein* | *Nucleotide Change* | *Effect* |
|---|---|---|---|
| *CYP1A1* | | | |
| *CYP1A1*2A* | CYP1A1.1 | T3927C | – |
| *CYP1A1*2C* | CYP1A1.2 | A2455G | I462V |
| *CYP1A1*4* | CYP1A1.4 | C2453A | T461N |
| *CYP1B1* | | | |
| *CYP1B1*2* | CYP1B1.2 | C142G; G355T | R48G; A119S |
| *CYP1B1*3* | CYP1B1.3 | C4326G | L432V |
| *CYP1B1*4* | CYP1B1.4 | A4390G | N453S |
| *CYP2E1* | | | |
| *CYP2E1*5B* | CYP2E1.1 | C-1053T | - |
| *CYP2E1*6* | CYP2E1.1 | T7632A | - |
| *CYP2C19* | | | |
| *CYP2C19*2* | – | G681A | Splicing defect |
| *CYP2C19*3* | – | G636A | Premature stop codon |
| *CYP2D6* | | | |
| *CYP2D6*4* | – | G1846A | Splicing defect |
| *CYP2D6*10* | CYP2D6.10 | C100T; G1661C; G4180C | P34S; S486T |
| *CYP2A6* | | | |
| *CYP2A6*1B* | CYP2A6.1 | A-1013G | gene conversion in the 3' flanking region |
| *CYP2A6*4C* | – | CYP2A6 deleted | CYP2A6 deleted |

Variant allele of Arg48Gly (*CYP1B1*2*) polymorphism of *CYP1B1* exhibit high frequency in Oriental, Caucasian and Indian populations (Sasaki *et al.*, 2003; Long *et al.*, 2006; Singh *et al.*, 2008). LD studies of two polymorphisms (Arg48Gly and Ala119Ser) of *CYP1B1* also showed that these SNPs exhibit complete LD in Caucasian, Chinese and Indian populations but not in Japanese population (Sasaki *et al.*, 2003;

Long *et al.*, 2006; Singh *et al.*, 2008). The C4326G transition (*CYP1B1*3*) leading to the corresponding amino acid transition (L432V) is associated with increased catalytic activity of the CYP1B1 enzyme. The A4390G polymorphism (*CYP1B1*4*) leading to the corresponding amino acid transition (N453S) is not associated with catalytic changes in the protein product but has been associated with increase in the CYP1B1 degradation rate. *CYP1B1*4* (Asn453Ser) was not found to exist in Oriental population but was found in European population (Sasaki *et al.*, 2003; Long *et al.*, 2006). Entire functionally important SNPs of *CYP1B1* gene exist in all Indian sub-populations irrespective of the stratification based on linguistic lineages and geographical zones. A study has also shown statistically significant genotypic distribution of the *CYP1B1*2* in the patients suffering from head and neck cancer and in controls. A higher frequency of variant genotypes of *CYP1B1*2* was also reported in laryngeal cancer patients. Similar results have also been reported in case control studies with breast, lung and prostate cancer. Although the mechanism by which Ala119Ser site of *CYP1B1* plays a role in carcinogenesis is not fully understood, polymorphic variants of *CYP1B1* showed much stronger hydroxylation activities compared to wild type. Multivariate logistic regression models after adjustment for age, sex, smoking and alcohol revealed no significant association of *CYP1B1*3* or *CYP1B1*4* genotypes with relation to HNSCC risk (Singh *et al.*, 2008). These results are consistent with the earlier report indicating no association between *CYP1B1*3* genotypes and risk to HNSCC (Li *et al.*, 2005) though Ko *et al.* (2001) reported a significant association of variant genotypes of *CYP1B1*3* with smoking induced HNSCC. Other studies also showed no association between *CYP1B1*3* genotypes and lung cancer.

An increase in HNSCC risk associated with variant genotypes of *CYP2E1*, particularly *CYP2E1*5B*, has been reported in several case-control studies. The increased risk could be attributed to enhanced gene transcription that may lead to increased metabolic activation of carcinogens and consequent initiation of malignancies (Watanabe *et al.*, 1994). However, studies such as that conducted by Soya *et al.* (2008) reported no association of *RsaI* and *DraI* polymorphisms of *CYP2E1* with risk of upper aerodigestive tract cancers among South Indians. The reason for this finding may be the diverse regions associated with upper aerodigestive tract cancer such as nasal cavity, nasopharynx, oropharynx, hypopharynx, oral cavity, esophagus, etc., some of which may or may not be significantly associated with polymorphism in *CYP2E1*. There are also reports indicating a lack of association of variant *CYP2E1* genotypes with risk of head and neck cancer, lung tumors and other malignancies, and several studies have even reported an association of the wild type genotype (c1c1) of *CYP2E1* with esophageal, liver and lung cancer (Harth *et al.*, 2008). Likewise, Uematsu *et al.* (1994) reported that the frequency of the variant allele (C) of the *DraI* polymorphism was significantly lower in lung cancer cases when compared to healthy controls. These inconsistencies could be attributed to the genotype distribution frequency in different ethnic groups and specific exposures associated with these polymorphisms.

As opposed to that observed for other tobacco metabolizing CYPs, *CYP2A6* has a protective effect against HNSCC as has been reported in a study by Ruwali *et al.* (2009). The study observed an increase in frequency of variant genotypes of *CYP2A6* containing at least one deletion allele of *4C amongst controls when compared to the cases. The decreased risk associated with *CYP2A6* gene deletion could be attributed to decrease in the production of carcinogens from tobacco-specific nitrosamines leading to decreased DNA damage as *CYP2A6* is known to be involved in the activation of several of the nitrosamines such as N-nitrosodimethylamine (NDMA), N-nitrodiethylamine (NDEA) and 4-(methylnitrosoamino)-1-(3-pyridyl)-1-butanone (NNK). Individuals with deleted *CYP2A6* gene (*CYP2A6*4C*) have several fold less or completely absent proportion of urinary cotinine excretion when compared to individuals with wild type genotype. A decrease in risk was also observed for oral cancer risk in betel quid chewers in Sri Lanka (Topcu *et al.*, 2002), for lung cancer in various studies in Japanese (Miyamoto *et al.*, 1999; Ariyoshi *et al.*, 2002) and Caucasian populations. There are also conflicting reports which report an increase in risk for cancers of lung and esophagus, and nasopharynx with *CYP2A6* gene deletion (Tiwawech *et al.*, 2006) which could be attributed to the fact that *CYP2A6* gene deletion leads to an increase in levels of procarcinogens (tobacco-specific nitrosamines) in affected organs where polymorphism in other phase-I enzymes like *CYP2E1*, *CYP1A2* and *CYP3A4* may promote carcinogenesis by activating procarcinogens leading to increased DNA damage and ultimately cancer of the affected organ (Tiwawech *et al.*, 2006). No association of *CYP2A6* conversion type allele (*1B) with HNSCC risk was observed in the study by Ruwali *et al.* (2009) which was consistent with other studies which also reported a lack of association of *1B with oral cancer (Topcu *et al.*, 2002) and lung cancer (Miyamoto *et al.*, 1999) though an increased risk associated with *1B conversion type allele was reported for nasopharyngeal carcinoma (Tiwawech *et al.*, 2006) and lung cancer (Ariyoshi *et al.*, 2002).

## Genetic Variations in Drug Metabolizing Enzymes and their Association with HNSCC Treatment Response

CYPs in families 1-3 are responsible for about 75 per cent of all phase I metabolism of clinically used drugs. In clinical terms, the important candidate genes for the study of association of genetic variations and treatment response are *CYP2D6*, *CYP2C9* and *CYP2C19* (Evans and Relling, 1999). Genetic polymorphisms have been reported for these drug metabolizing enzymes which results in variability in the expression, activity, function, and stability of the enzyme and to its responsiveness to regulators leading to differences in treatment outcome. Table 20.3 lists the CYP enzymes involved in the metabolism of anticancer drugs.

### CYP 2C Subfamily

The important members of the CYP2C subfamily are *CYP2C8*, *CYP2C9*, *CYP2C18*, and *CYP2C19*, which collectively represent about 20 per cent of total CYP enzymes expressed in the liver. The *CYP2C* enzymes are involved in the metabolism of about 20 per cent of currently used drugs with *CYP2C8* playing an important role in the metabolism of antidiabetics, antimalarials, paclitaxel, and retinoic acid. *CYP2C9* is responsible for the metabolism of warfarin, phenytoin, and the nonsteroidal anti-inflammatory drugs and *CYP2C19* is shown to catalyze the metabolism of citalopram, cyclophosphamide, diazepam, and omeprazole.

**Table 20.3: CYP Enzymes Involved in the Metabolism of Anticancer Drugs**

| Drug | CYPs | Effect |
|---|---|---|
| Cyclo-phosphamide | CYP2A6 | 4-Hydroxylation, minor contribution |
| | CYP2B6 | 4-Hydroxylation, activation |
| | CYP2C8 | 4-Hydroxylation, minor contribution |
| | CYP2C9 | 4-Hydroxylation, minor contribution |
| | CYP2C19 | 4-Hydroxylation, minor contribution |
| | CYP3A4 | 4-Hydroxylation, activation<br>N-dechloroethylation, inactivation |
| Docetaxel | CYP 1A1 | minor contribution |
| | CYP 1A2 | minor contribution |
| | CYP 2A6 | minor contribution |
| | CYP 2B6 | minor contribution |
| | CYP 2C8 | minor contribution |
| | CYP 2C9 | minor contribution |
| | CYP 2D6 | minor contribution |
| | CYP 2E1 | minor contribution |
| | CYP3A4 | Low-Km |
| | CYP3A5 | 10x Higher Km than 3A4 |
| Gefitinib | CYP2D6 | Minor pathway |
| | CYP3A4 | O-demethylation, major pathway |
| | CYP3A5 | O-demethylation, <3A4 |
| Ifosfamide | CYP2A6 | 4-Hydroxylation, minor contribution |
| | CYP2B6 | N-dechloroethylation |
| | CYP2C8 | 4-Hydroxylation, minor contribution |
| | CYP2C9 | 4-Hydroxylation, minor contribution |
| | CYP2C19 | 4-Hydroxylation, minor contribution |
| | CYP2A4 | N-dechloroethylation |
| Imatinib | CYP1A2 | Capable of metabolizing, role unclear |
| | CYP2C9 | Capable of metabolizing, role unclear |
| | CYP2C19 | Capable of metabolizing, role unclear |
| | CYP 2D6 | Capable of metabolizing, role unclear |
| | CYP3A4 | N-demethylation to active metabolite |
| | CYP3A5 | N-demethylation to active metabolite |
| Paclitaxel | CYP2C8 | Major pathway |
| | CYP3A4 | Major pathway |
| Tamoxifen | CYP2B6 | 4-Hydroxylation, activation |
| | CYP2C9 | 4-Hydroxyl, activation |
| | CYP2C19 | 4-Hydroxylation, activation |
| | CYP2D6 | 4-Hydroxylation, activation |
| | CYP3A4 | N-demethylation, inactivation |
| | CYP3A5 | N-demethylation, inactivation |

## Cytochrome P450 2C9 (*CYP2C9*)

*CYP2C9* is part of the *CYP2C* gene cluster located on chromosome 10q24 along with three other *CYP2C* genes. It spans over 50 kb and consists of nine exons and large intronic regions. The *CYP2C9* gene contains a number of variant alleles, both in the exons and the 5´-flanking region. Most of the *CYP2C9* alleles result in decreased enzyme activity including the most common and most studied variant alleles, the *CYP2C9**2 and *CYP2C9*3*. The *CYP2C9**2 variant codes for a cysteine instead of an arginine at position 144, while the *CYP2C9**3 variant results in substitution of isoleucine with leucine at position 359.

*CYP2C9**2 (Arg144Cys) and *CYP2C9**3 (Ile359leu) are the two most common variant alleles in Caucasian subjects though they are less prevalent in African Americans and Asians (Kirchheiner and Brockmoller, 2005). Other variant alleles such as *CYP2C9*5*, *CYP2C9*6*, *CYP2C9*8* and *CYP2C9*11* were also reported to be present and identified at a low frequency in only African-derived populations while *CYP2C9*4* has been reported in a Japanese patient. These alleles, except for the *CYP2C9*6*, encode enzymes with single amino acid substitutions and reduced activity while *CYP2C9*6* contains an adenine base pair deletion at nucleotide 818, which results in a premature stop codon and a truncated inactive protein (Lee *et al.*, 2002). Most of the mutations located in the 5´-flanking region of the gene are linked to either the *CYP2C9**2 or *CYP2C9**3 alleles, and are part of different haplotypes (Veenstra *et al.*, 2005). The influence of these mutations on gene expression is probably minor, although some reports claim that they may cause reduced enzymatic activity on their own (Shintani *et al.*, 2001). Since none of the 5´-flanking mutations are localized at any transcription factor binding sites, their potential influence on enzyme activity may not be present (King *et al.*, 2004). Though there are several polymorphisms within the *CYP2C9* gene, there is still no explanation for variation of the enzyme activity among individuals who are genotyped with wild type genotype (Shintani *et al.*, 2001; King *et al.*, 2004).

## Cytochrome P450 2C19 (*CYP2C19*)

*CYP2C19* gene is also located in the same *CYP2C* gene cluster as *CYP2C9*. It is quite a large gene covering a 90-kb genomic region that consists of nine exons and large intronic regions. Amongst the drug metabolizing enzymes, *CYP2C19* is one of the most important enzymes contributing to the metabolism of clinically used drugs though it is expressed at low levels and predominantly in the liver and the small intestine. Several mutant alleles of *CYP2C19* have been reported out of which *CYP2C19*2*, *CYP2C19*3*, *CYP2C19*4*, *CYP2C19*5A*, *CYP2C19*6*, *CYP2C19*7* and *CYP2C19*8* are examples of some of the important variations (Inoue *et al.*, 1997). *CYP2C19*2*, which shows a single base mutation (G to A) in exon 5, has been shown to be present in both Japanese and Caucasian populations (Xie *et al.*, 2001). The other important mutation, the *CYP2C19*3*-type, consisting of a premature stop codon (G to A) in exon 4, is found only in Orientals, including both Japanese and Chinese populations. As compared with Caucasian, poor metabolizers (PMs) are very common in Orientals with frequency as high as 20 per cent. It has been possible to predict nearly 100 per cent of the PMs observed among Japanese using two defective alleles of *CYP2C19*2* (*m1*) and *CYP2C19*3*(*m2*). A common novel variant, *CYP2C19*17*, associated with ultrarapid drug metabolism was also described (Sim *et al.*, 2006).

## Cytochrome P450 2D6 (*CYP2D6*)

The *CYP2D6* gene spans a 4.2-kilobase (kb) region located on chromosome 22q13 and is part of the *CYP2D* cluster together with highly homologous *CYP2D8P* and *CYP2D7P* pseudogenes (Kimura *et al.*, 1989). Like other members of the *CYP2* gene family, the *CYP2D6* gene consists of nine exons and eight introns. *CYP2D6* exhibits highest genetic diversity among a set of 11 genes coding for DMEs (Solus *et al.*, 2004). Several genetic variants of *CYP2D6* have been described which include point mutations, single or multiple base insertions and deletions, gene conversions, and whole-gene deletion and duplication. The wild type allele *CYP2D6**1 has a normal enzyme activity (Kimura *et al.*, 1989) while the *CYP2D6**2 allele carries two mutations, 2850C>T and 4180G>C, leading to amino acid changes R296C and S486T, respectively. So far, only minor phenotypic consequences of these two nonsynonymous SNPs have been described. When recombinant 2D6*2 (R296C, S486T) was tested, it appeared to have somewhat lower activity as 2D6*1 (Zanger *et al.*, 2001) but studies in human liver tissues revealed higher expression of 2D6*2 versus 2D6*1 at the protein level (Zanger *et al.*, 2001) though 2D6*1 and 2D6*2 alleles were discovered to have very similar activities in vivo (Zanger *et al.*, 2004). In Caucasians the 2D6*2 allele has a frequency of about 20-30 per cent (Zanger *et al.*, 2001). Allele *CYP2D6*35*, a variant form of the *2 allele, carries the additional 31G>A mutation leading to Val11Met change and having functional activity and expression level comparable to wild type.

Individuals with extremely high hydroxylation capacity were found to have multiple copies of apparently functional *CYP2D6* as a result of unequal crossover events and other mechanisms. The *2D6xN* (N including *1, *2, *35, *41) alleles were shown to occur with frequencies of 1 to 2 per cent in northern European populations, whereas much higher frequencies were observed in Ethiopians and in Saudi Arabians and an intermediate frequency was found in Spanish populations (Agundez *et al.*, 1995). *CYP2D6*9* has a 2613-15delAGA in exon 5 which results in the lack of codon 281 and low amount of an enzymatically functional variant CYP2D6 protein. The *CYP2D6*9* allele has a frequency of about 1 to 2 per cent in Caucasians. *CYP2D6*10* has a C100T mutation in exon 1 leading to Pro34Ser which is the functionally dominant mutation. The consequences of this mutation are decreased CYP levels but residual enzymatic activity (Zanger *et al.*, 2001). In Caucasians, the *CYP2D6*10* allele has a frequency of about 2 per cent and accounts for 10 to 20 per cent of individuals with IM (Intermediate Metabolizer) phenotype. In Oriental populations the frequency of 2D6*10 exceeds 50 per cent and is responsible for a shift in the median metabolic ratio of debrisoquine/sparteine to higher values compared to Caucasians (Raimundo, 2004). Null alleles of *CYP2D6* do not encode a functional protein product and there is no detectable residual enzymatic activity. Only null alleles cause the PM phenotype if present in homozygous or compound heterozygous compositions. Different mechanisms were found to lead to the total loss of function. Several alleles have single base pair mutations or small insertions/deletions that interrupt the reading frame or that interfere with correct splicing leading ultimately to prematurely terminated protein products (Kagimoto *et al.*, 1990). *CYP2D6*4* is the most frequent null allele in Caucasians which occurs with an allele frequency of about 20-25 per cent and is responsible for 70 to 90 per cent of all PMs. The 1846G>A key mutation causes a shift of the consensus acceptor splice site of the third intron by one base, which results in a spliced mRNA with one additional base that has an altered reading frame and premature stop codon.

## Association of Genetic Variations in Drug Metabolizing Enzymes with Treatment Response in HNSCC

Pharmacogenetics has a great relevance in cancer therapy because chemotherapeutic agents are, in general, unspecific and with narrow therapeutic indexes that result in frequent severe or even fatal toxicities. With respect to anticancer therapy, CYP enzymes play a role in the metabolism of several drugs used in standard chemotherapy for the treatment of different forms of cancers. Gene polymorphisms of the enzymatic isoforms *CYP2C8, CYP2C9, CYP2C19, CYP2B6* and those belonging to the *CYP3A* subfamilies (3A4, 3A5, 3A7 and 3A43) could be involved in modifying the metabolism of the cytotoxic drugs metabolism. *CYP2C8* mediates the major pathway of paclitaxel metabolism (Kumar *et al.*, 1994). Irinotecan, cyclophosphamide, tamoxifen, taxanes and doxorubicin are substrates for the *CYP3A* subfamily. *CYP2B6* as well as *CYP2C9* and *CYP2C19* have an important role in the metabolism of several drugs, particularly cyclophosphamide.

A recent study by Yadav *et al.* (2014) has shown that chemotherapeutic response is modified in patients with poor metabolizer (PM) genotypes of *CYP2C9* in HNSCC cases. The poor response rate in the patients with PM genotypes and treated with radio and chemotherapy regimen could probably be a result of decreased availability of metabolites in PMs. It has been reported that the dose and therapeutic response for several drugs in the cases vary according to the genotypic status of *CYP2C9*. *CYP2C* subfamily (including *CYP2C9* and *CYP2C19*) is involved in the metabolism of commonly used anticancer drugs such as ifosfamide, cyclophosphamide, thalidomide and could have an impact on the efficacy and toxicity of chemotherapeutic agents and other drugs used in standard oncology (Ando *et al.*, 2002). In vitro studies have demonstrated that wild type *CYP2C9* was more efficient than the mutant allele *CYP2C9*3* in cyclophosphamide 4-hydroxylation and ifosfamide 4-hydroxylation. Kinetic studies have revealed a 3-fold lower intrinsic clearance (Vmax/Km) for cyclophosphamide in a yeast expression system with recombinant CYP2C9*2 and CYP2C9*3 protein when compared to CYP2C9*1 protein (Griskevicius *et al.*, 2003). Another study (Yadav *et al.*, 2008) showed that the treatment response in the HNSCC patients with variant genotypes of *CYP2C19* (*CYP2C19*2* and *CYP2C19*3*) and treated with radio and chemotherapy regimen which included combination of drugs, was poor as judged clinically, particularly in the cases with *CYP2C19*2* genotype. High percentage of the non-responders in the patients carried the variant genotypes of *CYP2C19*. It has been reported that a decrease in CYP2C19 activity could have an impact on the efficacy and toxicity of chemotherapeutic agents and other drugs used in standard oncology. PMs of *CYP2C19* were reported to exhibit little response to the respective chemotherapy than the normal genotype (Ando *et al.*, 2002). *CYP2C19**2 was found to reduce the metabolic activation of cyclophosphamide and ifosfamide, thereby lowering the risk of toxicity but worsening the therapeutic response.

A study has also shown that the chemotherapeutic response is modified in the cases carrying PM genotypes of *CYP2D6* (Yadav *et al.*, 2010). *CYP2D6* is reported to be involved in the metabolism of anti-cancer drugs such as cyclophosphamide, tamoxifen and gefitinib. The prevalence of non-responders in the cases with PM genotypes of *CYP2D6* could possibly be attributed to the overall poor drug metabolizing capacity of the PMs. Drug metabolizing CYPs have been reported to have an impact on the efficacy and toxicity of chemotherapeutic agents and other drugs used in standard oncology. Sailaja *et al.* (2007) have shown that the leukemia patients with PM genotypes of *CYP2D6* did not respond to the treatment of imatinib leading to drug resistance and poor prognosis. Accumulation of the drug was found to occur in the cases with PM genotypes of *CYP2D6*. Another interesting finding has been that much higher prevalence of non-responders in the cases with combination of PM genotypes of *CYP2D6*, *CYP2C9**2 and *CYP2C19**2 (Yadav *et al.*, 2010). Though cisplatin is not reported to interact with *CYP2D6* and majority of other CYPs, except *CYP2C9* to a small extent, these drug metabolizing CYPs are involved in the metabolism of several of the supportive care drugs used in chemotherapy. Polymorphisms in these CYPs may, therefore, confer a high risk of clinically significant drug–drug interactions. As the cytotoxic drugs used in chemotherapy generally have a narrow therapeutic index, a slight increase or decrease in cytotoxic activity due to drug–drug interaction may result in excessive toxicity or reduced efficacy and may be particularly important in the management of chemotherapy in HNSCC cases.

## Interactions of Genetic Variations with Environment and their Role in HNSCC

### Environmental Risk Factors and HNSCC

About 90 per cent of all cancers are related to environmental factors with tobacco smoke, alcohol and diet being the main causative exposures (Brennan and Boffetta, 2004). In South-Asia, including India, the main etiological factors for oral cancer have been smoking and chewing of tobacco alone or together with betel quid (BQ), cigarette smoking and heavy alcohol consumption (La Vecchia *et al.*, 1997). The following sections discuss tobacco and alcohol as the main causative factors for HNSCC and later on the interaction of these causative factors with genetic polymorphisms.

#### Tobacco and HNSCC Risk

Analysis of data from the 52nd round of National Sample Survey (NSS), a national household survey conducted across India (Neufeld *et al.*, 2005), revealed the following trends: men are reported to smoke tobacco more than chewing, while the reverse trend is seen in women; regular smoking has been found to be greatest in fifth decade for men, while in eighth decade for women; and the greatest prevalence of tobacco chewing is seen in seventh decade amongst both genders. Table 20.4 lists major components reported in cigarette mainstream smoke. Approximately 75 per

**Table 20.4: Major Components in Cigarette Mainstream Smoke**

| *Class* | *Compound Name* | *Molecular Formula* |
|---|---|---|
| | **Hydrocarbons** | |
| Aliphatic | Isoprene | $C_5H_8$ |
| | 4-Vinylcyclohexene | $C_8H_{12}$ |
| Polycyclic | Benzo[*a*]pyrene | $C_{20}H_{12}$ |
| | Benzo[*b*]fluoranthene | $C_{20}H_{12}$ |
| | Benzo[*j*]fluoranthene | $C_{20}H_{12}$ |
| | Benzo[*k*]fluoranthene | $C_{20}H_{12}$ |
| | 5-Methyl-chrysene | $C_{19}H_{14}$ |
| | Dibenzo[*a,e*]pyrene | $C_{24}H_{14}$ |
| | Dibenzo[*a,h*]pyrene | $C_{24}H_{14}$ |
| | Dibenzo[*a,i*]pyrene | $C_{24}H_{14}$ |
| | Dibenzo[*a,l*]pyrene | $C_{24}H_{14}$ |
| | Indeno [1, 2, 3-cd] pyrene | $C_{22}H_{12}$ |
| Aromatic | Ethyl benzene | $C_8H_{10}$ |
| | Styrene | $C_8H_8$ |
| | **Nitrogen containing components** | |
| Aromatic amines | *o*-Anisidine | $C_7H_9NO$ |
| | *p*-Chloroaniline | $C_6H_6C_lN$ |
| | 2,6-Xylidine (2,6-dimethylaniline) | $C_8H_{11}N$ |
| Aza-arenes | Dibenz[*a,h*]acridine | $C_{21}H_{13}N$ |
| | Dibenz [*a,j*]acridine | $C_{21}H_{13}N$ |
| | Dibenzo[*c,g*]carbazole | $C_{20}H_{13}N$ |
| Nitrosamines | *N*-Nitrosodi-n-butylamine (NDBA) | $C_8H_{18}N_2O$ |
| | *N*-Nitrosodi-n-propylamine (NDPA) | $C_6H_{14}N_2O$ |
| | *N*-Nitrosoethylmethylamine (NEMA, MEN) | $C_3H_8N_2O$ |
| | *N*-Nitrosodiethanolamine (NDELA) | $C_4H_{10}N_2O_3$ |
| | 4-(*N*-Nitrosomethylamino)-1-(3-pyridinyl) -1-butanone (NNK) | $C_{10}H_{13}N_3O_2$ |
| | *N'*-Nitrosonornicotine (NNN) | $C_9H_{11}N_3O$ |
| | *N*-Nitrosopiperidine (NPIP, NPP) | $C_5H_{10}N_2O$ |
| | *N*-Nitrosopyrrolidine (NPYR, NPY) | $C_4H_8N_2O$ |

cent of oral cavity and pharyngeal cancers are attributed to the use of smoked and smokeless tobacco and the rate increases with the amount smoked or chewed and the duration of use. In India, the per capita consumption of cigarettes has been increased by 2 per cent over the last decade (Sankaranarayanan *et al.*, 1998; Sankaranarayanan *et al.*, 2000). Estimates for the annual number of persons developing disease attributable to their tobacco habit in mid eighties included 1,08,000 annual incident cases of cancer (Sankaranarayanan *et al.*, 1998; Sankaranarayanan *et al.*, 2000; Brennan and Boffetta, 2004). The chances of smokers to develop oral cancer are six times more likely than nonsmokers. Cigarettes are the predominant form of tobacco used; however, other forms of tobacco, including smokeless tobacco, cigars, and pipes, are also used (Iribarren *et al.*, 1999). The quality of smoked cigarette and quitting smoking also have a role in oral cancer risk as has been shown in a study that those who smoke filtered cigarettes and those who quit smoking are at a substantially lower risk of oral cancer than are users of other forms of tobacco and current smokers (Kabat *et al.*, 1994).

Cigarette smoke contains about 4,000 chemical agents, including more than 60 carcinogens. Cigarette smokers have a lifetime increased risk for head and neck cancers that are 5 to 25 fold increased over the general population (Andre *et al.*, 1995). Those who quit smoking are at the same level of risk as of a non-smoker twenty years after smoking cessation. Risks of lung cancer and of cancers of the upper respiratory and upper digestive tract increase with increased consumption and duration of smoking. An incidence of laryngeal cancer is strongly connected with exposure to tobacco smoke containing dozens of carcinogens. Genotoxic agents such as PAHs present in tobacco smoke are responsible for lesions of structure of DNA and formation of DNA adducts by metabolically activated intermediates. Smokeless tobacco products which include chewing tobacco and sucked (moist oral) tobacco are the other tobacco related risk factors involved in HNSCC (Accortt *et al.*, 2005). Chewers of betel nuts who also smoke and drink are 195 times more likely to develop cancer of the esophagus than those who do not. Stepanov *et al.* (2005) observed the highest level of tobacco-specific nitrosamines to be present in 'khaini' that is a mixture of tobacco, lime and menthol or aromatic species.

The mode of tobacco processing, favoring the reduction of nitrate to nitrite (and other nitrosating agents) seems responsible for the high levels of nitrosamines as has been seen in 'khaini', which probably has the highest amount of nitrites amongst the smokeless tobacco (Hoffmann *et al.*, 1995). Hence, as khaini is usually placed in the mouth and retained there, an extremely high amount of nitrite as well as additional amounts of N-nitroso compounds formed endogenously will impact the oral mucosa. Increased levels of tobacco specific nitrosamines in saliva of khaini-tobacco chewers have been reported (Stitch *et al.*, 1992). Zarda, another smokeless tobacco product, contains relatively high tobacco-specific nitrosamine levels and is produced by boiling tobacco leaves in water with lime and spices until evaporation. Gutka contains less nitrosamine, while Supari does not contain tobacco and hence undetectable nitrosamines.

**Alcohol and HNSCC Risk**

Epidemiological studies show that the consumption of alcohol increases the risk of developing cancer of the regions of head and neck such as mouth, pharynx and larynx. Alcohol is a major risk factor for HNSCC which has been supported from findings of increased prevalence of HNSCC among populations with a higher alcohol intake, from cohort studies in which incidence of cancer have been observed in men and women who have been previously classified according to the amount of alcohol intake, a large number of case–control studies and from observations of increased incidence of cancer among certain occupational groups with increased alcohol intake (Viswanathan and Wilson, 2004). When ethanol is administered to animals along with other chemical carcinogens, ethanol shows a co-carcinogenic effect (Hsu *et al.*, 1991). Thus the effect of a given alcohol intake on the absolute risk of cancer of the head and neck is influenced by the presence of other risk factors especially smoking. However, Schlecht *et al.* (1999) in 784 cases of cancers of the mouth, pharynx, and supraglottis reported the joint effects of alcohol and moderate smoking to be more than multiplicative in pharyngeal cancers. The study clearly showed that alcohol may act not only synergistically with tobacco but also as an independent risk factor with 9.2 times greater risk of cancers of mouth, pharynx and supraglottis among heavy drinkers who never smoked compared with never drinkers. Alcohol may, thus, act as a solvent for other carcinogens or perhaps generate and promote coincident inflammation, producing significant reactive oxygen species.

There are site specific differences in susceptibility for cancers of different regions of head and neck upon exposure to alcohol. All forms of alcohol have been linked to oropharyngeal cancer and alcohol use may be a stronger risk factor for cancer of the oral cavity than for cancer of the larynx and pharynx. Alcohol is a weaker risk factor for laryngeal cancer than is cigarette smoking; however, cancer of the oral cavity has an increased incidence among those who smoke and abuse alcohol (Choi and Kahyo, 1991). The incidence of oral cancer may remain high several years after alcohol cessation. The exact mechanism of action of alcohol's carcinogenicity has not been determined; however, it may be related to the carcinogenic effect of acetaldehyde, the first metabolite of ethanol (Harty *et al.*, 1997). Another study suggested that *p15*-gene methylation can be induced by chronic smoking and drinking and may have a role in the early stages of HNSCC (Chang *et al.*, 2004). Alcohol and tobacco are likely to be synergistic in causing HNSCC. The high prevalence of tobacco and alcohol use worldwide and the high association of these cancers with these substances make them ideal targets for enhanced cancer prevention.

## Gene-Environment Interactions and HNSCC Risk

Tobacco and alcohol are the main causative factors for HNSCC. Two individuals having the same amount and duration of usage of tobacco and alcohol may have differences in susceptibility to HNSCC. One of the reasons for these differences could be the gene-environment interactions in which the genetic make-up of an individual decides the outcome of exposure to either tobacco or alcohol or both. Following sections discuss examples of the interaction of genetic risk factors with tobacco and alcohol and modulation of HNSCC risk.

### Interaction of Genetic Risk Factors and Tobacco

A significant increase in the risk in cases who were regular tobacco users (in the form of smoking or chewing) and carriers of variant genotypes of *CYP1A1* had demonstrated that tobacco use significantly influences the susceptibility to HNSCC (Singh *et al.*, 2009) though Rothman and Keller (1972) and Blot *et al.* (1988) have shown that alcohol and tobacco have both independent and synergistic roles in the genesis of HNSCC. There are also differences in the relationship with the amount of tobacco use with a study in Japanese population reported that the risk associated with *CYP1A1*2A* genotype is inversely proportional to the amount of tobacco use (Sato *et al.*, 1999; Tanimoto *et al.*, 1999). However, Sato *et al.* (2000) observed that the risk was highest in the group with the lowest cigarette dose level. Even though tobacco chewing is the major risk factor in enhancing susceptibility to HNSCC, a much greater increase in risk in cases with variant genotypes of *CYP1A1* (*CYP1A1*2A* or *CYP1A1*2C*) and who were cigarette smokers when compared to such cases with regular tobacco chewing habit could be attributed to the differences in the efficiency of CYP1A1 to metabolize the reactive chemicals generated from tobacco/cigarette smoke and tobacco chewing respectively.

A significant interaction between the variant genotypes of *CYP1B1*2* and cigarette smoking was found in the patients who had a history of smoking when compared to the controls while no such significant interaction was reported with *CYP1B1*3* or *CYP1B1*4* (Li *et al.*, 2005; Singh *et al.*, 2008). In contrast to the study, Ko *et al.* (2001) reported that genotypes of *CYP1B1*3* significantly interact with smoking and are a putative susceptibility factor in smoking related HNSCC. Their study showed that smokers with the at-risk genotype of *CYP1B1*3* were 20 times more likely to show evidence of p53 mutations than those with a *CYP1B1* wild type genotype. As observed with tobacco smoking, data clearly indicate a several fold increase in the risk of developing HNSCC with relation to the variant genotypes of *CYP1B1*2* and *CYP1B1*3* and tobacco chewing (Singh *et al.*, 2008). It is likely that these CYP1B1 variants in combination with the other genotypes of DMEs, may augment the risk in tobacco chewers for HNSCC.

Several fold increased risk to HNSCC in cases with variant genotypes of *CYP2C19* and those using tobacco (chewers or smokers) have suggested the possibility of interaction of drug metabolizing *CYP2C19* genotypes with these risk factors. A statistically significant increase in risk to HNSCC in tobacco chewing (approx. 13 fold) or cigarette smoking (approx. 9-fold) cases with variant genotypes of *CYP2C19* (*CYP2C19*2*) could be possibly attributed to the lower ability of PMs for detoxifying the carcinogens generated by tobacco chewing or smoking (Sugimoto *et al.*, 2005). In contrast, a slight increase in the risk in tobacco chewers or smokers in cases with variant genotypes of *CYP2D6*4* while no change or even a small decrease in risk in cases with *CYP2D6*10* allele when compared to non-tobacco users have suggested that *CYP2D6* genotypes do not appear to interact significantly with environmental risk factors in modifying the susceptibility to HNSCC. These observations are consistent with the previous studies though Kato *et al.* (1995) have reported reduced levels of DNA nitrosamine adducts in individuals carrying inactivating alleles of the *CYP2D6* gene. Caporaso *et al.* (2001) demonstrated that *CYP2D6* does not influence the disposition of nicotine or nicotine dependency and therefore this gene is not likely to have major affect on tobacco intake.

A study by Ruwali *et al.* (2009) suggested that polymorphic *CYP2E1* genotypes interact with tobacco and increase the HNSCC risk. A greater increase in risk observed in tobacco-chewers in this study is consistent with earlier reports showing that more of the tobacco-specific nitrosamines are formed with tobacco chewing [Hecht and

Hoffmann, 1988], which interact with *CYP2E1* genotypes in modifying the risk to HNSCC. Correspondingly, the greater magnitude of increased risk in smokers and tobacco chewers who carry the *RsaI* polymorphism of *CYP2E1* compared with *DraI* provides further evidence that the *RsaI* polymorphism leads to a greater increase in transcription or enzyme activity than the *DraI* genotypes [Uematsu *et al.*, 1994]. Soya *et al.*, [2008] also reported an interaction between *CYP2E1*6* (*DraI*) genotypes and tobacco that significantly increased the risk of upper aerodigestive tract cancers among South Indian tobacco users. In addition, Harth *et al.*, [2008] found an interaction of *CYP1B1* (Leu432Val) and *CYP2E1* (-70G>T) genotypes among smokers indicating the relevance of combined genotypes with exposure to tobacco smoke in significantly enhancing the HNSCC risk. In contrast, another study by Ruwali *et al.* (2009) reported a reduction for HNSCC risk in individuals who are exposed to risk factors such as tobacco consumption (in the form of smoking and chewing) when compared to individuals who are not exposed to such risk factors and carrying the variant genotypes of *CYP2A6*. This reduction in HNSCC risk may possibly be attributed to a higher number of cases with variant genotypes of *CYP2A6* to be non-tobacco users than tobacco users as variant alleles of *CYP2A6* have been reported to reduce the risk of tobacco consumption in the form of smoking and thus it is smoking that results in a decreased HNSCC risk.

**Interaction of Genetic Risk Factors with Alcohol and HNSCC Risk**

Similar to tobacco use, several fold increase in the risk has also been observed in the cases who were regular alcohol drinkers and carried variant genotypes of tobacco or drug metabolizing CYPs when compared to alcohol users with wild type genotype of these CYPs. Approximately 5-6 fold increase in the risk to HNSCC was observed in the patients with heterozygous or homozygous genotype of *CYP1B1*2* as compared to the controls. In cases who carried the combination of heterozygous and homozygous mutants and were regular alcohol drinkers, a much higher increase in risk was observed (approximately 20-fold) compared to alcohol users in the controls with wild type genotype, suggesting the interaction of alcohol with *CYP1B1* genotypes in enhancing the susceptibility to HNSCC (Singh *et al.*, 2008). Alcohol was, however, not found to interact with *CYP1A1* genotypes (Singh *et al.*, 2009). Another study has further shown that number of the individuals with variant genotypes of drug metabolizing *CYP2C19* (*CYP2C19*2*) and who were regular alcohol users were at more than 7-fold increase in the risk to HNSCC when compared to the controls, who were regular alcohol drinkers. A slight increase in the risk (approx. 3-fold), though not statistically significant, was also observed in cases amongst the alcohol users with variant genotypes of *CYP2C19*3* when compared to alcohol users in the controls (Yadav *et al.*, 2008). Alcohol use was found to slightly increase the risk in cases with variant genotypes of *CYP2D6*4* when compared to the controls who were also regular alcohol users (Yadav *et al.*, 2010). Similarly, alcohol consumption was also found to interact with variant *CYP2E1* genotypes to modify the HNSCC risk. The risk associated with *RsaI* variant genotypes and *DraI* variant genotypes was significantly higher in the cases who were regular alcohol users when compared with the controls (Ruwali *et al.*, 2009).

## References


1. Accortt NA, Waterbor JW, Beall C, Howard C. 2005. Cancer incidence among a cohort of smokeless tobacco users (United States). *Cancer Causes Control.* 16: 1107-1115.
2. Agundez JA, Ledesma MC, Ladero JM, Benitez J. 1995. Prevalence of CYP2D6 Gene Duplication and Its Repercussion on the Oxidative Phenotype in a White Population. *Clin. Pharmacol. Ther.* 57: 265-269.
3. Altshuler D, Brooks LD, Chakravarti A, Collins FS, Daly MJ, Donnelly P. 2005. A haplotype map of the human genome. *Nature.* 437: 1299–320.
4. Anantharaman D, Chaubal PM, Kannan S, Bhisey RA, Mahimkar MB. 2007. Susceptibility to oral cancer by genetic polymorphisms at CYP1A1, GSTM1 and GSTT1 loci among Indians: tobacco exposure as a risk modulator. *Carcinogenesis.* 28: 1455-1462.
5. Ando Y, Fuse E, Figg WD. 2002. Thalidomide metabolism by the CYP2C subfamily, *Clin. Cancer Res.* 8 : 1964–1973.
6. Andre K, Schraub S, Mercier M, Bontemps P. 1995. Role of alcohol and tobacco in the aetiology of head and neck cancer: a case-control study in the Doubs region of France. *Eur. J. Cancer B. Oral. Oncol.* 5: 301-309.
7. Ariyoshi N, Miyamoto M, Umetsu Y, Kunitoh H, Dosaka-Akita H, Sawamura Y, Yokota J, Nemoto N, Sato K, Kamataki T. 2002. Genetic polymorphism of CYP2A6 gene and tobacco-induced lung cancer risk in male smokers. *Cancer Epidemiol. Biomarkers. Prev.* 11: 890-894.
8. Bandiera S, Weidlich S, Harth V, Broede P, Ko Y, Friedberg T. 2005. Proteasomal degradation of human CYP1B1: effect of the Asn453Ser polymorphism on the post-translational regulation of CYP1B1 expression. *Mol Pharmacol.* 67: 435-43.
9. Barbujani G, Goldstein DB. 2004. Africans and Asians abroad: genetic diversity in Europe. *Annu. Rev. Genomics Hum. Genet.* 5: 119-50.
10. Bielawski JP, Dunn KA, Yang Z. 2000. Rates of nucleotide substitution and mammalian nuclear gene e.ution. A.roximate and maximum-likelihood methods lead to different conclusions. *Genetics.* 156: 1299-1308.
11. Blot WJ, McLaughlin JK, Winn DM, Austin DF, Greenberg RS, Preston-Martin S, Bernstein L, Schoenberg JB, Stemhagen A, Fraumeni J.F. Jr. 1988. Smoking and drinking in relation to oral and pharyngeal cancer. *Cancer Res.* 48: 3282-3287.
12. Brennan P, Boffetta P. 2004. Mechanistic considerations in the molecular epidemiology of head and neck cancer. *IARC Sci. Publ.* 157: 393-414.
13. Brookes AJ. 1999. The essence of SNPs. *Gene* 234: 177-186.
14. Caporaso NE, Lerman C, Audrain J, Boyd NR, Main D, Issaq HJ, Utermahlan B, Falk RT, Shields P. 2001. Nicotine metabolism and CYP2D6 phenotype in smokers. Cancer Epidemiol. *Biomarkers Prev.* 10: 261-263.

15. Cascorbi I, Brockmoeller J, Roots IA. 1996. C4887A polymorphism in exon 7 of human CYP1A1: population frequency, mutation linkages, and impact on lung cancer susceptibility. *Cancer Res.* 56: 4965–4969.
16. Chamary JV, Parmley JL, Hurst LD. 2006. Hearing silence: non-neutral e.ution at synonymous sites in mammals. *Nat Rev Genet.* 7: 98-108.
17. Chang HW, Ling GS, Wei WI, Yuen AP. 2004. Smoking and drinking can induce p15 methylation in the upper aerodigestive tract of healthy individuals and patients with head and neck squamous cell carcinoma. Cancer. 101: 125-132.
18. Chang L, Zheng SL, Isaacs SD, Turner A, Hawkins GA, Wiley KE, Bleecker ER, Walsh PC, Meyers DA, Isaacs WB, Xu J. 2003. Polymorphisms in the CYP1A1 gene are associated with prostate cancer risk. *Int. J. Cancer.* 106: 375–378.
19. Choi SY, Kahyo H. 1991. Effect of cigarette smoking and alcohol consumption in the aetiology of cancer of the oral cavity, pharynx and larynx. *Int. J. Epidemiol.* 20: 878-885.
20. Crofts F, Cosma GN, Taioli E, Currie DC, Toniolo PT, Garte SJ. 1993. A novel CYP1A1 gene polymorphism in African-Americans. *Carcinogenesis.* 14: 1729–1731.
21. Drake JA, Bird C, Nemesh J, Thomas DJ, Newton-Cheh C, *et al.*, 2006. Conserved noncoding sequences are selectively constrained and not mutation cold spots. *Nat. Genet.* 38: 223–27.
22. Evans WE, Relling MV. 1999. Pharmacogenomics: Translating Functional Genomics into Rational Therapeutics. *Science.* 286: 487–491.
23. Fredman D, White SJ, Potter S, Eichler EE, Den Dunnen JT, Brookes AJ. 2004. Complex SNP-related sequence variation in segmental genome duplications. *Nat. Genet.* 36: 861–66.
24. Genomes Project Consortium, Abecasis GR, Altshuler D, Auton A, Brooks LD, Durbin RM, Gibbs RA, Hurles ME, McVean GA. 2010. A map of human genome variation from population-scale sequencing. *Nature.* 467: 1061-1073.
25. Graur D, LiW H. 2000. *Fundamentals of Molecular Nutrition.* Sunderland, Massachusetts: Sinauer Assoc.
26. Griskevicius L, Yasar U, Sandberg M, Hidestrand M, Eliasson E, Tybring G, Hassan M, Dahl ML. 2003. Bioactivation of cyclophosphamide: the role of polymorphic CYP2C enzymes. *Eur. J. Clin. Pharmacol.* 59: 103–9.
27. Harth V, Schafer M, Abel J, Maintz L, Neuhaus T, Besuden M, Primke R, Wilkesmann A, Their R, Vetter H, Ko YD, Bruning T, Bolt HM, Ickstadt K. 2008. Head and neck squamous-cell cancer and its association with polymorphic enzymes of xenobiotic metabolism and repair. *J. Toxicol. Environ. Health.* 71: 887-897.
28. Harth V, Schafer M, Abel J, Maintz L, Neuhaus T, Besuden M, Primke R, Wilkesmann A, Their R, Vetter H, Ko YD, Bruning T, Bolt HM, Ickstadt K. 2008.

Head and neck squamous-cell cancer and its association with polymorphic enzymes of xenobiotic metabolism and repair. *J. Toxicol. Environ. Health.* 71: 887-897.

29. Harty LC, Caporaso NE, Hayes RB, Winn DM, Bravo-Otero E, Blot WJ, Kleinman DV, Brown LM, Armenian HK, Fraumeni JF Jr, Shields PG. 1997. Alcohol dehydrogenase 3 genotype and risk of oral cavity and pharyngeal cancers. *J. Natl. Cancer Inst.* 89: 1698-1705.
30. Hashibe M, Brennan P, Strange RC, Bhisey R, Cascorbil I, Lazarus P, Oude Ophuis MB, Benhamou S, Foulkes WD, Katoh T, Coutelle C, Romkes M, Gaspari L, Taioli E, Boffetta P. 2003. Meta- and pooled analyses of GSTM1, GSTT1, GSTP1, and CYP1A1 genotypes and risk of head and neck cancer. *Cancer Epidemiol. Biomarkers Prev.* 12: 1509-1517.
31. Hayashi S, Watanabe J, Kawajiri K. 1991b. Genetic polymorphisms in the 5′-flanking region change transcriptional regulation of the human cytochrome P450IIE1 gene. *J. Biochem.*110: 559-565.
32. Hayashi SI, Watanabe J, Nakachi K, Kawajiri K. 1991a. PCR detection of an A/G polymorphism within exon 7 of the CYP1A1 gene. *Nucl. Acids. Res.* 19: 47-97.
33. Hecht SS, Hoffmann D. 1988. Tobacco-specific nitrosamines, an important group of carcinogens in tobacco and tobacco smoke. *Carcinogenesis.* 9: 875-884.
34. Hellmann I, Prufer K, Ji H, Zody MC, Paabo S, Ptak SE. 2005. Why do human diversity levels vary at a megabase scale? *Genome Res.* 15: 1222–31.
35. Hildebrand CE, Gonzalez FJ, Kozak CA, Nebert DW. 1985. Regional linkage analysis of the dioxin-inducible P-450 gene family on mouse chromosome 9. *Biochem. Biophys. Res. Commun.* 130: 396-406.
36. Hildesheim A, Anderson LM, Chen CJ, Cheng YJ, Brinton LA, Daly AK, Reed CD, Chen IH, Caporaso NE, Hsu MM. 1997. CYP2E1 genetic polymorphisms and risk of nasopharyngeal carcinoma in Taiwan. *J. Natl. Cancer Inst.* 89: 1207-1212.
37. Hinds DA, Stuve LL, Nilsen GB, Halperin E, Eskin E, *et al.*, 2005. Whole-genome patterns of common DNA variation in three human populations. *Science.* 307: 1072–79.
38. Hodgkinson A, Eyre-Walker A. 2010. Human triallelic sites: evidence for a new mutational mechanism? *Genetics.* 184: 233-241.
39. Hodgson J, Marshall A. 1998. Pharmacogenomics: Will the regulators a.rove? *Nature Biotechnology.* 16: 243-246.
40. Hoffman SM, Fernandez-Salguero P, Gonzalez FJ, Mohrenweiser HW. 1995. Organization and e.ution of the cytochrome P450 CYP2A–2B-2F subfamily gene cluster on human chromosome 19. *J. Mol. E.* 41: 894–900.
41. Hoffmann D, Djordjevic MV, Fan J, Zang E, Glynn T, Connolly GN. 1995. Five leading US commercial brands of moist stuff in 1994: assessment of carcinogenic N-nitrosamines. *J. Natl. Cancer Inst.* 87: 1862–1869.

42. Hsu T, Furlong C, Spitz MR. 1991. Ethyl alcohol as a cocarcinogen with special reference to the aerodigestive tract: a cytogenetic study. *Anticancer Res*. 11: 1097-1101.

43. Hu Y, Hakkola J, Oscarson M, Ingelman-Sundberg M. 1999. Structural and functional characterization of the 5′-flanking region of the rat and human cytochrome P450 2E1 genes: identification of a polymorphic repeat in the human gene. *Biochem. Biophys. Res. Commun*. 263: 286-293.

44. Hung RJ, Boffetta P, Brockmöller J, Butkiewicz D, Cascorbi I, Cla.er ML, Garte S, Haugen A, Hirvonen A, Anttila S, Kalina I, Le Marchand L, London SJ, Rannug A, Romkes M, Salagovic J, Schoket B, Gaspari L, Taioli E. 2003. CYP1A1 and GSTM genetic polymorphisms and lung cancer risk in Caucasian nonsmokers: A pooled analysis. *Carcinogenesis*. 24: 875-882.

45. Hurst LD. 2009. Fundamental concepts in genetics: genetics and the understanding of selection. *Nat Rev Genet*. 10: 83-93.

46. Ingelman-Sundberg M, Rodrquez-Antona C. 2005. Pharmacogenetics of drug-maetabolizing enzymes: implications for a safer and more effective drug therapy. *Phil Trans R Soc* B. 360: 1563-1570.

47. Inoue K, Yamazaki H, Imiya K, Akasaka S, Guengerich FP, Shimada T. 1997. Relationship between CYP2C9 and 2C19 genotypes and tolbutamide methyl hydroxylation and S-mephenytoin 4′-hydroxylation activities in livers of Japanese and Caucasian populations. *Pharmacogenetics*. 7: 103–113.

48. Iribarren C, Tekawa IS, Sidney S, Friedman GD. 1999. Effect of cigar smoking on the risk of cardiovascular disease, chronic obstructive pulmonary disease, and cancer in men. *N. Engl. J. Med*. 340: 1773–1780.

49. Kabat GC, Chang CJ, Wynder EL. 1994. The role of tobacco, alcohol use, and body mass index in oral and pharyngeal cancer. *Int. J. Epidemiol*. 23: 1137-1144.

50. Kagimoto M, Heim M, Kagimoto K, Zeugin T, Meyer UA. 1990. Multiple Mutations of the Human Cytochrome P450IID6 Gene (CYP2D6) in Poor Metabolizers of Debrisoquine. Study of the Functional Significance of Individual Mutations by Expression of Chimeric Genes. *J. Biol. Chem*. 265: 17209-17214.

51. Kato S, Bowman ED, Harrington AM, Blomeke B, Shields PG. 1995. Human lung carcinogen–DNA adduct levels mediated by genetic polymorphisms *in vivo*. *J. Natl. Cancer Inst*. 87: 902-907.

52. Kaur G, Rapthap CC, Kumar N, Kumar S, Neolia S, Mehra NK. 2007. Frequency distribution of cytokine gene polymorphisms in the healthy North Indian population. *Tissue Antigens*. 69: 113-120.

53. Kawajiri K, Eguchi H, Nakachi K, Sekiya T, Yamamoto M. 1996. Association of CYP1A1 germ line polymorphisms with mutations of the p53 gene in lung cancer. *Cancer Res*. 56: 72-6.

54. Kim S, Misra A. 2007. SNP genotyping: technologies and biomedical a.lications. *Annu Rev Biomed Eng*. 9: 289-320.

55. Kimura S, Umeno M, Skoda RC, Meyer UA, Gonzalez F.J. 1989. The human debrisoquine 4-hydroxylase (CYP2D) locus: sequence and identification of the polymorphic CYP2D6 gene, a related gene, and a pseudogene. *Am. J. Hum. Genet.* 45: 889-904.

56. King BP, Khan TI, Aithal GP, Kamali F, Daly AK. 2004. Upstream and coding region CYP2C9 polymorphisms: correlation with warfarin dose and metabolism. *Pharmacogenetics*. 14: 813-22.

57. Kirchheiner J, Brockmoller J. 2005. Clinical consequences of cytochrome P450 2C9 polymorphisms. *Clin. Pharmacol. Ther.* 77: 1–16.

58. Ko Y, Abel J, Harth V, Brode P, Antony C, Donat S, Fischer HP, Ortiz-Pallardo ME, Their R, Sachinidis A, Vetter H, Bolt HM, Herberhold C, Bruning, T. 2001. Association of CYP1B1 codon 432 mutant allele in head and neck squamous cell cancer is reflected by somatic mutations of p53 in tumor tissue. *Cancer Res.* 61: 4398-4404.

59. Kumar GN, Walle UK, Walle T. 1994. Cytochrome P450 3A-mediated human liver microsomal taxol 6 alpha-hydroxylation. *J. Pharmacol. Exp. Ther.* 268: 1160–5.

60. Kwok PY. 2002. SNP genotyping with fluorescence polarization detection. *Hum. Mutat.* 19: 315–23.

61. La Vecchia C, Tavani A, Franceschi S, Levi F, Corrao G, Negri E. 1997. Epidemiology and prevention of oral cancer. *Oral Oncol.* 33: 302-312.

62. Landegren U, Kaiser R, Sanders J, Hood L. 1988. A ligase-mediated gene detection technique. *Science*. 241: 1077–80.

63. Lee CR, Goldstein JA, Pieper JA. 2002. Cytochrome P4502C9 polymorphisms: a comprehensive review of the in-vitro and human data. *Pharmacogenetics*. 12: 251–263.

64. Li G, Liu Z, Sturgis EM, Chamberlain RM, Spitz MR, Wei Q. 2005. CYP2E1 G1532C, NQO1 Pro187Ser, and CYP1B1 Val432Leu polymorphisms are not associated with risk of squamous cell carcinoma of the head and neck. *Cancer Epidemiol. Biomarkers Prev.* 14: 1034-1036.

65. Li G, Liu Z, Sturgis EM, Chamberlain RM, Spitz MR, Wei Q. 2005. CYP2E1 G1532C, NQO1 Pro187Ser, and CYP1B1 Val432Leu polymorphisms are not associated with risk of squamous cell carcinoma of the head and neck. *Cancer Epidemiol. Biomarkers Prev.* 14: 1034-1036.

66. Long JR, Shu XO, Cai Q, Cai H, Gao YT, Jin F, Zheng W. 2006. Polymorphisms of the CYP1B1 gene may be associated with the onset of natural menopause in Chinese women. *Maturitas*. 55: 238-246.

67. Massingham T, Goldman N. 2005. Detecting amino acid sites under positive selection and purifying selection. *Genetics*. 169: 1753-1762.

68. McManus ME, Burgess WM, Veronese ME, Huggett A, Quattrochi LC, Tukey RH. 1990. Metabolism of 2-acetylaminofluorene and benzo(a)pyrene and

activation of food-derived heterocyclic amine mutagens by human cytochromes P-450. *Cancer Res.* 50: 3367–3376.

69. Miles JS, Bickmore W, Brook JD, McLaren AW, Meehan R, Wolf CR. 1989. Close linkage of the human cytochrome P450IIA and P450IIB gene subfamilies: Implications for the assignment of substrate specificity. *Nucleic Acids Res.* 7: 2907–2917.

70. Mittal RD, Srivastava DS, Mandhani A, Mittal B. 2005. Genetic Polymorphism of Drug Metabolizing Enzymes (CYP2E1, GSTP1) and Susceptibility to Bladder Cancer in North India. *Asian Pac. J. Cancer Prev.* 6: 6-9.

71. Miyamoto M, Umetsu Y, Dosaka-Akita H, Sawamura Y, Yokota J, Kunitoh H, Nemoto N, Sato K, Ariyoshi N, Kamataki T. 1999. CYP2A6 gene deletion reduces susceptibility to lung cancer. *Biochem. Biophys. Res. Commun.* 261: 658-660.

72. Nakajima M, Yokoi T, Mizutani M, Kinoshita M, Funayama M, Kamataki T. 1999. Genetic polymorphism in the 5'-flanking region of human CYP1A2 gene: effect on the CYP1A2 inducibility in humans. *J Biochem. (Tokyo).* 125: 803-8.

73. NCBI dbSNP Build 138, Accessed on April 25 2013. (http: / /www.ncbi.nlm.nih. gov/projects/SNP/snp_summary.cgi?view+sumary=view+summaryandbuil d_id=138).

74. Nebert DW, Russell DW. 2002. Clinical importance of the cytochromes P450. *Lancet.* 360: 1155–1162.

75. Nelson DR, Koymans L, Kamataki T, Stegeman JJ, Feyereisen R, Waxman DJ, Waterman MR, Gotoh O, Coon MJ, Estabrook RW, Gunsalus IC, Nebert DW. 1996. P450 superfamily: update on new sequences, gene making, accession numbers and nomenclature. *Pharmacogenetics.* 6: 1-42.

76. Nelson DR. 2006. Cytochrome P450 nomenclature, 2004. *Methods Mol Biol.* 320: 1-10.

77. Neufeld KJ, Peters DH, Rani M, Bonu S, Brooner RK. 2005. Regular use of alcohol, and tobacco in India and its association with age, gender and poverty. *Drug Alcohol Depend.* 77: 283–291.

78. Nordmark A, Lundgren S, Ask B, Granath F, Rane A. 2002. The effect of the CYP1A2*1F mutation on CYP1A2 inducibility in pregnant women. *Br J Clin Pharmacol.* 54: 504–510.

79. Oscarson M, McLellan RA, Gullsten H, Yue QY, Lang MA, Bernal ML, Sinues B, Hirvonen A, Raunio H, Pelkonen O, Ingelman-Sundberg M. 1999. Characterisation and PCR-based detection of a CYP2A6 gene deletion found at a high frequency in a Chinese population. *FEBS Lett.* 448: 105–110.

80. Pakstis AJ, Speed WC, Kidd JR, Kidd KK. 2007. Candidate SNPs for a universal individual identification panel. *Hum Genet.* 121(**3-4**): 305-17.

81. Raimundo S, Toscano C, Klein K, Fischer J, Griese EU, Eichelbaum M, Schwab M, Zanger UM. 2004. A Novel Intronic Mutation, 2988G>A, With High Predictivity for Impaired Function of Cytochrome P450 2D6 in White Subjects. *Clin. Pharmacol. Ther.* 76: 128-138.

82. Ramachandran S, Deshpande O, Roseman CC, Rosenberg NA, Feldman MW, Cavalli-Sforza LL. 2005. Su.ort from the relationship of genetic and geographic distance in human populations for a serial founder effect originating in Africa. *Proc. Natl. Acad. Sci. USA.* 102: 15942–47.

83. Rao Y, Hoffmann E, Zia M, Bodin L, Zeman M, Sellers EM, Tyndale RF. 2000. Duplications and defects in the CYP2A6 gene: Identification, genotyping, and in vivo effects on smoking. *Mol. Pharmacol.* 58: 747–755.

84. Rosenberg NA, Mahajan S, Ramachandran S, Zhao C, Pritchard JK, Feldman MW. 2005. Clines, clusters, and the effect of study design on the inference of human population structure. *PLoS Genet.* 1: e70.

85. Rothman K, Keller A. 1972. The effect of joint exposure to alcohol and tobacco on risk of cancer of the mouth and pharynx. *J. Chronic Dis.* 25: 711-716.

86. Ruwali M, Khan AJ, Shah PP, Singh AP, Pant MC, Parmar D. 2009. Cytochrome P450 2E1 and head and neck cancer: interaction with genetic and environmental risk factors. *Env. Mol. Mut.* 50: 473-482.

87. Ruwali M, Pant MC, Shah PP, Mishra BN, Parmar D. 2009. Polymorphism in cytochrome P450 2A6 and glutathione S-transferase P1 modifies head and neck cancer risk and treatment outcome. *Mutat Res.* 669: 36-41.

88. Sachidanandam R, Weissman D, Schmidt SC, Kakol JM, Stein LD, Marth G, Sherry S, Mullikin JC, Mortimore BJ, Willey DL, Hunt SE, Cole CG, Coggill PC, Rice CM, Ning Z, Rogers J, Bentley DR, Kwok PY, Mardis ER, Yeh RT, Schultz B, Cook L, Davenport R, Dante M, Fulton L, Hillier L, Waterston RH, McPherson JD, Gilman B, Schaffner S, Van Etten WJ, Reich D, Higgins J, Daly MJ, Blumenstiel B, Baldwin J, Stange-Thomann N, Zody MC, Linton L, Lander ES, Altshuler D; International SNP Map Working Group. 2001. A map of human genome sequence variation containing 1.42 million single nucleotide polymorphisms. *Nature.* 409: 928-933.

89. Sailaja K, Vishnupriya S, Surekha D, Rao DN, Rao DR. 2007. Association of CYP2D6*4 Polymorphism with Chronic Myeloid Leukemia. *Journal of Medical Sciences Research.* 1: 43-46.

90. Sankaranarayanan R, Masuyer E, Swaminathan R, Ferlay J, Whelan S. 1998. Head and neck cancer: a global perspective on epidemiology and prognosis. *Anticancer Res.*18: 4779-4786.

91. Sankaranarayanan R, Mathew B, Jacob BJ, Thomas G, Somanathan T, Pisani P, Pandey M, Ramadas K, Najeeb K, Abraham E. 2000. Early findings from a community based cluster randomized controlled oral cancer screening trial in Kerala, India. *Cancer.* 88: 664-673.

92. Sasaki M, Tanaka Y, Kaneuchi M, Sakuragi N, Dahiya R. 2003. Alleles of polymorphic sites that correspond to hyperactive variants of CYP1B1 protein are significantly less frequent in Japanese as compared to American and German populations. *Hum Mutat.* 21: 652.

93. Sato M, Sato T, Izumo T, Amagasa T. 1999. Genetic polymorphism of drug-metabolizing enzymes and susceptibility to oral cancer. *Carcinogenesis*. 20: 1927-1931.

94. Sato M, Sato T, Izumo T, Amagasa T. 2000. Genetically high susceptibility to oral squamous cell carcinoma in terms of combined genotyping of CYP1A1 and GSTM1 genes. *Oral Oncol*. 36: 267-271.

95. Schlecht NF, Franco EL, Pintos J, Negassa A, Kowalski LP, Oliveira BV, Curado MP. 1999. Interaction between tobacco and alcohol consumption and the risk of cancers of the upper aero-digestive tract in Brazil. *Am. J. Epidemiol*. 150: 1129-1137.

96. Serre D, Paabo S. 2004. Evidence for gradients of human genetic diversity within and among continents. *Genome Res*. 14: 1679–85.

97. Shah PP, Singh AP, Singh M, Mathur N, Pant MC, Mishra BN, Parmar D. 2008. Interaction of cytochrome P4501A1 genotypes with other risk factors and susceptibility to lung cancer. *Mutat. Res*. 639: 1–10.

98. Shen JC, Rideout WM 3rd, Jones PA. 1994. The rate of hydrolytic deamination of 5-methylcytosine in double-stranded DNA. *Nucleic Acids Res*. 22: 972–76.

99. Shintani M, Ieiri I, Inoue K, Mamiya K, Ninomiya H, Tashiro N, Higuchi S, Otsubo K. 2001. Genetic polymorphisms and functional characterization of the 5′-flanking region of the human CYP2C9 gene: *in vitro* and *in vivo* studies. *Clin. Pharmacol. Ther*. 70: 175–182.

100. Sim SC, Risinger C, Dahl ML, Aklillu E, Christensen M, Bertilsson L, Ingelman-Sundberg M. 2006. A common novel CYP2C19 gene variant causes ultrarapid drug metabolism relevant for the drug response to proton pump inhibitors and antidepressants. *Clin. Pharmacol. Ther*. 79: 103-113.

101. Singh AP, Shah PP, Mathur N, Buters JT, Pant MC, Parmar D. 2008. Genetic polymorphisms in cytochrome P4501B1 and susceptibility to head and neck cancer. *Mutat. Res*. 639: 11-19.

102. Singh AP, Shah PP, Ruwali M, Mathur N, Pant MC, Parmar D. 2009. Polymorphism in cytochrome P4501A1 is significantly associated with head and neck cancer risk. *Cancer Invest*. 27: 869-876.

103. Smith TJ, Guo Z, Guengerich FP, Yang CS. 1996. Metabolism of 4-(methylnitrosamino)-1-(3-pyridyl)-1-butanone (NNK) by human cytochrome P450 1A2 and its inhibition by phenethylisothiocyanate. *Carcinogenesis*. 17: 809–813.

104. Sokolov BP. 1990. Primer extension technique for the detection of single nucleotide in genomic DNA. *Nucleic Acids Res*. 18: 3671.

105. Solus JF, Arietta BJ, Harris JR, Sexton DP, Steward JQ, McMunn C, Ihrie P, Mehall JM, Edwards TL, Dawson EP. 2004. Genetic variation in eleven phase I drug metabolism genes in an ethnically diverse population. *Pharmacogenomics*. 5: 895-931.

106. Song N, Tan W, Xing D, Lin D. 2001. CYP 1A1 polymorphism and risk of lung cancer in relation to tobacco smoking: a case-control study in China. *Carcinogenesis*. 22: 11-6.

107. Soya SS, Padmaja N, Adithan C. 2005. Genetic Polymorphism of CYP2E1 and GSTP1 in a South Indian Population-Comparisons with North India, Caucasians and Chinese. *Asian Pac. J. Cancer Prev*. 6: 315-319.

108. Soya SS, Vinod T, Reddy KS, Gopalakrishnan S, Adithan C. 2008. CYP2E1 polymorphisms and gene-environment interactions in the risk of u.er aerodigestive tract cancers among Indians. *Pharmacogenomics*. 9: 551-560.

109. Stepanov I, Hecht SS, Ramakrishnan S, Gupta PC. 2005. Tobacco specific nitrosamines in smokeless tobacco products marketed in India. *Int. J. Cancer*. 116: 16–19.

110. Stitch HF, Parida BB, Brunnemann KD. 1992. Localized formation of micronuclei in the oral mucosa and tobacco specific nitrosamines in the saliva of reverse smokers. *Int. J. Cancer*. 50: 172–176.

111. Strachan T, Read AP. 1999. Human Molecular Genetics. New York: Wiley-Liss.

112. Sugimoto M, Furuta T, Shirai N, Nakamura A, Kajimura M, Sugimura H, Hishida A, Ishizaki T. 2005. Poor metabolizer genotype status of CYP2C19 is a risk factor for developing gastric cancer in Japanese patients with Helicobacter pylori infection. *Aliment Pharmacol. Ther*. 22: 1033-1040.

113. Suh Y, Vijg J. 2005. SNP discovery in associating genetic variation with human disease phenotypes. *Mutat Res*. 573: 41-53.

114. Syvänen AC. 2001. Accessing genetic variation: genotyping single nucleotide polymorphisms. *Nat Rev Genet*. 2: 930-42.

115. Tanaka K, Waki H, Ido Y, Akita S, Yoshida Y, Yoshida T. 1988. Protein and polymer analyses up to m/z 100,000 by laser ionization time-of-flight mass spectrometry. Rapid Commun. *Mass Spectrom*. 2: 151–53.

116. Tanimoto K, Hayashi S, Yoshiga K, Ichikawa T. 1999. Polymorphisms of the CYP1A1 and GSTM1 gene involved in oral squamous cell carcinoma in association with a cigarette dose. *Oral Oncol*. 35: 191-196.

117. Tiwawech D, Srivatanakul P, Karalak A, Ishida T. 2006. Cytochrome P450 2A6 polymorphism in nasopharyngeal carcinoma. *Cancer Lett*. 241: 135-141.

118. Topcu Z, Chiba I, Fujieda M, Shibata T, Ariyoshi N, Yamazaki H, Sevgican F, Muthumala M, Kobayashi H, Kamataki T. 2002. CYP2A6 gene deletion reduces oral cancer risk in betel quid chewers in Sri Lanka. *Carcinogenesis*. 23: 595–598.

119. Uematsu F, Ikawa S, Kikuchi S, Sagami I, Kanamaru R, Abe T, Satoh K, Motomiya M, Watanabe M. 1994. Restriction fragment length polymorphism of the human CYP2E1 (cytochrome P450IIE1) gene and susceptibility to lung cancer: Possible relevance to low smoking exposure. *Pharmacogenetics*. 4: 58–63.

120. Uematsu F, Kikuchi H, Motomiya M, Abe T, Sagami I, Watanabe M. 1991. Association between restriction fragment length polymorphism of the human

cytochrome P45011E1 gene and susceptibility to lung cancer. *Jpn. J. Cancer Res.* 82: 254–256.

121. Veenstra DL, Blough DK, Higashi MK, Farin FM, Srinouanprachan S, Rieder MJ, Rettie AE. 2005. CYP2C9 haplotype structure in European American warfarin patients and association with clinical outcomes. *Clin Pharmacol Ther.* 77: 353-64.

122. Viswanathan H, Wilson JA. 2004. Alcohol: The neglected risk factor in head and neck cancer. Clin. Otolaryngol. *Allied Sci.* 29: 295-300.

123. Wang DG, Fan JB, Siao CJ, Berno A, Young P, Sapolsky R, Ghandour G, Perkins N, Winchester E, Spencer J, Kruglyak L, Stein L, Hsie L, Topaloglou T, Hubbell E, Robinson E, Mittmann M, Morris MS, Shen N, Kilburn D, Rioux J, Nusbaum C, Rozen S, Hudson TJ, Lipshutz R, Chee M, Lander ES. 1998. Large –scale identification, ma.ing, and genotyping of single nucleotide polymorphisms in the human genome. *Science.* 280: 1077-10823.

124. Watanabe J, Hayashi S, Kawajiri K. 1994. Different regulation and expression of the human CYP2E1 gene due to the RsaI polymorphism in the 5′-flanking region. *J. Biochem.*116: 321–326.

125. Weinshilboum R. 2003. Inheritance and Drug Response. *New Engl J Med.* 348: 529-537.

126. Xie HG, Kim RB, Wood AJ, Stein CM. 2001. Molecular basis of ethnic differences in drug disposition and response. *Annu. Rev. Pharmacol. Toxicol.* 41: 815-850.

127. Yadav SS, Ruwali M, Pant MC, Shukla P, Singh RL, Parmar D. 2010. Interaction of drug metabolizing cytochrome P450 2D6 poor metabolizers with cytochrome P450 2C9 and 2C19 genotypes modify the susceptibility to head and neck cancer and treatment response. *Mutat. Res.* 684: 49-55.

128. Yadav SS, Ruwali M, Shah PP, Mathur N, Singh RL, Pant MC, Parmar D. 2008. Association of poor metabolizers of cytochrome P450 2C19 with head and neck cancer and poor treatment response. *Mutat. Res.* 644: 31-37.

129. Yadav SS, Seth S, Khan AJ, Maurya SS, Dhawan A, Pant S, Pant MC, Parmar D. 2014. Association of polymorphism in cytochrome P450 2C9 with susceptibility to head and neck cancer and treatment outcome. *Applied and Translational Genomics.* 3: 8–13.

130. Zanger UM, Fischer J, Raimundo S, Stüven T, Evert BO, Schwab M, Eichelbaum M. 2001. Comprehensive Analysis of the Genetic Factors Determining Expression and Function of Hepatic CYP2D6. *Pharmacogenetics.* 11: 573-585.

131. Zanger UM, Raimundo S, Eichelbaum M. 2004. Cytochrome P450 2D6: Overview and Update on Pharmacology, Genetics, Biochemistry. *Naunyn Schmiedebergs Arch. Pharmacol.* 369: 23-37.

*Section IX*

# Stem Cells and Nanotechnology

**2016, Biotechnology: Progress and Applications** *Pages* **505–524**
*Editors:* **Saif Hameed and Zeeshan Fatima**
*Published by:* **DAYA PUBLISHING HOUSE, NEW DELHI**

# *Chapter 21*

# Stem Cells of Skin and Hair Follicle and their Clinical Application

***Anil Kumar*[1]*, *Sujata Mohanty*[2] *and Somesh Gupta*[3]***

[1]*Amity Institute of Biotechnology. Amity University, Haryana, India*
[2]*Centre of Excellence for Stem Cell Research, All India Institute of Medical Sciences, New Delhi, India*
[3]*Department of Dermatology, All India Institute of Medical Sciences, New Delhi, India*

## ABSTRACT

Ever since the first seminal paper on the successful bone marrow transplantation by *Edward* Donnall Thomas was published in The New England Journal of Medicine in September 1957, the field of stem cells has progressed rapidly. He was awarded the Nobel Prize in Physiology or Medicine in 1990 for his development of bone marrow transplantation, which could cure patients with advanced leukemia.

Today we have identified the niche of stem cells in various organs of the human body and their clinical application in the treatment of various degenerative diseases. In tune with this one very interesting niche of stem cells has been found in epidermal tissues which comprises of skin and hair follicle. The skin and hair

* Corresponding Authors: E-mail: akumar8@ggn.amity.edu; someshgupta@hotmail.com

follicle tissue has attracted a lot of scientific and clinical group interest towards understanding their basic biology and clinical application as regenerative medicine.

Here have tried to summarize some of the milestone achievements in this area and the recent trend in their application in the surgical management of vitiligo.

## Historical Perspective of Depigmentation

Vitiligo is a disorder characterized by the loss of melanin skin pigmentation leading to severe psychological problems among the subjects. Melanin in humans is produced at the base of epidermis by melanocytes, which package granules of melanin into specialized cellular vesicles called melanosomes. Melanin granules are transported through dendrites to adjacent keratinocytes, where most pigment is found. Vitiligo is a disease that was observed very early in history. The history of the vitiligo has been addressed in elaborated manner by Rafael Falabella. The following historical excerpts has been adopted from his paper entitled "Vitiligo and The Melanocyte Reservoir" (Falabella 2009). Most ancient civilizations had some sort of reference about this disease (Lahiri 2009). Rig Veda of India's oldest sacred literature, has term "Kilas" to describe vitiliog, The Sanskrit word "kilas" is derived from Kil, meaning white as similar to a white spotted deer (Milington GW 2007). Two types of diseases characterized by discoloration of the skin were described by pharonic medicine in the "Ebers Papyrus" (1550 BC). The Ebers Papyrus in 1550 BC mentioned two forms of depigmentation one with swellings referring to leprosy and the other only macular and probably referring to vitiligo. The latter condition was regarded to be "treatable" (Nair 1978). By 1400 BC white leprosy spots were called Sveta khushtha in the Atharva Veda (Koranne and Sachdeva 1988) and in 1200 BC Japanese Shinto prayers described depigmentation in the Amarakosa (Nair 1978; Lahiri 2009). In the Buddhist sacredd book Vinay Pitak (624-544 BC) the term "kilas" was also used to describe the white spots on the skin. Men and women with kilas were not eligible for ordainment. In 200 BC, the Indian Manu Smriti described "Sweta Kushtha" meaning "white disease" probably referring to vitiligo (Koranne and Sachdeva, 1988). Herodotus (484-425), "The Father of History" claimed that foreigners "had sinned against the sun" and must leave the country (Goldman, Moraites *et al.*, 1966). The word "vitiligo" was probably first used in the first century AD (circa 30 AD) by the Roman physician Celsus in his Latin medical classic "De Medicina". It is still a debate whether vitiligo was derived from the Latin word "vitium" or "vitulum" respectively meaning "blemish" and "small blemish" or from the Latin word "vitelius" meaning "calf", referring to the white and glistening appearance of the skin of this animal (Kopera D. In: Hann SK 2000; Lahiri 2009). Many centuries went by and vitiligo continued to be one of the most important depigmentation ailments worldwide provoking discrimination, segregation or even ostracism in certain cultures where affected individuals are unable to get jobs or even become married most probably based upon ancient religious beliefs (Porter, Beuf *et al.*, 1978). World attention was grabbed by Vesalius Andreas (1533) who proposed about the skin having two layers (Vesalius 2007). Jean Riolan the Younger (1580- 1657), several decades later, separated the skin of a black subject into the upper black layer and the lower white layer "as snow". In

1665, Marcello Malpighi proposed that skin colour was mainly determined by the granules of stratum mucosum, not those of stratum corneum or dermis. Giosué Sangiovanni in 1819 was the first to describe melanocytes in the squid, which he termed as 'chromatophores'. In 1837, Friedrich Henle also identified pigment producing cells in human epidermis, as identical to pigment cells of the eye (Westerhof 2006). Bruno Bloch in 1917 with L-3,4-dihydroxy-phenylalanine (DOPA) reaction concluded the presence of tyrosinase enzyme within melanocyte for melanin synthesis (Boissy 2000). It took around 4000 years from the time man became aware of disturbing white spots on the skin until the melanocyte was finally identified as the responsible factor for pigmentary disorders (Falabella, 2009).

## Melanocytes Origin and their Role in Pigmentation

Melanocytes are originally derived from the neural crest, migrate and become localized in hair follicles as well as in the epidermis to pigment the hair and skin, respectively (Mayer 1973; Silvers 1979). The "follicular-melanin unit" resides in the immune-privileged hair bulb and consists of hair matrix melanocytes and keratinocytes supported by dermal papilla (DP) melanocytes. (reviewed in (Slominski, Paus *et al.*, 1994; Stenn and Paus 2001). The ratio of melanocytes to keratinocytes in hair follicle varies from 1:1 to 1:6 *i.e.*, one melanocyte for every five keratinocytes in the hair bulb whereas almost 1:1 in the basal epithelial layer next to the dermal papilla (DP) (Tobin and Paus 2001). In the epidermal melanin unit the ratio of melanocytes to keratinocyte is approximately 1: 36. Hair bulb melanogenic melanocytes differ from epidermal melanocytes by being larger, more dendritic, with more extensive Golgi and rough ER, and by producing larger melanosomes. Melanin granules transferred into hair cortical keratinocytes remain minimally digested, contrasting with the epidermis, where melanin is almost completely degraded in the differentiating keratinocytes (Bell 1967). Although hair follicle melanocytes remain much less completely characterized than epidermal melanocytes, considerable progress has been made recently (Slominski, Wortsman *et al.*, 2005).

Growth and maturation of melanocytes in the hair bulb is coupled with the hair follicle cycle (Slominski and Paus 1993). The hair follicle undergoes cyclic development throughout a finite period of hair shaft formation (anagen; ~3–5 y on average in human scalp), followed by a brief regression phase that results in the apoptosis-driven resorption of up to 70 per cent of the hair follicle (catagen ~1–2 wk in human scalp), and by a relatively quiescent period (telogen ~3 mo) (reviewed in Stenn and Paus, 2001). The hair bulb melanocyte system has, however, long been perceived as self-perpetuating whereby melanocytes involved in the pigmentation of one generation are also involved in the pigmentation of the next wave by surviving the apoptosis-driven regression of the hair bulb (Sugiyama, 1979); (Lindner, Botchkarev *et al.*, 1997). However, fully differentiated bulbar melanocytes undergo apoptosis during catagen and are removed from the regressing hair follicle, still, some less differentiated, hair bulb melanocytes appear to survive catagen (Tobin, Hagen *et al.*, 1998; Commo and Bernard 2000). Canities can affect individual hair follicles with either gradual loss of pigment over several cycles, gradual loss of pigment along the same hair shaft, or the hair fiber may grow fully de-pigmented. Pigment loss in graying hair follicles is due to a marked reduction in melanogenically active melanocytes in the hair bulb (Tobin and Paus 2001). True gray hairs show a much reduced, but detectable, DOPA oxidation reaction (indicator of tyrosinase activity), whereas white hair bulbs are broadly negative (Slominski, Wortsman *et al.*, 2005).

There appears also to be a specific defect of melanosome transfer in graying hair follicles, as keratinocytes may lack melanin granules despite their close proximity to melanocytes with melanosomes (Slominski, Wortsman *et al.*, 2005). The remaining hair bulb melanocytes in canities-affected anagen hair follicles often appear enlarged, although this may reflect a reduction in dendricity rather than an overall increase in cell volume. Ultrastructural analysis of the human gray hair matrix reveals melanocytes with heterogeneous degrees of melanogenesis (Tobin and Paus, 2001). Melanocytes of gray/white hair bulbs contain fewer and smaller melanosomes and less supporting organelles. Interestingly, the remaining melanosomes may be packaged within autophagolysosomes suggesting that these melanosomes are defective, perhaps even leaking reactive melanin metabolites. Melanocytes in graying and white hair bulbs may be vacuolated, a common cellular response to increased oxidative stress, and may disappear very rapidly (Commo, Gaillard *et al.*, 2004). A parallel increase in dendritic cells (including Langerhans cells) and their shift from upper to lower hair follicles may represent response to degenerative changes within melanocytes (Slominski, Wortsman *et al.*, 2005). Gray hair is often unable to hold a permanent or temporary set and may be more resistant to incorporating artificial color, suggesting reprogramming of matrix keratinocytes in aging hair follicles to alter cortical keratinocyte differentiation (Slominski, Wortsman *et al.*, 2005).

Melanocytes in the epidermis rarely proliferate and produce melanin pigments to protect the skin from ultraviolet light (UV), while melanocytes in hair follicles repeatedly proliferate and differentiate for hair pigmentation in every hair cycle (Aristidis Moustakas 2013). Once melanocytes colonize the hair matrix at the bottom of each hair follicle during development or during the hair cycle, they differentiate into fully mature melanocytes that produce melanin pigments in an organelle called a melanosome. Melanosomes are then transferred to adjacent precortical keratinocytes that produce the hair shaft, whereby during each hair cycle, the growing hair becomes pigmented (Slominski, Wortsman *et al.*, 2005; Van Den Bossche, Naeyaert *et al.*, 2006). Melanocytes die and disappear from the hair matrix when hair follicles regress, but reappear at this location during the growing phase of the hair follicle (Figure 21.2). How melanocytes are regenerated during each hair cycle has been an enigma for the last more than half-century (Falabella 2009; Tobin 2009).

## Concept of Melanocyte Stem Cells (MelSC) or Amelanotic Melanocytes

The major role of cell division in adult life is to maintain the number of differentiated cells at a constant level: to replace cells that have died or been lost through injury, *i.e.*, to maintain a overall homeostasis (Hall and Watt 1989). It was hypothesized that different organs of the animal system contain pocket of naive cells which are responsible for the homeostasis of routine wear and tear of the tissues or organs. These naive cells which do not express any lineage specific differentiation markers were given the name of stem cells in due course of development of the science. The first evidence for the existence of such a stem cell population came from experiments in which bone marrow cells from healthy mice were injected into mice that had received lethal doses of radiation (Ford, Hamerton *et al.*, 1956; Till and Mc 1961). The donor marrow was able to reconstitute all the differentiated blood cell types in the host and, when low numbers of cells were injected, colonies derived from single marrow cells (CFU-S) were found to contain both myeloid and erythroid cells (Wu 1968).

The concept of stem cells in skin epidermis started with inter follicular epidermis (IFE), which generates the lipid barrier of adult skin, constantly renews its surface throughout the entire life of the animal and also undergoes re-epithelialization after wound injuries. These renewing and repairing activities of the skin epidermis imply the existence of stem cells to ensure these critical functions (Alonso and Fuchs 2003). It was initially hypothesized that the entire basal layer consisted of stem cells. Radiation dose–survival studies suggested that stem cells might comprise 2–7 per cent of basal layer cells (Potten, Schofield *et al.*, 1979). Histological analysis has shown that epidermis is organized in stacks of cells with a hexagonal surface area lying on a bed of ten basal cells (Mackenzie 1970; Potten 1974; Potten 1981).

The cellular reservoirs for epidermal and hair follicle melanocytes were originally considered to be distinct populations. In 1956, Montagna and Chase (Montagna and Chase 1956) reported the existence of a distinct cell population that does not contain visible melanin pigment but can be distinguished morphologically from surrounding keratinocytes in the outer root sheath (ORS) of human hair follicles 3 weeks following ionizing irradiation. In the clinic, skin repigmentation, in vitiligo, an acquired pigmentary disorder, usually begins at the orifices of hair follicles, then enlarges, and coalesces to cover the entire depigmented skin area. This phenomenon is called 'follicular repigmentation' and is characterized by repigmentation of an unpigmented skin area starting from the hair follicle orifices, suggesting the existence of a melanocyte 'reservoir' population in the skin (Staricco 1963; Ortonne, Schmitt *et al.*, 1980). Starricco studied the phenomenon extensively and found that the amelanotic ORS cell population, does not contain any melanin pigment and is negative for the L-DOPA reaction (Staricco 1959). Starricco subsequently found that this population can be activated after excision of the epidermis or by UV therapy (Staricco 1963). He assumed that the amelanotic melanocyte population provides DOPA-positive dendritic melanocytes that migrate into, and fill, the epidermis for follicular repigmentation. However, these amelanotic melanocytes were not always clearly distinguished from Langerhans progenitors because of the lack of specific lineage markers available at the time (Staricco 1963; Shukla 1969). Ortonne *et al.*, examined the unpigmented skin of vitiligo patients after treatment with UV therapy using scanning electron microscopy and DOPA staining. These authors found that the pigmented islands that appeared after the treatment had follicular cores containing melanocytes with large cell bodies that could be distinguished from other cells in the ORS. They also postulated that there is a melanocyte reservoir in human hair follicles (Ortonne, Schmitt *et al.*, 1980). Other studies also supported the view that ORS melanocytes are able to proliferate and to provide progeny to the epidermis in response to specific stimuli, such as UV exposure and epidermal wound healing (Cui, Shen *et al.*, 1991). Subsequently, the existence of melanocyte lineage cells with significant potential to proliferate was shown by Tobin *et al.* (Tobin, Colen *et al.*, 1995) with *in vitro* cultures of human hair follicles. In 1996, Horikawa *et al.*, reported that cells that stain with NK1beteb, a monoclonal antibody that recognizes the premelanosomal protein SILV D GP100 D PMEL17, are distributed in the ORS at the mid-portion of anagen hair follicles and are quite similar to 'amelanotic melanocytes' in their morphology and distribution (Horikawa, Norris *et al.*, 1996). Narisawa *et al.* (Narisawa, Kohda *et al.*, 1997) found that amelanotic melanocytes stained by NK1beteb possess limited numbers of dendrites and are distributed abundantly in the bulge area of human hair follicles. This series of studies demonstrated the lineage identity of 'amelanotic melanocytes' in human hair follicles and led to the view that 'amelanotic melanocytes' (melanoblasts) in the ORS represent a reservoir that can replenish melanocytes in the epidermis when necessary. Melanocytes originate from the neural crest and undergo a complex process of fate specification, proliferation, survival and differentiation, before finally residing in the epidermis (Le Douarin, 1999). The presence of a melanocyte stem cell niche has been shown in hair follicles of mouse and human skin (Nishimura, Jordan *et al.*, 2002; Yu, Fang *et al.*, 2006; Nishimura, 2011).

MelSC in hair follicles lose their self-renewal capacity with aging, leading to hair graying. By contrast, skin melanocytes sustain lifelong proliferative potential. Till recently no obvious spatially restricted niche of MelSC was reported in glabrous skin areas, which are abundant in melanocytes. Although it is a rare case, now studies show repopulation of melanocytes occurs in vitiliginous in glabrous skin areas after UVA treatment (Davids 2009; Davids LM. 2009). These data suggest that extrafollicular MelSC might exist, and could potentially function as a reservoir for melanocytes in postnatal epidermis (Li 2010; Li L 2012).

Melanocyte cell culture from the pigmented hair follicle gives rise to the two different kind of melanocytes one which are DOPA positive and the other population of DOPA negative cells. The DOPA positive cells which are pigmented and are not possible to be passaged further, whereas the second population of the cells are non pigmented or amelanotic melanocytes (AMMC), which can be expanded further in culture. These AMMC are now regarded as MelSC (Nishimura, 2011).

## Photobiology of the UV Radiation

UV radiation is electromagnetic radiation having wavelength between 100–400 nm. The biologically most important wavelength regions are UVA (320– 400 nm) and UVB (290–320 nm). UVB represents the most bio-reactive part of the spectrum both as inducer tanning. Our current conception is that UV radiation causes basal cell skin cancers, such as basal cell carcinoma and malignant melanoma, through its mutagenic effect on basal layer cells. Melanin has a remarkable capacity to absorb UV radiation and to reflect it at the shortest wavelengths (<300 nm) (Thingnes, Oyehaug *et al.*, 2009).

## The Tanning Respnose and Skin Physiology

Tanning is an acquired pigmentation process involving the synthesis of melanin by melanocytes in specialized organelles called melanosomes, which transport melanin to the extremities of their dendrites and distribute melanin to surrounding keratinocytes (Jablonski and Chaplin, 2000). The crucial role of skin pigmentation seems to be to absorb ultraviolet (UV) radiation, serving as a natural sun screen. UV radiation induces melanin production in melanocytes, and melanin acts as a physical barrier that scatters incident UV irradiation, thereby reducing penetration of UV irradiation through the epidermis (Furuya R, 2002). Pigmentation determines skin color and is the body's own protection against solar ultraviolet (UV) radiation. It consists of melanin which is a composition of mainly two substances, the black-brown eumelanin and the reddish-yellow pheomelanin. In melanocytes melanin is synthesized and stored in organelles called melanosomes. Melanocytes are located in the basal layer of the skin epidermis, in the so-called stratum basale. Melanocyte dendrites are responsible to spread melanosomes containing melanin within the skin by transferring them to keratinocytes (Kauser, Schallreuter *et al.*, 2003). Epidermal melanin has important evolutionary and physiological implications, particularly for unclothed humans. Thus high melanin content (racial pigmentation) protects the skin against ultraviolet (UV)-induced skin damage through its optical and chemical filtering properties (Ahene AB 1995). Indeed, skin pigment levels and anthropological origin are closely associated, with higher pigment amounts in regions of lower latitude and higher UV radiation levels. However, this connection may only be a recent human adaptation since early hominids may have possessed dark, dense, terminal body hair. A closely related primate, the chimpanzee, similar to most other nonhuman primates, exhibits white or lightly pigmented epidermis (Post, Daniels *et al.*, 1975). Interestingly, chimpanzees have active melanocytes only in the epidermis of those areas directly exposed to UV radiation, *e.g.*, face and friction surfaces (Montagna and Machida 1966). Without concomitant increase in epidermal melanization, the end result of reduced hair coverage in humans residing in high UV radiation areas would be direct exposure to the adverse effects of that radiation. These include sunburn damage to the sweat glands with resultant suppression of sweating and abnormal thermoregulation (Pandolf, Gange *et al.*, 1992), carcinogenesis, and nutrient inactivation by photolysis (Brandenberger, Tee *et al.*, 1997). Human populations living in areas with lower UV levels adapt with lesser pigmentation, which facilitates the cutaneous UV radiation-mediated conversion of 7-dehydrocholesterol to pre-vitamin D3 (Webb, Kline *et al.*, 1988.) In fact, if UV exposure of pigmented humans is limited in duration and/or intensity (*e.g.*, northern latitudes), vitamin D3 deficiency and its associated pathologies may result, as seen in Southern Asians living in northern European cities (Henderson, Dunnigan *et al.*, 1987). Nevertheless, the value of the melanin pigmentation as a truly effective sunscreen for seasonal tanning is debatable (Wong, Pawelek *et al.*, 1974). Additional properties of melanin may include a bactericidal potential via the production of orthoquinones (Riley 1997), and contribution to the tensile strength of hair via cross-linking with proteins. Hair color may have undergone a far more complex evolution than skin pigmentation. Although most humans are dark-haired and dark-eyed, melanization in skin, hair, and eyes do not closely correlate. Indeed, a large fraction of humans have dark eyes and hair but their skin would rate as "white," whereas in some western European populations, black hair commonly coexists with blue eyes. The occurrence of black scalp hair, a potent trap for radiant heat, may appear as a paradoxical development for primates and humans living in tropical climates; however, black scalp hair may provide some protection from sunstroke by helping with the salt balance through the highly efficient and fast ion exchange property by melanin (Slominski and Paus 1993; Wood, Jimbow *et al.*, 1999). In fact, the pigmented hair on the human scalp may have resulted from the littoral residence of *Homo sapiens* residing on sea coasts or riverbanks, with diet dominant in fish (many of which concentrate heavy metals). In this context, the capability to rapidly excrete toxic metals provided by the very high turnover of melanized cortical keratinocytes in the pigmented hair shaft confers a selection advantage (Bertazzo, Costa *et al.*, 1996). Thus the long, melanized scalp hair with its capability to trap and/or bind chemicals, toxins, and heavy metals would prevent their access to living tissues. Pigmented hair may also provide antioxidant defense for the skin and hair follicles due to the high capacity of melanin for binding transition metals. (Slominski and Paus, 1993).

## The Hair Cycle and Melanocytes

The hair follicle is a mini-organ composed of epidermis-derived follicular keratinocytes, mesenchymal cells, and neuroectodermal cells required for cyclic pigmented hair growth. The hair follicle repeats its cyclic regeneration and regression with alternating phases, anagen (growing phase), catagen (regressing phase), and telogen (resting phase), to regrow new hair (Fuchs, 2007). Any cellular structure that periodically renews itself must have been able to do so because of pool or niche of naïve cells or stem cells capable to differentiate into desired cell types, cells that retain the ability to divide over the lifetime of the animal and regenerate that structure. Hair follicle stem cells (HFSCs) are multipotent stem cells that are located in the bulge area of each hair follicle (Cotsarelis, 2006; Fuchs, 2007). This HFSC population normally contributes to regeneration of the cycling portion of hair follicles to grow hair at the anagen phase (Cotsarelis, 2006). The cycling D transient portion of each hair follicle preferentially dies by apoptosis and regresses up to the level of the permanent portion of the hair follicle during the catagen phase and is then maintained in a resting state during telogen. The regeneration of the cycling portion is repeated with every hair cycle. As manifest by its cycle, the hair follicle is a regenerating system (Silver and Chase, 1977); the inferior hair follicle dramatically reforms itself over the cycle but the upper, permanent, follicle undergoes substantial remodeling (Lindner, Menrad *et al.*, 2000). Hair follicle cycling is a developmental process, which occurs over the total lifetime of a mammal. The fates of follicular melanocytes, including their proliferation, differentiation, and death, all happen in synchronization with hair cycle (Slominski, Paus *et al.*, 1994; Tobin, Slominski *et al.*, 1999; Nishimura, Jordan *et al.*, 2002) This fact suggests that the follicular melanocytes regeneration is regulated tightly by hair follicle cycling (Stenn and Paus, 2001).

## Hair Follicle Pigmentation and Melanocyte Stem Cells

Follicular pigmentation is under complex genetic control, as determined from studies in mostly murine models. In the mouse, coat color is regulated by more than 150 alleles at over 90 loci (Hearing 1999; Nakamura, Tobin *et al.*, 2002). This organization allows control of melanin synthesis at all levels: cellular (follicular melanocyte), organ (hair follicle), and developmental steps. Cutaneous melanocytes originate from the neural crest where lineage commitment is determined by microphthalmia-associated transcription factor (MITF), fibroblast growth factor-2, endothelin 3 (ET3), and others (Dupin and Le Douarin 2003). Melanoblasts migrate from the neural crest and populate the basal layer of the epidermis and then non-randomly enter the developing hair follicles. Melanogenic melanocytes can be detected at all stages of hair morphogenesis, from hair germ stage onwards (Peters, Tobin *et al.*, 2002). Melanoblasts expressing c-Kit migrate into the stem cell factor (SCF)-supplying hair follicle epithelium and differentiated c-Kit-positive melanocytes target the bulb when "SCF-positive" (Peters, Tobin *et al.*, 2002). The importance of this signaling pair is evident from the ability of a Kit blocking antibody to induce apoptosis in murine follicular melanocytes (Ito, Kawa *et al.*, 1999). C-Kit-negative melanoblasts invade the outer root sheath and bulge region in the fully developed hair follicle (Peters, Tobin *et al.*, 2002). Post-natally, c-Kit is required during the hair growth cycle for activation of melanocyte, although in stem cell compartment appears to exhibit SCF/c-Kit independence (Botchkareva, Khlgatian *et al.*, 2001). The cellular reservoirs for epidermal and hair follicle melanocytes were originally considered to be distinct populations. In human, the former has been referred to as 'amelanotic melanocytes.' In 1956, Montagna and Chase reported the existence of a distinct cell population that does not contain visible melanin pigment but can be distinguished morphologically from surrounding keratinocytes in the outer root sheath (ORS) of human hair follicles 3 weeks following ionizing irradiation. In the clinic, skin repigmentation usually begins at the orifices of hair follicles, then enlarges, and coalesces to cover the entire depigmented skin area, such as that caused by vitiligo. This phenomenon is called 'follicular repigmentation' and is characterized by repigmentation of an unpigmented skin area starting from the hair follicle orifices, suggesting the existence of a melanocyte 'reservoir' population in the skin (Staricco 1963; Ortonne, Schmitt *et al.*, 1980). Starricco studied the phenomenon extensively and found that the amelanotic ORS cell population, which does not contain any melanin pigment and is negative for the DOPA reaction, contains nuclei that stain densely with toluidine blue and thionine (Staricco 1959). Starricco subsequently found that this population can be activated after excision of the epidermis or by UV therapy (Staricco, 1963). He assumed that the amelanotic melanocyte population provides DOPA-positive dendritic melanocytes that migrate into, and fill, the epidermis for follicular repigmentation. However, these amelanotic melanocytes were not always clearly distinguished from Langerhans progenitors because of the lack of specific lineage markers available at the time (Staricco 1963; Shukla 1969). Ortonne *et al.*, examined the unpigmented skin of vitiligo patients after treatment with UV therapy using scanning electron microscopy and DOPA staining. They found that the pigmented islands that appeared after the treatment had follicular cores containing melanocytes with large cell bodies that could be distinguished from other cells in the ORS. They and Cui *et al.*, also postulated that there is a melanocyte reservoir in human hair follicles (Ortonne, Schmitt *et al.*, 1980; Cui, Shen *et al.*, 1991). Subsequently, the existence of melanocyte lineage cells with significant potential to proliferate was shown by Tobin *et al.* (Tobin, Colen *et al.*, 1995) with *in vitro* cultures of human hair follicles. In 1996, Horikawa *et al.*, reported that cells that stain with NK1beteb, a monoclonal antibody that recognizes the premelanosomal SILV/GP100/PMEL17 protein are distributed in the ORS at the mid-portion of anagen hair follicles and are quite similar to 'amelanotic melanocytes' in their morphology and distribution and found that amelanotic melanocytes stained by NK1beteb possess limited numbers of dendrites and are distributed abundantly in the bulge area of human hair follicles (Horikawa, Norris *et al.*, 1996; Narisawa, Kohda *et al.*, 1997). This series of studies demonstrated the lineage identity of 'amelanotic melanocytes' in human hair follicles and led to the view that 'amelanotic melanocytes' (melanoblasts) or melanocyte stem cells in the ORS represent a reservoir that can replenish melanocytes in the epidermis when necessary. In 1968, Silver *et al.*, found a population of cells that reside in the secondary hair germ of telogen hair follicles, contain a clear cytoplasm when stained with alkaline toluidine blue, and can be morphologically distinguished from keratinocytes and called them 'clear cells' (Silver 1968). Because chronological analysis suggested that these cells subsequently produce melanin pigments in the hair matrix during the anagen phase, Silver *et al.*, assumed that the population in the hair germ represents melanocyte precursors. As these cells were found just above the basement membrane adjacent to the dermal papillae and below the bulge area, the population resides in the subbulge area of the hair follicles. In these studies, no marker for the melanocyte lineage was available, and the possibility that mature melanocytes dedifferentiate into immature melanoblasts during regression of the hair follicle remained (Sugiyama 1979). In 1998, Tobin *et al.*, demonstrated that mature melanocytes located in the hair matrix die by apoptosis during regression of the cycling portion of the hair follicle during catagen (Tobin, Hagen *et al.*, 1998). The dramatic loss of hair matrix melanocytes had to be replenished in the following hair cycles. A turning point in this series of studies occurred when Botchkareva *et al.*, provided evidence using an anti-kit blocking antibody that strongly supported the existence of a 'melanocyte reservoir' that is not dependent on Kit signaling (Botchkareva, Khlgatian *et al.*, 2001). It was not clear from their experiments, however, which population in the hair follicle is the real reservoir, but they reported that DCTD TRP2-expressing melanoblasts exist in the hair follicle bulge area and suggested that this population may represent MelSCs. As described earlier, the reservoir for human epidermal melanocytes and that for mouse hair matrix melanocytes have been studied independently as completely different cell populations, but both lines of studies eventually merged when MelSCs were identified (Nishimura, 2011).

## Mechanisms of Melanocyte Maintenance, Hair and Skin Pigmentation

Skin and hair phenotypes are powerful cues in human communication. They impart much information, particularly about our racial, ethnic, health, gender and age status. In the case of the latter parameter, we experience significant change in pigmentation in our journey from birth to puberty and through to young adulthood, middle age and beyond. The hair follicle pigmentary unit is perhaps one of our most visible, accessible and potent aging sensors, with marked dilution of pigment intensity occurring long before even subtle changes are seen in the epidermis. Research groups are actively pursuing the study of the differential aging of melanocytes in the hair bulb versus the epidermis and in particular are examining whether this is in part linked to the stringent coupling of follicular melanocytes to the hair growth cycle. Whether some follicular melanocyte subpopulations are affected, like epidermal melanocytes, by UV irradiation is not yet clear. A particular target of research into hair graying or canities is the nature of the melanocyte stem compartment and whether this is depleted due to reactive oxygen species-associated damage, coupled with an impaired antioxidant status, and a failure of melanocyte stem cell renewal (Tobin 2009). Our group has also worked a little bit in this area and have got clue regarding the role of stem cells in human hair graying, where we have reported a depletion of keratinocyte CD200+ cells in the grey hair follicle in comparison to pigmented hair follicle (Mohanty, Kumar *et al.*, 2013).

More than 120 loci and 800 alleles have been identified that explain the coat color diversity in mice, including pigmentation and spotting pattern diversity (Bennett 2003). A group of coat color mutants are born normally pigmented but turn gray prematurely after hair molting. If Melanocyte Stem Cells (MelSCs) in the bulge–subbulge area were the only source for mature functioning melanocytes in hair follicles, one would imagine defects in the population maintenance to result inevitably in the appearance of the hair-greying phenotype. Indeed, analysis of mutant mice revealed that the loss of MelSCs precedes the loss of hair matrix melanocytes and hair graying (Nishimura, Granter *et al.*, 2005). Induction of ectopic differentiation of MelSCs in the niche andD or the death of MelSCs resulting from deficiencies in a critical gene or genes cause depletion of MelSCs and, consequently, hair graying. MITF, the master regulator of melanocyte development, and its downstream target Bcl2 have been shown to play critical roles in MelSC maintenance (Nishimura, Granter *et al.*, 2005; Mak, Moriyama *et al.*, 2006). The surrounding microenvironment (the niche) also plays critical roles in fate determination of MelSCs (Nishimura, Jordan *et al.*, 2002). MelSCs are located at the basement membrane and have direct contact with HFSC marker expressing keratinocytes in the bulge area. Indeed, TGF-b secreted from the surrounding HFSCs is critical for the maintenance of MelSCs by promoting their quiescence and immaturity. Therefore, HFSCs provide a functional niche with MelSCs (Nishimura, Suzuki *et al.*, 2010; Tanimura, Tadokoro *et al.*, 2011). Notch signaling is also important both for melanocyte development and for MelSC maintenance by promoting their survival (Moriyama, Osawa *et al.*, 2006; Schouwey, Delmas *et al.*, 2007). The Wnt signaling pathway and the transcription factor Pax3, both located upstream of MITF, have also been suggested to be important for MelSC maintenance (Kubic, Young *et al.*, 2008), but we still await *in vivo* evidence supporting their critical role. Other signals critical to MelSC maintenance will no doubt be defined in the near future (Nishimura, 2011).

## Pigment Restoration

Both non-surgical and surgical therapies have been employed to restore pigmentation. Here the elaboration has been given keeping in view the vitiligo. Surgical therapies are useful when disease is recalcitrant to medical therapies and stable for at least one year. Tissue grafts can treat only a small area; therefore cellular grafts, like basal cell layer enriched epidermal suspension are preferred over tissue grafts when large area is to be treated. Tissue grafting procedures include transplantation of minigrafts, thin and ultra-thin split thickness grafts (STSG) and suction blister epidermal grafts (SBEG). Cellular grafting techniques include transplantation of non-cultured basal cell enriched epidermal suspension (NCES) and cultured melanocytes (CM) or cultured epidermis (CE) (Lontz, Olsson *et al.*, 1994; Gupta, Shroff *et al.*, 1999; Ongenae, van Geel *et al.*, 2001; Falabella 2005; Gupta, Goel *et al.*, 2006; Bahadoran P 2007). The major advantage of non- cultured cell suspension and culture expanded cell suspension is that they permit the treatment of the affected area many folds larger than the donor area (Olsson and Juhlin 1993; Pianigiani, Risulo *et al.*, 2005; Redondo, del Olmo *et al.*, 2008). However, culture techniques employed for the expansion of the cell number are time consuming, expensive and also require trained manpower and well-equipped tissue culture laboratory. In addition, there are concerns about the safety of the techniques owing to the xenobiotic properties of some of the additives in the culture medium (Geel, 2011).

These limitations of culture expanded cells were sorted out with the introduction of the non-cultured cellular grafting techniques in 1992 by Gauthier and Surleve-Bazeille (Gauthier and Surleve-Bazeille 1992) and followed by several other vitiligo surgery groups. Non-cultured cell transplant can cover larger vitiliginous areas (8-to10-fold size of donor skin) and the procedure can be completed in few hours on an outpatient basis. In summary, cultured cell transplants are technically more challenging and expensive than non-cultured epidermal cell suspension; hence the latter provides a simple, effective and aesthetically more suitable treatment option (Falabella, Escobar *et al.*, 1989; Gauthier and Surleve-Bazeille 1992; Jha, Pandey *et al.*, 1993; Olsson and Juhlin 1993; Gupta, Jain *et al.*, 1999; Gupta, Shroff *et al.*, 1999; Chen, Chang *et al.*, 2000; Chen, Yang *et al.*, 2001; Czajkowski 2004; Gupta, Goel *et al.*, 2006).

Vitiligo surgery using the hair follicle and the hair follicle ORS cell transplants are the recent trends. Ttransplantation of hair follicle in vitiligo uses the scalp biopsy from occipital area. The scalp biopsy tissue is cut into smaller pieces, each containing a single hair follicle unit. The individual hair follicle units are then transplanted onto the vitiligo patch. It leaves a visible scar on the donor site-scalp. The vitiliginous area covered is again very less due to transplantation of the intact hair follicle. These two limitations of the hair transplant in vitiligo have been overcome by the recently developed hair follicle ORS cell transplant. Vaenscheidt *et al.*, were the first group to use the hair follicle ORS cells for transplantation in vitiligo (Vanscheidt and Hunziker 2009). They used the plucked hair follicle ORS cells for transplantation in vitiligo. We observed that during follicular plucking CD200+ cells population is partially lost in comparison to the extracted hair follicle (Kumar, Gupta *et al.*, 2013). Based upon these observations (it was published later) we devised a novel method of surgical pigment restoration in vitiligo using non-cultured Extracted Hair Follicle Outer Rooth Sheath Cell Suspension (EH-ORS-CS) (Mohanty S., Kumar *et al.*, 2011; Kumar, Mohanty *et al.*, 2013). Our observation is further supported by the observation that transplantation of plucked hair doesn't result in hair growth, however transplantation of extracted follicular unit is promptly accepted by the recipient site with resulting hair growth (Gho and Neumann 2001; Gho, Braun *et al.*, 2004).

Literature search shows that, hair follicle ORS suspension is more enriched for pigment cells, if we compare it with the established method of vitiligo surgery-epidermal cell suspension transplantation. In follicular melanin unit, there is one melanocyte for every five keratinocytes in the hair bulb which is much higher than epidermal melanin unit, which has one melanocyte for every thirty-six keratinocytes (Tobin and Paus 2001; Tobin 2009). In comparison to epidermal melanocytes, anagen hair bulb melanogenic melanocytes are larger, more dendritic, with more extensive golgi and rough endoplasmic reticulum, and produce larger melanosomes (Tobin and Paus 2001). Hair melanocytes have remarkable synthetic capacity and a relatively small number of melanocytes can potentially produce sufficient melanin to pigment up to 1.5 m of hair shaft. (Slominski, Wortsman *et al.*, 2005). All these properties make hair melanocytes a more attractive source than epidermal melanocytes for cell based therapies in vitiligo.

In conclusion, though, this procedure can serve as a novel, minimally invasive, technically less challenging technique with good yield of melanocytes, melanocyte stem cells, and other stem cells (keratinocyte stem cells, and mesenchymal stem cells in the surrounding dermis). It results in excellent pigmentation with no visible residual scarring (Mohanty S., Kumar *et al.*, 2011).

Epidermal stem cells isolated from hair follicles can also serve a good source for restoration of unrelated tissue such as nerves and cornea (Krejci and Grim; Yang, Jiang *et al.*, 2009). Thus, cell suspension prepared from extracted hair follicles provides an excellent, relatively less invasive and technically less challenging source for stem cells and/or melanocytes for vitiligo surgery.

## Conclusion and Future Direction

Lot of research focus has shifted towards the human melancoyte biology. These are very valuable pigments as they provide the shades of skin colour ranging from white to black skin. This gradation of skin colour gives rise to a lot of potential for the billion dollar cosmetic market which is rising day by day. Tropical people are darker in skin color and people inhabiting near the poles are fairer. Though this gradation is due to the amount of UV radiation exposure in that particular habitat, in some counties white is considered beautiful and it has created a lot of billion dollar market to cater to the needs of the developing demands of cosmetic products to make the skin fairer. This has fueled a lot of research on melanocyte biology. Finding of stem cells in skin and hair follicle is another big impetus to the research on melanocyte biology in the area of skin repigmentation for various conditions which are less medical and more of social ailments.

This is a very good time to research in the area of melanocyte biology, as we are gifted with lot of knowledge base earned by the clinicians and researchers. Many recent developments have brought interdisciplinary people together and are opening a vista of research avenues and product development.

## References


1. Ahene AB, S. S., and Nacht S. (1995). "Photoprotection of solubilized and microdispersed melanin particles. In: Melanin, Its Role in Human Photoprotection, edited by Zeise L, Chedekel M, and Fitzpatrick TB." Overland Park, KS: Valendmar: p. 255-269.
2. Alonso, L. and E. Fuchs (2003). "Stem cells of the skin epithelium." Proc Natl Acad Sci U S A **100 Suppl 1**: 11830-5.
3. Aristidis Moustakas, K. M. (2013). "TGF-β in Human Disease." Springer Science and Business Media.
4. Bahadoran P, O. J.-P. (2007). "Classification of surgical therapies for vitiligo. In Surgical Management of Vitiligo (Gupta S, Olsson M, Kanwar A, Ortonne JP, Eds.)." Oxford: Blackwell Publishing, pp. 59-68.
5. Bell, M. (1967). "Advances in Biology of Skin.In: Montagna W, Dobson RL (eds).Ultrastructure of differentiating hair follicles." Oxford: Pergamon Press.
6. Bennett, D. C. (2003). "Human melanocyte senescence and melanoma susceptibility genes." *Oncogene* **22**(20): 3063-9.
7. Bertazzo, A., C. Costa, *et al.* (1996). "Determination of copper and zinc levels in human hair: influence of sex, age, and hair pigmentation." *Biol Trace Elem Res* **52**(1): 37-53.
8. Boissy, R. (2000). "Histopathology of vitiliginous skin. In: Hann SK, Nordlund JJ, editors. Vitiligo." Oxford: Blackwell Science Ltd: p. 23-34.
9. Botchkareva, N. V., M. Khlgatian, *et al.* (2001). "SCF/c-kit signaling is required for cyclic regeneration of the hair pigmentation unit." *Faseb J* **15**(3): 645-58.
10. Brandenberger, A. W., M. K. Tee, *et al.* (1997). "Tissue distribution of estrogen receptors alpha (ER-alpha) and beta (ER-beta) mRNA in the midgestational human fetus." *J Clin Endocrinol Metab* **82**(10): 3509-12.
11. Chen, Y. F., J. S. Chang, *et al.* (2000). "Transplant of cultured autologous pure melanocytes after laser-abrasion for the treatment of segmental vitiligo." *J Dermatol* **27**(7): 434-9.
12. Chen, Y. F., P. Y. Yang, *et al.* (2001). "Transplantation of autologous cultured melanocytes for treatment of large segmental vitiligo." *J Am Acad Dermatol* **44**(3): 543-5.
13. Commo, S. and B. A. Bernard (2000). "Melanocyte subpopulation turnover during the human hair cycle: an immunohistochemical study." *Pigment Cell Res* **13**(4): 253-9.
14. Commo, S., O. Gaillard, *et al.* (2004). "Human hair greying is linked to a specific depletion of hair follicle melanocytes affecting both the bulb and the outer root sheath." *Br J Dermatol* **150**(3): 435-43.
15. Cotsarelis, G. (2006). "Epithelial stem cells: a folliculocentric view." *J Invest Dermatol* **126**(7): 1459-68.

16. Cui, J., L. Y. Shen, *et al.* (1991). "Role of hair follicles in the repigmentation of vitiligo." *J Invest Dermatol* **97**(3): 410-6.
17. Czajkowski, R. (2004). "Comparison of melanocytes transplantation methods for the treatment of vitiligo." *Dermatol Surg* **30**(11): 1400-5.
18. Davids, L., du Toit, E., Kidson, S. H. and Todd, G (2009). "A rare repigmentation pattern in a vitiligo patient: a clue to an epidermal stem-cell reservoir of melanocytes?" *Clin. Exp. Dermatol* **34**: 246-248.
19. Davids LM., d. T., E., Kidson, S. H. and Todd, G (2009). "A rare repigmentation pattern in a vitiligo patient: a clue to an epidermal stem-cell reservoir of melanocytes?" *Clin. Exp. Dermatol* **34**: 249-258.
20. Dupin, E. and N. M. Le Douarin (2003). "Development of melanocyte precursors from the vertebrate neural crest." *Oncogene* **22**(20): 3016-23.
21. Falabella, R. (2005). "Surgical approaches for stable vitiligo." *Dermatol Surg* **31**(10): 1277-84.
22. Falabella, R. (2009). "Vitiligo and the melanocyte reservoir." *Indian J Dermatol* **54**(4): 313-8.
23. Falabella, R., C. Escobar, *et al.* (1989). "Transplantation of in vitro-cultured epidermis bearing melanocytes for repigmenting vitiligo." *J Am Acad Dermatol* **21**(2 Pt 1): 257-64.
24. Ford, C. E., J. L. Hamerton, *et al.* (1956). "Cytological identification of radiation-chimaeras." *Nature* **177**(4506): 452-4.
25. Fuchs, E. (2007). "Scratching the surface of skin development." *Nature* **445**(7130): 834-42.
26. Furuya R, A. S., Ideta R, Naganuma M, Fukuda M, and Hirobe T. (2002). "Changes in the proliferative activity of epidermal melanocytes in serum-free primary culture during the development of ultraviolet radiation B-induced pigmented spots in hairless mice." *Pigment Cell Res* **15**: 348-356.
27. Gauthier, Y. and J. E. Surleve-Bazeille (1992). "Autologous grafting with noncultured melanocytes: a simplified method for treatment of depigmented lesions." *J Am Acad Dermatol* **26**(2 Pt 1): 191-4.
28. Geel, N. v. (2011). "A Review of Non-cultured Epidermal Cellular Grafting in Vitiligo." *Journal of Cutaneous and Aesthetic Surgery* **4**(1): 17-22.
29. Gho, C. G., J. E. Braun, *et al.* (2004). "Human follicular stem cells: their presence in plucked hair and follicular cell culture." *Br J Dermatol* **150**(5): 860-8.
30. Gho, C. G. and H. A. Neumann (2001). "Hair transplantation of plucked hair biopsies." *Dermatol Surg* **27**(10): 913.
31. Goldman, L., R. S. Moraites, *et al.* (1966). "White spots in biblical times. A background for the dermatologist for participation in discussions of current revisions of the bible." *Arch Dermatol* **93**(6): 744-53.

32. Gupta, S., A. Goel, *et al.* (2006). "Autologous melanocyte transfer via epidermal grafts for lip vitiligo." *Int J Dermatol* **45**(6): 747-50.

33. Gupta, S., V. K. Jain, *et al.* (1999). "Suction blister epidermal grafting versus punch skin grafting in recalcitrant and stable vitiligo." *Dermatol Surg* **25**(12): 955-8.

34. Gupta, S., S. Shroff, *et al.* (1999). "Modified technique of suction blistering for epidermal grafting in vitiligo." *Int J Dermatol* **38**(4): 306-9.

35. Hall, P. A. and F. M. Watt (1989). "Stem cells: the generation and maintenance of cellular diversity." *Development* **106**(4): 619-33.

36. Hearing, V. (1999). "Biochemical control of melanogenesis and melanosomal organization." *J Investig Dermatol Symp Proc* **4**: 24-28.

37. Henderson, J. B., M. G. Dunnigan, *et al.* (1987). "The importance of limited exposure to ultraviolet radiation and dietary factors in the aetiology of Asian rickets: a risk-factor model." *Q J Med* **63**(241): 413-25.

38. Horikawa, T., D. A. Norris, *et al.* (1996). "DOPA-negative melanocytes in the outer root sheath of human hair follicles express premelanosomal antigens but not a melanosomal antigen or the melanosome-associated glycoproteins tyrosinase, TRP-1, and TRP-2." *J Invest Dermatol* **106**(1): 28-35.

39. Ito, M., Y. Kawa, *et al.* (1999). "Removal of stem cell factor or addition of monoclonal anti-c-KIT antibody induces apoptosis in murine melanocyte precursors." *J Invest Dermatol* **112**(5): 796-801.

40. Jablonski, N. G. and G. Chaplin (2000). "The evolution of human skin coloration." *J Hum Evol* **39**(1): 57-106.

41. Jha, A. K., S. S. Pandey, *et al.* (1993). "Inoculation of a cultured autologous epidermal suspension containing melanocytes in vitiligo." *Arch Dermatol* **129**(6): 785-6.

42. Kauser, S., K. U. Schallreuter, *et al.* (2003). "Regulation of human epidermal melanocyte biology by beta-endorphin." *J Invest Dermatol* **120**(6): 1073-80.

43. Kopera D. In: Hann SK, N. J., editors. Vitiligo. (2000). "History and cultural aspects of vitiligo." Oxford: Blackwell Science Ltd: p 13-7.

44. Koranne, R. V. and K. G. Sachdeva (1988). "Vitiligo." *Int J Dermatol* **27**(10): 676-81.

45. Krejci, E. and M. Grim "Isolation and characterization of neural crest stem cells from adult human hair follicles." *Folia Biol (Praha)* **56**(4): 149-57.

46. Kubic, J. D., K. P. Young, *et al.* (2008). "Pigmentation PAX-ways: the role of Pax3 in melanogenesis, melanocyte stem cell maintenance, and disease." *Pigment Cell Melanoma Res* **21**(6): 627-45.

47. Kumar, A., S. Gupta, *et al.* (2013). "Stem Cell Niche is Partially Lost during Follicular Plucking: A Preliminary Pilot Study." *Int J Trichology* **5**(2): 97-100.

48. Kumar, A., S. Mohanty, *et al.* (2013). "Extracted hair follicle outer root sheath cell suspension for pigment cell restoration in vitiligo." *J Cutan Aesthet Surg* **6**(2): 121-5.

49. Lahiri, K. (2009). "Evolution and evaluation of autologous mini punch grafting in vitiligo." *Indian J Dermatol* **54**(2): 159-67.

50. Le Douarin, N. M. (1999). *"The Neural Crest."* Cambridge, UK: Cambridge University Press.

51. Li L, F.-K. M., Herlyn M. (2012). "Isolation and cultivation of dermal stem cells that differentiate into functional epidermal melanocytes." *Methods Mol Biol.* **806**: 15-29.

52. Li, L., Fukunaga-Kalabis, M., Yu, H., Xu, Xiaowei., Kong, Jun., Lee, JT., Herlyn, M. (2010). "Human dermal stem cells differentiate into functional epidermal melanocytes " *Journal of Cell Science* **123**: 853-860.

53. Lindner, G., V. A. Botchkarev, *et al.* (1997). "Analysis of apoptosis during hair follicle regression (catagen)." *Am J Pathol* **151**(6): 1601-17.

54. Lindner, G., A. Menrad, *et al.* (2000). "Involvement of hepatocyte growth factor/ scatter factor and met receptor signaling in hair follicle morphogenesis and cycling." *FASEB J* **14**(2): 319-32.

55. Lontz, W., M. J. Olsson, *et al.* (1994). "Pigment cell transplantation for treatment of vitiligo: a progress report." *J Am Acad Dermatol* **30**(4): 591-7.

56. Mackenzie, I. (1970). " Relationship between mitosis and the ordered structure of the stratum corneum in mouse epidermis." *Nature* **226**: 653-655.

57. Mak, S. S., M. Moriyama, *et al.* (2006). "Indispensable role of Bcl2 in the development of the melanocyte stem cell." *Dev Biol* **291**(1): 144-53.

58. Mayer, T. C. (1973). "The migratory pathway of neural crest cells into the skin of mouse embryos." *Dev Biol* **34**(1): 39-46.

59. Milington GW, L. L. (2007). "The historical course of depigmentation." *Int J Dermatol* **46**: 900-45.

60. Mohanty, S., A. Kumar, *et al.* (2013). "Depletion of CD200+ Hair Follicle Stem Cells in Human Prematurely Gray Hair Follicles." *J Cutan Aesthet Surg* **6**(2): 90-2.

61. Mohanty S., A. Kumar, *et al.* (2011). "Noncultured extracted hair follicle outer root sheath cell suspension for transplantation in vitiligo." *Br J Dermatol* **164**(6): 1241-6.

62. Montagna, W. and H. B. Chase (1956). "Histology and cytochemistry of human skin." *Am J Anat* **99**(3): 415-45.

63. Montagna, W. and H. Machida (1966). "The skin of primates. XXXII. The Philippine tarsier (Tarsius syrichta)." *Am J Phys Anthropol* **25**(1): 71-83.

64. Moriyama, M., M. Osawa, *et al.* (2006). "Notch signaling via Hes1 transcription factor maintains survival of melanoblasts and melanocyte stem cells." *J Cell Biol* **173**(3): 333-9.

65. Nair, B. K. (1978). "Vitiligo: A retrospect." *Int J Dermatol* **17**(9): 755-7.
66. Nakamura, M., D. J. Tobin, *et al.* (2002). "Mutant laboratory mice with abnormalities in pigmentation: annotated tables." *J Dermatol Sci* **28**(1): 1-33.
67. Narisawa, Y., H. Kohda, *et al.* (1997). "Three-dimensional demonstration of melanocyte distribution of human hair follicles: special reference to the bulge area." *Acta Derm Venereol* **77**(2): 97-101.
68. Nishimura, E. K. (2011). "Melanocyte stem cells: a melanocyte reservoir in hair follicles for hair and skin pigmentation." *Pigment Cell Melanoma Res.*
69. Nishimura, E. K., S. R. Granter, *et al.* (2005). "Mechanisms of hair graying: incomplete melanocyte stem cell maintenance in the niche." *Science* **307**(5710): 720-4.
70. Nishimura, E. K., S. A. Jordan, *et al.* (2002). "Dominant role of the niche in melanocyte stem-cell fate determination." *Nature* **416**(6883): 854-60.
71. Nishimura, E. K., M. Suzuki, *et al.* (2010). "Key roles for transforming growth factor beta in melanocyte stem cell maintenance." *Cell Stem Cell* **6**(2): 130-40.
72. Olsson, M. J. and L. Juhlin (1993). "Repigmentation of vitiligo by transplantation of cultured autologous melanocytes." *Acta Derm Venereol* **73**(1): 49-51.
73. Ongenae, K., N. van Geel, *et al.* (2001). "Autologous cellular suspensions and sheets in the treatment of achromic disorders: the need for future controlled studies." *Dermatology* **202**(2): 158-61.
74. Ortonne, J. P., D. Schmitt, *et al.* (1980). "PUVA-induced repigmentation of vitiligo: scanning electron microscopy of hair follicles." *J Invest Dermatol* **74**(1): 40-2.
75. Pandolf, K. B., R. W. Gange, *et al.* (1992). "Human thermoregulatory responses during heat exposure after artificially induced sunburn." *Am J Physiol* **262**(4 Pt 2): R610-6.
76. Peters, E. M., D. J. Tobin, *et al.* (2002). "Migration of melanoblasts into the developing murine hair follicle is accompanied by transient c-Kit expression." *J Histochem Cytochem* **50**(6): 751-66.
77. Pianigiani, E., M. Risulo, *et al.* (2005). "Autologous epidermal cultures and narrow-band ultraviolet B in the surgical treatment of vitiligo." *Dermatol Surg* **31**(2): 155-9.
78. Porter, J., A. Beuf, *et al.* (1978). "Personal responses of patients to vitiligo: the importance of the patient-physician interaction." *Arch Dermatol* **114**(9): 1384-5.
79. Post, P. W., F. Daniels, Jr., *et al.* (1975). "Cold injury and the evolution of "white" skin." *Hum Biol* **47**(1): 65-80.
80. Potten, C. S. (1974). "The epidermal proliferative unit: the possible role of the central basal cell." *Cell Tissue Kinet* **7**(1): 77-88.
81. Potten, C. S. (1981). "Cell replacement in epidermis (keratopoiesis) via discrete units of proliferation." *Int Rev Cytol* **69**: 271-318.

82. Potten, C. S., R. Schofield, *et al.* (1979). "A comparison of cell replacement in bone marrow, testis and three regions of surface epithelium." *Biochim Biophys Acta* **560**(2): 281-99.

83. Redondo, P., J. del Olmo, *et al.* (2008). "Repigmentation of vitiligo by transplantation of autologous melanocyte cells cultured on amniotic membrane." *Br J Dermatol* **158**(5): 1168-71.

84. Riley, P. A. (1997). "Melanin." *Int J Biochem Cell Biol* **29**(11): 1235-9.

85. Schouwey, K., V. Delmas, *et al.* (2007). "Notch1 and Notch2 receptors influence progressive hair graying in a dose-dependent manner." *Dev Dyn* **236**(1): 282-9.

86. Shukla, R. C. (1969). "The origin of Langerhans' cell from the melanocyte." *Acta Anat (Basel)* **73**(3): 372-88.

87. Silver, A. F. and H. B. Chase (1977). "The incorporation of tritiated uridine in hair germ and dermal papilla during dormancy (telogen) and activation (early anagen)." *J Invest Dermatol* **68**(4): 201-5.

88. Silver, A. F., Chase, H.B., and Potten, C.S. (1968). "Melanocyte precursor cells in the hair follicle germ during the dormat stage (telogen)." *Experientia* **25**: 299-301.

89. Silvers, W. (1979). "The Coat Colors of Mice-A Model for Mammalian Gene Action and Interaction. The Coat of the Mouse and Its Development." Springer Verlag.

90. Slominski, A. and R. Paus (1993). "Melanogenesis is coupled to murine anagen: toward new concepts for the role of melanocytes and the regulation of melanogenesis in hair growth." *J Invest Dermatol* **101**(1 Suppl): 90S-97S.

91. Slominski, A., R. Paus, *et al.* (1994). "Melanogenesis during the anagen-catagen-telogen transformation of the murine hair cycle." *J Invest Dermatol* **102**(6): 862-9.

92. Slominski, A., J. Wortsman, *et al.* (2005). "Hair follicle pigmentation." *J Invest Dermatol* **124**(1): 13-21.

93. Staricco, R. G. (1959). "Amelanotic melanocytes in the outer sheath of the human hair follicle." *J Invest Dermatol* **33**: 295-7.

94. Staricco, R. G. (1963). "Amelanotic melanocytes in the outer sheath of the human hair follicle and their role in the repigmentation of regenerated epidermis." *Ann N Y Acad Sci* **100**: 239-55.

95. Stenn, K. S. and R. Paus (2001). "Controls of hair follicle cycling." *Physiol Rev* **81**(1): 449-494.

96. Sugiyama, S. (1979). "Mode of redifferentiation and melanogenesis of melanocytes in mouse hair follicles. An ultrastructural and cytochemical study." *J Ultrastruct Res* **67**(1): 40-54.

97. Tanimura, S., Y. Tadokoro, *et al.* (2011). "Hair follicle stem cells provide a functional niche for melanocyte stem cells." *Cell Stem Cell* **8**(2): 177-87.

98. Thingnes, J., L. Oyehaug, *et al.* (2009). "The mathematics of tanning." *BMC Syst Biol* **3**: 60.

99. Till, J. E. and C. E. Mc (1961). "A direct measurement of the radiation sensitivity of normal mouse bone marrow cells." *Radiat Res* **14**: 213-22.

100. Tobin, D. J. (2009). "Aging of the hair follicle pigmentation system." *Int J Trichology* **1**(2): 83-93.

101. Tobin, D. J., S. R. Colen, *et al.* (1995). "Isolation and long-term culture of human hair-follicle melanocytes." *J Invest Dermatol* **104**(1): 86-9.

102. Tobin, D. J., E. Hagen, *et al.* (1998). "Do hair bulb melanocytes undergo apoptosis during hair follicle regression (catagen)?" *J Invest Dermatol* **111**(6): 941-7.

103. Tobin, D. J. and R. Paus (2001). "Graying: gerontobiology of the hair follicle pigmentary unit." *Exp Gerontol* **36**(1): 29-54.

104. Tobin, D. J., A. Slominski, *et al.* (1999). "The fate of hair follicle melanocytes during the hair growth cycle." *J Investig Dermatol Symp Proc* **4**(3): 323-32.

105. Van Den Bossche, K., J. M. Naeyaert, *et al.* (2006). "The quest for the mechanism of melanin transfer." *Traffic* **7**(7): 769-78.

106. Vanscheidt, W. and T. Hunziker (2009). "Repigmentation by outer-root-sheath-derived melanocytes: proof of concept in vitiligo and leucoderma." *Dermatology* **218**(4): 342-3.

107. Vesalius, R. R. A. (2007). "Founder of the Modern Human Anatomy." *Int J Morphol*: 847-50.

108. Webb, A. R., L. Kline, *et al.* (1988). "Influence of season and latitude on the cutaneous synthesis of vitamin D3: exposure to winter sunlight in Boston and Edmonton will not promote vitamin D3 synthesis in human skin." *J Clin Endocrinol Metab* **67**(2): 373-8.

109. Westerhof, W. (2006). "The discovery of the human melanocyte." *Pigment Cell Res* **19**(3): 183-93.

110. Wong, G., J. Pawelek, *et al.* (1974). "Response of mouse melanoma cells to melanocyte stimulating hormone." *Nature* **248**(446): 351-4.

111. Wood, J. M., K. Jimbow, *et al.* (1999). "What's the use of generating melanin?" *Exp Dermatol* **8**(2): 153-64.

112. Wu, A. M., Till, J. E., Siminovitch, L. and Mcculloch, E. A. (1968). "Cytological evidence for a relationship between normal hematopoietic colony-forming cells and cells of the lymphoid system." *J. exp. Med.* **127**: 455-46.

113. Yang, K., Z. Jiang, *et al.* (2009). "Corneal epithelial-like transdifferentiation of hair follicle stem cells is mediated by pax6 and beta-catenin/Lef-1." *Cell Biol Int* **33**(8): 861-6.

114. Yu, H., D. Fang, *et al.* (2006). "Isolation of a novel population of multipotent adult stem cells from human hair follicles." *Am J Pathol* **168**(6): 1879-88.

**2016, Biotechnology: Progress and Applications** *Pages* **525–556**
*Editors:* **Saif Hameed and Zeeshan Fatima**
*Published by:* **DAYA PUBLISHING HOUSE, NEW DELHI**

# *Chapter 22*

# Nanotechnology: Size DOES Matter

***Nitai Debnath* and Sumistha Das***

*Amity Institute of Biotechnology, Amity University Haryana, Gurgaon (Manasar) – 122 413, Haryana, India*

## ABSTRACT

Nanotechnology is the understanding and control of matter at the nanoscale, approximately at dimensions between 1 and 100 nm, where unique phenomena enable novel applications. It is the science and technology of manipulating and rearranging individual atoms and molecules to create useful materials, devices, and systems. Though the term 'Nanotechnology' is a coinage of modern science, nanoparticles (NPs) have been used for a very long time in Indian medicine, Chinese porcelain and European church paintings. But Nanotechnology in its current form started in the 1980s with the enhancement in resolution of electron microscope and invention of newer microscopic tools which helped researchers to visualize and manipulate particles at nano domain. Among different NPs carbon nanomaterials in different form (fullerene, quantum dot, grapheme, nano tube etc.) got much attention due to their potential application in various fields. Oxide nanomaterials like silica and zinc oxide, titanium dioxide; magnetic NPs, gold and silver NPs also have wide spread application ranging from cosmetics to drug and nucleic acid delivery. In this chapter different physico-chemical tools for NP characterization, widely used NPs and their application in biosensing and biomedical science are elaborately discussed.

* Corresponding Author: E-mail: nitai.debnath@gmail.com

## Introduction

Nanotechnology refers to the art and science of manipulating and rearranging individual atoms and molecules to create useful materials, devices, and systems at nanoscale, dimensions between approximately 1 and 100 nanometers. At this length scale unique physicochemical properties enable novel applications. Understanding nanotechnology involves nano-material science, engineering, and technology which also include imaging, measuring, modeling, and manipulating matter at this length scale.

Nanoscale particles are not new in either nature or science. 'NPs' have long been used in ancient human civilizations in pottery and medicines. In India red colloidal gold was used as *"Swarna Bhasma"* and *"Makaradhwaja"* in Ayurveda - the traditional medicine (Bhattacharya and Mukherjee 2008). Even, there are plenty of nanoscale objects in nature. Fossil fuel exhaust derivatives, volcanic ash, sea foam, products of photoionization are some of the sources of atmospheric nano particulate matter. Living organisms possess millions of nano sized devices which perform various biological functions. DNA, RNA are nothing but biological NPs. Ribosomes, various enzymes like, ion pumps, the photosynthetic machinery in plants and bacteria are complex nano devices functioning inside the biological world. However, the recent leaps in areas such as microscopy have given scientists new tools to understand and take advantage of phenomena that occur naturally when matter is organized at the nanoscale. The ideas and concepts behind nanoscience and nanotechnology started with a talk entitled "There's Plenty of Room at the Bottom" by physicist Richard Feynman, often referred as father of the field of NP research, at an American Physical Society meeting at the California Institute of Technology (CalTech) on December 29, 1959, long before the term nanotechnology was used. The term 'nanotechnology' was defined by Norio Taniguchi and later the idea was vividly elucidated by K. Eric Drexler in the 1980s. However, the very fascinating qualities of size dependent optical, electronic, magnetic, and mechanical properties of NPs have engaged the attention of researchers from almost all branches of science including biology and medicine.

Materials exhibit novel physical, chemical, and biological properties at the nanoscale, differing largely from their properties of bulk counterparts and as well as from single atomic or molecular level. These phenomena are based on "quantum effects" and other simple physical effects such as increased surface area (more on these below). Compared to the bulk, NPs are inherently unstable due to their immense surface-to-volume ratio and the consequent high surface free energy. In general, they have to be stabilized by a matrix which, in the simplest case, may be a mono-molecular coating of organic molecules, such as dodecanethiol coated gold NPs (GNPs). The interaction between the functional group of these molecules and the NP surface passivates the surface energy of the latter, but in the process the entropy may not be maximized. In particular, the organic molecules may be constrained to a smaller set of conformational states than in the bulk. This lowered entropy produces a force which tries to oppose the lowering and acts as driving force for the interaction of the NPs with their surroundings. Therefore, a coated NP, as described above, when applied on cells and tissues, may interact strongly with such systems, leading to some very striking effects.

Now-a-days different types of engineered NPs are being used in industrial and biomedical application. The most popular of them is probably carbon nanotubes and buckyballs, which are members of fullerene structural class. Carbon nanotubes are the strongest material yet to be discovered. They can be made excellent electrical conductors or semi conductors just by controlling how they are rolled. They can be used in various nanoelectromechanical systems and even in biomedical applications like scaffold material for bone formation. Buckyballs can be used as drug-delivery agents. Graphene, one-atom-thick planar sheets of carbon atoms, hold the key to almost everything from super small computers to high-capacity batteries. Lipid based liquid crystals or liposomes are being extensively used in pharmaceutical industries for their ability to encapsulate molecules and then breaking down inside cells once their payload has been released. Gold NPs are being used for killing cancer cells (Chen *et al.*, 2007), intravenous contrast enhancers in medical imaging and also as delivery vehicles of drug (Han *et al.*, 2007). Silver NP, as an antibacterial component, has been used in the formulation of dental resin composites and in coatings of medical devices (Bosetti *et al.*, 2002).There are many oxide NPs available in the market, like titanium, zinc, aluminium, iron, silicon oxides to name a prominent few. Titanium dioxide NPs cause photocatalytic degradation of water contaminants; hence it can be used in water purification system (Abdel-Mottaleb 2009). Nano Zinc oxide has superior UV blocking capacity compared to bulk zinc oxide. So nano zinc oxide is often used in sunscreens. Moreover, it has excellent bactericidal properties (Yuvaraj *et al.*, 2010). NPs of iron oxide are biocompatible, non toxic and can be used as contrast agents in magnetic resonance imaging (Cheong *et al.*, 2011), in labeling cancerous tissues and localized thermotherapy. NPs made of semi conductor materials are called quantum dots. These nanostructures confine conduction band electrons, valance band holes or excitons in all three spatial directions. Quantum dots can be used as fluorescent biological labels. Several core shell NPs are also being used in bioimaging, cell labeling and drug delivery. Polymer NPs have potential applications in controlled drug release, targeted nucleic acid delivery (Davis 2009), and biological and antibody receptor mimics. Calcium phosphate NPs present a unique class of non-viral vectors, which can serve as efficient and alternative DNA carriers for targeted delivery of genes.

## Nanotechnology Timeline

This timeline features pre modern example of nanotechnology, as well as modern era discoveries and milestones in the field of nanotechnology (Table 22.1).

## Physico-chemical Characterization Tools for Nanoparticles

The nanotechnological revolution is very much associated with the advancement of microscopic and NP characterization techniques. Enhancement in resolution of electron microscope and invention of newer microscopic tools helped researchers to visualize and manipulate particles at nano domain. Atomic force microscope (AFM), scanning tunneling microscope (STM), magnetic resonance force microscope (MRFM) and high resolution electron microscopes, having resolution of 0.1 nm to 0.2 nm, enabled researchers to characterize materials at nanoscale, allowing individual atoms to be observed and analyzed.

**Table 22.1: Major Breakthroughs in the Field of Nanoscience and Technology (adapted from NNI)**

| *Timeline* | *Major Breakthrough in the Field of Nanoscience and Technology* |
|---|---|
| 4th century | The Lycurgus Cup (Rome) is an example of dichroic glass; made up of colloidal gold and silver painting |
| 9th-17th centuries | Glowing, glittering "luster" ceramic glazes used in the Islamic world was made from colloidal silver, copper and other materials |
| 13th-18th centuries | "Damascus" saber blades contained carbon nanotubes and cementite nanowires |
| 1857 | Michael Faraday discovered colloidal "ruby" gold |
| 1936 | Erwin Müller, invented the field emission microscope |
| 1947 | John Bardeen, discovered the semiconductor transistor |
| 1950 | Victor La Mer and Robert Dinegar developed the theory and a process for growing monodisperse colloidal materials |
| 1951 | Erwin Müller pioneered the field ion microscope |
| 1958 | Jack Kilby built the first integrated circuit |
| 1959 | Richard Feynman delivered the talk, "There's Plenty of Room at the Bottom" at American Physical Society meeting at Caltech |
| 1965 | Intel co-founder Gordon Moore described "Moore's Law" |
| 1981 | Gerd Binnig and Heinrich Rohrer at IBM's Zurich lab invented the scanning tunneling microscope |
| 1981 | Russia's Alexei Ekimov discovered nanocrystalline, semiconducting quantum dots in a glass matrix |
| 1985 | Harold Kroto, Sean O'Brien, Robert Curl, and Richard Smalley discovered the Buckminsterfullerene (C60)/Buckyball |
| 1985 | Bell Labs's Louis Brus discovered colloidal semiconductor nanocrystals |
| 1986 | Gerd Binnig, Calvin Quate, and Christoph Gerber invented the atomic force microscope |
| 1991 | Sumio Iijima of NEC is credited with discovering the carbon nanotube (CNT) |
| 1992 | C.T. Kresge and colleagues at Mobil Oil discovered the nanostructured catalytic materials MCM-41 and MCM-48 |
| 2000 | National Nanotechnology Initiative (NNI) was launched in the US |
| 2003 | Naomi Halas, Jennifer West, Rebekah Drezek, and Renata Pasqualin at Rice University developed gold nanoshells |
| 2007 | Angela Belcher and colleagues at MIT built a lithium-ion battery with a common type of non-pathogenic virus |
| 2009–2010 | Nadrian Seeman and colleagues at New York University created several DNA-like robotic nanoscale assembly devices |

## Electron Microscope

Electron microscopes (EM) are a kind of advanced version of microscopes which relies on beam of energetic electrons to probe and examine objects at very fine scale. It is a step ahead from conventional light microscopes which are limited in their resolution which relies on the physics of light. More than 10,000 X magnification can be achieved with this EMs to identify intracellular, intra-organelle structural differentiations. EMs have the advantage of having high magnification at high resolution, it has largely standardized technique for visualization of ultra structural features. Electron microscopes are of two types: Scanning Electron Microscope and Tansmission Electron Microscope.

## Scanning Electron Microscope

Scanning Electron Microscope (SEM) uses a focused beam of high-energy monochromatic electrons to generate a variety of signals (secondary electrons) at the surface of solid specimens that are detected by specialized detector and represented in the form of image. Basic components of this instrument consists of (i) electron gun (produces an intense beam of electron, (ii) lenses (produce clear and detail images), (iii) scanning coils (used to raster the electron beam across the sample surface), (iv) detectors (used the detect the secondary and backscattered electrons), (v) sample stage (to hold sample) and (vi) vacuum system (to avoid collision of electrons with gases and filament damage). Morphology of the NP surface can be detected by SEM.

## Transmission Electron Microscope

In Transmission Electron Microscope (TEM) a thin specimen is irradiated with an high velocity electron beam of uniform current density. Electrons are then emitted from the electron gun and illuminate the specimen through a two or three stage condenser lens system (electromagnetic bending of electron beam). Objective lens provides the formation of either image or diffraction pattern of the specimen. The electron intensity distribution behind the specimen is magnified with a three or four stage lens system and viewed on a fluorescent screen. Image formation generally is dependent on loss and scattering of electrons by individual parts of the specimen. Emergent electron beam is focused by objective lens. Final image forms on a fluorescent screen for viewing. TEM consists of following major components: Illuminating system, Image forming system, projective system and apertures. Apart from ultra small size sample imaging and structure enumeration, TEM is also being used in diagnosis and identification of tissue or cell features related to several diseases. Size of NP can be determined from TEM images.

## Atomic Force Microscope

Atomic Force Microscope (AFM) is kind of microscope which is generally being used to detect 3D profile of the surface of nanoscale material. It measures the forces between a sharp probe (< 10 nm) and sample surface very narrow distance (0.2-10 nm). Tip scans sample through up and down movement of tip recorded by position sensing photodiode. The microscope operates in two modes. In contact mode short range interactions (inter-atomic forces) are considered. This also can be performed in two ways: constant force (feedback system moves tip in z direction to keep force constant) and constant height (no feedback system needed, usually used when surface roughness is small and higher scan speed is possible). In tapping mode, tip vibrates (105 Hz) close to specimen surface (50-150 Å) with amplitude 10-100 nm. In AFM, cantilever is a kind of flexible string, on the end of which a $Si_3N_4$/Si based probe is supported on a flexible cantilever. The AFM tip gently touches the surface and records the small force between the probe and the surface. The amount of force between the probe and s is dependent on the spring constant (stiffness of the cantilever and the distance between the probe and the sample surface. Surface topology of NPs can be examined with the help of AFM.

## Dynamic Light Scattering

Dynamic light Scattering (DLS) deals with measuring the variation of light intensity with time and measure the variation in intensity with time (usually microseconds). The change of light scattering angle is responsible for variation in time and in turn is caused by diffusion of particles. Interaction of light with electric field around a small particle is the consequence of scattered light. Dynamic light scattering is a non-invasive technique for measuring the size of particles and molecular suspension. The technique of dynamic light scattering measures the speed of particles undergoing Brownian motion, the random movement of particles due to collisions caused by bombardment by the solvent molecules that surround them. The smaller the particle is, the more rapid the Brownian motion becomes. The reverse is true for the larger particles. The higher the temperature the more rapid the Brownian motion. DLS gives the measurement of particle size in terms of its hydrodynamic diameter, which is the diameter of a hard sphere that diffuses at the same speed as the particle or molecule being measured. The hydrodynamic diameter depends not only on the size of the particle "core", but also on any surface structure, as well as the type and concentration of any ions in the medium.

## X-Ray Diffraction

About 95 per cent of all solid materials can be described as crystalline. When x-rays interact with a crystalline substance (Phase), a diffraction pattern is developed. Based on these background information the concept X-ray diffraction has been developed as a physical characterization tool for materials crystal structure. The single crystal is composed of all unit cells aligned in a perfect extended pattern. The single crystal diffractometer and associated computer package is used mainly to elucidate the molecular structure of novel compounds, either natural products or manmade molecules. Powder diffraction is mainly used for "finger print identification" of various solid materials, *e.g.* asbestos, quartz. In powder or polycrystalline diffraction is important to have a sample with a smooth plane surface. Normally the sample is grinded down to particles of about 0.002 mm to 0.005 mm cross section. The ideal sample is homogeneous and the crystallites are randomly distributed. A beam of X-rays directed at a crystal interacts with the electrons of the atoms in the crystal, which oscillates under the influence of the incoming X-ray and become secondary sources of electromagnetic radiation in all directions. The waves emitted by the electrons have the same frequency as the incoming X-ray *i.e.* coherent in nature. The emission will undergo constructive or destructive interference with waves scattered from other atoms which gives the idea about the structure of the crystal very well.

### Ultraviolet-visible Spectrophotometer

UV-Visible absorption spectroscopy is involved with the wavelength range between 160 nm to 780 nm. It involves the measurement of transmittance and absorbance. Working principle is based on Beer's law. Instrument is consisting of a light source (Deuterium and hydrogen lamps, W filament lamp, Xe arc lamps), sample container, cuvettes. Application of UV-Vis spectrophotometer involves quantity (using a spectrum) and quality analysis for various purposes like: chemical analysis, food safety analysis, blood analysis, DNA/RNA conc. analysis, residual pesticide analysis. The surface Plasmon resonance (SPR) peak of noble metals in nanoform can be detected in it and this is a confirmatory detection test for detection of these NPs. Functional groups attached to NP surface can also be detected by this analytical tool.

## Different Nanostructures

For the unique property in the nanoscale various nanostructures of different size and shape with different surface functionality are being synthesized to evaluate their potential application in various arenas starting from electronics to biomedical fields. Among all the NPs probably carbon nanostructure is the most studied one. Nowadays silica NP has also obtained much attention as its surface can be capped with different groups, making it suitable for application in diversified fields. Nobel metals like gold, silver in their nano form and metal oxide like zinc oxide and titanium dioxide are also very important.

### Carbon Nanostructure

Carbon is an element with unique properties and is the lightest among all group IV elements of periodic table. A new era in carbon materials began when the family of buckminsterfullerenes (Buckyballs) was discovered in the mid-1980s. From then on different other carbon based nanomaterials like graphene, carbon quantum dots etc. were discovered.

### Fullerenes

Harold W. Kroto, Robert F. Curl and Richard E. Smalley won the Nobel prize in the year 1986 for their discovery of a new allotrope of carbon [Kroto *et al.*, 1985]. This new allotrope was named Buckminsterfullerene, after the architect Buckminster Fuller who designed geodesic domes in the 1960's. Buckminsterfullerene or Buckyball ($C_{60}$) is a molecule that consists of 60 carbon atoms, arranged as 12 pentagons and 20 hexagons as the basis of icosohedral symmetry closed cage structure (Figure 22.1a). The shape is the same as that of a soccer ball: the black pieces of leather can be considered as the pentagon, while the hexagons are white. Each carbon atom is bonded to three others and is $sp^2$ hybridized. Several fullerene derivatives are now commercialized like PCBM (Phenyl Cn Butyric Acid Methyl Ester), PCB-C12 and PCB-C18 etc. Fullerene can be used as organic photovoltaics. They are powerful antioxidants, reacting readily and at a high rate with free radicals. It can behave like a "radical sponge," as they can sponge-up and neutralize 20 or more free radicals per fullerene molecule. Scientists are trying to develop medicine for neurological damage

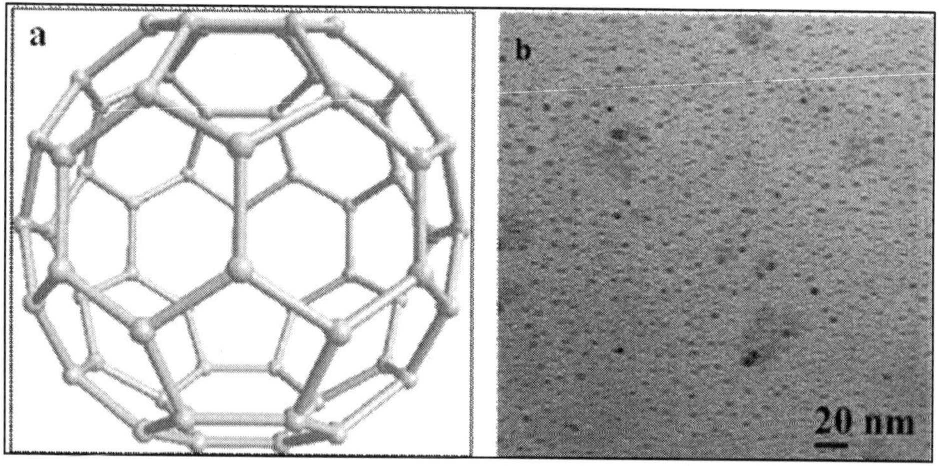


**Figure 22.1: (a) Buckminsterfullerene or Buckyball ($C_{60}$), (b) TEM image of Carbon Quantum Dot.**

of such diseases as Alzheimer's disease and Lou Gehrig's disease, which are a result of radical damage. Fullerene based drugs for atherosclerosis, photodynamic therapy, and anti-viral agents are also in development. Fullerenes are chemically reactive and can be added to polymer structures to create new copolymers with specific physical and mechanical properties and also, can be added to make composite.

## Carbon Quantum Dots

Carbon quantum dot (CQD, Figure 22.1b)) constitute an important class of nanomaterial for their unique fluorescence properties, resistance to photobleaching [Liu *et al.*, 2009], extremely small particle size for which they can enter the cells [Baker and Baker *et al.*, 2010], efficient light absorption throughout a wide spectrum and almost nontoxic characteristics. Recently CQDs have emerged great interest as fluorescent carbon family member for their potential application in bioimaging [Chandra *et al.*, 2011], drug delivery [Lai *et al.*, 2012] to biosensing [Liu *et al.*, 2012], catalysis [Yu and Kwak *et al.*, 2012], devising solar cells [Dong *et al.*, 2012], so on and so forth. Unique blend of nontoxicity and fluorescent property have made CQDs an efficient imaging probe. Numerous efforts have been made for the preparation of CQDs, such as microwave irradiation technique; combustion of carbon soot, activated carbons or carbon xerogels by nitric acid; laser ablation of graphite; hydrothermal treatment of protein and electrochemical route. Sometimes surface passivations are required to enhance the quantum yield (QY) of as prepared CQDs which have a poor QY. In most of the techniques CQDs are produced with hydrophilic property; in thrust of unique functional properties prompted the researchers to design hydrophobic or organophilic CQDs. Recently nature inspired hydrophobic surface tailoring is being pursued owing to their industrial applications, self cleaning property and ability to suppress chemical reaction on the surface.

## Graphene

Graphene is also a very popular carbon based nanostructure. This two-dimensional carbon nanostructure has attracted much attention from researchers due to its interesting mechanical, electrochemical and electronic properties. Graphene is a single atomic layer of $sp^2$-bonded carbon atoms tightly packed in a two dimensional (2D) honeycomb lattice. As a novel nanomaterial, graphene possesses unique electronic, optical, thermal, and mechanical properties [Balandin *et al.*, 2008]. Graphene and its derivatives have shown outstanding potentials in many fields such as nanoelectronics [Xuan *et al.*, 2008], engineering nanocomposite materials [Fan *et al.*, 2010], energy storage [Liu *et al.*, 2010], organic light emission diodes (OLED), sensors, catalysis, biosensor, biodevices, drug and gene delivery and cancer therapy etc.) [Lu *et al.*, 2009]. There are several methods to synthesize grapheme. Pristine graphene can be synthesized from the mechanical exfoliation of graphite using adhesive tapes. Single and few-layer graphene sheets on metal surfaces such as Ni and Cu can be grown by Chemical vapor deposition (CVD). High temperature annealing can also be used to obtain graphene layers by carbon segregation on carbon-containing substrates like SiC. Oxidation and exfoliation of graphite oxide followed by the chemical reduction has been used to prepare reduced graphene oxide or chemically functionalized graphene (CFG).

## Silica Nanoparticle

Silica nanoparticle (SNP) has gained much attention in scientific research because of its easy preparation and its wide range of industrial as well as biological application. The variety of chemical and physical modification possible with silica increases its versatility, and its biocompatibility makes it suitable for extensive biological application. For instance SNPs are being used as biosensors for assay of glucose, biomarkers for leukaemia cell identification using optical microscopy, drug delivery (Wang *et al.*, 2012;), DNA delivery (Gemeinhart *et al.*, 2005), enzyme immobilization cancer imaging, targeting and therapy. SNP has huge potential in agriculture and veterinary sector as well. It can be used as an entomotoxic agent (Debnath *et al.*, 2010; Debnath *et al.*, 2011), can control grasserie disease of silk worm (a viral disease caused by *Bombyx mori* nuclear polyhedrosis virus) and can also be applied as an anti-malarial drug in poultry farm (Rahman *et al.*, 2009).

Actually silica is the second most abundant compound in the earth's crust. Silicas can be amorphous or crystalline, porous or non-porous (dense), anhydrous or hydroxylated, regardless of their natural or synthetic nature. In a silica compound, the silicon atom is in tetrahedral coordination with 4 oxygen atoms. In crystalline silica the oxygen-sharing silicate tetrahedra are connected in an ordered fashion, whereas in amorphous form the tetrahedra are randomly connected. In nature, amorphous silica can have different origins. It can be condensed from vapors emitted in volcanic eruptions. Natural silica can also be deposited from supersaturated natural water or polymerized in living organisms (biogenic silica). These amorphous biogenic silicas can be found as isolated particles, skeletal structures or surface elements in different living organisms. Diatomaceous Earth (DE), which is siliceous deposits from skeletal remains of diatoms, is one of the major resources of amorphous natural silica.

There are two major synthetic routes for production of SNPs. It can either be synthesized by sol-gel method from an aqueous solution, with dissociated monomeric silicic acid, $Si(OH)_4$, or from a vapor of a silicon compound such as silicon tetrachloride. Stober *et al.* (1968) demonstrated a pioneering method for the synthesis of spherical and monodisperse SNPs from aqueous alcohol solution of silicon alkoxide in presence of ammonia as catalyst, and different sizes of SNPs were synthesized ranging from 50 nm to 1 μm with a narrow size distribution (Figure 22.2).

Lee and Choi (2002) synthesized SNPs of controlled morphology by using a $CO_2$ laser to re-heat flame synthesized SNP agglomerates, thereby sintering them in situ. Mdler *et al.* (2002) presented a detailed study on synthesis of SNPs by flame spray pyrolysis from hexamethyldisiloxane. Among other non conventional routes of SNP production, Guo *et al.* (1999) demonstrated synthesis of SNPs by using *Escherichia coli* as template. Amutha *et al.* (2010) extracted SNPs from rice husk, an agricultural waste. General scheme for synthesis of various silica materials is depicted in Figure 22.3.

## Gold Nanoparticle

Gold nanoparticles (GNPs) are of immense interest because of their broad chemical versatility and the resulting colloidal matrices. Gold colloid has long been used as therapeutic for unstable mental and emotional states, glandular problems, rheumatism, inflammatory skin disorders, due to its immense balancing and harmonizing effect on all levels of body (Higby *et al.*, 1982). Synthesis of GNPs can be divided into three major approaches.

### Chemical Method

Chemical synthesis is based on simple, economically cheap wet reduction methods. In the year of 1857, Michael Faraday first synthesized colloidal gold from reduction of gold chloride. Turkevich *et al.* (1951) first reported the synthesis of spherical monodisperse GNPs of 10 - 20 nm in diameter (Figure 22.4a). Later this method was modified by Frens in 1970s to obtain GNPs of pre chosen size (16 - 147 nm) by varying the molar ratio of the reactants and more controlled formation. In early 1990s, Brust developed a novel protocol for the formation of 5 - 6 nm of stable GNP in a non polar solvent. Spontaneous reduction of gold salts by inorganic matrix such as fumed silica is a novel method of GNP formation (Mukherjee *et al.*, 2002). These matrices serve the purpose of reducing substance and also provide a solid support to the resulting GNPs. The silica gold conjugates exhibit interesting catalytic properties.

### Physical Method

There are also several protocols for synthesizing GNPs using physical methods. This includes UV- irradiation, sonochemical methods, microwave irradiation, laser ablation, thermolysis, photochemical methods etc.

### Biological Method

Recently another approach, green synthesis, using biological organisms or routes for GNP synthesis is being widely researched. Here during synthesis three main factors are mainly considered: solvent choice, the use of an environmentally benign

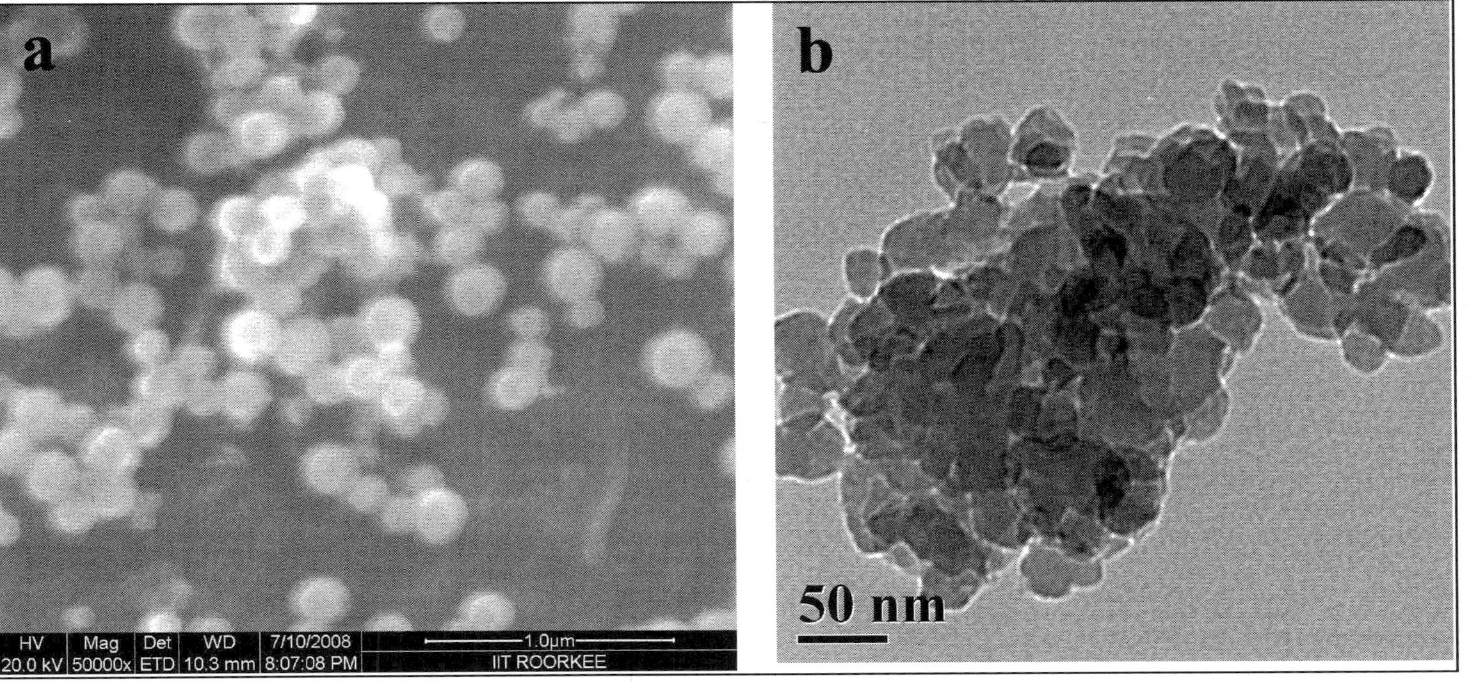


**Figure 22.2: (a) FESEM (b) TEM image of silica nanopaticle. All the particles are spherical in shape.**

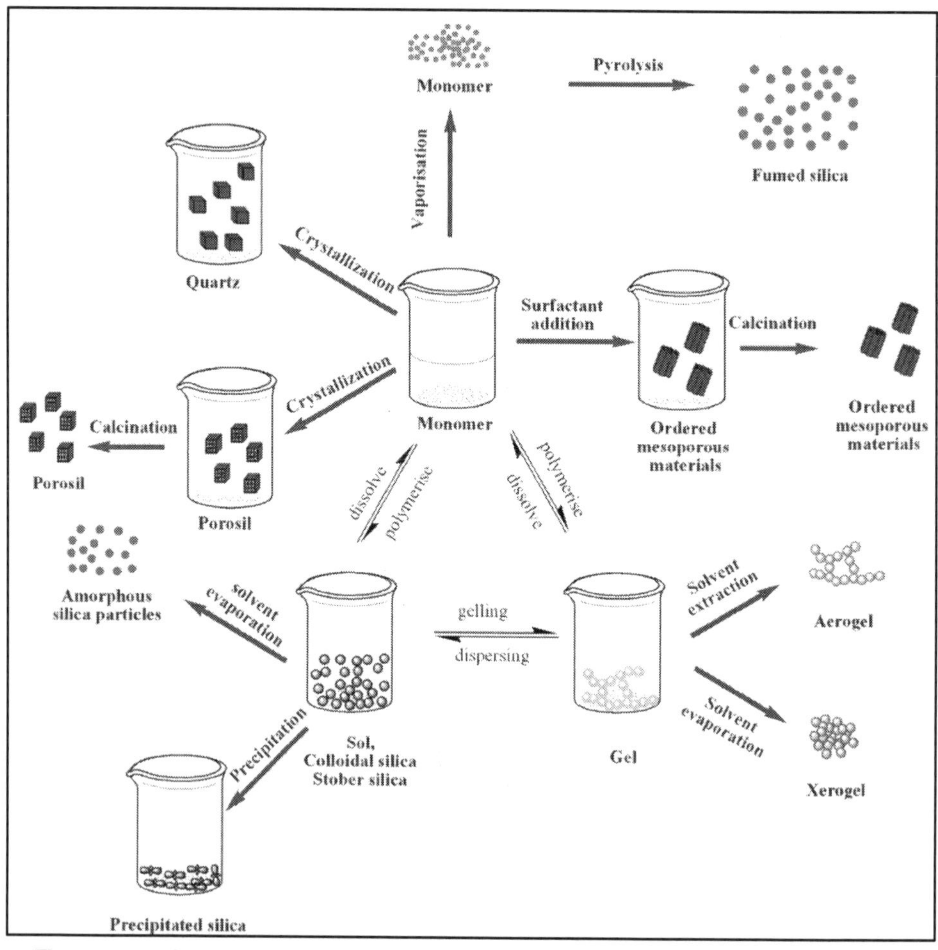


**Figure 22.3. General Scheme for Synthesis of Silica (Adapted and redrawn from Brinker CF, Schrerer GW: Sol-Gel Science. The Physics and Chemistry of Sol-Gel Processing. 2nd edition. London: Academic Press; 1990).**

reducing agent, and the use of a non-toxic material for NP stabilization. Phytochemicals such as tea extract is recently being used in synthesis of GNPs and AgNPs. Aloe vera and lemongrass (*Cymbopogon flexuosus*) extract are also popular constituents in synthesis of triangular-shaped GNPs (Vaseeharan *et al.*, 2010). Mukherjee *et al.* (2002) introduced another novel biological method for the synthesis of extra cellular and intracellular silver and GNPs of 107 - 109 nm size range using live microorganisms (*Verticillium, Fusarium oxysporum*).

Recent studies with GNPs revealed its potential possibilities in healing cancer. On irradiation with focused laser pulses of suitable wavelength, GNPs can kill cancer cells. Moreover, GNP based drug carriers and intravenous contrast enhancers in medical imaging are also being extensively researched. In spite of considerable progress in the field of synthesis, characterization, identification and biomedical

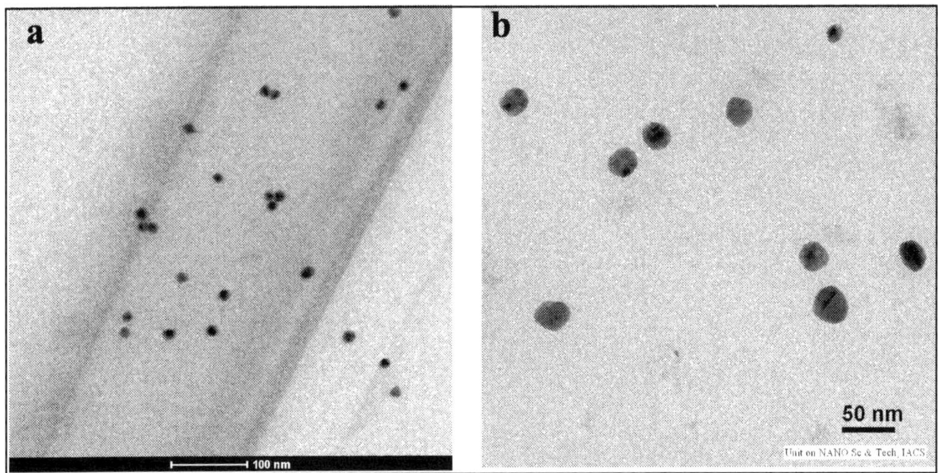


**Figure 22.4: TEM Image of (a) Gold Nanopaticle and (b) Silver Nanoparticle.**

application of GNPs, stability of these particles at high ionic strength and physiological conditions remains uncertain.

## Silver Nanoparticle

Silver nanoparticles (Ag-NPs) have attracted immense interest in researchers due to their unique physical, chemical and biological properties [Sharma *et al.*, 2009]. Ag-NPs have distinctive physico-chemical properties, including a high electrical and thermal conductivity, chemical stability, catalytic activity, surface-enhanced Raman scattering etc. Besides, Ag-NPs exhibit broad spectrum antimicrobial activity [Ahmed *et al.*, 2010] which has made them very popular in a diverse range of consumer products, including plastics, soaps, pastes, food and textiles, increasing their market value. Nanosilver technology is now being applied in a variety of manufacturing processes and end products. Colloidal nanosilver can be used in coating and spray; sometimes it is also present in shampoo. It is being utilized either in the textile industry by incorporating it into the fiber or utilized in filtration membranes of water purification systems. Mostly researchers are trying to store silver ions and incorporate a time-release mechanism so that a long-term protective barrier is created against bacterial and fungal pathogens. There are many published protocols for Ag-NP synthesis. Bovine serum albumin (BSA) capped water soluble Ag-NP can be synthesized by reducing aqueous silver nitrate ($AgNO_3$) solution with sodium borohydride ($NaBH_4$) and capping them with BSA under vigorous stirring condition. The molar ratio of Ag : BSA should be 28:1 and Ag : $BH_4^-$ is 1:1. The product should be filtered with cold ethanol. Similarly Ag-NP can be surface functionalized with Poly-*n*-vinyl-pyrrolidone (Figure 22.4b).

## Zinc Oxide Nanoparticle

Among different metal oxide NPs, zinc oxide NPs (ZNPs) have drawn much attention due to their potential applications in diversified fields, like cosmetics, antimicrobial agents, gas sensor, chemical sensor, bio-sensor, storage, optical and electrical devices, and drug-delivery [Baxter and Aydil, 2005]. ZNP is an attractive material for short-wavelength optoelectronic applications owing to its wide band gap 3.37 eV, large bond strength, and large exciton binding energy (60 meV) at room temperature. ZNP can be used in solid state blue to ultraviolet (UV) optoelectronics, including laser developments. In addition, due to its non-centrosymmetric crystallographic phase, ZnO shows the piezoelectric property, which is highly useful for the fabrication of devices, such as electromagnetic coupled sensors and actuators. Several ZnO nanostructures, including NPs, nanowires, nanorods, nanotubes, nanobelts, and other complex morphologies have been synthesized for various applications (Figure 22.5a). By different methods ZNPs can be synthesized. Sol-gel method is one of the popular methods for ZNP synthesis. Patra *et al.*, developed a simple microwave assisted route for the synthesis of ZNP. 20 ml 20 per cent aqueous TRIS solution was drop wise added into 25 ml of 0.05(M) zinc acetate dihydrate solution. The well stirred mixture was subjected to microwave irradiation at 300 watt for 3 minutes. Then this product was centrifuged at 10000 rpm for 10 minutes and washed several times with milli-Q water to remove excess of TRIS buffer followed by calcination overnight at 80°C.

## Titanium Dioxide Nanoparticles

Oxide NPs are matter of interest to a large number of researchers for their unique electrical, optical and magnetic properties. Amongst them titanium dioxide ($TiO_2$) NPs (Figure 22.5b) have achieved much of attention for their wide spread use in optical devices, sensors, and photocatalysis. $TiO_2$ belonging to the family of transition metal oxides, is present in 4 polymorphs in the environment as: anatase (tetragonal), brookite (orthorhombic), rutile (tetragonal), and $TiO_2$ (B) (monoclinic) In general sol-gel method and flame pyrolysis method are popular for nano $TiO_2$ synthesis. Various investigations have established that nano $TiO_2$ is very effective as a photocatalyst as the diameter of the crystallites of a semiconductor particle falls below a critical radius of about 10 nm; each charge carrier appears to behave as quantum mechanically. Due to this quantum confinement, the band gap increases and the band edges shift to yield larger redox potentials.

Wide application of of nano $TiO_2$ particles includes:

- ☆ Low cost, high efficiency, dye-sensitized solar cells exceptional stability
- ☆ Sensors of hydrogen, ethanol, humidity, oxygen, combustion fuel sensors
- ☆ As thin film capacitors, gate electrodes for MOS devices
- ☆ As interference filters, optical waveguides
- ☆ Photocatalysis for production of hydrogen and methane from ethanol and water water, air and wastewater treatment
- ☆ Coating and painting material
- ☆ Model system for the nanoporous materials research (electron transport, optical properties)
- ☆ In sunscreen, $TiO_2$ NPs are used to scatter ultraviolet light. The advantage of these over larger particles is that sunscreen containing NPs is transparent when applied. Even Surface sensitization of a wide band gap semiconductor

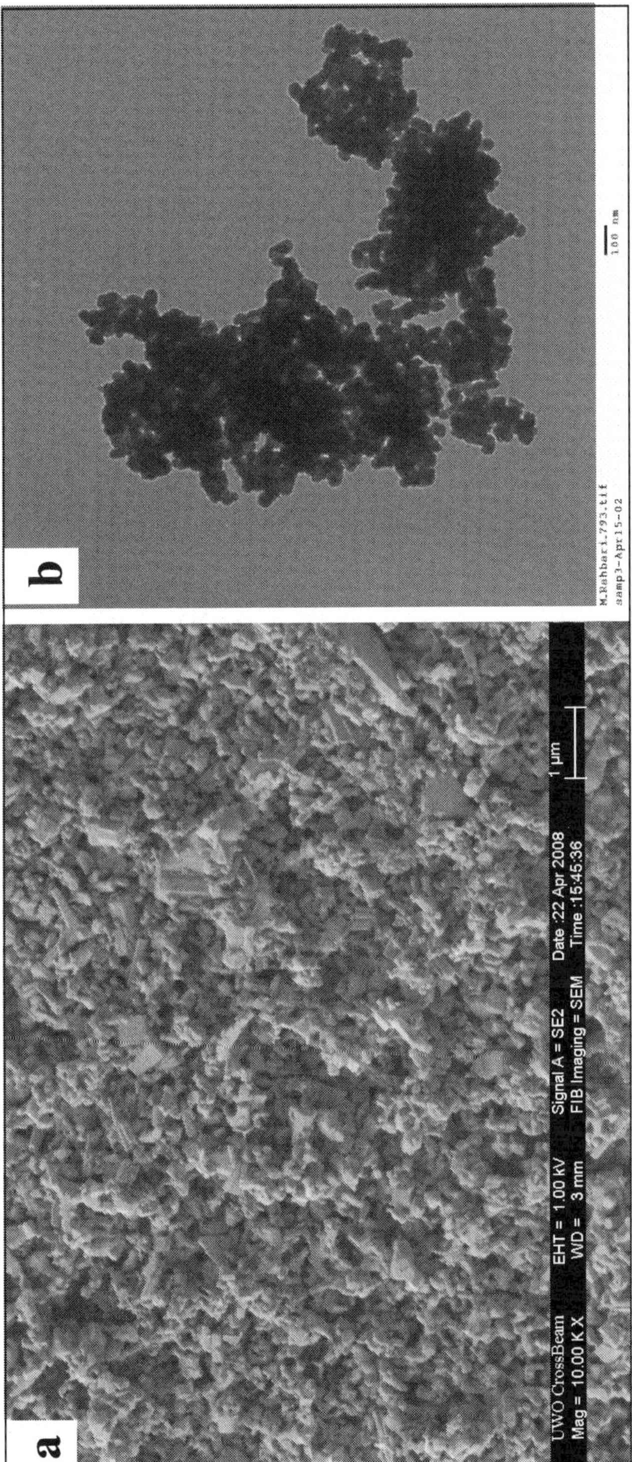


**Figure 22.5: (a) FESEM image of zinc oxide nanopaticle and (b) TEM image of titanium dioxide nanoparticle.**

☆ Photocatalyst $TiO_2$ can also be very good dye sensitizers. Phototoxic properties of Tio2 NPs have also been applied as for waste water treatment, as anti-cancer agent and ecotoxicological studies due to oxidative stress generation capacity.

## Nanotubes and Nanowires

Nanotubes are nanowires are two dimensional nanoscale materials consisting of well defined length and breadth. Nanowires are ultra fine wires or linear arrays of nano dots, formed by self-assembly. Nanowires are generally made from a wide range of materials such as semiconductor nanowires are made from silicon, gallium nitride and indium phosphide. Mostly nanowires are synthesized following two popular growth techniques such as: Chemical Vapor deposition (CVD) and electroplating. General characteristic features of Nanowires includes: Diameter in the range of 1-10 nm, single crystal formation (common crystallographic orientation along the nanowire axis), minimal defects within wire and minimal irregularities within nanowire arrays.

Nanotubes are long, slender tubular nanoscale structures where walls of the tubes are hexagonal and are often capped at each end. When buckyballs stretch themselves to form a long and narrow tubes of few nanometer diameter then it becomes a basic form of carbon nanotube (CNT). CNTs are broadly classified as single-walled nanotubes (SWNTs) and multi-walled nanotubes (MWNTs). Apart from applications of nanotubes in the fields nano-electronics, field effect emitters, and hydrogen storage, in recent years, efforts have also been devoted to explore the potential biological applications of NTs as motivated by their interesting size, shape, and structure, as well as attractive, unique physical properties. SWNTs are being utilized as optical tags or contrast agents for various biological imaging techniques. Surface functionalized NTs are able to enter cells without toxicity, shuttling various biological molecular cargoes into cells. Studies have also shown to use NTs for *in vivo* cancer treatment in mouse model system.

Semiconducting nanowires have the potential to act as highly sensitive detector elements. The optical and electrical properties of nanowires are expected to change in response to chemical and/or biological molecules, leading to new types of nanoscale biosensors. As for example nanofiber electrodes are being conjugated with biomolecules such as horesradish peroxidase to create new types of nanoscale biosensor systems. Magnetic nanowire arrays of cobalt, gold, copper and cobalt-copper are being used as storage device applications, Electrochemical deposition is the fabrication technique, as well as sensing devices. Nano sized sensing wires are laid down across a micro fluidic channel. As particles flow through the micro fluidic channel, the nanowire sensors pick up the molecular identifications of these particles and can immediately relay this information through a connection of electrodes to the outside world. They can detect the presence of altered genes associated with cancer and may help researchers pinpoint the exact location of those changes.

## Nano-scaffold

The term nano-scaffold can be defined as a three-dimensional structure composed of polymer fibers in nanometer scale. These new technology is now being applied in 'tissue engineering, to regrow tissue, bones, limbs and organs. synthetic and protein-based biocompatible polymer scaffolds, such as collagen, gelatin, elastin, proteoglycan, hyaluronan, laminin, silk fibroin, soybean, fibrinogen (Fbg), and fibrin (Fbn) have been widely used in tissue engineering purposes. During nano-scaffold fabrication of biomedical usage three considerations are taken into account:

- ✰ Scaffold should mimic the non-uniform tissue at tendon- bone interface
- ✰ Linear gradient of mineral deposition should be maintained onto nanofiber scaffold
- ✰ Varying orientation of the nanofibers.

Nanoscaffolds promote cell attachment and migration, deliver biochemicals to support cell growth and functionality and being biodegradable in nature ensures non-toxicity. As for example, Fibrinogen (Fbg)- and fibrin (Fbn)-based micro- and nanostructures can provide suitable natural matrix environments with advantages of consecutive chances of encapsulation and specific delivery, with multiple growth factors, proteins, and stem cells, in structures designed to aid in specific phases of the tissue regeneration process (Rajangam *et al.*, 2013).

## Nanogel

The term 'nanogel' can be defined as the nanosized particles formed by physically or chemically cross linked polymer networks that swells in a good solvent. Frequently they are composed of hydrogel – a crosslinked hydrophilic polymer network, synthetic or bio polymers which are cross linked with each other. These are now widely applied in biomedical usage specifically in drug delivery purposes (Sultana *et al.*, 2013). They are superior over other conventional delivery vehicles in respect to their:

- ✰ Flexibility in design with desired properties
- ✰ Controlled, sustained and target specific drug release
- ✰ High drug loading efficiency
- ✰ Able to solubilize hydrophobic drugs
- ✰ Simple fabrication method without need employing energy or harsh conditions such as sonication or homogenization
- ✰ High colloidal stability
- ✰ Due to tiny volume can reach smallest capillary vessels and can penetrate the tissues either through the paracellular or the transcellular pathway.

The route of administration of these smart tools follow general drug routes such as oral, pulmonary, nasal, parenteral, intra-ocular and topical. Nanogels are more commonly classified into two major types: Non-responsive (simply swell as a result of absorbing water) and stimuli-responsive microgels swell or deswell upon exposure to environmental changes such as temperature, pH, magnetic field, and ionic strength.

As for example, Doxorubicin has been successfully reported to be loaded in amphiphilic cross-linked on pluronic F127 and POEOMA based nanogels. Similarly proteins in cholesterol-modified pullulan nanogels, siRNA in HA nanogels, incorporation of hydrophobic molecules into nonpolar domains formed by hydrophobic chains present in selected nanogels have also been reported (Kabonov *et al.*, 2008). For example, prostaglandin E2 was solubilized in nanogels of cholesterol-modified pullulan nanogels.

## Nanobiosensor

A biosensor is an analytical device, used for the detection of an analyte that combines a biological component with a physicochemical detector. The biosensor has three basic components: biological element, transducer and the reader. Biologically sensitive element can be tissue, microorganisms, organelles, cell receptors, enzymes, antibodies, nucleic acids, etc. or a biologically derived material or biomimetic component that interact the analyte under study. The biologically sensitive elements can also be created by biological engineering. The transducer or the detector element transforms the signal resulting from the interaction of the analyte with the biological element into another signal that can be more easily measured and quantified. Biosensor reader is associated with the adjacent electronics or signal processors that are primarily responsible for the display of the results in a user-friendly way. Nanobiosensor can be potentially used in the following arena:

- ✰ Clinical diagnostics
- ✰ Food and agricultural processes
- ✰ Environmental (air, soil, and water) monitoring
- ✰ Detection of warfare agents.

The principle of detection by a nanobiosensor may be optical, electrochemical and piezoelectric. Optical biosensors generally measure changes in light or photon output. Electrochemical biosensors: which convert the biological binding events into useful electrical signals, whereas piezoelectric biosensors measure the mass change arising from the biological recognition process. In the nanoscale range, the sensitivity of signal detection is drastically improved due to nanoscale-specific physical properties [Cullum and Vo-Dinh, 2000]. For example, relatively weak optical signals such as surface plasmon resonance (SPR) and surface enhanced Raman scattering can be used as highly sensitive and specific signal in case of a biosensor made up of nano architectures [Erickson *et al.*, 2008]. Moreover, the miniaturized sensors may provide additional benefits such as parallel operation, portability, and low power consumption.

Rapid and convenient detection is essential for portable clinical monitoring systems, such as handheld point-of-care (POC) testing devices. In addition, miniaturization and conformal contact on curvilinear surfaces are necessary to allow the implantable microsized electronic system to function in real-time healthcare applications. The advent of nanofabrication technology has allowed the miniaturization of conventional benchtop instruments, resulting in devices such as the nano/microelectromechanical systems, microfluidic embedded sensor systems, and integrated systems such as the micro Total Analysis System, which can also called as lab-on-a-chip [Voldman *et al.*, 1999]. Microfluidics is an essential technology utilized in lab-on-a-chip used for medical diagnostics. In microfluidics some of the main challenges in traditional assays can be avoided by integrating sensing and sample pretreatment modules, reducing valuable reagent consumption, increasing the overall efficiency of the assays, and avoiding cross contamination. Hence, this type of portable microfluidic device can be used for fast, selective and portable POC diagnostics.

## Optical Nanobiosensor

Optical biosensors generally measure changes in light or photon output. The basic objective of optical biosensor is to produce an electronic signal which is proportional in magnitude or frequency corresponding to the concentration of a specific analyte or group of analytes, to which the biosensing element binds.

Surface plasmon resonance (SPR) based on the measurement of small changes in refractive index that occur in response to analyte binding at or near the surface of a noble metal (Au, Ag, Cu) thin film. SPR has been widely used to monitor a broad range of analyte-surface binding interaction including the adsorption of small molecules, ligand-receptor binding, protein adsorption on self-assembled monolayers, antibody-antigen binding, DNA and RNA hybridization, and protein-DNA interactions [Heaton *et al.*, 2001]. There are at least four different NP-based sensing mechanisms that enable the transduction of macromolecular or chemical binding events into optical signals based on changes in the localized SPR signal. These may be: (1) resonant Rayleigh scattering from NP labels like fluorescent dye labels, (2) aggregation of NP, (3) charge-transfer interactions at NP surfaces, and (4) changes in local refractive index. It was found that there is red-shift of the SPR wavelength of colloidal gold particles coated by a monoclonal antibody (Englebienne 1998). Englebienne demonstrated that this shift originated neither from agglutination nor aggregation, but was due to a change in the refractive index of the individual particles subsequent to further coating by the specific ligand when binding to the antibody. From then on, numerous researchers have been trying to correlate peak intensity and position of the surface plasmon absorbance of GNPs with the local refractive index of the surrounding medium. A fiber based biosensor for organophosphosphate pesticide detection was developed by Lin at al. by utilizing the SPR effect of GNPs. Self-assembling technique was used for immobilization of an acetylcholinesterase (AChE) layer on GNP layer by covalent coupling. When encountered with pesticides, the activity of AChE to hydrolyze acetylcholine chloride is reduced leading to the change of the light attenuation due to a local increase of the refractive index. The concentration of pesticide can be determined based on the correlation between inhibition rate and light attenuation. The sensitivity of the fiber is much enhanced in presence of GNP. When oligonucleotide probes are conjugated to colloidal gold particles for recognition of surface-confined target DNA via sequence-specific hybridization, the signal is 1000 times amplified in comparison with the probes where GNPs were not tagged [Nath and Chilkoti, 2004]. Even single nucleotide polymorphisms (SNPs) in genomic DNA can be analyzed by using SPR imaging. GNP-based SPR biosensors can also be fabricated into an array, which enables rapid, high-throughput screening of biomolecular interactions. This type of SPR resonance chip can be used for small molecules like dopamine (Matsui *et al.*).

## Electrochemical Nanobiosensor

Electrochemical nanobiosensor converts the biological binding events into useful electrical signals. NPs can be used in a variety of electrochemical biosensors.

*Nanomaterial-based enzyme electrodes*: Enzyme electrodes are often used to monitor a wide variety of clinically and environmentally important analytes. The main challenge in enzyme electrodes is the establishment of adequate electrical communication between active site of the enzyme and the electrode surface [Guindilis *et al.*, 1997].11 As the redox centre of the oxidoreductase enzyme is insulated by protein, it cannot be oxidized or reduced at an electrode at any potential. If the electrons can be transferred directly between the enzyme and the electrode surface, a reagentless biosensing device with superior sensitivity can be designed. Willner *et al.*, [Patolsky *et al.*, 2004] has shown that glucose oxidase (GOx) can be covalently aligned on single walled carbon nanotubes (SWCNT). Here SWCNT acts as molecular wire helping in electrical communication between underlying electrode and GOx. GNPs also can act as 'electrical nanoplug' in glucose sensor. Redox activity of hydrogen peroxide and NADH is greatly enhanced in CNT-modified electrodes.

*Nanomaterial based bioaffinity electrochemical sensors*: The development of electrical DNA hybridization biosensor for DNA sensing has attracted huge attention throughout the globe. Use of NPs in this type of sensors can amplify the sensitivity of oligonucleotide interaction. Colloidal gold can be used as tag for electrochemical detection of DNA. NP-induced changes in the conductivity across microelectrode gap can also be utilized for detection of DNA hybridization. When silver NP is used in DNA sensor, a conductive metal layer is formed leading to enhanced conductivity signal. This sensor can detect DNA even if it is present in concentration as low as 500 fmol. Bioelectronic measurement of proteins based on sandwich immunoassays and GNP tracer have also been reported. Different inorganic nanocrystals can be used in multi-target electronic detection of DNA or proteins. ZnS, CdS and PbS nanocrystals can be used to differentiate the signals of three proteins in a multi-antigen sandwich immunoassay. CNTs also can help in extremely sensitive detection of DNA hybridization [Wang *et al.*, 2004].

## Piezoelectric Nanobiosensor

Piezoelectric biosensors measure the mass change arising from the biological recognition process. Piezoelectric effect is basically the relation between a mechanical stress and an electrical change in solid. The quartz crystal microbalance (QCM) is one of the most important techniques based on piezoelectric effect. The sensitivity of QCM should be much enhanced in biosensing, as the target detection molecules are always present in trace quantity. GNPs can amplify the mass change on the crystals during the analysis. Numerous research groups are trying to improve the analytical sensitivity by coupling GNPs with the QCM sensing process. A sandwich-type complex consisting of an oligodeoxynucleotide (ODN) immobilized on a QCM electrode, a target DNA and a GNP modified DNA, in which the latter two oligonucleotides were both complementary to the ODN was fabricated. By studying the interaction of the ODN and target DNA complex with four different oligonucleotides (GNP modified DNA complementary to ODN, DNA complementary to ODN (no GNP), GNP modified DNA complementary to ODN with a five-base mutation, and GNP modified DNA not complementary to ODN), it was found that the GNP which was binding to the ODN served to enhance the signals (Zhou *et al.*, 2000).

It was considered that in comparison to the masses of the binding pair members themselves, the relatively large mass of each GNP contributed to amplify the frequency shift and substantially extend the limits of sensitivity of the QCM detection system.

## Nanotechnology in Biomedical Science

Enormous growth and development in the field of nanoscience and technology has opened up increasing optimism toward application of this high throughput technology to revolutionize the field of medical sciences. The term 'nanomedicine' refers to the application of nanotechnology in diagnosis and treatment of diseases. Broadly this new science deals with the interactions of nanomaterials or analytical nanodevices with human cells, tissue, body fluids. Nanomaterial based drugs are able to diffuse through the body wall. Altogether nanomedicine is opening an extremely large field of their application ranging from *in vivo* and *in vitro* diagnostics to therapy including targeted delivery and regenerative medicine. Engineered NPs can itself play the role of 'carrier' or drug and also can be fabricated in nanoscale which then can function at its own delivery mechanism. Engineered nanomedicines of biological origin mostly include proteins, lipids, phospholipids, chitosan, dextran and lactic acid etc as building stones. Whereas for no-biological ones, chemicals like polymers, metal and semiconductor NPs are broadly being used.

## Nanoparticles in Medicinal Application: Expected Features of Nanoparticles

"Clever" drug targeting system refers to the coordinated behavior of three components: the targeting moiety, the carrier and the therapeutic drug. First of all the nanomaterials should be small enough to reach cells. Secondly small size ensures high surface area to volume ratio from particle's bulk counterpart. High surface area indicates greater chance of interaction with biomolecules ensuring high reactivity. A special optical property that operates in small size range also provides added advantages for detecting diseases. Apart from these nanomedicine should exhibit:

- Higher intracellular uptake
- Tunable size, shape and surface properties
- Penetrate the sub-mucosal layers while the microcarriers are predominantly localized on the epithelial lining
- Target specific delivery
- Must be biocompatible
- Less toxic but more effective as therapeutic.

- ☆ Can be administered into systemic circulation without the problems of particle aggregation or blockage of fine blood capillaries.

The development of targeted delivery is firmly built on extensive experience in pharmaco-chemistry, pharmacology, toxicology, and nowadays is being pursued as a multi-and interdisciplinary effort. At present scenario more than 95 per cent drugs are having poor pharmacokinetic and biopharmaceutical properties. Nanomedicine based drug delivery system (Liu *et al.*, 2007) is expected to be target specific. Therapeutically active molecules allow to lower required doses of drugs and to increase their therapeutic indices and safety profiles. It is possible to fabricate NPs or nanocapsules with different properties as relates to drug encapsulation and release (Koo *et al.*, 2005). Interaction of nanostructures with plasma proteins and relation between protein adsorption and removal of nanostructures from the circulation by the reticulo-endothelial system is another important aspect regarding nanomedicine for their safe and non-toxic delivery. Adsorption of nanostructures within cells (in relation to the surface chemical characteristics, size and shape of the nanostructures) is of utmost importance for successful application. Uptake and recycling, trans-endocytosis and endosomal escape of nanostructures and safety evaluation in terms of *in vitro* and *in vivo* cytotoxicity, haemocompatibility, immunogenicity and genotoxicity testing, bio distribution and degradation: all these aspects needed to be taken care of during engeeneering of nanobiologicals for therapeutic application.

## Nanomaterials Used in Imaging, Biosensing, Treating and Preventing Diseases

The initial aim of "Nanobiopharmaceutics" is to provide breakthrough advances in novel biopharmaceutics delivery systems for the treatment of diabetes, cancer, AIDS, Alzheimer's disease, and other neurodegenerative diseases. Nanocarrier-based protein/peptide delivery systems for respiratory and oral delivery and blood brain barrier (BBB) crossing applications are also being developed within this project allowing a targeted and controlled release of the drugs. Huge amount of nanoscale systems in this particular aim has been investigated which include liposomes, dendrimers, quantum dots, nanotubes, polymeric biodegradable NPs and nanocapsules [Singh *et al.*, 2009].

## Liposomes

Liposomes are nano/micro-particular or colloidal carriers, usually with 80–300 nm size range, composed of phospholipids and steroids (*e.g.*, cholesterol), bilayers, or other surfactants by the process of sonication. Liposomes are very popular in nanomedicine due to its unique features that includes:

- ☆ Pharmacokinetic properties of the drug can be modulated by controlling the liposome size –smaller liposomes have reduced immunogenicity
- ☆ Lipid composition – give the membrane varied rigidity
- ☆ Suitable for delivery of hydrophobic, hydrophilic and amphipathic drugs and agents
- ☆ Chemically and physically well characterized entities

- Biocompatible
- Suitable for controlled release
- Suitable to give localized action in particular tissues
- Suitable to administer via various routes
- Stealth component –reduces the immunogenicity
- The drug enclosed so better chance of invading body immunologic responses
- Liposomes generally solubilizes drug to increase pharmacokinetic properties

Encapsulation of chemotherapeutics within stealth liposomes is a matter of research interest for long time in cancer treatment. Optimal antitumor activity occurs when cancer cells are exposed to moderate concentrations of a drug encapsulated in liposomes over an extended period of time. Doxil is the drug doxorubicin encapsulated in an antibody linked PEGylated liposome, composed of multiple monoclonal antibodies to target cancer cells. Here PEG (polyethylene glycol) makes the liposome less vulnerable to immune system. There are a lot of other drug examples apart from anticancerous ones based on liposomal formulations, *e.g.* neurotransmitters (serotonin), antibiotics, antirheumatic, and anti-inflammatory drugs [Metselaar and Storm, 2005]. Cationic liposmes are under wide application due to their efficiency to encapsulate and deliver nucleic acids through simple electrostatic interaction [Farhood *et al.*, 1995].

## Dendrimers

The term originates from two Greek word "Dendron" meaning tree and "Meros" meaning part. These are hyper-branched, globular, monodisperse, three dimensional nanoscale synthetic polymers, having very well defined size, shape and definite molecular weight [Cheng *et al.*, 2005], exhibiting characteristics features of both molecular chemistry and polymer chemistry. Dendrimer based therapeutic carrier systems are being developed for their features like:

- Easy to diffuse out of vasculature (<20 nm)
- Can recognize target cells and bind with high avidity and specificity to extracellular binding domain
- Internalize and intracellularly traffic to site of intended action
- Can avoid "normal" tissue
- Remain intact until reaching its intended site of action
- Stable and biologically inert

Dendrimers are ideal building block for creating a biologically active nanomaterial. High expectation starting from diseased cell recognition, reporting detection, diagnosis of disease state, reporting outcome of therapy and ultimately for drug delivery purposes are imposed to these smart dendrimer based systems [Mohammad and Antony, 2006]. Poly(amido amide) (PAMAM) is a dendrimer which has been intensively investigated for both the distribution of drugs or genes inside these molecules. Pure drug (Ibuprofen) encapsulated in dendrimers can enter into a cell very rapidly even less than an hour. Doxorubicin entrapped in dendrimers showed enhanced permeation and retention in tumour tissues to reduce cytotoxicity of anticancer drug and increased uptake by cancer cell lines. A number of drugs including methotrexate, 5-FU, and anti-inflammatory drugs, *e.g.* piroxicam, or indomethacin is being incorporated in PAMAM based dendrimers. Dendrimers can be utilized to regulate chemokine and cytokine release which inturn have great therapeutic potential.

Nano-vehicular targeted cancer therapy based on dendrimers are expected to be the next generation cancer treatment. DNA-linked dendrimers are providing versatility and controllability for making them strong candidates for targeting vehicles. Dendrimer-folate receptor mediated molecular recognition currently is the most promising oncological target. Development of targeted therapy using these dendrimers has proven to provide large returns on investment with the potential for extremely large revenue generation.

## Quantum Dots

Quantum dots (QDs) are nanometer scale semi-conductors that follow quantum mechanical principle of quantum confinement and exhibits energy band gap that determines required wavelength of radiation absorption and emission spectra. Requisite absorption and resultant emission wavelengths is dependent on particle size. The advantages of these very smart optically active QDs are having high potential for medical imaging and disease detection including:

- ☆ Set up in any arbitrary emission spectra to allow labeling and observation of detailed biological processes.
- ☆ QDs can be useful tool for monitoring cancerous cells and providing a means to better understand its evolution.
- ☆ QDs can also be armed with tumor-fighting toxic therapies to provide the diagnosis and treatment of cancer.
- ☆ QDs are much more resistant to degradation than other optical imaging probes such as organic dyes, allowing them to track cell processes for longer periods of time.
- ☆ QDs offer a wide broadband absorption spectrum while maintaining a distinct, static emission wavelength.

QD based sensing system surface modified with folate decorated nanomaterials of biocompatible polymers has been developed in the aim of breast cancer diagnosis and imaging by confocal laser scanning microscopy. QDs are now also being used for long-term labeling and imaging [Jaiswal *et al.*, 2003], optically guided surgery [Mihaljevic *et al.*, 2004], deep tissue cancer detection [Gao *et al.*, 2004], tracking cancer metastasis [Voura *et al.*, 2004], and optical barcodes. Quantum dots have been linked with protein transduction domains such as HIV, TAT, Pep-1, polyarginine, and SV40 T antigen to facilitate intracellular delivery. Orndorff *et al.* (2009) reported on high-affinity peptide neurotoxin QD nanoconjugates to image endogenous proteins in living cells and *ex vivo* tissue. Another wide application includes antibody targeting through peptide-functionalized QDs have been successfully used for targeting cellular proteins such as growth factor receptors, G protein-coupled receptors, integrins, and ion channels [Vu *et al.*, 2005]. QDs has become indispensable tool for biological research to elucidate the molecular mechanisms of protein binding, signaling, and regulation. The variety of QDs applied in biomedical application is wide and showing a bright path similar to its bright optical feature.

## Magnetic Nanoparticles

Magnetic NPs are also emerging out to be highly potential candidate for carrier based drug delivery purposes because of their following advantages:

- ✰ Tracking is easier under external magnetic field
- ✰ Passive and active drug delivary system can be developed based on these particles
- ✰ Easier cellular uptake ensures low dosage therapeutic potential.

There are several varieties of magnetic NPs present including pure metals such as cobalt, nickel, manganese, and iron, their alloys and oxides. But application in the field of nanomedicine is restricted with biocompatibility issue. According to Food and Drug Administration (FDA) approval iron oxide NPs are the most suitable one to be used in the field of food and drug development (Prijic and Sersa 2011). Two very common forms of iron oxide – magnetite and maghemite are naturally occurring substances in human heart which ensures their biocompatibility and nontoxicity at its physiologically relevant concentration.

For biosensing magnetic NPs are being utilized as contrast enhancing agents for MRI, where it reduces lesion detection as well as improve spatial resolution. Magnetic albumin microspheres containing low doses of doxorubicin administered intravenously resulted in total remission of 77 per cent (17/22) of tumors after only single dosage of the drug therapy [Widder *et al.*, 1983]. Recently there are several commercially available magnetic NPs (50 to 200 nm in diameter), *e.g.* CombiMAG, PolyMAG and TransMAGPEI, to be used in many *in vitro* and some *in vivo* studies of magnetofection (binding of the nucleic acids to magnetic NPs that concentrate and transfect cells in the area of interest by means of a magnetic field) [Krötz *et al.*, 2003]. The technology of magnetofection of therapeutic genes *in vivo* is under veterinary clinical trial consisting of dose-escalation of neoadjuvant gene therapy to surgery. Magnetic hyperthermia, involves local heat evolution due to either the relaxation losses for super paramagnetic particles (Neel relaxation) or the frictional losses of particles (Brownian motion) and causes reduction in the dose of chemotherapy, resulting lesser side effects. It also ensures that malignant cells are more sensitive to the rise in temperature than healthy ones.

## Nanoparticles of Noble Metal

From ancient era noble metals and their compounds as for example, gold and silver are being utilized in the field of medicine. Several diseases such as smallpox, skin ulcers, measles, and arthritis are being treated with gold. Metal NPs are attracting much interest because of their ultra small size generated optical properties (Surface Plasmon resonance), biocompatible nature, easier cell incorporation and simple synthesis procedure, easier characterization and surface modification with desirable functionalities etc. With time the wide application of noble metals at their nano-form, particulary NPs of gold and silver, in the field of diagnosis, treatment and healing made them more precious than bulk counterparts. This is because of the fact that metals at their nanoscale come out with newer properties which can apart from their smart entry to cells also ensures other added advantages like; sensing, heat generating, delivering (drugs, antibodies, genes, peptides, hormones, chemokines, cytokines etc) and ultimately curing life threatening diseases.

Recent studies with GNPs revealed its potential possibilities even in healing cancer. On irradiation with focused laser pulses of suitable wavelength, GNPs can kill cancer cells (Chen *et al.*, 2007). Moreover, GNP based drug carriers (Han *et al.*, 2007) and intravenous contrast enhancers in medical imaging are also being extensively researched. Cells can uptake small gold NPs without producing cytotoxicity. Gold-EGFR conjugates have been used to detect the cancer cells by exploiting the light scattering properties of gold nanoconjugates. Core shell nanocomposite based on gold-copper/gold-silver NPs are being used as efficient contrast enhancement agent in magnetic resonance imaging (MRI). Photothermal ablation generated by gold-silica coreshell nanoconjugates is useful in killing tumor cells. Similarly silver is known as one of the oldest antimicrobial agents. Silver ions are thought to affect bacterial enzymes and bind to DNA. Silver NP has also been used effectively against different bacteria, fungi and viruses [Stoimenov *et al.*, 2002]

### Polymer Nanoparticles

The Polymeric NPs (PNPs) including poly-e-caprolactone, polyacrylamide and polyacrylate, or natural polymers, *e.g.*, albumin, DNA, chitosan, gelatin are widely used in different biomedical purposes. Drugs can be immobilized on PNPs surface after a polymerization reaction and drugs may be released by desorption, diffusion, or NP erosion in target tissue. Cationic polymers such as PAMAM dendrimers, PEI and chitosan are also proven to be efficient gene delivery system [Fischer *et al.*, 1999] and also considered to be nontoxic after oral administration to humans. Drug-biodegradable polymeric nanocarrier conjugates can be used for drug delivery as they are stable in blood, non-toxic, and non-thrombogenic, non-immunogenic as well as non-proinflammatory, and they neither activate neutrophils nor affect reticuloendothelial system.

## Conclusion

Many researchers are working in the field of nanotechnology around the globe to decipher the noble properties at nanoscale. Many nanotech based products have already reached the market and many more are in the pipeline for commercialization. It should be mentioned that NPs can show toxicity in biological organisms if applied beyond a threshold level. Growth of prokaryotes was reported to be affected by stable aggregation of fullerene ($C_{60}$) in water (25 - 500 nm). Different NPs (fullerenes, carbon nanotube, QDs, zinc oxide NP) produce reactive oxygen species and thus regulate intracellular calcium concentrations, activate transcription factors, and induce cytokine production. Neuronal translocation of inhaled NPs, which reach different organs through lymphatic system, and circulatory systems, has also been reported. Even NPs can cross blood brain barrier. Although interaction between different types of NPs and cells with respect to their size, shape and surface properties has already been published, complete realization of *in vivo* and *in vitro* toxicity profiles of differently capped NPs of similar size and shape is yet inconclusive. So detailed *in vitro* and *in vivo* toxicity study of these NPs is necessary before large scale commercialization of NP based products.

## References

1. Abdel-Mottaleb MSA (2009). Duality of $TiO_2$ nanoparticles: applications in photocatalysis and dye-sensitised solar cells fabrication. *Int. J. Nanomanuf.* 4: 175-185.
2. Ahamed M, Alsalhi MS and Siddiqui MK (2010). Silver nanoparticle applications and human health. *Clin. Chim. Acta* 411: 1841-1848.
3. Amutha K, Ravibaskar R. Sivakumar G (2010). Extraction, synthesis and characterization of nanosilica from rice husk ash. *Int. J. Nanotechnol. Appl.* 4: 61-66.
4. Baker SN, Baker GA (2010). Luminescent carbon nanodots: Emergent nanolights. *Angew. Chem. Int. Ed.* 49: 6726-6744.
5. Balandin AA, Ghosh S,. Bao WZ, Calizo I, Teweldebrhan D, Miao F, Lau CN (2008). Superior thermal con-ductivity of single-layer graphene. *Nano Lett.* 8: 902-907.
6. Baxter JB, Aydil ES (2005). Nanowire based dye sensitized solar cells.*Appl. Phys. Lett.* 86: 53114-53116.
7. Bhattacharya R, Mukherjee P. 2008. Biological properties of "naked" metal nanoparticles. *Adv. Drug. Deliv. Rev.* 60: 1289-1306.
8. Bosetti M, Masse A, Tobin E, Cannas M (2002). Silver coated materials for external fixation devices: *in vitro* biocompatibility and genotoxicity. *Biomater.* 23: 887-892.
9. Chandra S, Das P, Bag S, Laha D, Pramanik P (2011). Synthesis, functionalization and bioimaging applications of highly fluorescent carbon nanoparticles. *Nanoscale.* 3: 1533-1540.
10. Chen YW, Hwang KC, Yen CC, *et al.* (2004). Fullerene derivatives protect against oxidative stress in RAW 264.7 cells and ischemia-reperfused lungs. *Am. J. Physiol. Regul. Integr. Comp. Physiol.* 287: 21–26.
11. Cheng Y, Tongwen X (2005). Dendrimers as Potential Drug Carriers. Part I. Solubilization of non-steroidal anti-inflammatory drugs in the presence of polyamidoamine dendrimers. *Europ. J. Med. Chem.* 40: 1188-1192.

12. Cheong S, Ferguson P, Feindel KW, Hermans IF, Callaghan PT, Meyer C, Slocombe A, Su CH, Cheng FY, Yeh CS, Ingham B, Toney MF, Tilley RD (2011). Simple Synthesis and Functionalization of Iron Nanoparticles for Magnetic Resonance Imaging. *Angew. Chem. Int. Ed.* 50: 4206–4209.

13. Cullum BM, Vo-Dinh T (2000). The development of optical nanosensors for biological measurements. *Trends Biotechnol.* 18: 388–393.

14. Davis ME (2009). The first targeted delivery of siRNA in humans via a self-assembling, Cyclodextrin polymer-based nanoparticle: From concept to clinic. *Mol. Pharmaceutics* 6: 659–668.

15. Debnath N, Das S, Brahmachary RL, Chandra R, Sudan S, Goswami A (2010). Entomotoxicity assay of silica, zinc oxide, titanium dioxide, aluminium oxide nanoparticles on *Lipaphis pseudobrassicae*. American Institute of Physics conference proceedings. 1276: 307-310.

16. Debnath N, Das S, Seth D, Chandra R, Bhattacharya SC, Goswami A (2011). Entomotoxic effect of silica nanoparticles against *Sitophilus oryzae* (L.). *J Pest Sci* 84: 99-105.

17. Dong J, Jia S, Chen J, Li B, Zheng J, Zhao J, Wang Z, Zhu Z (2012). Nitrogen-doped hollow carbon nanoparticles as efficient counter electrodes in quantum dot sensitized solar cells. *J. Mater. Chem.* 22: 9745-9750.

18. Erickson D, Mandal S, Yang A, Cordovez B (2008). Nanobiosensors: Optofluidic, electrical and mechanical approaches to biomolecular detection at the nanoscale. *Microfluid Nanofluidics*. 4: 33–52.

19. Englebienne P (1998). Use of colloidal gold surface plasmon resonance peak shift to infer affinity constants from the interactions between protein antigens and antibodies specific for single or multiple epitopes. *Analyst*. 123: 1599-1603.

20. Fan H, Wang L, Zhao K, Li N, Shi Z, Ge Z, Jin Z (2010). Fabrication, mechanical properties and biocompatibility of grapheme reinforced chitosan composites. *Biomacromolecules* 11: 2345-2351.

21. Farhood H, Serbina N, Huang L (1995). The role of phosphatidylethanolamine in cationic liposome mediated gene transfer. *Biochim.et. Biophys. Acta.*1235: 289-295.

22. Fischer D, Bieber T, Li Y, Elsässer HP, Kissel T (1999). A novel non-viral vector for DNA delivery based on low molecular weight, branched polyethylenimine: effect of molecular weight on transfection efficiency and cytotoxicity. *Pharm. Res.* 16: 1273–1279.

23. Gao X, Cui Y, Levenson RM, Chung LWK, Nie S (2004). In vivo cancer targeting and imaging with semiconductor quantum dots. *Nat. Biotechnol.* 22: 969-976.

24. Gemeinhart RA, Luo D, Saltzman WM (2005). Cellular fate of a modular DNA delivery system mediated by silica nanoparticles. *Biotechnol Prog* 21: 532–537.

25. Guindilis A. Atanasov P, Wilkins E (1997). Enzyme-catalyzed direct electron transfer: Fundamentals and analytical applications. *Electroanalysis* 9 : 661.

26. Guo Z, Wang C, Wang L, Cao C, Chen X, Sun C, Li T (1999). A New Biological Synthetic Route for Preparation of Silica Nano- Particles. *Moi Cryst. and Liq Cryst* 337: 249-252.

27. Han MS, Lytton-Jean AKR, Mirkin CA (2006). A gold nanoparticle based approach for screening triplex DNA binders. *J. Am. Chem. Soc.* 128: 4954-4955.

28. Heaton RJ, Peterson AW, Georgiadis RM (2001). Electrostatic surface plasmon resonance: Direct electric field-induced hybridization and denaturation in monolayer nucleic acid films and label-free discrimination of base mismatches. *Proc. Natl. Acad. Sci. U.S.A.* 98: 3701-3704.

29. Higby GJ (1982). Gold in medicine: a review of its use in the West before 1900. *Gold Bull* 15: 130-140.

30. Huang X, El-Sayed IH. Qian W, El-Sayed MA (2006). Cancer cell imaging and photothermal therapy in the near-infrared region by using gold nanorods. *J. Am. Chem. Soc.* 128: 2115-2120.

31. IJahnke A, Hirschberger J, Fischer C, Brill T, Köstlin R, Plank C, Ku¨chenhoff H, Krieger S, Kamenica K, Schillinger U (2007). Intratumoral gene d elivery of feIL-2, feIFN-gamma and feGM-CSF using magnetofection as a neoadjuvant treatment option for feline fibrosarcomas: a phase-I study. *J. Vet. Med.* A 54: 599-606.

32. Jaiswal JK, Mattoussi H, Mauro JM, Simon SM (2003). Long-term multiple color imaging of live cells using quantum dot bioconjugates. *Nat. Biotechnol.* 21: 47-51.

33. Kabanov AV, Vinogradov S (2008). Nanogels as Pharmaceutical Carriers, Multifunctional Pharmaceutical Nanocarriers, Springer Science, New York, 67-80.

34. Koo OM, Rubinsterin I, Onyuksel H (2005). Role of nanotechnology in targeted drug delivery and imaging: a consice review. *Nanomedicine: Nanotechnology, Biology and Medicine.* 1: 193-212.

35. Kroto HW, Heath, JR, O'Brien SC, Curl RF, Smalley RE (1985). C60: Buckminsterfullerene. *Nature.* 318: 162-163.

36. Krötz F, Sohn HY, Gloe T, Plank C, Pohl U (2003). Magnetofection potentiates gene delivery to cult ured endothelial cells. *J. Vasc. Res.* 40: 425-434.

37. Lai CW, Hsiao YH, Peng YK, Chou PT (2012). Facile synthesis of highly emissive carbon dots from pyrolysis of glycerol; gram scale production of carbon dots/ m$SiO_2$ for cell imaging and drug release. *J. Mater. Chem.* 22: 14403-14409.

38. Lee D, Choi M (2002). Coalescence enhanced synthesis of nanoparticles to control size, morphology, and crystalline phase at high concentrations. *J Aerosol Sci* 33: 1 –16.

39. Liu Y, Miyoshi H, Nakamura M (2007). Nanomedicine for drug delivery and imaging: A promising avenue for cancer therapy and diagnosis using targeted functional nanoparticles. *Int. J. Cancer,* 120: 2527-2537.

40. Liu R, Wu D, Liu S, Koynov K, Knoll W, Li Q (2009). An aqueous route to multicolor photoluminescent carbon dots using silica spheres as carriers. *Angew. Chem. Int. Ed.* 48: 4598-4601.

41. Liu C, Alwarappan S, Chen ZF, Kong XX, Li CZ (2010). Membraneless enzymatic biofuel cells based on graphene nanosheets. *Biosens Bioelectron.* 25: 1829-1833.

42. Liu S, Tian J, Wang L, Luo Y, Sun XA (2012). general strategy for the production of photoluminescent carbon nitride dots from organic amines and their application as novel peroxidase-like catalysts for colorimetric detection of $H_2O_2$ and glucose. *RSC Adv.* 2: 411-413.

43. Lu CH, Yang HH, Zhu CL, Chen X, and Chen GN (2009). A graphene platform for sensing biomolecules. *Angewandte Chemie Int. Edn.* 48: 4785-4787.

44. Matsui J, Akamatsu K, Hara N, Miyoshi D, Nawafune H, Tamaki K, Sugimoto N (2005). SPR sensor chip for detection of small molecules using molecularly imprinted polymer with embedded gold nanoparticles. *Anal Chem.* 77: 4282-4285.

45. M•dler L, Kammler HK, Mueller R, Pratsinis SE (2002). Controlled synthesis of nanostructured particles by flame spray pyrolysis. *J Aerosol Sci* 33: 369 –89.

46. Metselaar JM, Storm G (2005). *Liposomes in the treatment of inflammatory disorders.* 2: 465-476.

47. Mihaljevic T, Laurence RG, Dor DM, Cohn LH, Bawendi MG, Frangioni JV (2004). Near-infrared fluorescent type ii quantum dots for sentinel lymph node mapping. *Nat. Biotechnol.* 22: 93-97.

48. Mohammad N, Antony D (2006). Crossing cellular barriers using dendrimers nanotechnologies. *Current Opinion in Pharmacology.* 6: 522-527.

49. Mukherjee P, Ahmad A, Mandal D, Senapati S, Sainkar SR, Khan MI, Ramani R, Parischa R, Ajayakumar PV, Alam M, Sastry M, Kumar R (2001). Bioreduction of AuCl–4 ions by the fungus, verticillium sp.and surface trapping of the gold nanoparticles formed. *Angew Chem* 40: 3585-3588.

50. Mukherjee P, Senapati S, Mandal D, Ahmad A, Khan MI, Kumar R, Sastry M (2002). Extracellular synthesis of gold nanoparticles by the fungus Fusarium oxysporum. *Chembiochem* 3: 461-463.

51. Nath N, Chilkoti A (2004). Label-Free Biosensing by Surface Plasmon Resonance of Nanoparticles on Glass: Optimization of Nanoparticle Size. *Anal. Chem.* 76: 5370–5378.

52. Orndorff RL, Rosenthal SJ (2009). Neurotoxin quantum dot conjugates detect endogenous targets expressed in live cancer cells. *Nano Lett.* 9: 2589–2599.

53. Patra P, Mitra S, Debnath N, Goswami A. (2012). Biochemical-, biophysical-, and microarray-based antifungal evaluation of the buffer-mediated synthesized nano zinc oxide: An *in vivo* and *in vitro* toxicity study. *Langmuir.* 28: 16966–16978.

54. Pan J, Feng SS (2009). Targeting and imaging cancer cells by Folate-decorated, quantum dots (QDs)- loaded nanoparticles of biodegradable polymers. *Biomat.* 30: 1176-1183.

55. Patolsky F, Weizmann Y, Willner I (2004). Long-Range Electrical Contacting of Redox Enzymes by SWCNT Connectors. *Angew Chem. Int. Ed.* 43: 2113-2117.

56. Prijic S, Sersa G (2011). Magnetic nanoparticles as targeted delivery systems in oncology. *Radiol. Oncol.* 45: 1-16.

57. Stoimenov PK, Klinger RL, Marchin GL, Klabunde KJ (2002). Metal oxide nanoparticles as bactericidal agents. *Langmuir.* 18: 6679–6686.

58. Rahman A, Seth D, Mukhopadhyaya SK, Brahmachary RL, Ulrichs C, Goswami A (2009). Surface functionalized amorphous nanosilica and microsilica with nanopores as promising tools in biomedicine. *Naturwissenschaften*, 96: 31–38.

59. Rajangam T, Soo S, An A (2013). Fibrinogen and fibrin based micro and nano scaffolds incorporated with drugs, proteins, cells and genes for therapeutic biomedical applications. *Int. J. Nanomedicine.* 8: 3641–3662.

60. Hussain M, Ceccarelli R, Marchisio DL, Fino D, Russo N, Geobaldo F (2010). Synthesis, characterization, and photocatalytic application of novel TiO2 nanoparticles. *Chem. Eng. J.* 157: 45–51.

61. Singh R, Lillard GWG (2009). Nanoparticle-based targeted drug delivery," *Experimental and Molecular Pathology*, 86: 215- 223.

62. Sharma VK, Yngard RA, Lin Y (2009). Silver nanoparticles: green synthesis and their antimicrobial activities. *Adv. Colloid Sur. Interface Sci.* 145: 83-96.

63. Stober W, Fink A (1968). Controlled growth of monodisperse silica spheres in the micron size range. *J of Colloidal and Interface Sci* 26: 62-69.

64. Sultana F, Md. Imran-Ul-Haque M, Arafat M, Sharmin S. (2013). An Overview of Nanogel Drug Delivery System. *J. Appl. Pharm. Sci.* 3: S95-S105.

65. Turkevich J, Stevenson PC, Hillier J (1951). A study of the nucleation and growth processes in the synthesis of colloidal gold. *Faraday Soc* 11: 55–75.

66. Vaseeharan B, Ramasamy P, Chen JC (2010). Antibacterial activity of silver nanoparticles (AgNps). synthesized by tea leaf extracts against pathogenic *Vibrio harveyi* and its protective efficacy on juvenile *Feneropenaeus indicus. Lett Appl Microbiol.* 50: 352-356.

67. Voldman J, Gray ML, Schmidt MA (1999). Microfabrication in biology and medicine. *Annu Rev Biomed Eng.* 1: 401–425.

68. Voura EB, Jaiswal JK, Mattoussi H, Simon SM (2004). Tracking metastatic tumor cell extravasation with quantum dot nanocrystals and fluorescence emission scanning microscopy. *Nat. Med.* 10: 993-998.

69. Vu TQ, Maddipati R, Blute TA, Nehilla BJ, Nusblat L, Desai TA (2005). Peptide-conjugated quantum dots activate neuronal receptors and initiate downstream signaling of neurite growth. *Nano Lett.* 5: 603–607.

70. Wang J, Liu G, Jan M (2004). Ultrasensitive Electrical Biosensing of Proteins and DNA: Carbon-Nanotube Derived Amplification of the Recognition and Transduction Events. *J. Am. Chem. Soc.* 126: 3010-3011.

71. Wang C, Li Z, Cao D, Zhao YL, Gaines JW, Bozdemir OA, Ambrogio MW, Frasconi M, Botros YY, Zink JI, Stoddart JF (2012). Stimulated Release of Size-Selected Cargos in Succession from Mesoporous Silica Nanoparticles. doi: 10.1002/anie.201107960 (Article first published online: 13 APR 2012).

72. Widder KJ, Morris RM, Po ore GA, Howard DP, Senyei AE (1983). Selective targeting of magnetic albumin microspheres containing low-dose doxorubicin: total remission in Yoshida sarcoma-bearing rats. *Eur. J. Cancer. Clin. Oncol.* 19: 135-139.

73. Xuan Y, Wu YQ, Shen T, Qi M, Capano MA, Cooper JA, Ye PD (2008). Atomic-layer-deposited nanostructures for graphene-based nanoelectronics. *Appl Phys Lett.* 92: 013101-1-013101-3.

74. Yu BY, Kwak SY (2012). Carbon quantum dots embedded with mesoporous hematite nanospheres as efficient visible light-active photocatalysts. *J. Mater. Chem.* 22: 8345-8353.

75. Yuvaraj D, Kaushik R, Rao KN (2010). Optical, Field-Emission, and Antimicrobial Properties of ZnO Nanostructured Films Deposited at Room Temperature by Activated Reactive Evaporation. *Appl. Mater. Interfaces* 2: 1019-1024.

76. Zhou XC (2000). Amplified microgravimetric gene sensor using Au nanoparticle modified oligonucleotides. *Chem Commun.* 11: 953-954.

# Index

## B



## C

## *Q*



## *R*